Fourth Edition

# Biology

## *God's Living Creation*

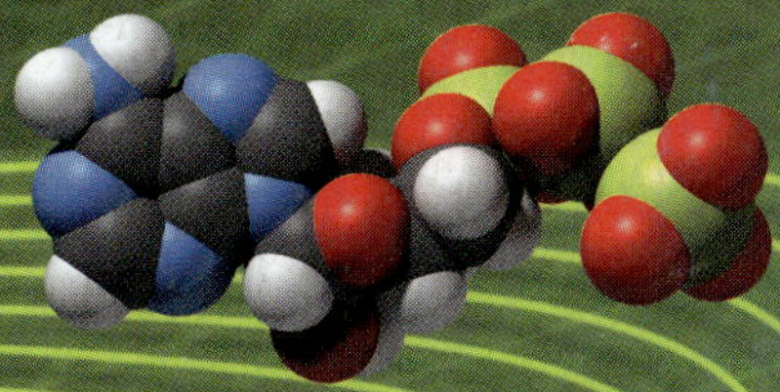

abeka
Pensacola, FL 32523-9100
an affiliate of PENSACOLA CHRISTIAN COLLEGE®

# *Biology: God's Living Creation*

Fourth Edition

**Staff Credits**

*Authors:* Keith Graham, Gregory Parker, John Durrand, Carlos Alvarez
*Managing Editor:* Corinne Sawtelle
*Edition Editor:* Jeremy Foster
*Designer:* Mitch Mann
*Production Artists:* Kristen Gherardini, Haley Cobb
*Illustrators:* Jesse Brady, Bobby Dalrymple, Jeremy Gorman, Tim Hansen, Ben Hensley, Brian Jekel, Audrey Karres, Jamie Kaufman, and Abeka Staff



Cataloging Data
Biology: God's Living Creation / Keith Graham, [et al.]—4th ed.
[Abeka Book science series]
xiii, 551 p. : col. ill. ; 26 cm.
Includes indexes.
1. Biology-Study and teaching (Secondary). II. Graham, Keith E. (Keith Ellis)
III. Abeka Book, Inc.
Library of Congress: QH 308.2 .B575 2015
Dewey System: 570

# Preface

Life is God's most marvelous and complex creation. *Biology: God's Living Creation* presents life as God created it and now controls it. In this textbook, the study of life is presented in a traditional manner as it was discovered by the great naturalists of the past, a large majority of whom revered the biblical account of Creation. Once students have acquired their bearings in the living world by studying the familiar flowering seed plants—how they grow and multiply and how they are important to man—students learn about the plants that are less familiar and more difficult to identify. The study of the human body is given prime emphasis as the temple of God and an integral part of God's creation. This sequence also allows for a more relevant, detailed, and understandable treatment of the animal kingdom. The more technical aspects of cellular biology are presented last in the text so that they can be presented in detail to help students understand why the "simple" cell of the evolutionists is simply nonexistent.

The Christian teacher will find that the unique Abeka approach to biology eliminates the conflict that results when evolutionary philosophy is combined with truth. Evolution is presented for what it is—a retreat from science. Students and teachers alike will feel more comfortable when they realize that it is not biology that is in conflict with Scripture, but rather the theory of evolution.

Since the day that Darwinism invaded the classrooms, God's glory has been hidden from students. Now there is an opportunity in the Christian classroom to declare that glory with *Biology: God's Living Creation.*

# Contents

## Unit 2 Human Anatomy & Physiology: Purpose and Design

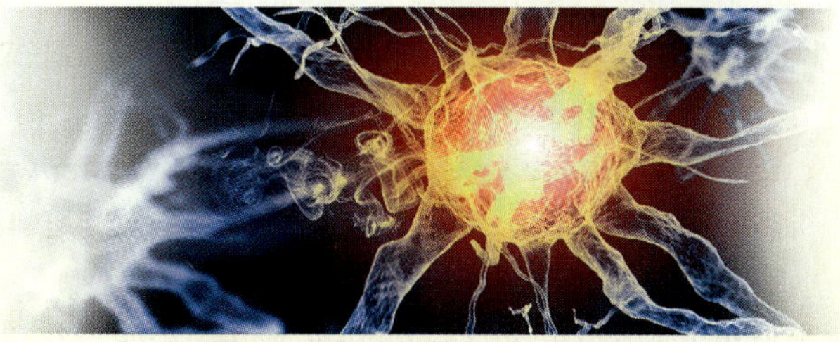

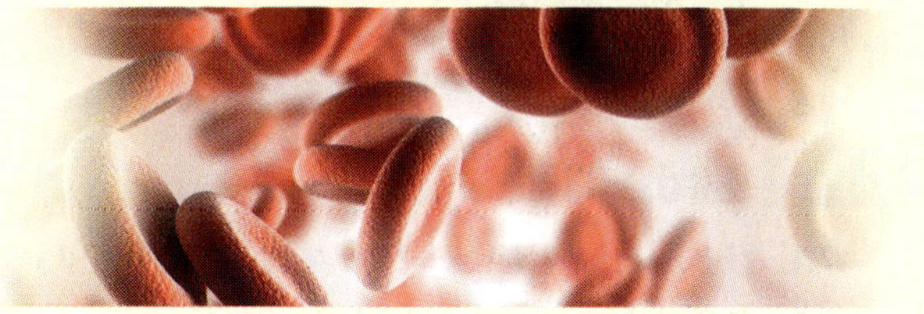



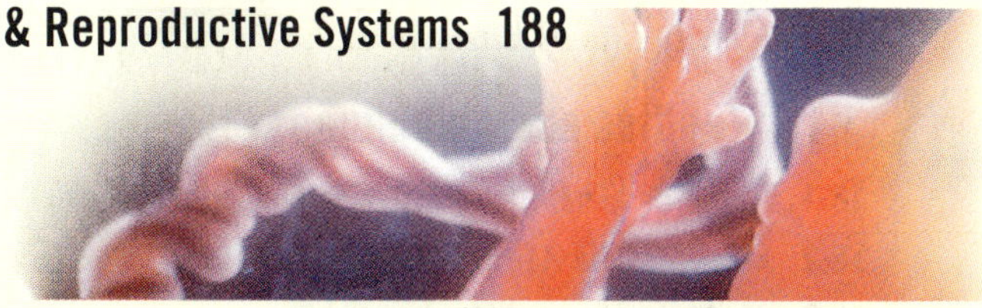



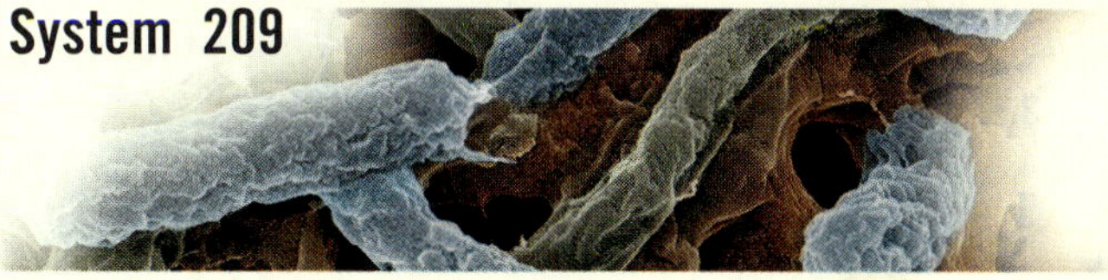

# How to Use This Text

Welcome to *Biology: God's Living Creation.* This unique textbook has many features designed to make your study of biology interesting and successful. Here are a few of them.

## Interesting Content

For an overview of what you will study this year, turn to the Table of Contents. At the beginning of the year, you will combine fieldwork, text work, and lab work to study the familiar plants around you and some plants that are less familiar. You will begin early to use the microscope and do dissections as you study the Creator's provision for living things.

The anatomy (structure) and physiology (function) of the human body are usually favorite topics with high school students. The Trans-Vision™ plates (between pp. 184 and 185) and Glossary of Medical Meanings (p. 529) will enhance your study of this unit.

You will study basic philosophical questions, such as who we are, why we are here, and where we are going, and will learn their relationship to biology. Following that is a brief chapter discussing the interrelationships between organisms that God has placed in His creation.

Your study of the fascinating and amazing world of animals will give you a greater understanding of man's unique place in creation. You may have the opportunity to study some of these animals by dissecting them or observing them under the microscope.

Finally, you will learn some of man's latest discoveries about the most complicated structure in all of creation: the living cell. You will be amazed both by the intricate detail that God has built into living things and by His master plan for transmitting information within an organism and from one generation of organisms to the next. You will also realize just how far man still has to go to gain a complete understanding of God's living creation.

## Study Aids

As you study, pay special attention to words printed in **bold.** Mastery of these key terms will help you to absorb the concepts of the chapter and to prepare for quizzes and tests. Some words are printed in *italics* to help you follow the logic of the section or to highlight connections to other chapters. An entire sentence or phrase printed in *italics* is an important concept or definition that you may want to mark or remember because you may be tested on it. Unusual or unfamiliar words are pronounced in the text the first time they appear. For the key to these **pronunciations,** see p. 500. Key terms are defined in the **Glossary** (p. 500). To find all the references in the book to a certain topic, use the **Index** (p. 530).

## Boxed Feature Articles

A number of articles giving fascinating facts or helpful observations about the living creation are included throughout the text. These articles are an important part of the book and need to be studied carefully.

## Section and Chapter Reviews

A **section review** follows each numbered section of the text. After you have read a section of a chapter, try to answer the questions and define the terms in the section review. Your teacher will assign portions of the section reviews for homework. The comprehension questions at the beginning of the review check your understanding of the material. The *Application* sections often require you to draw from your own experience or connect information presented in previous chapters to the present chapter. The *Critical Thinking* sections are designed to help you develop your higher thinking skills. *Observe* sections appear in some section reviews. These sections provide you with an opportunity for hands-on learning. Your teacher will tell you which of these activities should be done for homework.

Every chapter has a **chapter review** that will be helpful as you study for tests. Each chapter review consists of four sections: the *Define* section requires you to define relevant terms in the chapter; the *Identify* section includes important characteristics and lists; the *Explain* section asks questions about various terms and concepts; and the *Apply* section includes thinking questions that generally require a paragraph or short essay answer.

## Accurate, Detailed Illustrations

Every photograph and diagram in this book has been selected for a purpose: to enhance your interest in and understanding of the written text. As you read each section of the text, pay close attention to the illustrations. Match them up with the text material and read the captions and labels. If the illustration is showing a process, such as the life cycle of a fern or the movement of blood through the heart, trace the process as you read the text.

## Christian Perspective

The most important feature of this biology textbook is that it is written from the Christian perspective. This means that the authors give God the credit for all the wonders of the living creation and believe that the world and all things in it were created exactly as explained in the Bible. Most of the world's greatest scientists—men like Galileo, Sir Isaac Newton, Robert Boyle, Lord Kelvin, Michael Faraday, Gregor Mendel, Louis Pasteur, Johannes Kepler, and Joseph Lister—also worked from this perspective, so you will be in good company as you study this book. It is the authors' desire that as you gain a deeper knowledge of the living creation you will be drawn closer to the Creator.

# UNIT 1 | Botany

Structure and Function of Plants

# CHAPTER 1 The Flowering Seed Plants

**Key Concepts**

- Introduction to biology
- Parts of a green plant
- Woody and herbaceous plants
- Families of angiosperms
- Monocots and dicots
- Broadleaf trees

## 1.1 The Study of Living Things

### Introduction to Biology

**Biology,** the study of living things, has fascinated mankind since Creation. The word comes from the Greek root *bios*, which means "life," and the Greek suffix *-logy*, which means "science of" or "study of." You may see a field with numerous kinds of plants and small animals, a mountain stream, or an ocean with vast numbers of creatures. This is how biology began. God gave Adam and Eve the responsibility to subdue the earth and use it for their benefit. From the beginning, man has been naming, using, and studying living things.

In this text you will study many of the **organisms** (living things) that God created, often in their natural **habitats** (regions where particular organisms normally live). Your study will involve field observations, reading, using the microscope, and dissection of organisms. You will learn about your own body. The study will conclude with a detailed look at the intricate workings of living cells.

### Major Fields of Biology

Traditionally, the earth's living things are divided into three major groups—plants, animals, and people. These groups correspond to three major fields of biology—**botany,** the study of plants; **zoology,** the study of animals; and **human anatomy and physiology**

Fig. 1.1 **Fields of biology**

[fĭz′ē·ŏl′ə·jē], the study of the physical structure *(anatomy)* and function *(physiology)* of the human body.

As you become more familiar with organisms living near you, you will realize how important they are to nature and to you. The most obvious truths your study will reveal is that God used design in the creation of life and that all of nature fits together according to His plan.

> For the invisible things of him from the creation of the world are clearly seen, being understood by the things that are made, even his eternal power and Godhead; so that they are without excuse. —Rom. 1:20

Man's role in God's creation is explained in Genesis 1:28.

> And God blessed them, and God said unto them, Be fruitful, and multiply, and replenish the earth, and subdue it: and have dominion over the fish of the sea, and over the fowl of the air, and over every living thing that moveth upon the earth.

God instructed man to have dominion over His creation.

### The Importance of Botany

The study of botany is important to man, for all the food eaten by man comes directly or indirectly from green plants. Consider the foods you had for breakfast this morning: cereals and toast come directly from plants; and milk, eggs, and meat are produced by plant-eating animals. *About two thirds of the food we eat comes directly from plants. The other third comes from animals that eat plants.*

Green plants are biological food factories. When exposed to the sun, they carry on photosynthesis to make their own food, an ability no animal possesses. And all animals and humans depend on this for survival. Because plants are the ultimate source of man's food, botany is related to the science of agriculture.

A second reason botany is important is that many drugs and antibiotics used in medicine are derived from plants. And many plants and plant products are vital in industry. Products like lumber, paper, and clothing fibers (such as cotton, acetate, and rayon) are derived from plants.

### Section Review 1.1

1. Explain the origin of the word *biology.*
2. What are the three major fields of biology as presented in the text?
3. Name three reasons why the study of botany is important.

#### Critical Thinking

4. List the foods that you had for lunch today and explain how each one came from plants.

#### Identify

biology, habitat, botany, zoology

## 1.2 Collecting and Observing Seed Plants

### Parts of a Plant

A most basic activity of all scientists is collecting and observing data. Our study of botany begins with the collection and observation of plants and plant organs.

When observing plants, take a small notebook to note many distinguishing characteristics of each plant you study; you should include in your notebook a sketch or photograph of each plant. Most green plants possess roots, stems, leaves, flowers, and fruits. Your observation will be heightened as you look for similarities and differences of these parts in a variety of plants.

**Flowers.** Note color, type, and arrangement of flowers and flower parts. Flowers are important to plants; they help the plants to reproduce after their kind. Reproductive cells in the flowers unite to form embryos. These embryos are located within the seeds, which are located in the fruit. Under proper conditions, the embryo in each seed will develop into a mature plant. *Most plants with flowers will produce fruits and seeds.* When observing a plant that has lost its flowers, you may notice fruits or seeds.

**Leaves.** Note the shape, size, arrangement, and number of leaves. Leaves are vital to green plants because most of the plant's food is manufactured in the leaves through photosynthesis. In studying the leaves of different plants, you will notice tremendous variety. All are alike in producing food for themselves, animals, and man; yet each is unique in some way that makes it special among living things. The great variety of leaves testifies to the intelligence of the Creator.

**Stems.** Leaves and flowers are held up by the stem, which also transports substances and stores food. Many plants have peculiar stems that identify them. Note the appearance of the stems and the way they feel. Land plants are commonly classed as trees, shrubs, herbs, or vines according to size and type of stem. Trees and shrubs are **woody plants** that live several years and develop a large amount of wood. A plant with a single tall, woody stem capable of standing erect without support is a **tree.** A plant with several low, woody stems near the ground and able to stand erect without support is a **shrub** or *bush.*

Fig. 1.2
**Parts of a tomato plant**

Nonwoody plants are referred to as **herbaceous** [hûr·bā′shəs] **plants** or **herbs** [ûrbz]. To many people, an herb is any plant used as a medicine, seasoning, or flavoring; but to a botanist, an herb is any plant whose stem remains relatively soft. Herbaceous plants include familiar plants like daisies, mints, tomatoes, grasses, and daffodils.

The fourth kind of plant, the **vine,** can be either woody or herbaceous. A vine is any plant with a long, thin stem that grows along the ground or has tendrils that twine around a wall or other support. Herbaceous vines include cucumbers, beans, and morning glories. Grapes are a woody vine.

**Roots.** Carefully pull a plant out of the soil and note its root structure. The root anchors the plant to the soil, absorbs water and minerals, and stores food. Some roots spread out to make new plants. If you examine a dandelion plant and a grass plant, you see two main kinds of root systems. The dandelion has a **taproot system,** one in which the main root (primary root) grows straight down and remains larger than the secondary roots that branch off from it. Taproots often store large amounts of food. Grass has a **fibrous root system,** one in which the primary root remains small and many slender secondary roots grow from it in all directions. The tangled mass formed by grass roots serves to hold soil particles together and prevents soil erosion.

Fig. 1.3
**Types of stems**

Fig. 1.4
**Root systems**

## Section Review 1.2

1. Why are flowers important to plants?
2. How is the scientific definition of *herb* different from its everyday meaning?
3. How are taproot systems different from fibrous root systems?

### Application

4. If you were given one branch from a plant, describe the characteristics that you could look for that would help you to identify the type of plant from which the branch came.

### Critical Thinking

5. Why would you usually find fruits and seeds on a plant *after* its flowers have withered away?
6. Why would grass be superior to dandelions in preventing erosion of topsoil?

### Identify

woody plant, tree, shrub, herbaceous plant, vine

### Observe

- Bring in a branch with leaves and perhaps flowers or fruit and be prepared to explain what kind of plant it comes from—tree, shrub, herb, or vine.

# 1.3 Some Common Flowering Seed Plants

If you list the first thirty plants that come to your mind, the majority would probably be *flowering seed plants* or **angiosperms,** whose flowers produce seeds covered by fruit. All flowering plants, including broadleaf trees, grasses, garden flowers, weeds, and vegetables, are angiosperms. *About half of the plants known to man are angiosperms.* There are about 140 families of angiosperms, which are further divided into 250,000 species.

Fig. 1.6
**Plants of the composite family**

## The Composite Family

**Disks and rays.** One of the largest families of angiosperms is the **composite family,** also called the *sunflower family.* Roadside flowers such as the aster, goldenrod, daisy, dandelion, and sunflower are examples of composite flowers. These flowers superficially resemble a single flower, but each is in reality a cluster of many separate flowers referred to as a **head.**

In many composite flowers, there are two types of flowers in each head. The ones in the center of the head are called **disk flowers.** Usually they are the only seed-producing structures. Arranged around the disk flowers are the showy **ray flowers,** which are often mistaken for the petals of the flower. The flower heads of some composites, such as dandelion, consist only of ray flowers, each of which produces one seed.

Fig. 1.5
**Parts of a composite flower (sunflower)**

**Familiar composites.** One of the best-known and most beautiful composite flowers is the *daisy.* The *oxeye daisy,* probably the most common, has white ray flowers and yellow disks. Several types of plants, including some *asters,* have flowers with a color pattern similar to that of daisies but produce a branching stalk with several flower heads. (The daisy has only one head on each stalk.) Other similar flowers include the *black-eyed Susan* and the *common sunflower.*

The *goldenrods* are composite flowers that are easily recognized by their clumps of bright golden-yellow flowers. Each tiny flower has its own circle of yellow rays. There are about one hundred kinds of goldenrods in North America.

The *dandelion,* a common composite with bright yellow ray flowers, has edible leaves. Other composites with edible parts include the *globe artichoke* and *lettuce.*

### The Mint Family

When you hear the word *mint,* you probably think of the pleasant flavor of *peppermint* or *spearmint.* These flavors are produced by the oils of the peppermint and spearmint plants, important members of the **mint family.** The names of other members of the mint family may be familiar to you also. *Oregano, basil, sage, lavender, marjoram, rosemary, thyme,* and *balm* are used in cooking. You might find decorative *coleus* [kō′lē·əs] or *salvia* in the landscaping around your home or school. Mint plants have a square stem and are often very aromatic.

Fig. 1.7
**Plants of the mint family**

### The Parsley Family

The flowers of most plants in the **parsley family** are extremely complex in design; if you use a hand lens to observe them, you will marvel at their beauty. The "flower" of *Queen Anne's lace* is really a very large cluster of over two thousand individual white flowers. Many members of the parsley family are used as foods or as seasonings for foods, including *carrot, celery, parsley,* and *dill.* Several poisonous plants belong to the parsley family as well, including *poison hemlock* and *water hemlock.*

Fig. 1.8
**Plants of the parsley family**

## *Annuals, Biennials, and Perennials*

Understanding the LIVING CREATION

**P**lants are commonly classified according to their growth cycles. A plant that develops from a seed and produces new seeds in a single growing season is called an **annual.** These short-lived plants, which include peas, beans, grains, and many other plants, are among the most useful to mankind.

Some plants live through two growing seasons to complete their life cycles. These plants, which include beets, cabbage, turnips, carrots, and a great many others, are called **biennials.** During the first year of growth, a biennial does not grow very large but devotes its resources to storing food in its enlarging taproot. During the second year, the plant uses its reserves of stored food to produce flowers and seeds. Much of the time, biennials are harvested after the first year to obtain the stored food, and thus we do not see their flowers.

The final group of plants according to life cycle are the **perennials.** These are the plants that live from year to year and bloom each season. Perennials may be either herbaceous (soft-stemmed) or woody. In temperate regions like the United States, the shoots of herbaceous perennials die during the winter, but in the spring of the next growing season, a new shoot grows from underground roots or stems. The spring wildflowers, desert flowers, peonies, and asparagus plants are examples of herbaceous perennials. Trees and shrubs are woody perennials.

## The Rose Family

You are probably familiar with more members of the **rose family** than you realize. This family includes the plants with showy flowers and thorny stems known as *garden roses.* It also includes the *wild roses,* with their open, five-petal design. Stop and think of the plants you have seen that have blooms similar to those of a wild rose. Many of these plants, such as *blackberries, raspberries,* and *strawberries,* are in the rose family. A number of fruit trees, such as the *apple, apricot, cherry, nectarine, peach, pear,* and *plum,* as well as the *almond* tree, also belong to this family.

Fig. 1.9 **Plants of the rose family**

## The Pea Family

Think of as many plants as you can that have podlike fruits. These plants are probably members of the **pea family,** one of the largest families of flowering plants. Several members of the pea family, such as *English peas, beans* of all kinds, *peanuts,* and *lentils,* provide us with food. Others, such as *alfalfa, clover,* and *vetch,* provide forage for farm animals and help to prevent soil erosion. Some members of the pea family, such as the fragrant *wisteria* vine and the beautiful *redbud, locust, laburnum,* and *mimosa* trees, help to decorate our yards. A type of pea that you do not want to find in your yard is *kudzu* [ko͝od′zo͞o], a rapidly growing vine that was imported from China and Japan to the southern United States. Kudzu was first used to enrich the soil and prevent erosion in fields, but it escaped from many fields and became a nuisance by overtaking vacant lots and roadsides in much of the South.

Fig. 1.10 **Plants of the pea family**

## The Mustard Family

Many of our vegetables come from the leaves, stems, or roots of plants in the **mustard family.** We eat the leaves and stems of *cabbage, collards, kale,* and *turnips.* The turnip, *radish,* and *rutabaga* supply us with edible roots. We eat the flower buds of the *broccoli* plant and the actual flowers of the *cauliflower* plant. *Brussels sprouts* are the young leafy sprouts of the Brussels sprouts plant. The *mustard* plant itself is a dark green, leafy plant, of which there are several varieties. Black mustard grows to a height of about six feet and is the largest variety. Its seeds are used in the production of commercial mustard products.

Fig. 1.11 **Plants of the mustard family**

## Legumes, Bacteria, and the Nitrogen Cycle

Understanding the LIVING CREATION

Members of the pea family are commonly referred to as **legumes** [lĕg′ yo͞omz′]. The identifying feature of this family is its fruit, which takes the form of a pod called a *legume.* Clovers, vetches, peas, alfalfa, beans, and other legumes are important crop plants because they actually enrich the soil as they grow.

The element nitrogen is an essential part of proteins, but plants cannot directly use nitrogen out of the air to make protein. They can use only nitrogen that has been combined into special compounds of nitrogen and oxygen called **nitrates. Nitrogen-fixing bacteria** growing on the roots of legumes convert the nitrogen from air into ammonia and nitrates, which the plant absorbs through its roots to convert to proteins. By extracting nitrogen from the air, legumes can grow in poor soils that lack essential nitrogen compounds. When the plants die, their nitrogen-containing compounds are incorporated into the soil, making it more fertile for other plants. Soil nitrates also come from a few other sources. **Nitrifying bacteria** provide nitrates to the soil from decomposing plants and animals, and lightning produces small amounts of nitrates in the soil from nitrogen in the air. The main source of nitrates for cultivated plants comes from synthetic (manmade) fertilizers.

The movement of nitrogen from the air, into the soil, and back into the air is called the **nitrogen cycle.** The nitrogen cycle is essential to life on earth because plants, animals, and humans all require nitrogen for survival. **Denitrifying bacteria** in the soil convert unused nitrates back into atmospheric nitrogen to complete the nitrogen cycle.

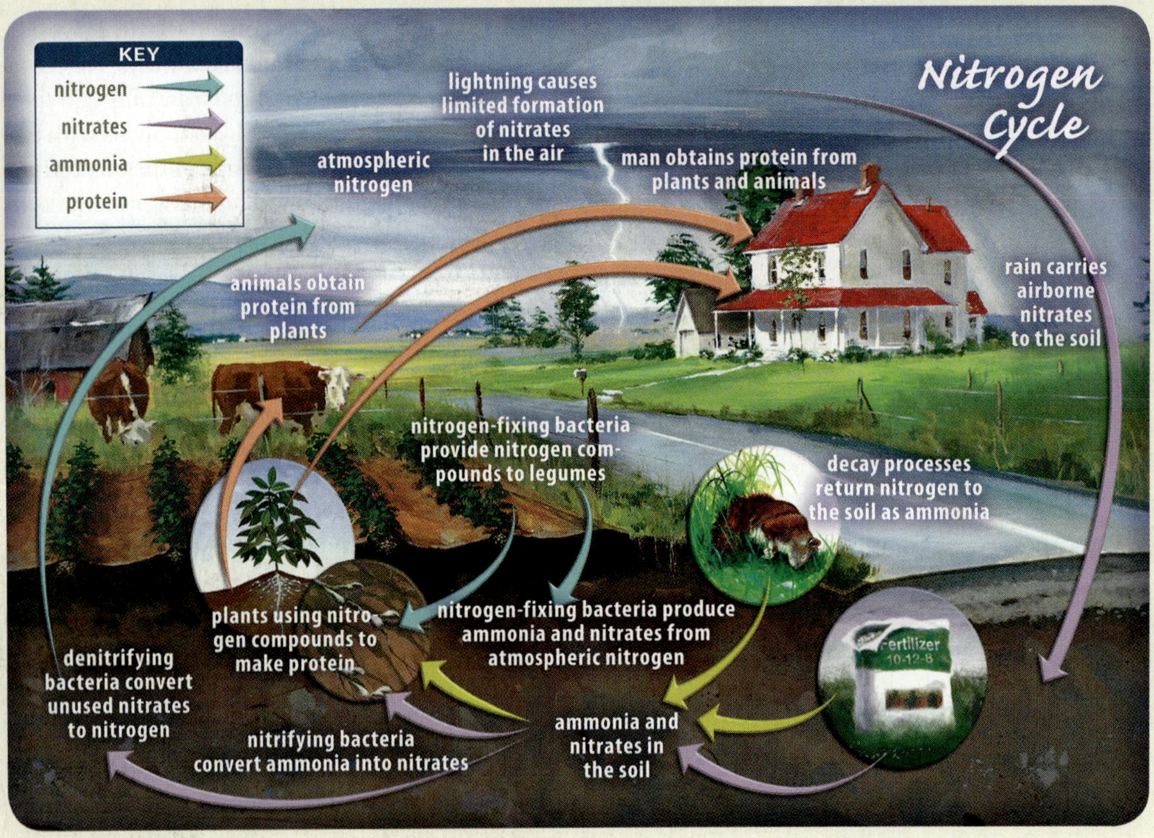

## The Lily Family

The **lily family** is a diverse plant family with some surprising members. You are probably familiar with *day lilies,* which grow in gardens, along roadsides, and in meadows. Their funnel-shaped flowers are open for only one day before they die and are replaced by other flowers. While day lilies bloom in summer, many members of the lily family bloom in early spring, such as the *tulip, trillium,* and *hyacinth.* In spite of its cactuslike appearance, *aloe vera* is also a lily; its juice is used in soothing skin lotions. The *yuccas,* including the *Joshua tree* and *Spanish bayonet,* are other cactuslike members of the lily family.

Fig. 1.12
**Plants of the lily family**

## The Cashew Family

The most familiar member of the **cashew family** is the *cashew tree,* with its delicious nuts. The cashew family also has the less pleasant distinction of being the group to which the three most common poisonous plants belong. *Poison sumac, poison ivy,* and *poison oak* are not ivies or oaks at all; they are all members of the cashew family. These three plants are not the most life-threatening among poisonous plants but are probably the greatest nuisance. The leaves contain *urushiol,* an oily substance that can produce a severe rash that lasts for days.

Fig. 1.13
**Plants of the cashew family**

## The Nightshade Family

The plants of the **nightshade family** include many important food plants like *tomatoes, Irish potatoes, peppers,* and *eggplants.* Many other members of the nightshade family are poisonous, including the *deadly nightshade* or *belladonna,* the *woody nightshade,* and *tobacco.* The *petunia,* a common ornamental flower, is also a member of the nightshade family.

Fig. 1.14
**Plants of the nightshade family**

## Monocots and Dicots

**T**he seed of an angiosperm plant contains special structures called **cotyledons** [kŏt′l·ēd′nz], which are used to store food until the seed begins to grow. All the families of angiosperm plants can be placed into two main categories, **monocots** and **dicots,** based on the number of cotyledons within each seed. Monocot plants have only one cotyledon per seed, whereas dicot plants have two cotyledons per seed. Most angiosperm families are dicots.

Understanding the LIVING CREATION

Monocots and dicots are distinguished by other basic characteristics. *Three characteristics of dicots are that their flower parts are usually arranged in groups of four or five, their leaves are usually broad and flat with a branching vein pattern, and their roots take the form of a taproot system.* Familiar dicot families are listed on pp. 10–11.

By contrast, *three characteristics of monocots are that their flower petals are always arranged in groups of three, their leaves are usually long and narrow with a parallel vein pattern, and their roots take the form of a fibrous root system.* Familiar monocot families are listed below. You will learn more about the characteristics of monocots and dicots in later chapters.

## Table 1.1 Some Major Families of Angiosperms

### Monocot Families

***Identification:*** *flower parts in groups of three; leaves usually long and narrow with veins running parallel to each other; root system usually fibrous*

**Amaryllis** (Amaryllidaceae) amaryllis, daffodil, garlic, narcissus, onion
**Arum** (Araceae) arum, caladium, elephant's ear, jack-in-the-pulpit, philodendron
**Banana** (Musaceae) banana
**Cattail** (Typhaceae) cattail
**Grass** (Poaceae or Gramineae) barley, bent, bluegrass, bamboo, corn, crabgrass, millet, pampas grass, rice, rye, sugar cane, timothy, quack grass, wheat
**Iris** (Iridaceae) crocus, gladiolus, iris
**Lily** (Liliaceae) aloe, asparagus, catbrier (greenbrier), day lily, Easter lily, hyacinth, Joshua tree, Spanish bayonet, tiger lily, trillium, tulip, yucca
**Orchid** (Orchidaceae) lady's slipper, orchid, vanilla
**Palm** (Arecaceae or Palmae) cabbage palm, coconut palm, date palm, ivory palm, palmetto
**Sedge** (Cyperaceae) bulrush, papyrus, sedge

### Dicot Families

***Identification:*** *flower parts usually in multiples of four or five; leaves usually broad and flat in shape with a branching vein pattern; usually have a taproot*

**Beech** (Fagaceae) beech, chestnut, oak
**Birch** (Betulaceae) birch, alder
**Cactus** (Cactaceae) cactus, prickly pear
**Cashew** (Anacardiaceae) cashew, mango, poison ivy, poison oak, sumac
**Composite** (Asteraceae) artichoke, aster, black-eyed Susan, daisy, dandelion, goldenrod, lettuce, marigold, ragweed, sunflower, thistle
**Dogwood** (Cornaceae) dogwood
**Elm** (Ulmaceae) elm, hackberry
**Gourd** (Cucurbitaceae) cucumber, gourd, pumpkin, squash, watermelon
**Heath** (Ericaceae) azalea, blueberry, brier, heath, mountain laurel, rhododendron
**Holly** (Aquifoliaceae) holly

Table 1.1 *Dicot Families (cont.)*

**Magnolia** (Magnoliaceae) magnolia, tulip tree (yellow poplar)
**Maple** (Aceraceae) box elder, maple
**Mint** (Lamiaceae) balm, basil, catnip, coleus, lavender, marjoram, peppermint, rosemary, sage, salvia, spearmint, thyme
**Mustard** (Brassicaceae) broccoli, Brussels sprouts, cabbage, cauliflower, collard, horseradish, kale, mustard, radish, rutabaga, turnip
**Nightshade** (Solanaceae) eggplant, jimsonweed, pepper, tobacco, tomato, white potato
**Olive** (Oleaceae) ash, jasmine, lilac, olive
**Parsley** (Apiaceae) anise, caraway, carrot, celery, coriander, dill, hemlock, parsley, parsnip, Queen Anne's lace
**Pea** (legume) (Fabaceae) alfalfa, bean, beggarweed, clover, English pea, indigo, kudzu, laburnum, lentil, licorice, locust, mesquite, mimosa, peanut, redbud, vetch, wisteria
**Pineapple** (Bromeliaceae) cultivated pineapple, Spanish moss, wild pineapple or air plant
**Pink** (Caryophyllaceae) baby's breath, carnation, pink, sweet William
**Plane tree** (Platanaceae) sycamore
**Rose** (Rosaceae) almond, apple, apricot, blackberry, boysenberry, bramble, bridal wreath, cherry, nectarine, peach, pear, plum, raspberry, rose, strawberry
**Rue** (Rutaceae) citron, grapefruit, kumquat, lemon, lime, orange, tangerine
**Walnut** (Juglandaceae) bitternut, butternut, hickory, pecan, walnut
**Willow** (Salicaceae) aspen, cottonwood, poplar, willow

## The Grass Family

**Characteristics of grasses.** One of the best areas for your study of angiosperms may be your own backyard. Your yard need not be stocked with roses, lilies, or fruit trees for you to have an abundance of angiosperms to study, nor do you need to have a vegetable garden—all you need is a lawn and the multitude of grasses it contains. **Grasses,** or *graminoids*, are among the most important groups of flowering seed plants, yet most people have never stopped to realize that grasses even have flowers. The flowers on most grass plants are small and inconspicuous, lacking large, colorful petals, yet they perform the same important function as their beautiful and fragrant counterparts.

In a temperate central North American meadow, as many as seventy species of grasses may be flowering at one time. An average lawn contains a variety of species as well.

Fig. 1.15 **Cereals**

**Cereal crops and forage grasses.** *The grasses may be considered the most important group of plants on Earth.* Many land animals depend on grasses, directly or indirectly, for their food. Man has taken the grasses and developed from them high-yielding **cereal** crops, including *wheat, barley, rye, oats, rice,* and *corn. Wheat, corn, and rice are the three most important cereal crops.* We also benefit indirectly from grasses when we eat meat because livestock are raised on grains and **forage grasses.**

**Turf grasses.** **Turf grasses** are those used as coverings for lawns, athletic fields, golf courses, and playgrounds. Common turf grasses include *Kentucky bluegrass, bentgrass, Bermuda grass, ryegrass, fescue, centipede grass, zoysia,* and *Bahia grass.* Two turf grasses, crabgrass and quack grass, are often considered weeds because they grow wild and tend to overrun the other, more beautiful, turf grasses planted by man. *Quack grass* has two- to three-foot-tall stems topped by a spike of grainlike fruits that is three to eight inches long. *Crabgrass* has leaves that are three to six inches long and hairy at the base. The flowers are in spikes that are clustered in a whorl at the end of the stem, somewhat like the ribs of an umbrella on the handle. Because of this arrangement, crabgrass is sometimes called umbrella grass.

Fig. 1.16 **Turf grass**

**Other grasses.** Ornamental grasses such as *pampas grass* beautify gardens, parks, and lawns. *Bamboo*, the most common of the *woody grasses*, is sometimes used in landscaping as well. Its hollow, woody stems may grow as tall as 120 feet and as wide as one foot. *Sugar cane*, another large grass plant, has stems 7 to 15 feet tall. The sugary juice from its stems is extracted and refined for use as table sugar.

Fig. 1.17 **Other grasses**

## Things to Notice about a Grass Plant

Observing the LIVING CREATION

Grass is one of those seemingly insignificant plants that we tend to take for granted and largely ignore. Perhaps you have never really taken time to observe a single grass plant before. Carefully uproot a grass plant. You already know that grass has *fibrous roots* rather than a taproot. Sometimes up to 90% of the grass's weight is in the roots, where vast amounts of water and food can be stored. Your grass plant may also have a **stolon,** a creeping stem that grows above the ground.

The *stem* is round and mostly hollow except at the *node*, or joint, where the leaf attaches. Unlike most plants, the leaves of grass plants have a **sheath** that seems to wrap partially around the stem at the base of the leaf. The sheath is open at least part of the way down. *The plant grows from the base of the sheath;* thus, if most of the leaf gets eaten or mowed, the grass plant can keep growing. Most plants grow from the tips of their stems and would be killed by constant mowing or grazing, but God designed grass to just keep growing.

The *flower cluster* is small and inconspicuous. From the flower the fruits, or **grains,** grow. Grain is high in starch content, making it an excellent source of energy for people and animals. Humans cannot digest the leaves and stems of grasses, but many animals can.

*Typical grass plant*

## Section Review 1.3

1. What two types of flowers make up a composite flower?
2. List several plants from each of the following families: composite, mint, parsley, rose, pea, mustard, lily, cashew, and nightshade.
3. What major structural characteristic do all members of the pea family have in common?
4. What are four differences between dicots and monocots? List several examples of both types of plant.
5. What are the three most important cereal crops?

**Application**

6. Would a pine tree be classified as an annual, a biennial, or a perennial?
7. Are grasses monocots or dicots? How can you tell?

### Critical Thinking

8. Describe the nitrogen cycle and explain the role of the pea family in it.
9. List as many characteristics of plants as you can think of that give evidence of their being designed by a Person with intelligence, imagination, creativity, and an interest in variety. Explain your conclusions.
10. Explain how grass reveals God's care and provision for people and animals. Read Romans 1:20. What does grass (a thing that is made) reveal about God's eternal power? Why does its widespread presence make all men responsible to God and without excuse?

### Identify

angiosperm, annual, biennial, perennial, legume, nitrates, denitrifying bacteria, graminoids, cereals, forage grasses, turf grasses, stolon, sheath, grain

### Observe

- Bring to class a member of one of the plant families described in this chapter and be able to describe its features according to the information given in the text.
- Uproot a grass plant and identify its parts.

## 1.4 Broadleaf Trees—The Largest Angiosperms

### Observing Broadleaf Trees

*Trees*, the largest of all plants, are another excellent subject to observe in your study of botany because of their size and their longevity. *As perennials, trees may live for hundreds or even thousands of years and continue to grow throughout their lives.*

One of the first things that you will notice in your study of trees is their leaves. Examine the shape of the leaves, their arrangement on the stem, the pattern of veins within each leaf, and the formation of the leaf margins (edges). As you observe trees, determine if they have broad and flat leaves, long and narrow leaves, or leaves that are needlelike or scalelike. In this section we will concentrate on trees having broad, flat leaves. These trees are referred to as **broadleaf trees.** *Nearly all broadleaf trees are flowering seed plants, or angiosperms, and all of those that are angiosperms are dicots rather than monocots.* Some of them belong to angiosperm families studied in section 1.3.

Because most broadleaf trees (*beeches, maples, oaks,* and others) have dense, hard wood, the term **hardwoods** is also applied to all broadleaf trees. However, a few broadleaf trees such as *basswoods* and *cottonwoods* actually have soft, lightweight wood. In addition to having broad, flat leaves and hard wood, most broadleaf trees in temperate regions are **deciduous** [dĭ·sĭj′o͞o·əs]; that is, they lose their leaves each fall and are bare all winter. **Evergreen trees,** by contrast, retain their leaves throughout the winter.

### Characteristics of Broadleaf Trees

The leaves of a tree form its bushy head, or **crown.** In broadleaf trees, the **trunk** (main stem) does not reach the top of the tree but divides into spreading branches, giving the crown a rounded shape. Certain broadleaf trees, when grown in the open where they are not cramped by other trees, develop distinctive crown shapes. *Oaks* have a wide-spreading crown. *Elms* are vaselike, and *poplars* have a tall, narrow crown. The trunk of the tree supports the leafy crown and holds it up to the sunlight.

All trees, including broadleaf trees, are considered to be woody plants because their trunk and branches are composed of wood. The purpose of wood is to strengthen and support the tree.

The trunk, branches, and certain roots of trees are covered by a rough protective layer of outer bark. Certain trees, such as the *paper birch*, can be identified by the unusual appearance of their bark.

If you have observed broadleaf trees very carefully in the spring, you may have noticed that they produce flowers. This is why they are included among the angiosperms, the flowering seed plants. Decorative broadleaf trees, including the *dogwood*, *magnolia*, and *laburnum* [lə·bûr′nəm] are prized for their beautiful flowers. Most of us are familiar with the strikingly beautiful blossoms of common fruit trees such as the *cherry* and *apple*. Most broadleaf trees do not have large, showy flowers, however, but tiny, drab flowers. The flowers of some trees are actually disguised. They appear as long, tassel-like clusters of tiny grayish- or yellowish-green flowers. These flowers-in-disguise, called **catkins,** can be seen in early spring on the branches of *oaks*, *willows*, *cottonwoods*, *poplars*, and *birches*.

# Guide to Familiar American Broadleaf Trees

## Maples

Maples are among the most popular trees of residential areas. Their distinctive fruits have tiny wings to catch the breeze so that the seeds will be efficiently scattered. This form of fruit is called a **samara** [săm′ər•ə]. The *sugar maple* of the northern forest is the source of maple syrup. It presents itself as a brilliant blaze of yellow and orange in the fall.

## Magnolias

Magnolias are a symbol of the South—their fragrant blossoms and shiny, dark green leaves beautify many yards and parks. The *bigleaf magnolia* has blossoms that measure up to ten inches in diameter, giving it the distinction of being the native American tree with the largest blossoms. The *tulip tree* bears small, greenish-yellow flowers shaped like tulips.

## Birches

Birches, especially the *paper birch*, are known for their thin, smooth bark that peels off in layers. Indians used this bark for their famed birch-bark canoes, and pioneers used it as roofing material and writing paper. Bark should never be forcibly peeled from a birch tree, as this results in permanent bark loss in that area of the tree. The pure white wood of the paper birch is valuable for small wood products. Alders are also members of the birch family.

## Beech family

The two most important nut trees in the beech family are the *American beech* and the *American chestnut;* however, the chestnut is rarely seen growing in the wild. In the 1890s a disease called chestnut blight, probably imported from Japan, destroyed nearly all of the wild chestnut trees in the United States. The species is not extinct, however; chestnut trees thrive under the care of horticulturists, who grow them to produce chestnuts for you to buy at the market. *Oaks* (p. 16) also belong to the beech family.

## Elms

Among the more than 15 known varieties of elm, the *American elm* is by far the most familiar because of its popularity as a shade tree and street border. It is known for its exceedingly graceful and symmetrical shape. Many elms have lived as long as 200 years.

**Black walnut**
*Juglans nigra*
Deciduous
Height: 60–90 ft

**Shagbark hickory**
*Carya ovata*
Deciduous
Height: 70–80 ft

**Pecan hickory**
*Carya illinoinensis*
Deciduous
Height: 100–180 ft

## Walnut family

The walnut family includes the *walnuts, pecans,* and *hickories,* of which there are many specific varieties. (Pecans are actually a type of hickory.) All members of the walnut family have compound leaves that are very similar in shape. The *black walnut* is valuable not only for its nuts but also for its high-quality lumber.

**Northern red oak**
*Quercus rubra*
Deciduous
Height: 60–80 ft

**Live oak**
*Quercus virginiana*
Evergreen
Height: 40–50 ft

**California white oak**
*Quercus lobata*
Deciduous
Height: 30–90 ft

## Oaks

Oaks, one of our most familiar and useful groups of trees, are known for their wood and for the acorns that they produce. About 60 varieties of oaks grow in the United States. The beautiful grain of oak wood produces fine furniture, and the strength of the wood makes it useful in the construction industry. Oaks are members of the beech family (p. 15).

**Weeping willow**
*Salix babylonica*
Deciduous
Height: 30–70 ft

**Quaking aspen**
*Populus tremuloides*
Deciduous
Height: 20–60 ft

**Black willow**
*Salix nigra*
Deciduous
Height: 30–40 ft

## Willows

Willows are distinguished by their long, narrow, graceful leaves with finely toothed margins. Willows often grow along rivers or in other damp areas. The young shoots of the shrublike *basket willow* are used in the making of wicker furniture and baskets. The twigs and furry catkins of the *pussy willow* are used as decorations. *Poplars* and *aspens* are also members of the willow family (although the yellow poplar is not). The *quaking aspen* gets its name from the trembling motion of its leaves.

## Sycamores

The *sycamore* is a majestic shade tree that can be recognized by its broad leaves with heavily toothed margins. The trunk has a mottled appearance that occurs as the dark brown outer bark flakes off to expose patches of white or yellowish-green inner bark. The sycamore, America's largest deciduous tree, is sometimes called the *button tree* because the fruits, which often hang on the tree throughout the winter, resemble one-inch balls.

## Ashes

Sixteen varieties of ash grow in the United States. Of these, *white ash* and *red ash* are the most common. All varieties of ash are recognized by their compound leaves, each normally consisting of six to nine leaflets. Ashes are members of the olive family.

## Hollies

Holly trees are usually associated with Christmas because of the widespread use of their shiny leaves and brilliant red fruits to form wreaths and other decorations. Holly trees are differentiated into male and female. The fruits are found only on female trees. The two most common varieties are *English holly* and *American holly.*

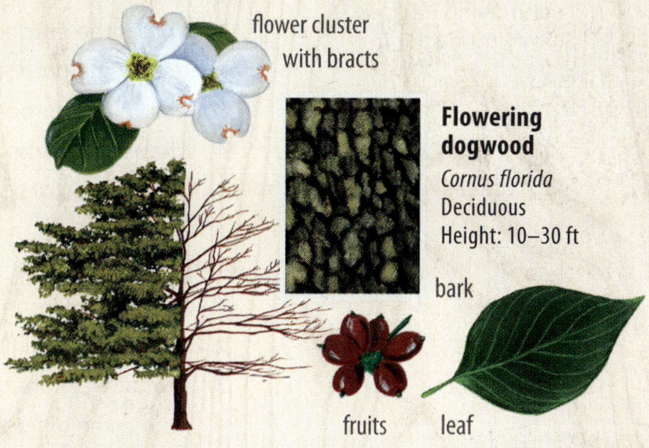


## Dogwoods

The *flowering dogwood* is widely used as an ornamental tree throughout the eastern half of the country. The *Pacific dogwood* is common on the West coast. The flowers of the dogwood are small, buttonlike clusters with tiny petals. The beautiful creamy white structures that surround the flowers like petals are actually modified leaves called **bracts.**

## Section Review 1.4

1. Why are broadleaf trees referred to as hardwoods?
2. What is the difference between a deciduous tree and an evergreen tree?
3. What group of trees is known for thin, smooth bark that peels off in layers?
4. Name two trees that produce flowers.
5. Which group of trees grows in damp areas?
6. What are two types of trees, in addition to willows, that are members of the willow family?

**Identify**

deciduous, evergreen, crown, catkin, samara, bract

**Observe**

- Begin a collection of tree leaves. Use a dichotomous tree key to identify the leaves in your region by scientific name and common name.

## Chapter 1 Review

**Define**

1. biology
2. organism
3. habitat
4. botany
5. zoology
6. woody plant
7. herbaceous plant
8. angiosperm
9. disk flower, ray flower
10. legume
11. nitrogen-fixing bacteria, denitrifying bacteria, nitrifying bacteria
12. dicot, monocot
13. cereals
14. forage grasses, turf grasses
15. stolon
16. sheath
17. hardwood
18. deciduous
19. catkin, bract

**Identify**

1. the study of living things
2. the study of plants
3. the three broad divisions of biology
4. the importance of plants
5. how land plants may be grouped according to the size and type of the stem
6. the two main kinds of root systems
7. the difference between annuals, biennials, and perennials
8. three major families of angiosperms and two members of each
9. the three most important cereal crops
10. the tree family that includes oaks and chestnuts

**Explain**

1. What is the purpose of flowers?
2. What is the purpose of leaves?
3. What is the purpose of stems?
4. What is the purpose of roots?
5. Which group of North American trees are known for distinctive fruits called samaras?

**Apply**

1. Why is the nitrogen cycle important?
2. What are four differences between monocots and dicots?
3. In what ways are grasses important to man?
4. What is the difference between deciduous trees and evergreen trees?

# The Structure and Function of Leaves

CHAPTER 2

**Key Concepts**

- Systems, organs, tissues, and cells in plants
- External structure of leaves
- Arrangement of leaves on stems
- Types of plant tissues
- Internal structure of leaves
- Structure of plant cells
- Photosynthesis
- Fall coloration of leaves
- Special leaves
- Transpiration and wilting

## 2.1 The External Structure of a Leaf

### Systems and Organs

Plants, like all living things, were created for a purpose and were designed according to the plan that would best fulfill that purpose. To perform their God-given purpose, plants are equipped with specially designed *systems,* which are composed of specialized *organs* made up of specialized *tissues* formed by specialized *cells.* This organizational design is found in almost all *multicellular organisms,* living things composed of more than one cell.

**Root system and shoot system.** A **system** is a group of structures designed to function together as a unit to perform a particular job for an organism. Your own body is provided with about eleven systems, including muscular, skeletal, circulatory, respiratory, and others. A typical flowering plant has just two systems: the root system and the shoot system. The **root system** is the part of the plant that is ordinarily underground, anchoring the plant, and that functions in absorption of nutrients and storage of food. The **shoot system** is the part of the plant that is usually found above ground, holding the leaves toward the sun for the manufacturing of food and providing for the production of flowers, fruits, and seeds. Leaves are part of the plant's shoot system.

**Root, stem, leaf, and flower.** Each root, stem, leaf, and flower of a plant may be considered an **organ,** a structure within a system that has a definite form and performs a definite function or functions for the system. For example, roots support and anchor plants, store food, and absorb water and minerals from the soil. Stems support leaves and conduct materials to all parts of plants. Leaves make food for plants. Flowers allow plants to reproduce.

Fig. 2.1 **Raffia palm and duckweed**

## Variety among Leaves

As organs, most leaves have the same basic function of producing food, but not all leaves have the same appearance. Leaves come in a tremendous variety of shapes, sizes, and textures, testifying to the imaginative powers of the Creator. *As you observe and collect plants, you will find three basic leaf shapes.* Some leaves are *broad and flat,* such as those of a rosebush, a dandelion, or a maple tree. Others are *long and narrow,* such as grasses, lilies, onions, and palms. Yet others are *needle-like or scalelike,* like those found on pines, firs, cedars, and other cone-bearing trees and shrubs.

You will also discover a variety of sizes among the leaves you observe. The giant taro, or elephant ear, has leaves as large as 6 feet long and 4 feet wide. Some palm trees have enormous fan-shaped leaves; the leaves of the raffia palm can grow to be well over 50 feet long. Other plants have small leaves, like duckweed, an aquatic plant whose tiny round leaves are seen on ponds.

Leaves also vary in texture and in other ways. They may be smooth and glossy, like the leaves of a magnolia tree, or hairy, like those of an African violet. Some are paper-thin, like the leaves of a maple tree; and others are relatively thick, like the leaves of the aloe vera plant. Leaves may also be leathery, fleshy, prickly, or as soft as velvet.

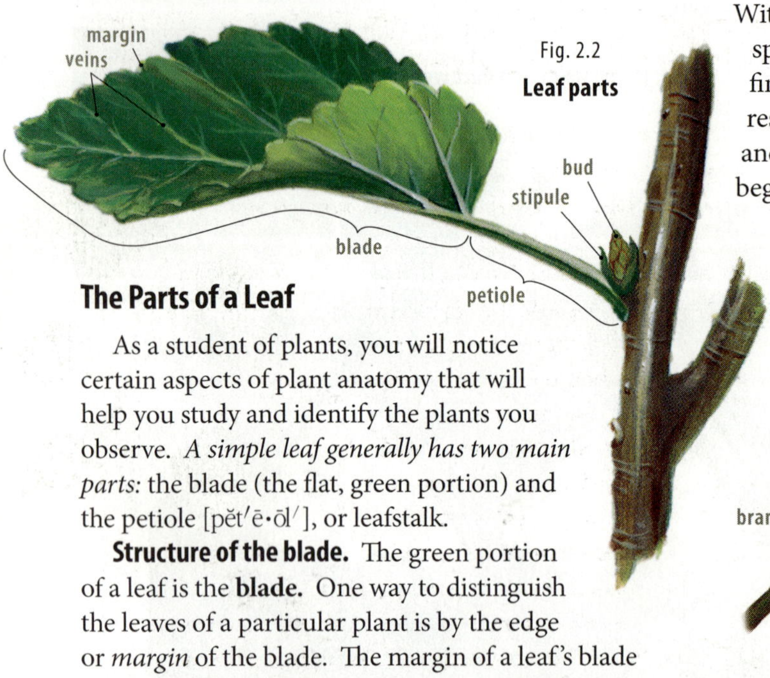


Fig. 2.2
**Leaf parts**

## The Parts of a Leaf

As a student of plants, you will notice certain aspects of plant anatomy that will help you study and identify the plants you observe. *A simple leaf generally has two main parts:* the blade (the flat, green portion) and the petiole [pĕt′ē·ōl′], or leafstalk.

**Structure of the blade.** The green portion of a leaf is the **blade.** One way to distinguish the leaves of a particular plant is by the edge or *margin* of the blade. The margin of a leaf's blade may be entire (smooth), toothed, or lobed. *Veins* are the pipelines that carry food and water (sap) throughout the blade; many leaves have a large vein called the *midrib* running down the center of the blade. Besides transporting food and water, the stiff, tough veins form a skeleton for the blade. This helps the leaf keep its form and prevents it from collapsing or tearing easily.

**Structure of the petiole.** The **petiole** is the second major part of a typical leaf. It attaches the blade to the stem and contains vascular tissues that conduct substances between the blade and the stem. The leaves of some plants do not have petioles. Leaves that lack a petiole are referred to as **sessile** [sĕs′ĭl′] leaves. Grasses and certain other monocots have sessile leaves that attach to the stem by a *sheath* that seems to wrap around the stem.

**Between petiole and branch.** If you examine the point where the petiole joins the branch, you may find stipules and buds. *Stipules* are small leaflike or scale-like structures that helped to cover the leaf when it was developing. The stipules of a rose look like pairs of tiny leaves attached to the base of each petiole. Some plants, such as the black locust, have spinelike stipules. In most plants, however, the stipules are so small that you must look very closely to see them at all.

*Buds* contain developing leaf or stem structures. In the cold winter months, the buds remain dormant and are covered by small appendages called bud scales. With the warm sunshine in springtime, new leaves spring forth from the buds. In certain trees, you can find *leaf scars* and *bud-scale scars* on the stems; these respectively show where leaves had previously grown and where one season's growth ended and the next began. (Fig. 2.3)

Fig. 2.3
**Stipules, buds, and scars in twigs**

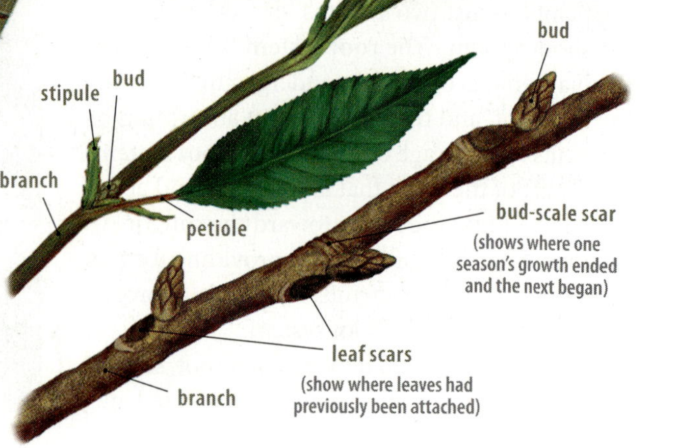

## Simple and Compound Leaves

Leaves may be classified as either simple or compound. **Simple leaves** have one blade on every petiole. **Compound leaves** have more than one blade on every petiole. Each small blade on a compound leaf is referred to as a **leaflet.** The petiole can be identified by the presence of a bud at the end where it attaches to the branch. The other end of the petiole will be attached to one or more leaflets. There are no buds at the point of attachment of each leaflet. If each petiole is attached to one blade, the leaves are simple. If each petiole is attached to several leaflets, the leaves are compound.

Fig. 2.4
**Simple and compound leaves**

## *The Arrangement of Leaves on Stems*

Understanding the LIVING CREATION

The arrangement of leaves on stems varies a great deal, but it is never haphazard. The leaves are placed with mathematical precision at definite locations along the branch. *This precise positioning ensures that all leaves on a stem will have maximum exposure to the life-sustaining sunlight.* The points at which leaves grow from the stem are called **nodes.** The various arrangements of leaves at the nodes have been given names that describe the arrangement. When two leaves grow from each node, the arrangement is referred to as **opposite.** The leaves of opposite-leaved plants are oriented so that they are attached to the stem at right angles to the leaves directly above and below them. Maples, coleus, and a number of members of the mint family have an opposite leaf arrangement. In a great number of plants, only one leaf grows from each node in an alternating pattern up the branch. Such an arrangement is referred to as **alternate.** Trees such as apple, oak, cherry, elm, birch, beech, and poplar have an alternate leaf arrangement. Many of the grasses, including corn and

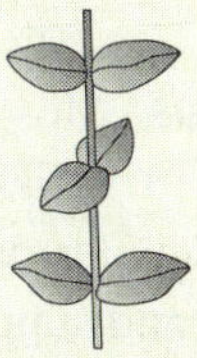

red maple

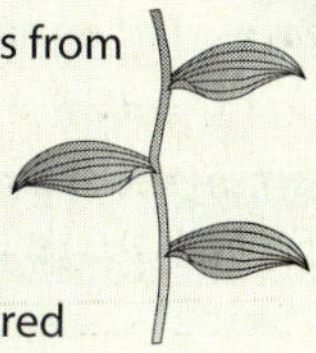

basswood

(cont.)

*(cont. from previous page)*

wheat, also have alternate leaf arrangements. All leaves have maximum exposure to the sunlight in this arrangement because no leaf is directly above another. When there are three or more leaves growing from each node, the arrangement is called **whorled** [wôrld]. Several members of the lily family have a whorled leaf arrangement.

loosestrife

Plants such as the dandelion that have a cluster of leaves growing around the base of the plant are said to have a **rosette** arrangement (also called a *basal* arrangement). This arrangement is given the name rosette because the leaves are situated like the petals of a rose.

adobe lily

When we observe the leaf arrangement of plants, we usually think only of the arrangement of simple leaves, but compound leaves also have a definite arrangement, which usually is either opposite or alternate.

Regardless of the pattern of arrangement, individual leaves orient themselves so that each blade is at approximately a 90° angle to the source of light. The petiole can twist and turn to maintain this orientation to the sun and thus expose the blades of each leaf to maximum illumination. The twisting of the petiole occurs because the cells on the side of the petiole away from the light are stimulated to grow more quickly than the cells on the lighted side of the petiole. More rapid growth on one side of the petiole than the other bends the petiole, thereby orienting the leaves to face the sun. This movement of the petiole is an example of **phototropism**, the growth response of a plant stimulated by light.

## Section Review 2.1

1. List the three basic leaf shapes. In what other ways do leaves vary from plant to plant?
2. Distinguish between opposite, alternate, whorled, and rosette leaf arrangements.
3. Explain how leaves orient themselves so that each blade is at about a 90° angle to the source of light.

**Application**

4. In a paragraph, write a complete explanation of the method of distinguishing between simple and compound leaves. Define all terms that you use.

**Identify**

system, root system, shoot system, organ, blade, petiole, margin, vein, midrib, sessile leaf, leaflet, node, phototropism

**Observe**

- Try to classify the leaves in a herbarium or outside garden as simple or compound.
- Find leaves that exhibit the opposite, alternate, whorled, and rosette arrangements.

## 2.2 The Internal Structure of a Leaf

### Three Kinds of Tissue

Living material that is constructed in such a way as to perform a particular task for the organs of an organism is called **tissue.** *A typical plant has three distinct kinds of tissue*: structural, vascular, and meristematic [mĕr′ĭ·stə·măt′ĭk].

**Structural tissue.** Most of the body, or structure, of the plant is **structural tissue.** *There are four types of structural tissue:* epidermal tissue, cork, parenchyma [pə·rĕng′kə·mə], and support and strengthening tissue. **Epidermal tissue** protects and covers leaves, roots, stems, and other exposed areas and prevents excessive

## Table 2.1 Important Plant Tissues

### Structural Tissue — forms the body, or structure, of the plant

| | Function | Structure | Location |
|---|---|---|---|
| **Epidermal tissue** | protects and covers plant parts | thin layer of cells normally covered by a cuticle (waxy layer) | outside covering of all plant parts except the stems and certain roots of woody plants |
| **Cork** | protects and waterproofs plant parts | rectangular cells that die and fill with a fatty substance soon after they form | forms the outer bark that covers the stems and certain roots of woody plants |
| **Parenchyma** | makes and stores food | cells of various shapes containing large vacuoles and often containing chloroplasts | roots, stems, leaves, and fruits |
| **Support and strengthing tissue** | provides support for leaves, stems, and roots | fibers composed of thick-walled cells | roots, stems, and leaves (veins and petioles) |

### Vascular Tissue — conducts materials throughout the plant

| | Function | Structure | Location |
|---|---|---|---|
| **Xylem** | transports water and dissolved minerals upward | complex tissue composed primarily of elongated cells called tracheids and long tubes called vessels | roots, stems, and leaves (veins and petioles) |
| **Phloem** | transports food downward | complex tissue composed primarily of elongated cells called sieve tubes | roots, stems, and leaves (veins and petioles) |

### Meristematic Tissue — produces new tissues of all kinds for growth and repair

| | Function | Structure | Location |
|---|---|---|---|
| **Apical meristems** | longitudinal plant growth | thin-walled, rapidly dividing cells | root and stem tips |
| **Vascular cambium** | makes new xylem and phloem tissue | thin-walled, rapidly dividing cells | layer between the xylem and phloem in roots and stems |
| **Cork cambium** | makes new cork tissue | thin-walled, rapidly dividing cells | layer just underneath the outer bark in woody plants |

water loss and injury by outside agents. Epidermal tissue of the shoot system is normally covered by a **cuticle** layer, which consists of a waxy substance that helps to seal in water. The epidermal tissue contains specialized cells like the *root hairs* of young roots, which absorb water and nutrients for the root. Generally, epidermal tissue is very thin; many times it is only one cell layer thick.

In woody plants, the epidermal tissue of the stem is split open and eventually lost as the young stem begins to grow in diameter. A layer of **cork tissue** grows in its place and forms a layer of outer bark. Cork tissue protects and waterproofs the stems and certain roots of woody plants. It is not found in herbaceous plants.

**Parenchyma,** found in roots, stems, leaves, and fruits, makes and stores food. **Support and strengthening tissue** consists of fibers that run through roots, stems, and large veins of leaves. The cells that form support and strengthening tissues live only a short time, but the thick walls of these cells remain as support structures long after the cell itself dies.

**Vascular tissue. Vascular tissues,** composed of elongated, tubular cells, are the sap-conducting tissues. They can be compared to your own circulatory system.

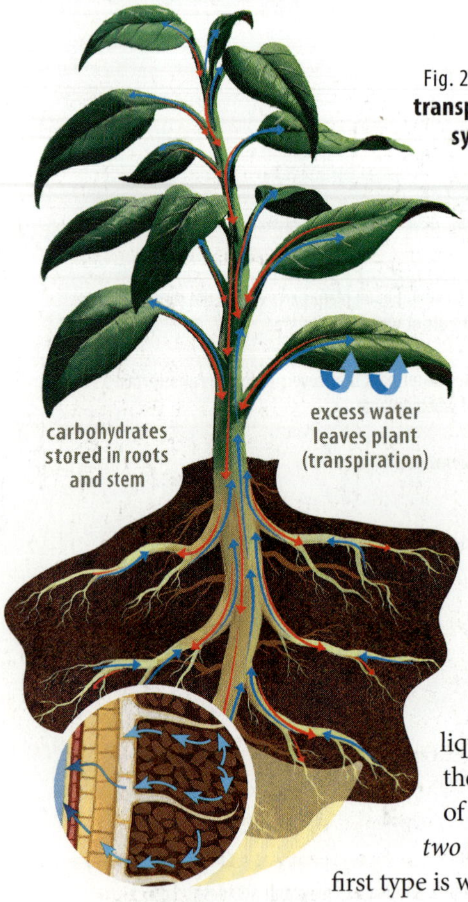


Fig. 2.5 **Plant transportation system**

Vascular tissue is found in the stems and roots and in the veins and petioles of leaves.

**Sap** is the sweet liquid found within the vascular tissue of plants. *There are two types of sap.* The first type is water with dissolved minerals that moves upward from the roots to the leaves. The other type of sap consists of food materials that have been made in the leaves and dissolved in a small amount of water to be transported downward to the stem and roots.

The two types of vascular tissue are xylem [zī′ləm] and phloem. **Xylem** transports water and dissolved minerals upward from the roots to the leaves, and **phloem** transports food manufactured in the leaves downward.

**Meristematic tissue.** The purpose of **meristematic tissue** is the growth and repair of plants. Meristematic tissue consists of rapidly growing and dividing cells that differentiate to produce new tissues during the plant's growing season or when the plant needs to repair or replace worn-out tissue. *When these new tissues mature, they form the various types of structural and vascular tissue that compose the plant's roots, stems, and leaves.* Meristematic tissue is not spread evenly throughout the plant but is present in definite growth regions, including root tips, stem tips (including buds), and the internal **cambium** layers. Meristematic tissue in root and stem tips is external and enables roots and stems to grow in length *(longitudinal growth),* whereas **cambium tissue** is internal and enables stems and roots to grow in diameter *(lateral growth).* Cambium tissue is one of the most important types of meristematic tissue. **Vascular cambium** produces new vascular tissue (xylem and phloem), and **cork cambium** produces new cork tissue.

## Leaf Structures

**The epidermis.** The **epidermis** provides covering and protection for both the upper and lower leaf surfaces. The waxy *cuticle* helps the epidermis to prevent water loss from the leaf.

The lower epidermis of leaves contains tiny openings called **stomata** [stō′mə·tə; sing. stoma] or *leaf pores;* many plants also have them in the upper epidermis. Stomata may be very abundant. An apple tree leaf may have 47,000 stomata per square inch of leaf surface, and an oak leaf may have over 100,000 of these tiny openings per square inch. The main purpose of stomata is to allow air to move into and out of the leaves. Oxygen is needed for respiration, and carbon dioxide is needed for photosynthesis. Stomata also regulate **transpiration,** the loss of water vapor from leaves and stems.

Stomata are not merely fixed openings that cannot be controlled; they are like doors that can be opened and shut. Two crescent-shaped **guard cells** expand and contract to open and close each stoma.

The epidermis of certain plants has hairlike structures called **leaf hairs** covering it. These hairs give the leaves of such plants a somewhat woolly appearance. Leaf hairs have a variety of functions. They are useful, for instance, in retarding the loss of water by reducing the speed of the wind that passes directly over the stomata. They are also useful as defense against insects and in many cases they are involved in the secretion of toxic substances.

**The mesophyll.** The middle portion of the leaf, which forms the bulk of the leaf, is called the **mesophyll** [mĕz′ə·fĭl′]. *It is in the mesophyll that most of the photosynthesis takes place.* Parenchyma is the most abundant type of tissue in the mesophyll.

In a typical dicot leaf, the mesophyll is divided into layers. The *palisade mesophyll* is located toward the upper side of the leaf and consists of elongated, column-like cells, which resemble a palisade, or row of fence posts. There may be several layers of palisade mesophyll. Below the palisade layer is the *spongy mesophyll.* This layer consists of large, irregularly shaped cells separated by large air spaces. The leaves of some dicots and most monocots contain only spongy mesophyll.

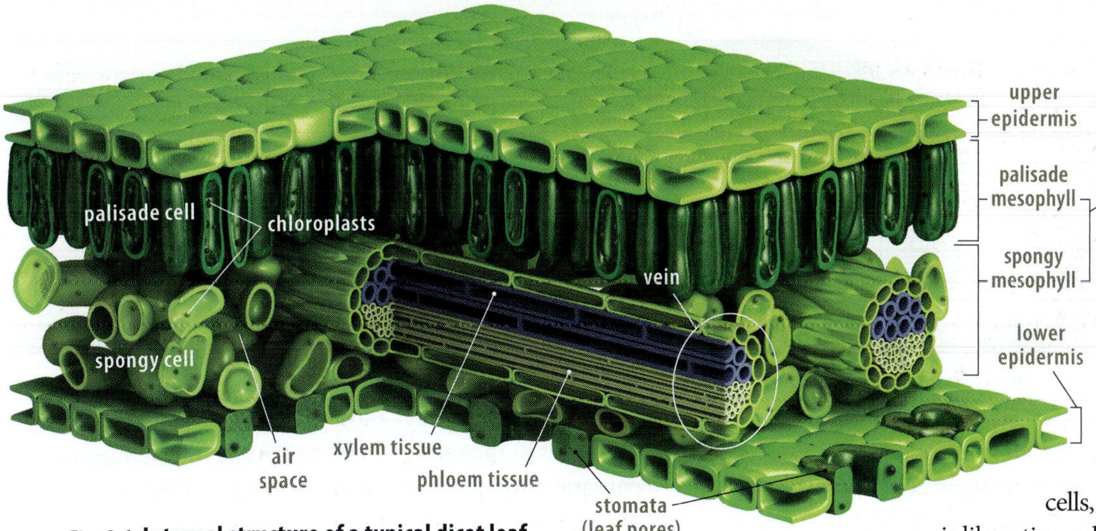


Fig. 2.6 **Internal structure of a typical dicot leaf**

### Inside a Plant Cell

The basic structural unit of all living things is the **cell.** A single mature leaf of a pear tree contains about 50 million cells, every one of which is like a tiny, self-contained city with its own power plants, chemical factories, food warehouses, waste disposal facilities, and transportation networks, all directed automatically by programs in the cell's control center. The complexity of each living cell and the tremendous variety among cells even in the same organism are two of the greatest wonders of creation.

The air spaces between the spongy mesophyll cells are called *intercellular air spaces.* There are also many smaller air spaces between the palisade cells. These air spaces form a system of passages throughout the leaf that permits air to come in contact with individual cells so that each cell can obtain needed oxygen and carbon dioxide.

**Veins.** The **veins** of the leaf run through the mesophyll layers and contain the vascular tissue. Each vein contains xylem that brings water and mineral-containing sap into the leaf and phloem that transports sap consisting of manufactured food out of the leaf. The larger veins enter the blade at its base from the petiole and connect the blade of the leaf with the stem of the plant. Smaller and smaller veins divide and subdivide in a network pattern until strands extend to all parts of the blade.

**Parts of the cell.** Biologists recognize three main parts of every plant and animal cell. The **cell membrane** surrounds the cell, separates it from its environment, and regulates what enters and leaves the cell. The **nucleus**, a spherical body often located near the center of the cell, controls the cell's activities. The **cytoplasm** serves as the fluid medium for the many molecules and *organelles* ("little organs") that make up the cell. In addition to these parts, plant cells also have a **cell wall.** The cell wall is found around the outside of the cell membrane.

## Venation: The Patterns of Leaf Veins

Understanding the LIVING CREATION

Veins are arranged in a definite pattern within a leaf. The pattern of the veins within leaves is called **venation** [vē·nā′shən]. Different plants have different venation patterns.

There are three main types of venation—parallel, pinnate, and palmate. In **parallel venation**, which is characteristic of monocots, the veins are usually parallel to each other along the length of the leaf.

arrowhead

In **pinnate venation**, the veins are in a branching pattern, characterized by one major vein (a **midrib**) with smaller veins extending outward from it. The word *pinnate* means "resembling a feather," and this is an excellent description of pinnate venation.

oak

**Palmate venation** is characterized by two or more major veins extending outward from one point like the fingers extending from the palm of a hand.

silver maple

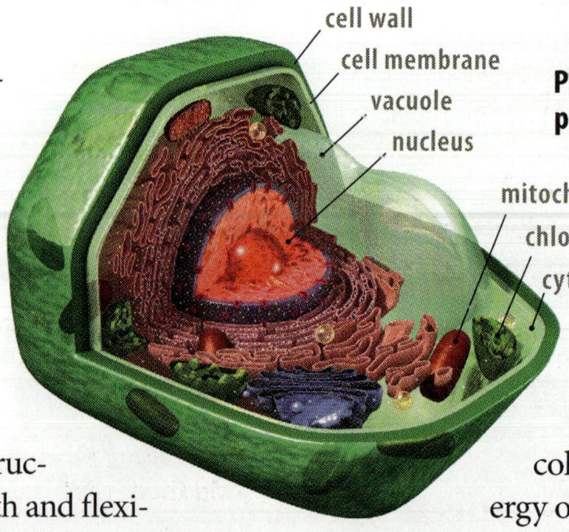


Fig. 2.7
**Parts of a plant cell**

**The cell wall.** Have you ever wondered how the stems and leaves of plants are able to stand erect with no bones or skeleton for support? The answer lies primarily in the structure of plant cell walls. The cell wall is made of **cellulose,** a complex carbohydrate made up of long chains of glucose (a simple sugar) molecules. This long chain structure gives cellulose fibers great strength and flexibility and provides the cell with support and protection.

In addition to cellulose, some plant cell walls also contain *lignin*, a substance that adds stiffening to the walls of the woody cells in trees and shrubs. Herbaceous plants do not have lignin in their cell walls but depend much more on the pressure of the fluid inside their cells to help keep their cells rigid. When a lack of water causes a drop in this internal pressure, herbaceous plants droop and wilt as their cells become limp.

Not only is cellulose important in supporting the plant, but it is also important to animals and man. Man uses cellulose in the production of many manufactured substances, including paper, clothing, foods, and industrial chemicals. Cellulose is also present in all plant foods, such as vegetables and fruits. Although man's digestive system is not equipped to break down cellulose, this complex carbohydrate plays a vital role as bulk and roughage to keep food substances moving through the digestive system. The digestive systems of cows and other grazing animals are equipped to digest cellulose. When we eat beef, we are taking advantage of the nutrients that the cow obtained from cellulose.

**Chloroplasts.** Within the cells of all green plants are organelles called **chloroplasts** (Fig. 2.8), which are tiny chemical "factories" that use light to manufacture food. Chloroplasts contain **chlorophyll,** the green pigment that gives plants their color and enables them to capture the energy of light. Chloroplasts are located in the cells of the mesophyll and are most numerous in the palisade mesophyll, where most of the photosynthesis occurs. Chloroplasts are also found in the guard cells and often in the stem and in unripe green fruits.

**Vacuoles.** Because plants produce large amounts of food and other substances, their cells must be equipped for storage. The storage structures in cells, called **vacuoles** [văk′yo͞o·ōlz′], store food materials, fluid substances, and minerals. Each plant cell usually has only one or two large vacuoles. Animal cells also have vacuoles; but because animals have less need for storage, their cells usually have several small vacuoles.

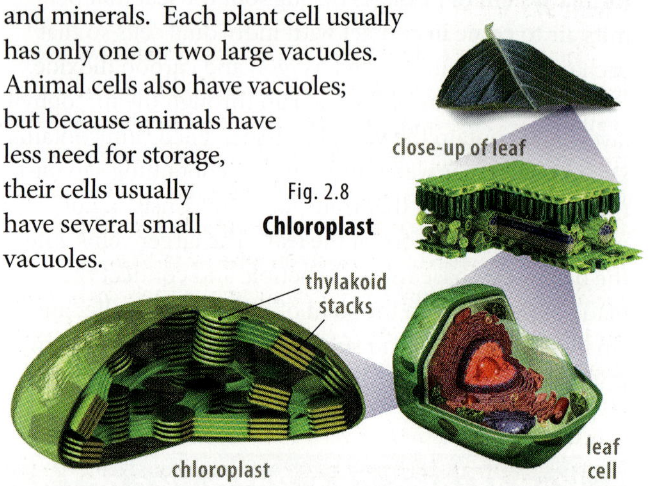


Fig. 2.8
**Chloroplast**

## *Plants and Food Storage*

Design in the LIVING CREATION

**P**lants are wonderful storehouses. They store sugars in the form of starch or fat. They also store vitamins and proteins. Plant cells are especially well equipped for storage because they are supplied with large vacuoles.

Man takes advantage of this food storage, and he harvests many of these stored foods for his own use. When you eat potatoes, you are taking advantage of starch that the potato plant stored in special underground stems. Storage roots in carrots, beets, radishes, and other plants also supply us with food. Of course, you can probably think of several fruits, seeds, and leaves that are used for food by man. The food storage ability of plants is another evidence of the loving provision of God for His creation.

## Section Review 2.2

1. What is the function of xylem and phloem?
2. What kind of tissue forms the majority of a leaf?
3. Broadly speaking, what are the three main parts of every cell?
4. Describe parallel, pinnate, and palmate venation.
5. What is the purpose of cell membranes and cell walls?

### Application

6. Describe the structure and function of each type of structural, vascular, and meristematic tissue in a plant.
7. How does the similar composition of the cells of plants, animals, and people show the great forethought and provident care of the Creator?

### Critical Thinking

8. Explain why a cactus in the desert might open its stomata at night and keep them closed during the day.

### Identify

tissue, cuticle, cambium tissue, epidermis, transpiration, midrib, cellulose, chloroplast, chlorophyll, vacuole

### Observe

- Collect three leaves to illustrate the three kinds of venation. Add them to the classroom herbarium.
- Use a hand lens to examine the external structure of a leaf, and identify the parts. Try to find a leaf that has an obvious cuticle layer.

# 2.3 Photosynthesis and the Leaf

## Photosynthesis

The intricate machinery of chloroplasts enables the leaves of a plant to transform radiant (light) energy, usually from the sun, into the chemical energy of food molecules. Plants are the first link in a very complex chain that ultimately supplies all living things with their food. Organisms, such as green plants, that can make their own food are called **autotrophs** [ô′tə·trŏfs′] or *producers*. Organisms that cannot make their own food and must obtain it from other organisms are called **heterotrophs** [hĕt′ər·ə·trŏfs′] or *consumers*. You will learn more about the relationships between producers and consumers in chapter 15.

The process whereby a plant's chloroplasts capture the radiant energy of light and convert it into the chemical energy of food is called **photosynthesis** ("putting together with light"). This amazing process is one of the most important chemical manufacturing processes on earth, for it produces a tremendous amount of food. It is estimated that green plants anually produce about 300 billion tons of food in the form of sugars and starch.

Photosynthesis occurs within the chloroplasts, which contain stacks of specialized disks called **thylakoids.** Chlorophyll molecules located on the surface of the thylakoid stacks absorb light energy to produce food in the form of sugars. Photosynthesis is a very complicated process, involving dozens of chemical reactions in the chloroplasts.

Traditionally, the entire process of photosynthesis has been divided into two phases: one that requires light directly (light reactions) and one that does not (dark reactions). You should realize that both phases occur together and that once the light is removed all reactions, light and dark, cease. The overall result of this process can be summarized by the chemical equation in Figure 2.9.

**Light reactions.** Carbon dioxide and water are needed for photosynthesis. The plant obtains water ($H_2O$) from the soil. The water molecules pass into the root cells,

Fig. 2.9 **Chemical equation of photosynthesis**

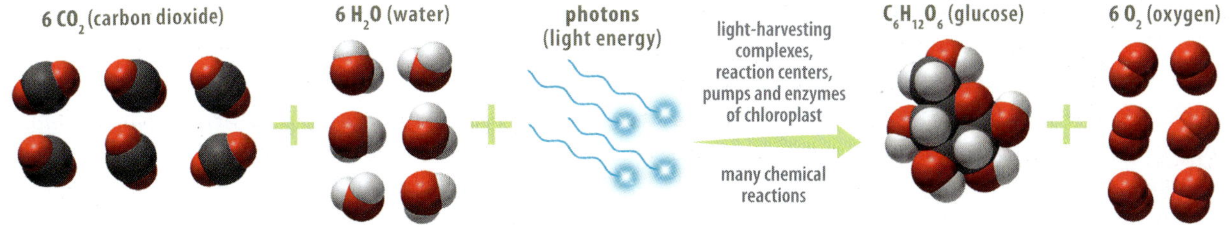

Fig. 2.10 **Photosynthesis**

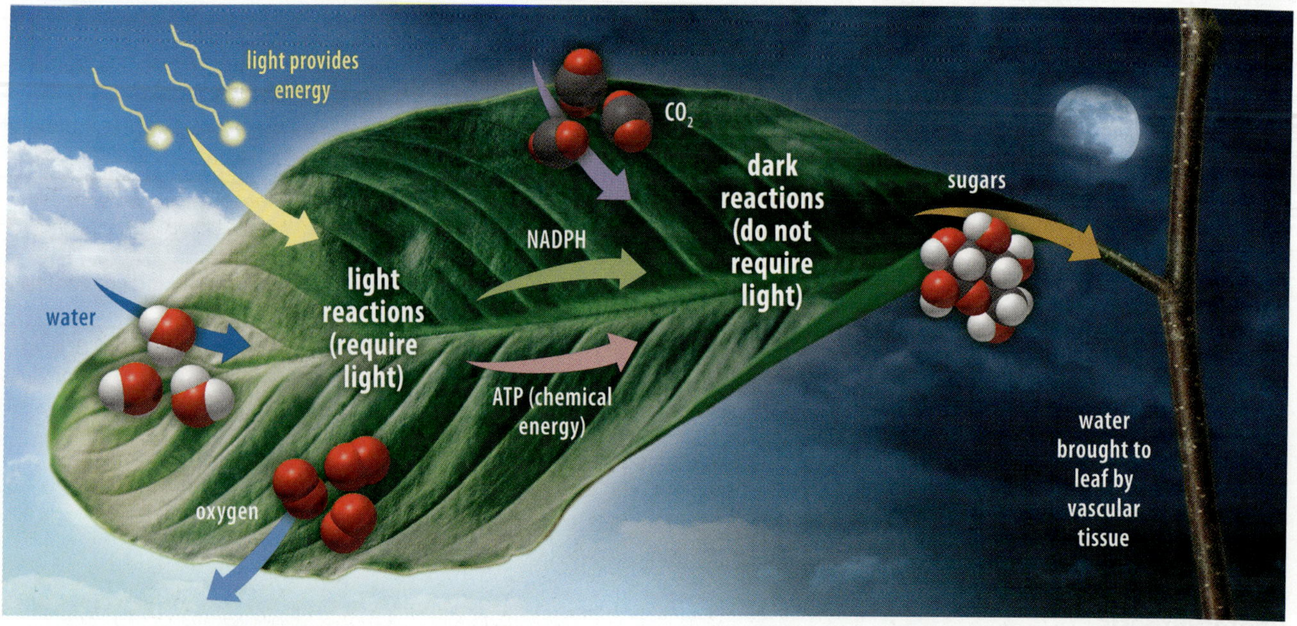


up through the vascular tissue in the stem, and into the leaves. The carbon dioxide ($CO_2$) is obtained from the air. Air passes through the stomata and into the air spaces of the spongy mesophyll cells. In the first step of photosynthesis, the plant uses the energy of light to split the water into hydrogen ions and oxygen; the oxygen is released into the air as a product, while the hydrogen is used to make two special, high-energy compounds called *ATP* and *NADPH*.

**Dark reactions.** Through a subsequent series of steps, which do not require light directly, carbon dioxide molecules from the air are combined to produce **glucose** ($C_6H_{12}O_6$), a simple sugar. The energy required to synthesize glucose is supplied by breaking down the ATP and NADPH produced in the light reactions. Thus the green plant can synthesize sugar from raw materials it acquires from both the soil and the air.

## Products of Photosynthesis

**Energy for the cell.** Plant cells, like all cells, need energy to stay alive. They obtain this needed energy by *oxidizing* sugars, or combining them with oxygen in a chemical reaction roughly comparable to burning. This process, called **cellular respiration,** enables the plant to "burn" sugars as fuel for energy. Both plants and animals perform cellular respiration.

Although all cells of the plant need energy to stay alive, only the cells containing chlorophyll can produce their own glucose through photosynthesis. Thus, much of the glucose produced is converted into **sucrose,** or table sugar, for transport to the rest of the plant. The presence of sucrose in plant sap accounts for the sweet taste of the sap of many plants.

Usually, the plant produces far more food than it needs right away; it stores this food for later use. Most plants store excess sugars by converting them to **starch,** a long, chainlike molecule consisting of thousands of glucose molecules linked together (Fig. 2.11).

**Materials for growth.** Plants use the sugars that they make not only for energy but also as building materials for construction of new cells, allowing the plant to grow. The plant obtains the proteins, fats, vitamins, and cellulose it needs by further modifying the glucose produced in photosynthesis. It obtains other materials it needs from the soil, including minerals and nitrogen (an ingredient of proteins).

## Factors That Influence Photosynthesis

Factors that affect the rate of photosynthesis include the amount of available carbon dioxide, the temperature, the amount of available water, and the amount of available light. Because carbon dioxide is an essential raw material of photosynthesis, *a decrease in the amount of available carbon dioxide will decrease the rate of photosynthesis.* Similarly, an increase in the amount of available carbon dioxide increases the rate of photosynthesis and often makes the process as a whole more efficient. Also, the plant wastes much less water when more carbon dioxide is available.

*In most plants, photosynthesis occurs most efficiently at temperatures between 65°F and 85°F.* Outside this range, photosynthesis slows; if the plant is too hot or too cold, photosynthesis may stop completely.

Water directly affects the rate of photosynthesis only slightly; however, *a shortage of water indirectly slows photosynthesis.* To conserve water, the guard cells close the stomata, shutting off the intake of carbon dioxide; the lack of ventilation can also cause the leaf's internal temperature to rise above the optimum range.

*The single most important factor affecting photosynthesis is the intensity of the light striking the leaf.* Plants vary greatly in the amount of light they require. Some require large amounts of light, but other plants, such as ferns and mosses, thrive in the shade.

Fig. 2.11
**Glucose, sucrose, and starch**

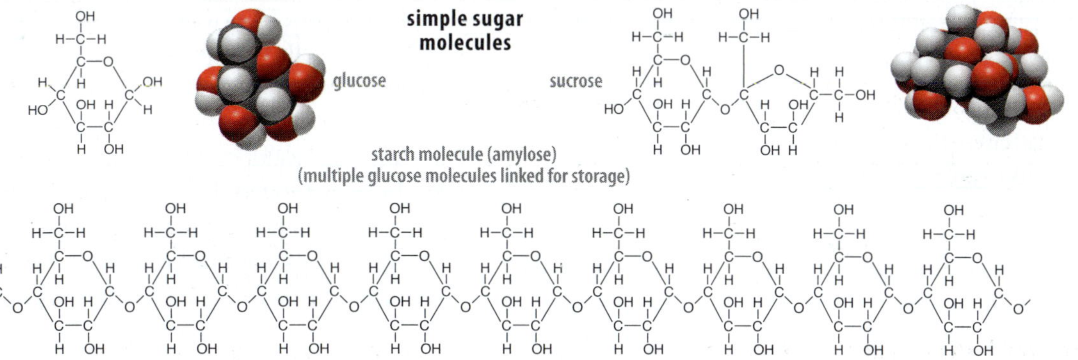


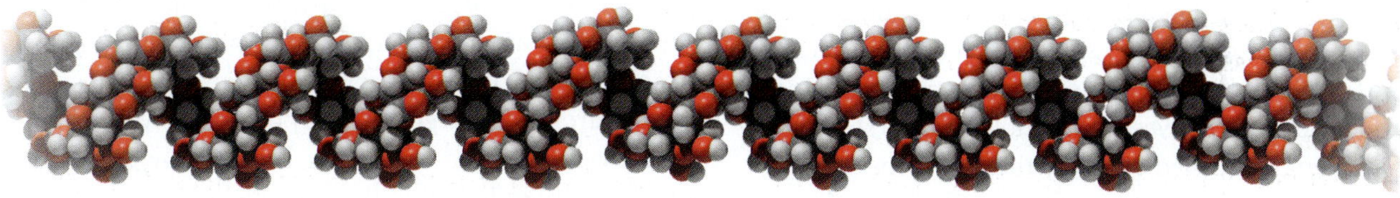

## Section Review 2.3

1. Describe the process of photosynthesis.
2. In what form is glucose usually transported throughout a plant?
3. In what form do plants store excess sugars?
4. What are the factors that affect the rate of photosynthesis? Which is considered the most important?

### Application

5. Why is photosynthesis important to life on earth?
6. Where does the oxygen released during photosynthesis come from?
7. Explain the main reactions that take place during the light and dark phases of photosynthesis.
8. Explain how the action of the guard cells can affect the rate of photosynthesis.

### Critical Thinking

9. Why would you expect to find a great deal of parenchyma tissue in the roots of plants such as the carrot, beet, and radish?

### Identify

autotroph, heterotroph, photosynthesis, glucose, cellular respiration, sucrose, starch

## 2.4 Fall Coloration and Wilting

### Fall Coloration of Leaves

In temperate parts of the world, as the weather turns cooler in the fall, the leaves of many deciduous trees change color. Most of the broadleaf trees that we have studied are deciduous and undergo this color change. If you have ever witnessed this beautiful display, you know that the leaves take on a range of colors, painting the landscape in shades of red, yellow, and orange. Shortly after the color changes occur, the leaves begin to turn brown and fall to the ground.

**Deterioration of chlorophyll.** Leaf coloration and fall is triggered by the changing lengths of the day and night as winter approaches. As the days grow shorter in autumn, the leaves of deciduous trees cease to form new chlorophyll, and the old chlorophyll they contain begins to deteriorate. As this deterioration progresses, other pigments, which are ordinarily hidden by the more abundant chlorophyll, become visible. Some of these pigments and the colors they produce include the following:

**xanthophyll** [zăn′thə·fĭl′]—produces yellowish colors
**carotene**—produces yellowish-orange colors
**anthocyanin** [ăn′thō·sī′ə·nĭn]—produces bright red, blue, and purple colors

**Formation of the abscission layer.** The change in daylight and temperature also causes the leaves to fall from trees in the autumn and winter months. Technically, the leaves do not just fall off; they are actually cut from the stem by an **abscission** [ăb·sĭzh′ən] **layer.** This special layer of cells forms at the base of the petiole, severing the connections between the petiole and the stem. A special enzyme called **cellulase** then weakens the cell walls of the abscission layer, allowing the leaf to break off under its own weight.

Either before or after the leaf falls, a protective layer of cork cells forms at the petiole's base. When the leaf falls, a mark called a *leaf scar* remains on the stem at the point where the petiole was attached. Leaf scars are most visible in winter when all the leaves have fallen. The pattern and arrangement of leaf scars is a characteristic that can be used to identify a plant in winter.

Fig. 2.12
**Fall leaves**

**Colors throughout the year.** In a number of plants, the leaves are multicolored throughout the year rather than just in autumn. These leaves exhibit xanthophyll, carotene, and anthocyanin year round. These pigments are also present in the flowers, fruits, and roots of some plants. For example, carrots contain large amounts of one form of carotene called *beta carotene*, giving their storage roots their characteristic yellowish-orange color. The human body converts beta carotene to vitamin A, an important nutrient. The roots of beets and the fruits of the plum tree get their purple and red colors from anthocyanin.

## Water and Wilting

All the processes that occur in a plant's cells require water, and cells that do not have an adequate supply of water function poorly. Most of the water that is lost by a plant is lost through *transpiration*, the process by which a plant loses water vapor through its leaves and stems. It has been estimated that a single corn plant may transpire more than four liters of water in a day. *Factors affecting the rate of transpiration include the humidity of the air, the temperature, the intensity of the sunlight, the wind speed, the amount of carbon dioxide in the air, and the amount of soil water available.* The guard cells in the leaves are important for controlling the rate of transpiration. The water within the guard cells produces a pressure called **turgor** [tûr′gər] **pressure**. When the guard cells are filled with water, they are **turgid** [tûr′jĭd: swollen, distended], and the stomata are open. When water moves out of the guard cells faster than it can be replaced, turgor pressure is reduced and the inner walls of the guard cells move together to close the stomata. When the stomata close, the rate of transpiration is reduced but not stopped completely.

A high rate of transpiration may produce *wilting*. **Temporary wilting** occurs when transpiration takes place faster than the roots can absorb soil water, causing the plant cells to lose their turgor. When this happens, leaves that have wilted in the daytime will recover during the night, when transpiration is reduced. **Permanent wilting**, resulting in the loss of leaves or even the death of the plant, can occur when there is a drought and there is not sufficient soil water to replace that lost by transpiration.

## Special Leaves

Some leaves have special functions, and thus their structure may be different from ordinary leaves. A **special leaf,** sometimes called a *modified leaf,* is one that has a special design for a special task. These leaves may be of particular value to man. For example, the upper green tubular leaves of the onion die after a season of growth, but the fleshy basal leaves remain wrapped around the short stem, storing food for later use. This is the onion bulb that we eat.

Botanists list several reasons why tendrils, bud scales, spines, fleshy storage leaves, and other special leaves are considered to be leaves. Here are two of those reasons:

- Buds often form at the place where special leaves are attached to the stem, just as they form where ordinary leaves are attached to the stem.
- Even though their external appearance differs, special leaves have the same basic tissue structure as foliage leaves.

Careful study is necessary to determine whether an unusual structure is a special leaf or an outgrowth of other tissue. **Tendrils** are special leaves (or stems) of vines that have the unique function of holding the vine upright by coiling around some support structure. The **spines** of cactuses are special protective leaves that have no chlorophyll.

A striking form of special leaf can be seen in the plants such as the *Venus flytrap,* the *sundew,* the *pitcher plant,* and the *bladderwort.* These unusual leaves can trap and digest small insects. Because of this ability of their special leaves, these plants are called **insectivorous** [ĭn′sĕk·tĭv′ər·əs] **plants.** Insectivorous plants contain chlorophyll to manufacture their own food but can obtain nitrogen (for building proteins) from the bodies of their victims, allowing the plants to grow in nitrogen-poor soil.

**ice plant**
(leaves store water)

**barrel cactus**
(spines protect plant)

**pitcher plant**
(leaves trap insects)

## Section Review 2.4

1. List the names and colors of three common leaf pigments normally hidden by chlorophyll.
2. What are five factors that can affect the rate of transpiration?
3. How are a leaf's stomata opened and closed?
4. Explain how a shortage of water causes a plant to wilt.
5. Name four insectivorous plants.

### Critical Thinking

6. Explain why deciduous trees may lose their leaves in autumn even if temperatures remain fairly warm.
7. Explain why the formation of the abscission layer causes sugar to accumulate in leaves.
8. Explain why it would be detrimental to desert plants to have a large number of stomata on their leaves.

### Identify

abscission layer, turgor pressure, turgid, tendrils, spines, insectivorous plants

### Observe

- Find as many examples of special leaves as you can.
- Throughout the fall season, add a variety of colored leaves to the classroom herbarium.

## Chapter 2 Review

### Define

1. system, organ, tissue
2. root system, shoot system
3. blade, petiole
4. leaflet
5. nodes
6. sap
7. xylem, phloem
8. vascular cambium, cork cambium
9. stomata, guard cells, leaf hairs
10. transpiration
11. mesophyll
12. cell
13. cytoplasm
14. organelles
15. venation
16. midrib
17. chloroplasts, chlorophyll
18. autotroph, heterotroph
19. photosynthesis
20. glucose, sucrose, cellulose, starch
21. cellular respiration
22. abscission layer
23. turgor pressure
24. wilting

### Identify

1. four ways in which leaves may be arranged on a stem
2. the three main types of plant tissue
3. four types of structural tissue
4. two types of vascular tissue
5. one of the most important types of meristematic tissue
6. the parts of a plant cell that separate its contents from the surrounding environment
7. the part of a plant cell that controls the cell's activities
8. three types of venation
9. the structures in a plant cell that carry out photosynthesis

### Explain

1. How are sessile leaves different from ordinary leaves?
2. How are simple leaves different from compound leaves?
3. What is the purpose of cork tissue?
4. What is the purpose of parenchyma tissue?
5. Explain the function of xylem and phloem tissue.
6. Explain the purpose of a leaf's epidermis.
7. State the purpose of vacuoles.
8. Give a simple overview of the process of photosynthesis.
9. What are some factors that can affect the rate of photosynthesis? Which can be considered the most important?
10. What causes leaves to turn different colors in autumn?
11. How do guard cells help conserve water?

### Apply

1. Why is the concept of the "simple cell" not valid?
2. Explain why a leaf would not function correctly without veins.
3. List two examples of special leaves and explain how they show their plants' design for their habitat.

# Flowers, Fruits, and Seeds

CHAPTER 3

**Key Concepts**

- Flower structure
- Types and arrangement of flowers
- Factors affecting flowering
- Pollination
- Fertilization
- Formation, types, and function of fruits
- Structure of seeds
- Germination

## 3.1 The Structure of Flowers

**Flowers,** with the seeds and fruits they produce, make up the reproductive parts of flowering seed plants. Like leaves, flowers vary greatly from plant to plant. They range in size from the almost microscopic flowers of certain aquatic plants; to the small, inconspicuous grass flowers; to the very large rafflesia [ră·flē′zhə] flower of Malaysia, which may be over three feet in diameter. Some plants produce a single flower, and others produce many hundreds of thousands of flowers per plant. Regardless of their size and number, *all flowers have one primary function for the plant: to produce seeds for reproduction.* Seeds are the way that angiosperms and certain other plants bring forth new plants after their kind to provide for life on earth.

Fig. 3.1 **Rafflesia flower**

Although flowers exhibit great variety in appearance, they all are similar in basic structure. The parts of a flower include the sepals, petals, stamens [stā′mənz], and pistils. A flower is usually formed on the enlarged end of a special stem, called the *flower stalk* or **pedicel.** The floral parts (sepals, petals, etc.) are attached to the end of the pedicel, called the **receptacle,** which is designed to hold the developing seeds.

### Flower Parts

**Sepals and petals.** The **sepals,** which are usually green in color, are the leaflike structures attached to the edge of the receptacle. The sepals enclose and protect the flower's **petals,** the most conspicuous part of a flower. Collectively, the sepals are called the flower's **calyx** [kā′lĭks: Latin, "cup"], and the petals are called the **corolla** [kə·rō′lə: Latin, "little crown"]. The petals of some flowers are fused together so that the corolla consists of a single trumpet-like structure. Some flowers have a fused calyx as well. Besides the beauty that they lend to flowers, petals play an important role in pollination by attracting insects with their showy colors and sweet nectar.

**Stamens.** Within the corolla are the **stamens,** the organs that produce the **pollen.** The pollen contains the male reproductive cells, or **sperm cells,** of the flower. Each stamen consists of two parts: a slender, elongated stalk called the **filament** and an enlarged structure at the tip of the filament, the **anther.** The anther produces the pollen. Though pollen grains appear as tiny specks, each type of pollen has a definite shape and surface texture. Each plant species has a unique pollen structure.

Fig. 3.2 **Parts of a flower**

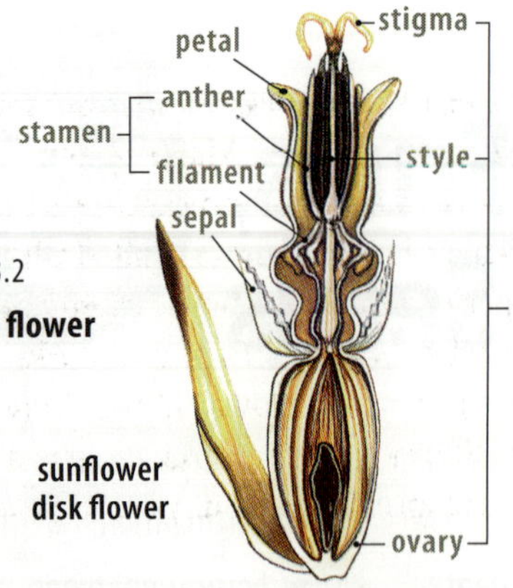


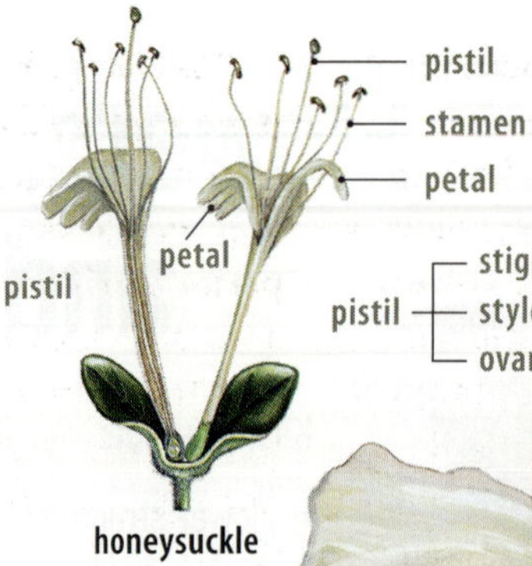


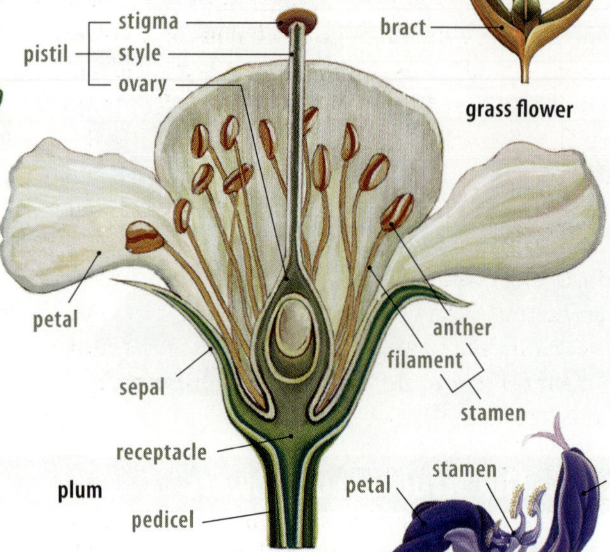


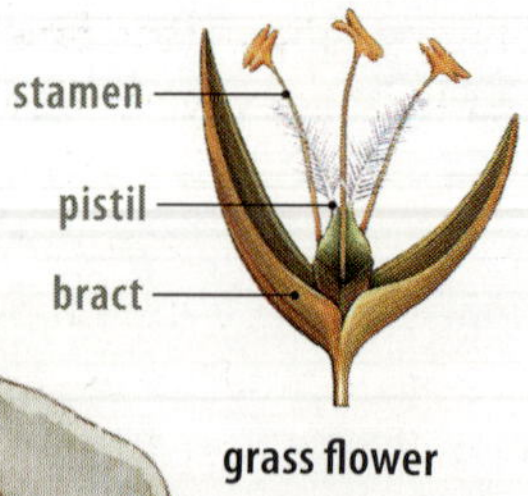


**Pistils.** The elongated, vase-shaped structure of a flower is the **pistil.** It is the central structure in a flower. Although several pistils may be present, most flowers have only one. A typical flower's pistil is divided into three parts: the stigma, the style, and the ovary. The swollen base of a pistil is called the **ovary.** Within the ovary are located **ovules** [ŏv′yo͞olz], or future seeds, which contain the female reproductive cells, called **egg cells.** When the egg is fertilized, the ovules will develop into seeds to produce new plants. The stalklike structure that connects the ovary to the tip of the pistil is the **style.** The tip, or **stigma,** of the pistil is the part of the flower that receives the pollen grains at the time of pollination. The stigma is sticky to aid in pollination.

## Variety in Structure

**Complete and incomplete flowers.** A **complete flower** is one that has sepals, petals, stamens, and pistils; an **incomplete flower** lacks one or more of these four basic flower parts. An incomplete flower that lacks petals is the *poinsettia* [poin·sĕt′ē·ə]. The bright red structures that appear to be its petals are actually special leaves called **bracts.** The actual poinsettia flowers are quite inconspicuous amid the colorful bracts. The flowers of the *dogwood* are also surrounded by bracts, but dogwood flowers are not considered to be incomplete flowers because the bracts do not replace the petals; each flower has its own tiny petals, sepals, stamens, and pistils. All of the plants of the *grass family* produce incomplete flowers that have bracts instead of petals and sepals.

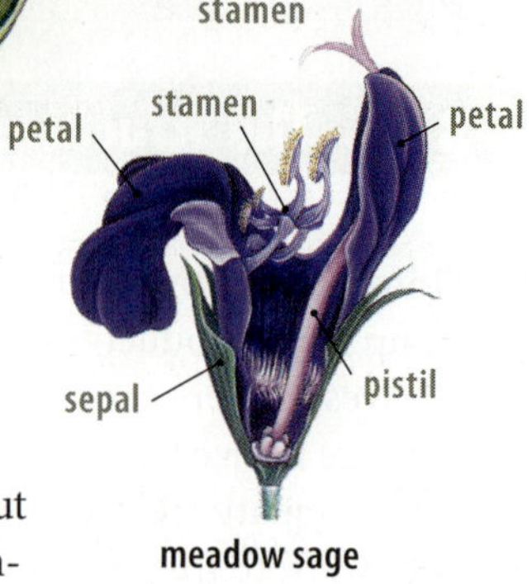


Fig. 3.3 **Bracts**

**Staminate and pistillate flowers.** Specific terms are applied to incomplete flowers that lack either stamens or pistils. Flowers that lack pistils and bear only stamens are called **staminate** [stā′mə·nĭt], or male, flowers; **pistillate** [pĭs′tə·lāt′], or female, flowers lack stamens and bear only pistils. The tassels on a *corn* plant are examples of staminate flowers, while the ears of the corn plant consist of many pistillate flowers. Each ear is composed of the silk, which consists of the style portions of the pistils, and the kernels, each of which is an ovary. Each corn plant produces both staminate and pistillate flowers. This characteristic identifies corn as a **monoecious** [mə·nē′shəs] plant, any plant on which both staminate and pistillate flowers are produced in the same plant. Oaks, walnuts, hickories, and birches are also monoecious plants. Willows, poplars, and the American holly tree are referred to as **dioecious** [dī·ē′shəs] plants. These plants produce staminate and pistillate flowers just as monoecious plants do, but the two types of flowers are borne by separate plants. For example, each individual *American holly* tree will produce either staminate flowers or pistillate flowers but not both. The trees that produce staminate flowers are known as male trees, while the trees that produce pistillate flowers are known as female trees. The familiar red fruits of the holly will form only when both male and female trees are growing in the same vicinity. Even under these circumstances, the fruits form on only the female trees.

**Arrangement on stem.** Flowers vary in the way they are arranged on the plant stem. The rose bears each flower individually upon a single stem, but most plants bear flowers in clusters called **inflorescences** [ĭn′flə·rĕs′əns·ĕz].

## Factors Affecting Flowering

What determines when a plant will burst into flower? Flowering in plants is not accidental; it is controlled by many factors. Among the factors that control flowering are light, temperature, and soil fertility.

*The chief factor controlling flowering in most plants is the length of daylight and night.* Many plants require a definite period of light and darkness before they will flower. This phenomenon is called **photoperiodism** [fō′tō·pēr′ē·ə·dĭz′əm]. Some plants can accurately measure the length of light and darkness to within minutes so they will flower at precisely the right time of year. Plants, like humans, use the sun "for signs, and for seasons, and for days, and years" (Gen. 1:14).

In addition to day length, *temperature and soil fertility also affect flowering.* Some plants produce flowers only when the temperature is relatively low, while others flower only when the temperature is high. Flowers bloom better in rich soil than in soil that has been depleted of necessary plant nutrients. Crowding, sudden drops in temperature, and drought can also affect the formation of flower buds.

Knowledge of how flowering occurs is very practical and profitable to **horticulturists,** agricultural technicians who specialize in growing flowers, fruits, vegetables, and shrubs.

## Things to Notice about Flowers

Observing the LIVING CREATION

### Symmetry

**radial symmetry** *(regular flower)*—If cut across the middle in any direction, both halves will be exactly alike.

**bilateral symmetry** *(irregular)*—Similar halves can be produced only if the flower is cut lengthwise.

### Flower Parts

**multiples of three** *(monocots)*

**multiples of four or five** *(dicots)*

## Section Review 3.1

1. What is the primary purpose flowers have for a plant?
2. What is the difference between complete and incomplete flowers?
3. How are staminate flowers different from pistillate flowers?

### Application

4. Explain the structure and function of each of the four main flower parts.

### Identify

flower, pedicel, receptacle, calyx, corolla, bract, monoecious, inflorescence, photoperiodism, horticulturalist

### Observe

- Add examples of complete and incomplete flowers and staminate and pistillate flowers to the classroom herbarium.

# 3.2 The Development of Fruits and Seeds

## Summary of Fruit Formation

As you have already learned, the purpose of flowers is to produce seeds. The seeds develop from ovules within the ovary, the swollen base of the pistil. As the fertilized ovules develop within the ovary, the ovary is said to be ripening. When the ovary is fully ripened, it is called a **fruit.** A fruit consists of a ripened ovary together with other flower parts that may adhere to it. *Fruits function as the seed-bearing structures of flowering plants.* They provide protection for the seeds while they are forming; and after the seeds have formed, the fruits help to distribute them.

A pear, a cucumber, an acorn, a tomato, an orange, a grain of wheat, and a milkweed pod are all fruits. These examples may seem unusual because in common usage the term *fruit* refers only to sweet, matured ovaries, such as peaches, apples, and oranges, and the term *vegetable* is applied to those fruits that are not necessarily sweet, such as peas, beans, and tomatoes. To a botanist, however, *any ripened ovary is called a fruit whether it is sweet or not and whether it is edible by humans or not.*

Before most plants can begin to form fruits, pollen grains must reach the pistil of the flower and then fuse with the ovules in the ovary. Pollen reaches the pistil through a process called *pollination* [pŏl′ə·nā′shən]. *Fertilization* occurs when pollen fuses with the ovules.

## Pollination

**Methods of pollination.** The transfer of pollen from an anther to the stigma of a pistil is **pollination.** Pollination often occurs within the same flower; this type of pollination is called **self-pollination. Cross-pollination** occurs when the pollen from an anther of one plant is transferred to a stigma of a flower on another plant. Bees, butterflies, and moths are the most important agents of cross-pollination; but wind, other insects, small animals, rain, birds, and even snails may be involved with the process.

Fig. 3.5 **Pollination**

Flowers can be artificially cross-pollinated by gathering pollen from a particular flower and using a special brush or blower to place the pollen on the stigma of another flower. If cross-pollination occurs between plants

Fig. 3.4 **Fruit formation**

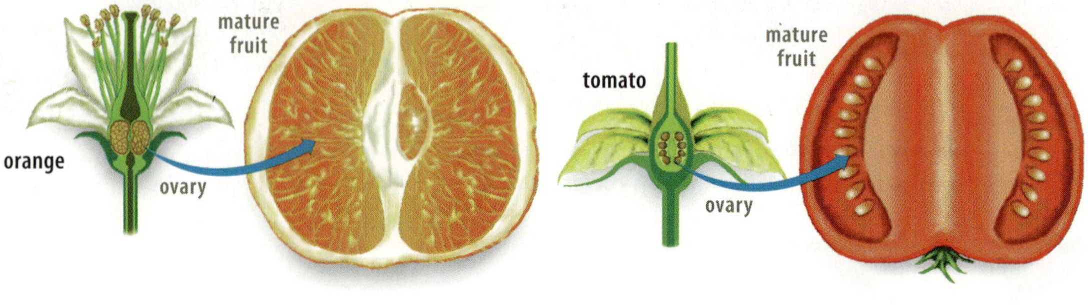

belonging to different species of the same plant, the result is a *hybrid.* Many modern food crops and garden vegetables are hybrids produced by plant breeders. They are characterized by improved nutritional quality and also improved resistance to pests and diseases.

**Designed for pollination.** The wide variety of ways in which pollination is ensured in flowering plants shows the ingenuity and provident care of the Creator. Showy, fragrant flowers usually have nectar-producing glands to attract pollinating insects. **Nectar** is a sweet-tasting, watery liquid produced by plants. It is used for food by butterflies, moths, hummingbirds, and other organisms. Bees use nectar to make honey. The nectar glands are helpful in pollination because their location, usually at the base of the petals on the inside surface, forces insects to brush against the pollen to reach the nectar glands. Color and odor are also important in attracting insects. One group of flowers, called *carrion flowers,* have a red-and-white mottled coloring that gives them the appearance of raw meat. They also produce the odor of decaying meat to attract flies that normally lay their eggs in raw meat.

Fig. 3.6 **Yucca flowers**

Flowers that are pollinated by insects have many other ingenious designs to ensure that insects will pick up and deposit the pollen. These special structures include leverlike stamens, filaments that have a springlike action to dust pollen on insects, and anthers located so that visiting insects must brush against them. Some flowers have stigmas situated so that visiting insects must brush against them and leave pollen behind.

Some flowers can be pollinated only by certain insects. For example, *yucca flowers* are pollinated only by yucca moths, and fig flowers by fig wasps. Certain *orchids* can be pollinated only by a certain type of moth. The nectar glands of these orchids lie at the base of long tubes, and only the *Morgan's sphinx moth* has a proboscis (sucking mouthpart) as long as the nectar tube. Only this moth feeds on the nectar, and it is the only insect to pollinate the orchid.

Wind-pollinated flowers, such as the grasses, produce large amounts of pollen, and their stigmas are designed to catch wind-carried pollen. Some of these stigmas are hairy, some are sticky, some are branched, and some have other special designs to ensure pollination. The flowers in wind-pollinated plants are usually small and inconspicuous. Many grass flowers are exquisite, but they are so small that they are seldom noticed by man.

## People and Pollen

Understanding the LIVING CREATION

Spring, summer, and fall are the peak seasons for plant growth and flower formation. During these seasons the air is often laden with pollen from the flowers of vast numbers of plants. When this pollen is inhaled, it causes an allergic reaction in some people that is called *hay fever* (also known as *allergic rhinitis*). The symptoms of hay fever include a runny nose; itchy, watery eyes; and sneezing. The symptoms of hay fever are triggered when pollen within the respiratory system causes the release of a chemical messenger called *histamine* [hĭs′tə·mēn′] into the delicate linings of the breathing passageways. Drugs taken to reduce the level of histamine are called *antihistamines* [ăn′tē·hĭs′tə·mēnz′].

In the United States, the most important type of pollen that triggers hay fever is that of *ragweed,* a common member of the composite family. Ragweed pollen is abundant in late summer or fall in most areas of the country. Hay fever can also be caused by the pollen of grasses, trees, and weeds, and even by airborne fungi and molds.

ragweed pollen — snapdragon pollen — pine pollen

## Fertilization

Pollination alone does not usually result in the formation of fruits and seeds. Ovules begin developing into seeds only after a sperm cell brought by a pollen grain fuses with the egg cell within the ovule through the process of **fertilization.** Fertilization ensures that the characteristics of the new plant that grows from the seed are determined by its parents. In other words, *the new plant receives a "blueprint" for growth from the sperm cell and egg cell that unite to form it.* The sperm cell and egg cell are called **gametes,** or reproductive cells. Reproduction in which sperm and egg unite is **sexual reproduction.**

The actual process of fertilization is quite complex. Following successful pollination, the pollen grain begins to form a **pollen tube** from the stigma, down through the style of the pistil and into the ovary. Sometimes this tube can be 900 times as long as the pollen grain is wide. Often many pollen tubes form because each pollen grain that reaches the stigma produces its own pollen tube. As the pollen grain produces the pollen tube, it also undergoes internal changes in which the one sperm cell that it contains divides into two. When the pollen tube reaches the ovary, it seeks out an ovule and grows into it, carrying the two sperm cells with it. One of the sperm cells unites with the egg cell, forming the embryo; the other sperm cell fuses with cells called *polar nuclei* to form the *endosperm* [ĕn′də·spûrm′], a nutritional tissue that surrounds the fertilized egg and provides nourishment to the growing embryo. This process of *double fertilization,* in which two sperm cells are required for reproduction, is unique to angiosperms.

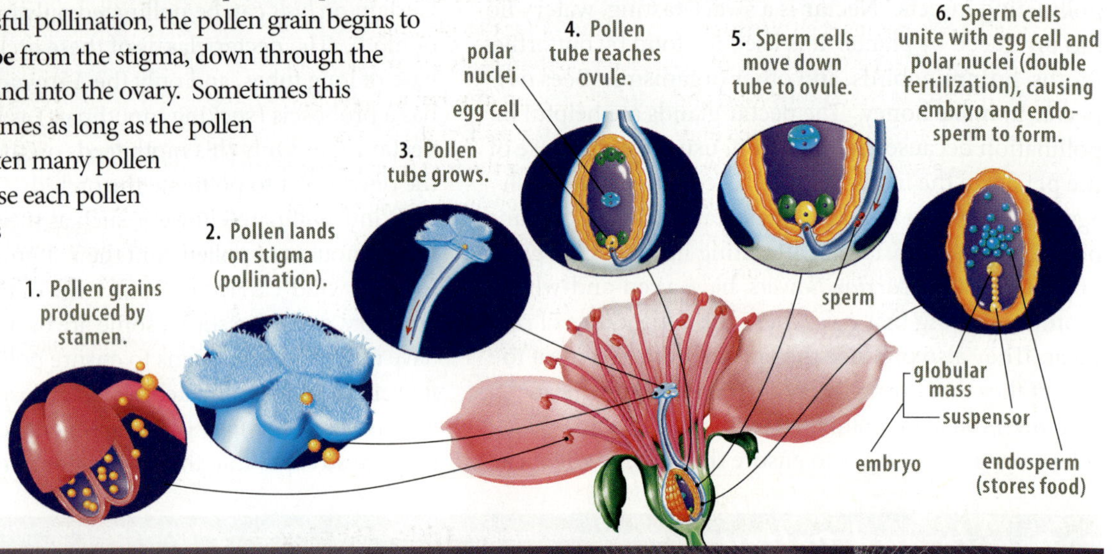


Fig. 3.7 **Pollination and fertilization**

## The Formation and Ripening of Fruits

*Understanding the* **LIVING CREATION**

The formation of fruits and seeds is fascinating. In most plants, it is triggered by pollination and fertilization, which cause the ovules to begin developing into seeds; however, a few plants are able to produce fruits without pollination, fertilization, or seed development. Certain seedless fruits such as navel oranges, bananas, pineapples, and some varieties of apples and pears are produced in this way. In most other plants, the slow development of the ovules into seeds results in the formation of **hormones** (chemicals produced in plants or animals to control or stimulate specific processes) that cause the ovary to begin **ripening,** growing larger and developing into a fruit.

The size increase in the ovary is dramatic. The one-eighth-inch ovary of a peach flower develops into a fruit four inches in diameter. The ovary of a watermelon flower is about the same size as that of the peach, yet it develops into a fruit that may weigh as much as twenty pounds. The number of seeds contained in fruits also varies depending on the number of ovules. Each fertilized ovule develops into a seed. The peach ovary contains only one ovule, while the ovaries of geranium and bean flowers contain several. One type of orchid flower was found to contain more than 3 million ovules.

As the ovary ripens, most of the other flower parts wither; some even fall away. A developing bean pod illustrates this process. The pod is the ovary of the bean flower. As the pod enlarges, the petals fall away while the stigma and style shrink in size. When the pod is fully ripened, the stigma and style are quite small. The mature bean seeds occupy most of the space within the pod. Occasionally a bean pod will contain small, undeveloped ovules that were never fertilized. The stringlike structure that joins the two sides of the pod is actually a strand of vascular tissue that brings food and water to the developing seeds and fruit. In certain fruits such as apples, pears, and strawberries, a portion of the receptacle (the base of the flower to which the other parts are attached) forms the fleshy portion of the fruit.

Most fruits are green when they are small and change colors as they ripen. The green color is the result of chlorophyll. Other pigments, such as carotene and anthocyanin, produce the reds, oranges, yellows, and purples associated with ripened fruits. Some fruits remain green even when they are ripe, yet they usually become softer and acquire a pleasant scent. As fruits change color, their taste also changes from sour or bitter to a pleasant flavor. Fully ripened fruits are prepared for picking by an **abscission layer** that forms in the stalk of the fruit and begins to cut the fruit from the stem. The abscission layer also blocks the vascular tissue in the stalk, often causing an accumulation of sugar in the fruit just at the time that it becomes fully ripe.

1. Flower pollinated by insect.
2. Ovules fertilized; seeds begin to develop.
3. Petals fall away; ovary enlarges; seeds and fruit begin to form.
4. Fruit ripens; abscission layer severs fruit from plant.

## Section Review 3.2

1. How are fruits and seeds formed?
2. What is the difference between cross-pollination and self-pollination?

### Application

3. Distinguish between pollination and fertilization and tell how each is involved in the reproduction of flowering plants.

### Critical Thinking

4. Describe the divine design and provision evident within the process of pollination.

### Identify

fruit, nectar, hay fever, gamete, sexual reproduction, pollen tube, endosperm, hormone, ripening, abscission layer

## 3.3 Fruits

### Simple Fruits

Most fruits are considered **simple fruits** because they form from one flower that has only one pistil. In contrast, *aggregate* [ăg′rĭ·gĭt] *fruits* form from one flower that has several pistils, and *multiple fruits* form from several flowers. Tomatoes, peaches, beans, and walnuts are all simple fruits. Botanists have placed simple fruits into categories according to their structures. The various simple fruits may be referred to as berries, drupes, pomes, legumes, samaras, nuts, achenes [ā·kēnz′], or grains. Certain of these fruits, such as the legumes and the grains, are so characteristic of a particular group of plants that the fruit itself was given the same name as the plant group.

Fig. 3.8
**Simple fleshy fruits**

**Berries.** Simple fruits in which the entire ovary is fleshy and juicy throughout are **berries.** The outer covering of these fruits may be soft, leathery, or somewhat hard. Examples of berries are *tomatoes, grapes, cucumbers, watermelons,* and *oranges.* Raspberries and strawberries are not classified as berries because they are aggregate fruits.

**Drupes. Drupes** are also fleshy and juicy, but not throughout the entire fruit. Drupes have an outer fleshy layer and an inner woody layer, called the **stone,** that surrounds the seed. *Peaches, plums, cherries, olives, apricots, mangoes,* and *walnuts* are drupes.

**Pomes.** Simple fruits with an outer fleshy layer and an inner papery core are called **pomes.** Pomes are unusual in that the outer fleshy layer is not actually formed from the ovary but is a part of the receptacle of the flower. The inner papery core was the ovary of the flower. *Apples* and *pears* are pomes.

**Legumes. Legumes** are simple fruits that consist of a pod enclosing several seeds. They differ from berries, drupes, and pomes in that they are not fleshy and juicy. When legumes are fully ripe, the pod dries out and splits open to release the seeds. *Peas, beans,* and other members of the pea, or legume, family all produce this type of fruit. *Peanuts* are especially interesting legumes. Their thin shells are actually pods that cover the seeds. Peanuts form from flowers, as do all fruits. After the flowers are fertilized, the ovary of each flower produces a stalklike structure that grows into the ground, where ripening of the fruit occurs.

Fig. 3.9 **Peanut plant**

Next to the cereal grains, legumes are the most important fruits to man. They are high in protein and are often canned or dried. The nitrogen-fixing bacteria that live in the nodules on the roots of the legumes are important to the nitrogen cycle.

**Samaras. Samaras** consist of small dry seeds with one or more winglike structures attached to them. When the seed falls from the tree, its wing spins like the rotor of a helicopter, greatly slowing the fall of the seed and allowing it to land much farther away from the tree. *Maples, ashes* and *elms* produce samaras.

Fig. 3.10 **Samara falling**

**Nuts.** Simple dry fruits consisting of a seed enclosed in a hard covering, or shell, are **nuts.** Examples of nuts are *chestnuts, hickory nuts, hazelnuts,* and *acorns.* Almonds, walnuts, and pecans are often referred to as nuts but are actually a type of drupe in which the stone forms a shell around the seed, which we eat.

**Achenes. Achenes** are simple fruits consisting of a seed and a shell. The structure of achenes is similar to that of nuts except that achenes have much thinner shells. The members of the composite

Fig. 3.11 **Simple dry fruits**

family are known for the multitude of achenes that are produced by the disk flowers in their flower heads. The seeds within sunflower achenes are used as food by man and animals.

**Grains.** The fruits of the grass family are called **grains.** Examples of grains important to man are *barley, oats, rye, corn,* and *rice.* These and other edible grains are called the **cereal grains.** Though grains are usually thought of as seeds, *each kernel of grain is actually a complete fruit consisting of a seed and a shell.* The shell of grains is quite thin, but unlike that of achenes, it is attached directly to the seed. It cannot be removed except by a special milling process. In wheat, this shell-like fruit covering is called the **bran.**

Among the fruits used by man, the cereal grains rank as most important. Rice is the cereal grain that feeds more people than any other. Wheat is the most important cereal in the United States and Europe. Cereal grains are valuable food sources because they are rich in starch. In addition to starch, oats are rich in protein, being the highest-protein cereal grain. Corn contains much oil. Cereal grains also supply essential mineral elements such as phosphorus, sulfur, calcium, potassium, magnesium, and sodium. This is especially true when the whole grain is used. When cereal grains are grown in proper soil, they are easily cultivated and produce plentiful yields. After the harvest, cereals are easily stored for considerable amounts of time because the low water content of the grain inhibits spoilage.

**Simple fleshy and simple dry fruits.** As you have probably noticed, the simple fruits that we have studied have either fleshy, juicy ovaries or dry ovaries. In fact, *all fruits are considered to be either fleshy or dry.* **Simple fleshy fruits** include berries, drupes, and pomes. **Simple dry fruits** include legumes, samaras, nuts, grains, and achenes. Dry fruits are often further divided into those that split open upon ripening, such as the legumes, and those that do not split open upon ripening, such as samaras, nuts, achenes, and grains.

## Aggregate Fruits and Multiple Fruits

Unlike the simple fruits, which form from one flower having only one pistil, **aggregate fruits** form from one flower that has several pistils. Two aggregate fruits that are similar in design to drupes are *blackberries* and *raspberries.* Each of the tiny structures on the surface of the *strawberry* that we commonly refer to as seeds is a fruit formed from a single flower ovary and contains its own very tiny seed. The fleshy part of the strawberry that we eat is part of the receptacle of the flower.

**Multiple fruits** form from several flowers. Examples are the *fig* and the *pineapple.* Pineapple fruits develop from individual flowers that grow together on a central, fleshy, stemlike structure. As each flower ovary develops into a small fruit, the entire mass of fruits fuses into the one large structure that we know as a pineapple. Pineapples grown for food are produced without pollination or fertilization of the flowers; therefore, the pineapple fruit has no seeds within it.

Fig. 3.12 **Aggregate and multiple fruits**

## Wheat: The Staff of Life

*Understanding the* LIVING CREATION

For many people around the world, wheat is the "staff of life," the main source of their daily bread. Every year, farmers harvest billions of bushels of wheat to feed the world.

Young, green wheat plants look similar to lawn grass. However, they grow as tall as five feet and turn golden brown when ripe. The kernel itself may be purple, red, white, or yellow. Although it is only $\frac{1}{8}$ to $\frac{1}{4}$ inch long, a single kernel of wheat is a complete fruit with four distinct parts: husk, bran, endosperm, and germ. The endosperm and the germ are the two parts that form the actual wheat seed. The **husk,** also called the *chaff,* is removed quite easily and has no value. Removal of the husk exposes the **bran,** the shell-like fruit covering that can be removed only by milling. Bran is a valuable source of dietary fiber, or roughage.

The starchy **endosperm,** which makes up about 85% of the kernel, is the part that is used in making white flour. The **germ** is the embryo, the young wheat plant. If the grain is planted, the germ begins to grow, producing a new plant. Since the germ spoils more quickly than the rest of the kernel, this most nutritious element of wheat is often removed before flour is made.

The endosperm alone is used to make white flour, while the entire wheat kernel except the husk is used to make whole-wheat flour. Because of the processing and bleaching and because the germ is taken out, white flour usually has less nutritive value than whole-wheat flour. Because it contains bran, whole-wheat flour also has more dietary fiber than white flour. To help make up for the nutrients lost during processing, white flour is *enriched* by adding vitamins and minerals to the flour.

We can illustrate how important bread is for human sustenance when we look at how Christ compared Himself to bread. As bread is the sustainer of the physical body, so Christ is the source and sustainer of spiritual life.

> *Then Jesus said unto them, Verily, verily, I say unto you, Moses gave you not that bread from heaven; but my Father giveth you the true bread from heaven. For the bread of God is he which cometh down from heaven, and giveth life unto the world.*
> —*John 6:32–33*

### Fruits and Seed Dispersal

Fruits are very important to humans and animals as sources of food. *The primary function of fruits is to scatter seeds.* Seed dispersal is of great importance because only seeds that come to rest in a favorable environment while the seed is still **viable** [vī′ə·bəl: able to germinate and grow] can germinate.

Fruits help scatter seeds in several ways. Animals eat fleshy fruits and the seeds they contain; some of these seeds pass safely through the animal's digestive tract and are expelled in the animal's wastes. Certain dry fruits, such as the beggar's ticks and various burs, have hooks. Seeds with hooks cling to animals and are carried long distances by the animals. Such a type of seed dispersal, in which an outside agent carries the seeds, is **agent dispersal.** Man, animals, wind, and water are the most important outside agents involved in agent dispersal. The wind disperses fruits equipped with wings, such as samaras; those with plumes, such as the *milkweed;* and the achenes of *dandelions* and *thistles.* Some seeds form in dry, podlike structures from which they are plucked by the wind and carried away. When the seeds of the *tumbleweed,* a plant of the Western plains, are ready to be dispersed, the entire plant withers and appears to die. When the wind blows, the plant stem breaks and the plant rolls across the plains, scattering seeds as it goes. Many seeds use ditches, rivers, and even oceans as agents to carry them to new locations for sprouting. Some of these water travelers, such as the coconut, even have air-filled chambers that help them to float.

Some fruits scatter their seeds by bursting open when they are ripe and catapulting the seeds to a new location. This is called **mechanical dispersal.** Examples of plants with mechanical dispersal are *violets, jewelweeds (touch-me-nots),* and *witch hazel.*

Fig. 3.13 **Seed dispersal**

## Section Review 3.3

1. Explain the characteristics of simple fruits, aggregate fruits, and multiple fruits and give an example of each.
2. Why are the cereal grains considered the most important fruits?
3. Make a chart showing the characteristics of berries, drupes, pomes, legumes, samaras, nuts, grains, and achenes.

### Critical Thinking

4. In 1963, a volcano produced a new island (Surtsey) in the North Atlantic. Although the island consisted at first of barren rocks, in just a few years it had become populated with several types of land plants. How might the plants have come to be there?

### Identify

stone, achene, grain, bran, fleshy fruit, dry fruit, husk, endosperm, (wheat) germ, viable

### Observe

- Collect examples of the various kinds of simple fruits. Observe their inner structure and make drawings of what you see.
- Collect kernels of wheat, corn, or other cereal grains and observe their structure using a hand lens. Make a drawing of what you see and label the parts.
- Collect several examples of fruits that scatter seeds by agent or mechanical dispersal and observe how each is specially designed.

# 3.4 Seeds

## Seed Structure

**Seeds** are the result of the process of flower and fruit formation. Seeds form from ovules within flower ovaries and are defined simply as fully developed and matured ovules that can produce a new plant.

Seeds have three main parts: the young **embryo plant,** called the *germ* in wheat; the **endosperm,** which provides nutrition to the growing embryo; and the **seed coat,** which provides a protective covering for the seed. Some plants have a double seed coat consisting of two layers. A scar, or *hilum* [hī′ləm], can be seen on the surface of the seed coats of many seeds. The hilum marks where the seed was attached to the ovary wall during its development.

The embryo is the living part within a seed. All embryos contain what they need to produce an adult plant of their kind, whether that plant be a small clover or a huge oak. Each embryo consists of a **plumule** [plo͞om′yo͞ol], a tiny shoot that will develop into the stem and leaves of the plant; a **radicle,** which will develop into the root system of the plant; and one or two **cotyledons.**

The cotyledons contain stored food that the embryo will use to supply the energy for sprouting. *Dicots are named for the fact that their seeds have two cotyledons; the seeds of monocots contain only one cotyledon.* The two cotyledons of a bean seed are easily observed after the seed has been soaked in water to remove the seed coat. The cotyledons are quite large and compose the majority of the bean embryo. Some seeds, especially those of certain monocots such as the cereal grains, have very small cotyledons.

## Seed Germination

The sprouting of a seed is called **germination.** Normally, a mature seed is dormant until it germinates. **Dormancy** [dôr′mən·sē] is a period of inactivity, which helps prevent seeds from sprouting when conditions are unfavorable for the growth of young plants. For example, many seeds are dormant in winter and do not germinate until the warm days of spring arrive. There is great variation in the viability of seeds. Some seeds, like those of the orchid, remain viable for only a few days or weeks; other types of seeds may retain their ability to germinate after being dormant for years. Seeds of some plants, such as the *Indian lotus* and the *date palms,* are apparently capable of germinating even after lying dormant for hundreds of years.

A common bean seed provides a good example of the processes that occur during germination. When a bean is planted and begins to germinate, it absorbs a great deal of water. The presence of water within the seed triggers chemical changes that cause the tissues of the embryo to swell and grow. The water also softens the seed coat so that the tiny root and shoot can push through the coat and begin growing into the soil. The embryo obtains energy for its rapid growth by "burning" the stored food in its cotyledons by the process of *cellular respiration.* After the young root and shoot have pushed out into the soil, the root grows quickly to anchor the young plant. The shoot then begins to grow. It produces a stem that arches up through the soil bearing the two cotyledons, which still contain some stored food for the plant. At the tip of the stem is a small bud that will produce the plant's first leaves. The tough cotyledons help to shield this delicate bud from receiving too much sun. The cotyledons also continue to provide food for the plant until the leaves are large enough to begin carrying on photosynthesis; then they shrivel and fall away.

Some plants do not display their cotyledons above ground as the bean does. The plant may not have large cotyledons, as is the case with the cereal grains, or the cotyledons may remain in the ground, as is the case with sweet peas.

Germination is a critical time in a plant's life. If the newly germinated plant encounters harsh conditions immediately during or following germination, it will not survive. Factors that favor germination include an adequate supply of moisture, an adequate supply of oxygen, a favorable temperature, proper soil conditions, and sufficient sunlight.

Before germination, a seed contains only about 5% to 10% water by volume. During germination, the seed

Fig. 3.14 **Monocot seed (corn)**

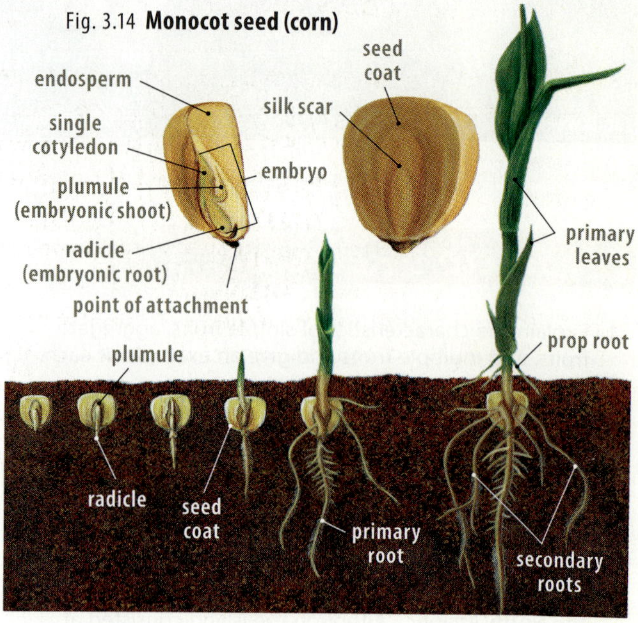


Fig. 3.15 **Dicot seed (bean)**

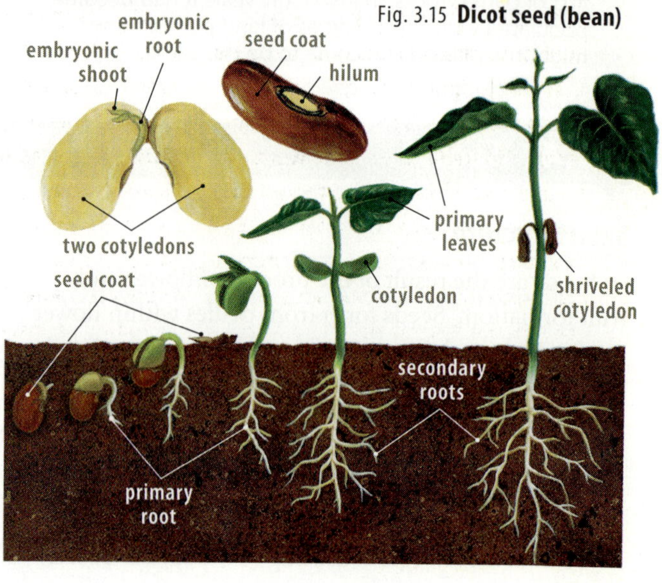

absorbs a large amount of water and increases its volume of water by as much as 200%. Without this extra water, the seed could not germinate.

The germinating seed must have oxygen to carry on cellular respiration and use the food stored in the endosperm or cotyledons. *Most of the food stored in the cotyledons is in the form of starch.* A special group of enzymes, the amylases [ăm′ə·lās′ĕz], convert the starch to sugar, and the embryonic cells absorb this sugar. This change explains why sprouting seeds such as bean sprouts have a sweet taste. Seeds germinate best when temperatures are between 10°C and 45°C (50°F and 113°F). Seeds usually must be planted in some type of soil before they will germinate. Some seeds will germinate in very poor soil, and others will germinate when completely out of the soil if other conditions for germination are met. Certain types of seeds also need light in order to germinate. These seeds usually lie dormant in winter, when days are short, and sprout in spring, when days become longer.

## George Washington Carver: Pioneer Plant Scientist

Explorers of the LIVING CREATION

One of America's greatest plant scientists was Dr. *George Washington Carver* (c. 1864–1943). For 47 years, Dr. Carver taught classes and directed research in agriculture at Tuskegee Institute (now Tuskegee University) in Alabama. His goal in his teaching and research at Tuskegee was to help the farmers of the South rise out of poverty. At that time, Southern cropland was almost useless. Its nutrients had been exhausted by the constant planting of cotton, and in many areas the land was barren, eroded, and sun parched. To help the farmers who could not attend classes at Tuskegee, Carver equipped a wagon as a traveling farmers' school. Every Friday after his regular classes, he went out into the countryside and taught the farmers how to restore their land by rotating crops and planting legumes to return nitrogen to the soil. He also taught them to diversify and plant other crops besides cotton.

It was Dr. Carver's dream that each farm be self-sufficient, able to produce food and household items as well as a cash crop to replace cotton. His desire was to find one or two plants that would grow well in the South and from which hundreds of practical products could be produced. The first plant that he successfully tested was the sweet potato. From it, he developed over one hundred products, including flour, glue, dyes, ink, paint, and candy.

Still not satisfied, Dr. Carver continued to search for a plant that he could exploit more fully. He settled on the lowly peanut, which grew wild in the South and was considered a nuisance by farmers and fit only to be fed to pigs. Patiently, Dr. Carver analyzed the peanut, broke it down into compounds, and began to put these compounds back together in various combinations until he had produced more than 300 products—including an instant coffee substitute, ink, soap, shampoo, shaving cream, meat sauces, and a milk substitute. George Washington Carver had at last achieved his dream. From the peanut crop, farmers could make many needed household items. (Dr. Carver wrote pamphlets explaining how to make the items.) Farmers could also sell peanuts for necessary cash. Thanks to the work of George Washington Carver, the multimillion-dollar Southern peanut industry had begun.

Dr. Carver's success lay in his familiarity with the natural world that God created. Believing that God made plants for man's delight and use and that it was man's duty to discover as many of those uses as possible, Dr. Carver observed and collected plants to obtain the knowledge and raw materials for his lab work. This philosophy, and the dedication with which he pursued it, made him a success.

Every day Dr. Carver rose at four to take a morning stroll. During these walks, he prayed for guidance and help for the day's work. When success came, he gave God the glory. He once said, "Without God to draw aside the curtain, I would be helpless."

## Section Review 3.4

1. Describe the structure of a seed and an embryo.
2. How are the seeds of monocots different from the seeds of dicots?
3. How are the embryo and the endosperm formed?
4. What purpose does the endosperm serve in the seed?

### Application

5. Write a complete description of the germination of a bean seed, including the factors that affect seed germination.

### Critical Thinking

6. Explain how Dr. George Washington Carver exemplified a scriptural attitude toward science.

### Identify

seed coat, hilum, plumule, radicle, cotyledon, germination, dormancy, viable

### Observe

- Collect some corn and lima bean seeds to illustrate monocots and dicots. Observe their inner and outer structure, make drawings of what you see, and label the parts of each seed.
- Soak some bean seeds in water for several hours and place them in a beaker lined with wet blotting paper. Place some wet (but not too wet) absorbent cotton in the center of the beaker to keep the blotter from drying. Set the germinating seeds in a dark place and moisten and observe daily for five days. On the fifth day, carefully remove the seedlings from the beaker and place them on a paper towel. Observe and measure the growth of the stems and roots (the roots should be about one inch long).

## Chapter 3 Review

### Define

1. sepals, petals, calyx, corolla
2. stamen, filament, anther
3. pollen
4. pistil, ovary
5. stigma, style
6. bract
7. inflorescence
8. photoperiodism
9. horticulturist
10. fruit
11. pollination
12. hybrid
13. nectar
14. fertilization
15. sexual reproduction
16. legume
17. bran
18. chaff
19. endosperm
20. germination
21. dormancy

### Identify

1. the primary purpose of flowers
2. another name for a flower stalk
3. the chief factor affecting when most plants flower
4. eight groups of simple fruits
5. five important cereal grains
6. the four parts of a wheat kernel
7. the three main parts of a seed
8. three parts of the embryo
9. five factors that favor seed germination

### Explain

1. What is the difference between complete and incomplete flowers?
2. What is the difference between staminate and pistillate flowers?
3. Explain the difference between monoecious plants and dioecious plants.
4. Distinguish between self-pollination and cross-pollination.
5. How does pollination lead to fertilization and the development of a seed?
6. Explain the difference between simple fruits, aggregate fruits, and multiple fruits.
7. Describe the stages of germination of a bean seed.

### Apply

1. Why does whole-wheat flour tend to be more nutritious than white flour?
2. What is the primary function of fruits?
3. How do agent dispersal and mechanical dispersal help to scatter seeds?
4. Which method of seed dispersal (agent or mechanical) can spread seeds farther from the plant? Explain your answer.
5. What is meant when a seed is said to be viable?

# Stems & Roots

CHAPTER 4

**Key Concepts**

- Stem structure
- Vegetative reproduction
- Special stems
- Plant hormones and tropisms
- Root structure and growth
- Root function

## 4.1 External Structure of Woody Stems

The principal parts of a green plant are the *leaves, stems,* and *roots.* As you know, the leaves manufacture most of the plant's food. By design, the leaves are supported by a *stem* to provide sunshine to as many leaves as possible. The food manufacturing process of photosynthesis also requires water and minerals found in the soil. *Roots* are designed to absorb necessary nutrients from the soil and to anchor the plant to the soil. Roots and stems contain transport, or vascular, tissues, which carry nutrients to the leaves and food to the roots. In this chapter, we will look more closely at the design and function of stems and roots.

### Examining a Dormant Twig

To begin our study, let us look at a twig similar to one you could snip from a nearby deciduous tree. We will examine its external design, noting each part and its purpose. Then we will look at the internal design to see different kinds of cells and tissues and why God placed them there.

**Buds and bud scales.** There is much to be seen on the outside of a twig, or small branch, of a woody stem. **Buds** are the parts of the stem that allow it to grow in length or to develop new stems, flowers, or leaves. The large bud at the end of the twig in Fig. 4.1 is called a **terminal bud** or *apical* [ā′pĭ·kəl] *bud.* The terminal bud allows the stem to grow in length. Those buds located along the sides of the twig are called **lateral** or *axillary* [ăk′sə·lĕr′ē] **buds.** Lateral buds, which grow in the space formed between the leaf petiole and stem, develop during the spring and summer very soon after the leaves reach maturity. Their purpose is to allow growth of new stems from the sides of the main stem. Lateral buds are responsible for a tree's breadth and branching. Each bud is covered with thick **bud scales,** which protect the bud in winter from frost injury and from drying out. If you take a bud scale off, you will also find under it a stem tip and tiny leaf parts. (You will probably need a hand lens to see this.) In some buds you would also be able to observe flower parts.

Fig. 4.1
**Dormant woody stem**

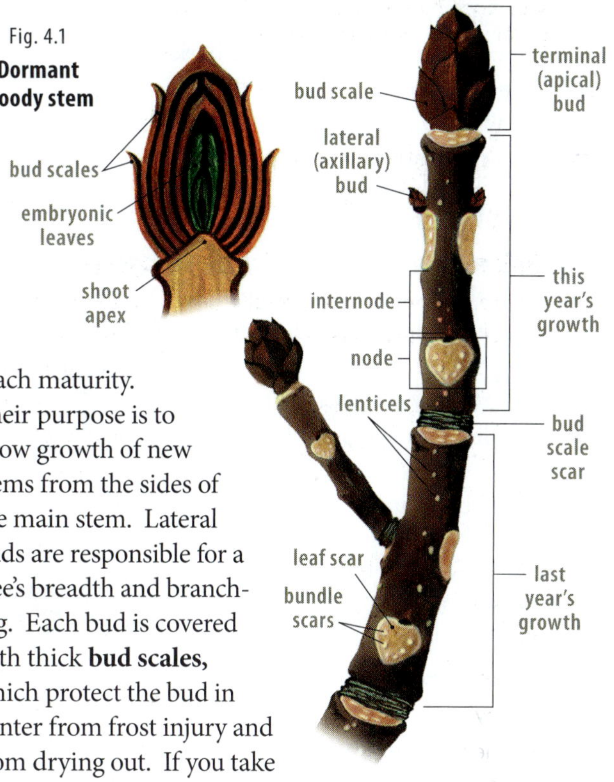


**Nodes, scars, and lenticels.** Rings that circle the stem, **bud-scale scars,** indicate where last year's growth started. Each year, growth begins at the bud. The age of the twig can easily be determined by counting the sets of bud-scale scars. Growth continues in the terminal region of the stem.

When the season of growth is complete, the new twig will have *nodes* at intervals along its length. The nodes are areas where leaves are growing or have grown. Botanists call the section of the twig between nodes the **internode. Leaf scars** at a node reveal where leaves were attached to the stem in previous years. One leaf scar at each node tells us that the tree has alternate leaf arrangement; two leaf scars show opposite leaf arrangement; and three or more indicate a whorled arrangement.

As you look closely at the leaf scars, you will notice tiny markings, or **bundle scars,** which indicate the passage points of vascular tissue from the stem to the petiole of the leaf. The shape and arrangement of leaf scars and bundle scars give botanists keys to identifying the plants, especially if the twig is dead or dormant. Other markings on the bark are **lenticels** [lĕn′tĭ·sĕlz′], openings in the bark that allow air to enter the stem, providing a means of respiration.

## Growth in Length

*Stems grow in length at the tips of their branches.* This growth is called **primary growth.** New tissues are formed by cell division of meristematic cells in the stem tip. When warm weather comes, the leaf, stem, and flower parts in the bud begin to grow, and the bud scales fall off. The stem tip grows longer, and the leaf and flower parts grow larger. Each summer, new terminal and lateral buds are formed and become dormant; each spring, they begin again to grow. Remember that a tree or shrub grows in height only at the terminal area. Suppose you were to measure a young tree and find that the first branch is exactly three feet above the ground. If you were to return in ten years, the same branch would still be only three feet above the ground, even though the tree itself may have more than doubled in height. Not all lateral buds begin to grow in the spring; some remain dormant and eventually fall off. It is not practical for every bud to grow; if it did, there would be a branch wherever there was a leaf. *The terminal bud of a stem produces hormones that control the growth of the lateral buds.*

Trees with strong terminal buds on the main vertical stem grow tall and straight with branches coming from a large central shaft. These trees, such as pines, firs, hemlocks, redwoods, and cypresses, are described as having **spirelike branching.** Trees such as willows, cottonwoods, and elms have strong lateral buds and thus develop several main branches, usually rather close to the ground. This growth pattern is described as **spreading branching.**

Fig. 4.2
**Types of branching**

## Section Review 4.1

1. What is the purpose of bud scales?
2. Describe the process by which stems grow in length.
3. Why do not all buds grow? What determines which buds grow and which do not?
4. What are two branching patterns?

### Critical Thinking

5. If you carved your initials on a tree trunk and came back to look at them many years later, you would find that, though the tree is much taller, your initials are the same height that you carved them. Explain why.

### Identify

terminal bud, lateral bud, bud-scale scars, internode, leaf scars, bundle scars, lenticels, primary growth

# 4.2 Internal Structure of Woody Stems

When examined in cross section, a woody dicot stem is easily divided into three areas: the *bark*, the *wood*, and the *pith*. If you have a freshly cut, growing woody stem, you can easily peel the bark off. Underneath the bark, the stem is hard and white. This is the wood, which botanists call *xylem*. The tiny tubes formed by the xylem

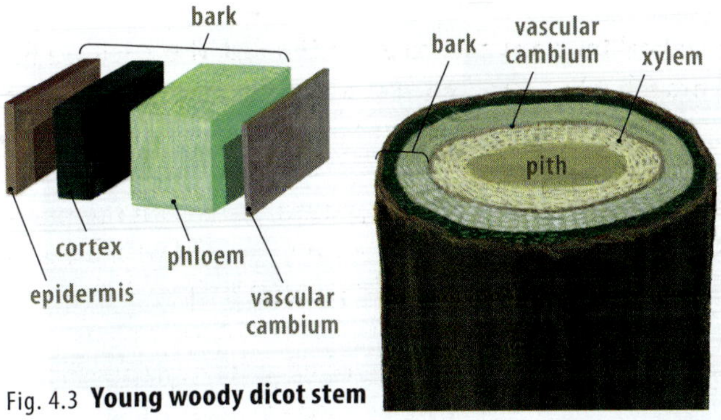

Fig. 4.3 **Young woody dicot stem**

carry water and minerals (sap) upward from the roots to the leaves. Between the bark and the wood is a thin layer of slimy material. This is the *vascular cambium*, a layer of meristematic tissue that produces new bark and new xylem and is responsible for the secondary growth of woody stems. The soft tissue at the center of young stems is *pith*.

## Bark

The outer section of a woody stem, known as **bark,** provides protection for the stem and helps transport food throughout the plant. Bark is made of several layers of tissues. In very young woody stems, the outermost layer of the bark is known as the *epidermis*. The epidermis of a woody stem, which is only one cell thick, helps protect the stem from drying out and from being destroyed by harmful parasites. As the stem grows older, the epidermis is replaced by a stronger protective tissue whose most noticeable part is the **cork.** Some trees, like the *cork oak*, produce so much cork that it is possible to extract it commercially without killing the tree, for use in bottle corks, cork tiles, etc. Young woody stems have a second layer of tissue underneath the epidermis called the *cortex*, which stores food for the stem. In very young stems, some cortex cells may develop chlorophyll, giving these twigs a green color. *Lenticels* in the bark allow the tissues in the stem to obtain necessary oxygen from the air.

The innermost layer of the bark, the *phloem,* carries sugars and other foods made in the leaves downward to the growing parts of the stem and roots. Sap-conducting phloem cells are in the form of thin-walled, hollow tubes with perforated plates (called *sieve plates*) at each end. Phloem cells are connected end to end at their sieve plates to form **sieve tubes,** which conduct food throughout the plant. Liquids containing dissolved food materials can pass through a sieve tube from one cell into the next through the sieve plates. Because the sap is under pressure, phloem cells are reinforced by neighboring *companion cells,* which strengthen the phloem tissue. New phloem cells are formed by the thin layer of *vascular cambium* cells that lie between the bark and the wood.

## Wood

The strong, resilient inner section of a woody stem, known as **wood,** provides strength and support for the stem and helps transport water and minerals from the roots to the leaves. Wood consists mostly of long, narrow, tubular cells with thick cellulose walls stiffened by a substance called *lignin*. These cells, which are arranged in long, water-conducting pipes, compose the *xylem* of the tree (the tissues that conduct sap upward from the roots to the leaves). Only the younger wood is able to conduct sap.

The wood of *conifers* (cone-bearing trees), such as pines and firs, contains only a simple system of water-conducting cells called **tracheids** [trā′kē·ĭdz]. Tracheids are long, narrow, rod-shaped cells that often have tapered ends. They have very strong, thick walls that impart great strength and a limited amount of flexibility to the tree. When a tracheid cell matures, it is designed to cease functioning and break down, leaving its sturdy walls behind as an empty tube that can then freely conduct water. Each tracheid has several dozen small holes in its walls, allowing water to flow in and out. By flowing to the top of one tracheid, then out the side into the bottom of an adjoining tracheid, and so on, water climbs from the roots of the tree, up through the stem, and into the leaves.

The wood of the angiosperms has a more complex structure. Like pines and firs, *broadleaf trees,* such as oaks and maples, contain tracheids that give strength to the tree and conduct water from the roots to the leaves. However, unlike the conifers, broadleaf trees also contain a second type of xylem tube called **vessels.** Vessels are considerably larger than tracheids; whereas tracheids are generally visible only under a microscope, vessels can sometimes be seen with the naked eye as tiny pores in a piece of cut wood. Vessels are composed of relatively short, wide, often barrel-shaped cells joined end to end like concrete pipes. Like tracheid cells, vessel cells are designed to cease functioning and break down when mature, leaving their thick walls behind to form a segmented pipe. The large diameter and straight walls of vessels allow water to be transported through them

Fig. 4.4 **Mature woody dicot stem**

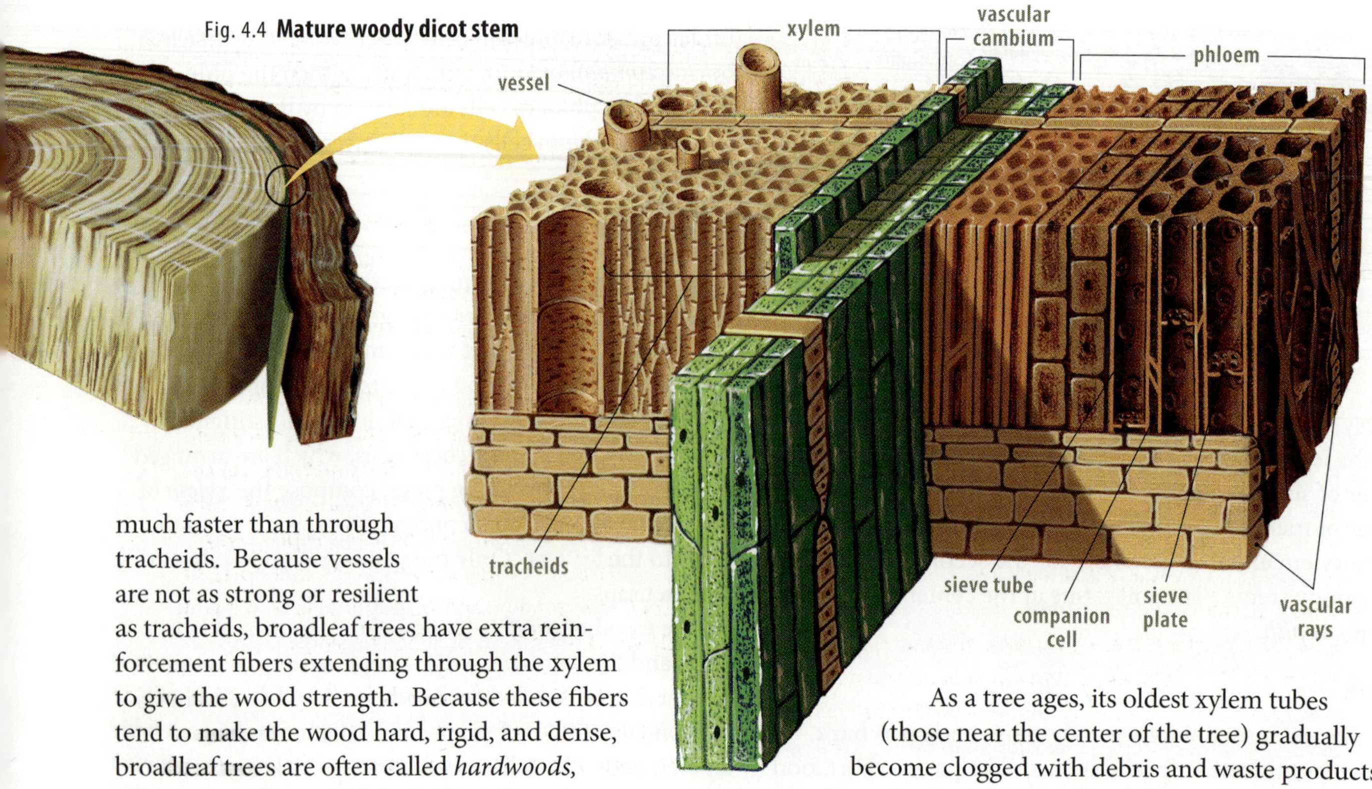


much faster than through tracheids. Because vessels are not as strong or resilient as tracheids, broadleaf trees have extra reinforcement fibers extending through the xylem to give the wood strength. Because these fibers tend to make the wood hard, rigid, and dense, broadleaf trees are often called *hardwoods,* whereas conifers, which lack these fibers, are called *softwoods.*

Tracheids, vessels, and the other parts of wood all make up the xylem tissue. Tracheids and vessels are types of xylem tubes. Together, the xylem and phloem cells are the *vascular tissue* that forms the circulatory system of the plant.

Some woody stems contain additional specialized transport structures. Most mature trees contain *vascular rays,* which radiate outward from the center of the trunk like the spokes of a bicycle wheel. Vascular rays transport water, food, and waste products between the center and edges of a tree. Many softwood trees also contain *resin ducts,* special tubes lined with resin-producing cells. If the bark of the tree is damaged, the resin seeps out of the wood and forms a protective seal against insects and fungi.

## Pith, Heartwood, and Sapwood

At the center of a woody stem in a young plant is the **pith.** This area of the stem is softer than wood and is made up of thin-walled cells. *Pith cells are packed loosely together and function mostly as water storage cells.* The pith becomes less important as the plant grows; it is hardly noticeable in the cross section of a mature tree and sometimes disappears altogether.

As a tree ages, its oldest xylem tubes (those near the center of the tree) gradually become clogged with debris and waste products from the tree's living cells. When xylem tubes become clogged, they are sealed off and no longer conduct sap from the roots to the leaves. This older, inner wood, known as **heartwood,** is then useful only as support for the stem. The accumulation of resins and oils often gives the heartwood a darker color than the rest of the wood and sometimes makes it aromatic.

As a result of this process, in a tree that is several years old only the younger xylem (that farthest from the center of the tree) conducts water and minerals from the roots to the leaves. This younger, functioning xylem tissue is called **sapwood.** Sapwood is usually lighter in color than the heartwood.

## Annual Rings

New wood is formed by the vascular cambium layer in the same manner that new phloem is produced. As can be expected, the production of new wood leads to an increase in the thickness of the tree trunk; this growth in width is called **secondary growth.** Cells of the vascular cambium are very active when growing conditions are good but slow down during droughts, cold weather, and whenever else conditions are not conducive to growth. This is particularly evident in temperate regions, where the cambium cells of trees become less active in the fall, remain dormant over winter, and

## *The Power to Change Matter* by Alfred M. Rehwinkel

Plants are the bridge between the inorganic and the organic, and unite the two in one magnificent system. Plants create food for man and animals. Without plants we could not exist. Neither man nor animal is able to absorb the inorganic and convert it into nourishment for itself. The plants perform this service for them. The mechanism by which this miracle is performed are the roots and the green leaves of the plant. These two, operating together, form a perfectly unified food-manufacturing system. . . .

Design in the LIVING CREATION

This is the most efficient manufacturing system known to man. And the wonder is that its operation is absolutely silent; there is no wear and tear on its mechanism; it replaces its own machinery; and it has continued its operation without interruption and with unfailing efficiency since it was set in motion on the third day of creation.

In addition, plants also maintain the delicate balance between the gases in the atmosphere by inhaling carbon dioxide which is exhaled by man and animals, and exhaling oxygen which again is essential for human and animal life.

Plants provide a protective covering for the soil and preserve it from wind and water erosion; they influence climate and beautify the landscape. They provide a shelter for animals and birds and make the earth a delightful habitation for man.

Plants are the source of all the coal on the earth. Without coal, our present industrial and economic progress would have been impossible. The great coal deposits in the bowels of the earth are the fossil remains of the luxuriant antediluvian plant life which was wiped away and buried in the Great Flood. Every plant, therefore, is a wonder of God's creation, and we ought to look at it with reverent awe as the poet does when he sings:

***There is not a plant or flower below***
***But makes Thy glory known.***
***And clouds arise and tempests blow***
***By order from Thy throne.***

*Alfred M. Rehwinkel,* The Wonders of Creation, *Bethany House Publishers, a division of Baker Publishing Group, Copyright ©1974, Used by permission.*

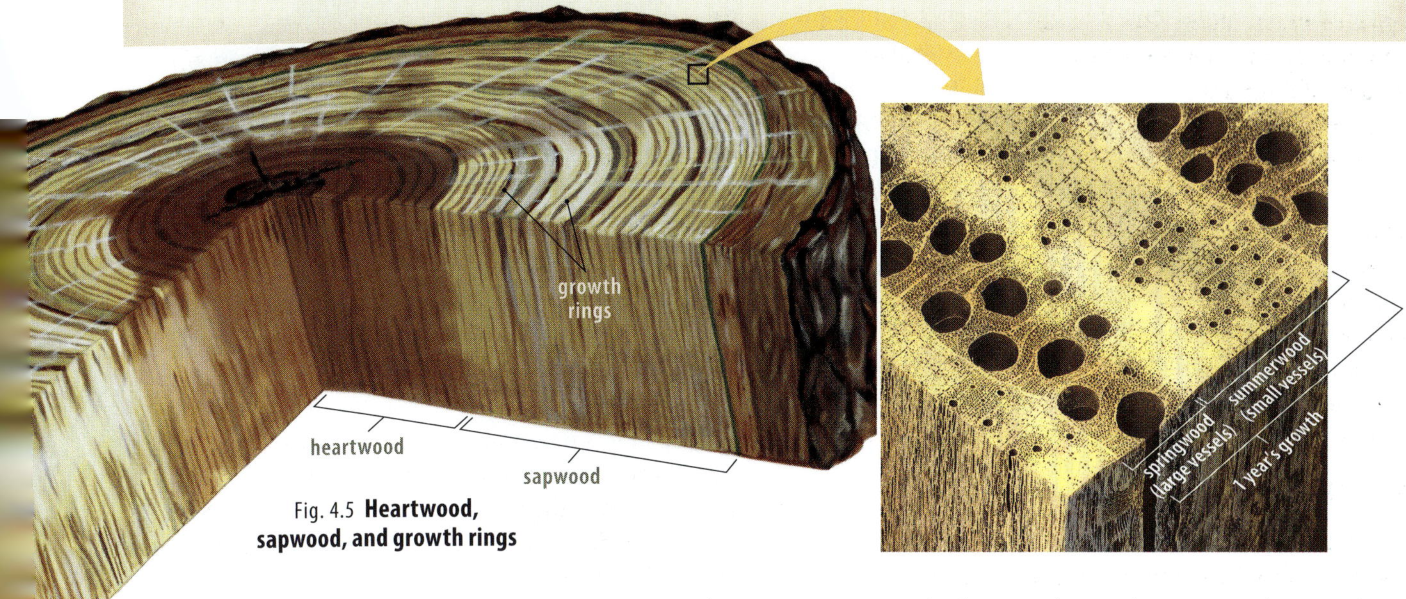


Fig. 4.5 **Heartwood, sapwood, and growth rings**

start to grow again in the spring and summer. Periods of faster and slower growth result in distinct layers of xylem, which are called **annual growth rings.** Much new lighter wood is formed in the spring; less but denser wood is formed in the summer. *One layer of spring- and summerwood taken together makes up a single annual growth ring.* Because of the clear difference between the denser summerwood and the lighter springwood of two consecutive years, the age of the trees can be estimated by counting the growth rings. In tropical regions, where trees grow year-round, this system of counting years is difficult to use.

Springwood of a given annual growth ring is located toward the center of the stem; wood toward the outside of a ring is summerwood. *Springwood is light in color* because xylem vessels are large to conduct the large amount of water necessary for spring growth. *Summerwood is darker in color* because the xylem vessels are smaller. These two areas make up the **grain** of the wood; when wood is stained, the two parts absorb the stain differently, causing the light and dark streaks (the beautiful grain of the wood) that appear in finished wood.

A tree does not have to be cut down for its rings to be examined. Foresters use a device called an *increment borer* to remove wood from a tree for examination. This device is a hollow drill that can bore into the center of a tree to collect a thin, rodlike plug of wood that shows the tree's growth rings.

## Section Review 4.2

1. Into what three major portions can a young woody dicot stem be divided?
2. How are vessels and tracheids different? What types of vascular tissue do they constitute?
3. What is the difference between sapwood and heartwood?
4. How are annual growth rings formed?

### Application

5. Describe the two kinds of vascular tissue in a woody stem. Give the name and location of each and tell the kind of sap each transports and the direction of transport.
6. List some of the commercial applications of cork.

### Identify

sieve plates, sieve tubes, wood, vascular rays, pith, secondary growth, grain

## 4.3 Herbaceous Stems

Unlike woody stems, herbaceous stems have soft tissues and generally live for only one growing season. Herbaceous plants tend to be smaller than woody plants because herbaceous stems do not experience significant secondary growth and therefore cannot grow thick. Herbaceous plants are vitally important to man. Field crops such as wheat, corn, and soybeans are examples of herbaceous plants that directly provide food for people. Alfalfa, clover, and pasture grasses are fed to animals used as food for man.

### Herbaceous Dicots

Some herbaceous plants are *dicots,* based on their seed structure. Examples of herbaceous dicots are clover, beans, and peas. These plants possess an internal design similar to that of the woody dicot, except there is only primary growth. Because the herbaceous stem usually dies after one growing season, the cambium cells do not produce annual ring patterns. The xylem and phloem cells form **vascular bundles,** which are arranged in a ring around the *pith,* the center of the stem. In each bundle, the *xylem* cells are located toward the center of the stem, and the *phloem* toward the outside. Segments of *vascular cambium* lie between the xylem and the phloem and produce new cells of each. Both the **cortex** (the tissue in which the vascular bundles are embedded) and the pith are used for structural strength and food storage. The entire stem is surrounded by an *epidermis,* which protects the delicate tissues inside the stem.

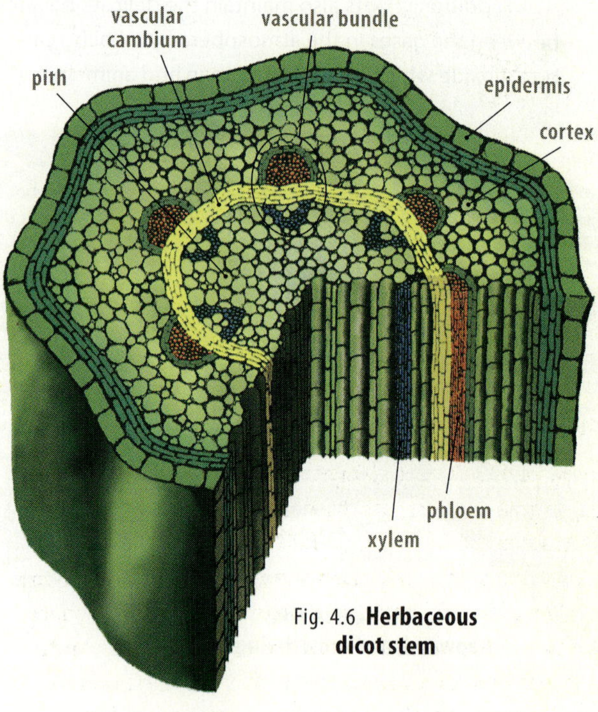


Fig. 4.6 **Herbaceous dicot stem**

### Herbaceous Monocots

The vascular bundles of herbaceous monocot stems are scattered throughout the stem rather than being arranged in a ring. The older and larger vascular bundles are near the center of the stem. The water-carrying *xylem* is toward the inside of each bundle, and the *phloem* is toward the outside.

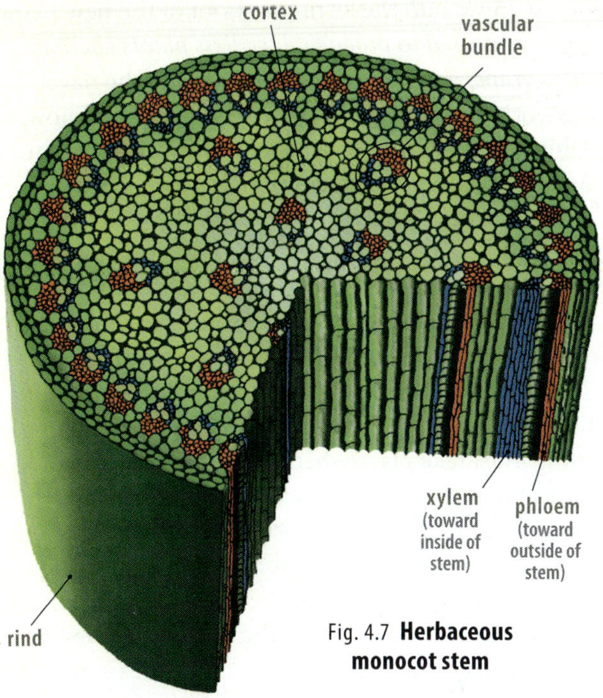


Fig. 4.7 **Herbaceous monocot stem**

The outer covering of a monocot stem is called the **rind.** The rind, which often contains the mineral *silica* to make it hard and strong, supports the plant. The thick outside wall of a cane fishing pole is an example of a rind that is hard. The rind of the bamboo stem is so thick and hard that it can be used for construction.

A monocot stem's vascular bundles usually do not possess cambium. In grasses and some other monocots, the stem is hollow and the vascular bundles are arranged around the edge of the stem. In this type of monocot stem, the vascular bundles exit the stem at the nodes and enter the leaves as large leaf veins.

## Section Review 4.3

1. How are herbaceous dicots different from woody dicots? How are they similar?
2. Compare and contrast the design of a herbaceous dicot stem with that of a herbaceous monocot stem.

#### Critical Thinking

3. Why is there no secondary growth in herbaceous plants?

#### Identify

vascular bundles, cortex, rind

#### Observe

- Make a cross section of a herbaceous dicot stem and a herbaceous monocot stem. Using a hand lens, study your cross sections and make drawings of what you see. Try to label your drawings with the different parts of the dicot and monocot stems.

## 4.4 Vegetative Reproduction and Special Stems

### How Stems Reproduce a Plant

New plants do not always have to grow from seeds; many plants can also produce new plants from sections of stems, leaves, or roots. When a new plant starts to grow from a stem, leaf, or root, **vegetative reproduction** has taken place. Vegetative reproduction occurs when some part other than the seed grows into a new plant, and is one type of **asexual** [ā·sĕk′shōō·əl: no union of sex cells] **reproduction.** Strawberry plants undergo vegetative reproduction when they send out runners, which develop roots and grow into new plants. In nature, stems of most woody and herbaceous perennials may undergo vegetative reproduction by either forming roots from the stems that are touching the ground or by sending up shoots from roots close to the surface.

### Vegetative Propagation

When a person uses vegetative reproductive methods to start a new plant from a root, stem, or leaf, the process is called **vegetative propagation.** There are several types of vegetative propagation, including the use of cuttings or slips, layering, grafting, budding, and tissue culturing.

**Cuttings.** A **cutting,** or *slip,* is a piece of a stem or root that can grow into a new plant. When the yellow-flowering shrub *forsythia* is trimmed, a new shrub may be started by simply sticking a trimmed stem into the ground. This is a common practice because of the advantages it offers. In the first place, when a plant is produced from a cutting, it is certain that *the new plant will be identical to the plant from which the cutting was taken.* Secondly, *using cuttings saves much time* compared to the time it would take to grow the plant from a seed. Cuttings are possible because certain types of plants are able to sprout special types of roots at the cut end of their stems. Roots that sprout from the bottom of a cut stem are examples of **adventitious** [ăd′vĕn·tĭsh′əs] **roots,** roots that grow from an unexpected region of the plant, such as from a stem or leaf.

**Layering. Layering,** like cutting, takes advantage of the fact that certain plants will sprout adventitious roots, and it provides the same advantages as the use of cuttings. A frequent method of layering involves fastening a container of soil around the nodes of a stem to induce the stem to develop roots. When the roots are well developed, the stem is cut at a point below the node and planted.

**Grafting. Grafting** is the process of transplanting living tissue from one plant to another. This method of propagation allows growing two kinds of apples or other fruit on the same tree or growing a more desirable fruit on a tree that is already growing. This is a great timesaver for fruit farmers, because by grafting young stems on an old tree they can start to produce a more desirable fruit in a short time. It saves years of waiting for the more desirable tree to mature and begin to bear fruit.

The grafting procedure is rather simple. The person who is doing the graft cuts a small branch that has several buds on it from the tree he wishes to propagate. This branch, which is to be grafted onto a stem growing on another tree, is called a **scion** [sī′ən]. The tree that receives the new stem is called the **stock.** The end bud of the scion grows into the terminal shoot of the new plant.

*Grafting is used to propagate seedless plants* such as seedless *grapes* and seedless *navel oranges.* The navel oranges originated in Brazil, where they are still known as Bahia oranges. In 1870, several were brought to the United States and propagated by grafting. All of our current navel orange trees, particularly those in California and Florida, descend from those few trees imported from Brazil.

**Budding. Budding** is a kind of grafting in which the scion is a bud. Native apple trees, whose fruit is often undesirable for eating, are made to bear desirable fruit by budding. To grow better fruit on a native apple tree, budding must take place when the tree is a seedling. A single leaf bud is cut from a desirable apple tree and placed in a slit in the bark of the seedling native tree. If the budding process has been done correctly, the bud will grow and become attached to the stock upon which it has been budded. The rest of the seedling above the bud is cut off, and the native tree will then bear only the desirable apples.

## Tropisms and Plant Hormones

Understanding the LIVING CREATION

Although plants lack nervous systems, God has designed them to be able to respond to their environment in many ways. The simplest response to a stimulus occurs when an organism moves either toward or away from a stimulus. This type of response is called a **tropism** [trō′pĭz′əm]. You may have noticed that a plant left near a window for several days without being turned will often "bend" toward the window. This occurs because the plant senses the window as a source of light and grows toward the light (a process known as **phototropism**).

Plants can also respond to gravity (a process called **geotropism**). Two types of geotropism can occur in the same plant. The roots grow downward, toward earth's gravity

(positive geotropism), while the stem grows upward, away from earth's gravity (negative geotropism). Other tropisms in plants include *chemotropism,* growing toward chemicals; *hydrotropism,* growing toward water; and *thigmotropism,* growing toward touch.

hydrotropism

Many aspects of plant growth and development, including most tropisms, are regulated by internally produced chemical messengers called *hormones.* Five main types of plant hormones are presently known: auxins [ok′sĭnz], cytokinins [sī′tə·kī′nĭnz], gibberellins [jĭb′ə·rĕl′ĭnz], abscisic acid, and ethylene [ĕth′ə·lēn′]. Very minute amounts of these chemicals can have drastic effects on plant growth.

thigmotropism

**Auxins,** the first plant hormones to be discovered, are growth hormones secreted by the tips of shoots. Auxins are involved in many tropisms, including phototropism and thigmotropism; auxin causes the stem or shoot of the plant to grow faster on one side than the other, causing the shoot as a whole to slowly bend. In addition, auxins cause the apical stem to remain dominant by preventing lateral buds from developing into new shoots. Auxins also stimulate root growth and fruit formation.

A second group of plant hormones is the **cytokinins.** Like auxins, cytokinins are growth hormones that can affect stem growth; cytokinins are especially important in promoting cell division, the process by which new plant cells are produced from existing cells. In addition, cytokinins stimulate lateral buds to grow into new shoots (the opposite effect of auxin). Thus, auxins and cytokinins work together to keep the plant's systems in balance.

Like auxins and cytokinins, **gibberellins** are growth hormones that help to regulate normal plant growth; a plant that manufactures too little gibberellin will be much smaller than a normal plant. Gibberellins also trigger germination of dormant seeds and are thought to play an important role in the formation of flowers.

A fourth plant hormone is known as **abscisic acid,** or *ABA.* Unfortunately, its name is something of a misnomer, because it does not actually trigger leaf or fruit abscission as was once thought. Instead, ABA signals plant tissues to remain dormant, such as in a tree that is shut down for the winter or a seed that is waiting for spring. Because its effects are in many ways opposite to those of gibberellins, ABA and gibberellin balance each other to ensure that the plant becomes dormant and active again at the appropriate times.

The simplest of the main plant hormones is **ethylene,** a simple hydrocarbon gas with the formula $C_2H_4$. Ethylene is the hormone that signals leaf abscission when a tree prepares for winter. Ethylene also triggers green fruits to ripen; the higher the concentration of ethylene, the more quickly the fruit becomes ripe. Because it is a gas, ethylene diffuses easily through all plant tissues, and even through the air around a plant or fruit, to trigger abscission or ripening. (For this reason, keeping fresh fruits in an enclosed bag or container will cause them to ripen, and possibly spoil, much more quickly than if they were kept in an open container.) In addition, ethylene seems to play a role in the changing color of leaves in the autumn by stimulating the breakdown of chlorophyll.

## Main Types of Plant Hormones

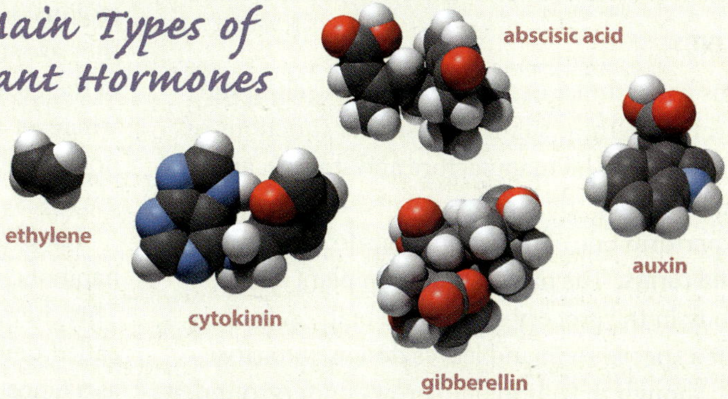

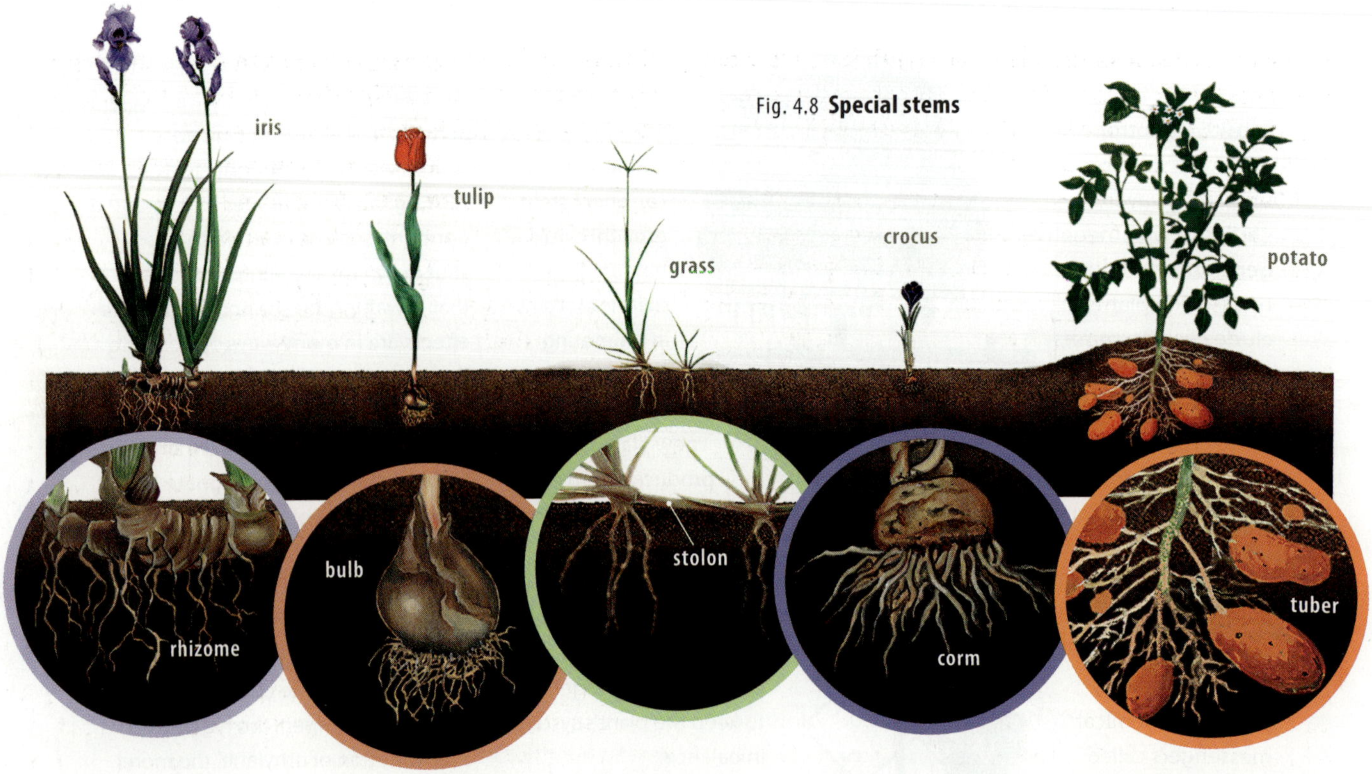

Fig. 4.8 **Special stems**

**Tissue culturing.** Modern technology has introduced an additional method of vegetative propagation known as **tissue culturing,** which allows seedlings to be produced from individual plant cells grown in the laboratory. In this technique, individual cells from a plant are separated and placed in small containers of a special nutrient solution, where the cells multiply to form small masses of millions of identical cells. Plant hormones are then applied to the cell cultures, causing leaves, stems, and roots to appear. The seedlings produced in this manner are then transferred to the soil to grow. Such techniques are especially helpful when working with plants that are not easily propagated by traditional methods or that are heavily infested by viruses that cause decreases in yield.

## Special Stems

All stems have similar design, as we have seen in the previous sections. These stems have similar functions of support, transport, food manufacture and storage, and protection. However, some plants have stems especially created to perform one of these functions to a high degree.

**Bulbs and corms.** The part of the *onion* plant that you generally buy in the grocery store is the bulb. A **bulb** is made up of a special stem and leaves designed to store food. Even though the onion bulb grows underground, it is nevertheless part of the shoot system. Pull an onion out of the ground, and you will see the roots branching from the bottom of the bulb. Some plants have swollen stems called **corms.** Botanists distinguish a corm as having a thicker stem and thinner leaves than a bulb. The *gladiolus* plant possesses corms.

**Rhizomes.** Thick stems that grow horizontally under the ground providing food storage and a means of vegetative reproduction are called **rhizomes** [rī′zōmz′]. The *iris* is a plant that depends on rhizomes. Irises grow in clumps and tend to spread out as they reproduce additional plants. A gardener occasionally needs to thin them out by digging up the rhizomes. By planting rhizomes in new areas, new plants are established.

**Stolons.** *Strawberry* plants are examples of plants with **stolons** or *runners.* Runners are well named because they are special stems that grow quickly along the surface of the ground. Their function is to produce new plants wherever a node touches the soil. Roots grow from the nodes, quickly establishing new plants in the area. Some grasses, such as *centipede grass,* also have stolons.

**Tendrils.** *Grape* plants and *Boston ivy* plants have a habit of climbing along fences or up the side of a wall. This characteristic is made possible by special stems called **tendrils.** When the stem contacts a surface such as a fence or wall, hormones stimulate cells on the opposite side of the stem to grow more rapidly than cells on the contact side of the stem, causing the stem to wrap itself around the object. This clinging supports the plant as it continues to grow in length. (Some tendrils are special leaves rather than stems.)

**Tubers.** Almost everyone is familiar with the *Irish potato,* which has a type of special stem called a **tuber.** Because it grows underground, the tuber is often thought of as a root, but a botanist looks much more closely at the design. The **eyes** of a potato are actually nodes consisting of buds and small leaves. That is why it will grow a new plant when you cut out the eye and plant it. Tubers provide the plant with food storage and a means of reproduction, while man benefits from their good nutrition.

**Thorns.** An example of special protection can be seen in the special stems called **thorns.** The *honey locust* tree possesses these kinds of thorns. Animals that otherwise would chew on the bark of the tree are discouraged by these long, sharp special stems. People as well as animals are discouraged from using plants with thorns. In Genesis 3:18 is recorded the origin of thorns as a result of man's sin and God's curse upon the earth. Botanists distinguish thorns (special stems) from *spines,* which are a type of special leaf; which a structure is can be determined by studying its design.

## Section Review 4.4

1. What are two advantages of producing a new plant by vegetative propagation instead of growing the plant from a seed?
2. Briefly describe the functions of the major plant hormones.
3. What are some examples of special stems used to store food for the plant?
4. How do tendrils help certain plants to "climb" objects?

### Critical Thinking

5. Describe as many tropisms as you can think of that would be characteristic of an ivy plant.

### Identify

asexual reproduction, vegetative propagation, cutting, adventitious roots, scion, stock, tropism, bulb, corm, rhizome, stolon, tuber

# 4.5 Roots: The Plant's Anchors

## Kinds of Root Systems

Root systems are of two basic designs, taproot systems and fibrous root systems. **Taproots,** which are commonly found in dicots, penetrate the soil with relatively little branching. They can reach deep into the ground where moisture is higher, and they are often used as a storehouse for food. Taproots can also be called **primary roots** because they grow directly from the plant embryo. Taproots that enlarge to store large quantities of food are called **fleshy roots.** Examples of plants with fleshy roots are *carrots, beets,* and *dandelions.*

**Fibrous roots,** like those of grasses and most other monocots, have no main section but spread out with very thin roots, forming a tangled mass in shallow soil. In these plants, the primary root dies very early and the fibrous root system consists of *adventitious roots* that arise from the stem. This mass of roots holds soil particles together and prevents the soil from being washed away. For this reason, plants with fibrous roots are planted on hillsides to hold the soil in place during heavy rains.

Fig. 4.9 **Root hairs of radish seedling**

Root systems, although not seen because they are underground, are quite large. They play an important role in the maintenance of the soil by breaking up compacted ground and adding organic material.

## Root Structure

The outer part of a root is covered by a protective layer called the *epidermis,* which helps protect the plant and prevent water loss. In a young root, the epidermis consists of a single layer of cells. Some of the epidermal cells develop into long, fingerlike projections called **root hairs.** These structures greatly increase the water-absorbing surface area of the root.

Root hairs are very delicate and live for only a short time. Root hairs are found near the root tip and in abundance on rootlets.

Inside the epidermis is a region of the root where food is stored, called the **root cortex.** The cortex surrounds the root's **central vascular cylinder,** which is also called the *stele* [stē'lē]. The vascular cylinder contains *xylem* tissues, which carry water to the stem, and *phloem* tissues, which carry food back to the root.

In the vascular cylinder, the xylem conducts water and dissolved minerals upward to the stem and leaves; the phloem conducts food made in the leaves, and sometimes in the stem, downward to all the cells that need it. Phloem cells, which contain cytoplasm but no nuclei, are connected end to end to form sieve tubes, just as in the phloem of the stem. Some mature roots contain pith cells at the center, which are used to store water.

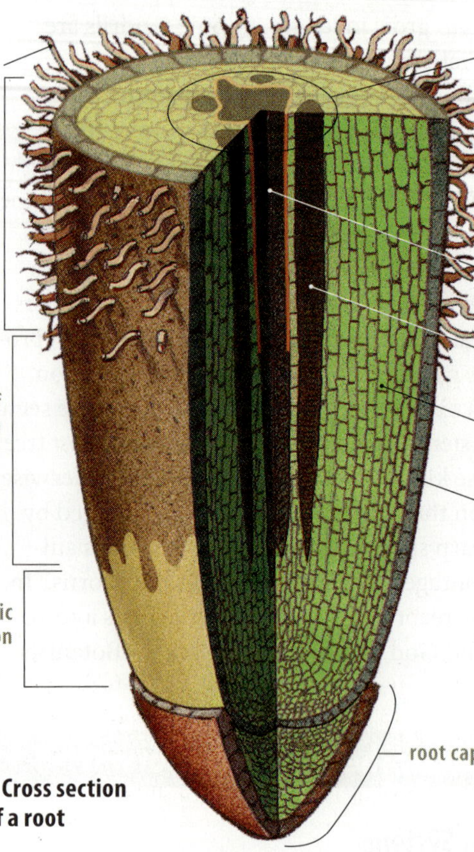


Fig. 4.10 **Cross section of a root**

## Root Growth

Near the tip is the main growing region of the root, the *apical meristem.* (Recall that meristematic tissue produces new tissues for growth and repair.) The apical meristem is responsible for forming the epidermis, the cortex, and the vascular cylinder. This growth is accomplished by cells undergoing cell division to increase the number of cells in the root. This increase of meristematic cells results in growth in length of the root *(primary growth).*

The increase in length of the root is observed in three different regions of the root tip. The region closest to the end is called the **meristematic region.** Here cell division takes place, adding new cells to increase the root's length. Just above the meristematic region is the **region of elongation.** In this region, cells increase in length, pushing the root further into the soil. The third region of growth is the **maturation region,** in which cells organize into the vascular cylinder. Root hairs grow from the epidermal cells in this region.

At the tip of the root providing a covering is the cone-shaped **root cap,** which consists of several layers of tough cells. The root cap protects the tender root tip as it pushes its way through the soil; it also secretes a slimy substance that allows the root to push through the soil more easily. As cells in these growth regions are very active, they give off significant amounts of carbon dioxide as a byproduct of their life processes. The carbon dioxide gas combines with water in the soil to form carbonic acid. This weak acid helps dissolve the minerals in the soil and make it easier for the root tip to move through the ground. The root cap also contains special gravity-sensing cells that guide the root through the soil in the proper direction.

As the roots of trees and shrubs mature, an additional layer of cells develops. It is a growth layer of meristematic cells located between the xylem and phloem called the *vascular cambium.* Vascular cambium cells carry on cell division and make it possible for this type of root to grow in diameter *(secondary growth).* Secondary growth accounts for the large size of tree roots. In later stages of secondary growth, a *cork cambium* is formed beneath the epidermis. It replaces the epidermis and forms a tough, protective layer of cork that also prevents the roots from absorbing water. The responsibility for water absorption remains with the younger roots that still possess root hairs.

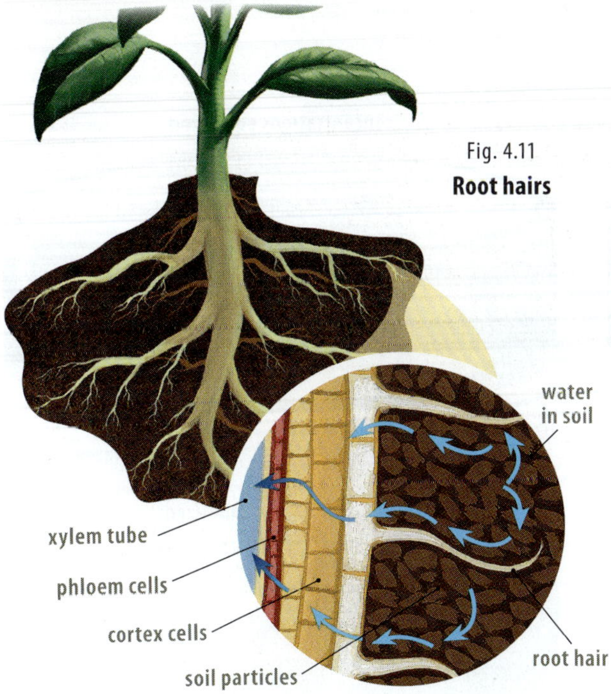


Fig. 4.11 **Root hairs**

## Lateral Roots

Unlike the branches of a stem, **lateral roots** do not develop from buds but from the central vascular cylinder of a root. Lateral roots, or *secondary roots,* originate from the primary or tap root; they have the same structural characteristics.

## Absorption and Transportation

Root hairs absorb water and minerals by allowing these substances to pass through the cell membrane (the thin covering of the cell) and enter the cell. All cell membranes are *permeable* [pûr′mē·ə·bəl] to certain substances. This means that certain substances are allowed to pass through the membrane.

**Diffusion.** You may recall that the molecules of matter are in constant motion. If you have ever added a drop of food coloring to a glass of water, you have seen one effect of this random molecular motion. The dye gradually spreads throughout the water because the dye molecules collide with the water molecules until the dye permeates the water. This process of mixing molecules of one substance through another by random molecular motion is called **diffusion.** Diffusion always occurs from areas of higher concentration to areas of lower concentration; that is, molecules always diffuse toward areas where there are fewer molecules of that type.

**Osmosis.** The cell membrane of a living cell is *semipermeable;* that is, it allows some substances to diffuse through it but not other substances. Small molecules such as water can pass from one side of the membrane to the other by diffusion, but larger particles, such as dissolved salt or sugar, cannot pass through. Such one-way diffusion through a semipermeable membrane is called **osmosis.**

Like ordinary diffusion, osmosis occurs from higher concentrations to lower concentrations. For example, if salt water and fresh water are separated by a semipermeable membrane, water will diffuse through the membrane by osmosis from the freshwater side, where the concentration of water is higher, to the saltwater side, where the concentration of water is lower.

Osmosis occurs in all plants and animals. Most nutrients and gases needed by living cells are in the form of solutions, water containing dissolved substances (solutes). The concentration of solutes in the water surrounding a cell determines the rate and direction of osmosis through the cell membrane. During osmosis, water moves toward the area of highest solute concentration, where the concentration of water molecules is the lowest.

Root hair cells acquire water from the soil by osmosis. If there is a higher concentration of solutes on the inside of the root hair than on the outside, water will move by osmosis from the outside of the root hair cell through the semipermeable membrane to the inside of the root hair cell. As the concentration of water in the root's surface cells increases, it exceeds that in neighboring cells

Fig. 4.12 **Diffusion in a liquid**

Fig. 4.13 **Diffusion and osmosis**

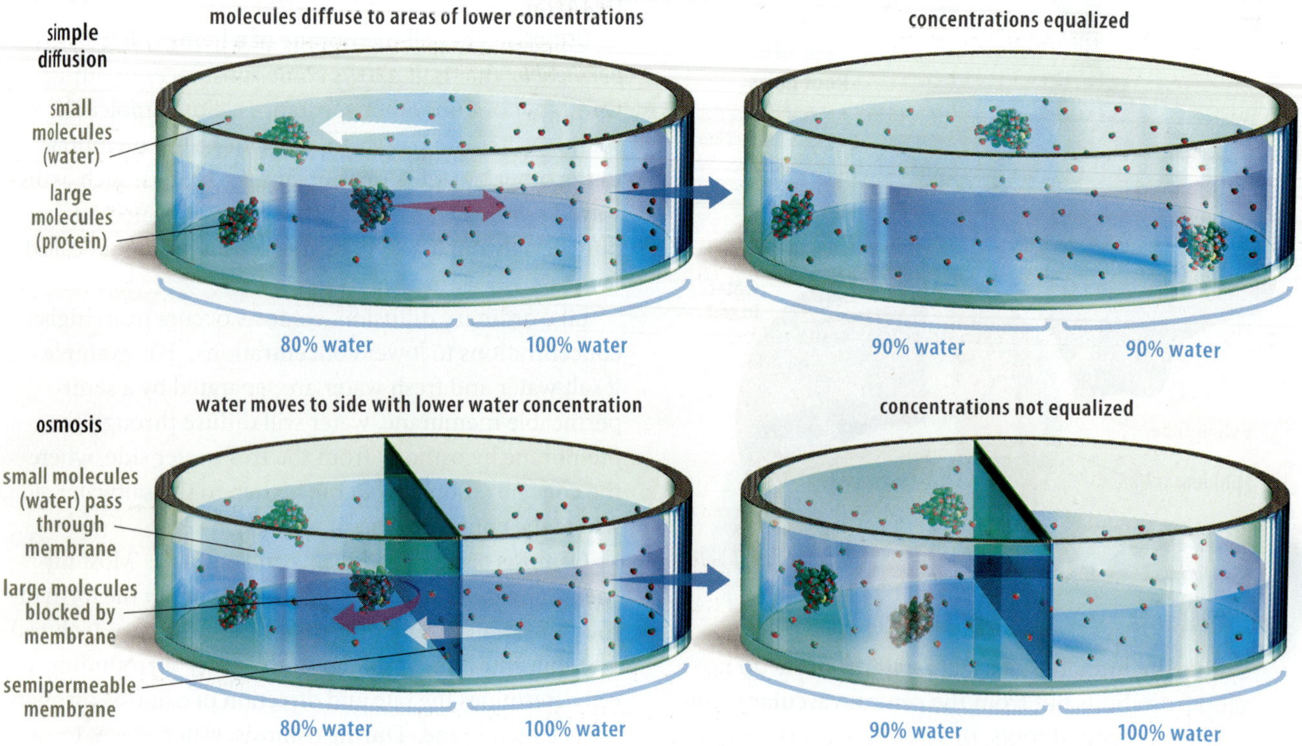


farther inside of the root. As water moves out of the surface cells and into these neighboring cells by osmosis, the concentration of water in the surface cells decreases, enabling more water to enter the surface cells from the soil. In this way, water continuously enters the root and moves toward its core so that it can be transported upward into the stem.

This osmosis of water into a root can also be reversed. If enough salt or some other soluble mineral (fertilizer, for example) is poured on the soil, the concentration of dissolved substances could become higher on the outside of the root cells than on the inside of the cells. This could be sufficient to cause water to move by osmosis out of the root. If this outward osmosis were to continue for very long, the plant's root cells could lose so much water that they would dry out and die. Gardeners refer to this situation as "burning" a plant when it occurs as a result of excessive fertilizing.

Plants can also be overwatered. If the soil around the root system is saturated with too much water, the naturally occurring air pockets in the soil become flooded. To carry on their life processes, root cells need oxygen, which is normally obtained from the air in the soil air spaces. If water displaces the air, the supply of oxygen is lost and the cells die.

**The sap stream.** After root hairs absorb water by osmosis, the water and absorbed minerals must flow up through the xylem tubes; this upward flow is the **sap stream.** Several factors drive the flow of the sap stream. *Root pressure,* the osmotic force that helps move water to the vascular cylinder, also helps transfer water up the stem. Another factor is *capillarity* [kăp′ə·lăr′ĭ·tē; also called capillary action], the upward movement of liquids through a very narrow tube. Capillarity occurs in plants because the attraction of water molecules to the surface of the xylem tubes *(adhesion)* is greater than their attraction to each other *(cohesion).*

Root pressure and capillarity by themselves cannot supply the force needed to raise water high into a plant or tree. The chief force that draws water up the stem is believed to be **transpiration pull** exerted by the leaves. As water evaporates from the leaves by transpiration, the leaf cells become deficient in water and tend to pull water into themselves in an attempt to establish a water balance; this pull is strong enough to draw the sap stream from the roots to the top of the tree.

## Section Review 4.5

1. Describe how roots grow in length.
2. What is the function of root hairs?
3. How can you tell a root hair from a lateral root?
4. How can a plant be killed by overwatering?
5. Name the four main functions of roots discussed in the text.

**Application**

6. What features of internal design are common to roots, stems, and leaves?
7. Explain the difference between osmosis and diffusion.

**Identify**

taproots, fleshy roots, fibrous roots, root cortex, lateral roots, capillarity, transpiration pull

**Observe**

- Collect as many different kinds of roots as possible from the two categories of root systems. Keep in your notebook sketches of these various roots. Note similarities and differences among roots of the same category.

## Chapter 4 Review

**Define**

1. terminal bud, lateral bud
2. internode
3. leaf scar
4. lenticel
5. primary growth, secondary growth
6. bark
7. sieve tubes
8. wood
9. pith
10. heartwood, sapwood
11. annual growth rings
12. vascular bundles
13. rind
14. vegetative reproduction
15. vegetative propagation
16. adventitious roots
17. grafting
18. scion, stock
19. tropism
20. taproots, fleshy roots, fibrous roots
21. root cortex, vascular cylinder
22. root cap
23. lateral roots
24. diffusion, osmosis
25. sap stream
26. transpiration pull

**Identify**

1. the two primary functions of a tree's bark
2. five types of vegetative propagation
3. the five types of tropism and the stimulus that causes each
4. five important plant hormones
5. seven examples of special stems

**Explain**

1. What are the differences between tracheids and vessels?
2. Describe the purpose of pith in a young stem.
3. How does the internal structure of herbaceous monocots differ from that of herbaceous dicots?
4. How are seedless plants propagated?
5. How does a root grow longer and thicker?
6. What are the purposes of the root cap?

**Apply**

1. Why are the terms *xylem* and *wood* sometimes considered synonymous?
2. Describe the microscopic featues that would allow a botanist to distinguish a spine from a thorn.
3. How do root hairs help a plant to absorb water and nutrients from the soil?
4. How does osmosis differ from simple diffusion?

CHAPTER 5

# Variety in the World of Plants

**Key Concepts**

- Classification
- Scientific names
- Conifers and other gymnosperms
- Ferns, club mosses, and horsetails
- Mosses and liverworts
- Algae
- Fungi
- Slime molds and lichens

## 5.1 Classification of Living Things

Biologists classify living things by arranging organisms into groups and subgroups. This process of arranging organisms into groups based on similarities is called **taxonomy** [tăk·sŏn′ə·mē]. A standardized taxonomy is important to biologists because it allows them to clearly describe similarities between organisms and to clearly distinguish similar organisms.

**Table 5.1 Classification of Various Plants**

| |  Rose |  Ponderosa pine |  Bracken fern |
|---|---|---|---|
| Kingdom | Plantae | Plantae | Plantae |
| Phylum | Anthophyta | Coniferophyta | Filicophyta |
| Class | Dicotyledonae | Pinopsida | Filicopsida |
| Order | Rosales | Pinales | Filicales |
| Family | Rosaceae | Pinaceae | Dennstaedtiaceae |
| Genus | *Rosa* | *Pinus* | *Pteridium* |
| Species | *chinensis* | *ponderosa* | *aquilinum* |
| Scientific name | *Rosa chinensis* | *Pinus ponderosa* | *Pteridium aquilinum* |

### The Basis for Modern Classification

In the mid-1700s, the Swedish naturalist **Carolus Linnaeus** (1707–1778) devised a simple, practical method of scientific classification. We use his general method today, with some changes. Linnaeus, an outstanding naturalist, was also an excellent teacher and a dedicated and outspoken Christian. Because he accepted the biblical account of Creation, Linnaeus concluded that God created distinct types of plants and animals.

Linnaeus developed his classification system using some ideas of **John Ray** (1627–1705), an English clergyman who devoted much time to natural studies and became well known for his work with plants. It was Ray who introduced a precise concept of *species,* which he defined as a group of individuals derived from similar parents and themselves capable of reproducing their kind.

*Linnaeus, observing that certain plants and animals often had similar characteristics, set up a classification system based on basic body structure of animals and on arrangement of reproductive organs in plants.* He grouped plants and animals like the topics in an outline. In the years after Linnaeus, biologists added more groups and subgroups to the Linnaean system, eventually resulting in the system of classification we use today.

In this system, all living things are divided into main groups called **kingdoms.** Traditionally, only two kingdoms were recognized—kingdom **Plantae** (the plant kingdom) and kingdom **Animalia** (the animal kingdom)—but it is known now that many organisms cannot be classified into two simple kingdoms. A set of body structure characteristics is used to further divide each kingdom into a few main groups called **phyla** [fī′lə; sing. phylum; in botany, phyla are also called *divisions*]. Each phylum is divided according to a different set of characteristics into **classes;** the classes are divided into **orders,** and the orders are divided into **families.** The families are divided into **genera** [jĕn′ər·ə; sing. genus], which are divided into one or more **species.**

The outline moves from general to specific. The kingdoms are general groups that include many different kinds of animals or plants; the species are specific groups that usually include organisms with very similar characteristics. The system can be illustrated with the classification of a strawberry plant (Table 5.2).

It is important to realize that what the Bible means by "created kinds" is not necessarily identical with what biologists classify as a "species." A *species* is a group of very similar organisms that can interbreed freely in nature, whereas the biblical **kind** (Hebrew *min*) means a group of similar organisms that are all descended from a single group of originally created organisms and may

## Table 5.2 Strawberry Classification (*Fragaria ananassa*)

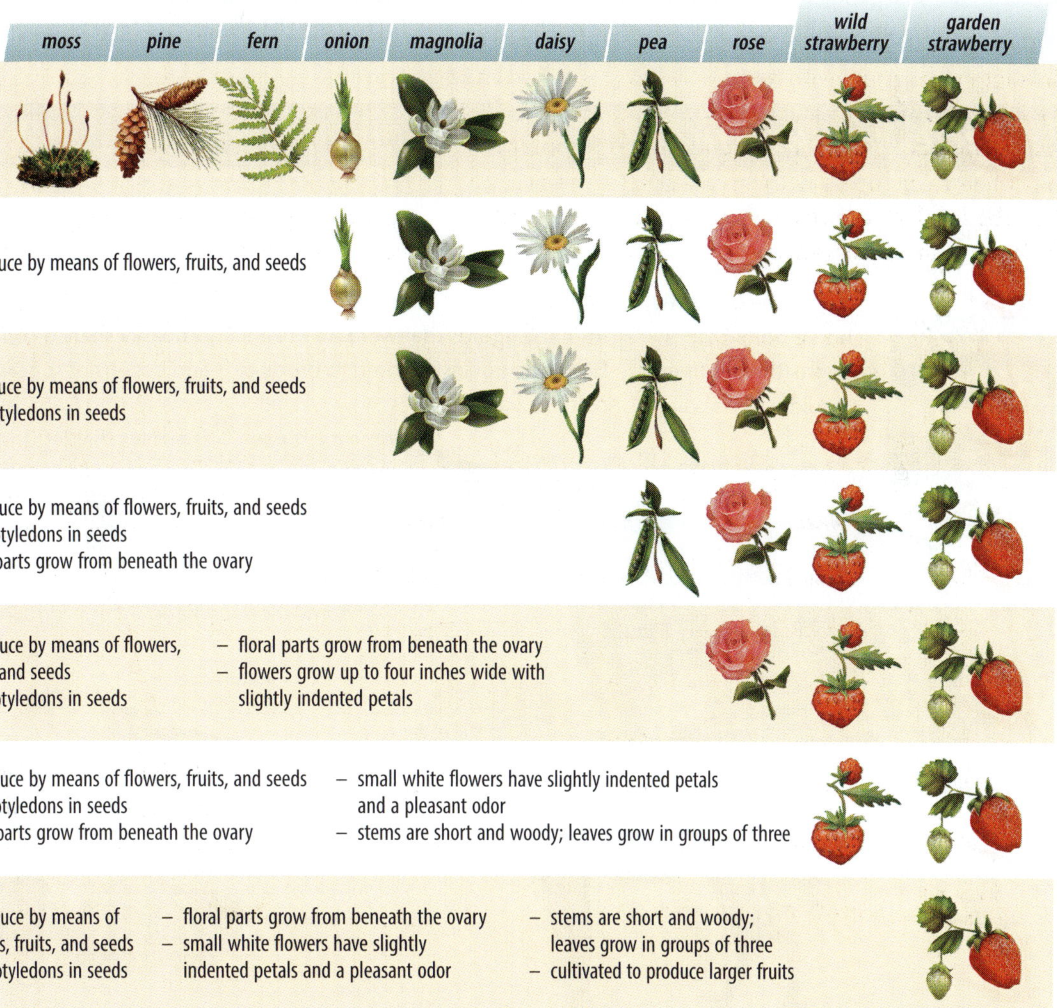


| | |
|---|---|
| **Kingdom** Plantae | – plants |
| **Phylum** Anthophyta | – reproduce by means of flowers, fruits, and seeds |
| **Class** Dicotyledonae (dicots) | – reproduce by means of flowers, fruits, and seeds<br>– two cotyledons in seeds |
| **Order** Rosales | – reproduce by means of flowers, fruits, and seeds<br>– two cotyledons in seeds<br>– floral parts grow from beneath the ovary |
| **Family** Rosaceae | – reproduce by means of flowers, fruits, and seeds<br>– two cotyledons in seeds<br>– floral parts grow from beneath the ovary<br>– flowers grow up to four inches wide with slightly indented petals |
| **Genus** *Fragaria* | – reproduce by means of flowers, fruits, and seeds<br>– two cotyledons in seeds<br>– floral parts grow from beneath the ovary<br>– small white flowers have slightly indented petals and a pleasant odor<br>– stems are short and woody; leaves grow in groups of three |
| **Species** *ananassa* | – reproduce by means of flowers, fruits, and seeds<br>– two cotyledons in seeds<br>– floral parts grow from beneath the ovary<br>– small white flowers have slightly indented petals and a pleasant odor<br>– stems are short and woody; leaves grow in groups of three<br>– cultivated to produce larger fruits |

refer to a species, a genus, or a family. Organisms that are biologically capable of producing offspring (fertile or not) are the same kind, but some members of a kind may have lost the ability to reproduce with each other, making them separate species.

## The Scientific Name

Linnaeus used Latin, the scientific language of his day, to name organisms. His system for assigning scientific names is called **binomial nomenclature** [bī·nō′mē·əl nō′mən·klā′chər] because he used two words for each scientific name. The first word indicates the genus, and the second word the species. For instance, the red wolf is *Canis rufus,* the gray wolf is *Canis lupus,* and the coyote is *Canis latrans.* The first part of each scientific name tells that they all belong to the genus *Canis;* the second part distinguishes them from other species of that genus. The scientific name is always either italicized or underlined. The genus name is always capitalized, but the species name is not capitalized. *For scientific purposes, an organism's scientific name is preferred over its popular name because many plants or animals have different popular names in different areas and languages.* The Linnaean system of nomenclature is still in use today and is little changed from the basic system he worked out in the 18th century.

Scientific names are often easy to learn and use because they tell some outstanding characteristic of the plant or animal. For example, the scientific name for the rice plant is *Oryza sativa.* The genus name *Oryza* is Latin for "rice," and the species name, *sativa,* is Latin for "cultivated." Thus, the scientific name of the rice plant tells us what type of plant it is and that it is the species of rice grown for food.

## Modern Taxonomy

Linnaeus built his system of classification upon the evidence of a world designed by God. His system of identification was based upon observable traits in each organism. Through his careful investigations, he

### *Species versus Kinds*

Understanding the LIVING CREATION

The biblical kind is usually broader than our modern species. For example, the domestic horse, donkey, and zebra (three different equid species) can interbreed to produce hybrid species and therefore all belong to the same biblical kind and are all descended from the equids that were on the ark. Domestic horses are bred to produce certain characteristics; however, all domestic horse breeds are part of the same species. Likewise, domestic dogs, wolves, jackals, and coyotes are in the same kind; the domestic cat and several species of wildcats probably share a common lineage; and dozens of species of sparrows have sprung from the three pairs of sparrows that left the ark. Creationists recognize that the origin of new species within a kind does occur. The origin of new species within a kind, however, is not the same as changing one kind into another.

attempted to identify each species according to its uniquely created characteristics.

As new species were discovered, it became apparent that not all living organisms easily fit into the two original Linnaean kingdoms. God's creation is more intricate than first perceived. In the 20th century, scientists discovered that some groups of organisms even have fundamental differences in their cellular structure and processes; based on these differences, organisms are divided into **eukaryotes** [yo͞o·kăr′ē·ōts], which have membrane-bound nuclei and membrane-enclosed organelles, and **prokaryotes** [prō·kăr′ē·ōts′], which do not have membrane-bound nuclei. Today, living things are commonly divided into the following six kingdoms:

**Plantae**—green plants; multicellular eukaryotes that have cell walls and usually perform photosynthesis

**Fungi**—all fungi, including yeasts, molds, and mushrooms; multicellular eukaryotes that have cell walls and do not conduct photosynthesis

**Animalia**—multicellular animals; eukaryotes without cell walls; usually can move and collect food

**Protista**—eukaryotes that are not green plants, fungi, or multicellular animals; mostly unicellular organisms, such as amoeba, euglena, and paramecium

**Eubacteria**—true bacteria, including common disease-causing bacteria; prokaryotes with cell walls made of the substance peptidoglycan [pĕp′tĭ·dō·glī′kən]

**Archaebacteria**—prokaryotes that do not contain peptidoglycan in their cell walls and have other key differences from true bacteria; many live in extreme environments (e.g., very cold habitats or highly saline, acidic, or alkaline water)

These six kingdoms are sometimes divided into three *domains* based on differences in cell chemistry: domain *Eukarya* (kingdoms Plantae, Fungi, Animalia, and Protista), domain *Bacteria* (kingdom Eubacteria), and domain *Archaea* (kingdom Archaebacteria).

## Evolutionary Misconceptions

Unfortunately, taxonomy was one of the first fields affected by modern evolutionary ideas in the late 1800s. Eager to apply this philosophy to the classification of organisms, scientists began to devise a host of hypothetical evolutionary relationships. Biological classification has begun to reflect philosophical suppositions invented to show alleged evolutionary relationships between organisms or *phylogeny* (the supposed development of a species, genus, or group over time). Instead of using common traits to classify organisms from general to specific for identification, evolutionists attempt to classify them according to presumed evolutionary ancestry. *Phylogenetic trees* are drawn to depict all forms of life coming from a single life form. The major weakness of such classification is the complete lack of convincing intermediate forms. Both in the fossil record and in the living creation today, birds are birds, reptiles are reptiles, plants are plants, and so on.

How should a Christian view taxonomy? The answer is simple. Use it as a tool to describe the wisdom of the Creator in nature, as John Ray and Linnaeus did. Use it as a key to identify the many plants and animals that you see, as Linnaeus did. Recognize that taxonomy is a manmade system of organization that aids in better understanding God's living creation.

## Section Review 5.1

1. What are the seven main categories of the classification system, from most general to most specific?
2. What are the two kingdoms that have been traditionally recognized since Linnaeus's time?
3. What are the two parts of an organism's scientific name?
4. Why have other kingdoms and domains been proposed?
5. How has the acceptance of evolution affected the science of taxonomy?

### Critical Thinking

6. According to Genesis, did God create each individual species? Explain.
7. Although they are in different genera, the common bottlenose dolphin *(Tursiops truncatus)* and the false killer whale *(Pseudorca crassidens)* share many physical characteristics and can interbreed to produce fertile offspring. From a biblical perspective, do these two animals belong to the same kind or to different kinds? Explain your answer.
8. How is the fact that organisms can be divided into distinct categories evidence against the evolutionary idea that all organisms developed from a common ancestor by gradual changes?

### Identify

taxonomy, Carolus Linnaeus, John Ray, kind, binomial nomenclature, eukaryote, prokaryote

# 5.2 Conifers and Other Gymnosperms

## Vascular and Nonvascular Plants

In chapters 1–4, we discussed the angiosperms, plants that reproduce by seeds formed in flowers. Angiosperms make up the phylum **Anthophyta** and are part of the **vascular plants** *(tracheophytes)*. The vascular plants also include gymnosperms [jĭm′nə·spûrmz′] (pines, firs, cedars, etc.), as well as the ferns and several other spore-producing plants. All vascular plants have vascular tissue for conducting water and minerals from the roots to the leaves and sugar from the leaves to all parts of the plant. *All plants are either vascular plants or nonvascular plants* (bryophytes [brī′ə·fītz′]); the majority of the familiar plants are vascular plants.

There are two groups of vascular plants—**seed plants,** which includes angiosperms and gymnosperms, and **nonseed plants,** which includes ferns, mosses, and others. In this chapter, we will study the vascular plants that reproduce by seeds that are not formed in flowers (gymnosperms) and the vascular plants that reproduce by spores rather than by seeds. The spore-producing vascular plants include ferns, horsetails, and club mosses.

We will end the chapter with a discussion of some of the **nonvascular plants,** including mosses and liverworts. We will also discuss fungi and algae, which are traditionally studied with plants but are no longer considered part of the plant kingdom.

### Table 5.3 A Survey of Plants, Fungi, and Algae

| | | Typical characteristics | Examples |
|---|---|---|---|
| Vascular Plants | ***Seed plants*** | | |
| | **angiosperms** (Phylum Anthophyta)<br>- monocots (Class Monocotyledonae)<br>- dicots (Class Dicotyledonae) | vascular tissue, reproduce by seeds formed in flowers | dandelions, lilies, grasses, oaks |
| | **gymnosperms**<br>- cycads (Phylum Cycadophyta)<br>- ginkgoes (Phylum Ginkgophyta)<br>- conifers (Phylum Coniferophyta) | vascular tissue, reproduce by seeds formed in cones | pines, firs, hemlocks, spruces, redwoods |
| | ***Nonseed plants*** | | |
| | **ferns** (Phylum Filicophyta)<br>**horsetails** (Phylum Sphenophyta)<br>**club mosses** (Phylum Lycopodiophyta) | vascular tissue, reproduce by spores | oak fern, tree fern, equisetum; club moss (sometimes called ground pine) |
| Nonvascular Plants | **mosses and liverworts** (Phyla Bryophyta and Hepaticophyta) | no true vascular tissue, reproduce by spores | sphagnum moss, haircap moss, *Marchantia* |
| | **fungi** (Kingdom Fungi; many phyla) | no vascular tissue, reproduce by spores or budding, no chlorophyll | shelf fungus, mushroom, mold, yeast |
| | **algae** (Kingdom varies*; many phyla) | mostly unicellular or colonial, contain chlorophyll, reproduce by fission or conjugation | spirogyra, kelp, chlorella |

Note: Because scientists disagree on how organisms should be divided, other texts may use different groupings or names.
*Prokaryotic blue-green algae are placed in kingdom Eubacteria. Other algae are typically placed in Protista, but some phyla are sometimes placed in Plantae.

## Gymnosperms

Plants that do not produce flowers but do produce seeds are called **gymnosperms.** The term *gymnosperm* means "naked seed," and *gymnosperms produce seeds not covered by the walls of an ovary.* Instead of flowers or fruits, most gymnosperms produce cones or conelike structures for the protection of the seeds. We will describe three of the main groups of living gymnosperms based upon their characteristic structure—the *conifers,* the *cycads* [sī′kădz′], and the *ginkgoes* [gĭng′kōz].

## Conifers

**Conifers** (phylum Coniferophyta) are the very familiar cone-bearing gymnosperms. Pines, spruces, firs, cedars, cypresses, hemlocks, larches, and redwoods

Fig. 5.1 **Types of cones**

Fig. 5.2 **Giant sequoia**

are examples of conifers. The term *conifer* refers to the cones that are produced by these trees and shrubs. The *yew* and *juniper* are conifers, but their cones are not obvious because they form a berrylike structure. Juniper "berries" are cones with fleshy scales fused together, while yew seeds are surrounded by a fleshy cup called an *aril*. Some conifers are shrubs, but most are tall, straight trees with branches growing nearly horizontally from a single large stem.

**Reproduction.** *Conifers produce two types of cones*, which are usually produced on separate shoots of the same tree. The pollen-producing cones are called **staminate cones** (like the stamen of the flower, the part that produces pollen). These cones are small and inconspicuous and are shed after the pollen season is over. **Ovulate** [ŏv′yə·lāt′] **cones** (or *seed cones*) are the seed-producing cones. They produce the ovules, which are fertilized by the pollen from staminate cones, resulting in seed formation. Ovulate cones are larger than staminate cones and range in size from less than an inch to more than two feet in length, depending upon the species. Ovulate cones are woody structures, each consisting of layers of cone scales attached to a central stalk. The seeds develop between the cone scales. Ovulate cones remain on the tree for many months while the seeds develop. The seeds produced by conifers are generally winged, and they are produced in pairs. Each pair of seeds develops on a cone scale; when the cone is mature, the scales spread open to disperse the seeds into the wind. Conifers depend upon the wind for pollination and reproduction, and our Creator-God has specifically designed each species' cones to capture its own kind of pollen. The cone's shape directs pollen from one scale to the next, increasing the chance of pollination.

**Some noteworthy conifers.** Conifers are among the largest trees in the world. The *Douglas fir* is one of the biggest trees, reaching a height of almost 300 feet. It is native to the western United States and Canada, where it forms great forests. The *giant sequoia* [sĭ·kwoi′ə] in central California is another huge conifer; many of these trees are estimated to be between 2,000 and 3,500 years old. Even older are the gnarled *bristlecone pines* of California's White Mountains—some of the oldest living things on earth—which were already seedlings when Abram left Ur of the Chaldees to go to the Promised Land. The *coast redwood* is very similar to the giant sequoia and also grows in California and Oregon. These redwood trees are the tallest known living things—some are 370 feet tall. The conifers furnish us with softwood lumber, the chief source of building materials for houses, doors, frames, panels, boxes, posts, planks, and beams. Most softwood lumber comes from pines, Douglas fir, Western firs, hemlock, and redwood.

Some of the most common conifers in temperate latitudes are the *pines,* which are evergreens that have needles two or more inches long found in clusters of two to five. Commercially, the most important pine in the southeastern United States is the large *loblolly pine,* which has two or three long needles in each cluster, although the *longleaf pine* and *slash pine* are also widespread in this region. An important pine found in the western United States is the *ponderosa pine,* also known as the *western yellow pine.*

Sometimes *spruces* are confused with pines. However, the single needles of the spruce protrude all around the twig, giving it a bristly appearance.

**Provisions for conifer survival.** In God's marvelous provision, several features enable conifers to flourish in harsh climates and poor soils. Their needles retain moisture well during arid summers, in which early-morning fog condenses on the needles, drops to the ground, and waters the tree. During the winter, when deciduous trees are barren and leafless, conifers still carry on photosynthesis and produce food. The tiny area and waxy covering of pine needles keeps snow from sticking, and any snow that does accumulate soon falls to the ground as the branches dip and shrug it off. The "steeple" shape of such conifers as firs and spruces also keeps snow from accumulating even as it helps the tree absorb the rays of the low winter sun. The root systems of conifers seem to be efficient at extracting nutrients from barren soils, and evergreen conifers are even able to use moisture condensed from clouds on mountaintops.

**Bunya-bunya, kauri, and Wollemi pine.** Three peculiar conifers are known to the people of the South Pacific. The *bunya-bunya,* which grows in Australia, has bright green leaves that are broad for a conifer. In New Zealand, the *kauri* [kou′rē] is even more unlike the usual pine tree. You would hardly recognize it to be a conifer, because its leaves are very broad. From these trees is extracted a resin used to make varnishes, adhesives, and linoleum.

In 1994, a grove of unusual conifers was found growing in a rain forest in the deep recesses of Australia's Wollemi National Park. In the fossil record, such *Wollemi pines* are often found in the same strata as the dinosaurs. Wollemi pines have unusual knobby-textured bark and waxy, fernlike leaves. In size, they are comparable to other conifers; the largest living Wollemi pine discovered so far is about 130 feet tall.

Fig. 5.3
**Tree height comparison**

Fig. 5.4
**Loblolly pine**

## *Variety of Plants* by Alfred M. Rehwinkel

Design in the LIVING CREATION

A remarkable aspect of plant life is the millions and billions of forms and species in which plants appear, from the microscopic diatom and plankton of the sea to the majestic giants of the redwoods of California. No two plants are exactly alike, whether shrubs or trees, flowers or weeds, grains or vegetables, and even the leaves of the same species are not mere duplicates one of the other. These marvels become even more spectacular when one compares the vegetation of one climatic zone with that of another, or the plants of one altitude and latitude with that of another, and desert plants with those of the rainy regions of our world.

This wonderful variation in the plant world is one of the things that makes traveling in foreign lands and climates so interesting. Every continent has its own unique kind of plant life. The magnificent vegetation of India and Ceylon [now Sri Lanka] is quite distinct from the equally fascinating vegetation of South America. And the native trees, shrubs, flowers, and grasses of Australia are quite different from those of Africa and North America.

I recall an exciting experience I had while traveling in Brazil. We were driving through a typical tropical rain forest and a new, wonderful world of God's creation was opened up to me. While we happened to be right in the midst of this tropical wilderness, our faithful Volkswagen suddenly developed tire trouble, causing a considerable delay. While my companions were busy correcting this problem, I used the time to examine more closely the vegetation surrounding us on all sides of our narrow trail. I was, first of all, profoundly impressed with the density of the vegetation. I found that the trees and shrubs and countless other varieties of plants were so closely knit together and interwoven that they formed a solid wall, so dense that neither man nor beast could penetrate them, except possibly a snake. And yet, I noticed that every plant had retained its own peculiar characteristics. None was confused with the other, and it was evident that even here every plant had propagated its own kind as the Creator had ordained it. . . . Every plant had remained strictly distinct from every other plant, though the roots, branches, and flowers were closely intertwined. I had somewhat similar experiences in the jungles of India and Africa.

All plants are wonderful, and the more carefully one examines them the more mysterious they become. A student of botany in a well-equipped botanical laboratory has a wonderful opportunity to observe this when examining these wonders. If he has eyes to see, he will discover that it is utterly absurd and irrational to assume that these wonders could have just happened, when everywhere he finds careful design and meticulous laws to govern them.

*Alfred M. Rehwinkel,* The Wonders of Creation, *Bethany House Publishers, a division of Baker Publishing Group, Copyright ©1974, Used by permission.*

### Unusual Gymnosperms

**Cycad.** The **cycads** (phylum Cycadophyta), which resemble palm trees, grow today only in certain tropical and subtropical regions. *Zamia* is a genus of cycads that are native to the Americas and includes the only cycad species native to the United States. The most common cycad in cultivation is the Asian *sago* [sā′gō] *palm (Cycas revoluta).* Fossil remains show that many varieties of cycads once lived that are not found today. As with many other kinds of plants and animals that lived in the past, they were destroyed by a major worldwide catastrophe such as the Flood of Noah's time. The environment was apparently so changed by this event that most cycads were unable to reestablish themselves and became extinct.

**Ginkgo.** The **ginkgo,** or *maidenhair tree (Ginkgo biloba),* is an oriental gymnosperm with two-lobed, fan-shaped leaves. No other tree has leaves remotely resembling those of the ginkgo; and unlike most other gymnosperms, which are evergreens, ginkgoes shed their leaves in the fall. This tree, which is native to China, is often used in parks and along streets in large cities as a decorative tree. The ginkgo is considered a "living fossil" because it is the only living species in phylum Ginkgophyta. As with the cycads, most ginkgo species apparently did not survive the post-Flood environment.

Fig. 5.5 **Sago palm (cycad)**

Fig. 5.6 **Ginkgo tree and leaf**

## Section Review 5.2

1. Write a brief overview of plant classification.
2. How are the reproductive structures of gymnosperms different from those of angiosperms?
3. How are staminate cones different from ovulate cones?
4. What evidence suggests that cycads and ginkgoes were once more widespread than they are now?

**Identify**

conifers, Wollemi pine, cycads, ginkgo

**Observe**

- Collect some samples of staminate and ovulate cones and locate the developing seeds between the cone scales on the ovulate cones.

## 5.3 Vascular Spore Plants

### Ferns

**Ferns** (phylum Filicophyta) are nonflowering vascular plants with spore-bearing leaves and horizontal underground stems. Ferns are similar to gymnosperms in that they do not form flowers. But, whereas gymnosperms form seeds, *ferns reproduce by spores.* **Spores** are tiny, one-celled reproductive structures that can grow into distinct or independent organisms under the proper conditions. (*Seeds,* in contrast, are reproductive structures of more than one cell.) Most ferns are only a foot or two tall, but some *tree ferns* in tropical regions grow as tall as 80 feet and have leaves 12 to 14 feet long. Other types of ferns hang from the branches of trees.

Fern stems, except for those of tree ferns, grow just below or along the soil's surface. The roots grow downward, and large leaves, called **fronds,** grow upward by unrolling from *fiddleheads*. The above-ground "stem" of a fern plant is actually a petiole. On their leaves, ferns produce small, brownish dots called **sori** [sôr′ī: sing. sorus; Greek, "a heap"], that are really groups of minute spore cases called **sporangia** [spə·răn′jē·ə: sing. sporangium]. Underneath the soil are the fern's underground stems, or **rhizomes.**

A careful study of fossil remains shows that many regions of the earth were at one time lush with giant ferns. Some scientists believe that different atmospheric conditions existed in the pre-Flood world. Coal, petroleum, and natural gas appear to have been formed from masses of ferns and other plants (and great numbers of marine creatures), which may have been swept together and crushed under the weight of the earth's surface during the worldwide Flood of Noah's day. Little, if any, of these fuels are being naturally produced today.

The life cycle of a fern involves both *asexual* (no union of sex cells) and *sexual* (union of sex cells) reproduction; this marvelous process is called **alternation of generations.** The two generations are the sporophyte [spôr′ə·fīt′] and the gametophyte [gə·mē′tə·fīt′]. The **sporophyte** generation is the stage that you see and think of as a fern; as its name implies, it is an asexual stage that produces spores. The **gametophyte** generation is much smaller and is the sexual stage that produces *gametes* (reproductive cells).

A new fern's life cycle typically begins in July when the brownish sori appear on the undersurface or along the margins of the sporophyte's fronds. The sori may be round, kidney shaped, oblong, linear, curved, or star shaped; the sori of each fern species have a unique shape—an aid to identification and another example of the wonderful variety in God's living creation. The sori consist of dense clusters of minute sporangia, each containing between 48 and 64 microscopic spores. When the spores ripen, the sporangia burst open, and millions of spores are released.

Fig. 5.7 **Fern life cycle**

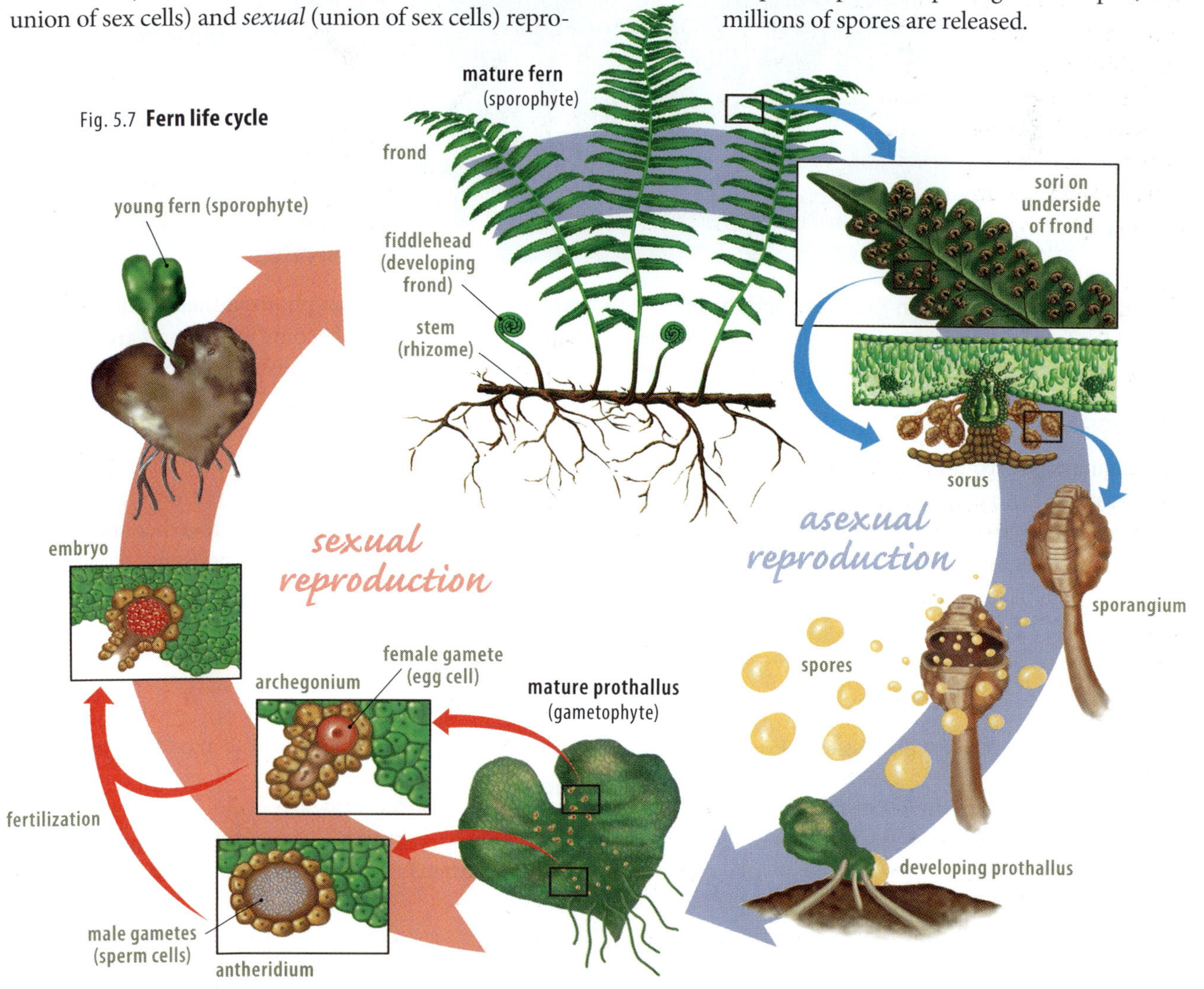

When the humidity of the soil is adequate, each spore develops into a tiny, green, heart-shaped *prothallus* [prō·thăl′əs: pl. prothalli], which is quite different from the original fern plant and is the beginning of the gametophyte generation. Because prothalli are typically only ¼ inch across and live for only three to seven weeks, they are seldom noticed. The tiny prothallus produces special reproductive organs (a male *antheridium* [ăn′thə·rĭd′ē·əm: pl. antheridia] and a female *archegonium* [är′kĭ·gō′nē·əm: pl. archegonia]), which in turn produce the reproductive cells. When the prothallus is fully developed, male cells swim through the moisture on the plant and fertilize the egg cell. A young fern plant (sporophyte stage) develops from the fertilized egg, and it usually requires between three and seven years to reach reproductive maturity. Then the sori begin to appear, and the life cycle of the fern begins again. Fern plants (sporophytes) live for several years and produce new fronds each year.

Fig. 5.8 **Lycopodium**

## Club Mosses

**Club mosses** (phylum Lycopodiophyta) are creeping vascular plants with erect stems that bear spores in club-shaped, conelike structures. Club mosses are also called *ground pine* because they grow along the ground in temperate regions and superficially resemble miniature pines. The stems of club mosses are slender, and their leaves are scalelike. They reproduce by spores.

**Lycopodium** [lī′kə·pō′dē·əm] is a typical club moss, with horizontal stems that trail along the ground or grow just below the surface. Upright branches arise from the stem. Lycopods produce spores in rather large, clublike sporangia called *strobili* (sing. strobilus) growing at the tip of the plant. When the sporangia crack open, the spores are dispersed as a yellow dust. A powder of dried lycopodium spores will explode and produce a brilliant white flash if ignited; therefore, lycopodium powder was used in camera flashes in the 1800s.

## Horsetails

**Horsetails** (phylum Equisetophyta) are vascular plants with unique hollow, jointed stems and very small leaves present only at the joints. The stems carry on photosynthesis. All but one genus of horsetails have become extinct and are known only from fossil remains. Many of these extinct horsetails grew as big as trees.

The epidermis, or outer layer of cells, of horsetails contains particles of the mineral *silica*, making the plant feel very rough to the touch. Until recent times, dried horsetails were commonly bunched together and used for scouring pots and pans and for polishing metals. For this reason, horsetails are often called *scouring rushes.*

## Section Review 5.3

1. How are spores different from seeds? How are they similar?
2. Name a well-known example of a club moss.
3. What gives horsetails their rough texture?

### Application

4. In a paragraph, describe alternation of generations in ferns.

### Identify

fern, spore, frond, sori, rhizome, club moss, horsetail

### Observe

- Find a fern frond that is a fiddlehead and another that has sori. Observe the rolled-up structure of the fiddlehead and look at the sori under a hand lens.

# 5.4 Mosses and Liverworts

## Mosses

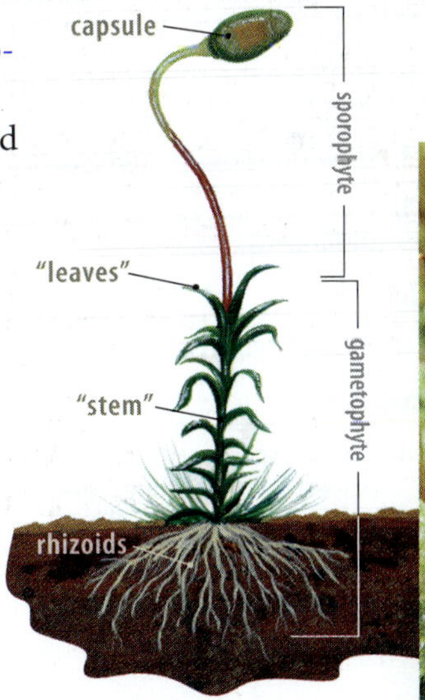

Fig. 5.9 **Structure of a moss plant**

**Mosses** (phylum Bryophyta) are nonvascular plants that lack xylem and phloem, the tissues that conduct water and food in vascular plants. Their simple leaflike, rootlike, and stemlike parts are not true leaves, roots, and stems because they lack vascular tissue. Botanists call mosses and liverworts **bryophytes,** which literally means "mosslike plants."

Where conditions are shady and moist, as in a forest, the ground will often be covered with a thick, velvety green carpet of mosses. These plants can be found in all the climatic regions of the world, from the poles to the tropics and at all altitudes. Mosses can also grow on moist rocks and trees.

Each individual moss plant in a carpet of moss has a slender stalk and leaflike structures often only a single cell thick. Instead of roots, a moss has tiny hairlike threads called **rhizoids** [rī′zoidz′], which grow into the soil to absorb water and minerals.

Fig. 5.10 **Peat moss *(Sphagnum)***

Several organisms commonly referred to as mosses are actually not mosses. *Spanish moss,* for example, a plant that hangs from trees, is not a moss but a flowering plant and a member of the pineapple family. *Reindeer moss* is actually a lichen [lī′kən].

**Benefits of mosses.** Mosses help to replenish the soil by secreting acids that gradually break down the minerals in rocks, a process called *chemical weathering.* Mosses also help prevent soil erosion in the places where they grow. A large, branching type of swamp-dwelling moss called **peat moss** (also known by its genus name ***Sphagnum*** [sfăg′nəm]) is especially helpful in this regard. The "leaves" of peat moss plants contain numerous nonliving water-storage cells, which give the moss a spongy texture and the ability to hold water. These spongy cells absorb rainwater and prevent much of the runoff that causes soil erosion. Peat moss also has antiseptic qualities and has occasionally been used for bandaging wounds in emergencies.

**Peat.** Decaying peat moss is the main constituent of a substance known as **peat.** Peat is a compressed accumulation of various partially decayed plant materials that collect in swamps and marshes *(peat bogs)* over a period of time. As plants, herbs, and mosses die and begin to decay, the weight of water and other plants compresses them into layers. The acid secreted by the plants keeps the accumulating peat from decaying completely. In countries where peat is plentiful, such as Ireland, peat is cut and dried and then used as fuel to heat homes. In the United States, gardeners use peat as a fertilizer to enrich the soil and as a mulch to keep soil water from evaporating.

**Moss reproduction.** Like ferns, moss plants experience *alternation of generations;* but unlike ferns, the familiar form of mosses is the gametophyte. A tuft of moss consists of male and female gametophytes. After an egg cell is fertilized by a sperm cell, the sporophyte grows from the top of the female gametophyte. The sporophyte, which is completely dependent on the "mother" plant for nutrition, consists of a tiny stalk bearing a *capsule,* or spore case. When a spore from the capsule falls in a moist place, it begins to grow by sending out a mass of fine threads. Buds form on this mass of threads and grow into the familiar male and female moss plants (gametophytes).

### Liverworts

Like the mosses, **liverworts** (phylum Hepaticophyta) are nonvascular plants (bryophytes); they lack true leaves, roots, and stems and grow along the surface of the soil, attached by rhizoids. The plants often look like tiny leaves. The male and female reproductive structures look like little umbrellas that rise about an inch above the flat body of the plant. Liverworts generally grow in moister places than mosses.

Fig. 5.11 **Liverworts**

## Section Review 5.4

1. Why are the leaflike and stemlike structures of mosses not considered true leaves or stems?
2. How are the rhizoids of a moss plant different from roots?
3. How do mosses reproduce?
4. How are mosses and liverworts alike? How are they different?

### Identify

moss, bryophyte, peat moss, peat, capsule, liverwort

### Observe

- Collect specimens of mosses and liverworts and observe the leaflike, rootlike, and stemlike parts of these nonvascular plants.
- Try germinating some wet bean seeds in a mass of peat. Does the peat retain the water so that the seeds germinate?

## 5.5 Algae

**Algae** (sing. alga), the smallest of the green, chlorophyll-containing organisms, live nearly everywhere. Unlike plants, most algae do not have stemlike, rootlike, or leaflike structures. Most algae live in water, though some species live in moist areas on land. If you have ever taken a walk by a pond, lake, or ocean, then you have probably seen algae in the form of water moss, pond scum, or seaweed. You can see algae attached to rocks or floating near the shore of almost every body of water. They often infest swimming pools and water purification plants. Snow in some areas of Yellowstone Park is sometimes green because of green algae. In addition to the green pigment chlorophyll, many algae contain other pigments, some of which give them a definite color other than green. In fact, *one common and simple way of classifying algae is by their color:* green, yellow, brown, red, and sometimes blue-green.

*Algae are extremely important because they are the major producers of food in aquatic (water) environments.* In fact, some scientists estimate that the algae in the oceans carry on about 70% of the photosynthesis on the earth. They are part of the **plankton,** microorganisms that float near the surface of the water and provide food for larger organisms.

### Green Algae

The 12,000 species of **green algae** (phylum Chlorophyta) form the largest group of algae. Most of the green algae are grass-green, but some are yellowish or brownish and a few are blue-green. Although most are microscopic, they may often be seen as a green scum on a pond or lake because of their huge quantities. Green

Fig. 5.12 **Algae on surface of pond**

algae may live as single cells or in one of a variety of different groups or kinds of groups. Some single-celled green algae clump together in differently sized and shaped groups called **colonies.** Some algae live in organized groups in which individual cells are joined end to end, forming a long chain of cells. Botanists refer to these hairlike structures as **filaments.**

**Unicellular green algae.** The unicellular green alga *chlorella* is one of the most important green algae because it has been widely used in the study of photosynthesis. *Chlamydomonas,* a unicellular green alga that lives in both soil and water, is unusual in that it possesses two distinctly animal-like characteristics: **flagella** [flə·jĕl′ə: whiplike structures that allow locomotion; sing. flagellum] and a red **stigma,** or *eyespot.* With these structures, the plant responds much like a one-celled animal.

Another type of unicellular green algae, the spherical ***Protococcus*** [prō′tō·kŏk′əs], often grows on the trunks of trees and on unpainted buildings and fenceposts. It is easy to see in wet weather and generally grows on the side of the tree trunk where less sunlight strikes, existing in a moister environment. During dry weather, these algal cells may be carried from tree to tree by birds or by the wind. Since they are photosynthetic organisms and make their own food, they do not harm the tree on which they grow.

The **desmids** [dĕs′mĭdz] are a large group of freshwater green algae. A typical desmid lives as a single cell with a beltlike constriction about its middle dividing it into two identical semicells. The nucleus is in the center, between the semicells. Many desmids are exquisitely patterned with geometric designs. Some are shaped like stars or crescents and have spines. Desmids form a significant part of the plankton of wetlands, and some can be found in ponds as well.

Fig. 5.13 **Unicellular green algae**

**Colonial green algae.** **Volvox** [vŏl′vŏks′] is an example of unicellular algae that live in colonies of hundreds or thousands of individual cells. These aquatic organisms live in pale green globular colonies, which are just barely visible to the unaided human eye. Each colony is hollow, with a watery fluid filling the inner cavity. With a microscope, one can see daughter colonies within the parent colony; in fact, the daughter colonies themselves may contain developing daughter colonies.

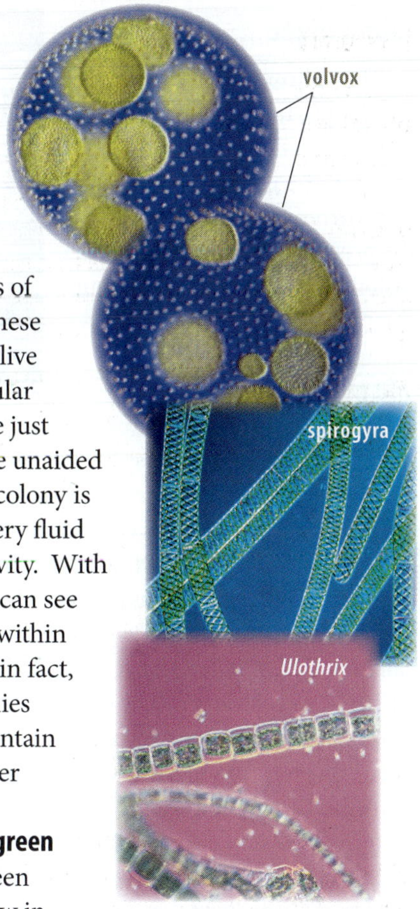

Fig. 5.14 **Colonial green algae**

**Filamentous green algae.** The green algae that grow in filaments, or threadlike strands, may be either branched or unbranched. One such organism, *Ulothrix,* grows along the seashore and in ponds. Its slender filaments grow attached to rocks and debris; the attachment is made by a special type of anchoring cell called a **holdfast,** which is found in certain types of algae.

**Spirogyra** [spī′rə·jī′rə] have been used in biological laboratories for many types of scientific investigations. The cells of these filamentous algae contain one or more ribbonlike chloroplasts arranged in a spiral. Attached to the chloroplasts are small protein structures that are surrounded by a layer of starch and that act as food reserves for the cell. Spirogyra can be found in almost any quiet pond in the spring and fall, and they resemble masses of green thread. Because a thin gelatinous sheath surrounds each individual cell, spirogyra have a slippery feel.

## Yellow Algae

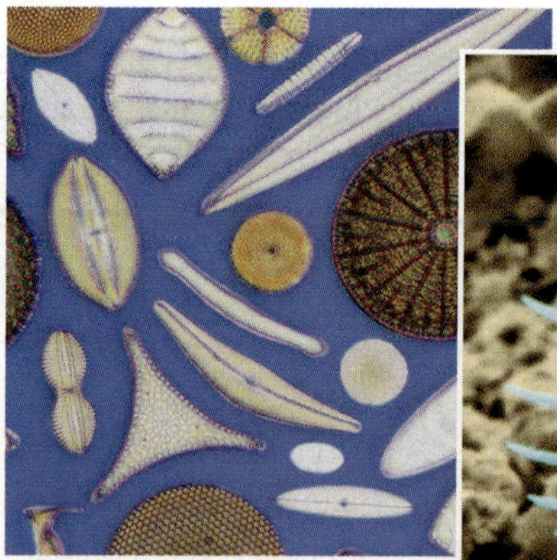

Fig. 5.15 **Diatoms**

The **yellow algae** (phylum Chrysophyta) are, in reality, yellow-green or golden-brown in color, the result of a yellow or brown pigment that masks the green chlorophyll. Most yellow algae are unicellular, but they often live in colonies or clusters. The most numerous and most interesting of the yellow algae are the **diatoms** [dī′ə·tŏmz], which are possibly the most abundant life form on earth except for bacteria. Diatoms have rigid cell walls composed of *silica* ($SiO_2$: the hard substance in sand and glass). The silica walls act as microscopic shells or "glass houses" for the tiny organisms. The cell wall is made of two almost equal halves called *valves.* The two halves give the diatom its name, which in Greek means "cut in two." Diatoms are found in more places than any other algae. You may have noticed some as brown, slimy coatings on stones or piling in fresh water or salt water. The microscopic shells of diatoms have an almost infinite variety of shapes and sizes and intricate designs—an example of the matchless beauty of God's living creation.

When diatoms die, their soft parts decay and their inner cell walls disintegrate, leaving the clear outer walls of silica to settle to the bottom of a body of water. Under certain conditions, the shells of diatoms may accumulate and become packed together to form layers of **diatomaceous** [dī′ə·tə·mā′shəs] **earth,** or *diatomite.* Many of these layers are now on dry land. One such deposit in California, which may have been formed during the Flood of Noah's day, is more than 495 yards thick. Diatomaceous earth is used in toothpastes, scouring powders, filters, bricks, tiles, and ceramics. Swedish chemist *Alfred Nobel* (1833–1896), creator of the Nobel prize, discovered that diatomaceous earth would absorb three times its own weight of nitroglycerin. By mixing diatomite with nitroglycerin he invented dynamite, which is much more stable than pure nitroglycerin and therefore much safer to handle. Diatomite is now widely used in the manufacture of explosives.

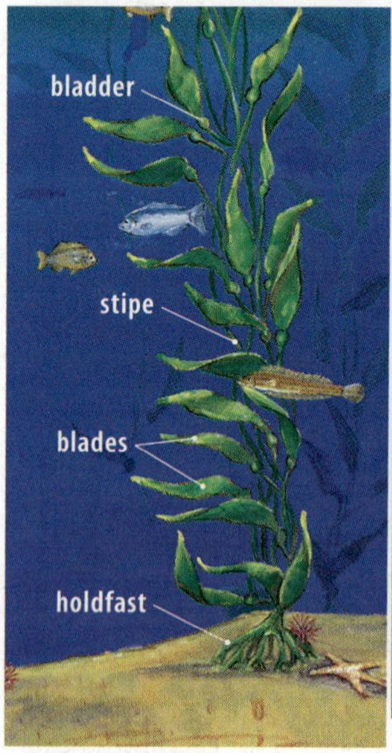


Fig. 5.16 **Giant kelp (brown algae)**

## Brown Algae

Like all algae, **brown algae** (phylum Phaeophyta) contain green chlorophyll, but it is masked by the presence of the same golden-brown pigment found in yellow algae. Virtually all the plants that belong to this phylum are marine organisms, and the **seaweeds** (large saltwater algae) of colder ocean waters are mostly brown algae. Seaweeds that are brown algae include the kelps, rockweeds, and gulfweeds, all of which are multicellular and have some cell specialization and tissue development.

**Kelps,** the largest of the brown algae, grow in both the Atlantic and Pacific oceans and may be over 200 feet long. Because of the constant motion of ocean waves and currents, kelps are attached to the ocean floor by large holdfasts. **Algin** [ăl′jĭn], a gummy substance derived from kelp, can hold several different liquids together in an emulsion. It is used in the manufacture of a variety of foods and drugs, including chocolate milk, ice cream, mayonnaise, puddings, salad dressings, aspirin, beauty creams, and skin lotions. In ice cream, algin thickens the mixture and keeps the water in the milk from forming crystals. Kelps are a regular part of the human diet in China and Japan. They also serve as fertilizer and as an important source of the element iodine.

Fig. 5.17 **Brown algae**

**Rockweeds,** a type of brown algae, are attached to rocks along the seashore by holdfasts. These brown algae are designed to survive in the harsh conditions of a changing environment that is submerged at high tide and above water level at low tide. Gas-filled air bladders serve as floats for the rockweeds. Rockweeds, like kelps, are a source of algin.

East of Florida are large masses of floating **gulfweeds** of the genus *Sargassum,* which give the *Sargasso Sea* in the Atlantic Ocean its name. The Sargasso Sea is an area of slow ocean currents filled with patches of gulfweeds, which remain afloat because of berrylike air sacs. The gulfweeds have a main stem with flattened, leaflike outgrowths and branches that contain the air sacs.

Fig. 5.18 **Sea grapes (red algae)**

## Red Algae

Although most **red algae** (phylum Rhodophyta) live in the sea like the brown algae, a few species grow in fresh water. These algae vary in color from pink and red to purple and even black. This great color range is caused by red and blue pigments. Some marine forms of red algae grow at depths of more than 800 feet; very little sunlight reaches these depths. The seaweeds in warmer ocean waters are mainly red algae. One species of red algae, commonly known as *Irish moss,* is the source of **carrageenan** [kăr′ə·gē′nən], a gelatinous substance used in many food products. Carrageenan is the gel that makes commercial ice cream smooth and creamy. The gelling agent **agar** [ā′gär′], which laboratories use as a culture medium for growing bacteria, is also derived from red algae. Several edible species of red algae, such as *dulse* [dŭls] and *nori,* are used in the traditional cuisine of several parts of Asia, Europe, and North America.

## Dinoflagellates

The **dinoflagellates** [dī′nō·flăj′ə·lĭtz: phylum Dynophyta] are microscopic, one-celled, aquatic organisms found in both fresh water and salt water. Dinoflagellates have two unlike flagella, which they use for locomotion. Most dinoflagellates live in the ocean, where they are a source of food for many other organisms. The best-known type of dinoflagellate is responsible for the

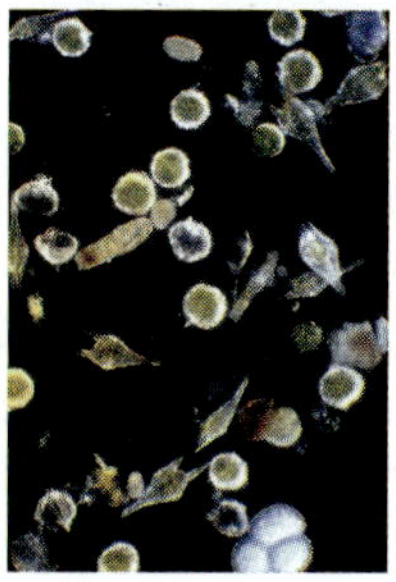

Fig. 5.19 **Dinoflagellates**

phenomenon commonly called **red tide.** A red tide occurs when this dinoflagellate blooms, or grows so rapidly that the water is tinged with a pinkish color. As the algae bloom, they produce great quantities of a nerve poison that is highly toxic to fish, small marine organisms, and humans. Scientists do not yet know exactly why red tide blooms occur. Some dinoflagellates can produce bioluminescence through a mechanism similar to that of fireflies. This phenomenon is generally observed at night in coastal waters, as the waves break on shore.

Fig. 5.20 **Red tide**

## Blue-Green Algae (Cyanobacteria)

Because *blue-green algae* (phylum Cyanophycota) commonly clump together to form colonies resembling those of unicellular algae, they have been traditionally studied with algae. But unlike other algae and true plants, *blue-green algae are prokaryotes and are thus classified with bacteria in kingdom Eubacteria.* Most blue-green algae have a slightly bluish tint because they contain blue and red pigments along with the green chlorophyll; they are called **cyanobacteria** because of their bluish appearance (*cyano-* means "dark blue"). However, some appear black, red-purple, yellow, or green. These organisms have a remarkable ability to thrive in environments quite hostile to other living things. Some live best in polluted water, even in manure-saturated barnyard pools. Others live in hot springs at temperatures up to 176°F. Many prefer water of high oxygen content. A common representative is **anabaena** [ăn′ə·bē′nə], a genus of nitrogen-fixing cyanobacteria whose cells look like tiny beads or barrels. Because anabaena can incorporate atmospheric nitrogen into its cells, it is not dependent upon dissolved nitrogen in soil or water. Anabaena is found as plankton in shallow water and moist soil and is important in food chains for fish and crustaceans.

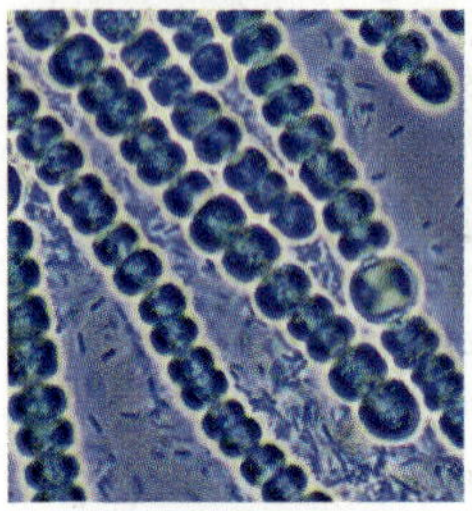
Fig. 5.21 **Cyanobacteria (blue-green algae)**

## Section Review 5.5

1. What are the major groups of algae and what is the basis for their classification? Give an example from each group.
2. What are some useful products derived from algae?
3. What is red tide? Why is it dangerous?
4. Why are blue-green algae classified as bacteria?

### Application

5. In a paragraph, explain why algae are very important to life on Earth.

### Identify

algae, plankton, colony, filament, flagellum, holdfast, silica, diatomaceous earth, seaweed

### Observe

- Obtain some diatomaceous earth and observe it with a hand lens. Notice the abrasiveness that makes it a common ingredient in such products as toothpaste and scouring powder.

# 5.6 Fungi

Toadstools in the front yard, shelf fungi on forest trees, green mold on bread or rotting oranges, itchy athlete's foot between your toes—all are examples of **fungi.** Fungi are nonvascular plantlike organisms that lack chlorophyll and depend upon other organisms for food. Unlike green plants and other *autotrophs,* which manufacture their own food, *fungi are heterotrophs,* organisms that must obtain their food from sources outside themselves. Some of the more familiar fungi are mushrooms and toadstools, molds, rusts and smuts, yeasts, and mildews.

Most fungi are multicellular and have a distinct threadlike structure; yeasts, however, are unicellular. The threadlike structures, or *filaments,* that form the body of a fungus are called **hyphae** [hī′fē]. The mass of hyphae forming the main body of a fungus is called the **mycelium** [mī·sē′lē·əm]. Scientists who study fungi are called **mycologists** [mī·kŏl′ə·jĭsts].

## Importance of Fungi

Most fungi are **saprophytes** [săp′rə·fīts′]; that is, they obtain their nutrition from dead organisms. Without fungi, bacteria, and other decay-producers, life on Earth as we know it would be impossible. When living things die, fungi and bacteria secrete digestive enzymes that break down the complex molecules of life into simpler ones that living organisms can use to build cells and tissues. Thus, the molecules of life are continuously recycled by saprophytic fungi and bacteria to make life possible on this present Earth.

Although some fungi are parasites and cause such diseases as athlete's foot, ringworm, Dutch elm blight, potato blight, wheat rust, and corn smut, other fungi are, in God's providence, extremely beneficial to mankind. Baker's yeast is a fungus used to make bread rise. Without certain fungi, many cheeses would lack their distinctive flavors. Countless human lives have been saved by penicillin and other powerful antibacterial drugs derived from fungi.

## Club Fungi

Fungi are classified according to several characteristics. *Some fungi can be classified by the shape of their spore-producing structures.* For instance, many fungi produce spores in microscopic club-shaped structures called **basidia** [bə·sĭd′ē·ə: sing. basidium]; these fungi are known as the **club fungi.** Examples of club fungi include mushrooms and toadstools, shelf fungi, rusts, and smuts. Nearly all club fungi are saprophytes; they feed upon such decaying organic matter as humus, rotten wood, and compost piles. Many club fungi are edible, but some are poisonous.

**Mushrooms and toadstools.** It is common practice to call inedible **mushrooms** toadstools and edible ones mushrooms, but this is no hard and fast rule because the two words are often used interchangeably. Mushrooms and toadstools consist of a mass of subterranean hyphae that spread through the soil but come together to form a main, compact mass, or mycelium, just under the surface. At reproduction, a stalk of tightly bound hyphae topped by an umbrella-shaped, spore-forming structure grows quickly into the air. This structure, called the **fruiting body,** contains many **gills** that radiate from the stalk like spokes on a wheel (Fig. 5.22). Every gill has hundreds of club-shaped basidia, each of which produces four spores. Because there are so many basidia, each fruiting body releases a small "cloud" of spores. *Puffballs,* for example, release millions of spores.

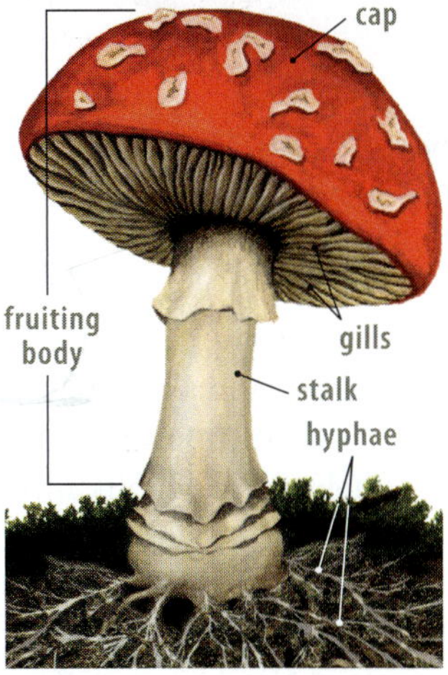

Fig. 5.22 **Parts of a mushroom**

Fig. 5.23 **Puffball**

Fig. 5.24 **Shelf fungi**

**Shelf fungi,** or *bracket fungi,* which are often found growing on trees or rotting logs, are in this same group of club fungi. Their mycelia spread throughout the log, digesting the wood and causing decay. Unlike mushrooms, shelf fungi grow slowly and do not have gills. Spores fall from holes on the undersurface of the fruiting body and are spread by the wind.

**Rusts.** Rusts and smuts are parasitic club fungi that do millions of dollars in damage to crops each year. (A **parasite** is an organism that lives on or in another living organism and derives its nutrition from that organism.) **Rusts** are named for the rusty-colored spores that they leave on their hosts. Many plants are hosts to rusts; these include wheat, oats, raspberries, melons, apple trees, and evergreen trees.

Most rusts are **heteroecious** [hĕt′ə·rē′shəs] **parasites,** which means that they must have two

Fig. 5.25 **Rusts and smuts**

hosts to complete their life cycle. Usually, one host is greatly harmed and the other host is not harmed at all by the parasite. **Wheat rust** is widespread where both its hosts, wheat and the barberry bush, are common. In the summer, the wheat rust fungus produces streaks of rust-colored spores on wheat stalks. Its hyphae so damage the wheat stalks that they fall and die before the grain is produced. When there is an outbreak of wheat rust, the damage may be widespread because the spores are scattered great distances by the wind.

At the end of the growing season, wheat rust fungi produce large, thick-walled spore cases, which appear as black patches on the wheat stubble. These spore cases allow the fungus to survive the winter; when spring comes, they release tiny spores, which will grow only on the common barberry bush, the rust's alternate host. The spores germinate on the barberry's leaves and send hyphae into the leaves. After a short period of growth, the rust forms cuplike clusters of spores on the underside of the host's leaves. When these spores are released into the wind, some are carried to wheat plants, and the cycle begins once again.

**Smuts. Smuts** produce masses of black spores on their hosts. Smuts commonly attack grains such as corn, oats, wheat, rye, and barley. The hyphae of smuts penetrate the tissues of their hosts and rob them of the nourishment needed for proper growth. Often smuts destroy the host plant's flowers so that no grains can be produced; at other times the smut causes the stalk to be much deformed and stunted.

**Corn smuts** often attack the tassels and ears of corn plants, forming large, shiny blisters. The blisters are spore cases, which eventually burst and spread millions of black spores into the air. These spores attack nearby plants, especially if the stems of those plants have been twisted by the wind or injured in some other way.

## Molds

**Bread molds.** Probably all of us are familiar with ordinary **bread mold.** One common type of bread mold is *Rhizopus stolonifer*. This mold is a miniature jungle of fine, shiny, white hyphae. When it is ready to reproduce, sporangia form on the tips of the mold's reproductive hyphae. In *Rhizopus,* each sporangium is a spore case that produces and disperses tiny black spores. When the hundreds of spores in a spore case are mature, the case bursts and discharges the spores into the air, where wind currents scatter them. If a spore settles on a piece of bread or some other particle of food, it may grow into a new mold. Conditions of temperature and moisture must be proper for a mold spore to grow, however. Environments of high humidity, warm temperatures, and darkness favor their growth. Molds of the genus *Aspergillis* are also common bread molds.

Fig. 5.26 **Bread mold**

**Cheese molds.** Cheese is made by using bacteria to help separate milk into the solid white *curd* and the watery *whey*. After this separation, most types of cheese are placed in a cool location to age. The distinct texture, color, and flavor of some cheeses come from **cheese molds,** which are allowed to grow on the cheese as it ages; the type of cheese that is formed depends partly on the kind of mold used and the rapidity with which the mold is allowed to grow. Roquefort cheese, for example, is made from sheep's milk and a dark green type of mold; this famous French cheese was originally produced in cool caves where the mold grew very slowly. Other blue cheeses are also aged with mold, as are Camembert and Brie cheeses.

**Parasitic molds.** Like bread and cheese molds, most molds are saprophytes, but some are parasites. Parasitic molds cause serious problems for man because they feed on important foods like potatoes and cereal grains. The *potato blight* and famine of 1845–1850 that caused the death of an estimated one million people in Ireland illustrates how serious a problem parasitic molds can be.

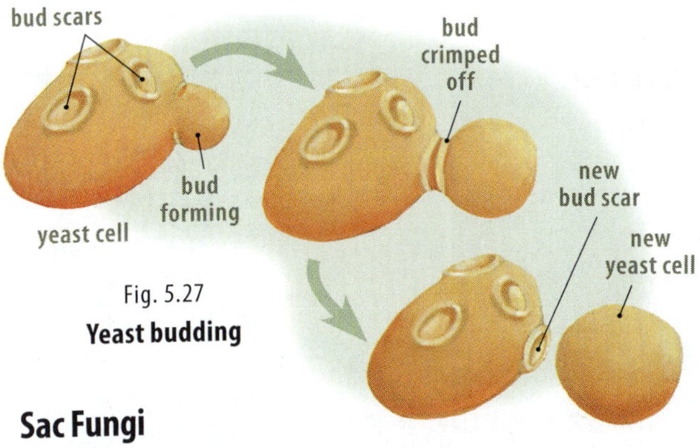


Fig. 5.27 **Yeast budding**

## Sac Fungi

**Yeasts.** Fungi with saclike spore-producing structures are called **sac fungi.** A well-known sac fungus is **yeast,** a small, colorless, single-celled organism that feeds on foods containing sugar, converting the sugar into alcohol and carbon dioxide in the process. When *baker's yeast* is added to bread dough, it causes the dough to rise. The dough rises because some of the starch in the dough is converted into sugar and the yeast ferments this sugar. The fermentation process produces bubbles of carbon dioxide, which become trapped within the dough. These bubbles raise the dough and make it light and porous. The bubbles of carbon dioxide are released into the air, and the alcohol is evaporated during baking. The tiny air spaces that form give the finished bread a light and tender texture.

When conditions favor the growth of yeasts, they reproduce rapidly by *budding.* A *bud* is an extension of a cell that breaks off and forms a new cell. When conditions are unfavorable for growth, each yeast cell forms four tiny spores in a sac called an *ascus* [ăs′kəs: pl. asci: ăs′ī′] within its cell wall. These spores may remain dormant for a long time until conditions again favor active growth. When the spores germinate, normal growth results.

**Mildews.** There are many kinds of **mildews** other than the ones that grow on wet clothes and shower stalls. Some mildews appear as a white, powdery substance on plants. Because they grow rapidly and reproduce by spores that are easily transported by the wind, these organisms are difficult to control.

Fig. 5.28 **Mildew**

## Slime Molds

**Slime molds** are the jellylike organisms often seen on the bark of fallen trees. These unusual organisms are often bright red, orange, or yellow, and as they grow, they ooze into their environment and engulf bacteria and other organic particles. Some slime molds are saprophytes and obtain their nutrition from dead organisms; others are parasites. Because their life cycle involves many changes in form, slime molds are of great interest to biologists and somewhat of a puzzle to taxonomists. They begin their life cycle as a slimy mass, which changes into a hardened spore-producing structure when the environment changes. Slime molds play an important part in the decay of old logs and dead leaves in a forest.

Fig. 5.29 **Slime mold**

## Lichens

Sometimes a fungus and an alga (either a green alga or a blue-green alga [cyanobacterium] colony) live together in a close relationship that combines the two organisms into a single unit called a **lichen.** The alga is surrounded by the mycelium of the fungus, which provides the alga with a secure environment. The alga contains chlorophyll and provides food for the fungus; the fungus, in turn, provides the water and minerals the alga needs to carry on photosynthesis. The partnership between the two organisms is an example of *symbiosis* (two different kinds of organisms living together for mutual benefit).

Lichens can be found growing on soil, rocks, and trees throughout the world, even in the harshest environments. The "moss" often seen growing on the sides of trees is often a lichen. Lichens also grow on sandy beaches and in the barren Arctic and Antarctic regions. Some grow as flat, crusty structures; others grow upward and have the appearance of a tiny shrub. Wherever they grow, lichens enrich the soil; and when they grow on rocks, they begin the process of breaking down the rock into soil. Many substances helpful to medical science have been derived from lichens.

Fig. 5.30 **Lichens**

Fig. 5.31 **Cup lichen**

**Careers in BIOLOGY**

### Botany

**Agricultural science** – developing better ways to grow cultivated plants

**Chemurgy** – developing new uses for plant products

**Horticulture** – commercially growing herbs, flowers, shrubs, etc.

**Marine botany** – analyzing the characteristics of and developing uses for aquatic plants

## Section Review 5.6

1. What characteristic do all club fungi have in common?
2. How are bacteria and molds used to make cheese?
3. How do yeasts make bread rise?
4. How do yeasts reproduce?

### Critical Thinking

5. Based on what you know of the life cycle of the wheat rust fungus, what might be one way to control the fungus?

### Identify

saprophyte, hyphae, mycelium, basidium, fruiting body, parasite, slime mold

### Observe

- Collect specimens of club fungi and identify the hyphae, mycelia, and fruiting bodies. Try to distinguish the individual basidia that make up the fruiting body.
- Expose a moist piece of bread for a number of days until mold is flourishing on it. Using a hand lens, observe the bread mold and try to distinguish the hyphae and sporangia.

## Chapter 5 Review

### Define

1. taxonomy
2. kind
3. vascular plant, nonvascular plant
4. gymnosperm
5. conifer
6. fern
7. spore
8. moss
9. bryophyte
10. liverwort
11. algae
12. plankton
13. diatom
14. dinoflagellate
15. cyanobacteria
16. fungus
17. club fungus, sac fungus
18. shelf fungus
19. slime mold
20. lichen
21. symbiosis

### Identify

1. the primary criterion by which Linnaeus classified organisms
2. the seven main categories of the classification system
3. three examples of spore-bearing vascular plants
4. a common club moss
5. two examples of bryophytes
6. three examples of seaweeds
7. the commercial products derived from yellow, brown, and red algae
8. the organism responsible for red tides
9. a common club fungus
10. a common sac fungus

### Explain

1. What is the difference between a taxonomic species and a biblical kind?
2. How do nonseed plants such as ferns reproduce?
3. Explain the difference between rhizomes and rhizoids.
4. Explain the difference between club mosses and true mosses.
5. Why are blue-green algae classified as bacteria?
6. Distinguish between saprophytes and parasites.

### Apply

1. Do most conifers exhibit spirelike branching or spreading branching? (See ch. 4.) Explain your answer.
2. Compare the life cycle of ferns and mosses.
3. Explain how saprophytes such as fungi are necessary, in God's providence, to life on the present Earth.

# UNIT 2 | Human Anatomy & Physiology

Purpose and Design

CHAPTER 6

# Fearfully and Wonderfully Made

**Key Concepts**

- Wonders of the human body
- Body cavities
- Body systems
- Tissues, membranes, fluids, cells, and matrix
- Vestigial organs

## 6.1 The Crown of God's Creation

> The human body is a veritable museum of science, full of God's creation wonders.
>
> —*Alfred M. Rehwinkel*

Man is the most magnificent part of God's creation—far more complex in structure and design than the earth or any heavenly body. Each of his body's 100 trillion cells is an enormously complex assembly of compartments, girders, factories, motors, sensors, and control systems; together, they contain roughly 60 billion miles of DNA representing 70 billion terabytes of information storage. Pound for pound, his intricately constructed, self-healing bones are as strong as reinforced concrete and much more resilient. With training, his muscles allow him to lift several times his own weight, sprint at nearly 40 feet per second, leap with the grace and agility of a gymnast, or push on for hour after hour in a grueling marathon. Yet his muscles are also capable of such precision that his hands can assemble a Swiss watch or repair a damaged eye.

A man's heart is an efficient four-chambered pump that beats, on average, a hundred thousand times per day, pumping blood through 60,000 miles of blood vessels to distribute food and oxygen throughout his body. The red blood cells that transport oxygen through this vast network are manufactured at a rate of 2.4 million per second. The oxygen the blood carries is supplied by 600 million gas-exchange chambers in his lungs, which have a total surface area the size of a volleyball court.

His intricately designed, autofocusing eyes provide him with extraordinarily high-resolution images of the world around him, from objects a few inches distant to giant galaxies a quintillion ($10^{18}$) miles away; they can adjust from the silvery outlines of a moonlit landscape to the millionfold greater brightness of a summer day. Unequalled image-processing equipment in his optic nerves and brain allow both rapid object recognition and the ability to almost intuitively analyze angles and speeds. His delicately crafted ears can detect the faint rustling of distant leaves yet allow him to analyze sounds that are ten billion times more intense. His sense of smell enables him to detect and distinguish ten thousand different chemicals, often in amounts so vanishingly small that they would have been difficult to detect by any other means until the 20th century.

All of this marvelous assembly is controlled and coordinated by the brain, a three-pound, ten-watt supercomputer that contains more than 100 quadrillion electrochemical connections. Not only is the human body the crowning glory of God's creation; it is superior to the best machines that modern technology can build.

The intricate workings of the human body point to God's wisdom in creation and cause us to exclaim with the Psalmist

> I will praise thee; for I am fearfully and wonderfully made: marvellous are thy works; and that my soul knoweth right well. —Psa. 139:14

## The Body Cavities

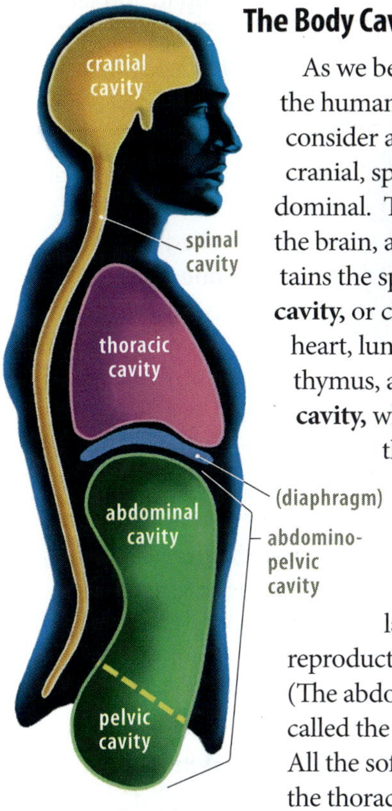


Fig. 6.1
**Major body cavities**

As we begin our examination of the human body, the first units to consider are the four body cavities: cranial, spinal, thoracic, and abdominal. The **cranial cavity** contains the brain, and the **spinal cavity** contains the spinal cord. The **thoracic cavity,** or chest area, contains the heart, lungs, esophagus, windpipe, thymus, and aorta. The **abdominal cavity,** which is separated from the thoracic cavity by the diaphragm, contains the liver, stomach, pancreas, spleen, kidneys, urinary bladder, small intestine, large intestine, and the reproductive organs, or **gonads.** (The abdominal cavity is sometimes called the *abdominopelvic cavity.*) All the soft, internal organs within the thoracic and abdominal cavities, such as the heart, lungs, and intestines, are called **viscera** [vĭs′ər·ə].

## Eleven Systems Functioning as One

*The body can also be divided into eleven functioning systems.* A **system** is a group of structures that function together as a unit to perform a definite job for the body. Each system is made of smaller units called *organs.* Organs are further divided into *tissues.* When tissues are observed under the microscope, we find that they are made of even smaller units—*cells.*

**Circulatory system.** The organs of the **circulatory system** are the *heart, arteries, veins,* and *capillaries.* These organs function together to bring food, water, and oxygen to the cells and to carry away waste products. The circulatory system is one of the most important systems of the body, because "the life of the flesh is in the blood" (Lev. 17:11). The rhythmic beating of the heart forces the blood throughout the body. Although it weighs only about eleven ounces, the heart can keep blood moving continuously through approximately 60,000 miles of blood vessels. *The arteries carry the blood away from the heart, and the veins return it to the heart.* The microscopic capillaries connect the arteries and the veins and provide for the exchange of nutrients and waste materials in every cell of the body.

**Immune system.** The human body is amazing in its ability to protect itself against disease-causing organisms such as bacteria and viruses. The **immune system** is the body's defense against these and other foreign invaders. The most important part of the immune system is a host of special cells *(white blood cells)* programmed to search out and destroy any organism that they do not recognize as part of the body. These cells are assisted by a complex system of vessels, glands, and fluids. An important part of the immune system is the *lymphatic* [lĭm·făt′ĭk] *system,* a network of vessels, similar in appearance to blood vessels, that collect lymphatic fluid *(lymph)* from the body's tissues. The lymph begins as a watery fluid that leaves the blood through the capillary walls. After passing through the body's tissues, lymph enters the lymph vessels and travels to special organs called *lymph nodes,* where white blood cells filter out harmful particles and bacteria that the lymph has picked up from the tissues; the lymph nodes also serve as "bases" from which white blood cells mount attacks against invaders. White blood cells are produced in the *bone marrow* and mature in the bone marrow and *thymus.* Other agents of the body's immune system include the *spleen* and the *tonsils.*

**Respiratory system.** The main function of the **respiratory system** is to supply body tissues with oxygen and to remove carbon dioxide. When the diaphragm contracts, creating a partial vacuum in the thoracic cavity, air passes through the *nose, nasal passages, pharynx, larynx* (voice box), *trachea* (windpipe), and *bronchial tubes* to the *lungs.* Oxygen and carbon dioxide pass between the blood and the air in the *alveoli* [ăl·vē′ə·lī′], the thin-walled air sacs at the ends of the smallest bronchial tubes in the lungs. The alveoli dissolve oxygen from the inhaled air and pass it to the capillaries, which begin the process of carrying it to every cell in the body. At the same time, carbon dioxide from the body's cells moves from the capillaries to the alveoli and is then expelled from the body by exhalation. Exhaled air contains 100 times more carbon dioxide than inhaled air.

**Digestive system.** The **digestive system,** which makes a path through the thoracic and abdominal cavities, is composed of a number of organs that make up a 20-foot-long tube, the **alimentary** [ăl′ə·mĕn′tə·rē] **canal,** that leads from the mouth to the *anus* [ā′nəs], the opening through which wastes are expelled from the digestive system. The function of the digestive system is to break down food into a form that can be absorbed by the bloodstream. The *mouth* and *teeth* chop the food to begin digestion. Enzymes produced in the mouth begin chemically digesting the food even before it reaches the stomach. The *esophagus* [ĭ·sŏf′ə·gəs] is the tube that transports food from the mouth to the stomach. The stomach continues the chemical breakdown of food, but its principal job is to store food until the small intestine is ready to receive it. In the *small intestine,* digestion is completed and nutrients are absorbed into the bloodstream. The *large intestine* absorbs water from undigested food before it is expelled from the body.

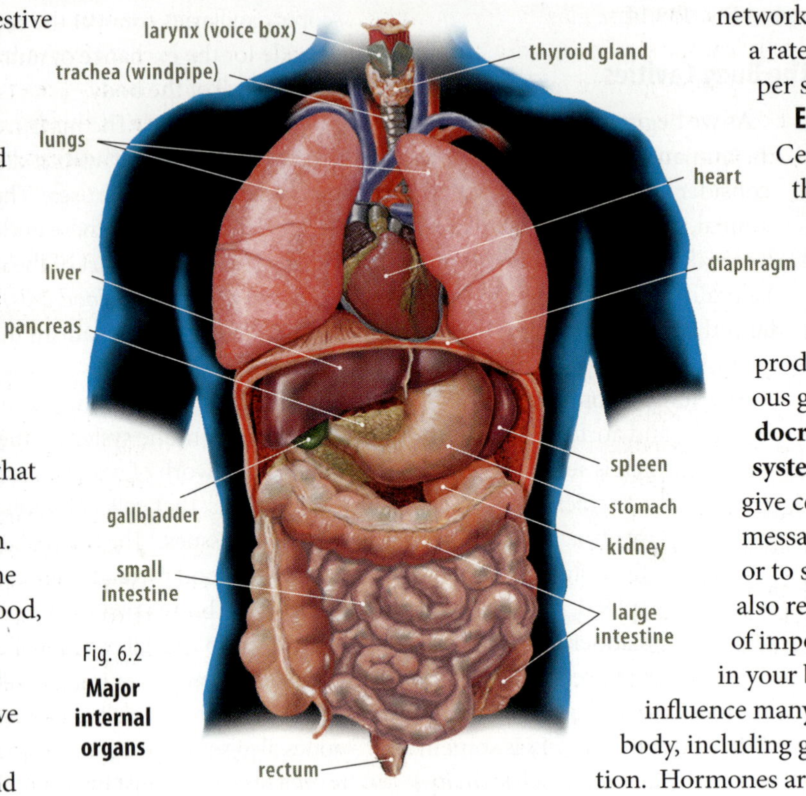


Fig. 6.2 **Major internal organs**

**Excretory system.** The **excretory system** provides a number of openings through which a variety of waste materials are excreted from the body. The *sweat glands* in the skin excrete urea and salts from the body. The *kidneys* filter the blood, removing excess water, urea, uric acid, and other waste materials. Liquid wastes from the kidneys are stored in the *urinary bladder* until they are excreted from the body. The tube connecting the urinary bladder to the outside of the body is the *urethra* [yo͞o·rē′thrə]. All the excretory organs except the sweat glands are located in the abdominal cavity.

**Nervous system.** The **nervous system** coordinates the activities of the body. The *brain,* the most important organ of the nervous system, can be considered the master control unit of the body. *Sense organs* provide the nervous system with information about the environment through such senses as sight, hearing, smell, taste, touch, pressure, and pain. *Nerves* connect all areas of the body to the brain and *spinal cord.* They carry information to and from the brain in the form of electrochemical signals, or *impulses.* When you touch something, impulses travel through the nerve network to your brain at a rate of up to 400 feet per second.

**Endocrine system.** Certain activities of the body are regulated by chemical messengers called **hormones,** which are produced by the various glands of the **endocrine** [ĕn′də·krĭn] **system.** Hormones give certain organs the message to slow down or to speed up; they also regulate the levels of important chemicals in your body. Hormones influence many activities of the body, including growth and digestion. Hormones are secreted into the bloodstream by the *endocrine glands.*

**Integumentary system.** Besides providing communication with the world around us through sensory receptors, the *skin,* which is the **integumentary** [ĭn·tĕg′yo͞o·mĕn′tə·rē] **system,** protects the body from its surroundings, providing protection from bacteria and other microbes.

**Skeletal system.** The **skeletal system** is made up of approximately 206 *bones* and numerous *ligaments.* Bones give support and rigidity to the body while they protect vital organs and produce blood cells. Bones also are the major storage area for calcium. The *cranium* is the part of the skull that provides a protective helmet for the brain. The *rib cage* protects the heart, lungs, and other vital organs.

**Muscular system.** The **muscular system** provides movement for the body. Muscles that are consciously controlled are called *voluntary muscles. Involuntary muscles* operate whether we think about them or not. The heart, or *cardiac muscle,* is an example of an involuntary muscle. The design of muscle tissue gives it great strength. If all the muscles in the body pulled together in one direction, they could lift 25 tons.

**Reproductive system.** Finally, the body system responsible for the continuation of the human race is the **reproductive system.** It consists of several organs influenced by hormones produced in endocrine glands. *A new living person begins at conception,* the time when the male gamete (sperm) unites with the female gamete (egg) in a process called *fertilization.* Before birth, development of the child takes place in the **uterus,** or *womb,* a muscular organ in the mother's abdomen.

**Homeostasis.** The body's systems work together to maintain **homeostasis,** a stable internal environment in the body; failure to maintain homeostasis results in disease or death. **Feedback mechanisms** help maintain homeostasis by detecting changes in the body and directing the body to respond appropriately. Most feedback mechanisms in the body are *negative feedback mechanisms,* which reverse the effect of a change; for example, a negative feedback mechanism lowers the blood pressure if it gets too high. *Positive feedback mechanisms,* which maintain or intensify the effect of a change, are relatively rare in the body but are involved in processes such as fever, inflammation, blood clotting, and childbirth.

## *The Wonders of Man's Body* by Alfred M. Rehwinkel

Design in the LIVING CREATION

The body of man is only the material and mortal side of his being, and yet, what a bundle of incomprehensible wonders it is! It is the most wonderful and the most complicated machine in the universe and has no parallel in the whole world of wonders. The most refined creations of man are crude compared to the human body.

The first thing to note is that the body of man is not made. It becomes. The human body is composed of the same materials found in the rest of the world all about us, but these materials are not put together by an outside agent, mechanic or builder, as men would construct a machine, make a watch, or build a house. It grows from within. . . .

So many of these wonders in the human body are taken for granted because everybody has them, and whatever has become common no longer arouses excitement. Take, for example, the beauty and graceful form of the human body. Consider the perfect harmony of all its parts: the lordly dignity of its upright stature, the marvelous mechanism of its bony framework, the miracle of its vital organs. The body grows—that is, the body itself collects the heterogeneous materials of which it is composed from outside of itself and manufactures them into itself. It follows a definite pattern in this process and always develops harmoniously in all its numerous parts. It has the ability to improve its own capacity and potentialities. It repairs itself when damaged anywhere in its members and reproduces by itself new models like itself, and yet sufficiently different to make each a distinct new model, a new and a different individual personality. Among the billions of human beings that have lived on the face of the earth or are living today, there have never been two people exactly alike.

The human body consists of thousands of individual parts, large, small, and microscopic; but all are so joined together that they function as a perfect, synchronized organism. The human body consists of billions and trillions of individual cells, and every cell is a distinct unit that functions in perfect harmony with every other cell.

Hence, the study of the human body reveals, more than all the other wonders of creation put together, the infinite wisdom, power, and goodness of the omnipotent Creator, who designed this masterpiece of His hand to be His vice-regent here upon earth.

*Alfred M. Rehwinkel,* The Wonders of Creation, *Bethany House Publishers, a division of Baker Publishing Group, Copyright ©1974, Used by permission.*

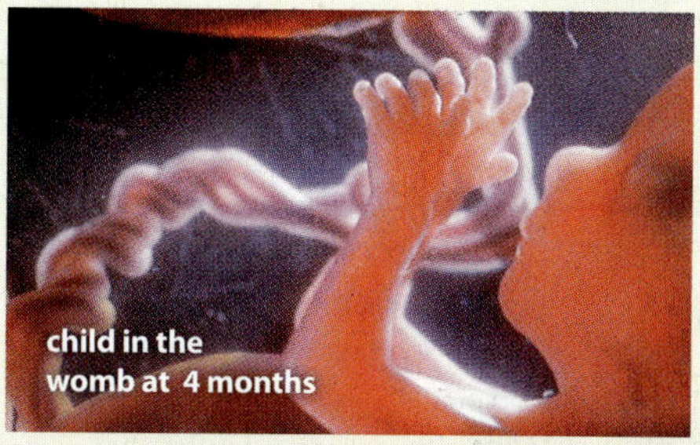
child in the womb at 4 months

## Section Review 6.1

1. List the location of each body cavity and name the organs it contains.
2. Identify the main function of each of the body's systems.
3. Explain the importance of the circulatory system.
4. Why is the immune system important?
5. Why is the brain considered the master control unit of the body?
6. Why must the body maintain homeostasis?

**Identify**

viscera, system, organ, alimentary canal, hormone, endocrine glands, uterus, feedback mechanism

# 6.2 To Everything a Purpose

## Tissues

**Four types of tissues.** As we have seen, each system has a job to do, and that job is accomplished by the coordinated efforts of the various organs that make up the system. Organs, in turn, are composed of various types of **tissue,** a group of similar cells that work together to accomplish the task of an organ. The tissues that compose organs are grouped into four broad categories: epithelial [ĕp′ə·thē′lē·əl] tissue, connective tissue, muscle tissue, and nerve tissue. For example, the stomach is an organ that is constructed of muscle tissue, lined with epithelial tissue, and surrounded by connective tissue. The epithelial tissue provides protection and produces secretions that aid in digestion. The connective tissue cushions and insulates the stomach.

**Muscle tissue** moves the body and its parts. **Nerve tissue** controls organs and coordinates all of their functions. **Epithelial tissues** line all body parts. They cover and protect every organ, regulate temperature, and regulate what enters and exits the organs. Epithelial tissues also may manufacture and secrete lubricants, and some have hairlike projections (cilia) that keep materials moving through the organs. **Connective tissues** connect the parts of the body. The tissues that make up bones connect and support the entire body. Tendons and ligaments help hold muscles and bones together. Blood and lymph can be thought of as liquid connective tissues that connect all the cells. Fat, which is also considered connective tissue, helps protect organs from injury and also stores energy.

Fig. 6.3 **Epithelial and connective tissue**

**Tissue fluids.** The composition of human tissues varies greatly, depending on gender, level of fitness, and other factors. A large percentage of the body tissue in an average, healthy adult is composed of water. Much of this water is found in body fluids such as *blood* and

### Table 6.1 Tissues of the Human Body

| | Examples | Function | Structure |
|---|---|---|---|
| Muscle | muscles | movement | designed to contract |
| Nerve | nerves | control body activities | long and thin |
| Epithelial | outer layers of the skin; lining of the alimentary canal and blood vessels | cover, protect, secrete, and regulate temperature | fit tightly together; reproduce rapidly |
| Connective | bones, tendons, ligaments, cartilage, fat, blood, lymph | hold the body parts together; store energy | varies according to functions; fat cells have large vacuoles for storage |

*cerebrospinal* [sĕr′ə·brō·spī′nəl] *fluid.* The area surrounding and between tissues (the interstitial [ĭn′tər·stĭsh′əl] area) contains **interstitial fluid.** Interstitial fluid that enters the lymphatic system is called *lymph.*

**Membranes.** Membranes are thin, flexible sheets of tissue that cover or line a part of the body. The body has four basic types of membranes. **Serous membranes,** which consist mostly of epithelial tissues, line enclosed body cavities (such as the thoracic and abdominal cavities) that do not open to the body's exterior. **Mucous membranes** are made of epithelial tissues that line openings to the outside. The throat, mouth, and nose are lined with mucous membranes. Our skin, or **cutaneous** [kyo͞o·tā′nē·əs] **membrane,** is composed of several types of tissue. The linings of the joint cavities are **synovial** [sĭ·nō′vē·əl] **membranes.** Synovial membranes are composed of a type of connective tissue.

## Cells and Matrix

When we take a closer look at tissues, we discover that they are made of individual *cells* and **matrix** [mā′trĭks], the material surrounding cells. The matrix of bone tissue includes materials in which calcium is stored. Cells of connective tissue secrete a matrix that holds tightly to neighboring cells. The matrix of blood tissue is called *plasma* and is a liquid. The matrix of cartilage, on the other hand, is flexible and rubbery. Much of the body volume is made up of matrix material.

## "Vestigial" Organs

At one time, there were about 180 parts of the body that evolutionists classified as **vestigial** [vĕ·stĭj′ē·əl], or serving no useful purpose today. Now scientists realize that there are no vestigial organs, only organs that we do not yet fully understand.

A few organs once thought to be vestigial include the parathyroid glands, the thymus, and the pineal gland in the brain. There are still a few parts of the body whose purpose has not yet been discovered, but that number is shrinking rapidly. One day perhaps we will know the function of all the vestigial organs. The fact remains that modern medicine does not know all that is to be known about our bodies. Mysteries abound. Some of God's designs are yet secret. There is plenty of work for man to do to fully understand God's greatest creation.

### Table 6.2 Some Important Human Organs Once Thought to Be Vestigial

| Organ | Known function(s) |
|---|---|
| appendix | contains much lymphatic tissue, which probably helps to protect the body against infection, especially in the early years of life |
| thyroid gland | produces hormones that are important in the growth and development of the body; regulates the amount of calcium in the blood; primary responsibility is to regulate metabolic rate |
| parathyroid glands | produce hormones that regulate the amount of calcium, magnesium, and phosphate ions in the blood |
| pituitary gland | the "master gland" of the body; regulates growth of the body and other bodily functions |
| pineal gland | produces the hormone melatonin, which regulates cycles of sleep and wakefulness |
| coccyx (tailbone) | muscles attached to it allow us to stand in an upright position; these muscles also aid in the elimination of wastes |
| tonsils | help promote good health by destroying harmful bacteria |
| third molar (wisdom tooth) | provides a surface to grind food into tiny bits to prepare it for proper digestion |
| cuspid (canine tooth) | tears large pieces of food into smaller pieces |
| "yolk sac" of embryo | serves as an early source of nutrients and blood cells; contains cells that later form into various parts of the embryo's developing reproductive system and GI tract |
| thymus | serves as the center where T cells (white blood cells that help fight bacteria and viruses) mature and are "trained" |

## Cells by Dr. Paul Brand and Philip Yancey

Design in the LIVING CREATION

I have closed my eyes. My shoes are kicked off, and I am wiggling the small bones in my right foot. Exposed, they are half the width of a pencil, and yet they support my weight in walking. I cup my hand over my ear and hear the familiar seashell phenomenon, actually the sound of blood cells rushing through the capillaries in my head. I stretch out my left arm and try to imagine the millions of muscle cells eagerly expanding and contracting in concert. I rub my finger across my arm and feel the stimulation of touch cells, 450 of them in each one-inch-square patch of skin.

Inside, my stomach, spleen, liver, pancreas, and kidneys, each packed with millions of loyal cells, are working so efficiently I have no way of perceiving their presence. Fine hairs in my inner ear are monitoring a swishing fluid, ready to alert me if I suddenly tilt off balance.

When my cells work well, I'm hardly conscious of their individual presences. What I feel is the composite of their activity known as Paul Brand. My body, composed of many parts, is one. And that is the root of the analogy we shall explore.

One drawer in my medical laboratory contains neatly filed specimens of an array of cells from the human body. Separated from the body, lifeless, stained with dyes and mounted in epoxy, they hardly express the churn of living cells at work inside me at this moment. But if I parade them under the microscope, certain impressions about the body take shape.

I am first struck by their variety. Chemically my cells are almost alike, but visually and functionally they are as different as the animals in a zoo. Red blood cells, disks resembling Lifesaver candies, voyage through my blood loaded with oxygen to feed the other cells. Muscle cells, which absorb so much of that nourishment, are sleek and supple, full of coiled energy. Cartilage cells with shiny black nuclei look like bunches of black-eyed peas glued tightly together for strength. Fat cells seem lazy and leaden, like bulging white plastic garbage bags jammed together.

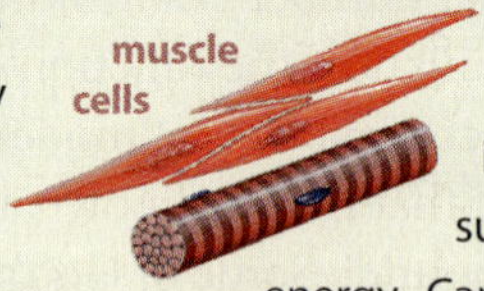


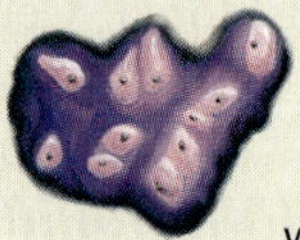


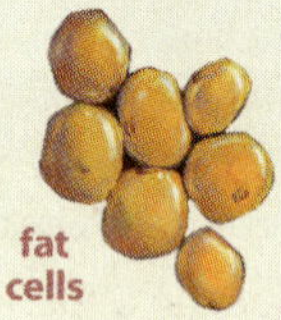


Bone cells live in rigid structures that exude strength. Cut in cross section, bones resemble tree rings, overlapping strength with strength, offering impliability and sturdiness. In contrast, skin cells form undulating patterns of softness and texture that rise and dip, giving shape and beauty to our bodies. They curve and jut at unpredictable angles so that every person's fingerprint—not to mention his or her face—is unique.

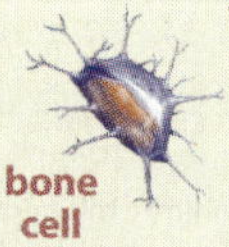


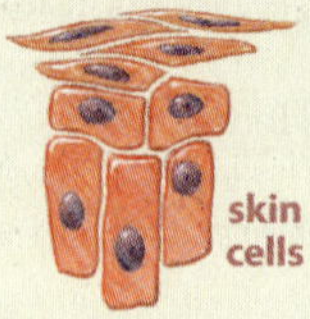


The aristocrats of the cellular world are the sex cells and nerve cells. A woman's contribution, the egg, is one of the largest cells in the human body, its ovoid shape just visible to the unaided eye. It seems fitting that all the other cells in the body should derive from this elegant and primordial structure. In great contrast to the egg's quiet repose, the male's tiny sperm cells are furiously flagellating tadpoles with distended heads and skinny tails. They scramble for position as if competitively aware that only one of billions will gain the honor of fertilization.

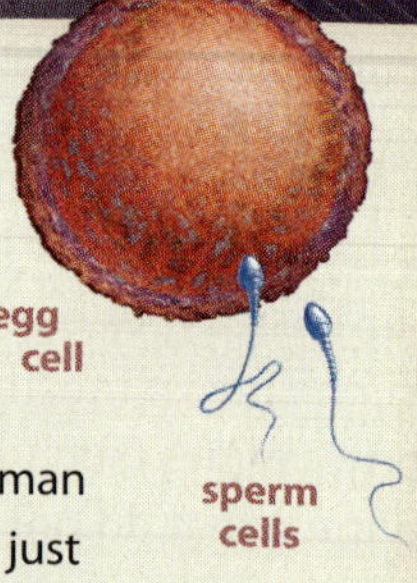


The king of cells, the one I have devoted much of my life to studying, is the nerve cell. It has an aura of wisdom and complexity about it. Spider-like, it branches out and unites the body with a computer network of dazzling sophistication. Its axons, "wires" carrying distant messages to and from the human brain, can reach a yard in length.

I never tire of viewing these varied specimens or thumbing through books which render cells. Individually they seem puny and oddly designed, but I know these invisible parts cooperate to lavish me with the phenomenon of life.

## An Imperfect Body

Though the design of the human body is a marvel of engineering, it has its imperfections. Nerve cells wear out and die, never to be replaced. Eventually so many are lost that muscle coordination and mental awareness fail, resulting in some of the symptoms of old age. Because of an accumulation of years of deposits, the blood vessels eventually narrow, diminishing the life-giving flow of blood. Eventually this well-designed machine, the human body, wears out.

> Then shall the dust return to the earth as it was: and the spirit shall return unto God who gave it. —Eccles. 12:7

God created for man a perfect body. Until Adam and Eve sinned, their bodies were without flaw. Ever since their sin, however, the human body has been subject to disease, decay, and death.

> And the LORD God said, Behold, the man is become as one of us, to know good and evil: and now, lest he put forth his hand, and take also of the tree of life, and eat, and live for ever: Therefore the LORD God sent him forth from the garden of Eden, to till the ground from whence he was taken. So he drove out the man; and he placed at the east of the garden of Eden Cherubims, and a flaming sword which turned every way, to keep the way of the tree of life. —Gen. 3:22–24

God promises to someday give the Christian a new body, which will be free from the imperfections brought upon us by sin.

> For our conversation is in heaven; from whence also we look for the Saviour, the Lord Jesus Christ: Who shall change our vile body, that it may be fashioned like unto his glorious body, according to the working whereby he is able even to subdue all things unto himself. —Phil. 3:20–21

## Section Review 6.2

1. Explain the purpose of tissues, fluids, membranes, cells, and matrix.
2. What are the four basic types of tissue?
3. What are the four main types of membranes found in the body?

### Critical Thinking

4. What do the supposedly vestigial organs tell you about the wisdom of God and the wisdom of man?

### Identify

interstitial fluid, serous membrane, cutaneous membrane, vestigial

## Chapter 6 Review

### Define

1. cranial cavity
2. spinal cavity
3. thoracic cavity
4. abdominal cavity
5. viscera
6. system
7. alimentary canal
8. hormone
9. endocrine gland
10. uterus
11. homeostasis
12. tissue
13. interstitial fluid
14. matrix
15. vestigial organ

### Identify

1. the eleven systems of the body and the function of each
2. four basic types of tissue and examples of each
3. four types of tissue fluids
4. four types of membranes and an example of each

### Explain

1. How do the circulatory, respiratory, digestive, and excretory systems work together to supply nutrients to the body's cells and to remove waste products?
2. Explain the importance of the immune system.
3. Why is the skin important?
4. How do the skeletal and muscular systems work together to allow movement?
5. List five important organs once thought to be vestigial and give their known functions.

### Apply

Why are there few positive feedback mechanisms in the body?

# CHAPTER 7 Bones & Muscles

## Key Concepts

- Protective design of the skeleton
- Structure of bones
- Bone growth and development
- Joints
- Types of muscle tissue
- Structure of skeletal muscles
- Muscles and exercise

The framework of the human body is its **skeletal system,** composed of approximately 206 bones that support the body and protect vital organs. The body's bones are masterpieces of structural design, comparable in strength to reinforced concrete, yet self-repairing and far more resilient. Precisely engineered joints between bones allow the body to move; the power for movement is provided by strong muscles attached to the bones by cablelike tendons. Besides their structural roles, bones also produce blood cells. This vital function is carried out by bone marrow inside the body's long bones, which manufactures new red blood cells at the astonishing rate of more than 2 million per second. The marrow also produces white blood cells, the agents of the body's immune system. Finally, bones store important minerals such as calcium and phosphorus. Bones account for about 20% of the body's weight.

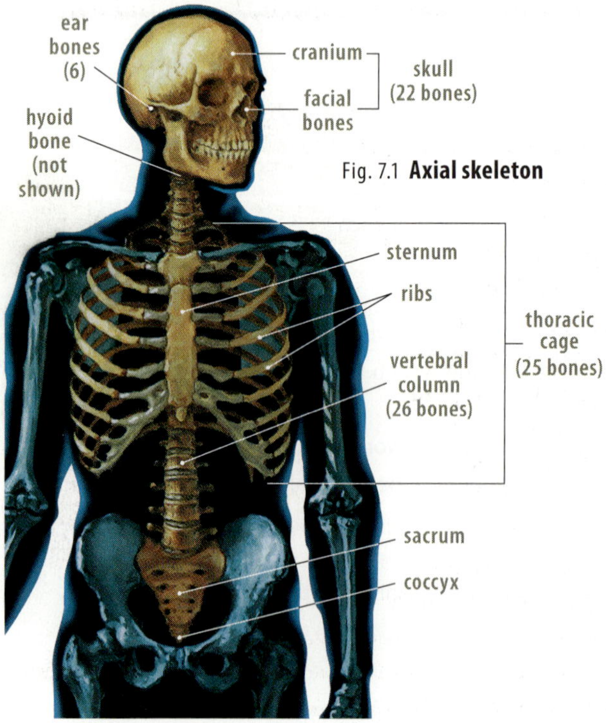


Fig. 7.1 **Axial skeleton**

According to God's precise engineering, the **muscular system** functions together with the skeletal system and the nervous system to make body movements possible. The individual organs of the muscular system are the skeletal muscles, which make up between 40% and 50% of the body's weight.

## 7.1 The Axial Skeleton

*The skeletal system consists of two broad divisions,* the axial [ăk′sē·əl] skeleton and the appendicular [ăp′ən·dĭk′yə·lər] skeleton. The **axial skeleton** is the "backbone" of the skeletal system; it is made up of eighty bones that constitute the head and spine, including the bones of the skull, the hyoid [hī′oid′] bone in the neck, the vertebrae [vûr′tə·brā′] of the neck and backbone, the sternum or breastbone, and the ribs. The *appendicular skeleton,* on the other hand, refers to the bones of the appendages (arms and legs), shoulders, and hips.

## Bones of the Head

**Cranial bones.** Your **skull** can be divided into two main parts, the cranium and the facial bones. The largest portion of the skull is the **cranium,** or *braincase,* which composes the top, sides, and rear of the skull. The cranium is made up of eight bones joined tightly together to form an armored shell for the brain. Important cranial bones include the *frontal bone* (forehead), the *parietal* [pə·rī′ĭ·təl] *bones* (top of the head), the *temporal bones* (sides of the head), and the *occipital* [ŏk·sĭp′ĭ·tl] *bone* (back of the head).

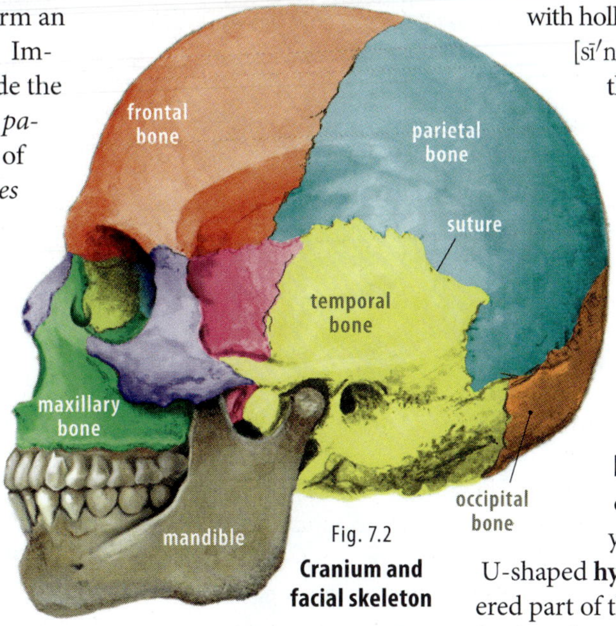

Fig. 7.2
**Cranium and facial skeleton**

Although these bones are fused tightly in an adult, they are joined only loosely in the skull of an infant. The bones of an infant's cranium are connected by **fontanels** [fŏn′tə·nĕlz′], tough membranes made of fibrous connective tissue; the fontanels are the *soft spots* on the top and back of an infant's head. The flexibility imparted by the fontanels allows the baby's skull to deform somewhat in order to fit through the narrow birth canal during the miracle of birth. The fontanels also enable the skull to grow larger as the child grows. The gaps between the infant's cranial bones gradually shrink as the fontanels are replaced by growing bone. Within a year or two, the fontanels disappear completely, and the cranial bones are joined tightly together at uneven lines known as **sutures** [so͞o′chərz]. The jagged sutures are another evidence of God's design; they give the skull great strength by helping to absorb some of the shock from a blow to the head.

**Facial bones.** The 14 bones of the **facial skeleton** serve as the framework of the face and jaw. The most important of the facial bones are the two *maxillary* [măk′sə·lĕr′ē] *bones,* which form the central portion of the face and serve as the attachments for your upper teeth. The maxillary bones, together with the *palatine* [păl′ə·tīn′] *bones* located behind them, also form the roof of the mouth, or *palate* [păl′ĭt]. The *mandible* [măn′də·bəl], which forms the lower jaw and holds the lower teeth, is the only movable bone of the skull and is held to the cranium by strong connective tissues known as **ligaments** [lĭg′ə·mənts: tissues that join bones to other bones].

**Sinuses.** Certain bones of the skull are designed with hollow spaces known as **sinuses** [sī′nəs·ĕz]. Large sinuses are located in the maxillary bones of the face and in the frontal and sphenoid bones of the cranium. When you speak, sound waves resonate within the sinuses, helping to give your voice its characteristic vocal quality. In addition, your hollow sinuses allow your skull to be lighter than if it were made of solid bone.

**The hyoid and middle ear bones.** An important bone hidden in the upper neck, just above your larynx (voicebox), is the U-shaped **hyoid bone.** Although not considered part of the skull, the hyoid bone serves as the foundation of many of the tongue muscles and of certain other muscles that allow you to swallow.

*The smallest bones in the human body are the three middle ear bones,* the *malleus* [măl′ē·əs: hammer], *incus* [ĭng′kəs: anvil] and *stapes* [stā′pēz: stirrup]. These structures, located in a hollow cavity inside each temporal bone, constitute a mechanical linkage between the eardrum and the cochlea (the snail-shaped structure of the inner ear that senses sound vibrations). This linkage amplifies the vibrations of the eardrum and passes them to the cochlea, enabling you to hear.

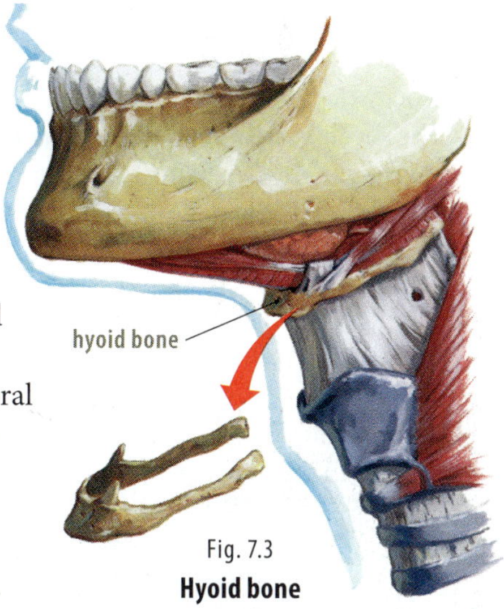

Fig. 7.3
**Hyoid bone**

## Bones of the Spine

The chief structural member of the body is the **vertebral column** (also known as the *spine, spinal column,* or *backbone*), a massive, columnlike structure that, in an adult, consists of 33 segments called **vertebrae** (sing. vertebra). The vertebral column is the support to which all the other parts of the skeleton are attached. Between the vertebrae are disks of flexible cartilage called **intervertebral** [ĭn′tər·vûr′tə·brəl] **disks,** which allow the spine to bend and twist somewhat; these disks also act as shock absorbers. (**Cartilage** is a tough, rubbery connective tissue that cushions the joints between bones.) Hollow cavities near the rear of each vertebra form a long vertical tunnel through the spine that encloses the spinal cord, protecting it from damage.

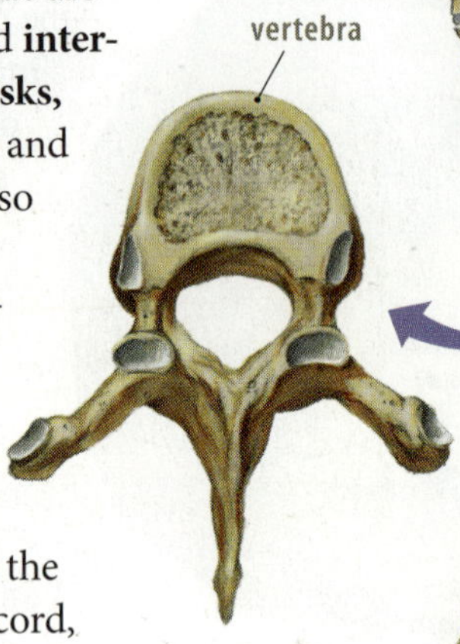


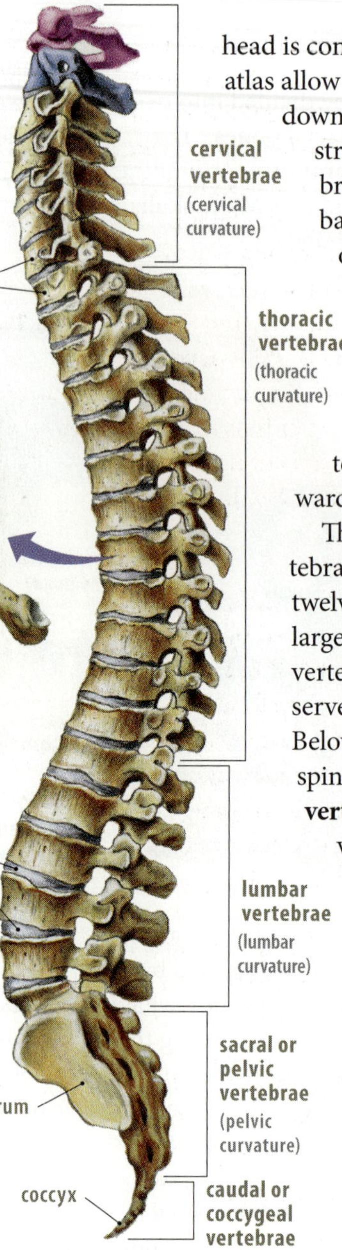


Fig. 7.4 **Vertebral column**

**Regions of the vertebral column.** The vertebral column is divided into five regions: cervical [sûr′vĭ·kəl: the neck], thoracic, lumbar [lŭm′bər: the lower back], sacral [sā′krəl: the sacrum] or pelvic [pĕl′vĭk: the pelvis], and caudal [kôd′l] or coccygeal [kŏk·sĭj′ē·əl: the coccyx].

The seven vertebrae that form the neck are called the **cervical vertebrae.** The uppermost of these, the *atlas*, is the mounting to which the head is connected. Special bearings on the atlas allow the head to rotate upward or downward while keeping your neck straight. The second cervical vertebra, the *axis*, is designed as the base for the atlas. Bearing surfaces on the axis allow the atlas to swivel left or right. The other five cervical vertebrae are joined together by intervertebral disks; the flexibility of these disks allows the head to tilt to the side and crane backward or forward.

The longest segment of the vertebral column is composed of the twelve **thoracic vertebrae.** Slightly larger and thicker than the cervical vertebrae, the thoracic vertebrae serve as attachments for the rib cage. Below the thoracic vertebrae, the spine widens into the five **lumbar vertebrae,** which are the largest vertebrae. The lumbar vertebrae form the lower back and support the weight of the entire upper body. When the lumbar vertebrae are engaged with the supporting muscles, they can bear very heavy burdens—another striking example of God's design and provision.

The pelvic region consists of the **sacral vertebrae** (or *pelvic vertebrae*), which are five separate vertebrae in infants but are fused into a single bone, called the *sacrum,* in adults. The sacrum is joined rigidly to the hip bones to form the foundation of the spine. Beneath the sacrum is the tiny **coccyx,** which consists of four separate vertebrae (the *caudal* or *coccygeal vertebrae*) in infants but are fused into a single bone in adults. Also known as the *tailbone*, the coccyx is an attachment for various muscles that allow us to stand in an upright position and aid in the elimination of wastes. The coccyx also helps absorb the shock to your spine when you sit down.

Fig. 7.5 **Atlas and axis**

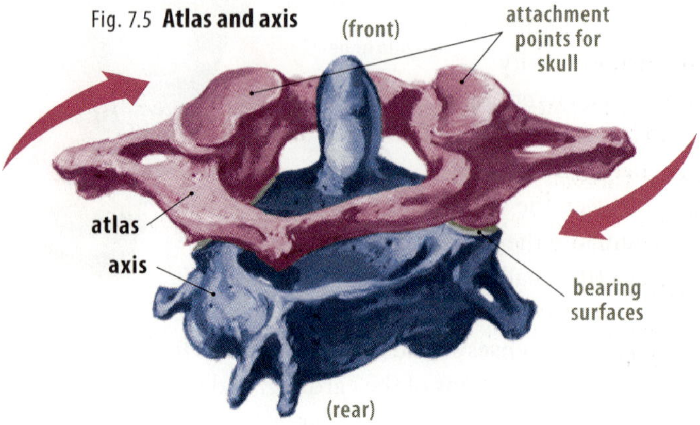

**Curvatures of the vertebral column.** The vertebrae do not form a straight, pillarlike column, but instead form four curvatures from front to back: the cervical curvature, the thoracic curvature, the lumbar curvature, and the pelvic curvature (Fig. 7.4). God designed these curves of the vertebral column to provide the spine with additional strength and flexibility, allowing it to absorb shocks and helping to protect it against fracture.

Normally, the spine also has a slight lateral (sideways) curve. But in some people, this curve is excessive, causing one shoulder blade to protrude abnormally and one side of the waist or hips to be higher than the other. A severe lateral curvature of the spine is known as **scoliosis** [skō′lē·ō′sĭs].

## Bones of the Chest

The vital internal organs of the thoracic cavity, such as the heart and lungs, are protected by the bones of the chest, which make up the **thoracic cage.** The most prominent bones of the thoracic cage are the **ribs**, long, curved bones that are attached to the thoracic vertebrae in the back and to the **sternum**, or *breastbone,* in the front. Although the ribs attach directly to the thoracic vertebrae, they are joined to the sternum indirectly by short segments of flexible cartilage called the *costal* [kŏs′təl: from Latin *costa,* "rib"] *cartilages.* Because the costal cartilages are not rigid bone but rather flexible cartilage, they allow the thoracic cage to expand and contract during breathing. Only the upper seven pairs of ribs are attached (via their costal cartilages) to the sternum itself; for this reason, the upper seven pairs of ribs are called *true ribs.* The rest of the ribs, which do not connect to the sternum itself, are known as *false ribs.* The bottom two pairs of false ribs are called *floating ribs* because they do not connect to the front of the thoracic cage at all; the remaining three pairs of false ribs are attached to the costal cartilages of the lowest pair of true ribs instead of to the sternum.

Fig. 7.6 **Thoracic cage**

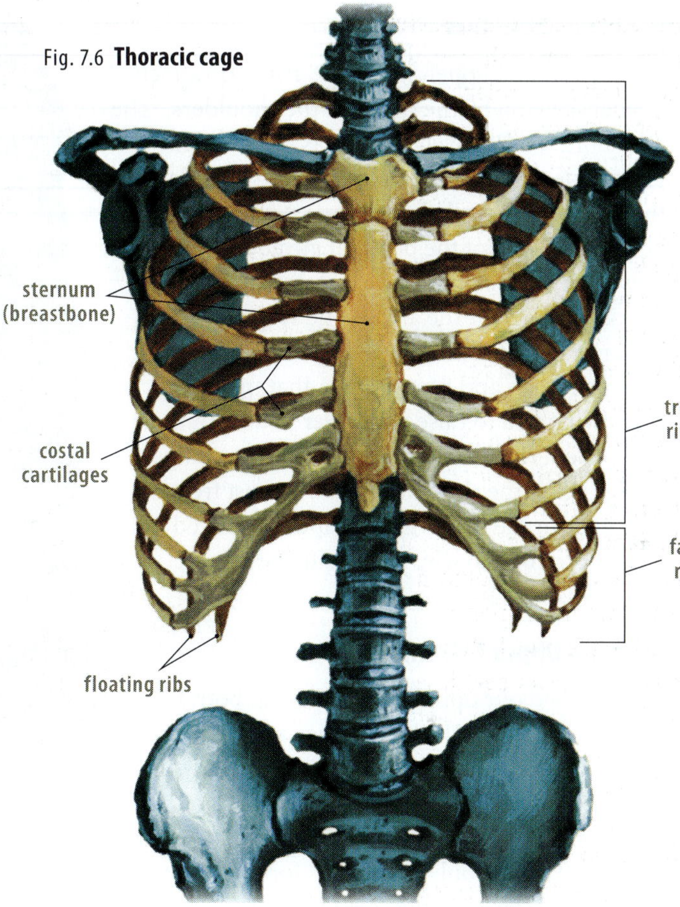


## Section Review 7.1

1. What are the two broad divisions of the skeleton?
2. Name the major bones of the head and chest.
3. Describe the functions of the cranium, fontanels, sutures, and sinuses.
4. Describe the functions of the hyoid bone and the middle ear bones.
5. Name the bones, regions, and curves of the vertebral column.
6. What characteristics distinguish true ribs from false ribs and floating ribs?

### Application

7. In a paragraph, explain how the structure of the skull is an example of God's design and provision for man.

### Critical Thinking

8. Men and women both have exactly 12 pairs of ribs. Why is this not a contradiction of Genesis 2:21–22?

### Identify

cranium, ligament, cartilage, cervical, thoracic, lumbar, scoliosis

## 7.2 The Appendicular Skeleton

The **appendicular skeleton** includes the 126 bones that form the pectoral girdle (the shoulders), the pelvic girdle (the hips), and the *appendages* (the arms and legs).

### Bones of the Shoulders

The body's upper appendages, the *arms,* are attached to the axial skeleton by the bones of the shoulders. The shoulder bones, collectively known as the **pectoral girdle,** include the shoulder blades (called scapulae) and the collarbones (called clavicles).

The largest bones of the pectoral girdle are the *shoulder blades,* or **scapulae** (sing. scapula), which contain the sockets for the arms. The scapulae are not fixed rigidly to the axial skeleton but are suspended by muscles and ligaments so that they can move, allowing you to raise and lower your shoulders and move them forward or backward. This arrangement helps make the shoulder joint the most freely movable joint in the body. Each scapula is braced by a collarbone, or **clavicle,** which attaches to the top of your sternum.

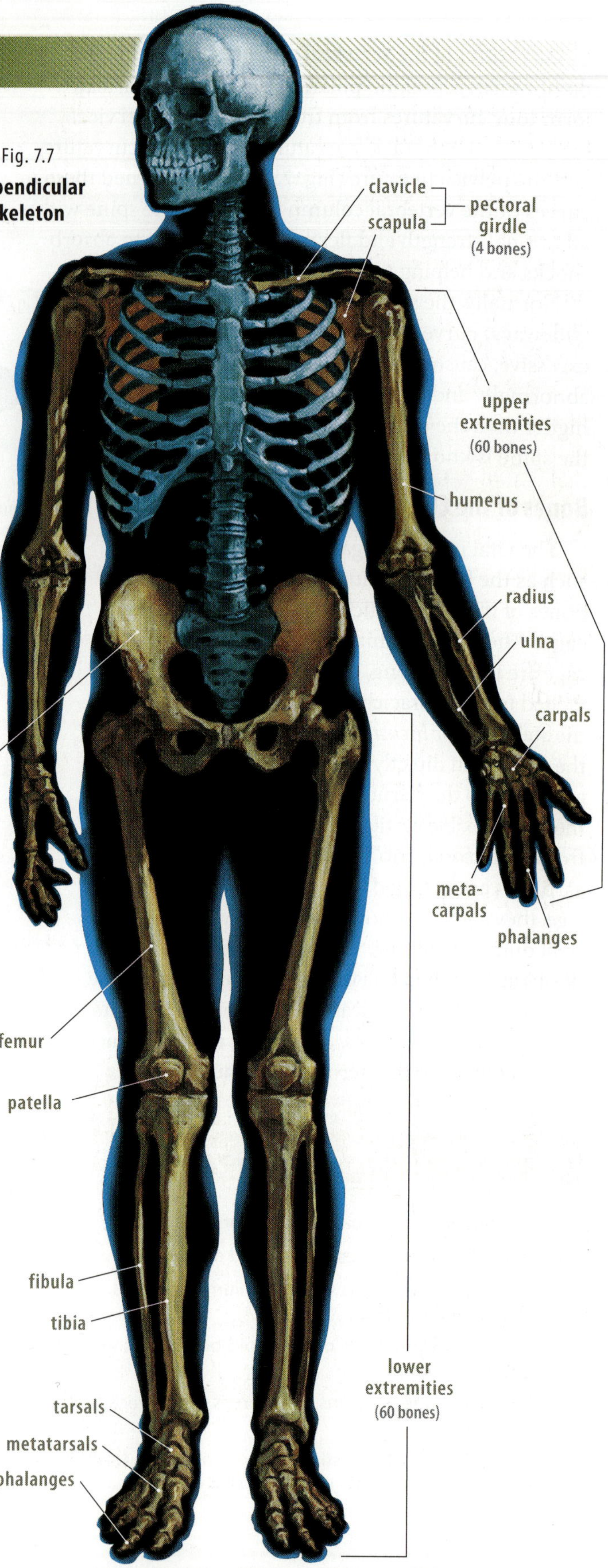


Fig. 7.7 **Appendicular skeleton**

### Bones of the Upper Extremities

God designed our upper extremities for both strength and dexterity; the large bones and muscles of the shoulders and arms provide great strength, while the intricate structures of the wrists and hands allow for precision movements. Each upper limb is composed of a total of 30 bones, only 3 of which form the arm itself; the remaining 27 bones form the wrist, hand, and fingers. The largest of the three bones in the arm is the **humerus** [hyo͞o′mər·əs], which makes up the upper arm; the humerus attaches to the scapula at the shoulder joint.

The forearm is composed of two bones, the ulna and the radius. This two-bone design allows you to twist your forearm clockwise or counterclockwise, such as when turning a key or using a doorknob. The **ulna,** the bone on the same side of your forearm as your little finger, is attached firmly to the humerus by a strong, hingelike joint. The **radius,** the bone on the same side as your thumb, is attached to the ulna and humerus by a weaker but more movable joint that allows it to rotate around the ulna (Fig. 7.8, p. 97). When you rotate your hand, the ulna does not rotate but remains fixed relative to the humerus; the radius and the hand rotate around pivots on the ulna. This wise design provides the forearm with great strength while allowing for rotation of the hand.

The hand is attached to the ulna and radius by the bones of the wrist, or *carpus.* These eight bones, called **carpals,** form an intricate assembly that allows the wrist to bend in various directions. Branching out from the wrist are five **metacarpals** [mĕt′ə·kär′pəlz], which compose the base of the thumb and the main part of the hand. The four metacarpals that compose the hand are bound together by strong ligaments, but the metacarpal that forms the base of the thumb is designed with a greater range of motion, allowing it to be pivoted outward from the hand. Attached to the metacarpals are fourteen **phalanges** [fə·lăn′jēz: sing. phalanx], which form the fingers and thumb. Each finger consists of three short phalanges, but each thumb contains only two.

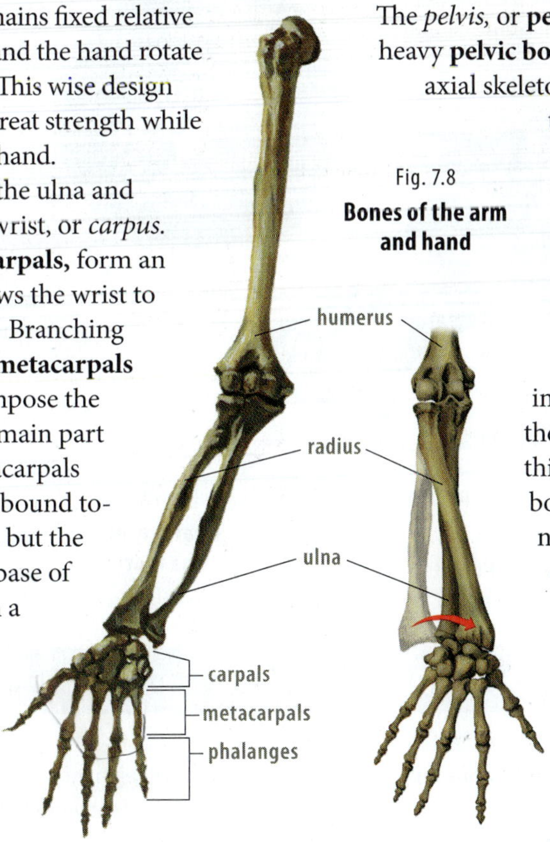


Fig. 7.8
**Bones of the arm and hand**

## Bones of the Pelvis

The body's lower appendages, the *legs,* are attached to the axial skeleton by the two bones of the pelvis.

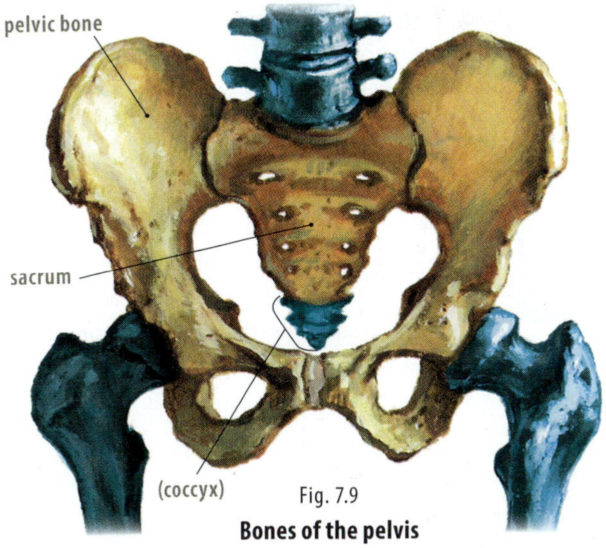


Fig. 7.9
**Bones of the pelvis**

The *pelvis,* or **pelvic girdle,** is composed of two large, heavy **pelvic bones** that attach to the sacrum of the axial skeleton, forming a rigid ring of thick bone that supports most of the body's weight. The legs attach to the pelvis at the hip joints, which are formed by large sockets on the side of each pelvic bone.

## Bones of the Lower Extremities

**Bones of the legs.** Although similar in basic layout to the upper extremities, the lower extremities are composed of thicker and stronger bones to support our bodies when standing, walking, running, and jumping. Each lower limb is composed of 30 bones. Four of these compose the leg itself; the remaining 26 bones form the ankle, foot, and toes. The largest bone of the leg (and the longest bone in the body) is the **femur** [fē′mər], which composes the thigh. The femur attaches to the pelvic girdle at the hip joint.

The lower leg is composed of two bones, the tibia and the fibula, which are similar to the ulna and radius of the forearm. The main weight-bearing bone of the leg is the *shinbone,* or **tibia,** which extends from the knee joint to the ankle. The other bone of the lower leg, the **fibula,** is much thinner than the tibia and does not attach to the femur at all. Rather, it is attached to the tibia, just as the radius is attached to the ulna in the forearm. However, the lower leg cannot be rotated like the forearm because the fibula cannot move freely around the tibia. Such rotation is not necessary, because you can rotate your foot side to side at the ankle joint and rotate your leg as a whole at the hip joint.

One additional bone found in the leg does not correspond to any bone in the arm. This is the *kneecap,* or **patella** [pə·tĕl′ə], which is found on the front of the knee. The patella serves as a brace for the tendon that straightens the leg. (A **tendon** is a strong "cable" of tough fibers that attaches a muscle to a bone.) By holding this tendon slightly out in front of the knee joint, the patella increases the leverage of the tendon, allowing you to exert more force when rising from a squatting position than would otherwise be possible. The patella also helps shield the knee joint from injury.

**Bones of the foot.** The foot is attached to the tibia and fibula by the bones of the ankle, or *tarsus.* These seven bones, known as **tarsals,** are somewhat similar to the carpals of the wrist but are thicker and stronger to support the weight of the body. But unlike the wrist, which does not allow twisting movements, the ankle allows the foot to be rotated left or right independently of the lower leg. One of the largest tarsals, the *talus,* is the only tarsal that is directly connected to the tibia and fibula; other, smaller tarsals form the rearmost part of the instep. Beneath the talus is the largest tarsal, called the *calcaneus* [kăl·kā′nē·əs], which forms the heel.

In front of the tarsals are the five **metatarsals** [mĕt′ə·tär′səlz], which correspond to the metacarpals of the hand. These bones, which form the framework of the instep and the ball of the foot, are arranged in curved arches that help make the foot stronger and more resilient. Attached to the metatarsals are fourteen *phalanges,* which correspond to those of the fingers and thumb.

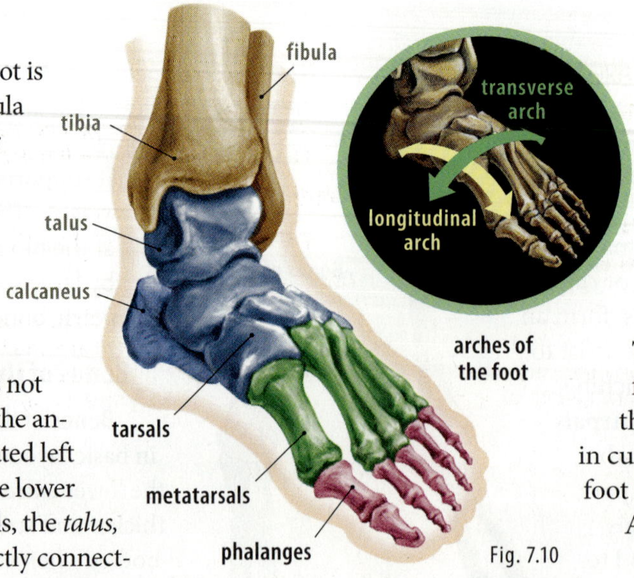


Fig. 7.10
**Anatomy of the foot**

## *The Bones of the Human Foot* by Alfred M. Rehwinkel

Design in the LIVING CREATION

The foot is the humblest member of man's anatomy. The human foot with its five toes is a miracle of construction. It consists of 26 separate bones of various sizes and shapes bound together by a system of ligaments. It is supported by a complex array of muscles and supplied with a network of fibers and blood vessels. The different bones articulate in gliding joints, giving the foot a degree of elasticity and a limited amount of motion. The arrangement of the bones is such as to form several arches, the most important of these being the long arch from the heel to the ball of the foot.

These arches are held in place and supported by a complex of strong muscles to carry the weight of the body just as the steel cables carry the load of a suspension bridge; the construction also gives elasticity to the foot, making walking and running and other movements possible. The ability to walk and to move from place to place and to balance the tall upright body on a comparatively small pedestal is itself a most remarkable feature of the human body. If the foot were flat and rigid, fixed at right angles to the bone of the leg, walking would be difficult or impossible. The elastic arches also serve as shock absorbers to soften the jar resulting from walking on a hard surface. The human foot is a masterpiece of engineering. It is a miniature suspension bridge but more complicated than an ordinary bridge. . . .

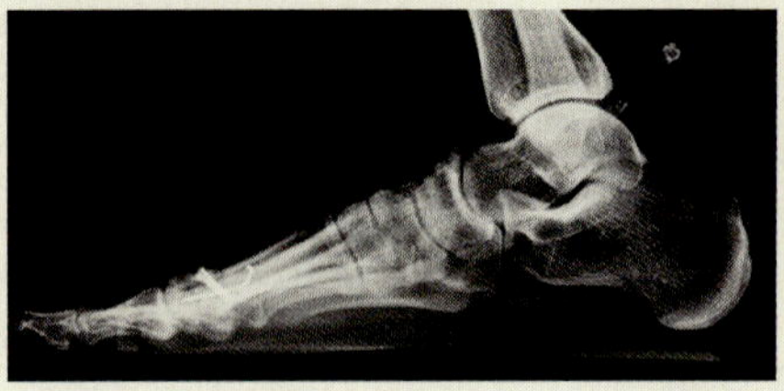

Would anyone say that the Golden Gate suspension bridge just happened? Of course not, if he were sane! But why do people assume that an even more intricate mechanism of the human foot could have just happened without design or the workmanship of a master engineer. To add to the wonders of the human foot we must also remember that it has been reproduced billions and billions of times in every human birth with exactly the same shape and form and with the same number of bones and tendons and nerves. And so we see that even the humble foot of man is one of the wonders of his body that glorifies the wisdom of the Creator who made it.

*Alfred M. Rehwinkel,* The Wonders of Creation, *Bethany House Publishers, a division of Baker Publishing Group, Copyright ©1974, Used by permission.*

## Section Review 7.2

1. What are the major parts of the appendicular skeleton?
2. Explain how the bones of the forearm allow the hand to rotate. Why is this a wise design?
3. Explain how the foot is a marvel of God's design and engineering.

### Critical Thinking

4. Professional football players have been known to play through an entire game with a cracked fibula and (except for the pain) run and walk normally. Use your knowledge of the lower limbs to explain how this is possible.

### Identify

pectoral girdle, phalanges, pelvis, pelvic girdle, patella, tendon

### Observe

- Using a skeleton or a picture, identify the parts of the appendicular skeleton and name the bones of the shoulders, limbs, and pelvis. Identify the bones of the wrist, palm, fingers, ankle, instep, and toes; notice the structure of the hands and feet.

# 7.3 Bones: Designed for Strength

## How Bones Are Classified

The 206 bones of a typical adult have a wide variety of shapes and sizes, from the long, straight bones of the legs and arms to the irregularly shaped bones of the face and spine. Bones are classified into four main categories based upon their shape. **Long bones** are longer than they are wide; most of them support the weight of your body and work with your muscles to provide movement. Long bones include the *humerus, ulna, femur,* and *phalanges.* **Short bones** are roughly cube shaped, nearly as wide as they are long; most of the *carpals* and *tarsals* are short bones. The *ribs* and *cranial bones* are **flat bones,** which often have the job of protecting vital organs. Any bone that cannot be easily classified into one of the first three categories is considered an **irregular bone;** the *vertebrae* are irregular bones.

Fig. 7.11 **Types of bones**

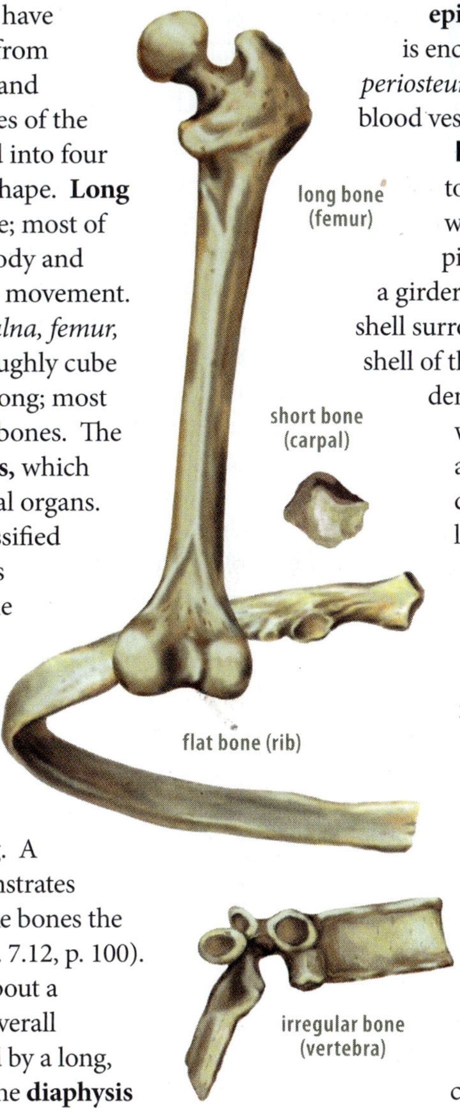


## The Design of a Bone

At first glance, a bone appears to be a relatively simple construction; but closer inspection reveals God's precision engineering. A long bone such as the femur demonstrates some of the design details that make bones the marvelous organs that they are (Fig. 7.12, p. 100).

The first thing that you notice about a long bone such as the femur is its overall shape—two bulged ends connected by a long, straight shaft. (The bone's shaft is the **diaphysis** [dī·ăf′ĭ·sĭs], while each bulged end is an **epiphysis** [ĭ·pĭf′ĭ·sĭs].) The bone as a whole is enclosed in a protective sheath called the *periosteum* [pĕr′ē·ŏs′tē·əm], which also contains blood vessels that supply the bone with nutrients.

**Engineered for strength.** If you were to cut the bone in half lengthwise, you would find that it is not solid like a piece of cement but is constructed like a girder, with a solid, weight-bearing outer shell surrounding a hollow interior. The outer shell of the bone is composed of a strong, dense substance called **compact bone,** which gives the bone great strength and rigidity. Inside the epiphyses, the dense compact bone is replaced by a lightweight, porous tissue called **spongy bone.** This name is something of a misnomer because spongy bone is not soft like a sponge; rather, this type of bone consists of an intricate network of tiny struts and girders that is porous but rigid. The spongy bone in the epiphyses allows the bone to be much lighter for its strength than if it were composed of solid compact bone. Besides reducing the bone's weight, the spongy bone provides tiny open spaces for cells of **red marrow,** special tissue that manufactures red blood cells and white blood cells for the body's circulatory and immune systems.

At the core of the bone's shaft is a hollow, cylindrical cavity, like a tunnel running the length of the bone. This cavity, called the **medullary cavity** or *marrow cavity,* houses red marrow in the bones of infants and children. In adults, the red marrow in the medullary cavity is replaced by a fatty substance called **yellow marrow,** which stores fats.

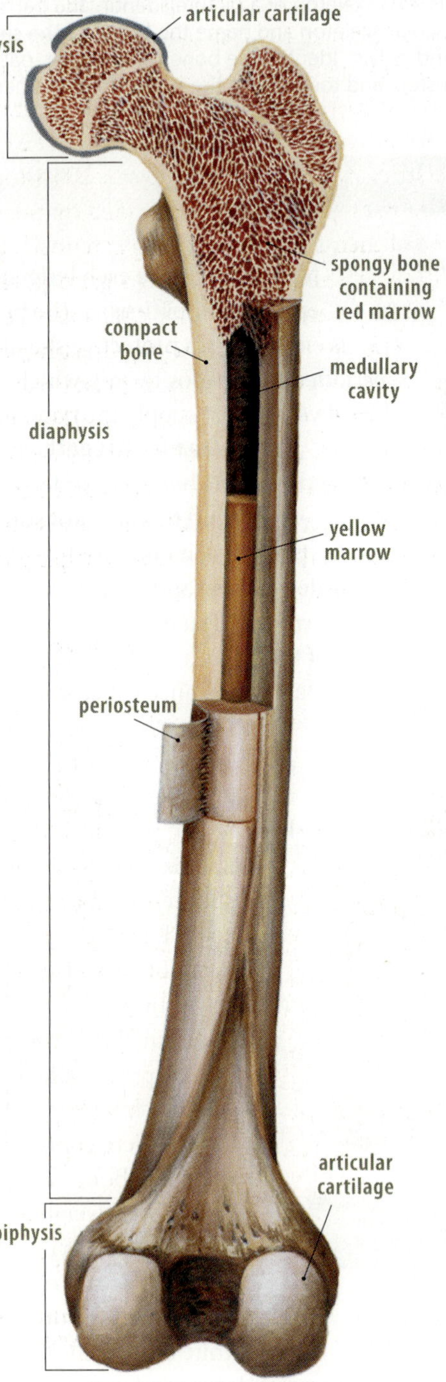


Fig. 7.12
**Structure of a long bone**

**A closer look.** If you look at a cross section of a bone under a powerful microscope, you will find even more design details that contribute to the bone's strength. The hardness of bone is due to microscopic crystals of a mineral substance called **hydroxyapatite** [hī·drŏk′sē·ăp′ĭ·tīt′: $Ca_{10}(PO_4)_6(OH)_2$], which contains much calcium and phosphorus. Hydroxyapatite is extremely rigid and allows bone tissue to withstand greater crushing forces than the finest reinforced concrete. Running between the crystals of hydroxyapatite are reinforcing fibers of **collagen** [kŏl′ə·jən], a tough, resilient protein that serves the same purpose as the steel rods in reinforced concrete, helping to prevent the mineral crystals from being pulled away from each other. These reinforcing protein fibers are so abundant in bone that they make up nearly a third of the bone's dry weight. The collagen fibers bind bone tissue together so strongly that bone can withstand pulling forces about as well as steel-reinforced concrete.

## The Life of a Bone

**Maintenance and renovation.** One of the wonders of bone tissue is that it is self-maintaining, self-repairing, and self-modifying. Because the collagen fibers and hydroxyapatite crystals of your bones are constantly deteriorating, God designed living bones to constantly renovate themselves, breaking down old collagen fibers and hydroxyapatite crystals and replacing them with newly manufactured materials to prevent the bones from becoming weak and brittle. This important job is performed by two types of cells. **Osteoclasts** [ŏs′tē·ə·klăsts′: literally, "bone breakers"] are special cells that constantly move through your bones, removing old materials to make room for new. As osteoclasts move through the bone, they carve tiny tunnels (called *Haversian canals*) lengthwise through the bone. Other cells called **osteoblasts** [ŏs′tē·ə·blăsts′: "bone growers"] move into the Haversian canals behind the osteoclasts, constructing new collagen fibers and hydroxyapatite crystals as they go; blood vessels also enter the canals, supplying the osteoblasts with nutrients. This constant process of renovation helps keep your bones strong and resilient. Unfortunately, due to the curse of sin upon God's creation, this process of bone renovation slows as we grow older, so that our bones generally become more brittle with age.

**Nutrition for healthy bones.** A proper diet is necessary for healthy bones. Important nutritional factors include the minerals calcium and phosphorus and vitamins D,

A, and C. *Calcium* (from milk, milk products, and some leafy green vegetables) and *phosphorus* (from meats, poultry, fish, cheese, nuts, whole grain cereals, milk, and legumes) form the inorganic structure of bones and teeth but cannot be properly absorbed by the body without the presence of *vitamin D*. In God's providence, skin produces needed vitamin D in the presence of sunlight; such foods as fortified milk, egg yolk, and fish-liver oils are also rich in vitamin D. Vitamin D deficiency in children causes **rickets,** a disease in which the bones are weak and deformed. In adults, the same malady is called *osteomalacia* [ŏs′tē·ō·mə·lā′shə] and results in a softening of the bones. A deficiency of *vitamin A* (found in liver, milk products, eggs, dark green vegetables, and deep yellow vegetables and fruits) may retard bone development. A deficiency of *vitamin C* (found in citrus fruits and juices, tomatoes, cabbage, potatoes, dark green vegetables, and fresh fruits) prevents osteoblasts from producing enough collagen in the intercellular matrix of bone tissue, resulting in abnormally slender and fragile bones.

**Exercise for healthy bones.** *Physical exercise* (or the lack of it) *also affects bone growth and development.* The osteoclasts and osteoblasts in your bones can somehow determine which portions of your bones undergo the most stress and strengthen each bone accordingly. *Julius Wolff* (1836–1902), a German orthopedic surgeon, formulated a law that describes the effects of exercise or *stress* on bones. According to **Wolff's law,** bones adjust their shapes to the physical stress placed upon them.

If you expose your bones to stress, such as by running, lifting weights, or other physical activity, your osteoclasts and osteoblasts will increase their activity and build additional bone to handle the extra strain. Thus, the bones of athletes are bigger, denser, and stronger than those of nonathletes because the bones of athletes are subjected to more physical stress. The opposite is also true: bones that are not exercised lose strength and mass. For example, astronauts tend to lose a quarter or more of their bone mass during extended space missions (because of lack of use in a "zero-gravity" environment) unless they undergo vigorous daily exercise while in orbit. The same kind of atrophy (wasting) occurs when a person is recovering from a broken bone. An arm or a leg in a cast may lose as much as 30% of its bone mass within only a few weeks. Of course, after the bone has properly healed, physical exercise can counteract the effects of inactivity. Physical exercise also helps reduce the effects of aging on your bones, helping to prevent them from becoming brittle.

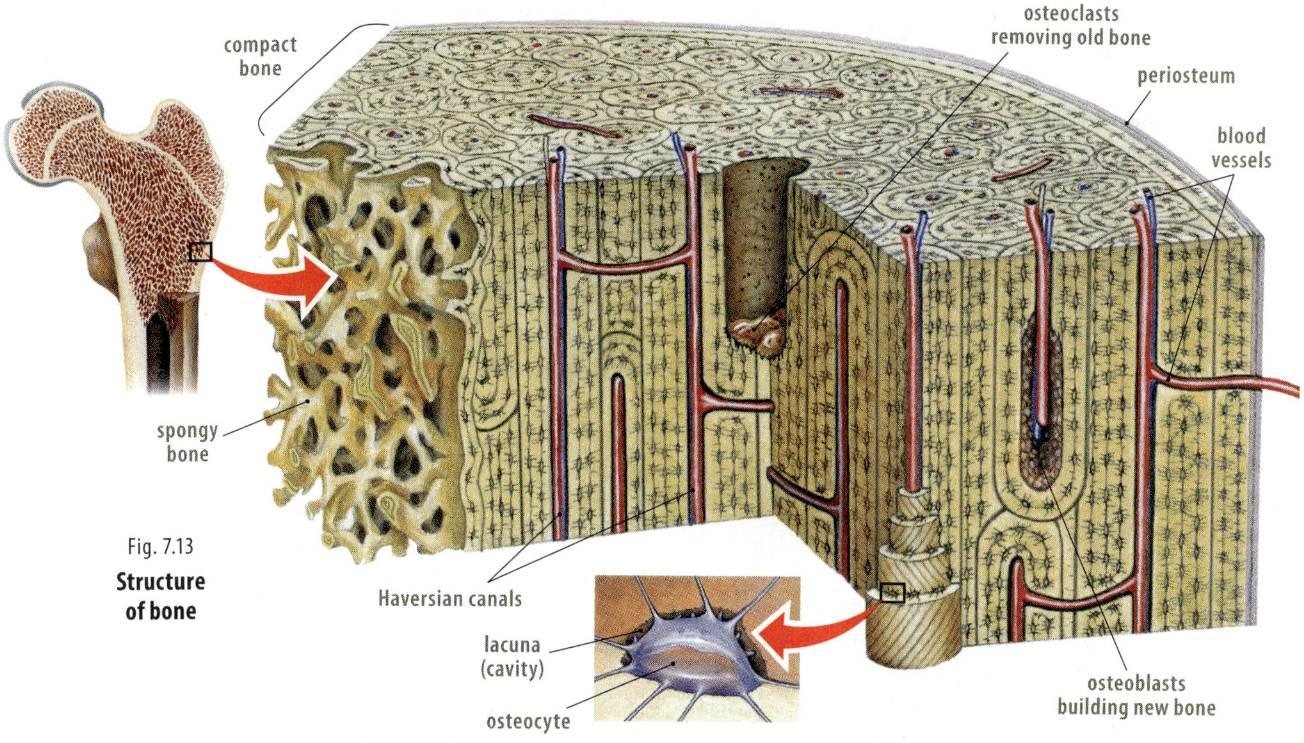


Fig. 7.13
**Structure of bone**

**Bones under construction.** Your osteoclasts and osteoblasts are well suited for maintaining and renovating your bones because they are the cells that built your bones in the first place. When you were developing in your mother's womb, special cells began to construct the forms of your body's bones out of cartilage or membrane material. When these cartilage forms were complete, osteoblasts began to move through them, replacing the soft, rubbery cartilage with collagen-reinforced hydroxyapatite. This process, called **ossification,** began only a few weeks after you were conceived. During your childhood, your bones grew in length as new cartilage was added within the ends of the bone and became ossified. In fact, for most high school students, the process of ossification is not yet finished; the last bone in your body to ossify, your sternum (breastbone), does not become completely ossified until about age 40.

> As thou knowest not what is the way of the spirit, nor how the bones do grow in the womb of her that is with child: even so thou knowest not the works of God who maketh all. —Eccles. 11:5

**Broken bones.** Although bones are extremely strong and resilient, accidents can occur that exert more force on a bone than the bone can handle, causing it to break. A break or crack in a bone is called a **fracture.** There are several types of fractures, based on how the bone breaks. A *transverse fracture* occurs when a bone breaks cleanly in two, while a *greenstick fracture* occurs when the bone cracks and bends but remains held together by its collagen fibers. A *spiral fracture* occurs when a bone is twisted or exposed to sudden impact at either end; and a *comminuted* [kŏm′ə·no͞ot′ĕd] *fracture* occurs when part of a bone is broken into multiple fragments.

Fractures may also be classified into two categories based on the severity of the break. A **simple fracture** *(closed fracture)* occurs when a bone cracks or breaks but does not pierce through the skin. A **compound fracture** *(open fracture)* occurs when the broken bone pierces through the skin. Compound fractures are more serious than simple fractures because heavy bleeding may occur, and the wound must also be treated carefully to prevent infection.

**Repairing a break.** When a concrete girder breaks, it will never again regain its strength. Fortunately, God has designed our bones so that they can repair themselves when broken. *Unlike most body tissues, bone heals with its own cells and not those of scar tissue.* Your osteoclasts and osteoblasts repair your bones in emergencies; they often perform their jobs so well that the bone is actually stronger after the break than before.

When a bone is broken, blood vessels and other tissues in the area are damaged, causing internal bleeding. Your body's damage-control systems attend to this problem first, shutting off the broken blood vessels using blood clots. Once the bleeding is stopped, your body forms new blood vessels to supply the region of the break with extra blood. The break soon becomes a busy construction site as your osteoclasts and osteoblasts move in and set to work.

Working outward from the new blood vessels, special cells begin to form cartilage and collagen in the gap between the broken ends of the bone, forming a mass of cartilage that eventually spans the break. Meanwhile, osteoclasts scour the area of the break for bone fragments, absorbing them to make room for the new construction. Osteoblasts then begin moving through the cartilage, replacing the rubbery substance with spongy bone, which connects the broken ends more firmly. Osteoclasts and osteoblasts then move together through the new spongy bone, replacing it with stronger compact bone where necessary and removing any excess. When the process is complete, the osteoclasts and osteoblasts return to their normal work of bone renovation.

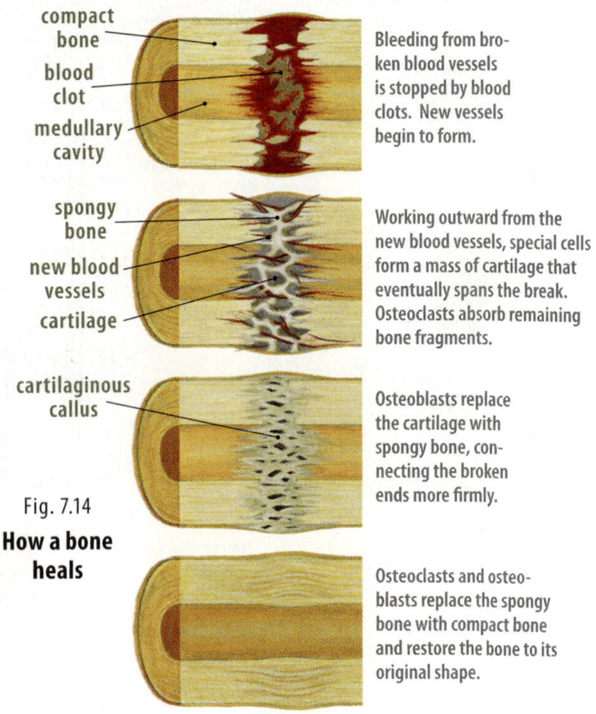


Fig. 7.14 **How a bone heals**

## Section Review 7.3

1. How can bones be classified by their shapes?
2. What is the difference between compact bone and spongy bone?
3. What is the purpose of bone marrow?
4. How do osteoclasts and osteoblasts keep your bones strong and healthy?
5. What vitamins and minerals are necessary for proper bone growth and development?
6. What effect does physical exercise have on bone strength?
7. Why is a compound fracture more serious than a simple fracture?

### Critical Thinking

8. Explain how the composition and structure of a bone reveal its design by an intelligent Creator.
9. Use Wolff's law to explain how appropriate physical exercise by older individuals can often reduce their risk of sudden bone fractures.

### Identify

diaphysis, epiphysis, hydroxyapatite, collagen, osteoclast, osteoblast, rickets, Wolff's law, ossification, fracture

### Observe

- Obtain a clean, uncooked long beef bone and examine its external structure. Identify such parts as the epiphyses, articular cartilage, diaphysis, and periosteum. Using a bone saw, cut the bone longitudinally (along its length) and use a hand lens to identify compact bone, spongy bone, and the medullary cavity containing marrow. (Caution: Do not use a knife or scalpel for this exercise.)

## 7.4 Joints of the Skeleton

The places where bones join, or articulate, are called **joints.** Although we normally associate joints with movement, not all joints in the body allow movement. For example, the joints of the cranium are known as *immovable joints* because they join the bones rigidly together. The joints between the vertebrae are examples of *slightly movable joints* because they make some bending and twisting movements possible. The most important joints in the body, however, are the **freely movable joints,** which allow a wide range of motion. Freely movable joints include the joints of the limbs and jaw.

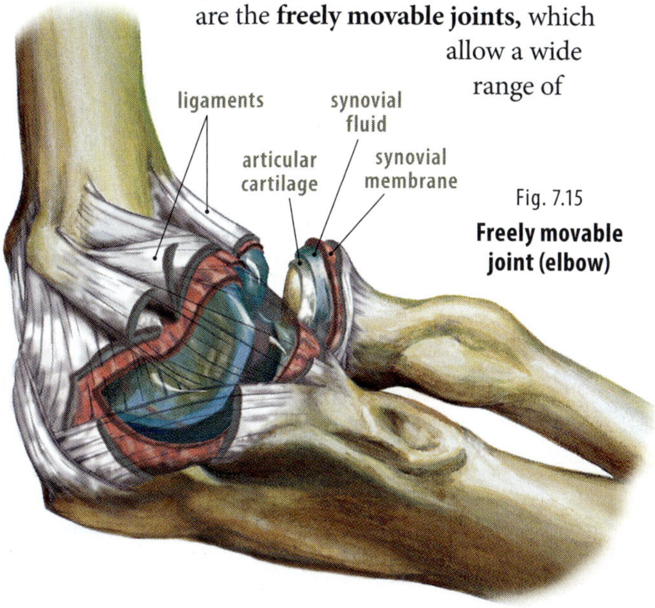


Fig. 7.15 **Freely movable joint (elbow)**

### Structure of a Freely Movable Joint

A good example of a freely movable joint is the joint of your elbow between the humerus and the ulna, which allows you to bend your arm. The bony ends of the humerus and ulna do not actually touch each other, because direct bone-to-bone contact would create grinding friction whenever you moved your arm. Instead, the ends of the bones are covered with tough, slippery *articular* [är·tĭk′yə·lər] *cartilage,* which serves as the "bearings" that allow the bones to glide smoothly over each other. The articular cartilage is lubricated with a clear, water-based lubricant known as **synovial fluid,** which resembles egg white in appearance and consistency. (Because of the synovial fluid, freely movable joints are also called *synovial joints.*) The bones are held together by strong bands of fibrous connective tissue called *ligaments,* which are arranged to completely encase and support the joint but allow it to bend freely. Together, the ligaments form an enclosure called the *joint capsule,* which also serves as a container for the synovial fluid. The synovial fluid is produced by the *synovial membrane,* which lines the inside of the joint capsule.

## Types of Freely Movable Joints

Freely movable joints may be further classified based on their structure. *Six types of freely movable joints are generally recognized:* the hinge joint, the ball-and-socket joint, the pivot joint, the ellipsoid joint, the gliding joint, and the saddle joint.

One of the most straightforward types of freely movable joints is the **hinge joint,** which allows a bone to move back and forth in a single plane, like a door swinging on its hinges. Examples of hinge joints include the knees and the joints of the fingers. The most freely movable joints in the body are **ball-and-socket joints,** in which the rounded head of a bone fits into a hollow socket in another bone. Not only does this design allow a bone to move in two planes, but it also allows the bone to rotate in place. The two shoulder joints and the two hip joints are the only ball-and-socket joints in the body.

An interesting type of joint that allows only rotating movement is the **pivot joint,** in which a bone rotates in place against another bone. The joint between the atlas and axis in the neck, which allows the head to swivel left or right, is an excellent example of a pivot joint. The only other pivot joints in the body are the joints just below each elbow between the ulna and the cylinder-shaped head of the radius. This joint allows the radius to rotate as you twist your forearm.

The joints between the metacarpals and phalanges, which allow you to move your fingers up, down, left, and right, are **ellipsoid joints** (also known as *condyloid* [kŏn′dl·oid′] *joints*). An ellipsoid joint consists of a bone with a convex (bulged) surface that fits into a concave (indented) portion of another bone. Ellipsoid joints allow movements in two planes but do not permit the bone to rotate in place.

One of the simplest types of joints is the **gliding joint,** in which one bone merely slides across the surface of another. Gliding joints are found between the carpal bones of your wrist, allowing these bones to slide over each other as you flex your wrist. Gliding joints are also found between the projections of your vertebrae, between your vertebrae and ribs, and between your clavicle and sternum.

The last type of joint, found only in your hands, is the **saddle joint.** In this type of joint, a saddle-shaped portion of one bone nestles into a saddle-shaped portion of another bone. A saddle joint is found in the lower palm of your hand, near the wrist, where the metacarpal bone of your thumb attaches to a carpal bone.

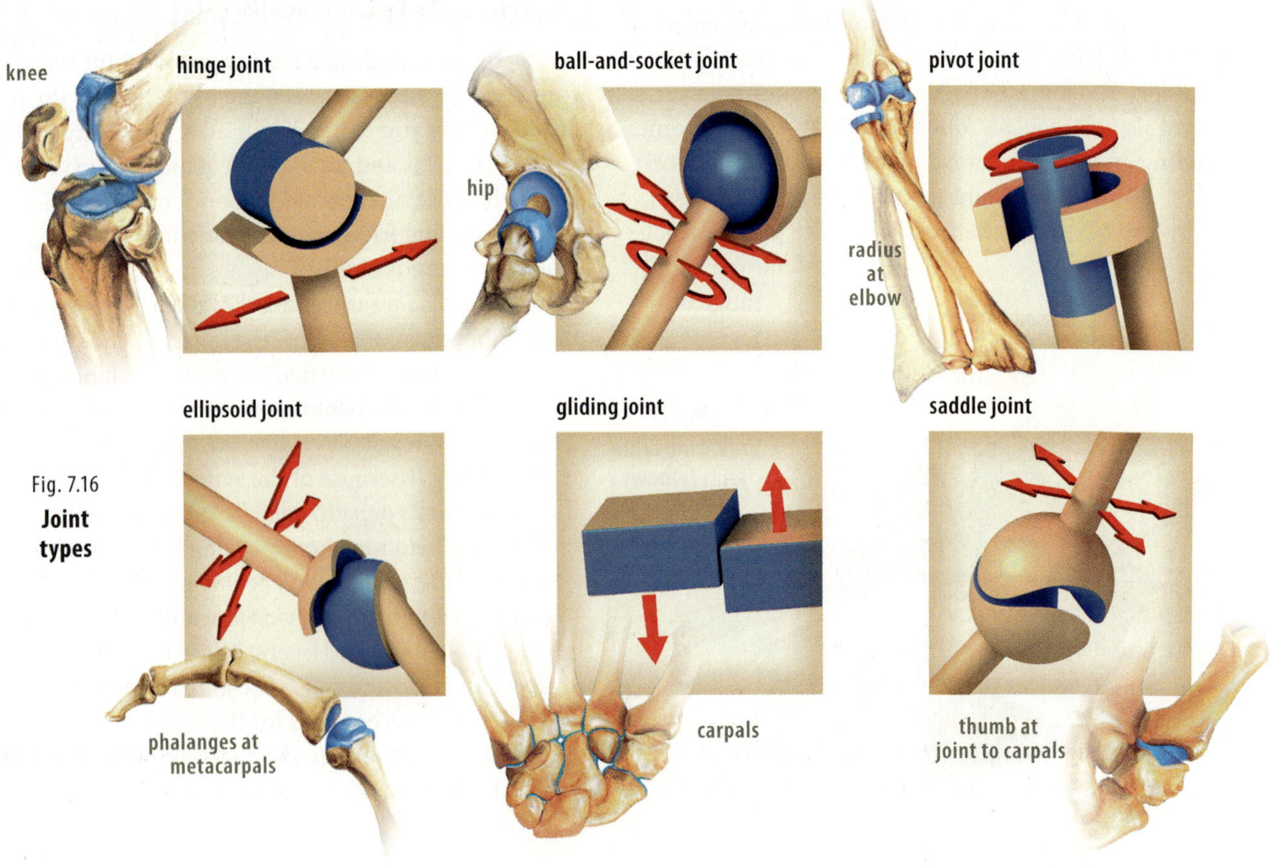


Fig. 7.16 **Joint types**

This saddle joint allows motion in two planes, allowing you to hold your thumb against the side of your hand, to stretch out your thumb to the side, and to rotate your thumb out in front of your palm in order to grasp something.

## Problems of the Joints

God designed our joints to last a lifetime, but one of the consequences of sin's curse is that our bodies will inevitably wear out. The joints are often affected by this process of deterioration. The most common degenerative conditions of the joints are the various forms of arthritis. Mechanical injuries to the joints include sprains, dislocations, and torn ligaments. A joint can also be injured by a fall or other accident that subjects it to more strain than it was designed to handle.

**Degenerative conditions.** A common degenerative condition of the joints is **arthritis,** or inflammation of the joints. The most common form of arthritis is *osteoarthritis,* a condition in which years of wear and tear on a joint cause the articular cartilage in the joint to deteriorate. Osteoarthritis is a common condition among the aged, but its severity varies greatly; its effects may range from occasional mild joint discomfort to severe pain and stiffness. It is most serious when it affects the hip joint.

A form of arthritis that is less common but generally more serious than osteoarthritis is *rheumatoid arthritis.* Rheumatoid arthritis occurs when the body's immune system attacks the tissues of a joint. Repeated attacks can devastate the articular cartilage, making movements difficult and painful.

**Joint injuries.** Joints can also be damaged by accidentally placing too much stress on the joint, as in a fall. Probably the most common type of joint injury is the **sprain,** which occurs when the ligaments of a joint are overstressed, causing them to become tender and inflamed; a severe sprain may also be accompanied by bleeding inside the joint from damaged blood vessels. The only sure treatment for a sprain is rest to give the body time to repair the strained ligaments. A more serious problem than a simple sprain is a **dislocation,** in which a joint is overstressed to the extent that a bone pops out of alignment and must be relocated. Accidents may also result in *torn ligaments,* which vary in severity. Sprains, dislocations, and torn ligaments are especially common among athletes as a result of sports injuries.

### Section Review 7.4

1. Give an example of a freely movable joint, a slightly movable joint, and an immovable joint.
2. Which type of freely movable joint allows the greatest range of motion?
3. What is the most common type of arthritis? Which type is most serious?

**Application**

4. In a paragraph, explain the design of a freely movable (synovial) joint and describe the types of movements for which each type of synovial joint is designed.

**Critical Thinking**

5. How does osteoarthritis help to reveal the importance of articular cartilage in freely movable joints?

**Identify**

synovial fluid, ligament, joint, arthritis, osteoarthritis, rheumatoid arthritis, sprain, dislocation

**Observe**

- Obtain a pork or beef shoulder and examine the structure of the joint. Notice the articular cartilage and the ligaments that hold the bones together.

# 7.5 Muscles: Designed for Motion

The skeleton forms the basic framework of the body, but the muscles give the body its shape and form. Nearly half of the body's weight is composed of muscles. Muscles consist of uniquely designed cells that contract when stimulated by nerve impulses, allowing the body to move with strength, speed, and grace. In addition to their functions in body movement, muscles serve vital roles in transporting materials around the body: they move blood through your arteries, food through your digestive system, and lymph through your lymphatic system. Without our carefully engineered muscles, our bodies would be completely unable to function.

## How Muscles Are Classified

Although the muscular system is composed only of skeletal muscles, there are two other major types of muscle tissue in the body—smooth muscle and cardiac muscle. Because skeletal muscles are generally under conscious control, they are also known as **voluntary muscles.** Smooth muscle and cardiac muscle are not under completely conscious, voluntary control and are therefore known as **involuntary muscles.**

**Skeletal muscle.** Over 600 muscles make up a large part of your chest, abdomen, neck, face, arms, and legs. Since their primary function is to move the parts of the skeleton, they are called **skeletal muscles.** These muscles are composed of individual cells, known as **muscle fibers.** Each skeletal muscle is composed of thousands or millions of individual muscle fibers bound together in parallel bundles.

Muscle fibers (muscle cells) are marvels of engineering. The contraction filaments inside skeletal muscle cells are arranged in rows that give skeletal muscle tissue a striped, or striated [strī′āt·ĕd], appearance when viewed under a microscope; thus, skeletal muscle is also called *striated muscle.* Most skeletal muscles are attached to bones by tendons, although the skeletal muscles of the face are attached directly to the facial bones. Skeletal muscles are designed to move quickly and powerfully.

**Smooth muscle.** Smooth muscle is designed quite differently from skeletal muscles. Whereas skeletal muscles consist of long, parallel fibers, **smooth muscle** consists of shorter, wider cells in a looser arrangement. The internal structure of a smooth muscle cell is also quite different from that of a skeletal muscle cell. Skeletal muscles are designed to contract and relax very quickly, but *smooth muscle is specialized for relatively slow, powerful, and prolonged contractions.* These characteristics perfectly suit the purposes for which God designed smooth muscle. Smooth muscle tissue is used in the body to operate certain valves (called *sphincters* [sfĭngk′tərz]), to adjust the tension in the walls of blood vessels, to squeeze food through the organs of the digestive system (such as the stomach and intestines), to adjust the focus of the lenses in your eyes, and even to help hold your eyelids open when you are wide awake. Contractions of smooth muscles in a mother's womb move the baby through the birth canal during the miracle of birth.

**Cardiac muscle.** *Cardiac muscle* is found only in the heart. Although striated and similar in structure to skeletal muscle, it is specially designed to contract over and over without tiring, allowing your heart to beat for a lifetime. Its intricate network of fibers is

Fig. 7.17 **Three types of muscle tissue**

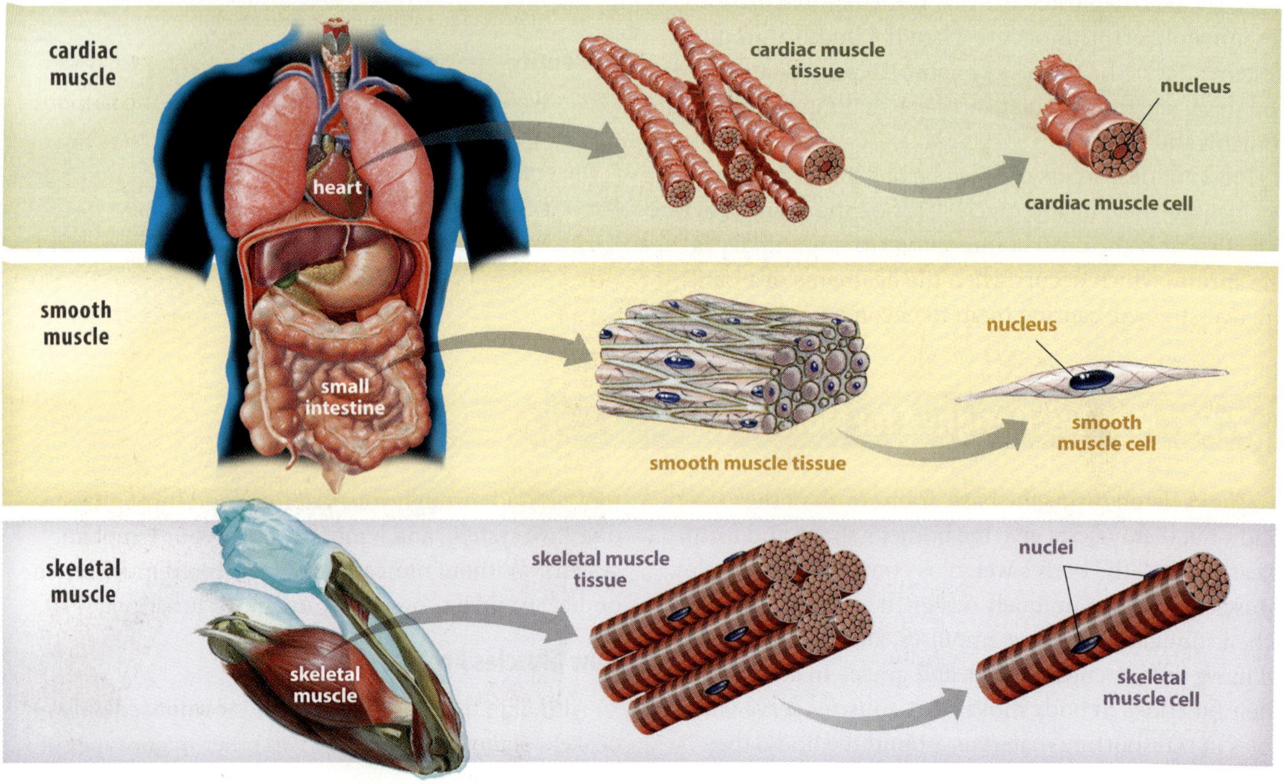

self-stimulating and contracts rhythmically to supply blood, the stream of life, to all parts of the body. The most efficient pump in the world, the heart begins beating four weeks after conception and continues to do its job until the body dies.

## Some Important Muscles

**Movements of the head and face.** Several important muscles are responsible for head movements, facial movements, and chewing. Some of the chief muscles responsible for moving the head are the two **sternocleidomastoid** [stûr′nō·klī′də·măs′toid] muscles, which connect the temporal bones of the skull to the sternum and clavicles. When both of these muscles contract simultaneously, they bend the head and neck forward, as when you bow your head. When they are contracted one at a time, they help turn the head from side to side. Many other muscles cooperate with the sternocleidomastoids to make head movements possible.

Some of the most important muscles of the face are those that move the lower jaw, as when chewing or talking. A **temporalis** [tĕm′pō·rā′lĭs] muscle and a **masseter** [mə·sē′tər] muscle are located on each side of the head; both sets of these muscles connect the mandible to the cranium and function to close the jaw, as when you bite a piece of food.

Many smaller muscles of the face are involved in facial expressions, allowing you to smile, frown, and raise your eyebrows. The muscles that move the lips are also important in speaking.

**Movements of the shoulders and ribs.** One of the most important muscles in shoulder movement is the **trapezius** [trə·pē′zē·əs].

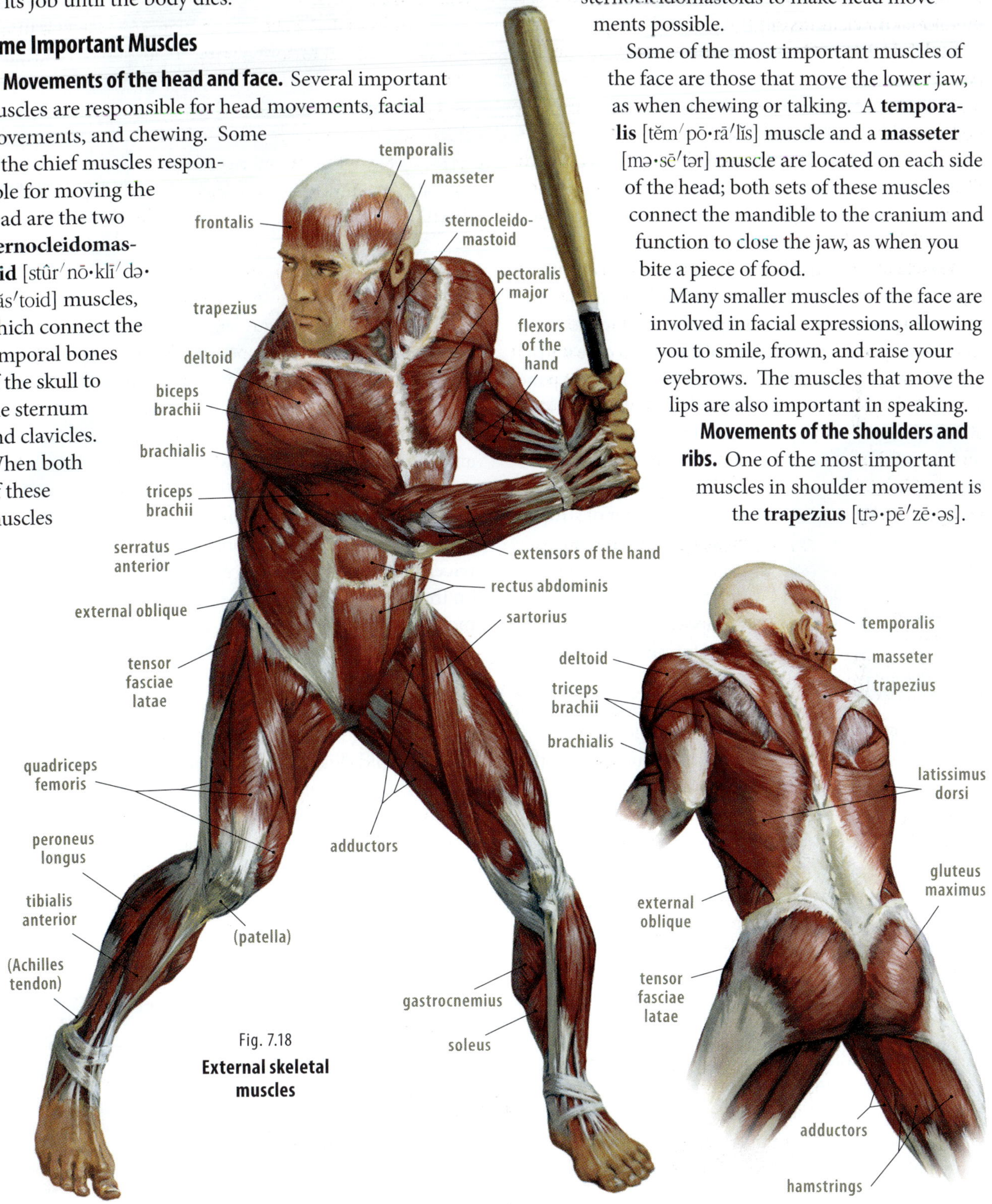


Fig. 7.18
**External skeletal muscles**

You have two of these muscles, which are located in your back on either side of your spine. The trapezius muscles permit you to pull your shoulders back or to shrug. They also tilt your head back, as when you look up at the ceiling, and work with the sternocleidomastoid muscles to help turn the head from side to side. The *serratus anterior* muscles, which attach to the side of each scapula and stretch forward around the sides of the rib cage, work with the *pectoralis minor* muscles of the chest to pull the shoulders forward.

The muscles between the ribs, known as the **intercostal** [ĭn′tər·kŏs′təl] **muscles,** are much smaller than most chest muscles but are very important because they help you breathe. When the intercostal muscles contract, they lift the rib cage and expand the thoracic cavity, allowing the lungs to expand. When they relax, they lower the rib cage and help to force air out of the lungs.

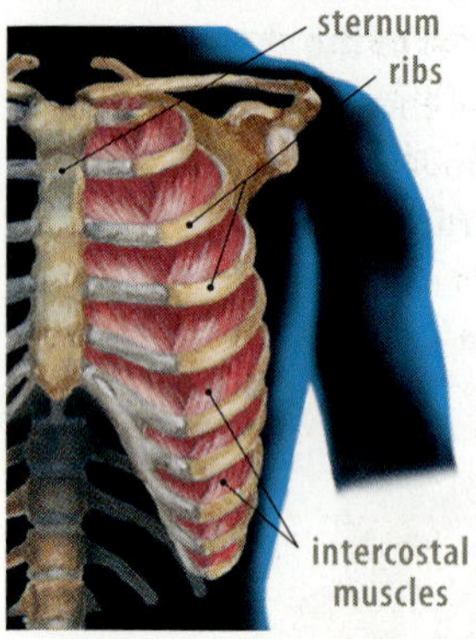

Fig. 7.19 **Intercostal muscles**

**Movements of the upper arms.** A number of strong muscles are involved in moving your upper arms. Some of the most important of these are the **pectoralis major** muscles, the largest of the chest muscles. Each pectoralis major extends from the sternum toward the outside of the chest, where it narrows to form a tendon that connects to the humerus. You use these muscles to shove your arms forward, such as when pushing away from an object or doing push-ups. Your **deltoid** muscles, which form the curves of your shoulders, lift the upper arms away from the body, such as when you raise your hands to the side. Your **latissimus dorsi** muscles connect each humerus to the lumbar region of the spine; they are responsible for drawing your arms to the rear, as when you pull on a rope or paddle a canoe.

**Movements of the lower arms.** Most of the muscles responsible for moving your forearms are located in your upper arms. Probably the best known of these are your **biceps brachii** [brā′kē·ī′] muscles, or *biceps.* Located on the front of each upper arm, the biceps muscles allow you to bend your forearms. The opposite job, that of extending or straightening the arms, belongs to the **triceps brachii** muscles *(triceps)* located on the backs of your arms. Several smaller muscles, mostly located in the forearm, allow you to twist your forearm to rotate your wrist. Other forearm muscles bend the wrists up, down, left, or right.

Fig. 7.20
**Muscles that move the forearm**

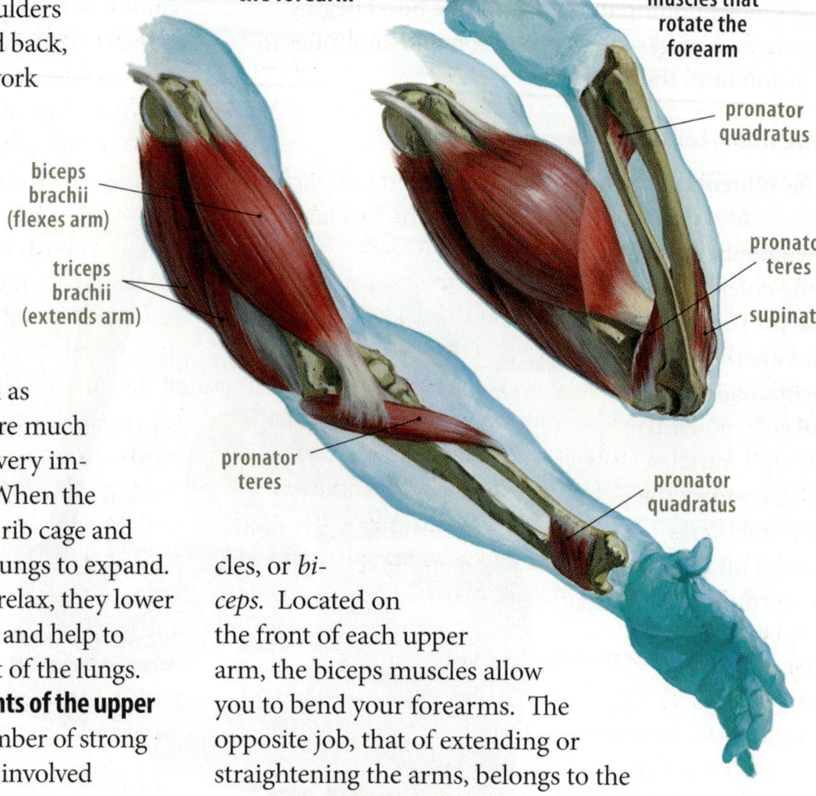


**Movements of the hands.** The human hand is a marvel of divine design and engineering. Most of the muscles that open and close the hand are actually located in the forearm; they operate the fingers through long tendons, like the strings of a marionette, that pass through special tunnels in the wrist. This wise arrangement allows much larger and stronger muscles to operate the hands than if these muscles were located in the hands themselves.

Fig. 7.21 **Muscles and tendons that move the hand**

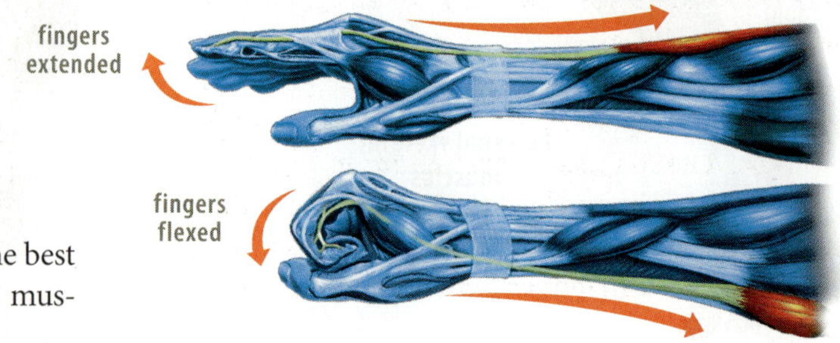

The muscles that are located in the hands are relatively small and serve primarily to move the fingers from side to side and to rotate the thumb. The carpenter, the automobile mechanic, the watchmaker, the concert pianist or violinist, the brain surgeon—all depend upon the precise movements of these muscles and bones to perform their tasks.

**Movements of the trunk.** Many muscles of the abdomen and back are involved in bending your trunk. These include the **rectus abdominis** muscles, which extend from the bottom of the sternum to the front of the pelvic girdle; you use these muscles whenever you sit up from a reclining position. Your **external oblique** [ō·blēk′] muscles also connect the ribs to the pelvis but extend more to the sides than do the rectus abdominis muscles. The abdominal muscles are vital to proper posture; a lack of exercise of these muscles can cause your abdomen to protrude and can also cause swayback. These muscles are also important in giving birth, in eliminating body wastes, and in heavy breathing during strenuous exercise.

Many muscles of the back are also involved in moving the trunk. Some of the most important of these are the **erector spinae** [spī′nē] muscles, which extend vertically along the spinal column from the upper back to the pelvis. These muscles straighten the back, allowing you to hold your body upright. Weak erector spinae muscles, like weak abdominal muscles, can lead to poor posture.

**Movements of the upper legs.** Several large muscles are involved in moving the upper leg. The largest and strongest muscles in your whole body are your two **gluteus maximus** [glo͞o′tē·əs măk′sə·məs] muscles, which pull your upper legs downward, away from the body and toward the rear. You use these muscles, which form much of the buttocks, whenever you jump, run, or even stand up. Several other muscles work together to pull the upper legs upward and forward or to the side.

**Movements of the lower legs.** The quadriceps femoris [kwŏd′rĭ·sĕps′ fĕm′ə·rĭs], hamstrings, and sartorius [sär·tōr′ē·əs] muscles are the primary muscles involved in moving your lower legs. Each **quadriceps femoris** is a group of four muscles, located in the front of each thigh, that are all connected to the tibia by the *patellar tendon* (the tendon that contains the kneecap). The quadriceps femoris muscles straighten the leg at the knee, as when kicking a ball or rising from a squatting position. The **hamstrings** are a group of three muscles in the back of each thigh that work together to bend the leg at the knee. They are called hamstrings because their cordlike tendons are easily felt in the back of the knee, especially when the knee is bent. Each **sartorius** muscle extends diagonally downward from the front of the pelvic bone (near the hip) and passes around the side of the thigh, eventually connecting to the tibia. The sartorius lifts the lower leg and turns it inward, such as when a person rests his ankle on his knee when seated or when a soccer player traps the ball in front of him with the inside of his foot.

**Movements of the feet.** Just as most of the muscles that operate the hands are located in the forearms, most of the muscles that move the feet are located in the lower legs (calves). The largest muscle of the calf is the **gastrocnemius** [găs′trŏk·nē′mē·əs], which forms the bulge on the upper back part of your calf; the main job of the gastrocnemius is to extend the foot downward by pulling up on the heel via the *Achilles tendon.* (You can feel this tendon as a cordlike ridge in the back of your ankle.) The gastrocnemius also helps the hamstrings bend the leg. The *tibialis* [tĭb′ē·ā′lĭs] *anterior* muscle performs the opposite job of the gastrocnemius, pulling the foot upward. Various other muscles allow you to tilt and turn your feet and curl or straighten your toes. Without these well-engineered muscles, walking would be impossible.

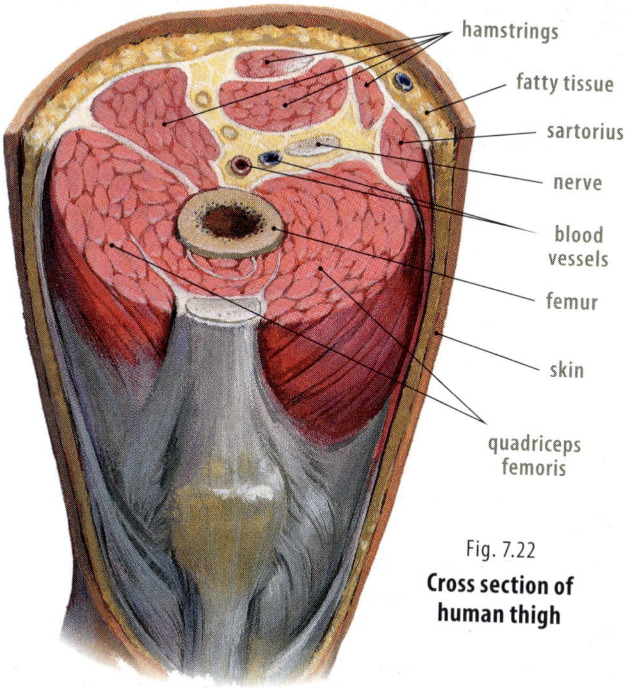


Fig. 7.22
**Cross section of human thigh**

## *The Human Hand* by Alfred M. Rehwinkel

Design in the LIVING CREATION

Ordinarily we give little thought to the mechanism of the hand. We simply take it for granted. And yet, the hand is one of the most wonderful features of the human body. The hand is so unique and so wonderful in its construction and function that it alone marks man as apart and distinct from every other creature. No other creature in the world, not even the ape, is equipped with an instrument comparable to the human hand. Normal human life without the hand would be inconceivable; without it man could never have become what he is or could have achieved what he has accomplished. Without it our civilization would be impossible. In everything that man has created, his hands in one way or another have been involved.

Like the foot, so also the hand consists of a bony framework: muscles, tendons, fats, and nerve fibers. There are a total of 27 individual bones in the hand; 8 of these are in the wrist, 5 in the palm, and 14 in the fingers. The hand is so constructed that it lends itself to an unlimited variety of functions. But every act of the hand sets in motion a whole series of operations of the human mechanism. For example, to close the fist or make a simple grasping motion involves an array of muscles, joints, and tendons, all the way from the shoulder to the fingertips. Doing such a simple thing as lifting a spoonful of soup to one's mouth involves more than 30 joints and 50 muscles, all of them functioning together in perfectly synchronized order.

The five fingers of the hand determine the variety and character of the manual manipulations. The most important of these is the thumb. It is the strongest and has a wider sphere of motion than the rest of the fingers. It provides the base against which the hand can perform a grasping or a holding operation. The hand without a thumb has lost most of its usefulness, but a thumb with only two or even one finger is still an efficient mechanism.

While the hand is of greatest importance to everyone in normal life, it is especially important to such people as musicians, artists, craftsmen, physicians, engineers, machine operators and, of course, many others. The hand can be developed and trained to perform astonishing feats. For example, it is estimated that a master pianist can strike over a thousand notes a minute. The well-trained typist will accomplish a similar feat. With the fingers of the hands, the skilled surgeon is able to perform a delicate operation which requires the precision of a mechanical instrument. The trapeze artist can develop the . . . muscles of his hand to such a degree that he is able to hold with his hands and fingers many times his own weight and perform the most breathtaking stunts.

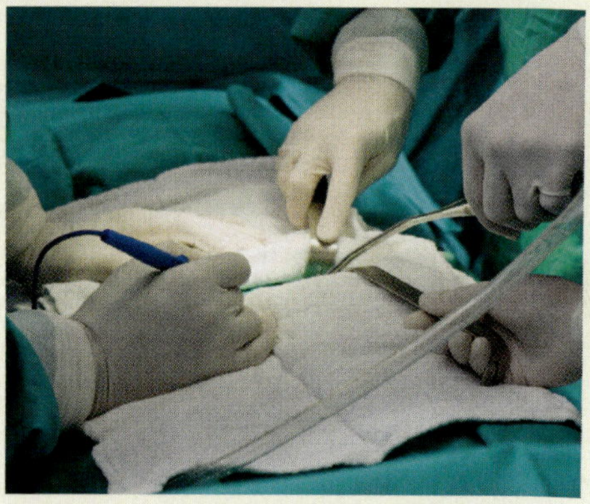

No other part of our body is so intimately associated with man's moods and general behavior. With our hands we work and play. We create and destroy, we sow and reap, we build and construct; with them we love and heal, we communicate and express our emotions, we write books and create works of art. With our clasped hands we express love, faith, and friendship, and with the clenched fist we threaten and attack. With the open hand we welcome friends and loved ones; the caressing hand expresses deep emotion and attraction, and with a handclasp we seal a solemn covenant. With our wringing hands we reveal the anguish of our soul, and with folded hands we express our devotion and reverence to our God.

The hand is a wonderful gift of the Creator to man. It is one of the seven wonders of the human body. Only an omnipotent and all-wise God could have designed and created the human hand.

*Alfred M. Rehwinkel,* The Wonders of Creation, *Bethany House Publishers, a division of Baker Publishing Group, Copyright ©1974, Used by permission.*

## Section Review 7.5

1. What are the three types of muscle tissue? Which type is voluntary?
2. How is the design of smooth muscle different from that of striated muscle?
3. Identify twelve important muscles of your body and the movements they make possible.

### Application

4. Explain how God's creative design of your hands makes them both strong and dexterous.

### Critical Thinking

5. List as many muscles as you can think of that are involved in throwing a baseball or football. Do not forget that other parts of your body are involved besides the throwing arm.

### Identify

voluntary muscle, involuntary muscle, muscle fibers, striated muscle, cardiac muscle

### Observe

- Holding your hand in front of you (palm outward), open and close your hand several times. Place your palm flat on a desk or table and raise and lower your thumb several times without lifting your palm. In both cases, observe the movement of muscles and tendons in your hand, wrist, and forearm. Use the fingers of your other hand to try to trace the moving tendons and muscles.

## 7.6 Engineered for Power: The Design of a Skeletal Muscle

Although we often think of our skeletal muscles as relatively simple masses of flesh, they are actually extraordinarily complex mechanisms that provide impressive evidence of God's design. Muscle tissue is unique in that it swiftly, silently, and powerfully contracts when activated. It is extremely reliable, requires no external maintenance, and can be strong enough to exert forces of hundreds or even thousands of pounds yet can be precise enough to position your eye or finger within a few thousandths of an inch of a desired position. A close look at a skeletal muscle reveals some of the design characteristics that help make muscles so useful.

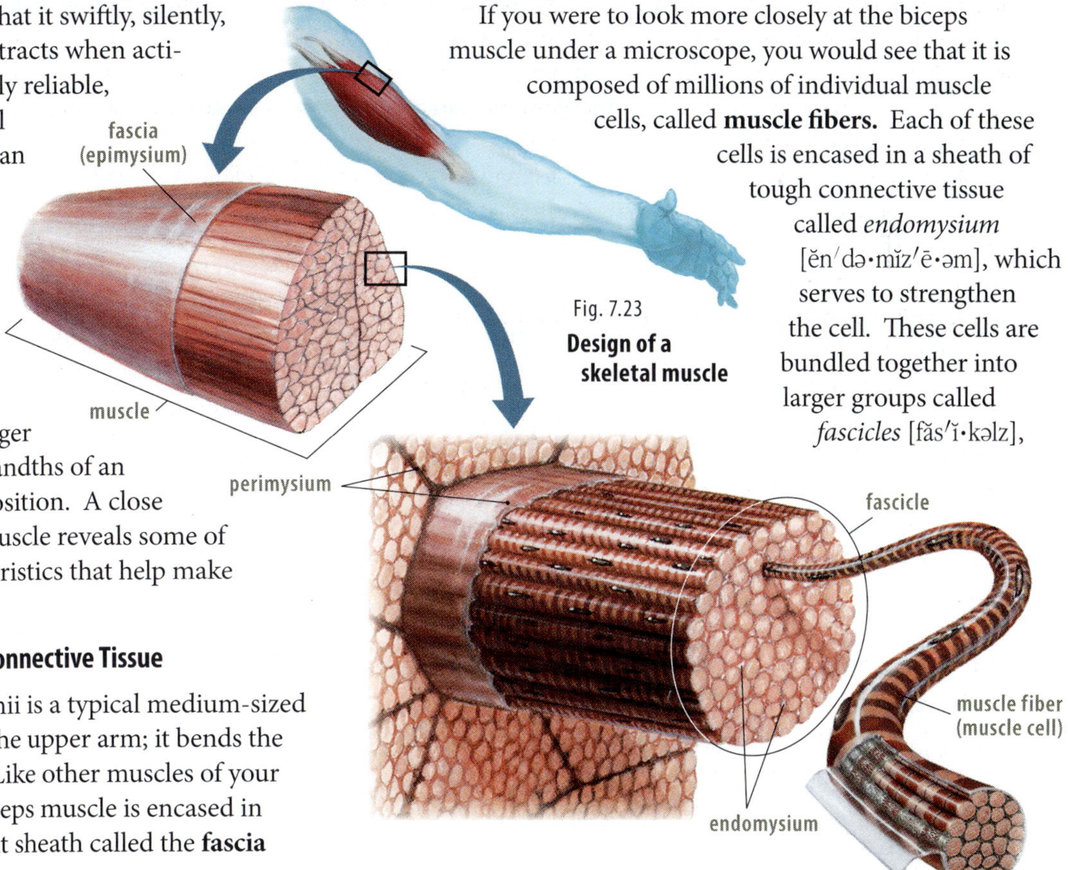


Fig. 7.23
**Design of a skeletal muscle**

### Fiber Bundles and Connective Tissue

The biceps brachii is a typical medium-sized muscle located in the upper arm; it bends the arm at the elbow. Like other muscles of your body, the entire biceps muscle is encased in a tough, translucent sheath called the **fascia** [făsh′ə], also known as the *epimysium* [ĕp′ə·mĭz′ē·əm], which binds the muscle together. At both ends of the muscle, the fascia tapers into strong *tendons* that fasten the biceps to the scapula above and to the radius below.

If you were to look more closely at the biceps muscle under a microscope, you would see that it is composed of millions of individual muscle cells, called **muscle fibers.** Each of these cells is encased in a sheath of tough connective tissue called *endomysium* [ĕn′də·mĭz′ē·əm], which serves to strengthen the cell. These cells are bundled together into larger groups called *fascicles* [făs′ĭ·kəlz], which consist of several dozen endomysium-wrapped fibers bound together by an additional sheet of connective tissue called *perimysium* [pĕr′ə·mĭzh′ē·əm]. Thus, a skeletal muscle is composed of thousands of fascicles bound together by the fascia (epimysium). Between the fascicles run the capillaries and nerves that supply the muscle with fuel and oxygen and signal it to contract.

Fig. 7.24
**Structure of a skeletal muscle cell**

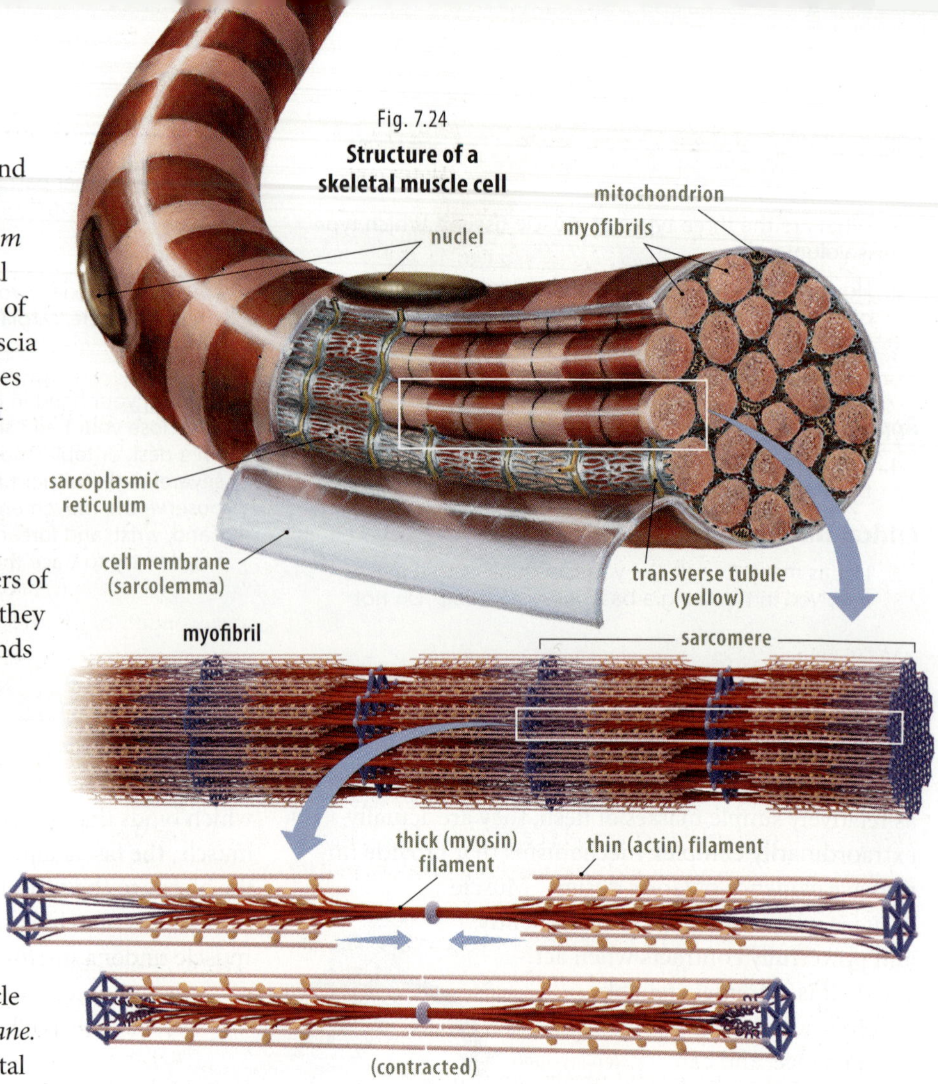


### Structure of Muscle Cells

As you examine the muscle fibers of a skeletal muscle, you will see that they are long, flexible cylinders, thousands of times longer than they are wide. Each individual muscle cell is a marvel of miniaturization—a tiny self-contained and self-repairing machine, only a few thousandths of an inch across, complete with its own fuel and oxygen storage systems, power plants, and contraction mechanisms.

Like other types of cells, a muscle cell is surrounded by a *cell membrane.* Unlike most cells, however, a skeletal muscle cell contains more than one *nucleus;* these nuclei are found at regular intervals along the length of the cell. The cell also contains an unusually large number of *mitochondria* [mī′tə·kŏn′drē·a; sing. mitochondrion], which serve as the cell's power plants.

Most of the volume of the muscle cell is devoted to the contraction machinery and the complex system of pipes and pumps that activate it. The contraction machinery itself is found inside long cylinders called **myofibrils** [mī′ə·fī′brəlz], which extend the length of the muscle cell. A single muscle cell may contain dozens or hundreds of myofibrils, which nearly fill the cell.

Each myofibril is composed of millions of overlapping protein fibers arranged in repeating units called **sarcomeres** [sär′kə·mērz′], which give skeletal muscle its banded, or striated, appearance. These fibers are *thick filaments,* made of the protein *myosin* [mī′ə·sĭn], and *thin filaments,* made of the protein *actin.* When the muscle is stimulated to contract, the myosin and actin filaments work together like winches and cables to shorten the cell.

Each myofibril is surrounded by an extensive network of pipes and reservoirs called the *sarcoplasmic reticulum.* Additional pipes called *transverse tubules* connect the sarcoplasmic reticulum with the surface of the muscle cell (the cell membrane). The sarcoplasmic reticulum and transverse tubules work together to trigger the myofibrils to contract.

### Nerve Impulse Reception

Muscle contractions are triggered by your nervous system, which transmits signals to individual muscle cells through *motor neurons.* (A *neuron* is a nerve cell.) The point at which a motor neuron connects to a muscle cell is called a **neuromuscular junction.** At the neuromuscular junction, the nerve triggers the muscle cell to contract by releasing a special chemical known as a *neurotransmitter* into the junction. When sensors in the membrane of the muscle cell detect neurotransmitter molecules in the junction, they generate electrical

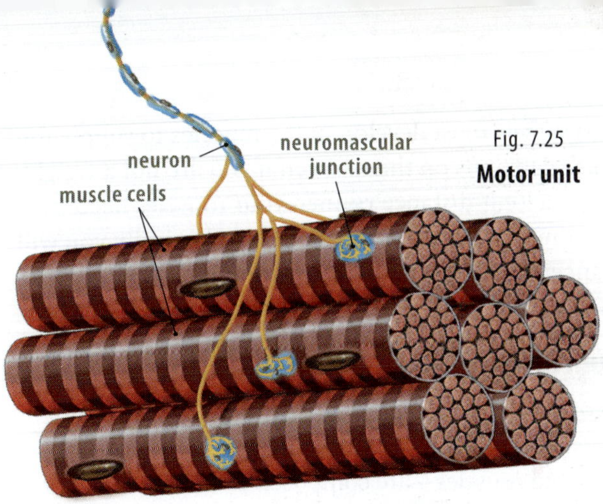


Fig. 7.25
**Motor unit**

signals that spread throughout the muscle cell and activate the contraction mechanisms.

### Precision Control

On average, a single motor neuron connects to about 100 muscle cells, which all contract as a unit when the motor nerve transmits an impulse. A motor neuron and the group of muscle cells it controls are a **motor unit.** The arrangement of muscle cells into motor units allows much fewer neurons to be used than if a separate neuron were required for every muscle cell. In general, muscles that must produce movements of great precision have fewer muscle cells per motor unit than muscles responsible for coarser movements. For example, certain tiny muscles of the larynx (voice box) and eyes have as few as three muscle cells in each motor unit, whereas certain muscles of the legs and abdomen have more than four hundred muscle cells per motor unit. Most large muscles, such as those in your arms, contain motor units with varying numbers of muscle fibers. When you need to accomplish a task requiring fine dexterity but relatively little strength (such as threading a needle or tying your shoes), your brain activates the motor units with the fewest fibers, allowing more control; when you need to accomplish a task requiring more strength than dexterity (such as lifting a heavy weight), your brain activates the larger motor units.

Each muscle has thousands of motor units. When you raise your arm, only a fraction of the motor units fire an impulse, causing some of the muscle fibers to contract while others remain relaxed. When a muscle fiber is stimulated to contract, it does so completely and then relaxes completely until another nerve impulse reaches it. This is the **all-or-none principle** of muscle contraction. Slow, steady body movements are accomplished as motor neurons within a muscle stimulate particular motor units to contract and allow others to relax. Within split seconds, contracted muscle fibers relax and relaxed fibers contract, balancing the work load among neighboring fibers. The coordination of muscle contractions, allowing you to move with precision, is the job of the central nervous system.

## Section Review 7.6

1. Why does the number of muscle fibers in a motor unit vary?
2. What is meant by the all-or-none principle of muscle cell contraction?

#### Critical Thinking

3. How do the structure of muscle fibers and the process of muscle contraction both support the idea that the human body did not come into being by a series of genetic mistakes but rather was carefully designed and engineered by an intelligent Creator?

#### Identify

fascia, muscle fiber, myofibril, sarcomere, neuromuscular junction, neurotransmitter, motor unit

#### Observe

- Examine a cut of uncooked round steak, noticing how the groups of fibers are bound together by connective tissue. Use a hand lens to observe this tissue more clearly.

## 7.7 Muscles in Strength and Weakness

### Exercise for Healthy Muscles

**Hypertrophy and atrophy.** Muscles perform more efficiently with use. Muscles that are exercised will enlarge and become stronger. Enlargement of muscles through use is **hypertrophy** [hī·pûr′trə·fē]. Muscles can also shrink through lack of use. Muscle degeneration is known as **atrophy** [ăt′rə·fē]. Some atrophy occurs naturally with time; in your late twenties or early thirties, a few of your muscle fibers will begin to shrink and will eventually be replaced by connective tissue. However, most muscle atrophy is caused by inactivity; our muscles are designed to be used. As a general rule, *skeletal* *muscles must be used or they will atrophy.* A person who leads a sedentary lifestyle will gradually lose strength and muscle mass as the body breaks down little-used muscle tissues. Fortunately, proper exercise can slow or reverse the process of atrophy and keep your muscles strong for many years.

**Red fibers and white fibers.** All muscle cells contain *mitochondria,* which serve as the "power plants" of the cell, and a protein called *myoglobin* [mī′ə·glō′bĭn], which stores oxygen supplied by the blood for the functioning of the muscle. However, muscle cells do not have equal amounts of mitochondria and myoglobin. Muscle fibers that contain many mitochondria and large amounts of myoglobin are called **red fibers,** because these structures give the cell a pronounced red color; muscle fibers containing fewer mitochondria and less myoglobin are known as **white fibers.**

White fibers contain more myofibrils than do red fibers. Because of this, white muscle fibers can contract more rapidly and more strongly than do red fibers; therefore, white fibers are commonly called *fast-twitch fibers.* Because white fibers have fewer mitochondria and less myoglobin than red fibers, they run out of energy (become fatigued) much more quickly. Red muscle fibers, on the other hand, contract more slowly and less strongly than white fibers; thus they are commonly called *slow-twitch fibers.* However, red fibers have much greater endurance than white fibers and can contract over and over for long periods without tiring. Most of the body's major muscles contain a mixture of both kinds of fibers. Generally, when you move a limb with little force, your brain activates mostly red muscle fibers; it brings the white muscle fibers into play only when rapid or forceful movement is required.

**Exercise and you.** Whether or not you are an athlete, you are responsible to nourish and exercise the body God has given you so that you can better use it in His service. An important part of taking proper care of your body is physical exercise; the size and efficiency of each muscle will not remain stable without physical activity. Exercise stimulates the development of muscle cells, the development of more blood vessels, and a greater blood supply within muscle tissue. If a muscle remains unused, the blood vessels that supply it begin to deteriorate, and little blood flows through that muscle. It takes time and effort to regain strength in muscles that have weakened, because new blood vessels must develop in the muscles to supply the greater demands on them. Even if you are not a weight lifter or a long-distance runner, your muscles need proper nutrition and exercise so that physically you will be "meet for the master's use, and prepared unto every good work" (2 Tim. 2:21).

## A Functioning Unit

We have considered the skeletal and muscular systems in this chapter. The organs of these body systems—bones and muscles—are composed of various kinds of tissues and cells that are constantly performing complex tasks that make body support and movement possible. However, the muscular and skeletal systems could not function together without the actions of another system—the *nervous system,* which we will consider in the following chapter. The skeletal system provides the support and attachment for the muscular system and also serves as a reservoir of calcium, an element that is important in muscular contraction. Electrochemical impulses from the nervous system spark muscle fibers into action, initiating a series of chemical reactions that culminate in muscle contraction and body movement. Body movements are coordinated because of the divinely engineered and designed

Fig. 7.26
**Physical exercise**

interaction between the muscular system and the nervous system. Special sensors in muscle tissue relay messages to the brain, informing it of the location and tension of each body part at a given time. This total awareness of body movements is known as **muscle sense** and is impossible without the complex and intricate cooperation of the muscular system and the nervous system—another striking example of God's marvelous design and precision engineering of the human body and its systems.

## Section Review 7.7

1. What is the greatest cause of muscle atrophy?
2. Why is proper exercise essential to the health of muscles?
3. How do red muscle fibers and white muscle fibers differ?

### Critical Thinking

4. Use your knowledge of red and white muscle fibers to explain why your legs tire quickly when sprinting or climbing stairs but tire very slowly or not at all when walking at a normal pace.

### Identify

hypertrophy, atrophy, muscle sense

# Chapter 7 Review

### Define

1. cranium
2. fontanel, suture
3. sinus
4. vertebral column
5. cartilage
6. coccyx
7. thoracic cage
8. pectoral girdle
9. phalanges
10. tendon, ligament
11. osteoclast, osteoblast
12. rickets
13. Wolff's law
14. ossification
15. fracture
16. joint
17. synovial fluid
18. arthritis, sprain, dislocation
19. muscle fiber
20. fascia
21. myofibril
22. neuromuscular junction
23. motor unit
24. hypertrophy, atrophy
25. red fiber, white fiber

### Identify

1. the two main divisions of the skeleton
2. the major bones of the head and spine
3. the five regions of the vertebral column
4. the major bones of the shoulders, hips, and limbs
5. four categories of bones based on their shape
6. the parts of a long bone
7. six types of freely movable (synovial) joints
8. two types of arthritis
9. three major types of muscle tissue
10. ten major muscles of the body

### Explain

1. Why is the thoracic cage important?
2. Explain how the design of the ulna and radius allows you to rotate your hand.
3. How do the structures of compact bone and spongy bone differ?
4. List the functions of bone marrow.
5. How does physical exercise affect bones?
6. Describe how bone cells repair a fracture.
7. Explain how a muscle fiber contracts.
8. What is meant by the all-or-none principle of muscle fiber contraction?
9. How does physical exercise affect muscles?

### Apply

1. Why does the spine of an adult have seven fewer bones than the spine of an infant?
2. How do the two primary substances that compose bone tissue provide the ability to resist both crushing and pulling forces?
3. How is the process by which fractures are repaired similar to the process by which bones are constructed in the first place?
4. Why do muscles in the eyes and larynx have a higher ratio of nerve fibers to muscle cells than do muscles of the limbs and back?
5. How can the arrangement of motor units in a muscle provide the muscle with both dexterity and strength?
6. Why is physical exercise vital for healthy muscles?

CHAPTER 8

# The Nervous System

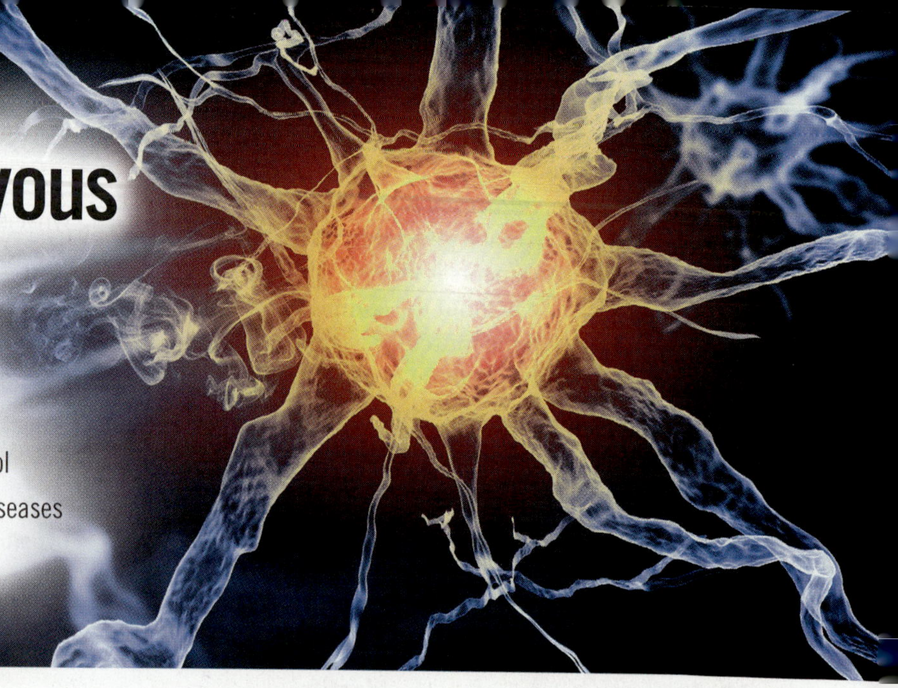

**Key Concepts**

- Divisions of the nervous system
- How nerves work
- Reflex action
- Parts of the brain
- Importance of avoiding alcohol
- Neurological diseases
- Sense organs

The human nervous system is one of the most amazing designs in God's living creation. Biologists consider it the most highly organized unit of any form of life. The nervous system is a superefficient communication system, coordinating physical activities and allowing you to control your body. Your nervous system gathers information about your surroundings. Some of this information is acted upon by your spinal cord, allowing you to react quickly to danger. Other information travels to your brain, which helps you understand situations, think about them, and solve problems. Then you can decide on the action you want to take and send messages along the nervous system to the other parts of your body, which will do what you tell them to do.

Although the nervous systems of animals and of humans were created according to similar design, there is no doubt that the human nervous system has distinct differences. Only people have the ability to use language, exercise judgment, solve difficult problems, produce creative objects and ideas, and transmit abstract notions from one person to another. The human nervous system is so complex that biologists have not been able to explore it fully. In this chapter, we will learn what scientists have discovered about our nervous system so that we may better understand the body in which God has placed us. A biologist who studies the nervous system is a **neurobiologist** (*neuro-* means "nerve"). A physician who specializes in disorders of the nervous system is a **neurologist.**

## 8.1 Outline of the Nervous System

Because the human nervous system is large and complex, neurobiologists have divided it into two major divisions according to its structure: the central nervous system and the peripheral [pə·rĭf′ər·əl] nervous system. The **central nervous system** includes the brain and the spinal cord, which are both encased in bone for protection. The *peripheral nervous system* consists of the nerves that branch from the brain and spinal cord.

### Central Nervous System

**The brain and spinal cord.** The **brain,** which is protected by the *cranium,* is the principal organ of the nervous system. It contains the higher centers of thought and is the chief organ of the mind. The brain of the average-sized adult weighs only about three pounds yet contains more knowledge than can be printed in several tons of books. It receives messages from the eyes, ears, and muscles to constantly monitor the location of body parts. From within, it receives messages from the blood and other body fluids and from the organs to keep the body in control. In some inexplicable way, the brain uses all of this information to allow you to form ideas, make decisions, and communicate your orders to your body.

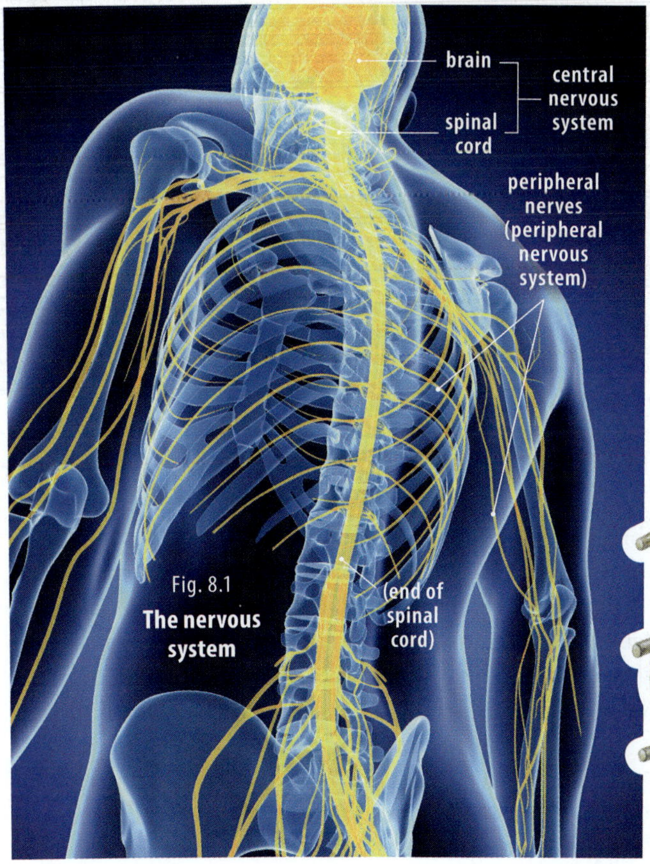

Fig. 8.1
**The nervous system**

The brain communicates with most of the body through the **spinal cord,** a thick bundle of nerve fibers located within the spinal cavity. At its upper end, the spinal cord merges with the brain stem, where its nerve fibers continue upward into the brain. At intervals along the spinal cord, pairs of nerves branch from the cord and extend out of the spinal cavity through gaps between the vertebrae. These thirty-one pairs of **spinal nerves** transmit nerve signals to and from the rest of the body. (Another twelve pairs of nerves, the **cranial nerves,** branch directly from the brain stem and transmit nerve signals to and from the eyes, ears, mouth, face, and scalp.)

The spinal cord does not extend all the way down your spine but abruptly tapers off in the upper lumbar region of your back, just below the first lumbar vertebra (see Fig. 8.1). Ten of the thirty-one pairs of spinal nerves continue past the end of the spinal cord, exiting the spinal cavity through gaps in the lumbar vertebrae and sacrum.

The spinal cord and brain are covered by a triple layer of protective tissues known as the **meninges** [mə·nĭn′-jēz]. The meninges are composed of an outer layer called the *dura mater* [dŏŏr′ə mā′tər], a middle layer called the *arachnoid* [ə·răk′noid′] *mater,* and an inner layer called the *pia* [pī′ə] *mater.* The dura mater, which is one of the strongest tissues of the body, serves primarily as a flexible protective layer. The arachnoid mater is composed of thin fibers, like the fibers of a spider's web, that form an intricate three-dimensional network around the brain. A clear fluid called **cerebrospinal fluid** circulates through the fibers of the arachnoid mater, serving to cushion the brain when you bump your head. Beneath the arachnoid mater is the pia mater, which forms a delicate covering that rests directly against the brain and spinal cord.

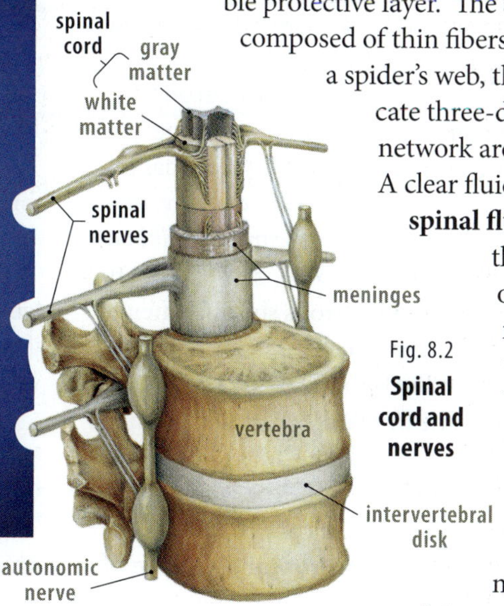

Fig. 8.2
**Spinal cord and nerves**

**Neurons.** The brain and spinal cord are both made of two basic kinds of cells: **glial** [glē′əl] **cells,** which support and insulate nerve tissue, and **neurons,** the actual nerve cells. Neurons are different from all other body cells in that long fibers branch out from their cell bodies. The **cell body** of each neuron contains a nucleus (which controls its metabolic activities) and most of the nerve cell's cytoplasm. Extending from the cell body of a typical neuron are hundreds or thousands of dendrites and a single axon. A **dendrite** is a short, branched extension of the cell that receives nerve impulses from other neurons and

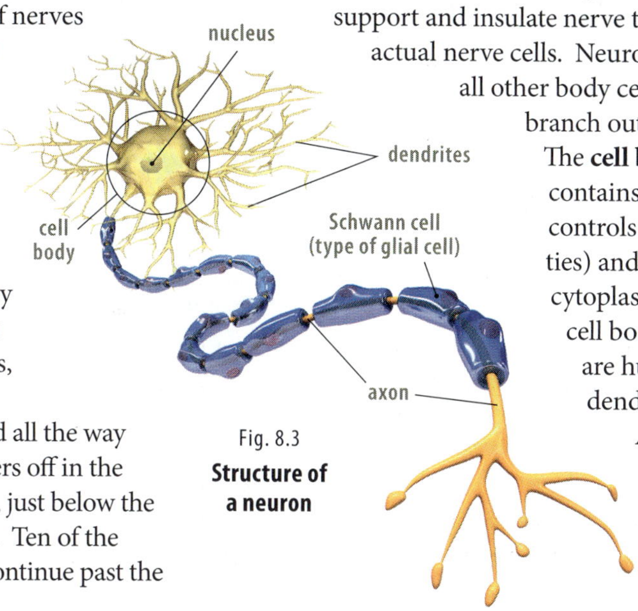

Fig. 8.3
**Structure of a neuron**

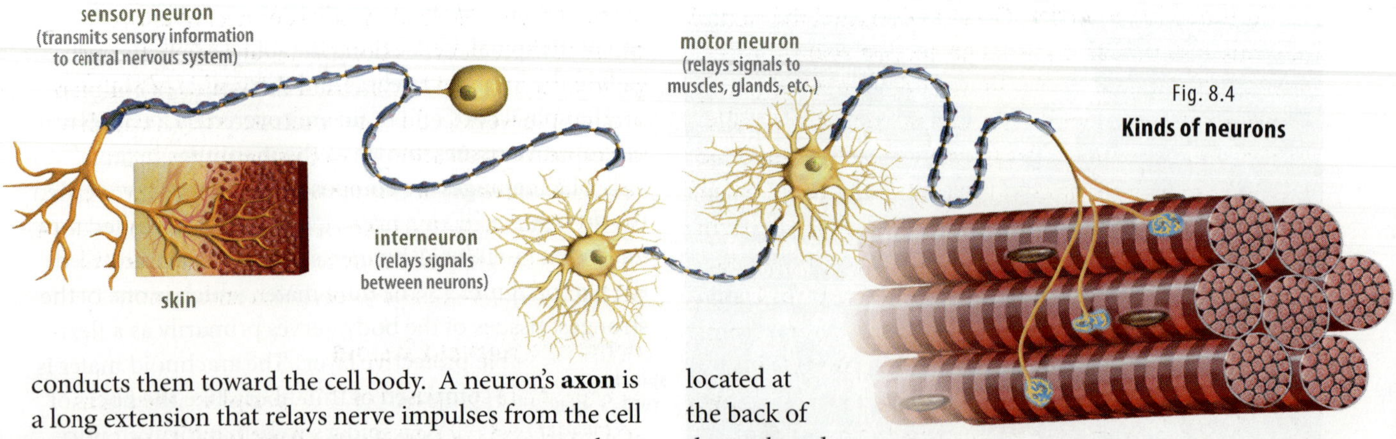


Fig. 8.4
**Kinds of neurons**

conducts them toward the cell body. A neuron's **axon** is a long extension that relays nerve impulses from the cell body to other neurons. (In most sensory neurons, the dendrites bypass the cell body and conduct the nerve signals directly to the axon, which relays the signal to the spinal cord.) *Nerve impulses travel in one direction in a nerve fiber—from the dendrites toward the cell body and from the cell body toward the axon.* The end of an axon is usually divided into numerous short filaments that attach to many other nerve cells.

The **gray matter** found within the brain and spinal cord consists largely of the cell bodies of neurons and is gray because the cell bodies lack the white, specialized covering known as **myelin** [mī′ə·lĭn]. The **white matter** of the brain and spinal cord, on the other hand, is composed of axons and glial cells that are white because of their myelin content.

Nerve cell bodies are found only in the brain, the spinal cord, and **ganglia** [găng′glē·ə; sing. ganglion], masses of cell bodies. Some ganglia are grouped together to form a large nerve mass called a **plexus.** The *brachial* [brā′kē·əl] *plexus,* for example, located at the back of the neck and shoulder, is the group of ganglia from which branch the median nerve and other arm nerves. The brachial plexus helps interconnect these nerves to allow better control of arm and hand muscles. A group of cell bodies in the brain or spinal cord is known as a **nerve center.** The nerve centers in the *medulla oblongata* [mĭ·dŭl′ə ŏb′lông·gä′tə] of the brain stem, for example, control your breathing.

Neurons can be classified into three broad groups according to their function. Neurons that transmit information to the central nervous system from the senses of sight, hearing, taste, touch, and smell, as well as those that transmit pain signals, are **sensory neurons.** Neurons that relay signals from the central nervous system to the other parts of the body are **motor neurons. Interneurons,** which are found only in the central nervous system, relay signals between neurons or groups of neurons and are responsible for the processing of information by the brain, like the logic circuits of a computer.

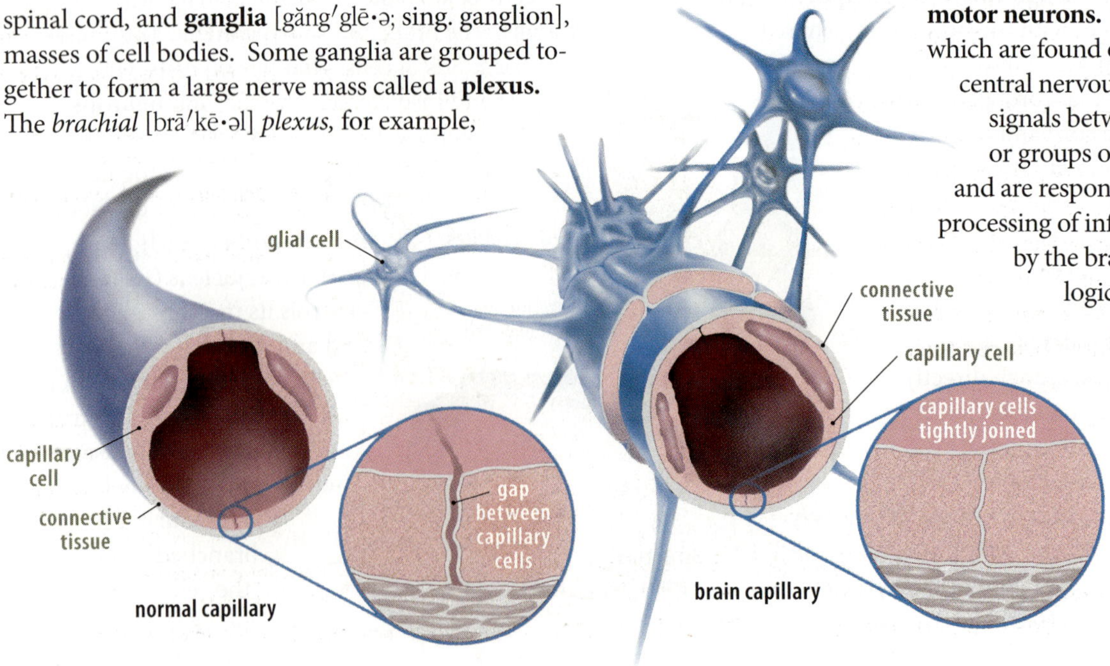


Fig. 8.5
**Blood-brain barrier**

**The blood-brain barrier.** God has designed the central nervous system with a special protection against infection. Unlike the capillaries in most tissues of your body, which are designed with microscopic gaps in the walls to allow nutrient molecules to easily enter the surrounding tissues, the capillaries of the central nervous system are tightly sealed to prevent bacteria, viruses, and harmful molecules from entering the brain and spinal cord. These tightly sealed capillary walls, known as the **blood-brain barrier,** protect the central nervous system from being permanently damaged every time you get sick. Under normal circumstances, not even your body's own white blood cells are allowed to cross the barrier. (Your brain contains its own white blood cells, called *microglia,* which help protect the brain from infection.) Even individual proteins and glucose molecules, which are tiny compared to cells, can cross the blood-brain barrier only through special "gates" or by special molecular shuttles.

**Meningitis.** Unfortunately, some bacteria and viruses have special mechanisms that allow them to slip through the blood-brain barrier. When invading microorganisms enter the nerve tissue and infect the meninges, they can cause a disease called **meningitis** [mĕn′ĭn·jī′tĭs]. When such an infection occurs, the blood-brain barrier must open slightly to allow additional white blood cells to squeeze through and fight the invaders, causing inflammation and swelling inside the skull. A person with meningitis generally suffers from headache, nausea, vomiting, and fever. The neck becomes stiff and sometimes cannot be bent forward. In severe cases, the pressure on the brain impairs its functioning, causing the patient to become delirious or slip into a **coma,** a state of prolonged unconsciousness; serious brain damage can also result. This disease, which can be fatal, is more likely to affect small children because their blood-brain barrier is not completely developed.

**Poliomyelitis.** A serious disease that attacks the spinal cord is **poliomyelitis** [pō′lē·ō·mī′ə·lī′tĭs], or *polio.* It is caused by a virus that enters the motor neurons of the spinal cord (the *anterior horn* cells) and destroys some of them. Skeletal muscles connected to the brain by these neurons become paralyzed. Although polio once afflicted hundreds of thousands of children around the world, efforts to eradicate polio through widespread vaccination programs have drastically reduced the threat of the disease. In 1988, there were over 350,000 reported cases of polio worldwide; in 2014, only 358 cases of polio were reported. Polio is still a problem, however, in parts of Africa and Asia.

## Peripheral Nervous System

**Cranial and spinal nerves.** The **peripheral nervous system** consists of **nerves,** which are bundles of nerve fibers (axons) branching from the brain and spinal cord and connecting the central nervous system to the extremities of the body. The structure of a nerve can be likened to that of a telephone cable constructed of thousands of tiny wires or optical fibers bound in a single sheath. Many of the axons in a nerve are as long as the nerve itself, perhaps several feet long. The *sciatic* [sī·ăt′ĭk] *nerve,* one of the longest nerves in the body, connects the leg extremities with the spinal cord in the lower back.

*There are twelve pairs of cranial nerves* (branching directly from the brain) *and thirty-one pairs of spinal*

Fig. 8.6 **Cranial nerves**

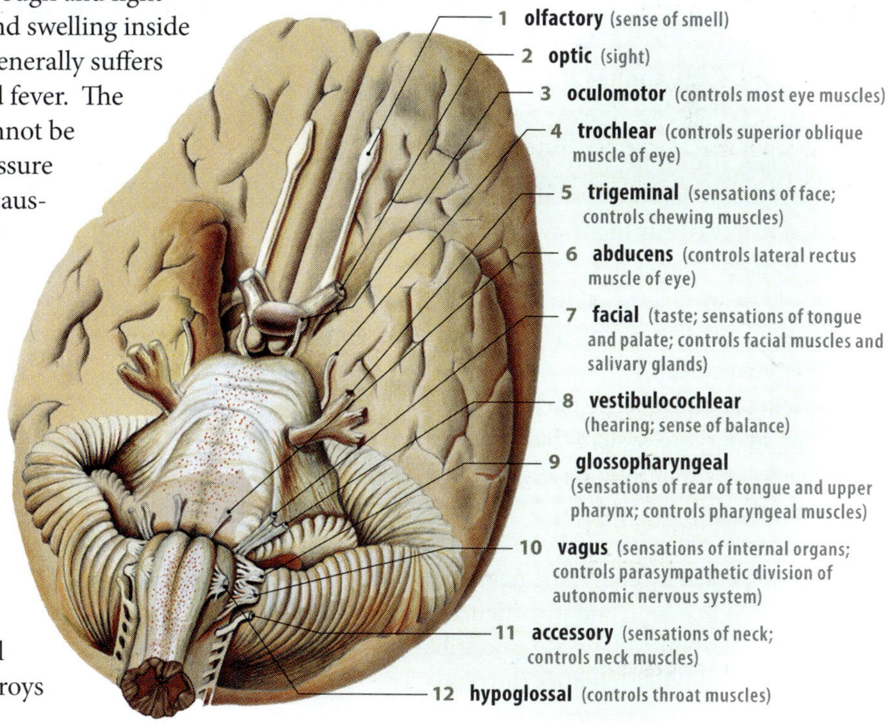

*nerves* (connected to the brain through the spinal cord). Each spinal nerve supplies a particular body part. Most nerves contain both sensory and motor fibers and thus are known as **mixed nerves.** A few nerves have only sensory fibers or only motor fibers. **Sensory nerve fibers** carry impulses from light, taste, sound, touch, and pain from other parts of the body to the spinal cord and brain for analysis. **Motor nerve fibers** carry impulses from the brain and spinal cord to produce action in muscles and organs. The cell bodies of the spinal sensory nerve fibers are in ganglia outside the spinal cord, while the cell bodies of motor nerve fibers are in the gray matter of the spinal cord. The sensory fibers connect to the spinal cord through the posterior (rear) nerve root, and the motor fibers leave the spinal cord through the anterior (front) nerve root. The peripheral nervous system can be likened to a two-way highway consisting of thousands of lanes in each direction, all of which are channeled to the ascending and descending nerve tracts of the spinal cord.

Fig. 8.7
**Spinal nerves**

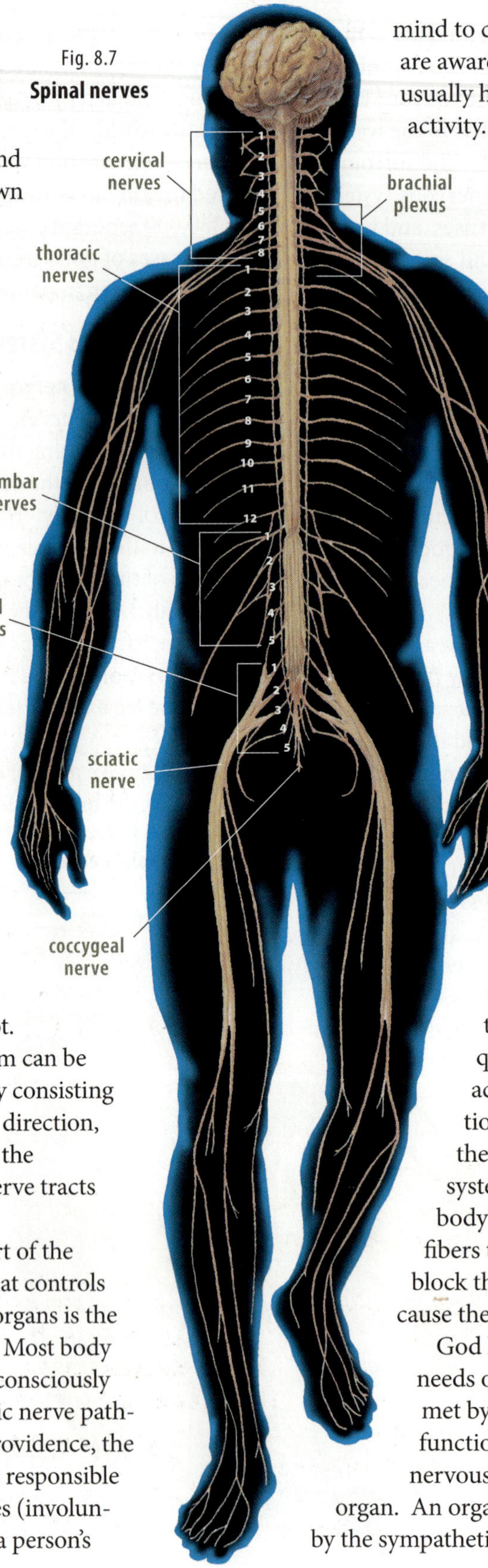


**Automatic control.** The part of the peripheral nervous system that controls the heart and other internal organs is the **autonomic nervous system.** Most body functions are not controlled consciously but are regulated by automatic nerve pathways. In God's design and providence, the autonomic nervous system is responsible for all routine bodily activities (involuntary functions), thus freeing a person's mind to concentrate on other matters. You are aware that your heart is beating, but you usually have no conscious control over its activity. The autonomic nervous system also controls the digestive system, various glands, and thousands of other vital functions.

Consider the importance of the autonomic nervous system. You will begin to see how very complicated it is for your nervous system to coordinate all of your body's organs and metabolic activities. Each organ or gland must be turned on at specific times and turned off at other times. This job is the task of the sympathetic and parasympathetic divisions of the autonomic nervous system, both of which are controlled by the *hypothalamus* of the brain. Each of your organs is supplied with sympathetic and parasympathetic nerve fibers.

The **sympathetic division** generally responds to your body's needs during increased activity and in emergency situations by causing heartbeat and breathing rates to increase, thus supplying more blood to your body so that you can respond quickly. The **parasympathetic division** acts as a balance to counteract the actions of the sympathetic division. When the emergency is over and your nervous system detects the need to slow down your body's metabolic activities, parasympathetic fibers transmit impulses to your heart that block the action of sympathetic fibers and cause the heart to slow down.

God has wonderfully provided for the needs of each of your body's organs to be met by the autonomic nervous system. The function of each division of the autonomic nervous system depends upon the individual organ. An organ such as the heart may be stimulated by the sympathetic division and inhibited by the parasympathetic division. Just the opposite is true of the stomach; stomach muscles are stimulated by the parasympathetic division and inhibited by the sympathetic division. Generally, the parasympathetic division stimulates organs while you rest, and the sympathetic division stimulates organs when you are active, in danger, or under stress.

Fig. 8.8 **Neurons and spinal cord**

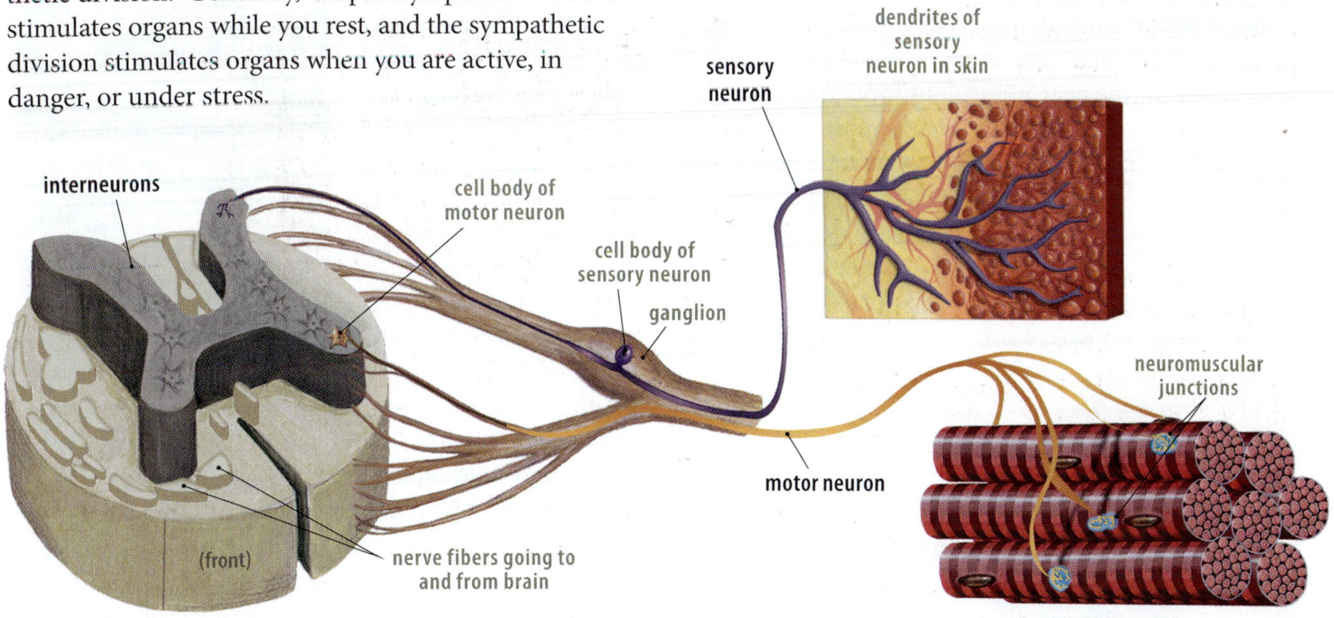


## Section Review 8.1

1. What are the main divisions of the nervous system? What organs belong to each?
2. Identify the main parts of a neuron and give the function of each.
3. What is the difference between gray matter and white matter?
4. What are the three broad categories of neurons?
5. What viral diseases affect the nerve cells of the spinal cord?
6. Explain how the sympathetic and parasympathetic divisions of the autonomic nervous system work together to regulate body activities.

### Identify

brain, spinal cord, spinal nerves, cranial nerves, meninges, cerebrospinal fluid, glial cells, neurons, myelin, ganglion, plexus, nerve center, blood-brain barrier, meningitis, coma, poliomyelitis, nerve

# 8.2 Nerves and How They Work

## Structure of a Typical Nerve

**The median nerve.** The median nerve, which controls the muscles of the forearm and the muscles and skin of the hand, is representative of how all peripheral nerves are designed. From sense receptors in the skin of the hand, messages travel along the median nerve's sensory nerve fibers to the spinal cord and brain, enabling you to sense temperature, pressure, or pain. From the brain and spinal cord, messages pass along the median nerve's motor nerve fibers to stimulate the muscles of the forearm and hand to move.

Like the other nerves of the peripheral nervous system, the median nerve consists of bundles of nerve fibers surrounded by connective tissues. Within these bundles lie the nerve fibers (axons), embedded among glial cells and protected by a myelin sheath. Each axon relays either a sensory or a motor impulse (never both). Blood and lymph vessels, along with cushioning fatty tissue, help nourish and protect the nerve fibers. Thus, the median nerve is well designed to function as a physical link between the central nervous system and remote (peripheral) body parts.

**Nerve cells.** Although most cells within nerve tissue are glial cells, the *neurons,* or nerve cells, are the ones responsible for the near-miraculous process of nerve impulse conduction. Neurons, unlike most other body cells, rarely reproduce. Therefore, when the body of a nerve cell dies because of injury or disease, it is not usually replaced. Whenever a person uses alcohol or other drugs, for example, he destroys irreplaceable nerve cells in the central nervous system.

Fig. 8.9
**Structure of a nerve**

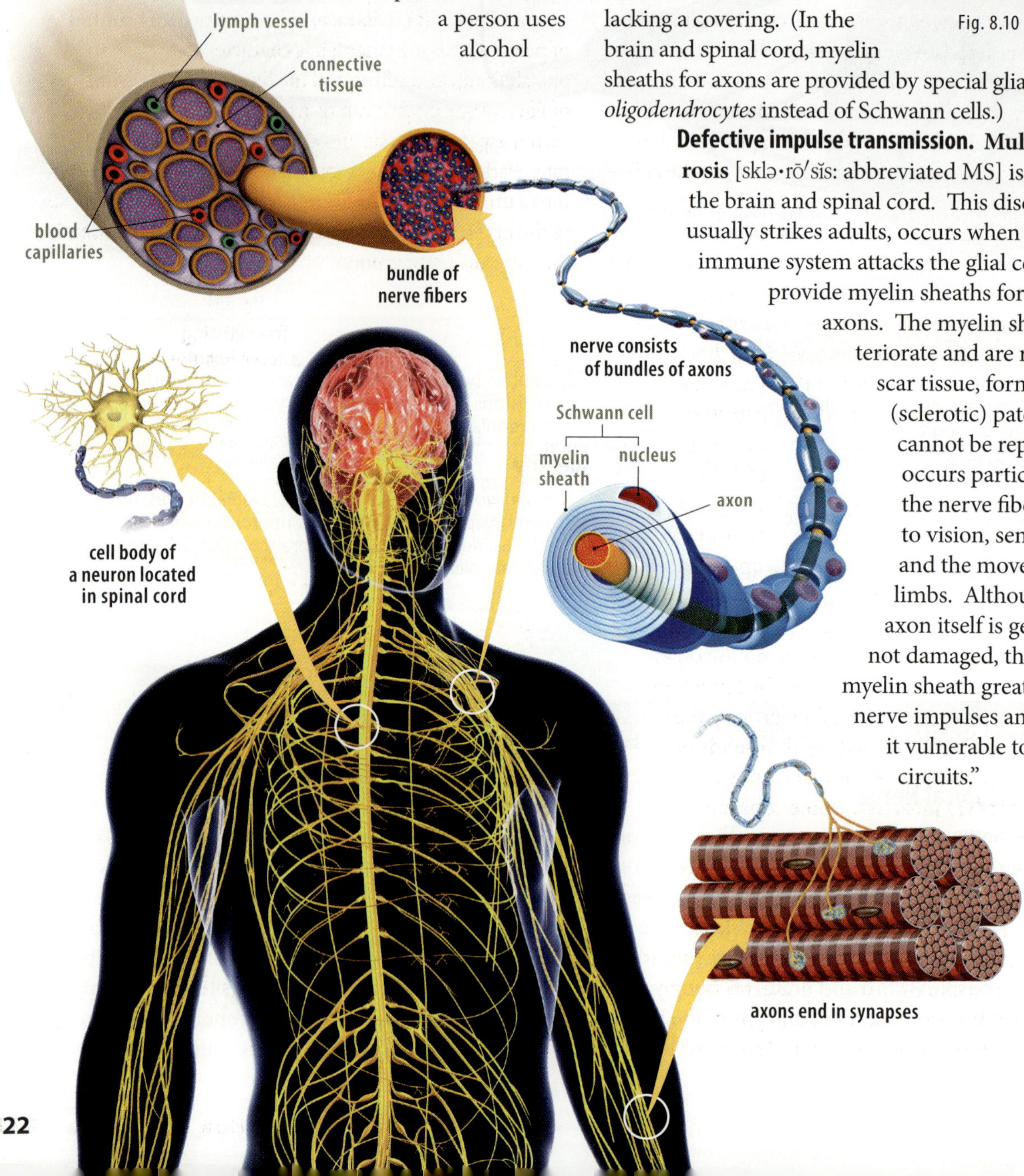


The axons extending from the nerve cell bodies of peripheral nerves are surrounded by special glial cells called **Schwann cells,** which produce layers of myelin sheathing that acts much like the insulation on an electrical wire. Myelinated axons transmit nerve impulses much faster than axons lacking a covering. (In the brain and spinal cord, myelin sheaths for axons are provided by special glial cells called *oligodendrocytes* instead of Schwann cells.)

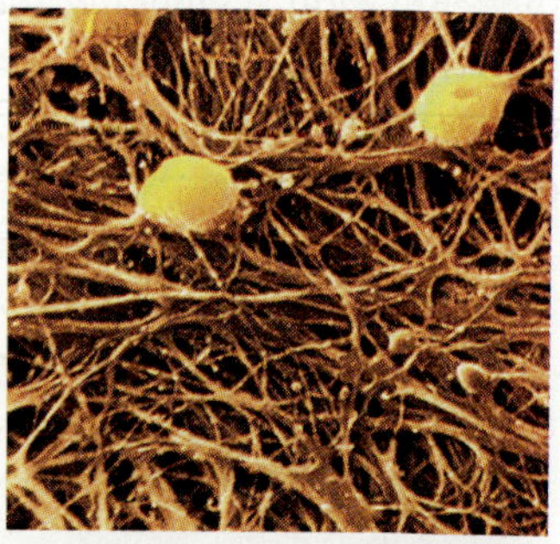

Fig. 8.10 **Nerve cells**

**Defective impulse transmission. Multiple sclerosis** [sklə·rō′sĭs: abbreviated MS] is a disease of the brain and spinal cord. This disease, which usually strikes adults, occurs when the body's immune system attacks the glial cells that provide myelin sheaths for nerve cell axons. The myelin sheaths deteriorate and are replaced by scar tissue, forming hard (sclerotic) patches that cannot be repaired. This occurs particularly in the nerve fibers related to vision, sensations, and the movement of limbs. Although the axon itself is generally not damaged, the loss of its myelin sheath greatly slows nerve impulses and makes it vulnerable to "short circuits."

Symptoms of blurred or double vision, disturbance of sensation, and muscle weakness may come and go. Early symptoms may disappear but recur at intervals of months or years as the nerves are repeatedly attacked. As time passes, permanent damage is more likely, leading to gradual **paralysis,** the inability of the muscles to move.

**How neurons work.** The transmission of a nerve impulse by an axon is a marvelous electrochemical process that shows the creative design and skillful engineering of its Creator. When a neuron is triggered to fire, a brief pulse of positive charge sweeps through the neuron and races down the axon like the fuse of a firecracker. This wave of electrical activity, called an **action potential,** is propagated by fast-acting, voltage-sensing ion gates that quickly open and close, allowing sodium and potassium ions to briefly flow into and out of the cell. After the action potential has passed, the cell membrane's ion pumps restore the cell's original ion balance, preparing the cell to fire again in as little as $\frac{1}{2000}$ second.

**Synaptic transmission.** The axon of a neuron ends at a **synapse,** an enclosed junction between two neurons or a neuron and another cell. (The *neuromuscular junction* is a synapse between a neuron and a muscle fiber.) When the action potential reaches the synapse, it causes a chemical known as a **neurotransmitter** to be released into the synapse. When sensors in the membrane of a receiving neuron detect the neurotransmitter, they generate electrical signals that trigger that neuron to fire. The second neuron may in turn relay the impulse to a third neuron, and so forth. In this way, peripheral nerves are constantly relaying impulses to and from the central nervous system. In the central nervous system, on the other hand, most neurons do not automatically fire when they receive an impulse. Rather, they process the incoming signals from hundreds or thousands of other neurons to determine whether or not to fire.

**Inhibitors of nerve impulses.** Some substances can inhibit the transmission of nerve impulses. The widely used local anesthetic procaine blocks nerve impulses in sensory nerves, a property that makes it a good local anesthetic. A potent inhibitor of impulse transmission in motor nerves is *botulinum* [bŏch′ə·lī′nəm] *toxin,* a powerful poison that is responsible for the most deadly type of food poisoning (*botulism* [bŏch′ə·lĭz′əm]). The most toxic substance known to man, this toxin is manufactured by the bacterium *Clostridium* [klŏ·strĭd′ē·əm] *botulinum* and is sometimes found in spoiled or improperly canned foods. The toxin is an enzyme that destroys a specific protein inside motor nerve synapses, preventing the motor nerve from discharging neurotransmitter molecules into the neuromuscular junction. The blocked nerve impulses cause paralysis death by suffocation. Very small injections of botulinum toxin can be used medically to relax muscles, such as in the treatment of migraines and certain eye conditions and in cosmetic surgery.

**Parkinson's disease.** One serious disease of the nervous system, **Parkinson's disease,** affects the patient's control of posture and movement; it is characterized by tremors (shaking) and stiffness of the limbs. Some victims of Parkinson's disease can move their limbs only at an extremely slow (sometimes almost imperceptibly slow) rate. Parkinson's disease appears to result from a lack of the neurotransmitter *dopamine* in the motor-control areas of the brain, caused by damage to a region of the midbrain that produces dopamine.

Fig. 8.11

**Transmitting a nerve impulse**

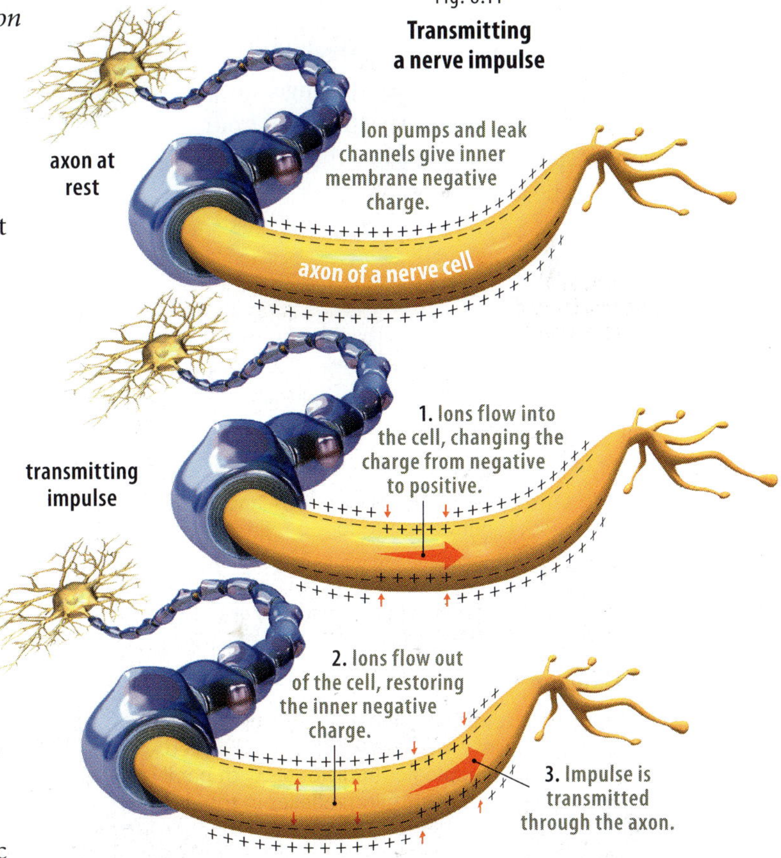

Usually, the patient's mind remains alert, although the patient is unable to respond. Some of the symptoms of Parkinson's disease, which usually occurs in people over 50 years of age, can be treated with drugs or surgery.

Fig. 8.12 **Reflex pathway**

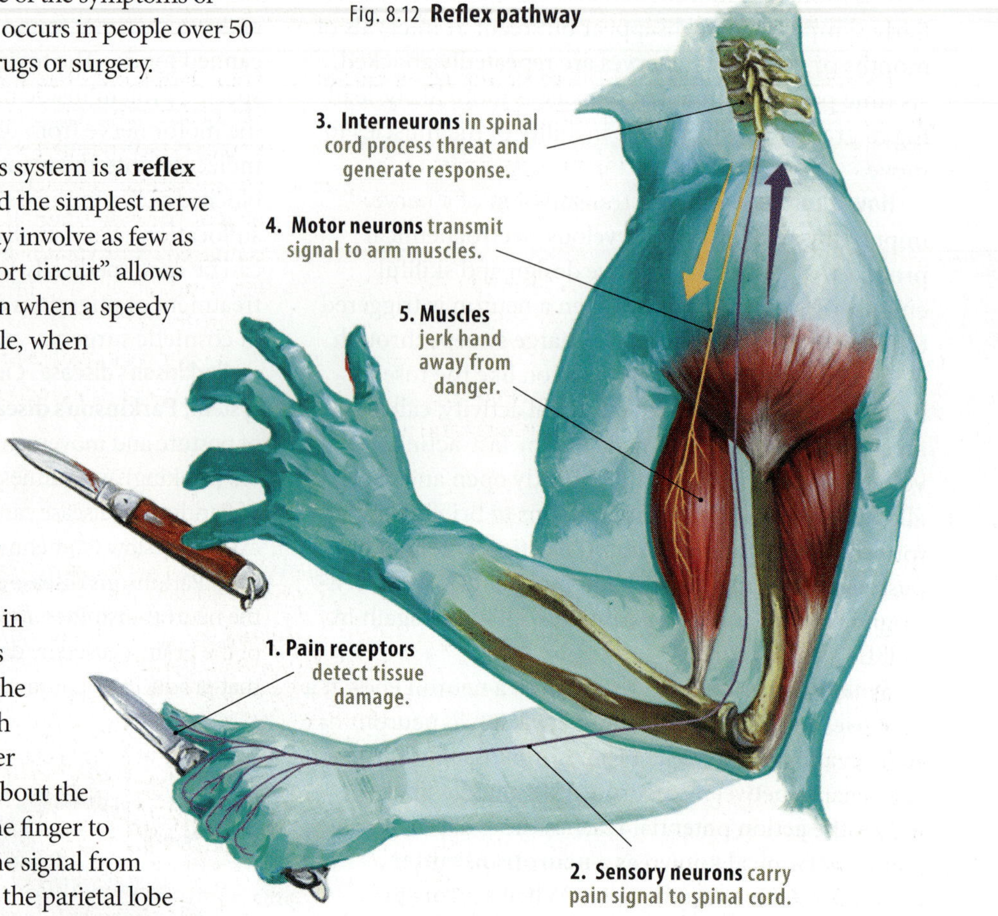


## Reflex Action

The simplest act of the nervous system is a **reflex** (a quick, automatic response), and the simplest nerve pathway is a **reflex arc,** which may involve as few as two or three nerve cells. This "short circuit" allows nerve impulses to bypass the brain when a speedy response is necessary. For example, when your finger touches a hot stove burner or is accidentally cut by a knife, a reflex arc helps prevent severe damage to the tissues. When your finger touches the object, sensory neurons in the skin transmit the sensation of pain to interneurons in the spinal cord. The interneurons transmit an emergency signal to the appropriate motor neurons, which cause the muscles to jerk the finger away from the burner or knife. About the same time the reflex arc caused the finger to jerk away, you felt pain because the signal from the pain receptors finally reached the parietal lobe of the brain, the area where pain is felt.

## Section Review 8.2

1. Describe the structure of a nerve.
2. What is the purpose of Schwann cells and oligodendrocytes in nerve tissue?
3. Name two diseases caused by brain cell damage.
4. How do sensory neurons, interneurons, and motor neurons work together to produce a reflex?

### Application

5. Write a summary of the activity involved in the transmission of a nerve impulse.

### Critical Thinking

6. An injury that severs the spinal cord in the neck or upper back causes a person's legs to be paralyzed. However, some paraplegics (individuals with paralyzed lower bodies) have functioning reflexes in the paralyzed limbs. How can this be possible?

### Identify

multiple sclerosis, paralysis, synapse, reflex, reflex arc

### Observe

- To observe the "knee-jerk reflex," an example of a reflex arc employing only two neurons, sit on a chair with your legs crossed at the knees and strike the top leg lightly with a reflex hammer or similar object just below the kneecap. As the ligament is struck, the quadriceps femoris muscle is pulled slightly, stimulating stretch receptors within the muscle. An impulse travels along sensory nerve fibers to the spinal cord, where the sensory neuron forms a synapse with a motor neuron. A motor impulse travels back to the quadriceps femoris, which contracts slightly, extending the lower leg. (This reflex helps a person maintain an upright posture by automatically adjusting the tension in the leg muscles.)

## 8.3 The Brain: Organ of the Mind

The **brain,** which is the largest, most important part of the nervous system, has three main parts:

- The *cerebrum* [sə·rē′brəm], the upper part of the brain, coordinates thought, memory, and learned behaviors.
- The *cerebellum* [sĕr′ə·bĕl′əm], the lower part of the brain, helps control balance and coordinate voluntary muscle activity.
- The *brain stem,* the part that connects to the spinal cord, controls the involuntary muscles and activities of the autonomic nervous system.

The human brain is so densely packed with thoughts, memories, and impulses that it would take a skyscraper full of computers to begin to come close to containing it. You may have a small computer that can always beat you at a chess game, but that same computer is totally helpless at problems that you consider quite simple. Some neurobiologists estimate that your brain, which weighs about three pounds, contains over 100 billion neurons, each directly linked to an average of 100,000 others and forming more than 10 quadrillion connections. Packing so many communication channels into such a small space could be compared to taking all of the telephone systems in the world and putting them into a thimble. God's design of the human brain goes beyond human comprehension.

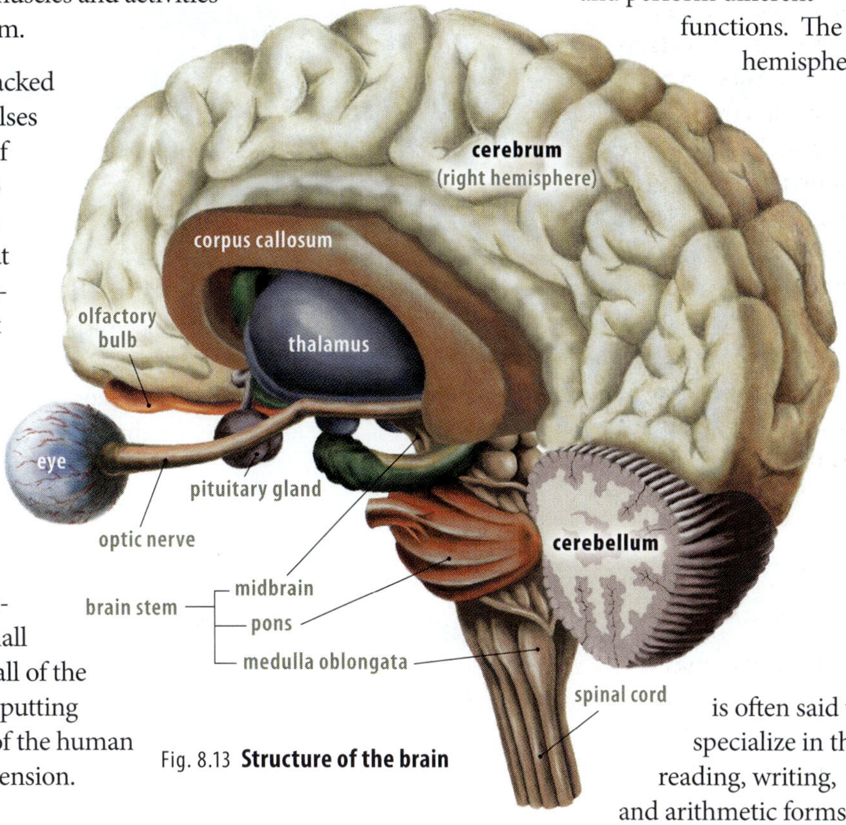


Fig. 8.13 **Structure of the brain**

### Cerebrum: Thought and Will

**Right and left hemispheres.** The **cerebrum** is the largest part of the brain and is the physical organ that helps you in the areas of consciousness, memory, voluntary actions, thinking, and intelligence. It consists of two halves called the right and left **hemispheres.** These halves are split by the *longitudinal fissure* [fĭsh′ər], a deep groove. *The right hemisphere is connected with the left side of the body and the left hemisphere is connected with the right side of the body;* therefore, if you touch something with your right hand, that information is sent primarily to your left brain. The two halves are not exactly equal. If you are right-handed, your left brain is slightly better at coordinating hand activity than your right brain. If the area of your left brain that controls your hand were to be damaged, your right hand would be paralyzed, but your left hand, which is controlled by the right brain, would be unaffected.

As odd as it may seem, your right and left hemispheres, even though they look alike, are not identical and perform different functions. The left hemisphere is often said to specialize in the reading, writing, and arithmetic forms of intellect, while the right hemisphere is said to specialize in artistic and geometric kinds of intelligence. However, this is something of an oversimplification. In language skills, for example, individual words and their sounds and meanings are processed by the left hemisphere; but humor, metaphor, and literary themes are processed by the right hemisphere. In general, *the left brain seems to specialize in details, known procedures, and formal logic, whereas the right brain specializes in the "big picture," creativity, and intuitive approaches to problems.*

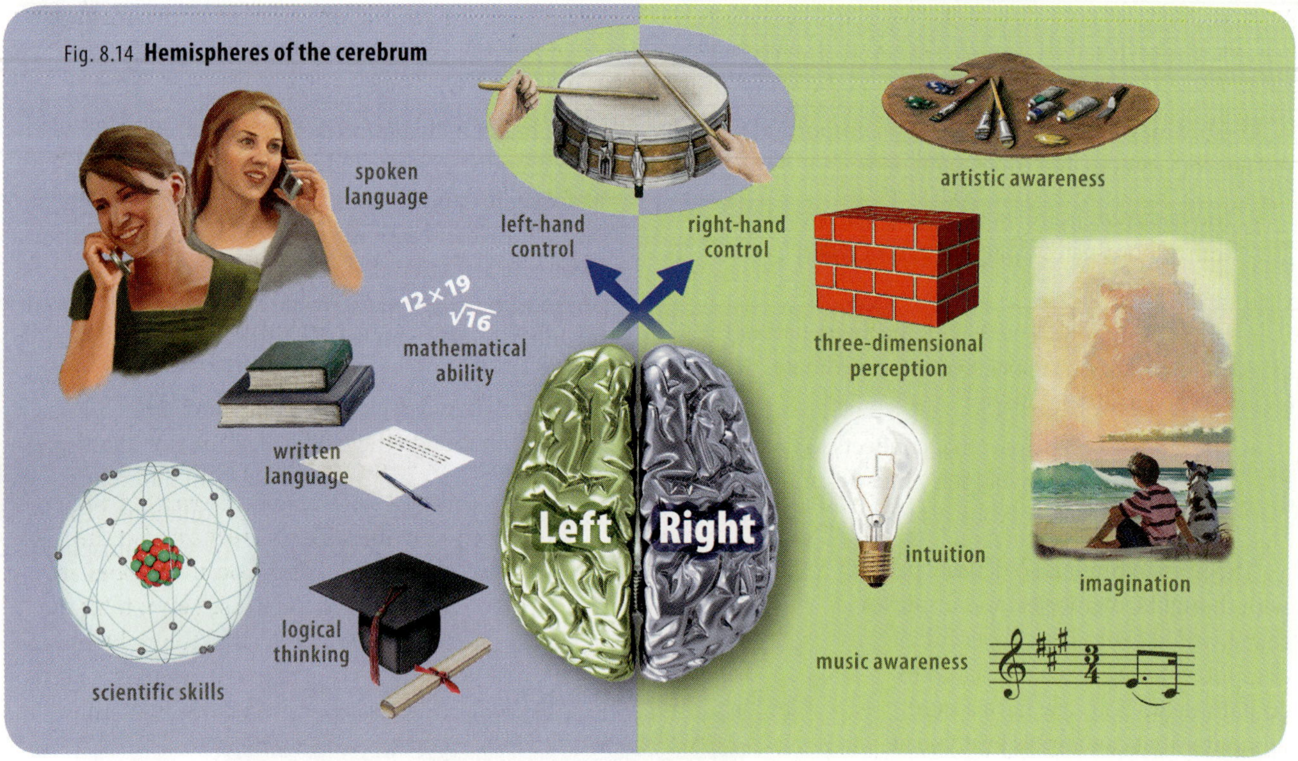


Fig. 8.14 **Hemispheres of the cerebrum**

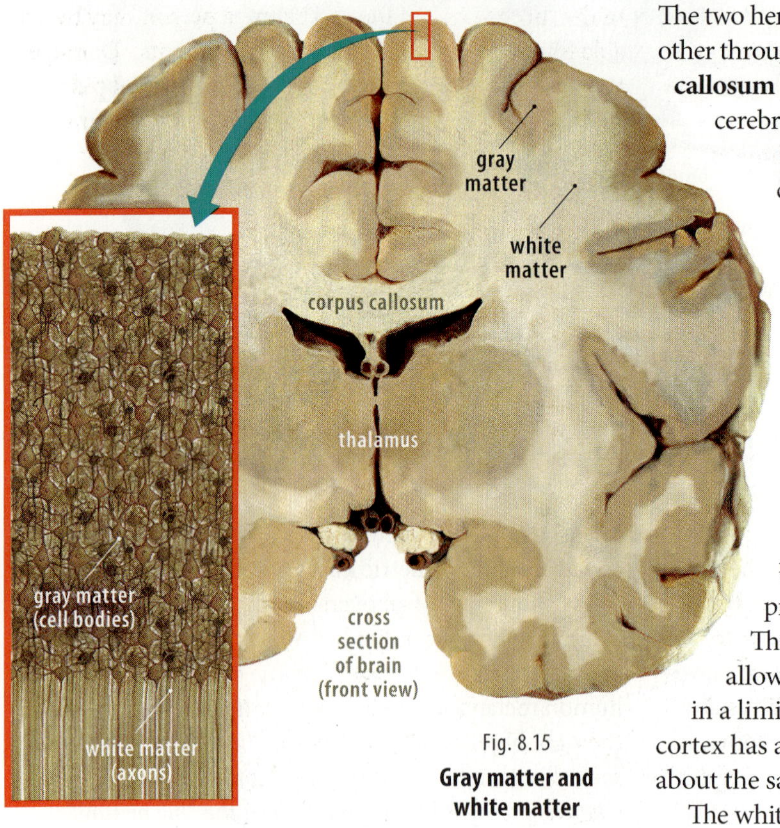


Fig. 8.15 **Gray matter and white matter**

The two hemispheres communicate with each other through a mass of nerve fibers called the **corpus callosum** [kôr′pəs kə·lō′səm], located at the base of the cerebrum.

**Gray matter and white matter.** The cerebrum consists of two layers of tissue: gray matter and white matter. As we have seen, *gray matter* is made largely of nerve cell bodies, which lack the fatty covering called myelin. *White matter* gets its color from the myelin covering of the nerve fibers. Much of the brain's gray matter is located in the **cerebral** [sə·rē′brəl] **cortex,** the outer layer of the cerebrum. The cerebral cortex consists mainly of nerve cell bodies, which are located in a thin layer less than ⅛ inch (3 mm) thick with their axons projecting toward the interior of the cortex. The cortex is deeply grooved, or *convoluted,* allowing a maximum amount of gray matter to fit in a limited space; it is estimated that your cerebral cortex has a total surface area of about 16 square feet, about the same as a typical beach ball.

The white matter of the cerebrum lies in the interior and consists largely of bundles of myelin-covered nerve fibers that extend from the cell bodies in the cerebral cortex. Messages generated in the cortex travel along these nerve fibers to other areas of the brain, to the spinal cord, or to the body parts. Also, impulses from sense organs travel along the white nerve fibers to the cerebral cortex, where they are perceived and acted upon.

**Cerebral lobes.** The convolutions of each of the cerebral hemispheres are divided by grooves into various regions, or **lobes,** that correspond to the major bones of the cranium. Each part of the body is controlled by a specific location on a specific cerebral lobe, and each hemisphere has its own set of lobes.

Fig. 8.16 **The brain at work**

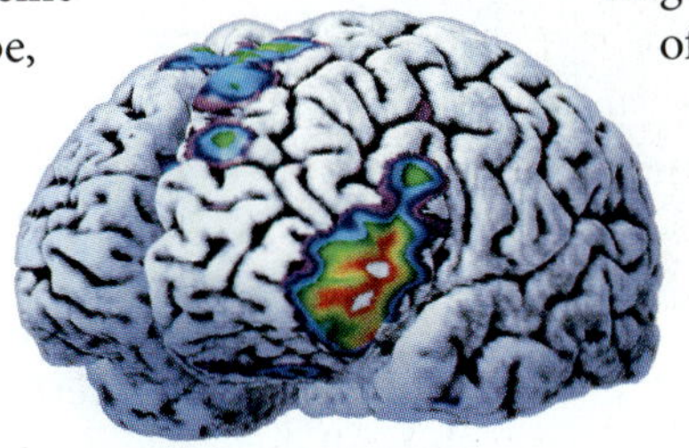

brain activity in a person reciting synonyms

Throughout most of the 1800s and early 1900s, man's study of specific brain locations for particular functions was limited to observations of people with brain damage. People suffering from injuries to specific areas of the brain caused by accidents or strokes developed paralysis of specific body parts or impairments of specific functions. The advent of neurosurgery in the mid-1900s led to additional knowledge; as neurosurgeons attempted to repair damaged brains or cut out brain tumors, they found that stimulating certain brain parts always caused certain actions. By combining the findings of many brain surgeons, neurobiologists were able to develop a rudimentary brain map to illustrate the function of the brain areas. More recent advances in technology have allowed neurobiologists to observe brain activity much more directly; these advances allow researchers to observe which locations in a person's brain become activated when the person performs specific tasks or answers various questions.

The forward portion of the **frontal lobes** is associated with personality, judgment, and self-control. The rear portion of the frontal lobes is called the **motor area** because it controls voluntary movement of the skeletal muscles. The movement of each part of the body is controlled by a specific area of this portion of the frontal lobes.

The left and right **parietal lobes** are located behind the frontal lobe. They analyze the senses on their respective sides of the body to allow us to feel sensations such as temperature, pressure, and pain and to make judgments about such things as shape and texture. The sensations for each part of the body are handled by a different spot in the parietal lobe. If you hit your finger with a hammer, you would become aware of the pain when the nerve impulses from your finger reached the area of the parietal lobe responsible for that particular finger.

The sense of vision is interpreted by the **occipital lobes,** which are located at the back of the brain. If the occipital lobes are damaged, a person cannot see even if his eyes are functioning perfectly. The right and left **temporal lobes,** which are located on either side of the brain, are associated with the senses of hearing, taste, and smell.

**Damage to the cerebrum.** When the brain is damaged in the motor areas of the cerebrum, a person may be unable to move or control his body movements. Damage to the cerebral motor area is known as **cerebral palsy.** This disorder manifests itself in many ways, ranging from inability to walk, to poor speech, to irregular

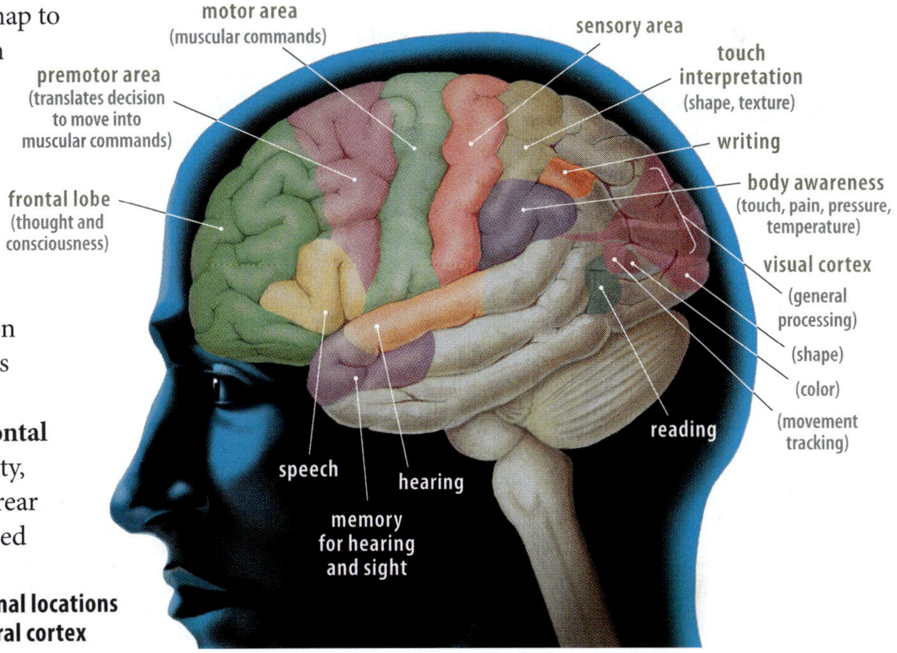


Fig. 8.17 **Functional locations of the cerebral cortex**

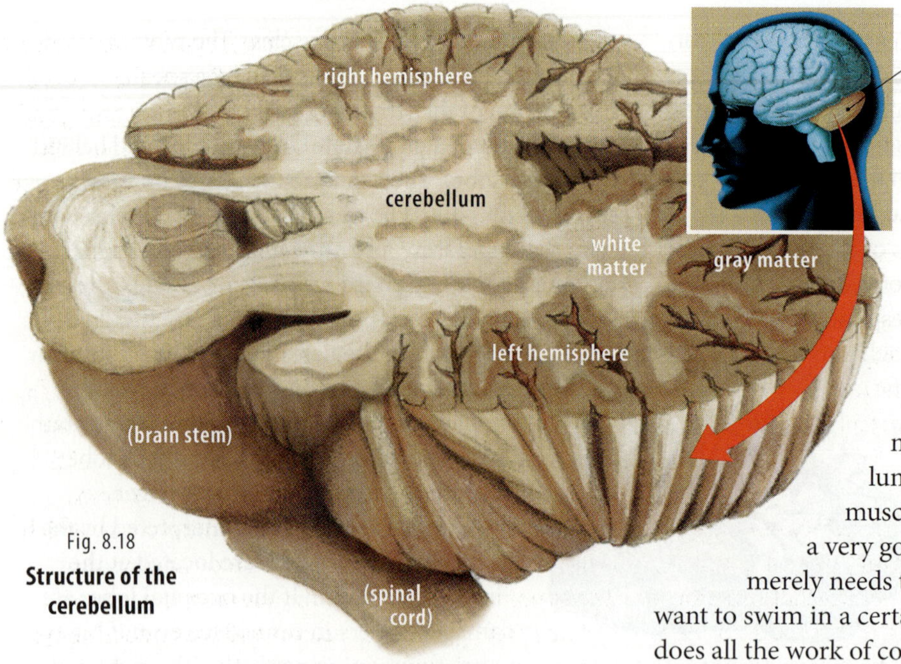

Fig. 8.18
**Structure of the cerebellum**

contractions of face, neck, or extremities; the intellect is not affected, however. The person's right side, left side, or both sides of the body may be affected, depending on how much of the motor area is damaged. Although medications can help relieve muscle spasms, the best help the patient can receive is love and understanding coupled with training to help him develop the sensory and motor abilities left to him.

## Cerebellum: Muscle Coordination

The **cerebellum** is the second-largest brain region. It lies behind the brain stem and just below the occipital lobe of the cerebrum. It is divided into right and left hemispheres and is highly convoluted. Like that of the cerebral cortex, the tissue of the cerebellum is organized as white matter and gray matter. The design of the cerebellum is very similar to that of the cerebrum except that the convolutions are much smaller, packing the cortex of the cerebellum even more densely than that of the cerebrum.

The function of the cerebellum is primarily to coordinate skeletal muscle activity. This relieves the cerebrum of much work because it means that you need not consciously think about which muscles to use when you are walking, driving, or brushing your teeth. To illustrate how the cerebellum operates, let us assume that you are not able to swim. Swimming demands a coordination of almost every skeletal muscle in the body. When you begin learning to swim, the cerebrum must think about every move. You concentrate on the muscle activity so that you will not drown. The messages from the cerebrum are also relayed to the cerebellum. As you learn to swim and your body moves with more coordination, the cerebellum is being trained to operate the muscles habitually. When you become a very good swimmer, your cerebrum merely needs to tell the cerebellum that you want to swim in a certain way and the cerebellum then does all the work of coordinating the muscles. *Most complex muscle coordination is handled by the cerebellum.* If the cerebellum is damaged, even a simple task like tying shoes becomes difficult.

## Brain Stem: Involuntary Functions

The **brain stem** is located between the cerebrum and the spinal cord. Because of its location, all nerve fibers connecting the brain and spinal cord must pass through it. The brain stem consists of the medulla oblongata, pons, and midbrain.

The lowest part of the brain stem is the **medulla oblongata.** It contains nerve centers that monitor and regulate breathing, heartbeat, blood pressure, and other vital body functions such as swallowing; it also is responsible for triggering you to sneeze.

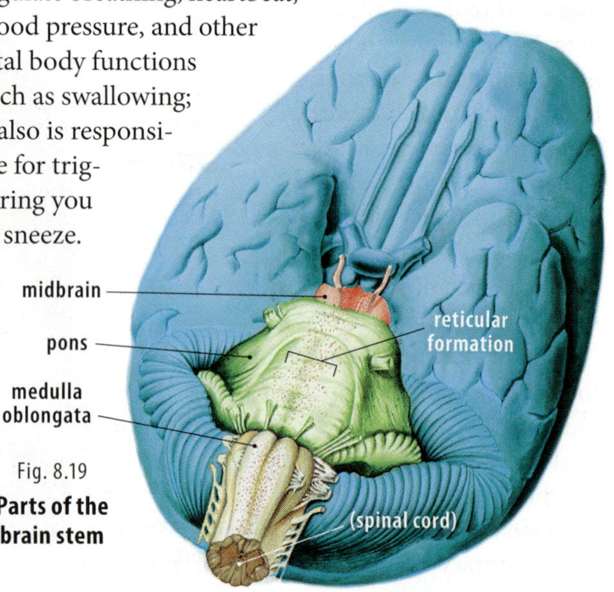

Fig. 8.19
**Parts of the brain stem**

All major sensory and motor pathways between the body and the cerebrum cross over in this part of the brain. That is why the left cerebral hemisphere controls the right hand instead of the left hand.

Just above the medulla oblongata is the **pons** (Latin for "bridge"), which links the cerebrum with the cerebellum. The pons assists the medulla oblongata in regulating breathing; it also helps coordinate some eye movements and facial expressions.

Above the pons lies the **midbrain.** Nerve centers in the midbrain help coordinate the movements of both eyes, adjust the size of the pupils in response to light, and operate the lens muscles to focus the eyes on the object of your attention.

Within the brain stem lies an intricate network of neurons called the **reticular formation,** from a Latin word meaning "small net." Its most important function is serving as the "master switch" of the cerebrum. When you wake up in the morning, it is the reticular formation that "switches on" the cerebral cortex to bring you to consciousness; when you drop off to sleep at night, it is the reticular formation that "shuts down" the cerebral cortex, causing you to sleep. The reticular formation is also actively involved during the day, constantly adjusting your alertness level in response to input from your senses and feedback from the cerebral cortex. The reticular formation activates your cerebral cortex to a state of high alert when you hear a loud or unexpected noise or feel pain, helping to protect you from danger. If the reticular formation is damaged, a person may slip into a *coma* (a state of prolonged unconsciousness). The reticular formation also serves to keep your body balanced and upright when you are not moving and plays a role in carrying out certain functions of the autonomic nervous system.

## Limbic System: Coordinating Emotions

A number of complex brain structures lie clustered around the brain stem at the core of the brain, surrounded by the cerebrum; neurobiologists sometimes refer to these structures collectively as the **limbic system.** The structures of the limbic system are involved in coordinating the activity of the different parts of the brain; they also seem to be important in generating and regulating emotions and desires in coordination with incoming sensory information and the powers of reason in the cerebrum.

**The thalamus and hypothalamus.** Two important structures of the limbic system are the thalamus and the hypothalamus, which are located above the midbrain, at the very core of the brain. At the uppermost end of the brain stem is the **thalamus,** which acts like a switchboard, routing activation signals from the reticular formation and sensory impulses from various parts of the body to the appropriate areas of the cerebral cortex.

Just below the thalamus is the **hypothalamus,** which is the control unit for your body's automatic systems. The hypothalamus controls the autonomic nervous system (through the brain stem) and the endocrine system (through the pituitary gland). Your hypothalamus monitors the temperature, pressure, and composition of your blood and makes adjustments through its control of sweating and shivering, heart rate, blood vessel tension, liver and digestive system activity, kidney activity, and waste management. *The hypothalamus is also responsible for the physical effects of emotions* such as rage and fear. When you become afraid, for example, your hypothalamus activates your sympathetic nervous system and directs your adrenal glands to release epinephrine (adrenaline). Your heartbeat, breathing rate, and cell activity increase; your pupils dilate; your muscles tense; your digestion slows; and other effects occur as your body readies itself for emergency action. Other important functions of the hypothalamus are generating physical desires such as hunger and thirst and (through the pituitary gland) regulating the growth of a person from early development in the womb through adulthood.

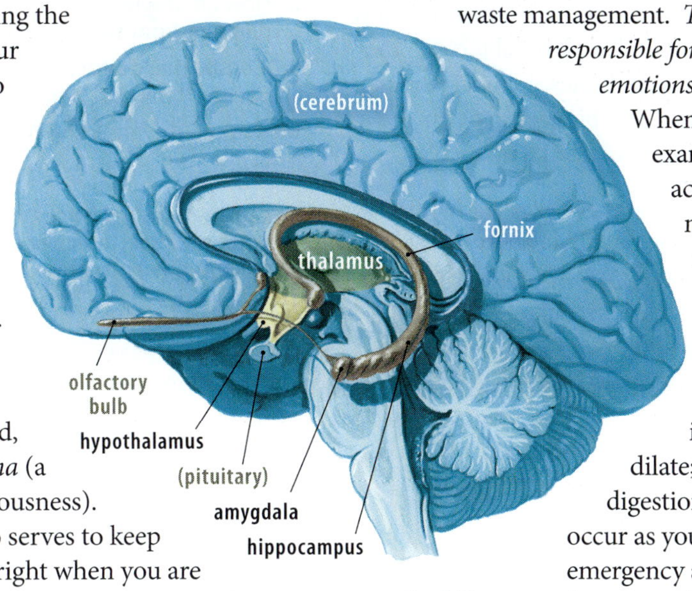


Fig. 8.20
**Limbic system**

**Other structures of the limbic system.** The other structures of the limbic system also have important functions. The *hippocampus* [hĭp′ə·kăm′pəs] processes factual memories for storage, while the *amygdala* [ə·mĭg′də·lə] helps generate emotions and processes emotional memories. Various other structures are thought to help coordinate many activities of the limbic system and the brain as a whole, although their specific functions are not completely understood.

### The Mind and the Brain

Because the mind is a very difficult reality to measure, some biologists have suggested that the behavior of a person is determined by his environment, not by his mind. According to this philosophy, known as **behaviorism,** a person's actions can be completely explained as responses to particular stimuli. Although some popular writers in the fields of psychology and progressive education today still subscribe to behavioristic thinking, behaviorism has been largely discredited in the field of neurobiology by more recent discoveries about brain function. As Christians, we know that the mind does not consist simply of neurons and electrochemical reactions. Although the brain can be thought of as the seat of the mind, *the mind and the physical brain are not the same.* God's Word makes it clear that man also has a spiritual component, a soul and a spirit, that cannot be explained in physical terms. The eminent neurobiologist Sir John Eccles, a Nobel laureate in medicine and physiology and a pioneer in brain research, had this to say of man's spiritual nature:

> Science has gone too far in breaking down man's belief in his spiritual greatness and has given him the belief that he is merely an insignificant animal who has arisen by chance and necessity on an insignificant planet lost in the great cosmic immensity.
>
> But that does not mean that religion and science are necessarily at odds. Max Planck, the great physicist, was a practicing Catholic. Albert Einstein believed in a God of the cosmos. Werner Heisenberg, the world famous physicist, also held religious views, though he was not a man who practiced religion. I, myself, am a practicing Christian.
>
> To hold views such as mine about the mystery of existence, you don't have to be a religious person. The great philosopher of science Sir Karl Popper, with whom I have written a book on this subject, holds similar beliefs—and he describes himself as an agnostic. Both of us recognize the great wonder of existence. We believe in both a material world and a mental-spiritual world.
>
> *"Science Can't Explain 'Who Am I? Why Am I Here?'"*
> U.S. News and World Report, *Dec. 10, 1984, 80.*
> Copyrighted 2015. U.S. News & World Report. 115414:0215BC

### Section Review 8.3

1. Identify the location and basic functions of the cerebrum, cerebellum, and brain stem.
2. How are the left and right cerebral hemispheres different from each other?
3. Identify the location and function of the frontal, parietal, temporal, and occipital lobes of the cerebrum.
4. Why are the brain and the mind not the same thing?

#### Identify

corpus callosum, cerebral palsy, medulla oblongata, pons, midbrain, reticular formation, limbic system, thalamus, hypothalamus, behaviorism

## 8.4 Neurological Health

### Caring for the Nervous System

Because your nervous system is an intricately designed and vitally important part of your life, you should know how to keep it in good working order. Although you cannot control all the diseases or injuries that might affect the nervous system, you can control your habits. *Some habits harm the nervous system from within,* such as smoking and the drinking of alcoholic beverages. Nutrition is also important; some substances found in the foods we eat, such as calcium, are vital to the proper function and maintenance of the nervous system. Others are harmful.

According to one study by the American Medical Association, a person can, on the average, live eleven years longer if he avoids cigarettes and alcohol and reduces his consumption of fatty foods. Equally important is regular and vigorous exercise.

*External factors can also affect the nervous system.* For example, repeated or prolonged exposure to excessively loud sounds results in a form of partial or complete hearing loss known as **sensorineural** [sĕn′sə·rē·no͝or′əl] **deafness.** Repeated exposure to sounds that exceed 90–100 *decibels* (a scientific measure of loudness) for more than a short time will cause serious hearing loss, but many teenagers and young adults routinely listen to music at levels of 100–120 decibels for hours at a time.

*Rest or sleep is another factor critical to neurological health.* Sleep gives the body an opportunity to rid itself of accumulated waste materials and to restore strength and energy to the nerves and muscles. Fatigue is a warning sign that waste materials are being formed faster than they are being eliminated from the body. Failure to pay attention to the warning signals of fatigue may lead to nervous problems.

Sleep is especially important for the brain, which requires sufficient sleep for proper functioning; if a person is deprived of sufficient sleep, his brain will become less capable of functioning properly. The brain's neurons synchronize their communication by firing in rhythmic pulses, which can be recorded as waves of electrical activity *(brain waves)* by special instruments. During the day, these pulses are short and fast (up to 50 per second). When a person passes from wakefulness to sleep, his brain produces slower but larger waves. As a person sleeps, the electrical activity of the brain goes through cycles. Periods of small, fast waves, similar to those of an awake person, occur periodically throughout the night. Because the sleeper's closed eyes show rapid eye movements as though he were watching the events of a dream, these periods are called **REM** (rapid eye movement) **sleep.** A person who is awakened during REM sleep will usually be able to recall details of the dream.

## Alcohol and the Nervous System

One of the most commonly abused drugs in the world today is **alcohol** *(ethanol).* It is estimated that alcohol claims the lives of up to 100,000 Americans each year through alcohol-related accidents, disease, and criminal behavior.

**Loss of inhibitions.** When alcohol is ingested, it quickly enters the bloodstream from the stomach and small intestine and is carried to every part of the body. When alcohol reaches the brain, it interferes with specific neurotransmitter systems in the thought and motor areas of the cerebral cortex. Alcohol acts as a *depressant* of the central nervous system, slowing the functions of the neurons that it affects. A small dose of alcohol begins by depressing the controlling centers of the brain, producing a general loss of inhibitions, or self-control. The person's behavior becomes unpredictable. He may do things he ordinarily would consider wrong. Motor neurons are also affected, resulting in clumsiness and lack of coordination. As more alcohol is consumed, the person's reactions become slower; his judgment and the ability to reason are further impaired; his thinking becomes increasingly confused; his speech becomes slurred; and his muscular control deteriorates. He becomes emotionally unstable and may act silly, irritable, angry, or depressed. Very large quantities of alcohol result in unconsciousness and may be fatal.

**Breaking down alcohol.** The liver, the only organ able to break down alcohol in the body, removes the alcohol from the bloodstream by slowly converting it into another poison, *acetaldehyde* [ăs′ĭ·tăl′də·hīd′]; this poison is responsible for the headaches and nausea (hangover) that follow alcohol consumption. The acetaldehyde is eventually changed to acetate and "burned" in cellular respiration. Because alcohol remains in the body until it is broken down by the liver, it may take the body hours to rid itself of only a few drinks.

**Long-term effects.** Studies have shown that heavy consumption of alcohol causes accelerated death of brain cells and a gradual, permanent decline in brain function; nervous disorders such as visual disturbances, tremors, and **neuritis** (degeneration and inflammation of nerves) may also result. Heavy drinking for long periods of time also greatly increases one's risk of chronic hepatitis (inflammation of the liver), cirrhosis, pancreatitis, kidney failure, heart disease, dementia, and certain types of cancer. In addition, heavy alcohol use lowers hormone levels in men, and women who drink during pregnancy can cause brain damage and birth defects in the developing child.

### Alarming Facts

- Alcohol is the most widely abused drug in the world.
- The average age of a person who starts drinking alcoholic beverages is about 14 years.
- About 80,000 alcohol-related deaths occur every year in the United States alone.
- Nearly a quarter of all suicides are alcohol related.
- There are perhaps 18 million alcoholics in the United States.

## Injuries to the Nervous System

At about two years of age, a child's nervous system is completely developed. From that point on, new cells will never be added to his brain, spinal cord, or nerves; *dead nerve cells cannot regenerate and existing nerve cells cannot reproduce.* Cell death occurs when the cell body (containing the nucleus) of the neuron dies. If nerves are damaged without cell death, the nerves can grow back, but damage resulting in cell death is permanent. If the axon of a peripheral nerve is damaged or severed, it can repair itself by regenerating the damaged part. This process takes much time because of the length of fiber to be regenerated. Unfortunately, nerve fibers in the central nervous system generally do not regenerate, so that injury to the brain or spinal cord often results in permanent damage.

**Spinal cord and nerve injuries.** Each year, between 7000 and 10,000 people in the United States suffer serious spinal cord injuries; nearly half suffer complete paralysis and loss of sensation in the affected areas. The vertebrae in the neck may be injured when a person's body is jolted by a sudden stop, such as in an automobile accident. If the spinal nerves protruding from the spinal cord in the neck are injured, *whiplash* has occurred, resulting in pain, muscle spasm, and limited motion. When intervertebral disks rupture, the vertebrae move closer together, pinching the spinal nerve that runs between them. **Sciatica** [sī·ăt′ĭ·kə], a condition characterized by pain in the lower back and the back of the thigh, is caused by a pinching of the sciatic nerve of the lower spinal cord.

**Brain injuries.** The most common serious injury of the nervous system is **stroke,** a leading cause of death throughout the world. This injury is not caused by outside forces but by an internal blockage or rupture of a blood vessel in the brain. When this happens, certain brain areas are not supplied with blood, resulting in their death. Stroke victims may suffer speech loss or paralysis, depending on the location and severity of the damage.

One of the most common causes of death among young adults is injury to the head, which damages the brain. Fortunately, most everyday bumps and knocks to the head do not cause any permanent damage, even when the blow causes a **concussion,** a period of paralysis of the central nervous system accompanied by a short period of unconsciousness (usually less than five minutes and rarely more than ten minutes). Varying amounts of **amnesia** [ăm·nē′zhə], a loss of memory, can result from a concussion, but recovery from amnesia is soon complete. Usually, time and rest give the body the opportunity to repair the nerve tissue in the brain.

*A more prolonged loss of consciousness may result from severe effects of injury, poison, or disease;* a *coma,* a prolonged unconscious state that may not be reversible, can result from damage to the nervous system (particularly damage to the reticular formation or thalamus). Some areas of the brain are affected, while others, such as the centers that control breathing and the heartbeat, continue to function.

## Exercising Your Brain

Understanding the LIVING CREATION

Neurobiologists have discovered that *the brain constantly optimizes its systems to most efficiently meet the demands placed upon it.* The first time you attempt to perform an unfamiliar task (such as a new athletic skill or a new type of math problem), your brain must devote considerable resources to the process. But after performing the task a few times, your brain begins to adjust the intricate patterns of its synapses and connections to meet the particular mental challenges posed by the task. After the brain has optimized its systems to perform that task, it can perform it much more quickly while expending much less energy. It is thought that this process of constantly adjusting and readjusting connections is how you learn and remember. The more areas in which you challenge your brain, the more areas in which your brain will become proficient. Thus, spending less time watching television and more time in wholesome reading, outdoor activities, family interaction, physical exercise, and hobbies can make your brain more capable and efficient in many areas.

## Diseases of the Nervous System

**Tetanus.** Several diseases of the nervous system have already been discussed, including meningitis, poliomyelitis, multiple sclerosis, and Parkinson's disease. Another serious disease that affects the nervous system is **tetanus,** commonly known as *lockjaw* because severe spasms of the jaw muscles make it hard to open the mouth. Tetanus is caused by a type of bacterium *(Clostridium tetani)* that dwells naturally in the soil. Tetanus bacteria can invade the body through a break in the skin or from dirt in a wound. Since tetanus bacteria do not need oxygen to live, they can multiply rapidly in wounds caused by a puncture from a thorn, rusty nail, or other dirty object because no air gets to the wound. Although tetanus bacteria cause little damage to the body directly, they are extremely dangerous because of a toxin they produce that damages certain motor nerves. Symptoms include restlessness, irritability, headache, and difficulty in opening the mouth and swallowing. Death results if the person is not treated with antitoxin injections. Tetanus can be prevented by periodic immunization with a small amount of deactivated tetanus toxin, which stimulates the body to produce antibodies that destroy the toxin; these antibodies remain in your bloodstream for a decade or so, protecting you from further exposure. Cleaning all wounds thoroughly is also important in preventing tetanus.

**Dementia.** One of the most common mental disorders in the elderly, **acute confusion,** is a disturbance of the brain caused by physical illness elsewhere in the body. The disorder that results from the slow process of natural nerve cell loss, **senile dementia** [sē′nīl′ dĭ·mĕn′shə], may become noticeable between the ages of 70 and 80. Although the dementia begins with recent memory, it may gradually continue until the person forgets his relatives and even his own name. Senile dementia affects only about 10% of people over 65 years of age. **Arteriosclerotic** [är·tēr′ē·ō·sklə·rŏt′ĭk] **dementia** is also caused by the death of brain cells; however, the cells die because blockage in the arteries restricts their blood supply. The risk of dementia can be reduced by exercising regularly throughout life.

**Alzheimer's disease. Alzheimer's** [älts′hī·mərz] **disease,** which has symptoms similar to dementia but more severe, may occur much earlier in life. As the disease progresses, Alzheimer's patients become physically as well as intellectually disabled. Forgetfulness, loss of coordination, and continuous repetition of a word or gesture are common. Degeneration of the frontal and temporal lobes of the brain eventually results in a loss of ability to speak, write, and walk, while damage to the hippocampus results in the loss of short-term and long-term memory. At present, there is no cure for Alzheimer's disease, which can progress quickly (in three to four years) or slowly (for as long as fifteen years), although there are medications that may slow its progression.

**Epilepsy.** One common neurological disorder that affects people of all ages is the condition known as **epilepsy.** Ordinarily, brain cells of the cerebral cortex fire in seemingly random patterns as they help process the normal thoughts and actions of life. With epilepsy, however, neurons occasionally malfunction and begin to fire together over and over, disrupting ordinary cerebral processing and causing unconsciousness. Motor neurons may also misfire, causing muscle convulsions. This rhythmic firing may continue for a few minutes, until the malfunctioning neurons tire and the brain is able to bring its cerebral processing centers back online. Epilepsy can usually be controlled with medication.

## Section Review 8.4

1. How can listening to loud music harm the nervous system?
2. Briefly explain the effects of alcohol on the nervous system.
3. Why is the death of nerve cells very serious?
4. Name three neurological diseases or conditions typically associated with aging.

### Critical Thinking

5. List two good habits to develop and two bad habits to avoid if you want good neurological health. Explain the importance of each item on your list.

### Identify

sensorineural deafness, REM sleep, depressant, whiplash, stroke, concussion, tetanus, epilepsy

# 8.5 The Senses and Sense Organs

## Sensory Receptors

Although your brain is a remarkable mechanism, it is not designed to function in isolation. Rather, large portions of your brain are devoted to processing information that your senses provide about the world around you and about the condition of your body at any given moment. This information is obtained and relayed to the spinal cord or brain by living sensors known as **sensory receptors.**

These specialized neurons can be divided into two broad groups. The first group includes neurons designed to detect pressure or temperature changes; these neurons are responsible for the **somatic senses,** which include all the senses of your skin (primarily touch and temperature). Other neurons in this group (the pain receptors) are designed to detect extreme temperatures or damage to body tissues, helping to warn you of danger or injury to your body. The **special senses**—sight, hearing, smell, taste, and balance—are provided by a second group of neurons located in intricately designed **sense organs.** Your eyes, ears, and taste buds are examples of sense organs.

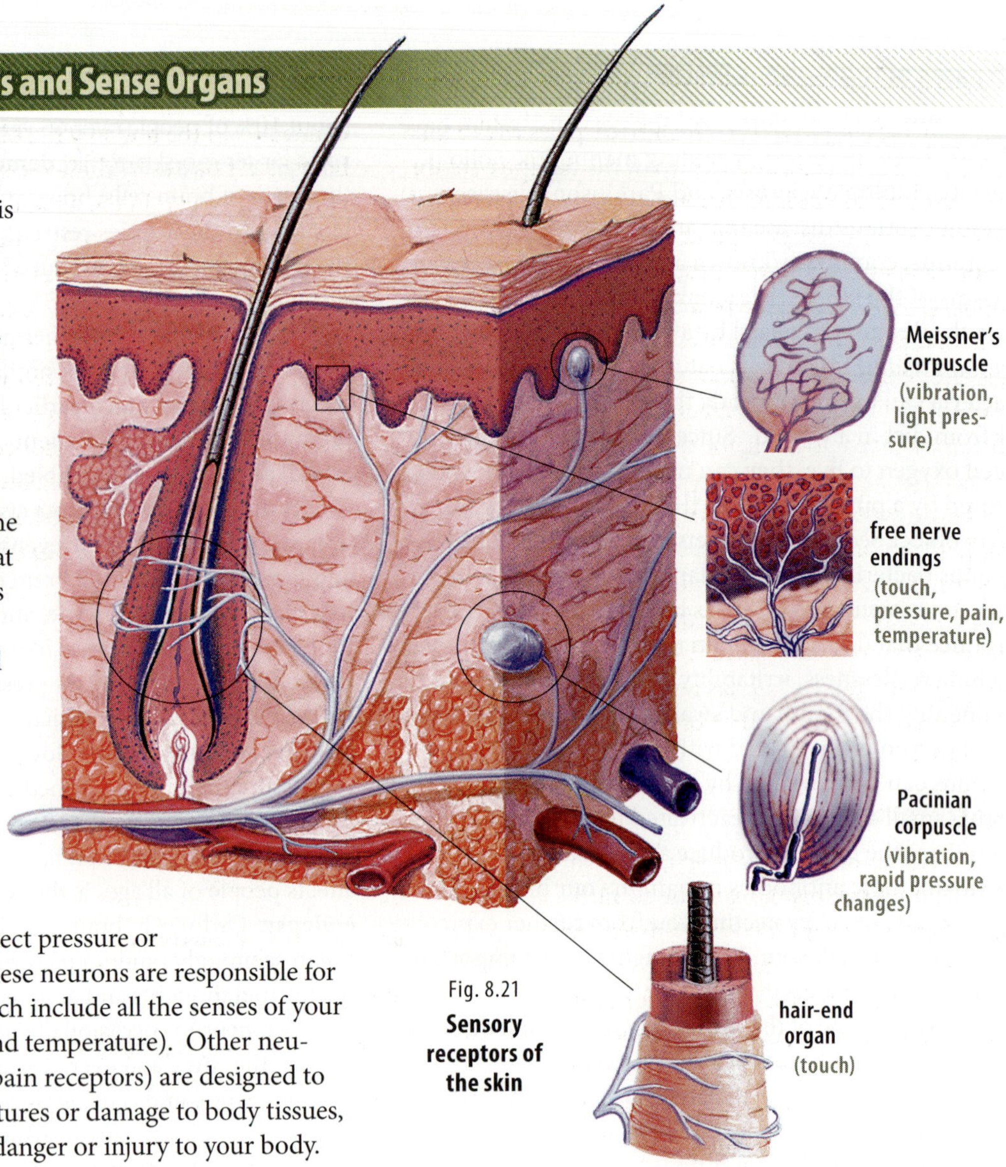


Fig. 8.21 **Sensory receptors of the skin**

### The Sensations of the Skin

The skin itself is equipped with several different kinds of sensory receptors, each of which is associated with one of the somatic senses. Each receptor is designed to respond to particular types of stimuli and relay the proper information to the central nervous system. The skin's sensory receptors may be classified in three categories: pain receptors, mechanoreceptors [měk′ə·nō·rĭ·sĕp′tərz], and thermoreceptors.

**Pain receptors.** Each of the skin's **pain receptors** is a bare dendrite that reacts to a certain type of strong stimulus. Some pain receptors become activated when extremes of heat or cold cause your skin temperature to rise or fall to dangerous levels. Other pain receptors are sensitive to excessive levels of mechanical stress, such as when you stub your toe or hit your thumb with a hammer. Still others are sensitive to chemical changes caused by damage to your body's cells. As a rule, extremes of temperature or mechanical injury produce a sharp, intense pain that subsides relatively quickly, while cell damage produces a dull, throbbing pain that lasts much longer.

**Mechanoreceptors.** The sensory receptors responsible for sensations of touch and pressure are called **mechanoreceptors;** these receptors lie at different depths in the layers of your skin. If you move the point of your pencil over your skin very lightly, you stimulate only the mechanoreceptors associated with touch, which lie close to the surface of your skin. Such body parts as the fingertips, the forehead, and the tip of the tongue contain abundant nerve endings that are sensitive to light touch or texture. Mechanoreceptors that react to pressure are found deeper in the skin of such body parts as the hands and feet. If you press the point of your pencil against your hand, you feel pressure as well as touch. Muscles and tendons also contain mechanoreceptors that monitor the length and tension of your muscles; these receptors are responsible for the *muscle sense* described in chapter 7.

**Thermoreceptors. Thermoreceptors,** as their name implies, respond to temperature changes. *Heat receptors* in your skin react to temperatures above normal body temperature; *cold receptors* are sensitive to temperatures below normal body temperature. God has provided thermoreceptors as another of the body's protective devices.

## The Chemical Senses

The senses of taste and smell, which are **chemical senses,** result from the stimulation of **chemoreceptors** [kē′mō·rĭ·sĕp′tərz] on the tongue and in the nose.

**Taste.** The organs responsible for taste are the **taste buds,** which are chemoreceptors on the back, sides, and front of the *tongue* that detect dissolved chemicals in the mouth. Different types of chemoreceptors identify substances in the mouth as sweet, sour, salty, bitter, or *umami* [o͞o·mä′mē: meaty]. Each taste bud consists of several *taste cells* surrounded by epithelial cells. Adults have some ten thousand taste buds, most of which are located in tiny projections called *papillae* [pə·pĭl′ē] on the tongue's surface. Taste cells are thought to be replaced about every ten days. As a person ages, taste cells are not replaced as rapidly; consequently, the ability to taste diminishes. This is one reason why as you get older you can eat foods that you once disliked. The senses of taste, touch, and temperature are picked up by different sensory receptors on the tongue. For this reason, the same food may taste somewhat different when dry, moist, hot, or cold. *Most of the flavors we taste are strongly affected by the odors we smell;* therefore, many foods taste bland and flavorless when you have a cold.

**Detecting odors.** The **nose's** sensory receptors are located in the upper part of the *nasal cavity* and are connected to the brain by the **olfactory nerve.** As you breathe, air passes by these receptors, which constantly sample chemicals in the air and transmit the information to the brain. As a rule, ordinary breathing does not cause as much stimulation as when we sniff, causing a larger volume of air to come into contact with the nose's chemoreceptors. Normally, if you are exposed to the same odor for more than a few minutes, your sense of smell will become temporarily accommodated to the odor and you will cease to notice it.

Some scientists think that seven different types of olfactory receptors are found in our noses, each of which is designed to recognize molecules of a certain shape (two types of these receptors also recognize electrical charge). As a result, there are thought to be seven *primary odors* that we can smell (Table 8.1). Combinations of the primary odors produce other familiar odors. For example, the odor of almonds is apparently produced by the stimulation of the camphorlike, floral, and minty olfactory receptors. It is estimated that humans can distinguish more than 10,000 different odors.

**Table 8.1 Primary Odors**

| Primary odor | Examples |
|---|---|
| camphorlike | camphor, mothballs |
| ethereal | ether |
| floral | flowers |
| minty | oil of peppermint |
| musky | musk |
| pungent | vinegar, some spices |
| putrid | rotten meat |

## The Ears: Organs of Hearing

References to the ears or to the sense of hearing abound in Scripture because the ears are a gateway to the mind. In many ways, we are what we see and hear. Hearing brings us into contact with our world through sound waves transmitted by air. Through hearing we communicate with other people, and others communicate with us. Our ears are among the most intricately designed organs in the human body.

**The outer ear.** The part of the ear that we see on the side of the head is only one component of the complex system that enables the brain to receive and interpret sound waves. Though the **outer ear** seems to be simply

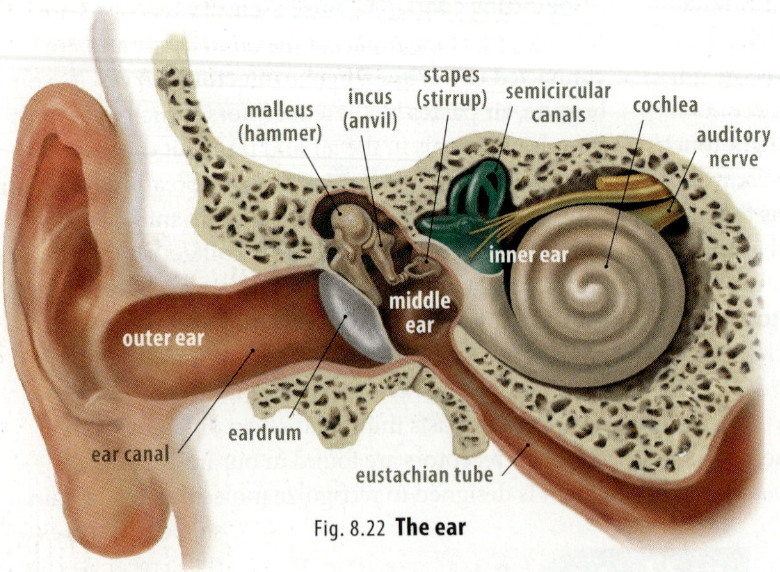

Fig. 8.22 **The ear**

a flap of cartilage and skin, it is a marvel of engineering, designed to collect sound waves and direct them into the ear.

From the outer ear, sound waves travel through the one-inch *ear canal,* which leads to the eardrum. Lining this canal are thousands of *hairs* and *wax glands* whose purpose is to trap dust, pollen, and other irritants and to guard against infection. At the end of this short canal lies the **eardrum,** a taut membrane stretched across the canal like the surface of a drum. Sound waves bouncing against the eardrum produce vibrations.

**The middle ear.** The vibrations of the eardrum are relayed to the inner ear by three tiny bones called the **malleus, incus,** and **stapes** (often called the *hammer, anvil,* and *stirrup*). These bones form a delicate structure designed to amplify the vibrations while preserving the quality of the sound (pitch, tone, and so forth). The chamber containing the malleus, incus, and stapes is the **middle ear.**

**The inner ear.** As the vibrations pass through the stapes, they enter the **inner ear,** the actual organ of hearing. The main component of this fluid-filled chamber is the **cochlea,** a coiled tube resembling a snail's shell. Thousands of hairlike nerve cells line the inside of the cochlea. As vibrations from the stapes enter the cochlea, they generate waves in the fluid, causing these nerves to wave. The waving of these cells sends electrical messages into the **auditory nerve,** which carries them to the brain, where they are translated into meaningful sounds—words, music, and so forth.

Also in the inner ear is an assembly of three fluid-filled tubes called **semicircular canals,** which serve as balance sensors. One loop detects forward motion, one side-to-side motion, and one up-and-down motion. As in the cochlea, hairlike nerve cells in the semicircular canals detect the motion of the fluid in these tubes and communicate that information to the brain. The brain, in turn, responds with messages for the muscles in the back and legs to balance the body.

The inner ear also receives vibrations from the jawbones. This type of hearing is known as **bone conduction.** When you speak, the vibrations of your voice travel to your inner ear in this way, which is why your voice sounds different on a recording. The voice you hear on the recorder is the voice that your listeners hear.

**Disease and damage.** We have already discussed sensorineural deafness and the damage that can be caused by exposure to sounds exceeding 90 decibels. *Eardrum punctures* are another threat to your sense of hearing. These injuries are usually treatable, however; in fact, the eardrum often heals itself. Many people suffer from a condition known as *tinnitus,* or ringing in the ears. This condition can also be remedied in most cases. Tinnitus may be caused by a number of factors, including fever, high blood pressure, tumors on the auditory nerve, drugs (including some antibiotics), and exposure to loud noises.

*Ear infections* are a common problem among infants and young children. Such infections occur when microbes from the throat travel up the **eustachian** [yo͞o·stā′shən] **tube** to the middle ear. The warm, moist environment of the ear is an ideal place for bacteria to grow. But the eustachian tube has the important purpose of balancing pressure in the middle ear when the outside air pressure changes. You may have experienced the painful effects of a change in air pressure when traveling by plane. To counter the pressure on your eardrums during takeoff or landing, you can yawn or swallow, opening up the throat and allowing air to travel up the eustachian tube to the middle ear.

## The Eyes: Organs of Vision

The **eye** is a fascinating organ. The more we learn about the structures and functions of this intricately designed instrument of sight, the more we appreciate the order and precision of God's creation and the more intelligently we can care for and use our eyes.

**Socket, eyelid, and tears.** The eye, a very delicate organ, is protected from injury by the socket and the eyelid. The **socket,** or *orbit,* provides a surrounding protection of heavy bone. The **eyelid** closes reflexively (blinks) to keep foreign objects from entering the eye. The blinking of the eyelid also cleans the surface of the eyeball and keeps it moist by distributing the moisture of the tear glands over the surface of the eye. During sleep, closed eyelids prevent the evaporation of tears.

**Tears** are produced by **lacrimal glands** (tear glands), which are located under the upper eyelid on the side away from the nose. Tears are more than 90% water. They also contain electrolytes in about the same concentration as those in blood plasma, giving them a salty taste. Another very important component of tears is the enzyme **lysozyme,** which destroys bacteria. Lysozyme (*lyse* means "to split") breaks a chemical bond in the cell walls of bacteria, causing the walls to split. Tears, therefore, help prevent bacterial infections of the delicate and wonderful eye.

**Eye movement.** Attached to each eye is a set of six muscles, the **extrinsic muscles,** which move the eye. These muscles work together as three matched pairs. One pair moves the eye to the right and left, and another pair moves the eye upward and downward. The third pair tilts the eye right or left, allowing your brain to keep the eye level even when your head is tilted slightly. When the extrinsic muscles are well coordinated and of equal strength, both eyes will turn in the same direction at the same time. If the muscles of the two eyes do not work together, the eyes may turn up, down, or out independently of each other. Such a condition is fairly common in very young children and frequently corrects itself.

**Eye structure.** The outer part of the eye is composed of three layers of tissue, each of which is made up of several parts. The outermost layer of the eye consists of the opaque *sclera* [sklĕr′ə], often called the white of the eye, and the transparent *cornea* [kôr′nē·ə]. The *uvea* [yo͞o′vē·ə], the middle layer, includes (in order from front to back) the *iris,* the *ciliary* [sĭl′ē·ĕr′ē] *body,* and the *choroid* [kôr′oid′]. The *retina* [rĕt′n·ə], the innermost layer of the eye, is itself composed of three main layers of cells.

The **sclera** is the tough outer layer of tissue that provides protection and the supporting frame for the interior of the eye. At the front of the eye, the opaque sclera is replaced by a transparent portion called the **cornea,** the area through which light enters the eye. Toward the back of the eye, the sclera continues to the point where the optic nerve leaves the eyeball. It then extends toward the rear and blends with the covering surrounding the *optic nerve.* The optic nerve passes through a hole in the skull in the back of the eye socket and connects with the thalamus at the base of the brain.

The largest portion of the **uvea,** the middle layer of tissue, is the **choroid,** a layer of connective tissue rich in blood vessels, which provides most of the eye's nourishment. Toward the front, the choroid becomes the **ciliary body.**

Fig. 8.23 **Tear gland and tears**

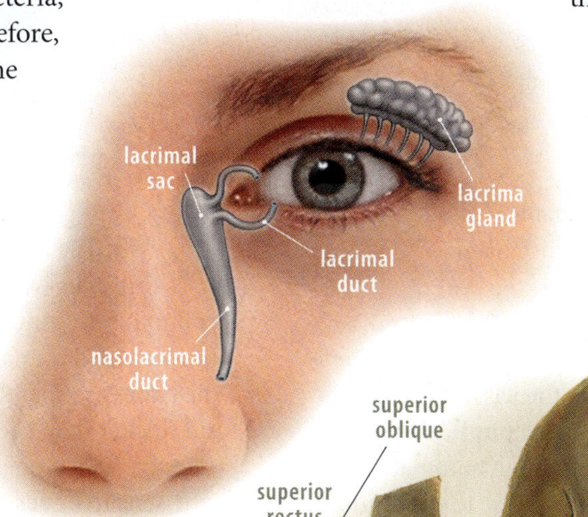


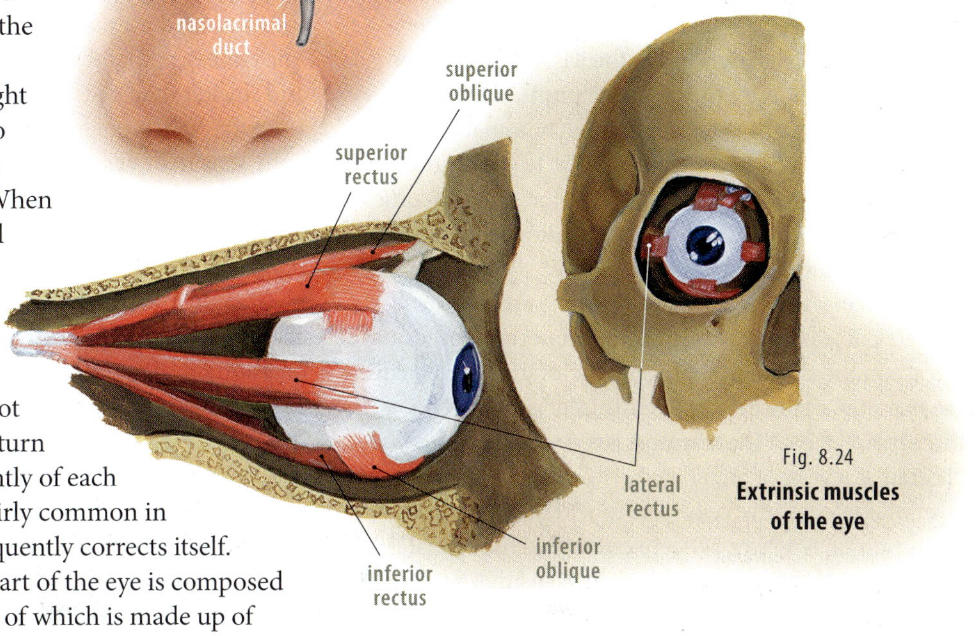


Fig. 8.24 **Extrinsic muscles of the eye**

Fig. 8.25 **Structure of the eye**

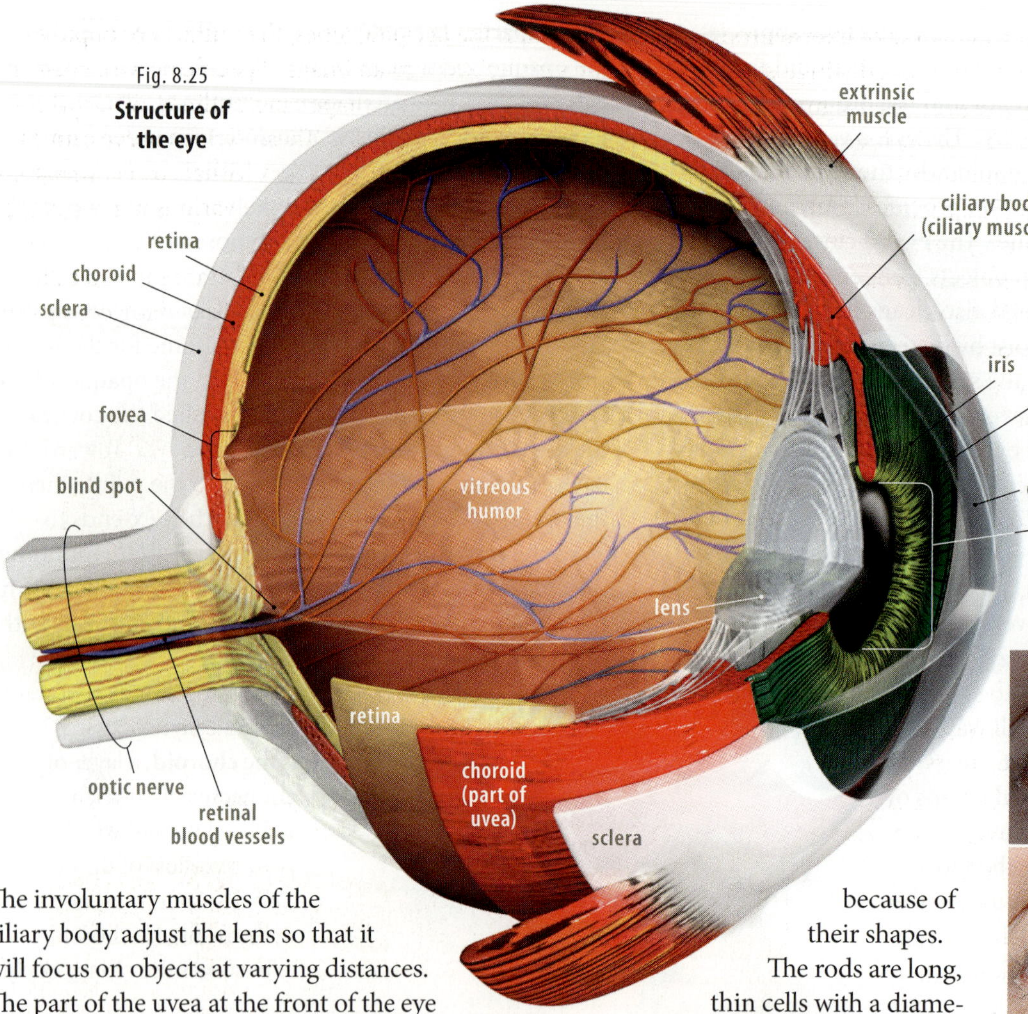


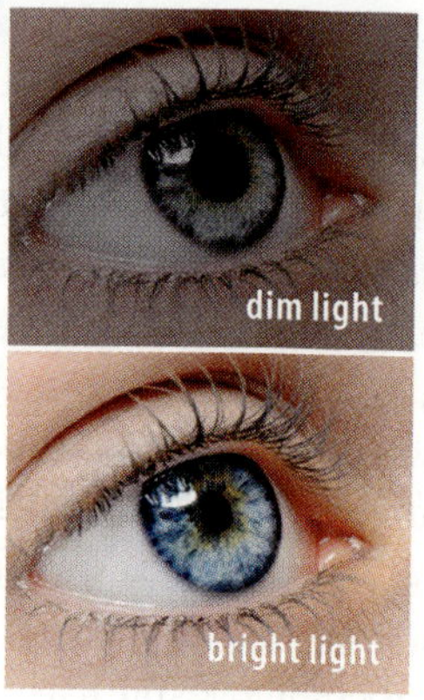


Fig. 8.26 **Iris adjustment**

The involuntary muscles of the ciliary body adjust the lens so that it will focus on objects at varying distances. The part of the uvea at the front of the eye is the **iris,** which is a continuation of the ciliary body. The iris is the colored portion of the eye. The more pigment cells there are in the iris, the darker the color. In the center of the iris is the **pupil,** the opening through which light enters. The brain automatically contracts or relaxes the iris in response to the amount of light entering the eye. In bright light, the pupil becomes small; in dim light, it becomes large, allowing about thirty times more light to enter the eye.

The **retina,** the innermost layer of the eye, is the most delicate part of the eye and the most important for vision. It rests on the choroid, and together with the choroid covers the rear 80% of the eye. The retina itself is composed of three main layers. The rearmost layer, which touches the choroid, is a layer of pigmented cells. The middle layer is a layer of light-sensitive cells. The inner layer, next to the vitreous humor, is a layer of nerve cells called the retinal layer.

**Function of the retina.** In the light-sensitive layer of the retina are two kinds of cells, known as **rods** and **cones** because of their shapes. The rods are long, thin cells with a diameter about one hundredth the thickness of a human hair. The cones are slightly shorter and thicker than the rods and have tips shaped somewhat like an ice-cream cone. There are over 100 million light-sensitive cells in the retina of each eye. God has designed the eye so that in the spot on the retina that is directly in line with the center of the cornea, the light-sensitive cells are much more densely packed than in other areas of the retina. The layer of nerve cells is also slightly thinner in this region, forming a small depression on the retina called the **fovea** [fō′vē·ə]. The fovea is the area that produces the clearest vision. Unlike the rest of the retina, the fovea contains a much higher concentration of color-sensing cone cells than rod cells.

The light-sensitive rods and cones are not connected directly to the optic nerve as one might expect. Rather, signals from the rods and cones are fed into sophisticated image-processing circuits in the middle layer of the retina, where networks of interneurons process the image to reveal motion, distinguish colors, and enhance contrast. The processed signals are then relayed to the visual centers of the brain via the **optic nerve,** which is not just one nerve but rather a large bundle of individual nerves, each of which carries nerve impulses from a specific region of the retina to the brain. At the point on the retina where the optic nerve leaves the eyeball is a **blind spot.** At that point there are no light-sensitive cells to receive visual messages.

As the optic nerves leave the retinas, they curve inward toward the center of the head where they meet at a point called the *optic chiasm* [kī′ăz′m], located just beneath the front of the hypothalamus. At the optic chiasm, some of the nerve fibers cross over to the opposite side; the fibers toward the outside of the head do not cross, but the fibers toward the inside of the head do cross. That means that each side of the brain receives impulses from both eyes; the right hemisphere receives signals from the left half of the field of view of both eyes, while the left hemisphere receives signals from the right half of the field of view of both eyes. As they leave the optic chiasm, the nerve fibers are grouped into bundles called *optic tracts* as they pass around the thalamus.

Just before the optic tracts reach the rear of the thalamus, they can stimulate other neurons to complete various visual reflexes, such as adjusting pupil size, turning the eyes toward sudden movement, or closing the eyelids to protect against an object flying at the eye. Most of the fibers continue to the rear of the thalamus and synapse with fibers that radiate to the occipital lobes of the cerebrum.

Within the occipital lobes, the millions of impulses are synthesized into an image. Specific nerve centers in this region recognize shapes and outlines, while others process colors or motion. These various nerve centers relay their results to the primary visual cortex, which integrates the information from the various nerve centers into a single image.

**The purpose of the rods.** The substance in rod cells that absorbs light is called **rhodopsin,** which consists of

Fig. 8.27 **Structure of the retina**

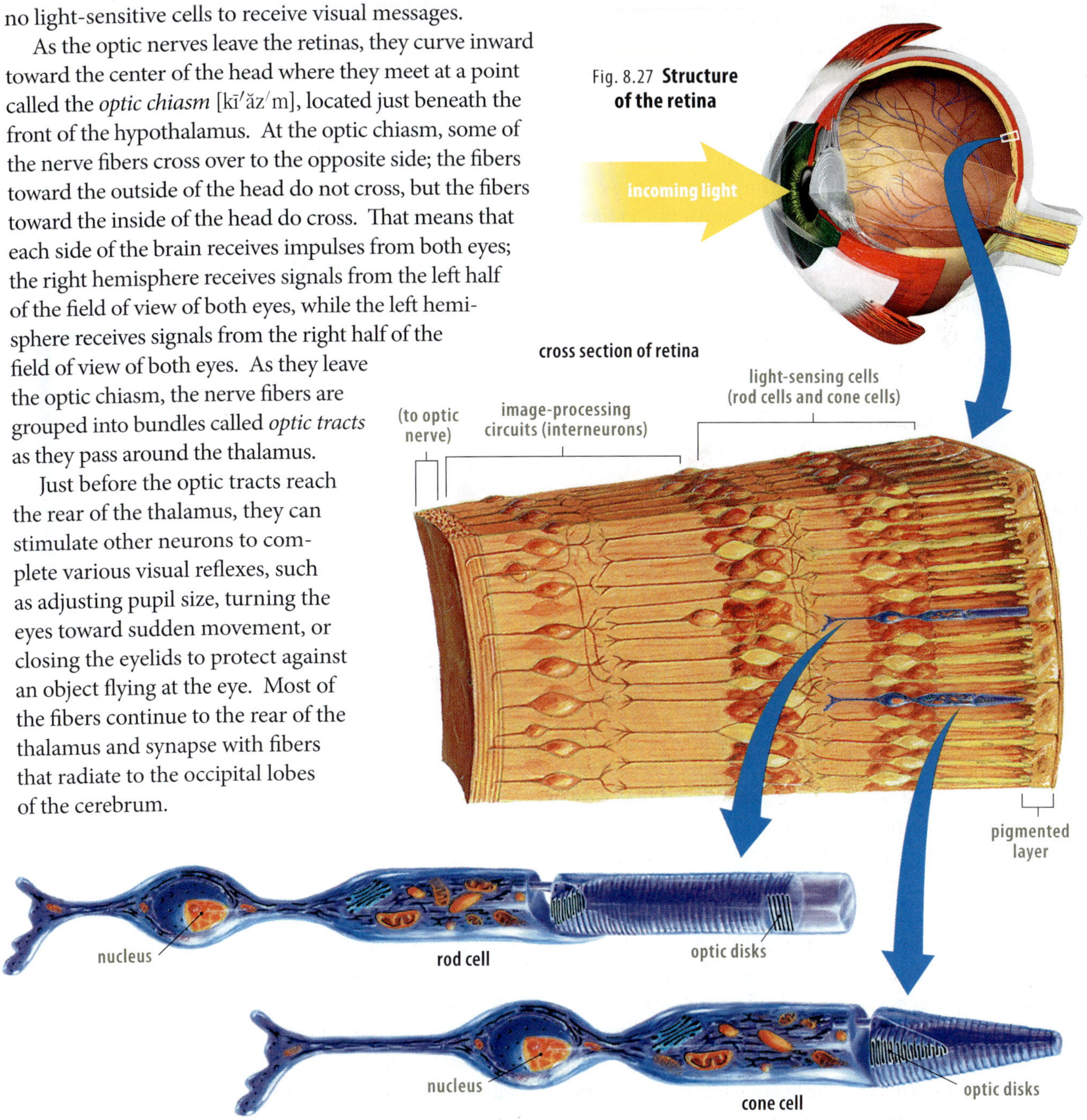

a molecule derived from *vitamin A* suspended in a protein framework. When a rhodopsin molecule is struck by a photon of light, it instantly twists into a new shape, activating a switch that causes the cell to emit a nerve impulse. Afterward, the rhodopsin molecule is reset to prepare it to detect another photon. Because rhodopsin is extremely sensitive to light (it can be triggered by a single photon), it enables us to see in near-darkness. But because *rod cells react indiscriminately to light of all colors,* they enable us to see only in shades of gray.

**Seeing color.** *The cones are responsible for our ability to see color.* The cones in the retina are of three different types, each containing molecules that are similar to rhodopsin but react most strongly to different wavelengths of light. One type of cone cell reacts most strongly to light from the red end of the spectrum, a second type reacts to the blue end of the spectrum, and a third type reacts to the middle (green) region of the spectrum. (The fact that we have three types of cones in our eyes is the reason that we see in three *primary colors.*) When red light enters our eyes, it stimulates the cones sensitive to red light to produce nerve impulses, which are transmitted to the brain along the optic nerve. The brain then recognizes the color of that light as red. The same is true for either blue or green light.

**Mixing colors.** Light of intermediate frequency stimulates more than one type of cone cell. For example, if pure yellow light enters the eye, it will stimulate both the cones that are most sensitive to red light and the cones that are most sensitive to green light. The combination of nerve impulses from the red-sensitive and green-sensitive cones is relayed to the brain, which interprets the color as yellow. (The brain will also interpret a mixture of red light and green light as yellow because the mixture stimulates both the red-sensitive and green-sensitive cones in the same fashion as pure yellow light.) Other colors result from unequal stimulation of the red, green, and blue cones. For example, if the green cones are stimulated only half as strongly as the red cones, the brain interprets the color as orange. When all three types of cones are stimulated equally, the brain recognizes the color as white.

Color vision is a rather complex process. But as our scientific knowledge of the eye steadily grows, our appreciation for God's handiwork increases, and the possible applications of science, especially medical science, multiply.

**Persistence of vision.** There is about a tenth of a second delay between the time that light from an object strikes the retina and the time that the brain receives the signal.

Fig. 8.29 **Color sensitivity of cone cells**

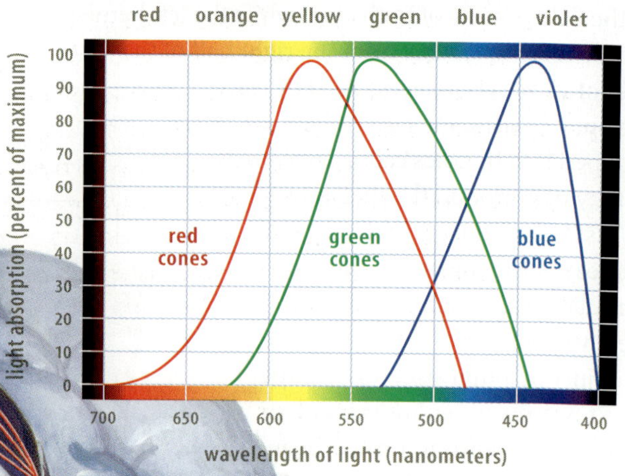


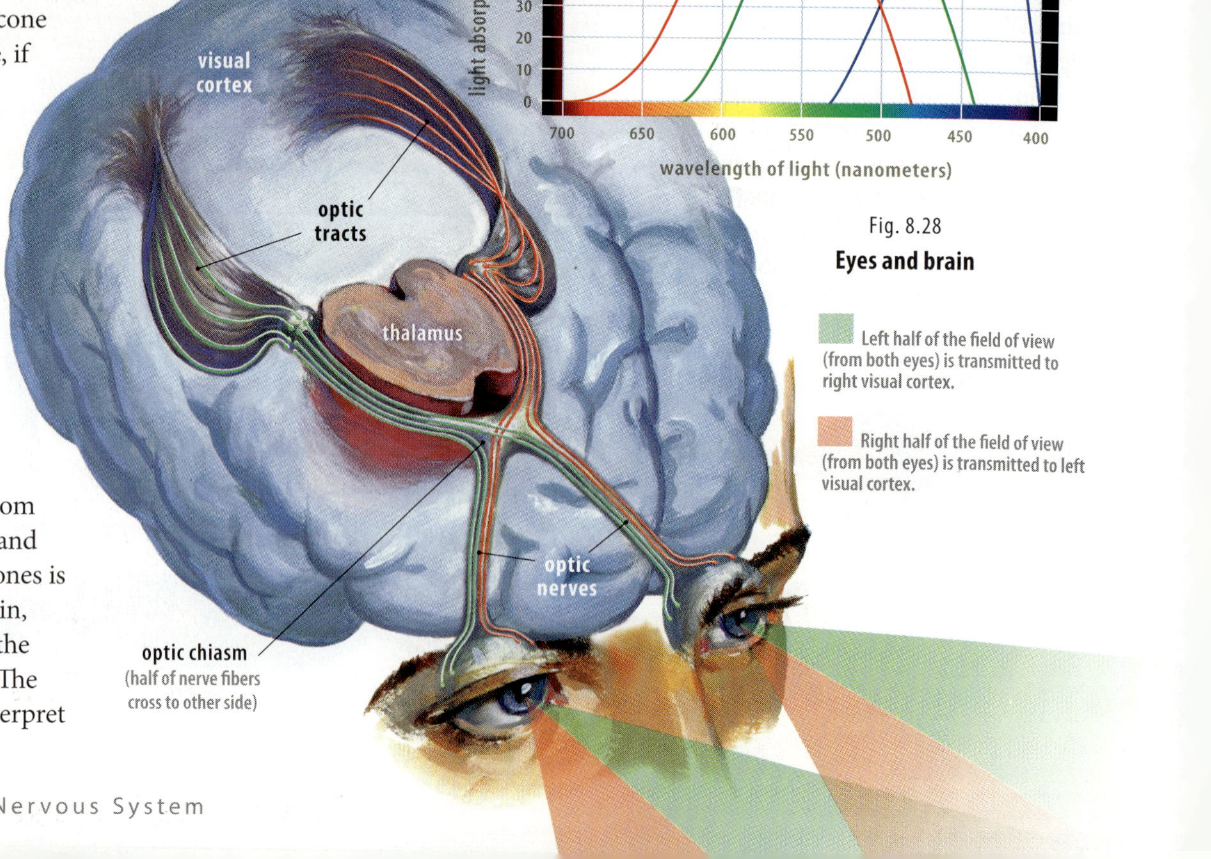


Fig. 8.28
**Eyes and brain**

Left half of the field of view (from both eyes) is transmitted to right visual cortex.

Right half of the field of view (from both eyes) is transmitted to left visual cortex.

The image or picture of the object on the retina persists about a tenth of a second after the object is gone. This phenomenon, known as **persistence of vision,** explains why a flashlight appears to leave a trail of light when moved rapidly in the dark. The ability to see a motion picture as continuous movement relies upon persistence of vision. About twenty-four individual pictures are projected onto the screen each second, but we see continuous movement because each picture is shown before the previous image has faded from the retina.

**The transparent interior.** Between the cornea and the iris and lens is a space filled with a clear fluid called the **aqueous** [ā′kwē·əs] **humor,** which is produced from blood circulating in the ciliary body. Its most important function is helping to keep the eyeball reasonably firm and in the proper shape. It also supplies the lens and cornea with necessary nutrients.

The part of the eye that focuses the rays of light passing through it is called the **lens.** The lens is connected to the ciliary body, just behind the iris. It consists of tightly packed transparent cells enclosed in a capsule of elastic tissue. Rays of light pass first through the cornea, aqueous humor, and pupil before reaching the lens. When these rays of light enter the lens, they are bent (refracted) by the lens to form an image on the retina at the back of the eye. Because focusing on nearby objects requires more curvature of the lens than is required for focusing on distant objects, some provision must be made for changing the curvature of the lens. This change in the curvature of the lens is accomplished by the contraction of the ciliary muscles. When the ciliary muscles are relaxed, the lens is partly flattened and the eye is adjusted for distant vision. When the ciliary muscles contract, the lens bulges, allowing the eye to focus clearly on nearby objects. This increase in the curvature of the lens to focus on nearby objects is called **accommodation.**

Between the lens and the rear of the eye is a space filled with **vitreous humor,** a transparent, jellylike substance that is about 98% water. Small amounts of certain kinds of protein and sugar are also present. The jellylike vitreous humor, along with the watery aqueous humor, gives the eye its semisolid structure. The vitreous humor also keeps the retina pressed against the choroid layer.

**Defective vision.** The eye may have certain defects that prevent a person from seeing normally. Some of these defects are inherited; others develop as a person grows older. The deeper we go in our scientific understanding of God's order and precision as it applies to the eye, the better we will be ready to remedy or prevent eye defects.

Sometimes the lens of the eye, in its relaxed state, is too thick and convex (curved outward on both front and back surfaces) to form a clear image on the retina; the

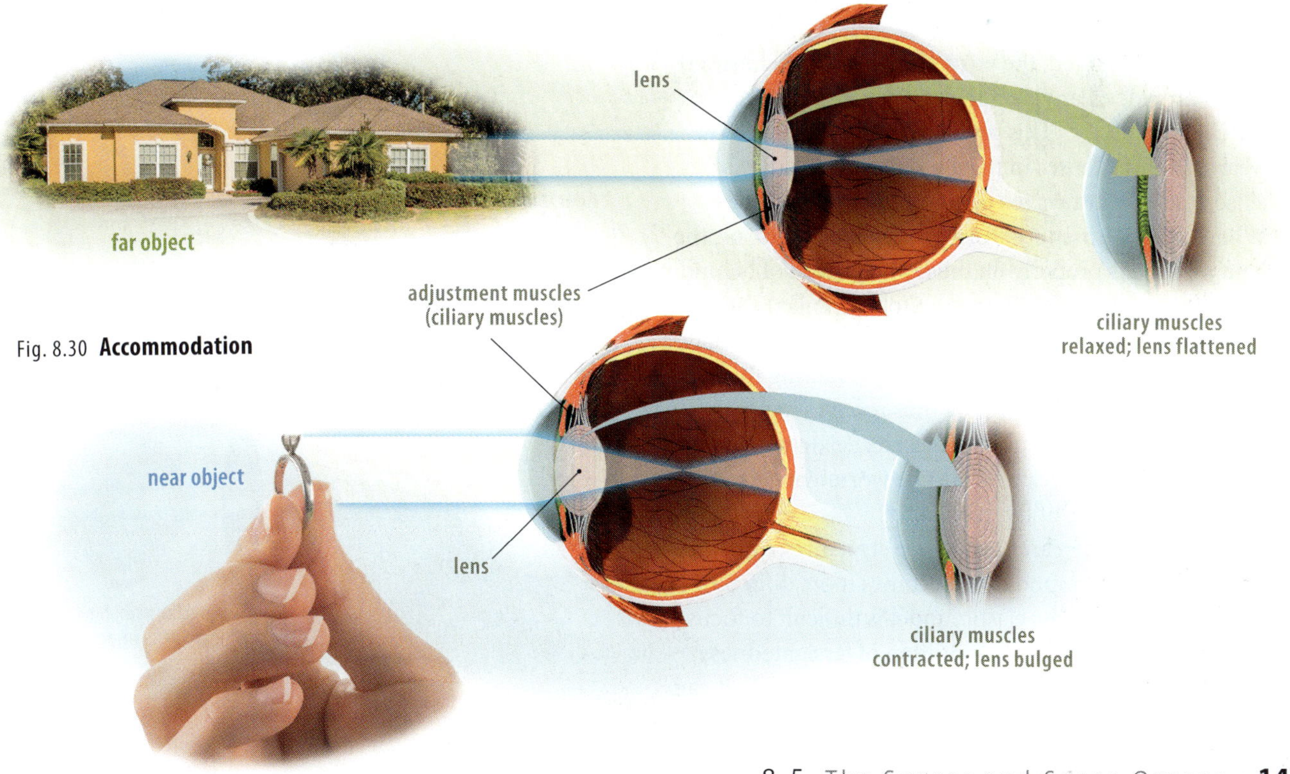


Fig. 8.30 **Accommodation**

same difficulty in focusing can also occur if the eyeball is too long from front to back. Such a condition results in **nearsightedness,** or *myopia* [mī·ō′pē·ə]—the ability to see only near objects clearly. In this condition, rays of light focus in front of the retina rather than on the retina. Since this improper focusing occurs when the ciliary muscle is relaxed, the eye cannot overcome the problem by accommodation: when the ciliary muscle contracts, additional curvature will be added to the lens, making the object focus even farther in front of the retina.

Nearsightedness can be corrected by wearing *concave* lenses in front of the eyes; these lenses are curved inward, toward the center of the lens. Such a lens causes light rays to diverge, or spread out, before they enter the eye; the focusing power of the eye will then bring these divergent rays to focus on the retina, and objects at a distance will be seen clearly. Nearsightedness can also be corrected surgically by reshaping the cornea using a computer-controlled laser.

When the eye is too short or the lens is too flat, objects are brought to focus at a point that would lie behind the retina. This results in **farsightedness,** or *hyperopia.* A person who is farsighted sees objects at a distance better than those that are nearby. Such a defect in the eye can be partly corrected by accommodation; the ciliary muscles are thus in use nearly all the time, when normally they would contract only to focus on nearby objects. Consequently, the eyes feel fatigued.

For the farsighted, the purpose of corrective lenses is less to provide clear vision than to reduce the strain on the ciliary muscles. This problem can be corrected by wearing *convex* lenses in front of the eyes. Convex lenses cause light rays to converge (come together). When convex lenses are worn by farsighted persons, the light rays focus properly on the retina instead of behind it.

Some people have a focusing problem called **astigmatism.** This is usually the result of an asymmetrically curved cornea, which causes light rays entering the eye to be bent incorrectly. The retina thus receives an image that cannot be properly focused. To correct astigmatism, a lens with special curvature must be worn that counteracts the incorrect curve of the cornea.

As a person increases in age, eye structures sometimes change, and accommodation becomes more difficult. The eye's lens becomes less elastic, and it is difficult to focus on nearby objects. Such a condition of farsightedness brought on by age is **presbyopia** [prĕz′bē·ō′pē·ə]. It can be corrected by wearing *bifocal* lenses—the upper portion for distant vision and the lower portion for near vision. Sometimes a person must wear trifocal lenses, each part of the lens for a different distance region.

People who cannot distinguish one or more of the primary colors of light (red, blue, or green) are said to be **colorblind.** This condition is about twenty times more common in men than in women. Colorblindness is usually caused by an inherited defect in certain types of cones in the retina but can also result from damage to the area of the brain where the sensations of color are received and interpreted.

Some people's eyes do not become adjusted to darkness, remaining almost totally blind in dimly lit places.

Fig. 8.31 **Correcting vision**

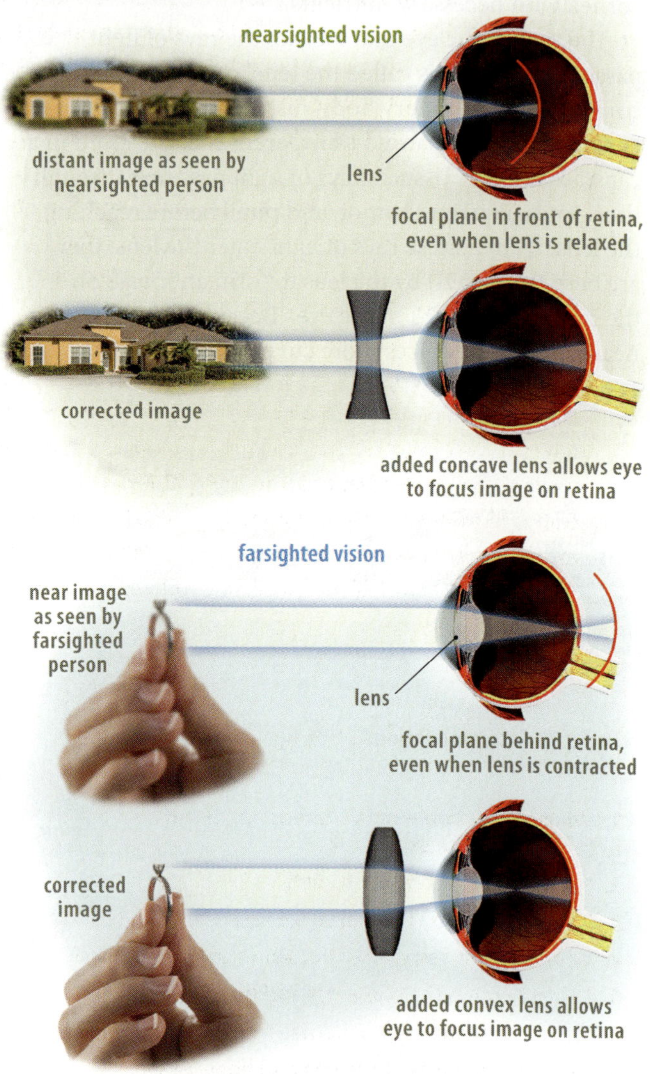

This condition is known as **night blindness.** Night blindness is usually caused by a lack of *vitamin A* in the diet. You will recall that vitamin A is needed to produce rhodopsin. Without enough rhodopsin, the rods cannot attain their maximum sensitivity. Since vision in dim light is due primarily to the action of the rods, inadequate vitamin A leads to night blindness. Night blindness can also be caused by certain hereditary eye defects.

A common cause of blindness is **glaucoma,** a condition in which the pressure of the fluid inside the eye becomes much higher than normal, causing permanent damage to the cells of the retina. If detected in time, glaucoma can be treated with drugs or surgery to prevent the person from becoming blind.

## Design in Nature *by Sir Isaac Newton (1642–1727)*

Thinking about the LIVING CREATION

Can it be by accident that all birds, beasts, and men have their right side and left side alike shaped, (except in their bowels,) and just two eyes, and no more, on either side of the face; and just two ears on either side [of] the head, and a nose with two holes; and either two fore-legs, or two wings, or two arms on the shoulders, and two legs on the hips, and no more? Whence arises this uniformity in all their outward shapes but from the counsel and contrivance of an Author? Whence is it that the eyes of all sorts of living creatures are transparent to the very bottom, and the only transparent members in the body, having on the outside a hard transparent skin, and within transparent humours, with a crystalline lens in the middle, and a pupil before the lens, all of them so finely shaped and fitted for vision, that no artist can mend them? *Did blind chance know that there was light, and what was its refraction, and fit the eyes of all creatures, after the most curious manner, to make use of it?* These, and suchlike considerations, always have, and ever will prevail with mankind, to believe that there is a Being who made all things, and has all things in his power, and who is therefore to be feared. . . .

We are, therefore, to acknowledge one God, infinite, eternal, omnipresent, omniscient, omnipotent, the Creator of all things, most wise, most just, most good, most holy. We must love him, fear him, honour him, trust in him, pray to him, give him thanks, praise him, hallow his name, obey his commandments, and set times apart for his service, as we are directed in the Third and Fourth Commandments, for this is the love of God that we keep His commandments, and His commandments are not grievous, 1 John 5:3.

From "A Short Scheme of the True Religion," quoted in Sir David Brewster, *Memoirs of the Life and Discoveries of Sir Isaac Newton* (Edinburgh: Thomas Constable and Co., 1855), 2:347–348. Emphasis added.

## Section Review 8.5

1. What are the three broad categories of sensory receptors in the skin?
2. How are pain receptors stimulated?
3. What are your two chemical senses?
4. What organ is responsible for your sense of balance?
5. Describe the parts of the ear and how we hear.
6. What is persistence of vision?
7. Describe the parts of the eye and how we see.

### Application

8. Why does a landscape look vividly colored in bright sunlight but appears only in shades of gray by moonlight?

### Critical Thinking

9. Many animals designed for night vision have only rod cells in their retinas and lack cone cells entirely. Why would this arrangement improve the animal's ability to see in near darkness? What disadvantages would such an arrangement have for seeing in daylight?

*(cont.)*

### Identify

sense organ, pain receptor, thermoreceptor, chemoreceptor, taste bud, primary odors, tinnitus, cornea, iris, pupil, fovea, rhodopsin, primary colors, persistence of vision, accommodation, astigmatism, glaucoma

### Observe

- Have someone blindfold you and touch your skin with two toothpicks about one inch apart in the following areas: back of the hand, palm of the hand, fingertip, back of the neck. Have your partner gradually shorten the distance between the two. Determine which of these body areas has the greatest concentration of touch receptors.

# Chapter 8 Review

### Define

1. neurobiologist, neurologist
2. meninges
3. cerebrospinal fluid
4. plexus, nerve center
5. blood-brain barrier
6. coma
7. synapse
8. cerebral cortex
9. limbic system
10. sensorineural deafness
11. REM sleep
12. whiplash
13. sciatica
14. stroke
15. concussion
16. taste buds
17. eardrum, cochlea
18. tinnitus
19. eustachian tube
20. cornea
21. choroid, ciliary body, iris, pupil
22. blind spot
23. rhodopsin
24. aqueous humor, vitreous humor
25. lens
26. accommodation
27. glaucoma

### Identify

1. two main divisions of the nervous system
2. the parts of a neuron
3. three types of nerve fibers
4. two divisions of the autonomic nervous system
5. three main portions of the brain stem
6. five diseases that affect the central nervous system
7. two broad groups of senses
8. three bones of the middle ear
9. three main layers of the eye
10. the area of the retina that produces the clearest vision

### Explain

1. Explain the difference between gray matter and white matter.
2. What is the function of each of the three types of neurons?
3. Briefly describe the process by which a signal travels from the dendrites of an interneuron to the dendrites of a motor neuron.
4. What are the purposes of the cerebrum, cerebellum, and brain stem?
5. How have the functions of specific parts of the cerebral lobes been discovered?
6. How do the cerebral hemispheres differ in their functions?
7. Why is the hypothalamus important?
8. Why does cell death in the brain or spinal cord cause permanent damage?
9. Why do foods sometimes seem bland when you have a cold?
10. What is the function of the semicircular canals?
11. Why are the lacrimal glands important?
12. What is the function of the auditory nerve? The olfactory nerve? The optic nerve?
13. Explain the differences between rod cells and cone cells.

### Apply

1. How is the design of having the cell bodies of the peripheral nervous system concentrated in ganglia near the spinal cord beneficial in case of injury to the extremities?
2. What problems would a person be expected to have if his body had no Schwann cells?
3. How do reflexes and pain receptors work together to help protect your body from damage?
4. Explain the difference between the brain and the mind; use Scripture to support your answer.
5. Why might alcohol be involved in large numbers of accidents and violent crimes?
6. How would a person's hearing be affected if he did not have earlobes?
7. Are the hairlike cells in the cochlea mechanoreceptors, thermoreceptors, or chemoreceptors? Explain your answer.
8. Why would a person with a punctured eardrum be advised against traveling by airplane?
9. How does the crossing of the optic nerves at the optic chiasm make binocular vision more efficient?

# Nutrition & Digestion

CHAPTER 9

**Key Concepts**

- Nutrients and nutrition
- Organs of digestion
- Processes of digestion

## 9.1 Nutrients and Nutrition

The complex machinery of our bodies' cells requires constant maintenance. Structural proteins and enzymes must be repaired or replaced; internal and external membranes must be maintained; the cell's transportation networks and structural supports must be rearranged to meet the changing needs of the cell; "power plants" and "factories" must be fueled and serviced; and errors in the cell's master program (DNA) must be found and repaired. In many areas of the body, entire cells must be continually manufactured to replace those that wear out. Together, all of these chemical processes of life are the body's **metabolism.** Metabolism requires a great deal of fuel and raw materials. We obtain fuel and raw materials from our food, which supplies the body with the elements and compounds it needs for energy, repair, and growth.

### Nutrients and Energy

The energy content of food is typically measured in **Calories** (with a capital *C*). One Calorie is the amount of energy needed to raise the temperature of one kilogram (or one liter) of water by one degree Celsius (the energy needed to raise the temperature of one pint of water by about four degrees Fahrenheit); this is one thousand of the calories (lowercase *c*) sometimes used in physics, or one *kilocalorie* (kcal). Rising from a chair, walking eight feet, and returning to the chair requires about one Calorie of energy.

If a person consumes more Calories than he needs, the excess energy-producing substances are converted to fat and stored in **adipose** [ăd′ə·pōs′: fat] **cells. Obesity** (being severely overweight due to excess body fat) increases a person's risk of diseases such as heart disease, diabetes, and some forms of cancer. *Generally, the only safe, healthy way to lose weight is to eat a healthy diet (including limiting Calorie intake) and exercise regularly.* Maintaining a healthy weight also requires proper eating and regular exercise.

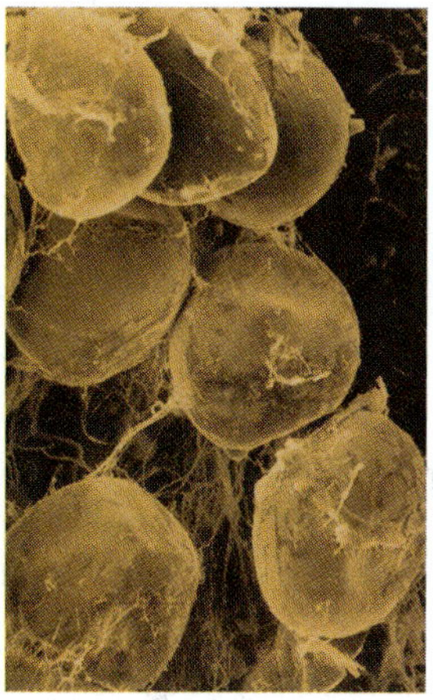

Fig. 9.1 **Adipose cells**

### Macronutrients

Nutritionists often divide the nutrients your body needs into two categories. Because they are needed in relatively large amounts, carbohydrates, proteins, and fats are considered **macronutrients.** Macronutrients are your body's primary source of energy and building materials. Although they are essential for proper body function, vitamins and minerals are needed in much smaller amounts and are therefore **micronutrients.** Your body also requires adequate *water.*

**Carbohydrates. Carbohydrates,** the most important energy-producing compounds in the cell, provide most of the energy for living things. All carbohydrates are organic compounds that consist of the elements carbon, hydrogen, and oxygen in about a 1:2:1 ratio. The molecules of most carbohydrates are like chains; each link in the chain is a simple sugar. Common **sugars**

## Table 9.1 Types of Nutrients

| Nutrient | Primary function |
|---|---|
| carbohydrates | provide energy |
| proteins | provide raw materials to build tissues |
| fats and oils | provide energy and raw materials to build tissues |
| vitamins | necessary for growth and body functions |
| minerals | regulate body functions and provide materials for bone and tooth construction and muscle contraction |
| water | provides a medium for chemical reactions and transport of materials |

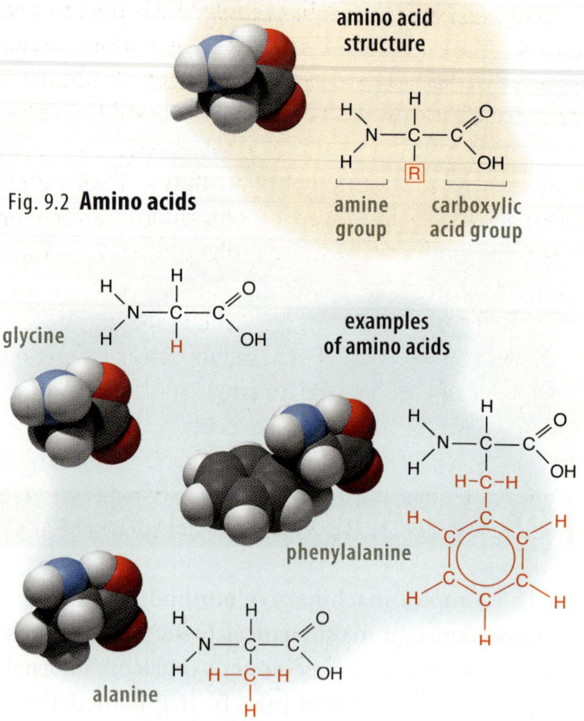


Fig. 9.2 **Amino acids**

include *glucose* ($C_6H_{12}O_6$), *fructose* (fruit and vegetable sugar), *lactose* (milk sugar), *maltose* (found in grains), and *sucrose* (table sugar, refined from sugar beets or sugar cane). Both *monosaccharides* [mŏn′ə·săk′ə·rīdz′: simple sugars] and *disaccharides* [dī·săk′ə·rīdz′: two monosaccharides combined] are **simple carbohydrates.**

*Polysaccharides* [pŏl′ē·săk′ə·rīdz′], formed from long chains of simple carbohydrates, are **complex carbohydrates** and are found in such foods as bread, cereals, pasta, potatoes, and rice. Two main types of polysaccharides are important nutrients. **Starch** is a mixture of polysaccharides manufactured by plant cells for glucose storage; the body breaks starch down into glucose for use as fuel. **Dietary fiber** is not digested or incorporated into the body. There are two types of dietary fiber. *Soluble fiber* turns into a gel during digestion, thereby slowing digestion and helping the stomach and intestines absorb nutrients. *Insoluble fiber* helps move undigested wastes along to keep the intestines clean and healthy. *Cellulose,* the primary component of plant cell walls, is a form of insoluble fiber.

**Proteins.** About half of the body's dry weight is composed of **proteins,** complex organic molecules used to build and maintain living cells. They are composed of carbon, hydrogen, oxygen, nitrogen, and sometimes other elements. Proteins serve many vital functions in the cell. Relatively simple proteins are used like cables or modular building blocks to construct the framework of the cell, while more complex proteins called *enzymes* control chemical reactions in the cell. Some of the most complex proteins serve as molecular machines—pumps, actuators, motors, sensors, and molecular assemblers that allow the cell to carry on the processes of life.

Proteins are made of simpler molecules called **amino** [ə·mē′nō] **acids.** A protein is a long chain of amino acids, usually folded into a three-dimensional shape. About 20 different amino acids are combined in various ways to produce different proteins. Because a single protein may be composed of hundreds or thousands of amino acids, proteins are among the most complex molecules in the cell.

Amino acids are all quite similar to each other, differing only in the type, number, and arrangement of a particular group of atoms in the molecule; this variable group is represented by the letter *R* in Figure 9.2. In the amino acid *glycine,* a single hydrogen atom (–H) is attached here. In *alanine,* a group of four atoms ($-CH_3$) is attached here. In other amino acids, the variable group is more complicated.

**Fats and other lipids.** Fats, oils, and fatlike substances are **lipids** [lĭp′ĭdz], a varied group of compounds characterized by their insolubility (inability to dissolve) in water. Lipids are used by the body primarily to store energy and to build cell parts; they also carry fat-soluble vitamins to all parts of the body. Some hormones, which carry chemical messages throughout the body, are lipids. All cell membranes are built mainly of lipids.

The simplest type of lipid molecule is the **fatty acid,** a molecule made of carbon and oxygen atoms attached to a long chain of carbon and hydrogen atoms. In **saturated fatty acids,** every possible bond to hydrogen is occupied; they are said to be saturated because they cannot hold any more hydrogen atoms. In **unsaturated fatty acids,** this hydrogen-carbon chain can form one *(monounsaturated)* or more *(polyunsaturated)* additional pairs of bonds to hydrogen.

In cells, fatty acids are combined into larger molecules called **fats.** Because fats consist of three fatty acid molecules attached to a single molecule called glycerol (Fig. 9.3), they are also called *triglycerides* [trī·glĭs′ə·rīdz′]. Animal fats, which are usually solid at room temperature, consist mostly of saturated fatty acids. Fats that are liquid at room temperature are called **oils.** Hydrogen is often added to vegetable oils to change them from liquid to solid or semisolid; such **hydrogenated oils** are used to make margarine and shortening. Most processed food, including fast food, bakery items, and snack foods, contain hydrogenated oils. Hydrogenated oils sometimes contain chemically altered *trans fatty acids,* which have been linked to heart disease and cancer.

Another lipid, **cholesterol,** is used by our bodies to make bile, vitamin D, and important *hormones* (chemical messengers); small amounts of cholesterol are also used in cell membranes. Your liver manufactures cholesterol if you do not consume it in your food. Cholesterol is an important nutrient, but too much of it is associated with *atherosclerosis* [ăth′ə·rō·sklə·rō′sĭs: a buildup of deposits inside blood vessels] and coronary heart disease. Although cholesterol is found only in animal fats, saturated plant fats can also cause high cholesterol levels by stimulating the body to produce cholesterol.

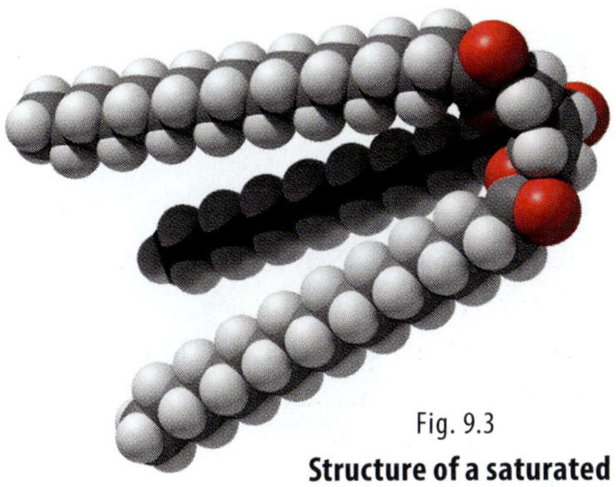

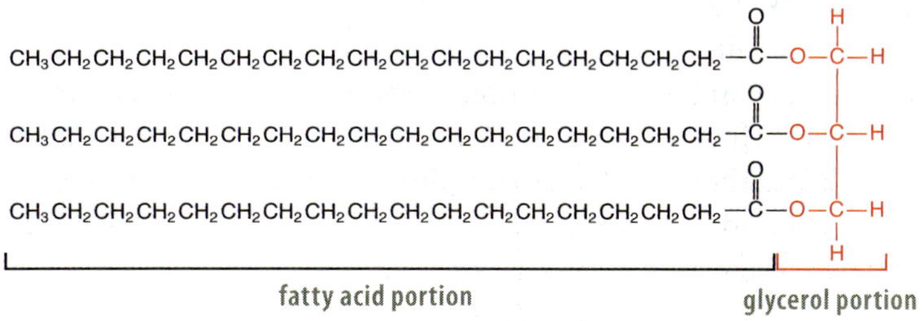


Fig. 9.3
**Structure of a saturated fat molecule (tristearin)**

## *ATP (Adenosine Triphosphate)*

Understanding the LIVING CREATION

Merely burning carbohydrates and fats to release heat would be useless for a cell because the cell needs chemical energy to drive its chemical processes. Instead, the cell breaks down its fuels in small steps, releasing a small, controlled amount of chemical energy at each step; this energy is used to synthesize a special phosphorus compound called *adenosine triphosphate* [ə·dĕn′ə·sēn′ trī·fŏs′fāt′], or **ATP.** ATP is the energy carrier of the cell and is a convenient form for the temporary storage of chemical energy. ATP molecules can, in a sense, be thought of as the "energy currency" of the cell.

ATP is a *nucleotide,* a molecule constructed of a nitrogen-containing base, a sugar called *ribose,* and three *phosphate* groups. The energy of an ATP molecule can be released by breaking a phosphate group off the molecule, forming *ADP* (adenosine diphosphate) and a free phosphate group.

When the cell needs energy for a process, it "plugs" an ATP molecule into a special "slot" on the enzyme or motor protein. The ATP molecule is then split, transferring its energy directly to the enzyme or mechanism, which carries out the desired process. After the ATP molecule is split, the resulting ADP and free phosphate group are ejected from the "slot" and transported back to a "power

*(cont.)*

(cont. from previous page)

plant" *(mitochondrion)* of the cell to be "recharged," or recombined into ATP, by *ATP synthase* [sĭn′thās′], a "generator" driven by the energy of glucose breakdown.

This cycle of ATP breakdown and recombination is vital to living things. ATP energizes muscle contraction, nerve signal transfer, DNA replication, protein manufacture, and nearly all the other functions of living cells in our bodies.

**formula of ATP**

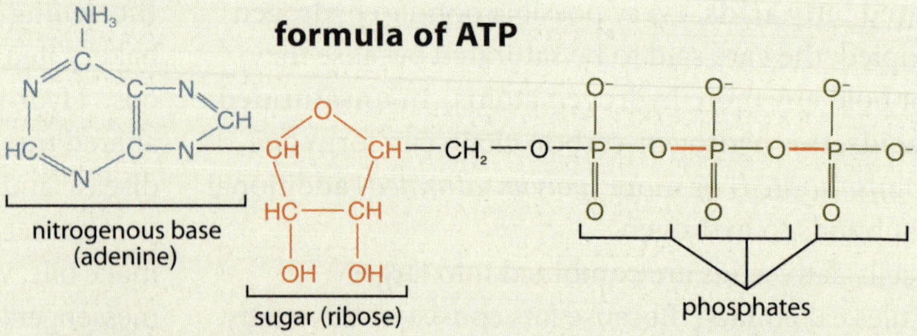


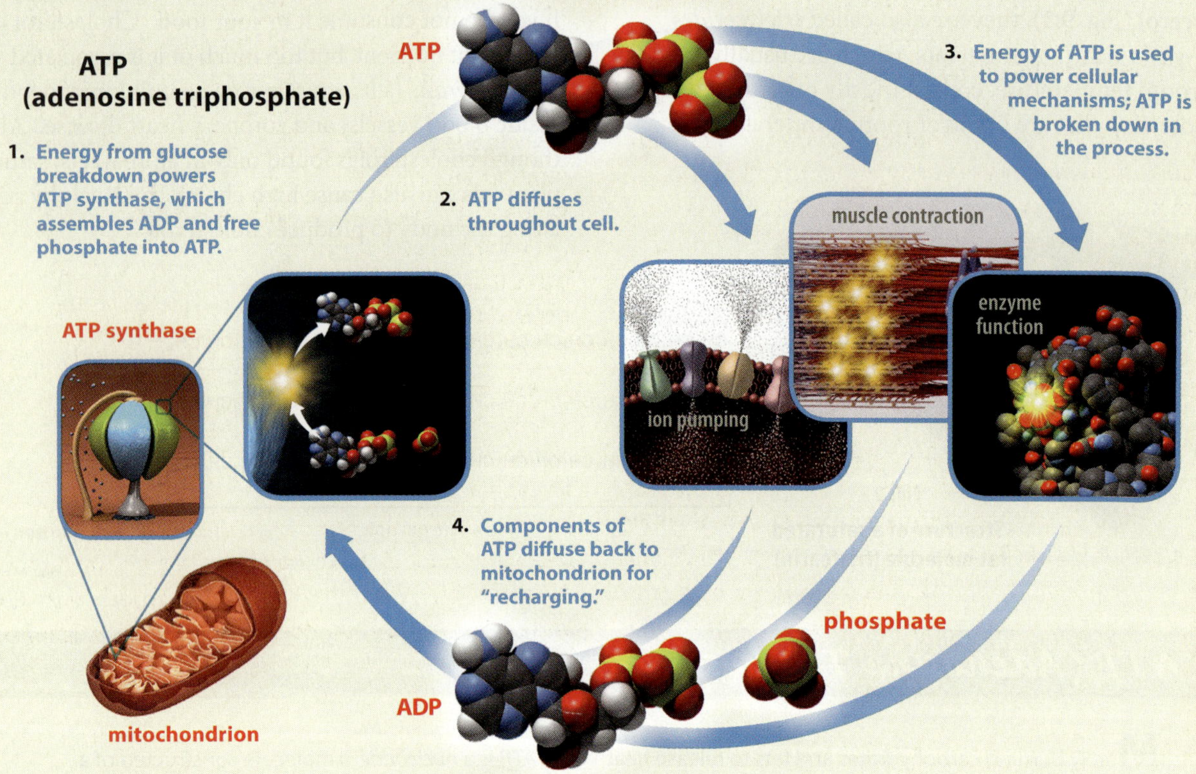


## Micronutrients

**Vitamins. Vitamins** are important organic substances found in plants and animals and foods made from them. Vitamins are needed in small amounts for the proper functioning of metabolism. Although only small amounts of vitamins are needed, they are just as important as the macronutrients. A deficiency of vitamins may cause illness or even death.

Vitamins can be divided into two groups. The **water-soluble vitamins,** which are easily eliminated from the body, are *vitamin C* and the eight *B-complex vitamins:* thiamin [thī′ə·mĭn: vitamin $B_1$], riboflavin [rī′bō·flā′vĭn: vitamin $B_2$], niacin (vitamin $B_3$), folate (also called folic acid), biotin, pantothenic acid, vitamin $B_6$ (pyroxidine), and vitamin $B_{12}$ (cobalamin). The **fat-soluble vitamins,** which can accumulate in the body if ingested in excess, are *vitamins A, D, E,* and *K.*

Most vitamins serve as **coenzymes,** molecules that assist a cell's enzymes in performing their jobs. Some vitamins (such as biotin) function as parts of enzymes, while others (such as niacin and vitamin $B_{12}$) function independently. Some vitamins (particularly vitamins C and E) are **antioxidants,** substances that neutralize harmful molecules called *free radicals,* which would otherwise damage important cell parts. Vitamin C neutralizes free radicals dissolved in water, while vitamin E neutralizes free radicals dissolved in fats and other lipids. Because free radicals are natural byproducts of cell processes, antioxidants such as vitamins C and E are an important part of our diet.

**Minerals. Minerals** are important inorganic nutrients needed for proper growth and repair of body tissues. *Calcium, phosphorus, iron,* and *iodine* are important minerals for building body tissues. Other important minerals include *sodium, potassium, magnesium, sulfur,* and *chlorine.* Iron and iodine are called **trace elements** because, though they are vitally important, they are needed in extremely small quantities. Other trace elements include zinc, copper, manganese, selenium, molybdenum, and chromium.

Like vitamins, small amounts of minerals are essential for good health. Atoms of some minerals, such as zinc, iron, magnesium, and iodine, are key parts of many proteins that your body produces. Iron, for example, is a part of *hemoglobin* [hē′mə·glō′bĭn], the protein that transports oxygen through your body. Some other minerals, such as sodium and potassium, are used to help regulate the balance of water in your body fluids, both inside and outside cells. Finally, several minerals, such as calcium and phosphorus, are used as raw materials in the production and repair of bones and teeth.

## Water

Normally, much of a person's body weight is **water.** The percentage of water in the body is related to the amount of fat in the body; a body that has more fat will have less water in its composition. Nearly 60% of the total body water is *intracellular* (inside the body's cells) *fluid.* The *extracellular fluid,* which surrounds the cells, is found primarily in blood plasma, lymph, cerebrospinal fluid, and the joints. All body fluids contain large amounts of water, and most of the chemical activities carried on by the body depend upon water. Water is necessary for proper digestion and absorption of food, for transporting nutrients to the body's cells, for constructing and repairing cells, for lubricating joints, and for regulating body temperature. When the body tissues absorb extra fluids and cannot maintain proper water balance, a condition called **edema** [ĭ·dē′mə] results.

## Section Review 9.1

1. How does the body respond when a person consumes more Calories than he needs?
2. What is the function of carbohydrates in our diet?
3. Explain why dietary fiber is necessary in the diet.
4. What is the relationship between amino acids and proteins?
5. What is the difference between fats and fatty acids?
6. What is the function of ATP in a cell?
7. List the water-soluble and fat-soluble vitamins.
8. Explain the importance of the minerals calcium, phosphorus, iron, iodine, sodium, potassium, and magnesium in body metabolism.
9. Why is water necessary for life?

### Application

10. Explain why daily intake of some vitamins is more important than daily intake of other vitamins.

### Identify

metabolism, Calorie, obesity, starch, cellulose, lipid, unsaturated fat, oil, hydrogenated oil, cholesterol, trace element, intracellular fluid, extracellular fluid, edema

### Observe

- Rub a piece of hamburger against a piece of brown paper. A greasy spot indicates the presence of fat in the meat.

# 9.2 The Beginning of the Digestive System

The complex nutrients we ingest must be broken down into smaller water-soluble substances before they can be absorbed and used by the body. The digestive system does this job: carbohydrates are converted into simple sugars, proteins into amino acids, and fats into fatty acids.

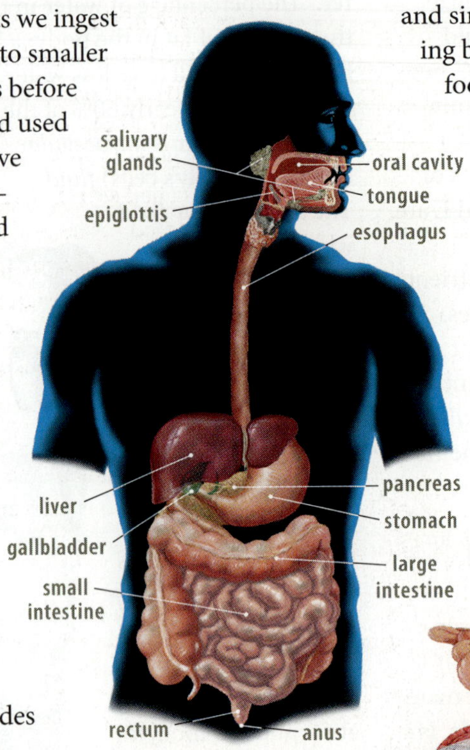


Fig. 9.4
**Organs of digestion**

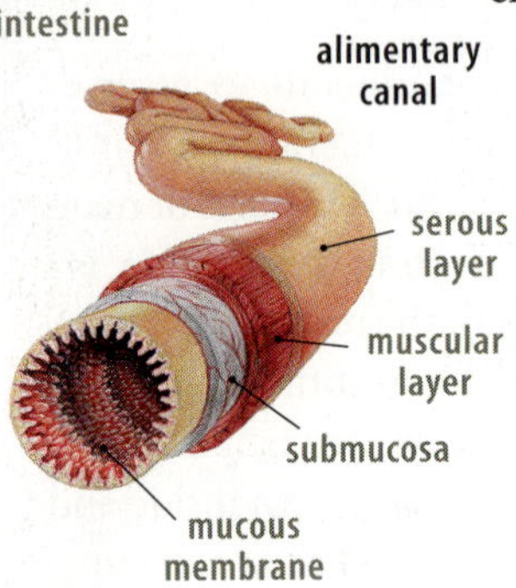


## The Alimentary Canal

The organs of the digestive system form a tube that extends from the mouth to the anus. This tube, called the **alimentary canal,** includes the oral cavity (mouth), esophagus, stomach, and the small and large intestines. Other organs and glands along the alimentary canal aid the digestive process. Four layers of tissue, shown in Figure 9.4, compose the alimentary canal. The purpose of the alimentary canal is to aid the body with the processing of food *(digestion),* the intake of nutrients *(absorption),* and the *elimination* of wastes.

## Digestion—A Chemical Process

**Digestion** is the changing of food substances with large, complex chemical molecules into substances with smaller, less complex molecules. The molecules of digested substances must be small enough physically to diffuse through cell membranes and simple enough chemically to serve as the "building blocks" of the cell. This changing of complex food molecules into simple food molecules by the digestive system is primarily a chemical process. Digestion is partly mechanical, however, in that food is broken down into small particles by the chewing action of the teeth; food is continually mixed by a muscular action called *peristalsis;* and fats are melted by heat and broken into smaller droplets *(emulsified)* by the action of *bile* (a soaplike substance formed by the liver). The chemical process begins in the mouth, where saliva starts to break down starch, and continues in the stomach, where proteins and starches are broken down into smaller parts. However, the majority of chemical digestion takes place in the small intestine.

Fig. 9.5
**Mechanical and chemical digestion**

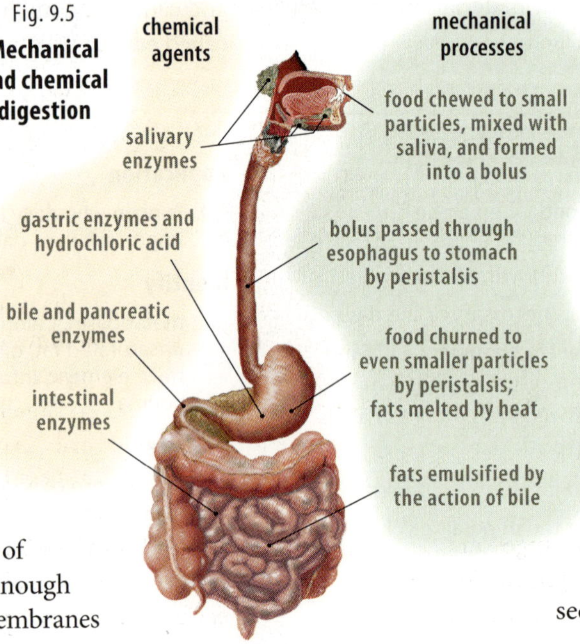


Digestion takes place in a series of small steps that are controlled by special proteins called **enzymes.** Enzymes are catalysts produced by living cells. (A **catalyst** is any substance that speeds up a chemical change but is not consumed in the process.) Many chemical changes (reactions) must take place before the large molecules are small enough and simple enough to be useful. Enzymes control the numerous step-by-step chemical changes involved in digestion.

The **digestive enzymes,** which control digestion, are produced by tiny secreting bodies called

**digestive glands.** These glands are located in the mouth, the stomach, the small intestine, and the pancreas. Although digestive glands synthesize and secrete digestive enzymes in small quantities, they secrete *digestive juices* in relatively large quantities. A digestive juice is mostly water, but it contains a small amount of one or more digestive enzymes and frequently certain other substances.

### Important Facts about Enzymes

- Enzymes are proteins produced by living cells.
- Enzymes speed up chemical changes in other substances.
- Enzymes are not consumed during the chemical reactions they control.

## The Oral Cavity

Inside the *mouth,* or **oral cavity,** food is lubricated and ground and digestion is begun. The mucous membrane, which lines the oral cavity, is kept moist and lubricated by a slimy substance called **mucus,** which is composed mainly of a compound of protein and sugar and is produced by the membrane's special secreting cells. Separating the mouth from the nasal passages is the **palate.** The palate is the roof of the mouth and works with the tongue, lips, and teeth to form words for speech.

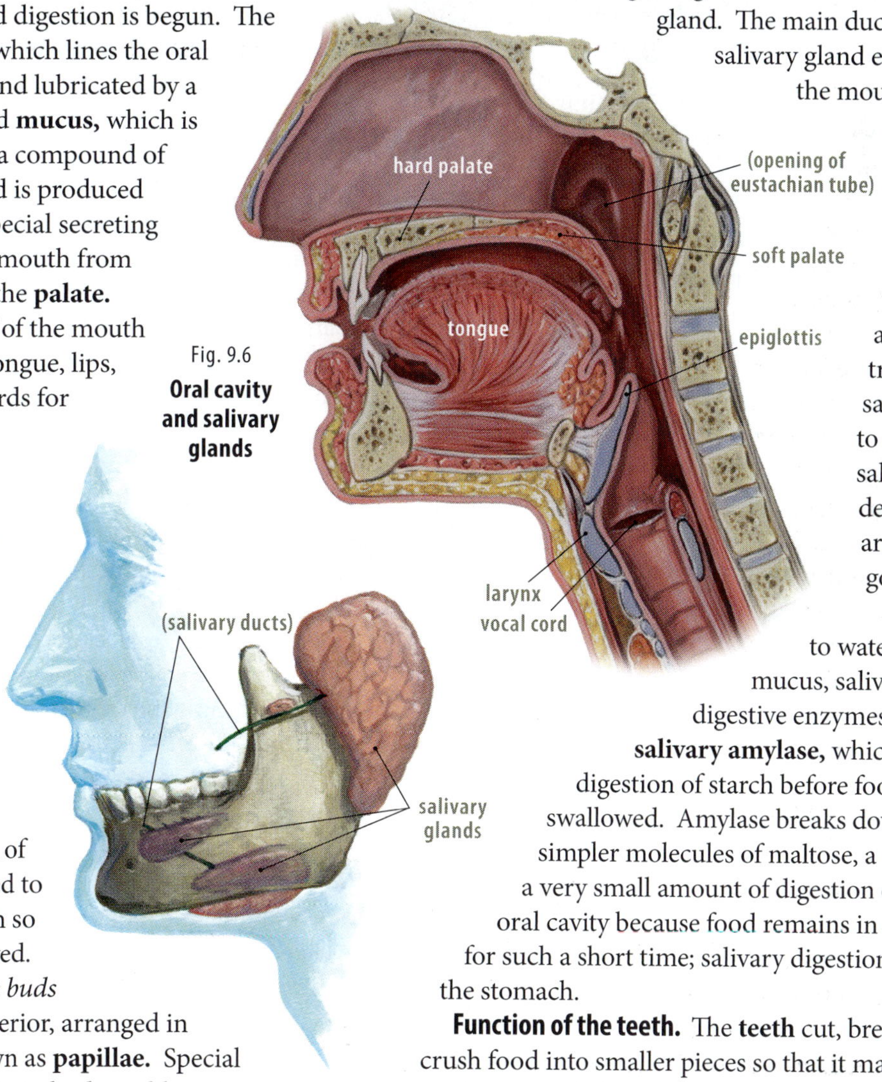


Fig. 9.6
**Oral cavity and salivary glands**

**Function of the tongue.** The **tongue** pushes food between the teeth and helps mix the food with mucus and saliva. Once the food has been broken down by the teeth, it is formed into a small ball, or **bolus** [bō′ləs], by the action of the tongue and pushed to the back of the mouth so that it can be swallowed.

Thousands of *taste buds* cover the tongue's exterior, arranged in tiny projections known as **papillae.** Special nerve endings in the taste buds enable us to taste the five fundamental tastes—sweet, sour, bitter, salty, and umami. The papillae also give the tongue a rough surface, allowing it to more easily move food to the back of the mouth.

**Function of the salivary glands.** The digestive glands of the mouth are the **salivary** [săl′ə·vĕr′ē] **glands,** which daily produce 1 to 1.5 liters (2 to 3 pints) of a sticky digestive juice called **saliva.** Six paired salivary glands are located in the oral cavity—two in the upper jaw, two in the lower jaw, and two under the tongue. Each salivary gland consists of many small clusters of secreting cells, which resemble clusters of grapes. Each cluster has its own small duct for carrying its saliva to a single large duct that serves the whole gland. The main duct from each salivary gland empties into the mouth. Saliva moistens food and makes it easy to swallow. Because foods that smell and taste good trigger the salivary glands to secrete more saliva, attractive, delicious foods are an aid to good digestion.

In addition to water and some mucus, saliva contains digestive enzymes such as **salivary amylase,** which begins the digestion of starch before food is even swallowed. Amylase breaks down starch into simpler molecules of maltose, a sugar. Only a very small amount of digestion occurs in the oral cavity because food remains in the mouth for such a short time; salivary digestion continues in the stomach.

**Function of the teeth.** The **teeth** cut, break, or crush food into smaller pieces so that it may be more easily digested. An adult normally has 32 teeth,

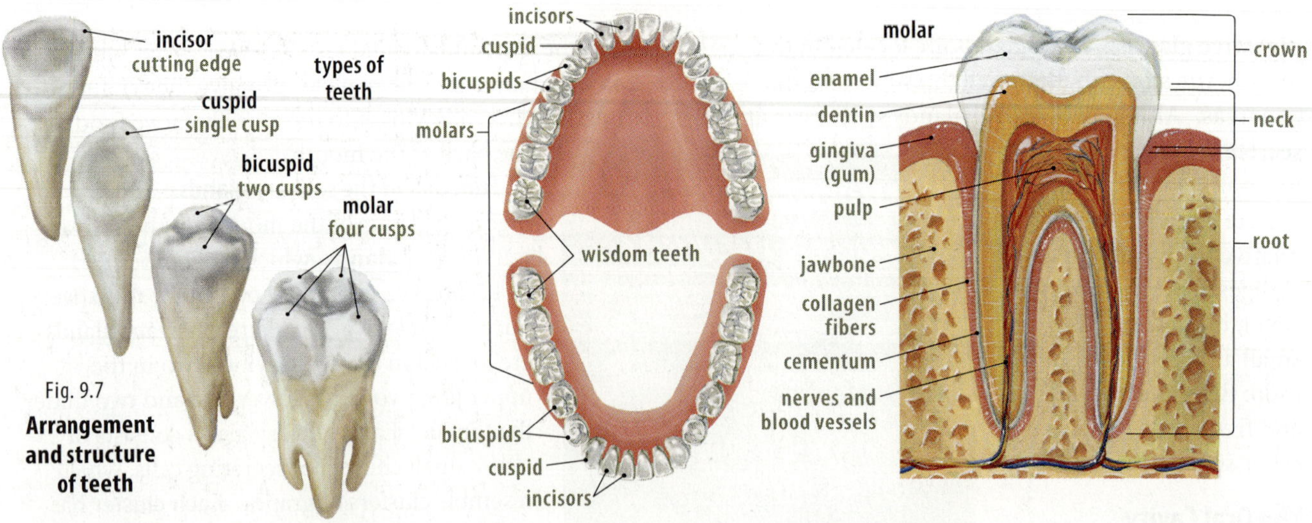


Fig. 9.7
**Arrangement and structure of teeth**

which are arranged in a semicircle on the upper and lower jaws. The number and kinds of teeth on each jaw are as follows:

**incisors** [ĭn·sī′zərz]—the four front cutting teeth;
**cuspids** (canines)—the two tearing teeth that flank the incisors;
**bicuspids** (premolars)—the four light-grinding teeth that are arranged as a pair on each side of the jaw next to the cuspids; and
**molars**—the six heavy-grinding teeth that are arranged three to a side at the back of the jaw.

The last molar on each side of the jaw is called a **wisdom tooth.** Wisdom teeth usually appear between the ages of seventeen and twenty-five; in some cases, they do not develop. Sometimes the wisdom teeth are too large for the space left on the jaw and have to be pulled. At other times, they may erupt too close to the adjacent molar and become stuck, so that they fail to grow up through the gums; wisdom teeth in this position are referred to as **impacted.**

Most of us have two sets of teeth during our lifetime. The first set, called the *primary teeth,* appear during the first four or five years of life and are lost by the age of fourteen. The roots of the primary teeth are slowly reabsorbed as the *permanent teeth,* which form on the jaw beneath the primary teeth, prepare to erupt.

Our teeth allow us to chew food and mix it with saliva. Because digestive enzymes must penetrate the food molecules, **mastication** [măs′tĭ·kā′shən: chewing], which breaks the food into small particles, increases the surface area of the food and thus aids digestion. Food that has been gulped down in chunks is more difficult for the digestive enzymes to break down than food that has been sufficiently chewed.

The general parts of a tooth are the **crown,** the **neck,** and the **root.** The crown is the part exposed above the gums, the neck is the part at the gum line, and the root anchors the tooth below the gum. In cross section, the tooth reveals the layers of materials of which it is constructed. The **pulp,** which consists mainly of blood vessels and nerves, is the inner part of the tooth. The **dentin** [dĕn′tĭn] is the bonelike layer that surrounds the pulp and forms the bulk of the tooth. The **enamel,** a mineral substance that covers the crown, is the hardest substance in the human body. The root of the tooth is fastened to the jaw by collagen fibers and a bonelike substance called **cementum.**

*Fluoride* is often added to drinking water and toothpaste to make the teeth more resistant to decay. Fluoride ions react chemically with the mineral crystals of the teeth, causing them to become roughly three times more resistant to decay. As a tooth develops in the gums, it can acquire fluoride in the enamel, dentin, and pulp, strengthening the entire tooth. After a tooth has erupted, only the enamel that comes in contact with the fluoride is strengthened and made more resistant to decay.

**The pharynx.** The *throat,* or **pharynx** [făr′ĭngks], is a funnel-shaped cavity at the back of the oral cavity. It has six openings: the mouth, the nasal passage, two *eustachian tubes,* the opening to the windpipe, and the opening to the esophagus. The eustachian tubes, which lead to the middle ears, help to equalize the pressure on the eardrum. It is to the pharynx that a food bolus is pushed by the tongue in preparation for swallowing.

# Life and Death of Teeth

Understanding the LIVING CREATION

God designed permanent teeth to last a lifetime; therefore, with proper nutrition and good dental hygiene, your teeth can remain strong and attractive. The ages from thirteen to eighteen are a critical time for oral health because decay, which occurs most often during these years, damages the permanent teeth. *Tooth decay,* or **dental caries** (commonly known as *cavities*), is the most common disease in man. The decay process begins when sugar mixes with **plaque,** a sticky, colorless film of bacteria (*Streptococcus mutans* and other types of bacteria) that is constantly forming on and between the teeth. Bacteria in plaque break down the sugars in foods to produce energy, forming *lactic acid* as a byproduct; the acid is trapped against the surface of the teeth by the sticky plaque. (Starches such as breads and crackers can also cause tooth decay because the mouth's digestive enzymes break down the starches into sugars that the bacteria can use.) After many acid attacks, tiny holes may be formed in the enamel, allowing the acid to attack the much softer dentin beneath the enamel. Once the protective enamel is penetrated, cavities may form relatively quickly. A toothache will occur when the decay reaches the nerves at the core of the tooth.

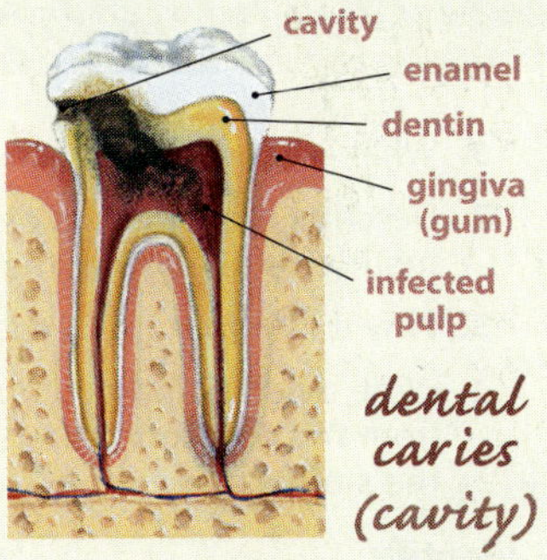


It is not primarily the amount of sugars and starches consumed that causes dental caries but rather how often they are consumed and how long they remain in the mouth. Each time such foods are consumed, more acid forms on the teeth. Sticky sweets such as soft candy; raisins and other dried fruits; mints; toffee and caramel desserts; and peanut butter and jelly sandwiches can be very harmful if they are allowed to remain on the teeth; acid could be attacking the teeth all day long. If sugar-sweetened soft drinks are sipped in small amounts throughout the day, the sugars may likewise allow bacteria to attack the tooth enamel all day.

Plaque is even more harmful if allowed to remain on the teeth for twenty-four hours; extended neglect or improper oral hygiene can result in *periodontal* [pĕr′ē·ə·dŏn′tl: gum] *disease.* Any plaque not removed by daily brushing and flossing begins to build up and form a hard deposit called **tartar** (known medically as *calculus*). When plaque and tartar accumulate, the gums become tender and inflamed and are likely to bleed easily when the teeth are brushed. This periodontal disease is **gingivitis** [jĭn′jə·vī′tĭs]; if left untreated, the infection can cause **periodontitis,** a disease in which the gums detach from the teeth and eventually the bones supporting the teeth are destroyed.

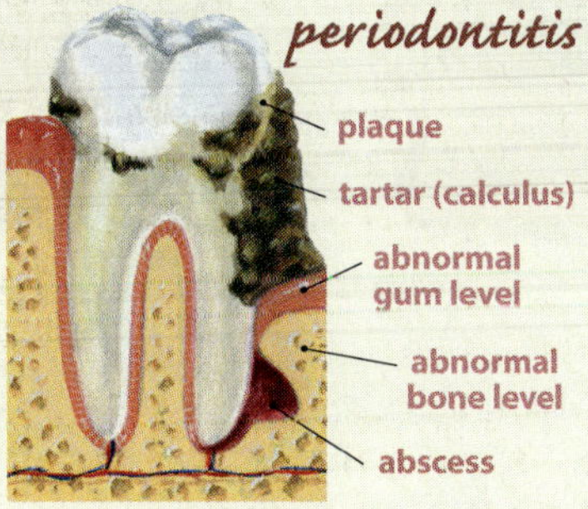


Healthy teeth and gums are important for good nutrition, and good nutrition is important for healthy teeth and gums. A poor diet can weaken the gums and bones supporting the teeth, making them more susceptible to periodontal disease. If sugar-rich foods are limited to mealtimes, the saliva that is flowing readily by the end of a meal helps to dilute the acid caused by bacteria. Brushing your teeth immediately after every meal to remove acid-forming plaque will also help preserve your teeth from decay; loss of teeth from periodontal disease can be prevented by regularly flossing your teeth and visiting the dentist periodically to have any tartar buildup removed. If you remove plaque daily, eat a balanced diet, and reduce or eliminate the consumption of sugar between meals, your teeth can last a lifetime.

dental hygiene

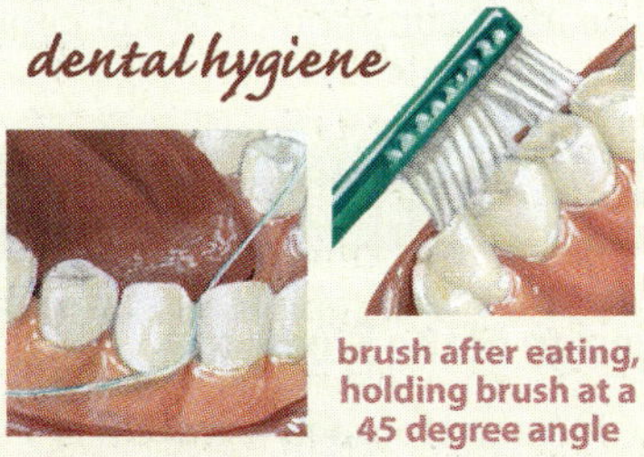

floss between teeth daily

brush after eating, holding brush at a 45 degree angle

Fig. 9.8 **Swallowing and peristalsis**

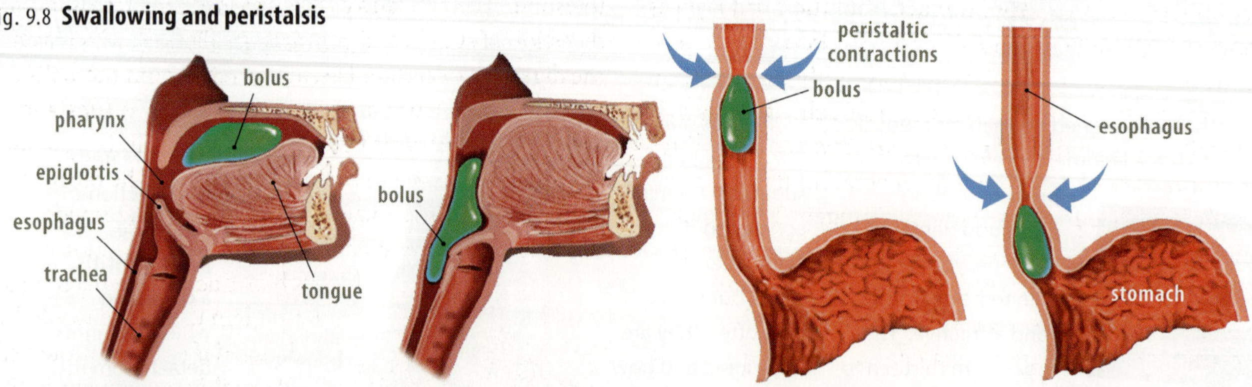


## The Esophagus

Once food has been lubricated with mucus and saliva, broken into small particles by chewing, and formed into a bolus and pushed to the pharynx by the tongue, it moves quickly to the esophagus. The **esophagus,** a foot-long muscular tube connecting the pharynx to the stomach, lies behind the trachea (windpipe). When you swallow, the esophagus takes over and you lose control of what happens to the food.

Two sets of involuntary muscles produce a rhythmic series of contractions that forces food down the esophagus toward the stomach. This rhythmic wave of muscular contractions of the esophagus and digestive tract is called **peristalsis** [pĕr′ĭ·stôl′sĭs]. Peristalsis is the method by which food is moved through all regions of the digestive tract. Before it reaches the stomach, the esophagus passes through the diaphragm, which divides the chest cavity from the abdominal cavity.

**Blocking the windpipe.** Swallowing is an automatic reflex response that occurs when chewed food is pushed into the pharynx by the tongue. A small flap of cartilage in the pharynx, the **epiglottis,** closes the top of the windpipe as you swallow to prevent food from entering the respiratory tract. When food is sucked into the windpipe instead of being swallowed, choking can occur. The most common causes of choking are laughing while eating, consuming excessive amounts of alcohol while eating, and allowing children to play with candy or small objects in their mouths. A choking victim cannot cough, speak, or breathe.

### Section Review 9.2

1. What are the three functions of the alimentary canal?
2. Describe the functions of the tongue, salivary glands, teeth, and pharynx in the process of digestion.
3. Why it is important to chew one's food well?
4. Write a paragraph describing the process of digestion from the time food enters the mouth until it reaches the stomach.

#### Identify

digestion, mucus, palate, bolus, mastication, gingivitis, periodontitis, esophagus, peristalsis, epiglottis

#### Observe

- Using a mirror, identify your incisors, cuspids, bicuspids, and molars. Count the number of teeth in each jaw and note whether or not your wisdom teeth have appeared yet.
- Slowly chew a soda cracker (saltine cracker). As saliva is mixed with the cracker, what do you think happens to the starch? Why?

# 9.3 The Stomach and Intestines

## The Design of the Stomach

The human **stomach** is a saclike, J-shaped storage structure that has a capacity of about three pints. It lies just below the diaphragm in the upper part of the abdominal cavity. Of all the sections of the gastrointestinal tract, the stomach is the largest in diameter. As part of the alimentary canal, the stomach is lined by the mucous membrane; however, instead of two layers of smooth muscle tissue, the stomach has three layers. One layer of muscle runs the length of the stomach, the second layer encircles the stomach, and the third layer is situated diagonally across the stomach. Peristaltic contractions of these muscle layers thoroughly mix the food and the digestive juices inside the stomach. Besides the peristaltic contractions, *hunger contractions* occur when the stomach has been empty for several hours; these are usually associated with a feeling of hunger. The major function of the stomach is to store food and release it to the small intestine at the required rate because we usually eat much more quickly than the intestines can digest and absorb food. The stomach also aids in digestion and absorbs water, glucose, and a few other substances.

## Digestion in the Stomach

As the stomach muscles continue peristaltic agitation, the food is broken down into smaller particles. Salivary amylase continues to digest starch until the gastric juices secreted by cells in the stomach walls penetrate the softened food mass, destroying the salivary amylase. Some cells in the stomach wall secrete mucus, some cells secrete hydrochloric acid, and some cells secrete gastric enzymes and other substances; these secretions combined form **gastric juice.**

**Hydrochloric acid** softens foods; kills bacteria; and reacts with certain minerals, making them soluble so they can better be absorbed through the walls of the small intestine. Hydrochloric acid also provides an acid medium that *pepsin,* the most important gastric enzyme, requires to begin the digestion of proteins. Pepsin breaks down proteins into shorter chains of amino acids; these chains are broken later, when the food reaches the small intestine. The enzymes *gastric amylase,* which aids in the digestion of starch, and *gastric lipase* [lĭp′ās′], which aids in the digestion of certain fats, are secreted from the walls of the stomach in lesser amounts than pepsin. A substance known as the *intrinsic factor* bonds to vitamin $B_{12}$, preparing it for absorption in the small intestine. The mucous membrane and the alkaline mucus it secretes provide protection to the stomach from the corrosive action of hydrochloric acid. (Stomach muscle is a protein tissue and would be digested if not thus protected.) Peristaltic movements in the stomach and gastric juice secretions reduce the semisolid bolus to a thick liquid called **chyme** [kīm].

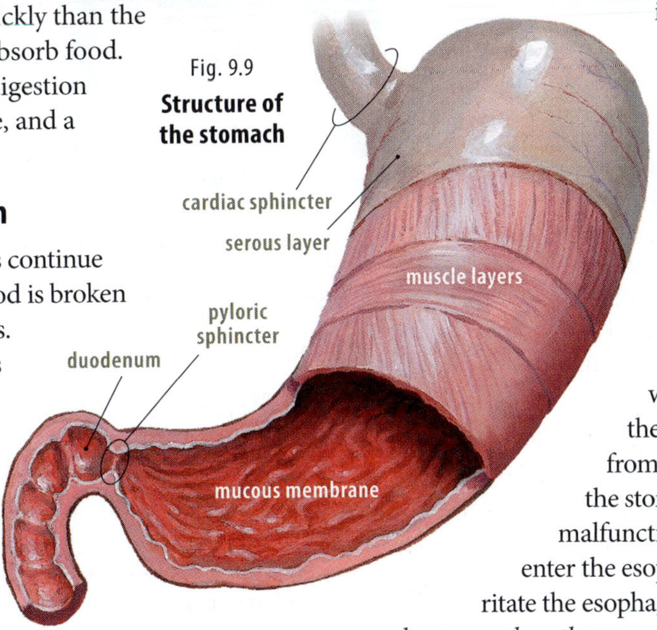

Fig. 9.9 **Structure of the stomach**

Where the esophagus joins the stomach is the *cardiac sphincter,* an area of thickened circular muscles. This valve, which allows food to pass into the stomach, also prevents food from returning to the esophagus from the stomach. Sometimes this valve malfunctions and allows gastric juice to enter the esophagus. The gastric secretions irritate the esophagus, causing a burning sensation known as *heartburn.*

The **pyloric** [pī·lôr′ĭk] **sphincter,** located at the lower end of the stomach, keeps food in the stomach until the food has been sufficiently mixed with gastric juice and is ready to enter the small intestine. Peristaltic contractions squirt the chyme through the pyloric sphincter. When a sufficient amount of chyme has passed through, the pyloric sphincter automatically closes. The average amount of time that food stays in the stomach is between 2 and 4 hours. Carbohydrate foods leave the stomach rather quickly, protein-rich foods take longer, and fatty foods take the longest.

Fig. 9.10 **Peristaltic contractions in the stomach**

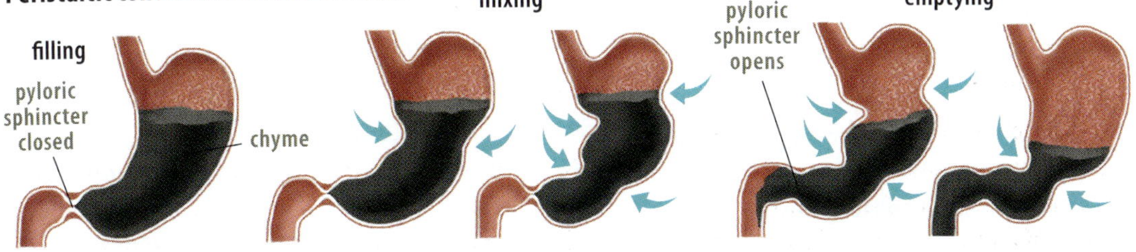

## The Liver and the Pancreas Aid Digestion

Two organs that are not part of the digestive tract but are important to its functioning are the liver and the pancreas. The **liver** produces a greenish-colored substance called **bile,** which is secreted from the liver and stored in the saclike **gallbladder** beneath the liver. **Bile salts,** the most important components of bile, help with the digestion of fats in the small intestine. One end of each bile-salt molecule is soluble in fat, and the other end is soluble in water. In the small intestine, fat molecules link themselves to the fat-soluble ends of the bile-salt molecules. This linkage breaks large droplets into smaller droplets, a process called **emulsification,** to expose a greater surface area for digestive enzymes to attack. The enzyme *lipase* then breaks down the fats into individual fatty acids. Bile also keeps the droplets separated and mixed with the watery digestive juices, and it aids in the absorption of fatty acids released during digestion. Interestingly, the gallbladder releases bile salts precisely when they are needed to digest fats. When fat is present, the duodenum releases a hormone that signals the gallbladder to release the bile salts. Bile enters the small intestine from the gallbladder through the **common bile duct.**

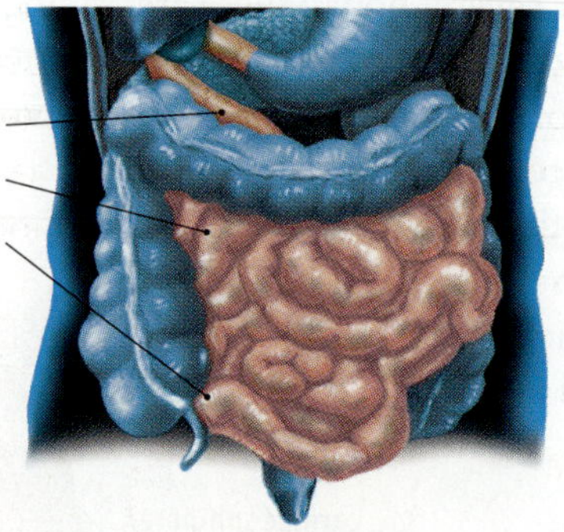

Fig. 9.11
**Parts of the small intestine**

The **pancreas** produces pancreatic juice and secretes this juice into the small intestine through the *pancreatic duct.* The common bile duct and the pancreatic duct merge to form a single duct that empties into the small intestine. **Pancreatic juice** contains several important enzymes, including *pancreatic lipase,* which digests fats; *pancreatic amylase,* which digests carbohydrates; and trypsin [trĭp′sĭn], chymotrypsin [kī′mə·trĭp′sĭn], and carboxypeptidase [kär·bŏk′sē·pĕp′tĭ·dās′], which act on proteins. The pancreas also produces dissolved **sodium bicarbonate** (the same chemical as household baking soda), which neutralizes the hydrochloric acid secreted by the gastric glands. The stomach acid must be neutralized before the pancreatic and intestinal enzymes can perform their work in the small intestine.

## The Structure of the Small Intestine

The **small intestine,** which is the primary organ of digestion and absorption, is a long tube (about 10 feet long in an average living adult) constructed of two layers of muscle; one layer of muscle runs lengthwise, and the other encircles the intestine. This ten-foot long, one-inch-diameter digestive tube is lined with a mucous membrane and small projections called *villi* [vĭl′ī], which give the inner surface of the small intestine a velvety appearance. To fit into the abdominal cavity, the small intestine is coiled; it fills most of the cavity below the stomach.

The first ten inches of small intestine is called the *duodenum* [do͞o′ə·dē′nəm]. The middle section, called the *jejunum* [jə·jo͞o′nəm], is somewhat longer than the duodenum. The longest part of the small intestine is the *ileum* [ĭl′ē·əm]. *The duodenum is where most chemical digestion occurs.*

## Digestion and Absorption in the Small Intestine

**Digestion.** The **intestinal glands** of the small intestine produce *intestinal juice,* which contains several enzymes, including intestinal amylase, intestinal lipase, peptidase [pĕp′tĭ·dās′], maltase, sucrase, and lactase. As the stomach contents enter the duodenum, the pancreas, liver, and gallbladder add their secretions to the chyme via ducts from the pancreas and gallbladder. Intestinal juice produced by glands in the wall of the duodenum is mixed with bile and pancreatic juice. Simultaneously,

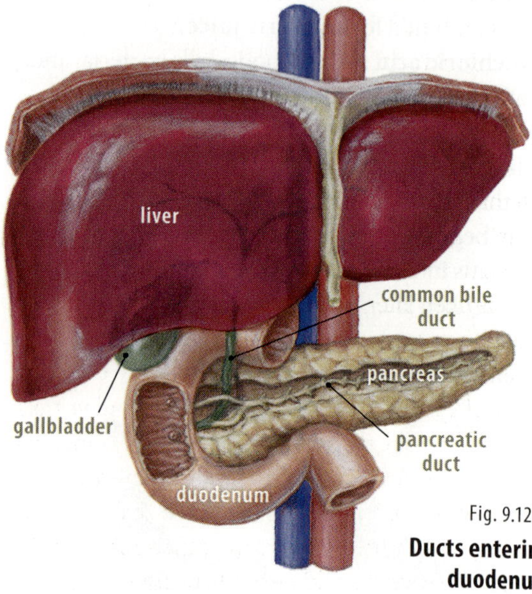

Fig. 9.12
**Ducts entering the duodenum**

the enzymes in pancreatic juice and intestinal juice begin the final phase of digestion. They complete the digestive processes started in the mouth and continued in the stomach; they also begin the digestion of foods that are not affected by digestive juices in the mouth or the stomach.

**Absorption.** Digestion is important, but it would do us no good to digest food if there were no way to make it available to the body. Food is made available to the body by **absorption,** the movement of water, digested food, and other dissolved substances into the bloodstream. Most absorption occurs in the jejunum, the middle section of the small intestine. Absorption is more than simple diffusion; absorption requires work. The cells doing the absorbing take an active part in the process. Biologists have determined that absorbing cells in the intestine use more oxygen and give off more carbon dioxide when they are doing the work of absorption than when they are not. This is evidence that absorption requires the expenditure of energy.

**Work of the villi.** The small intestine is well fitted for its work; its length and numerous villi provide a large surface area for absorption. **Villi** are tiny, hairlike projections of the small intestine whose primary function is to absorb digested food. The villi wave back and forth to come in contact with as many food molecules as possible.

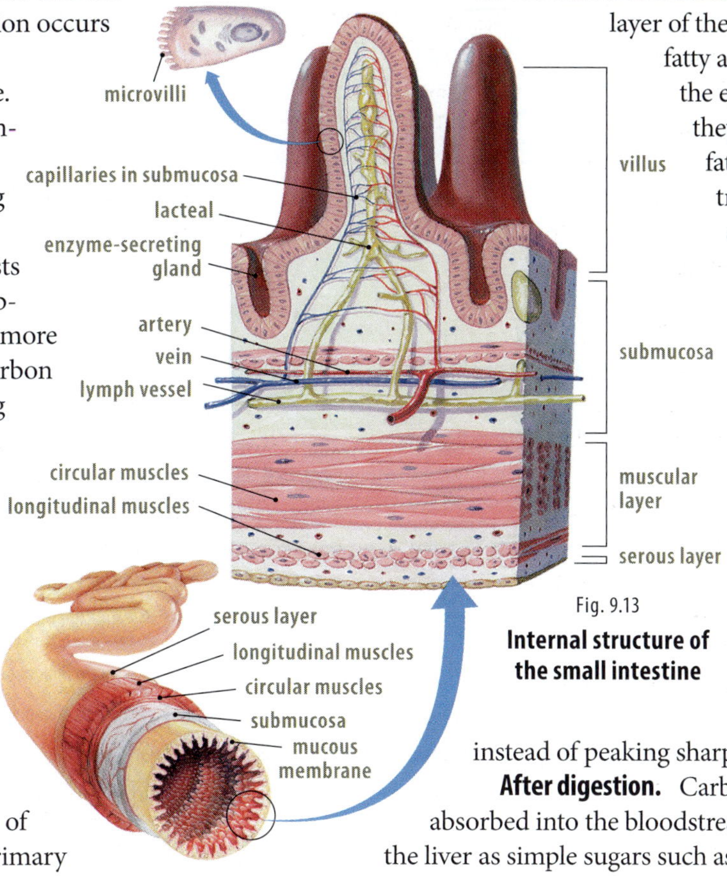


Fig. 9.13
**Internal structure of the small intestine**

Digested food molecules do not have to travel very far to enter the bloodstream. Each villus is covered by a single layer of epithelial cells, and the tiny blood vessels, called *capillaries,* that absorb food molecules lie just beneath this layer. On the free surface of the epithelial cell extending into the passageway of the digestive tract are fine extensions of the cell membrane called **microvilli.** These microvilli give the villi cells a brushlike border that greatly increases their surface area. Amino acids, glycerol, simple sugars, vitamins, and minerals move through the epithelial cells of the villi into the capillaries and are carried away by the bloodstream.

In the center of each villus is another type of tubelike vessel called a **lacteal** [lăk′tē·əl]. The lacteals, which are part of the *lymphatic system,* eventually empty into the bloodstream, but that occurs far from the small intestine. The lacteals serve an important purpose in the absorption and distribution of fats. Unlike other food substances, fatty acids do not immediately enter the bloodstream after crossing the epithelial layer of the villus. Instead, the fatty acids are retained by the epithelial cells, where they are reassembled into fats (triglycerides). As triglyceride molecules accumulate, they are packaged into spherical *lipoprotein* "containers," which are then transported out of the cell and released into the lacteal. Because fatty acid molecules enter lacteals instead of entering the bloodstream directly, the level of fats in the blood remains at a relatively steady level instead of peaking sharply after meals.

**After digestion.** Carbohydrate molecules absorbed into the bloodstream are carried first to the liver as simple sugars such as glucose. Much of this glucose is temporarily stored in the liver and muscles as *glycogen,* or animal "starch." The level of glucose in the blood is regulated by the pancreas. Whenever the level of glucose in the blood rises above normal (such as after you eat lunch), the pancreas releases a hormone called **insulin.** Insulin signals the liver and muscles to remove glucose from the blood and store it as glycogen. Whenever the level of glucose in the blood falls below normal, on the other hand, the pancreas releases a hormone called **glucagon** [glo͞o′kə·gŏn′], which signals the liver to convert glycogen back into glucose and release the glucose into the bloodstream. Together, insulin and glucagon help the body maintain a steady level of glucose in the blood, supplying cells with the energy they need for their activities.

Proteins, fats, and some carbohydrates are used to replace or repair tissues and cells. If more proteins are consumed than the body needs to build and repair its tissues, the liver converts the excess amino acids into fuels the body can use and a waste substance called **urea,** which is removed from the bloodstream by the kidneys. Digested fats, which are distributed to the tissues by the bloodstream, may be used immediately by body cells as energy-yielding food or stored in adipose cells for later use.

### The Large Intestine

Although the **large intestine** is larger in diameter than the small intestine, it is much shorter. The large intestine, which is about five feet long, lies on top of the small intestine and forms an inverted U in the abdominal cavity; it is divided into the cecum [sē′kəm], colon, and rectum.

The large intestine joins the small intestine from the side instead of end to end. At this junction is a

**Table 9.2 Chemical Processes of Digestion**

| Place of digestion | Glands | Secretion | Enzymes | Digestive function |
|---|---|---|---|---|
| **oral cavity** | salivary | saliva | amylase | breaks down starch into maltose, lubricates |
| | mucous | mucus | | lubricates |
| **esophagus** | mucous | mucus | | lubricates |
| **stomach** | gastric | gastric juice | pepsin | breaks down proteins into oligopeptides (short chains of amino acids) |
| | | | amylase | changes starch into maltose |
| | | | lipase | breaks down some fats into fatty acids |
| | | hydrochloric acid | | activates pepsin, dissolves minerals, kills bacteria |
| | mucous | mucus | | lubricates |
| **small intestine** | liver | bile | | emulsifies fats, activates lipase |
| | pancreas | pancreatic juice | trypsin | breaks down proteins into oligopeptides |
| | | | chymotrypsin | breaks down proteins into oligopeptides |
| | | | carboxypeptidases | break down oligopeptides into amino acids |
| | | | amylase | breaks down starch into maltose |
| | | | lipase | breaks down fats into fatty acids |
| | intestinal | intestinal juice | lipase | breaks down fats into fatty acids |
| | | | peptidases | break down proteins and oligopeptides into amino acids |
| | | | amylase | breaks down starch into maltose |
| | | | maltase | breaks down maltose into glucose |
| | | | lactase | breaks down lactose into glucose and galactose |
| | | | sucrase | breaks down sucrose into glucose and fructose |
| | mucous | mucus | | lubricates |
| **large intestine** | mucous | mucus | | lubricates |

pouch, called the **cecum,** possessing a small, fingerlike structure, the **vermiform appendix.** It is the appendix that sometimes becomes infected and causes **appendicitis,** a condition that often requires surgery. The appendix has no known function in digestion, but it does secrete mucus and may help the body fight infections. The junction of the small intestine, the cecum, and the appendix is in the lower right side of the abdominal cavity. The *ileocecal* [ĭl′ē•ō•sē′kəl] *valve* controls the peristaltic movement of materials from the small intestine into the large intestine (from the ileum to the cecum).

The **colon** is divided into ascending, transverse, descending, and sigmoid [sĭg′moid′] regions. Although the colon has no digestive function, it does help the small intestine absorb a certain amount of digested food. The primary function of the colon is to absorb water; much of the water used in digestion is recycled, preventing the body from becoming dehydrated. *Diarrhea* is the condition of expelling too much water when the colon is not working properly. A person with diarrhea needs extra water to prevent dehydration.

The **rectum,** joined to the end of the sigmoid colon, stores undigestible materials that pass through the digestive system. These wastes, called **feces** [fē′sēz], are semisolid and relatively dry by the time they finish their passage through the colon. Feces are stored in the rectum until they are passed from the body through the **anus.** The anus is a valve, composed of a pair of sphincter muscles, that closes the rectum and serves as the digestive tract's second opening to the exterior.

Feces contain many bacteria that live in the large intestine. Some of these bacteria are helpful because they produce vitamins. Significant amounts of vitamin K, for example, are produced by bacteria in the large intestine. The other bacteria help protect the body from disease by "crowding out" harmful bacteria that you ingest every day. The characteristic color of feces comes from the bile pigments that are byproducts of the liver's action. (Old red blood cells are broken down in the liver; although much of their parts are recycled, some pigments are excreted into the feces.) The characteristic odor of feces comes from bacterial action. As the bacteria live and reproduce, they expel waste products such as acids and gases. These processes are vital functions of the human body designed by a wise Creator. The large intestine is actually an active habitat for many species of living things. A proper balance of this habitat is important for good health.

Fig. 9.14

**Structure of the large intestine**

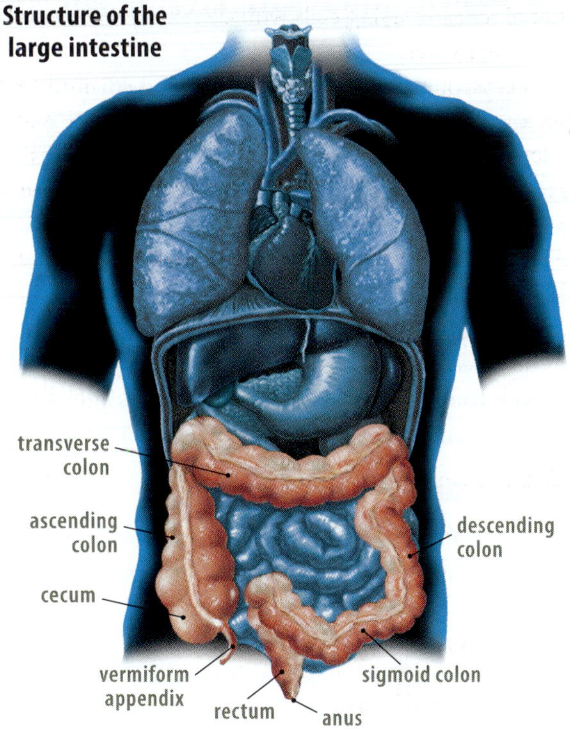


## Section Review 9.3

1. Explain how the stomach wall is protected from the acids and digestive enzymes inside the stomach.
2. How do hydrochloric acid and bile salts help digest food?
3. What is considered the primary organ of digestion and absorption?
4. What is the source and function of sodium bicarbonate in digestion?
5. In a paragraph, summarize the digestive processes in the stomach and small intestine.

### Application

6. How is the absorption of amino acids and simple sugars in the small intestine different from the absorption of fats?

### Critical Thinking

7. One form of diabetes is caused by an inability to produce the hormone insulin. What effect would a lack of insulin have on levels of glucose in the blood? What effect would too much insulin have?

### Identify

gastric juice, chyme, gallbladder, emulsification, duodenum, jejunum, ileum, lacteals, urea, large intestine, vermiform appendix, colon, anus, feces

To keep your digestive tract healthy and efficient, you must eat a balanced diet, chew your food properly, and drink enough fluids. The large intestine also requires a certain amount of dietary fiber to function properly; this indigestible fiber helps scour toxic waste products and dead cells from the walls of the large intestine as the fiber passes through. Nutritionists recommend eating 14 grams of fiber for every 1,000 calories in our diet. Your body also needs exercise to eliminate wastes properly. However, you should wait at least 30 minutes after a large meal before engaging in strenuous physical activity.

## Gastrointestinal Disorders

**Food poisoning.** A common form of **food poisoning** is produced by the bacterium *Staphylococcus aureus* [stăf′ə·lō·kŏk′əs ô′rē·əs], which grows in foods like precooked hams, milk, custards, cream fillings, and chicken salads. If these foods are not properly refrigerated, the bacteria in them grow rapidly. Symptoms of this poisoning—intestinal distress, nausea, vomiting, diarrhea, and weakness—occur from two to twelve hours after consuming food containing the bacteria.

An uncommon but very severe form of food poisoning is **botulism,** caused by the bacterium *Clostridium botulinum.* These bacteria grow well in areas lacking oxygen, such as sealed containers. These bacteria produce the most deadly neurotoxin known. Although most botulin-contaminated food shows no evidence of spoilage, the gas formed by the bacteria during fermentation may cause cans to bulge or leak. Therefore, food from cans that bulge or leak should be discarded. The botulinum toxin is up to one million times as deadly as the poison strychnine [strĭk′nīn′]; it is possible to receive a lethal dose from one contaminated string bean. Botulism is a special problem for home canners because the dormant bacteria *(endospores)* can withstand more than five hours of ordinary boiling. Fortunately, although the bacteria are difficult to kill, the poison they produce is easily destroyed by heat. As a result, boiling home-canned food for ten minutes just before eating it destroys any poison that may have accumulated in the can.

**Salmonellosis** [săl′mə·nĕ·lō′sĭs] is a common food infection caused when bacteria of the genus *Salmonella* [săl′mə·nĕl′ə] get into or on food. Meat, fish, poultry, eggs, dairy products, and shellfish are frequent causes of salmonellosis. A kitchen counter that has been in contact with contaminated meat is a source of contamination for any food placed on it; therefore, good sanitation in food preparation is extremely important. More serious than salmonellosis is **typhoid fever,** a disease rare in the United States; it is acquired from food or water contaminated with the bacterium *Salmonella* Typhi. Unlike the other *Salmonella* species, typhoid bacteria spread throughout the entire body.

**Dyspepsia.** *Gastric indigestion* is known as **dyspepsia** [dĭs·pĕp′sē·ə]. Symptoms of dyspepsia, such as heartburn, belching, nausea, and vomiting, may occur either during or following the ingestion of food. Dyspepsia may be *organic,* resulting from a disease of the organs of the gastrointestinal tract, or *functional,* usually resulting from nervous disorders and emotional tension. Other causes of functional dyspepsia include consuming too much food and drink, consuming excess fatty foods, and inadequate chewing of food.

**Peptic ulcers.** An **ulcer** is an open sore in the mucous membrane resulting from the breakdown of tissues. Ulcers often occur in the stomach; these ulcers are called *gastric ulcers.* Ulcers

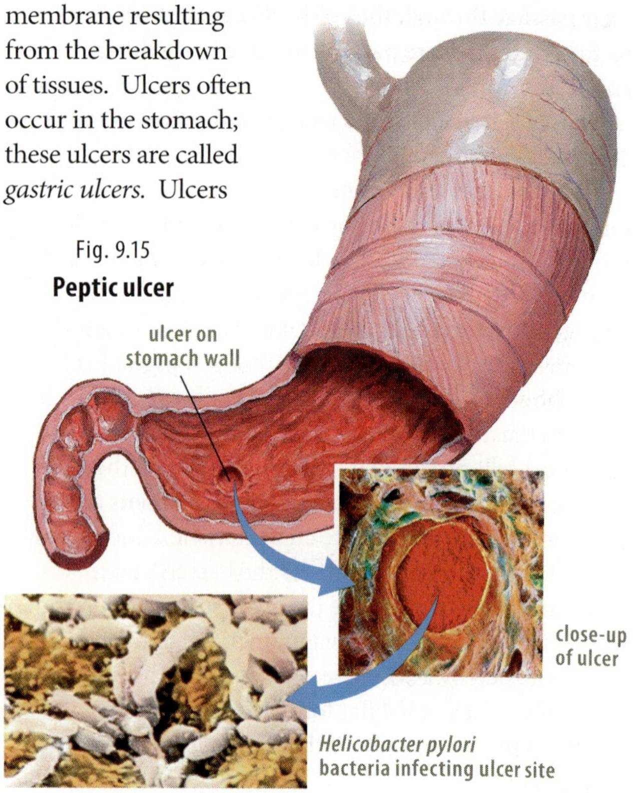


Fig. 9.15
**Peptic ulcer**

are also common in the duodenum. *Duodenal ulcers* occur where the duodenum is exposed to gastric juice as the stomach contents enter the small intestine. Since both types of ulcers are caused by the digestive action of pepsin, they are known as **peptic ulcers.** Although peptic ulcers were once thought to result from excessive secretion of gastric juice, scientists have discovered that nearly all peptic ulcers are caused by a type of bacteria *(Helicobacter pylori)* that infects the inner protective coating of the stomach and duodenum. As the bacteria multiply, they damage this inner lining, reducing its production of protective mucus. Stripped of its defenses, the wall of the stomach or duodenum is vulnerable to attack by its own digestive juices. Once an ulcer forms, it continues to be irritated unless it is treated; untreated ulcers may hemorrhage, or bleed, resulting in a need for hospitalization. Fortunately, ulcers caused by *H. pylori* can be treated with antibiotics, which kill the bacteria and allow the stomach wall to heal itself.

**Effects of alcohol.** Consumption of alcoholic beverages can be irritating to the mucous membrane of the gastrointestinal tract. Excessive alcohol consumption can cause diarrhea, nausea, and vomiting. Chronic drinking interferes with the body's absorption of essential nutrients, such as the B-complex vitamins, vitamin C, zinc, and magnesium. Medical studies indicate that prolonged heavy drinking increases the risk of cancer of the mouth, esophagus, larynx, and liver.

**Constipation.** A condition known as **constipation** results when the feces become so dry that defecation is difficult. Constipation can be caused by removal of too much water from the feces or by delaying defecation. Other factors relating to constipation include a low-fiber diet, insufficient intake of liquids, insufficient exercise, stress, and disease.

**Diarrhea.** When movement of feces through the colon is so rapid that insufficient water is removed, **diarrhea** results. The feces are soft and watery, and there is frequent defecation. Prolonged diarrhea may result in dehydration if the fluid is not replaced. Causes of diarrhea include bacterial and viral infections, various chemicals, and emotional factors. Cramping and pain in the abdomen, weakness, and sometimes fever and vomiting are symptoms of diarrhea.

**Dysentery.** A painful infection of the colon known as **dysentery** can be caused by bacteria, protozoa, or viruses. The symptoms include severe diarrhea and cramps accompanied by blood in the feces. Dysentery is spread by poor sanitation, often from drinking water that is polluted by sewage and other impurities or by eating fruit that has not been washed well. If you travel to parts of the world where the water is not purified, you need to be careful to drink and wash only with purified water. Even washing your hands with contaminated water and then touching your mouth can cause dysentery.

## Colon Cancer

Although **colon cancer** is one of the most common forms of cancer in America, it is up to 95% curable when detected early. Colon cancer most often affects people over forty years of age. Although highly curable if caught early, it is extremely difficult to treat once it has spread throughout the body.

Although a specific cause of colon cancer is not known, most scientists believe that diet is a significant factor. Colon cancer is much more prevalent in highly developed countries, such as the United States and United Kingdom, where consumption of fats is high and consumption of fiber is low. It is thought that people who consume ample dietary fiber have less chance of getting colon cancer because the indigestible carbohydrates in whole grains, raw fruits and vegetables, fruits with seeds, and nuts act as a scrub brush inside the colon, cleaning the surface and carrying the debris out of the body.

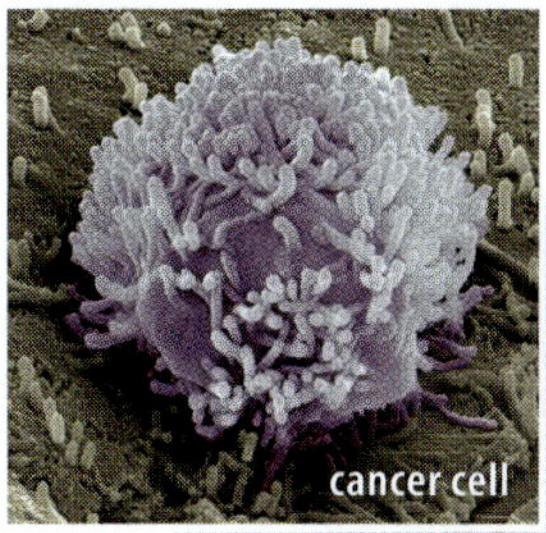


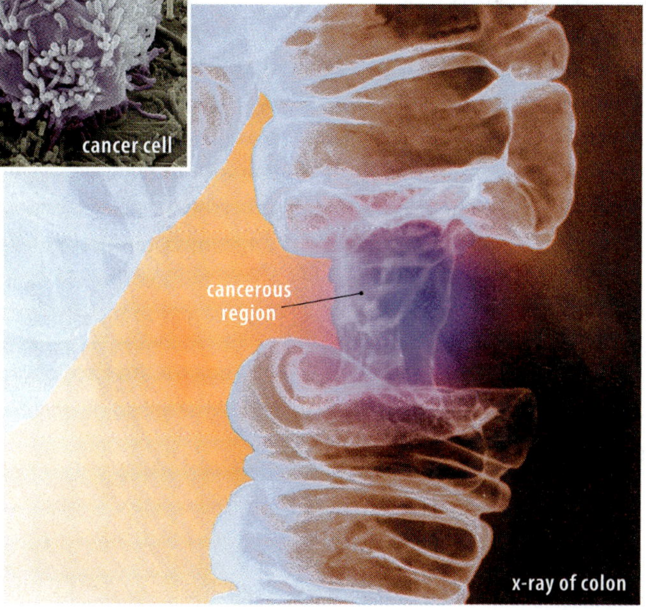


Fig. 9.16 **Colon cancer**

## Section Review 9.4

1. What gastrointestinal disorders can be caused by negative emotions?
2. Name three ways that good health practices can help prevent gastrointestinal disorders.
3. Describe the effects of alcohol on the gastrointestinal tract.

### Identify

botulism, typhoid fever, peptic ulcer, dysentery, colon cancer

# Chapter 9 Review

### Define

1. metabolism
2. Calorie, kilocalorie
3. adipose cell
4. obesity
5. macronutrients, micronutrients
6. starch
7. amino acid
8. lipid
9. fatty acid, fat, triglyceride
10. cholesterol
11. ATP
12. antioxidant
13. alimentary canal
14. digestion
15. oral cavity, pharynx, esophagus
16. dental caries, plaque
17. gallbladder, pancreas
18. villi, microvilli
19. salmonellosis, typhoid fever
20. dyspepsia
21. ulcer

### Identify

1. seven water-soluble vitamins
2. four fat-soluble vitamins
3. the major minerals and trace minerals needed by our bodies
4. four types of permanent teeth
5. the parts of a tooth
6. three divisions of the small intestine
7. two types of food poisoning

### Explain

1. What is the difference between saturated and unsaturated fats?
2. What is the purpose of saliva?
3. Why do wisdom teeth sometimes have to be removed?
4. Why is proper dental hygiene important?
5. Describe the stomach's role in digestion.
6. Explain how the liver aids digestion.
7. What are the purposes of the small intestine?
8. What are the purposes of insulin and glucagon?
9. Describe the function of each main segment of the large intestine.

### Apply

1. Which type of compounds in food are the main source of energy for doing things?
2. What is the purpose of eating fiber, since it cannot be digested by humans?
3. Which type of compounds in food are building blocks for the body?
4. Why is it important to drink sufficient water every day?
5. Why are enzymes vital to the process of digestion?
6. As part of his routine, an acrobat eats an apple while hanging upside down from a trapeze. How is the food able to get from his mouth to his stomach against the force of gravity?
7. What are some ways to prevent gastrointestinal disorders?

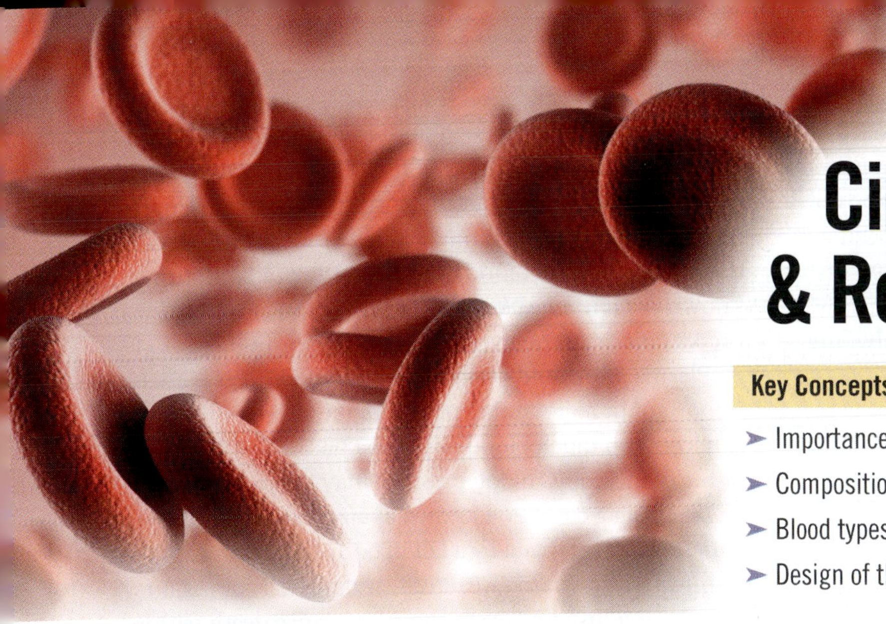

# Circulation & Respiration

CHAPTER 10

**Key Concepts**

- Importance of blood
- Composition of blood
- Blood types
- Design of the heart
- Blood circulation
- Cardiovascular health
- Function of respiration
- Respiratory health

## 10.1 Blood: The Life of the Flesh

The Bible tells us that "the life of the flesh is in the blood" (Lev. 17:11) and that at Creation God "breathed into [Adam's] nostrils the breath of life; and man became a living soul" (Gen. 2:7). Indeed, the circulatory and respiratory systems are absolutely vital to life. Ordinarily, a person can survive for days without water and weeks without food but will die within minutes if breathing or blood circulation ceases. A loss of blood flow to the brain for only five to ten seconds is enough to cause a person to lose consciousness.

### Importance of the Blood

The heart, the blood vessels, and the blood they contain all make up the **cardiovascular system,** also called the *circulatory system.* Blood permeates every part of the body, reaching the most remote areas through microscopic channels. Blood is pumped away from the heart through *arteries,* and it moves back to the heart through *veins.* The vessels linking these larger vessels are the microscopic *capillaries.* Every one of the trillions of cells of the body is close enough to a capillary to receive nourishment, oxygen, and chemical messengers. Blood carries away the wastes of the cell's activity and protects the cells from attack by microorganisms. It is completely impossible for the body to remain alive when the blood is removed. Scientists can remove individual cells and tissues and keep them alive in laboratory conditions, but human life requires blood.

In Scripture, God emphasizes blood more than any other body substance. Every year, scientists learn more

Fig. 10.1 **Cardiovascular system**

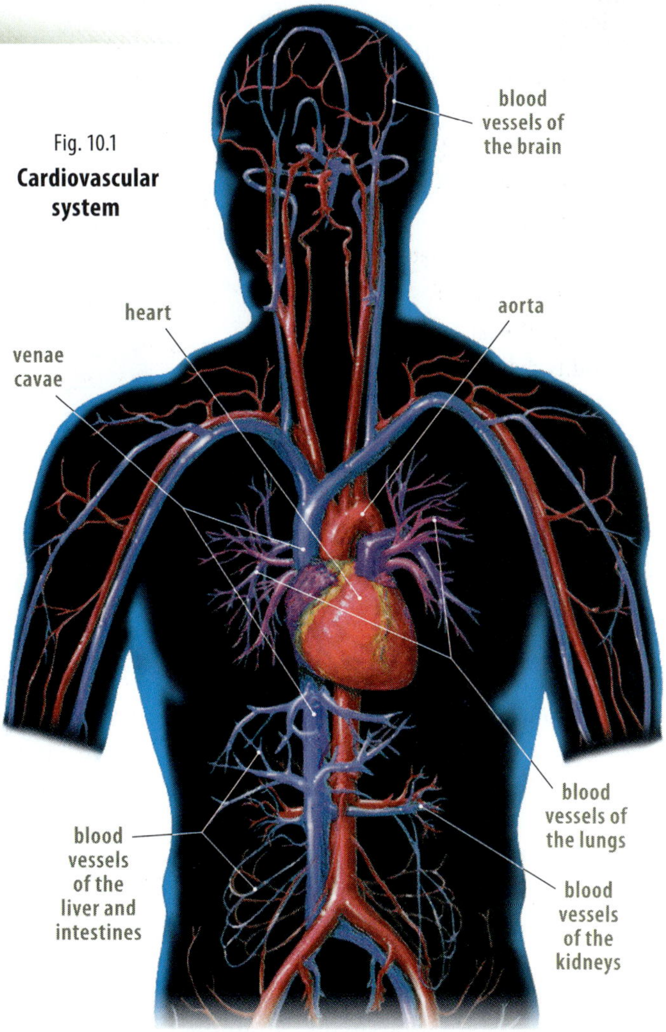

and more about this very complex material. What scientists once thought to be a simple liquid now appears to be a highly organized accumulation of thousands of different kinds of molecules, each with a very specific job. Today, biologists know that the blood is very complicated and that every person's blood is somehow different from anyone else's who ever lived. Truly, "the life of the flesh is in the blood."

## Components of Blood

**Blood** is a thick fluid containing many solid particles. About 55% of the total blood volume is a liquid called *plasma.* The rest of the blood (45%) consists of *red blood cells, white blood cells,* and *platelets.* The average person's circulatory system contains about 5 to 6 liters (1.2–1.5 gallons) of blood. Blood is red in color. The blood in the arteries is bright red because it is free of wastes and laden with oxygen. In the veins, blood is dull red because it has given oxygen to the cells and picked up waste materials from them. When seen through the pigment of the skin, the dull red blood in the veins commonly appears blue.

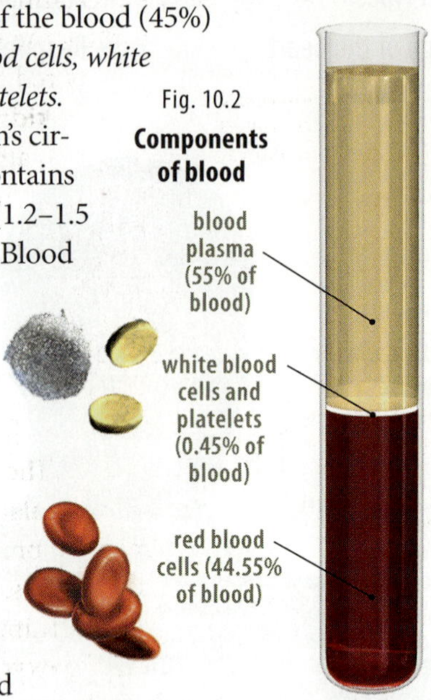

Fig. 10.2
**Components of blood**

## Plasma

**Plasma,** the straw-colored liquid part of the blood, is about 90% water and contains various dissolved substances, including digested food molecules in the form of glucose, fats, and amino acids; waste molecules; hormones; and many kinds of blood proteins.

There are three basic categories of blood proteins: albumin [ăl·byo͞o′mĭn], globulins [glŏb′yə·lĭnz], and fibrinogen [fī·brĭn′ə·jən], each having special jobs to do. **Albumin,** the most plentiful of the blood proteins, is manufactured in the liver and helps regulate the amount of water in the blood. **Globulins,** made in both the liver and lymph tissue, are a broad category of blood proteins that help transport fats throughout the body and help fight infections. (The disease-fighting proteins called *antibodies,* produced by white blood cells, are a type of globulin known as *gamma globulins.*) **Fibrinogen,** another blood protein made in the liver, helps in the clotting of blood.

Plasma, the liquid matrix of blood tissue, is extremely important in maintaining proper blood pressure. If a person loses too much blood or plasma because of an injury or burn, the body's blood pressure may fall to a critical low, causing a condition known as **circulatory shock.** Shock is characterized by a fall in body temperature, clammy skin, extreme weakness, and mental confusion or unconsciousness; severe shock can cause death within minutes from heart failure. Shock can be treated by prompt transfusion of whole blood or plasma. During war or natural disasters, when the wounded may sometimes be hours or even days away from a hospital, plasma transfusions may be given as treatment to keep the blood pressure high, fight disease, and allow blood clotting.

## Red Blood Cells

**Carriers of oxygen.** A single drop of blood contains about five million **red blood cells,** or *erythrocytes* [ĭ·rĭth′rə·sīts′]. It is estimated that the body contains 25 trillion red blood cells, enough to cover 3500 square yards; lined up end to end, the body's red blood cells would extend around the earth four times.

Fig. 10.3
**Structure of hemoglobin**
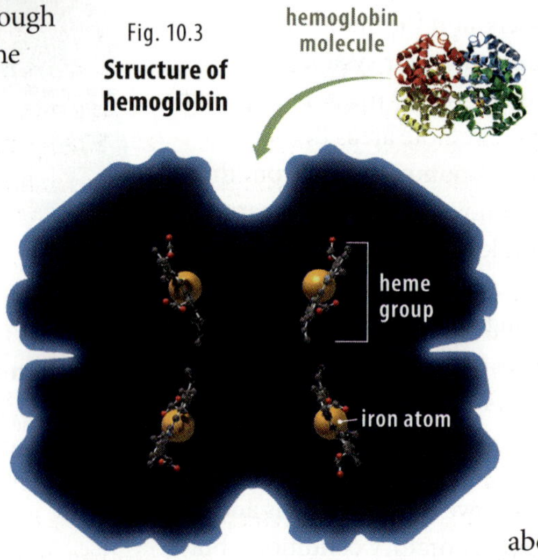


A red blood cell is a masterpiece of design for its job of carrying oxygen. Each red blood cell is a disk-shaped cell with a depression on each side. Red blood cells are smaller than most body cells and are extremely flexible, allowing them to easily pass through the smallest capillaries. A row of 3400 red blood cells would be only about an inch long. Unlike most other body cells, red blood cells lack nuclei; this architecture allows for more efficent use of the cell's volume.

The cytoplasm of red blood cells contains **hemoglobin,** a special iron-containing blood protein that binds easily to oxygen and gives blood its characteristic red color. The ability of red blood cells to unite easily with oxygen molecules and to give them up just as easily makes these cells useful. A mature erythrocyte can contain over 250 million molecules of hemoglobin, which compose about a third of the cell's volume. Each hemoglobin molecule contains four small *heme* [hēm] *groups* (Fig. 10.3). Central to each heme group is an atom of iron that can bind loosely to an oxygen molecule. When high levels of oxygen are present, such as in the lungs, the hemoglobin quickly becomes saturated with oxygen. This saturated hemoglobin, known as oxyhemoglobin, passes through body tissues low in oxygen, where the oxygen molecules break their bonds with the hemoglobin and move to the tissue, supplying the tissue with oxygen. After losing the oxygen molecules, the hemoglobin picks up carbon dioxide from the cells.

Fig. 10.4
**Life cycle of a red blood cell**

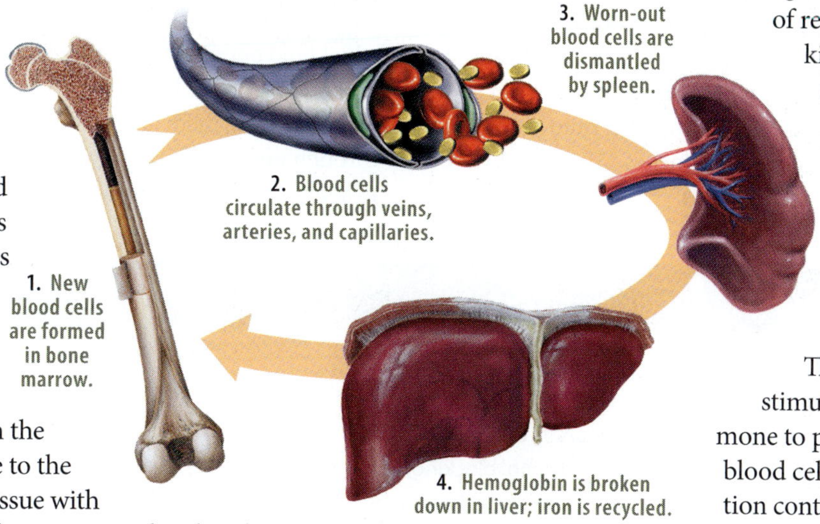


**Anemia** is a disorder resulting from too little oxygen being transported to body cells. Anemia can have several different causes, such as a lack of iron, a lack of hemoglobin, too few red blood cells, or a number of different diseases. A unique disease of the erythrocytes is **sickle cell anemia,** caused by a genetic defect in the hemoglobin molecules that makes them clump together into elongated crystals under certain conditions. When these crystals form inside a red blood cell, they cause the cell to become brittle and sickle-shaped instead of round, smooth, and flexible. Often, the sharp crystals destroy the red blood cell by puncturing its membrane. The sickle cells that survive tend to get stuck and block capillaries because of their stiffness and unusual shape; the kidneys are particularly affected.

**Production in bone marrow.** *Red blood cells and all other blood cells are produced in the bone marrow.* Bone marrow is found at the ends of long bones (femur and humerus) and in flat bones (ribs, cranial bones, sternum), vertebrae, and the pelvis. *Blood cells originate from special cells called stem cells, which can produce any of the different types of blood cells.* Just before the newly formed red blood cells mature and are released into the blood, the nuclei and certain other organelles are expelled. This allows the young red blood cells more space to carry hemoglobin molecules and therefore more oxygen, their primary job. The life expectancy of an erythrocyte is about 120 days. *Erythrocytes are replaced constantly at a rate of over 2 million each second.*

The body has an interesting way of controlling the production rate of red blood cells. The kidneys monitor the amount of oxygen in the blood. When this level drops, the kidney releases the hormone *erythropoietin* [ĭ·rĭth′rō·poi·ē′tĭn]. The bone marrow is stimulated by this hormone to produce more red blood cells. This production continues at a high rate until the oxygen deficiency has been relieved. When the kidney is satisfied that the blood is carrying enough oxygen, it reduces its production of erythropoietin, and the production of red blood cells slows.

**Storage in the spleen.** Surplus red blood cells are stored in the **spleen** until they are needed. The spleen, which is a hand-sized organ located in the abdominal cavity behind the stomach, supplies the body with stored red blood cells whenever it needs extra oxygen-carriers, such as during exercise or when moving from sea level to the mountains. The spleen, along with the liver, also destroys old, worn-out red blood cells. This function is important because red blood cells lose their elasticity as they age, becoming more fragile and less efficient in carrying oxygen. These worn-out cells may become damaged or rupture as they squeeze through tiny capillaries, especially in active muscles. Special cells in the spleen and liver detect these worn-out red blood cells and destroy them by engulfing and dismantling them. The bulk of the dismantled cells are used as a source of energy and repair for other body cells. The heme groups of the hemoglobin, which contain iron, are dismantled in a special way to recycle the iron, which is carried back to the bone marrow for use in new hemoglobin molecules. In this way, God ensured an adequate supply of iron for healthy blood.

Fig. 10.5
**Types of blood cells**

## White Blood Cells

The blood contains several kinds of **white blood cells,** or *leukocytes* [lo͞o′kə·sīts′], which are produced in the bone marrow and the lymph nodes. There is only about 1 white blood cell to every 600 red ones. Unlike red blood cells, white blood cells have nuclei, mitochondria, and all other normal cellular organelles. Since they have no hemoglobin, they are colorless.

**Designed for attack.** White blood cells function as part of the *immune system,* helping the body to fight disease-causing organisms. They are an important part of "the life of the flesh." There are different types of white blood cells, each designed for a specific function. Some white blood cells serve as mobile munitions factories, churning out billions of antibody molecules if they recognize bacteria or other foreign cells in the blood. Some white blood cells serve as hunter-killer systems, moving through the body and killing any body cells that have been infected by viruses or have become cancerous. These cells are not restricted to the blood vessels but can exit the bloodstream at the capillaries and move about in the interstitial fluid between the body cells. Other white blood cells cruise through the bloodstream, engulfing invading bacteria and tearing them apart. Still others are designed to sift through intracellular debris for traces of foreign invaders or to sound the alarm if tissue damage is detected.

### *Soldier Cells* by Dr. Paul Brand and Philip Yancey

Design in the LIVING CREATION

This morning I will examine a hibernating albino bat who sleeps in a box in my refrigerator. I rely on him to study how the body responds to injury and infection. I lift him carefully, lay him on his back, and spread his wings in a cruciform posture. His face is weirdly human, like the shrunken heads in museums. I keep expecting him to open an eye and shriek at me, but he doesn't. He sleeps.

As I place his wing under the microscope lens, again a new universe unfolds. I have found a keyhole. The albinic skin under his wing is so pale that I can see directly through his skin cells into the pulsing capillaries which carry his blood. I focus the microscope on one bluish capillary until I can see individual blood cells pushing, blocking, thrusting through it. Red blood cells are by far the most numerous: smooth, shiny disks with centers indented like jelly doughnuts. Uniform size and shape make them seem machine-stamped and impersonal.

More interesting are the white blood cells, the armed forces of the body which guard against invaders. They look exactly like the amoebae: amorphous blobs of turgid liquid with darkened nuclei, they roam through the bat's body by extending a finger-like projection and humping along to follow it. Sometimes they creep along the walls of the veins; sometimes they let go and free-float in the bloodstream. To navigate the smaller capillaries, bulky white cells must elongate their shapes, while impatient red blood cells jostle in line behind them.

Watching the white cells, one can't help thinking them sluggish and ineffective at patrolling territory, much less repelling an attack. Until the attack occurs, that is. I take a steel needle and, without waking the bat, prick through its wing, puncturing a fine capillary. An alarm seems to sound. Muscle cells contract around the damaged capillary wall, damming up the loss of precious blood. Clotting agents halt the flow at the skin's surface. Before long, scavenger cells appear to clean up debris, and fibroblasts, the body's reweaving cells, gather around the injury site. But the most dramatic change involves the listless white cells. As if they have a sense of smell . . . , nearby white cells abruptly halt their aimless wandering. Like beagles on the scent of a rabbit, they home in from all directions to the point of attack. Using their unique shape-changing qualities, they ooze between overlapping cells of capillary walls and hurry through tissue via the most direct route. When they arrive, the battle begins.

Lennart Nilsson, the Swedish photographer famous for his remarkable closeups of activity inside the body, has captured the battle on film as seen through an electron microscope. In the distance, a shapeless white cell, resembling science fiction's creature "The Blob," lumbers toward a cluster of luminous green bacterial spheres. Like a blanket pulled over a corpse, the cell assumes their shape; for awhile they still glow eerily inside the white cell. But the white cell contains granules of chemical explosives, and as soon as the bacteria are absorbed the granules detonate, destroying the invaders. In thirty seconds to a minute only the bloated white cell remains. Often its task is a kamikaze one, resulting in the white cell's own death. . . .

At the cellular level, massive warfare is a daily fact of life. Fifty thousand invaders may lurk on the rim of a drinking glass, and a billion can be found in a half-teaspoon of saliva. Bacteria enshroud my body—every time I wash my hands I sluice five million of them from the folds of my skin.

To combat these threats, some of the blood's white cells are specifically targeted to one type of invader. If the body has experienced contact with a severe danger, as in a smallpox vaccination, it imprints certain white cells with a death-wish to combat that signal danger. These cells spend their lives coursing through the bloodstream, waiting, scouting. Often they are never called upon to give battle. But if they are, they hold within them the power to disarm a foreign agent that could cause the destruction of every cell in the body.



**Sign of infection.** One way that doctors can detect an infection in your body is by doing a white blood cell count. If your blood contains more than the normal number, you probably have an infection. If a great excess is found and no apparent infection is found, a person may have **leukemia,** a form of cancer in which white blood cells (leukocytes) are produced in uncontrolled numbers. As with most cancers, the abnormal number of cells produced crowds out normally functioning cells, creating pain and weakness.

**Platelets.** The job of patrolling your 60,000 miles of blood vessels for leaks belongs to your **platelets,** tiny disk-shaped cell fragments produced in the bone marrow to serve as patch kits. Platelets originate from large cells produced in the red bone marrow; portions of these large cells break off to form platelets. Each platelet is a round disk, half the size of a red blood cell, that lacks a nucleus. Although they are much smaller than most cells and lack nuclei, platelets contain a host of cellular machinery devoted to blood vessel repair, including "power plants" (mitochondria), facilities for protein production (ribosomes), and a power-driven framework (cytoskeleton) that allows the platelet to change its size and shape. Platelets are the first participants in blood clotting. The average life span of a platelet is about two weeks, after which they shut down and are dismantled in the spleen.

## How Blood Vessels Are Healed

Understanding the LIVING CREATION

Blood vessels carry life-giving oxygen and nutrients to body tissues. If any of these blood vessels are damaged, precious blood can leak out of the circulatory system; the loss of 1–2 quarts can lead to irreversible shock and death. Thus, it is vital that damaged or severed blood vessels be quickly repaired; otherwise, you could bleed to death from a small cut or die of internal hemorrhaging from a bruise.

When blood starts to escape from a cut, millions of platelets attach

*(cont.)*

(cont. from previous page)

themselves to the rough edges of the wound, forming a *platelet plug*. In small or lightly injured blood vessels, the platelet plug may stop the bleeding entirely. In larger blood vessels or in areas of more extensive damage, a process known as **clotting** is used to seal the break. When platelets become activated, a protein called *prothrombin* [prō·thrŏm′bĭn], which is always present in the blood, becomes attached to the platelets, where it is converted into an enzyme called *thrombin* and released back into the blood. Thrombin rapidly converts another blood protein, *fibrinogen,* into an activated form called *fibrin* [fī′brĭn]. Molecules of fibrin link to form a web of threads that traps red blood cells, sealing the break. Motor proteins in the platelets then draw the fibrin fibers tight, pulling the edges of the break together to speed healing. If the blood clot is on the surface of the skin, a **scab** develops, and healing begins along the perimeter of the wound. As wounds heal, the fibrin web is dissolved by enzymes in the blood.

When a person is bleeding as a result of an injury, applying pressure to the wound helps to speed the clotting process. In years past, tourniquets were commonly used to shut off blood loss, but they are used only as a last resort today because the body part (such as a hand or foot) beyond the tourniquet can die from a total lack of blood. Instead, you may apply pressure to various pressure points (places where an artery passes over a bone) in order to slow the flow of blood to the wounded area. If the wound is on an extremity, keep it as high as possible. That will lower the blood pressure and limit blood flow. Most large open wounds need sutures (stitches) to heal properly.

Close-up of blood clot

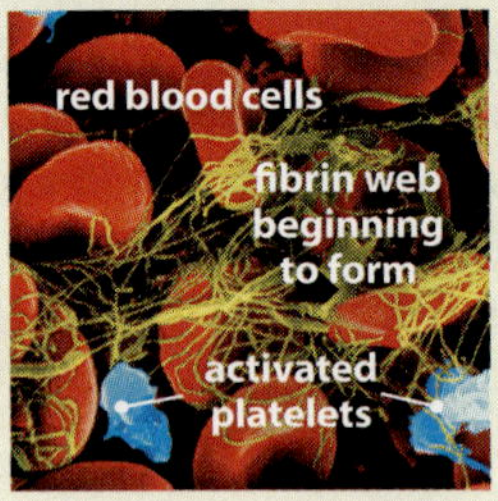


It is a marvelous wonder of God's design that all of the ingredients for blood clot formation are always present but do not operate until they are needed. If blood clots started on their own inside the blood vessels, they would cause instant death. If blood clots did not form quickly when blood vessels are damaged, you could bleed to death from a small cut.

An inherited disease called **hemophilia** [hē′mə·fĭl′ē·ə] results in the inability to form blood clots. Hemophilia is caused by a hereditary defect in one of several blood proteins involved in blood clot formation. Hemophiliacs frequently suffer from internal bleeding or may bleed to death from an otherwise minor accident.

God has designed a very delicate balance in your blood between a need to form clots and the necessity of keeping even the smallest of clots out of the blood flow. Considering the volume of circulation and the complex formulas involved, we must conclude that the human body is of supernatural design.

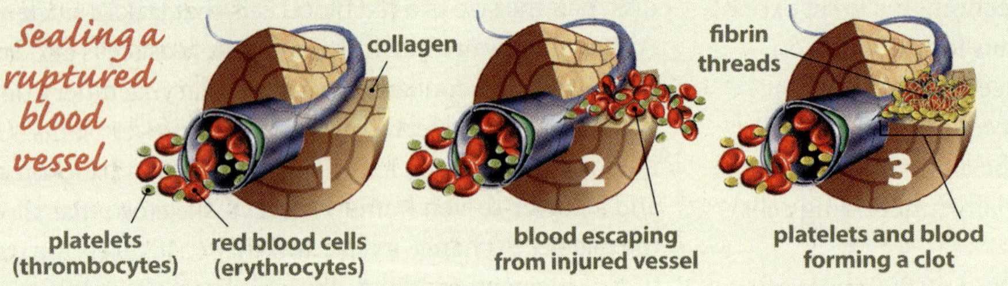


## Blood Types

Your immune system carefully guards your body against external invaders. This is possible because the surfaces of your cells contain protein or carbohydrate molecules, called **antigens,** that act as identification tags; your white blood cells attack and destroy any cells with different antigens than your natural cells.

This necessary vigilance can cause problems when physicians need to introduce foreign cells into the body. For example, a *blood transfusion* (a transfer of someone else's blood into your body) will fail if your immune system identifies the transfused red blood cells as foreign. The foreign red blood cells will immediately be attacked by *antibodies,* defense proteins in your bloodstream, causing the transfused cells to be clumped together (*agglutinated* [ə·gloōt′n·āt′əd]) and destroyed, often with fatal consequences.

Fortunately, successful transfusions are possible because red blood cells carry far fewer unique antigens than most other body cells. Although there are more than 20 antigen groups in red blood cells, the most important are the ABO and Rh groups.

## Table 10.1 Blood Types

| Type of antigen | Blood type | Type of antibodies |
|---|---|---|
| A | A | anti-B (against B) |
| B | B | anti-A (against A) |
| A, B | AB | none |
| none | O | anti-A, anti-B |

**The ABO blood group.** The **ABO blood group** includes two antigens, named A and B. A person can have type A blood (only antigen A present), type B blood (only antigen B), type AB blood (both antigens), or type O blood (neither antigen). He will have antibodies that attack cells with whichever antigens are not normally present in his blood.

Since people with type A blood have anti-B antibodies, they can receive transfusions of blood that does not have antigen B (types A and O) but cannot receive blood that contains antigen B (types B and AB). Similarly, those with type B blood can receive types B or O but not types A or AB. People with type AB blood are sometimes considered *universal recipients* because they have no ABO-group antibodies and can therefore receive any blood type. Because people with type O blood have both anti-A and anti-B antibodies, they can receive only type O blood; but since they can donate to people of any blood type, people with type O blood are sometimes considered *universal donors.*

Although the ABO blood group can be used to predict whether a transfusion will be compatible, any of the other antigen groups may also trigger an immune response. In practice, donor blood is tested before a transfusion to avoid mismatches.

**The Rh blood group.** The **Rh blood group** (named after the rhesus [rē′səs] monkey, in which scientists first identified this group) contains many antigens, but the most important is one called the **Rh factor.** People whose blood contains the Rh factor are considered *Rh positive,* and people whose blood lacks this factor are considered *Rh negative.* The Rh factor is usually typed and included with the A, B, and O types; for example, type A blood can be either A positive or A negative.

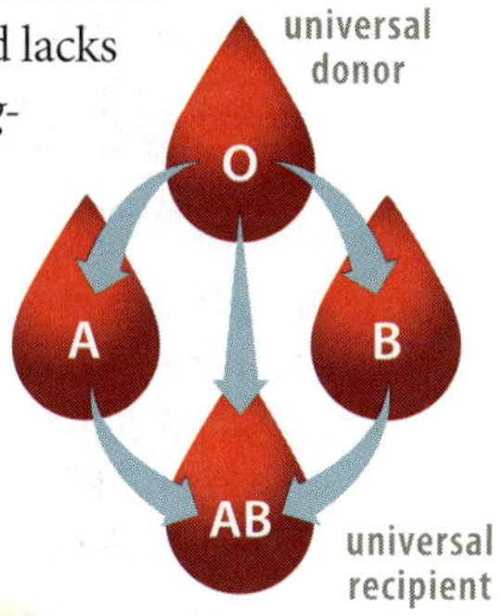


Fig. 10.6 **Blood types and compatibility**

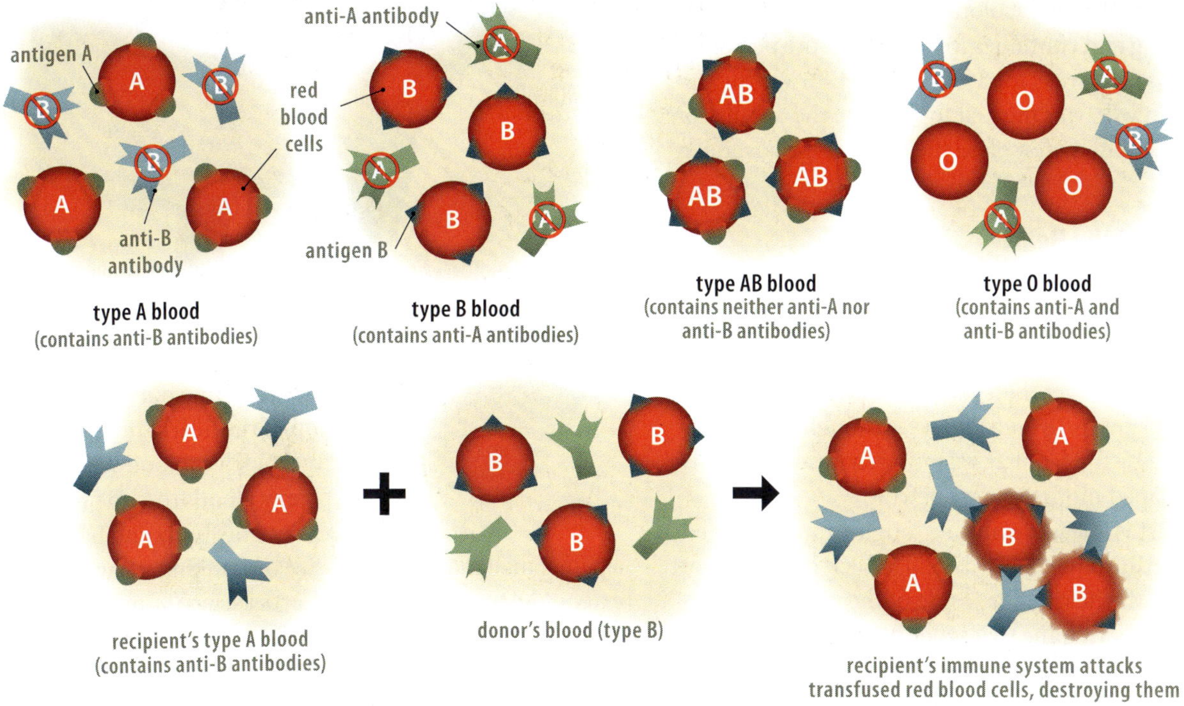

# Section Review 10.1

1. Name the primary components of the cardiovascular system.
2. What are the functions of albumin, globulins, and fibrinogen in the blood?
3. What are the functions of red blood cells, white blood cells, and platelets?
4. Where in the body are new red blood cells produced?
5. Name the four main blood types in the ABO group.

### Application

6. Why is proper blood typing important for safe blood transfusions? Under what circumstances is it safe for a person to receive blood of a different type than his own?

### Critical Thinking

7. Cite a variety of characteristics of blood that make it "the life of the flesh."

### Identify

arteries, veins, capillaries, circulatory shock, hemoglobin, anemia, sickle cell anemia, spleen, leukemia, hemophilia

## 10.2 The Heart: God's Handiwork

The **heart** is the world's most efficient pump, a marvel of divine design and precision engineering. Although your heart is only the size of your fist and weighs a mere eleven ounces, it squeezes your blood through an intricate network of 60,000 miles of blood vessels every moment of your life, pumping about 70 times per minute and 36 million times each year. The heart of the developing child begins as two tiny tubes surrounded by a cluster of muscle cells. Four weeks after conception, the tiny tubes miraculously fuse; the heart then starts beating and continues to do so until the person dies. *William Harvey* (1578–1657), the 17th-century Englishman who was the first to describe the circulation of blood accurately, described the heart as the sovereign of the body.

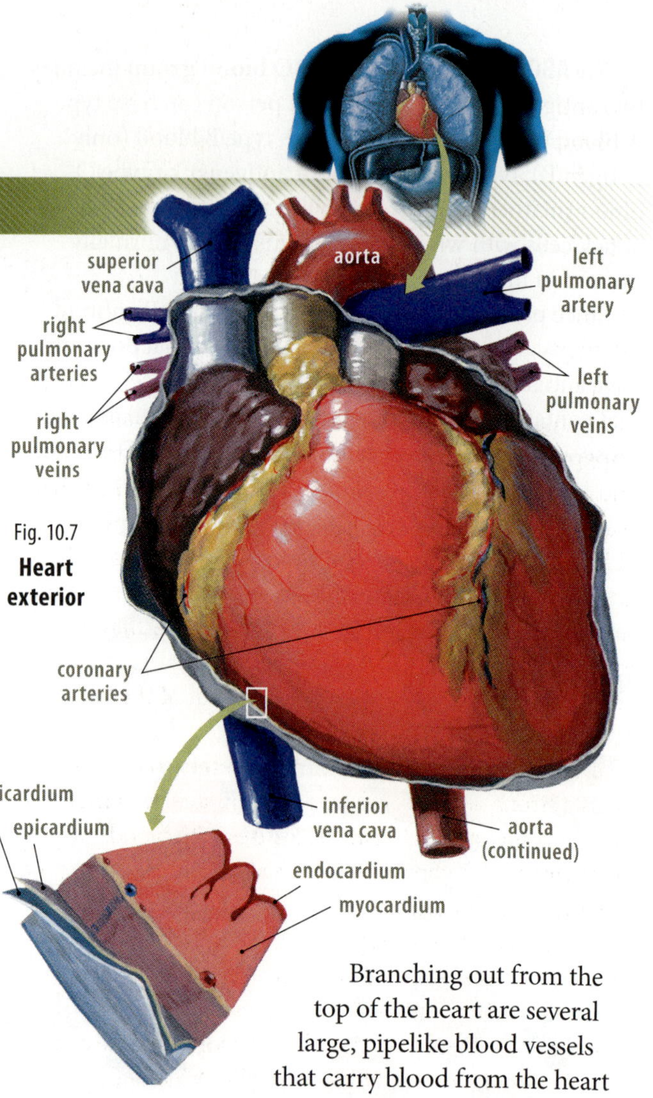


Fig. 10.7 **Heart exterior**

### The Design of the Heart

**External features.** Your heart is located in the center of your chest, just behind your sternum (breastbone), nestled in a hollow cavity between your lungs. Your heart is slightly larger than your fist and weighs between 250 and 350 grams (8 and 12 oz). Although commonly represented by the familiar valentine shape, the heart is actually shaped more like a grapefruit.

The heart is encased within a tough white sac called the **pericardium** [pĕr′ĭ·kär′dē·əm], which helps support and protect the heart. The inside of the pericardium is lined with a membrane that secretes a lubricating fluid; this fluid allows the pericardium to slide smoothly over the surface of the heart as the heart pumps.

Branching out from the top of the heart are several large, pipelike blood vessels that carry blood from the heart to the rest of the body and back again. Smaller arteries, the size of drinking straws, extend downward over the heart's surface to provide nourishment to the heart muscles.

**Internal features.** Internally, the heart's main features are its four hollow spaces, or **chambers.** The pumping

Fig. 10.8
**Heart interior**

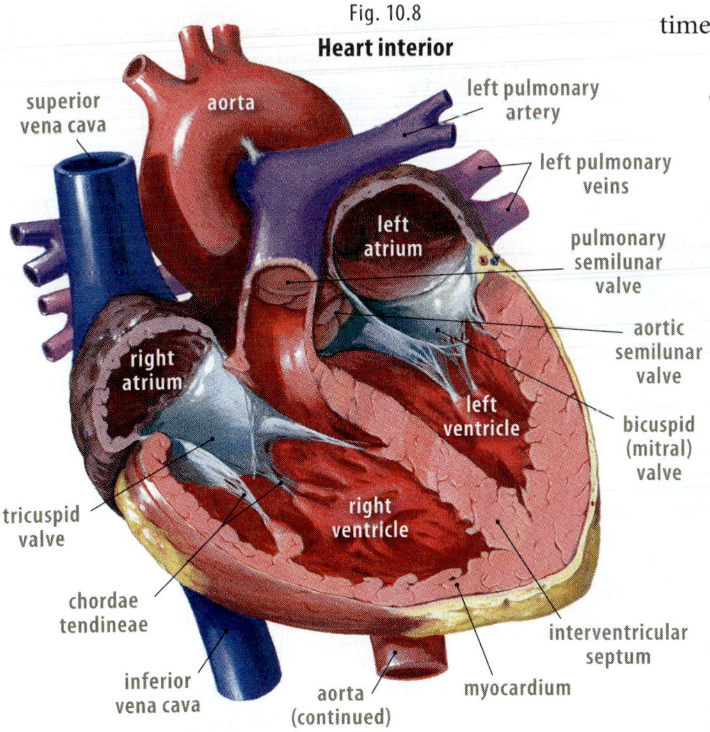


mechanism of the heart consists of two thin-walled upper chambers, called **atria** [ā′trē·ə: sing. atrium], and two thick-walled lower chambers, called **ventricles** [vĕn′trĭ·kəlz]. A thick vertical wall called the **septum** divides the left and right sides of the heart. The pumping action of the heart is provided by strong muscles in the walls of the heart chambers, which cause the chambers to contract forcefully when the heart beats. As the chambers contract, blood is squeezed out of the heart and forced through the body.

The chambers of the heart work together to make the heart a strong and efficient pump. Contraction of the atria helps inflate the thick-walled ventricles with blood. The ventricles, which contract after the atria, force blood out of the heart and through the body's blood vessels.

The walls of the heart's four chambers can be divided into three layers. The most important layer is the **myocardium** [mī′ō·kär′dē·əm], which contains the muscles of the heart. Lining the interior of the heart's chambers is the **endocardium** [ĕn′dō·kär′dē·əm], which provides a smooth coating for the heart's interior. The outside of the heart is coated with a layer of slippery tissue called the **epicardium,** which allows the heart to move easily within the pericardial sac as it beats. (Because the epicardium and the pericardium merge at their edges, the epicardium is sometimes considered part of the pericardial sac instead.)

Blood is prevented from flowing in the wrong direction by four **valves** that are strategically located in the heart. Two of these valves—the bicuspid valve and the tricuspid valve—are the **A-V valves** (atrioventricular valves) because they allow blood to flow from each atrium into the corresponding ventricle. The **bicuspid valve** *(mitral valve)* connects the left atrium and left ventricle, while the **tricuspid valve** connects the right atrium and right ventricle. The names of these valves refer to the number of cusps (flaps) that compose each valve: the bicuspid valve has two cusps, and the tricuspid valve has three.

The other two valves in the heart allow blood to flow to the body from the ventricles but prevent blood from flowing backward into the ventricles from the body. Both of these valves are known as **semilunar valves** because each of their three cusps is shaped somewhat like the crescent moon. The **aortic semilunar valve** allows blood to exit the left ventricle and enter the *aorta* (which carries blood to the body tissues), while the **pulmonary semilunar valve** allows blood to exit the right ventricle and enter the *pulmonary arteries* (which carry blood to the lungs to absorb oxygen). God has designed all four of the heart valves so that, unless they are damaged by injury or disease, blood will move in one direction only.

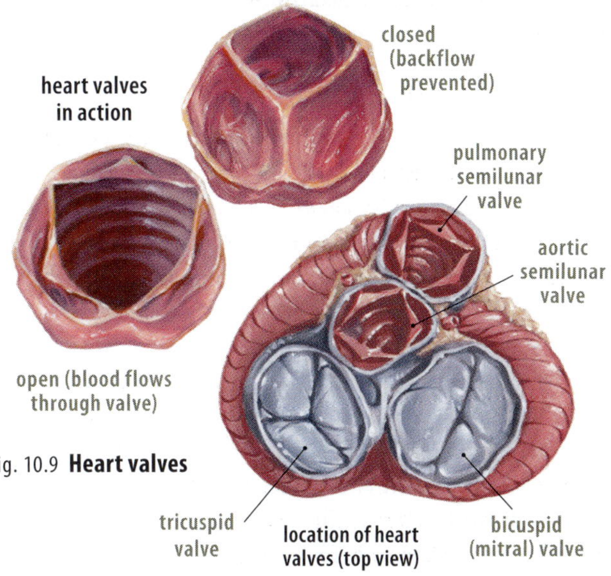


Fig. 10.9 **Heart valves**

The muscles and valves of the heart are supported by the heart's skeleton, which is composed of fibrous connective tissue. This strong but flexible skeleton supports the valves of the heart and provides a framework to which the muscles of the heart are attached.

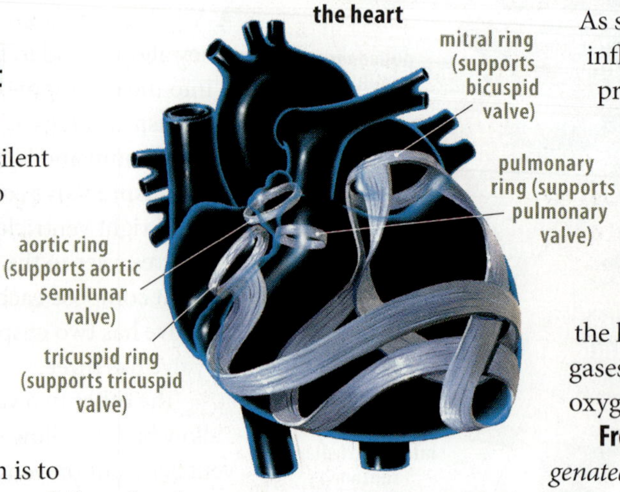


Fig. 10.10
**Fibrous skeleton of the heart**

## Blood Flow in the Heart

The pathway of blood through the heart bears silent yet eloquent testimony to the wisdom of its Designer. The heart can be regarded as a precision-engineered two-stage pump. The first stage of the heart consists of the two atria, whose sole function is to ensure that the ventricles are fully inflated with blood. When fully inflated, the ventricles contract to force blood out of the heart to the rest of the body.

The right side of the heart (right atrium and right ventricle) receives oxygen-depleted blood from the body and pumps it to the lungs. The left side of the heart (left atrium and left ventricle) receives oxygen-rich blood from the lungs and pumps it to the rest of the body. Thus, *each red blood cell passes through the heart twice on its way to the body tissues.*

**From the body to the right atrium and ventricle.** Blood returning to the heart from the tissues of the body has been depleted of oxygen and must be returned to the lungs to receive a fresh supply. This *deoxygenated blood* returns to the heart through the **venae cavae** [vē′nē kā′vē; sing. vena cava, vē′nə kā′və], the body's largest veins. The **superior vena cava** returns blood from the upper part of the body, and the **inferior vena cava** returns blood from the lower part of the body.

As the heart rests between beats, blood flows into the right atrium from the venae cavae; the open tricuspid valve allows much of this blood to flow directly into the right ventricle. The pressure of this blood "inflates" the relaxed heart like air inflates an empty balloon. The right atrium is filled to its normal capacity, but the right ventricle is inflated to only about 70% of its normal capacity because its thicker walls make it harder to inflate.

When the right atrium is filled with blood, its walls contract and squeeze the blood through the tricuspid valve into the right ventricle, inflating the ventricle another 30% so that it is fully filled.

**From the right ventricle to the lungs.** As soon as the right ventricle is fully inflated with blood, it contracts; the pressure of the blood shuts the tricuspid valve and forces open the pulmonary semilunar valve, allowing blood to pass through the pulmonary semilunar valve into the **pulmonary artery,** which transports blood from the heart to the lungs. In the lungs, an exchange of gases takes place as the blood receives oxygen and gives up carbon dioxide.

**From lungs to heart to body.** The *oxygenated blood* travels back to the heart from the lungs through four **pulmonary veins**—two for each lung—and enters the left atrium and left ventricle, inflating them. The left atrium is fully inflated, but the left ventricle is inflated to only 70% of its normal capacity. The left atrium then contracts, squeezing additional blood into the left ventricle until the ventricle is filled to its normal capacity. The left ventricle then contracts, closing the bicuspid valve and forcing blood through the aortic semilunar valve into the **aorta,** the body's largest artery. From the aorta and its branches, blood, the stream of life, is distributed to all parts of the body.

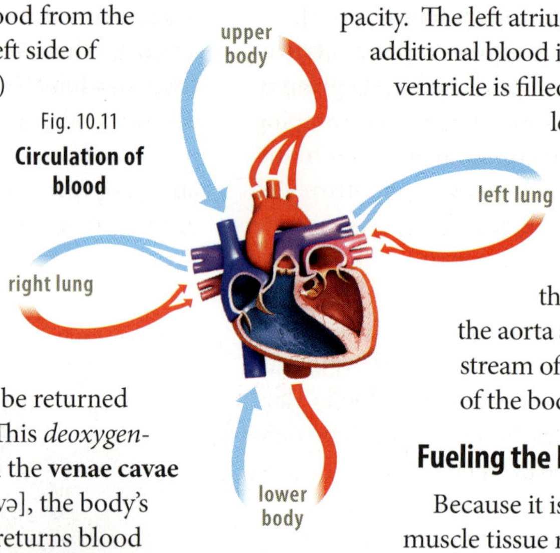


Fig. 10.11
**Circulation of blood**

## Fueling the Heart

Because it is continually active, cardiac muscle tissue requires a large amount of blood to supply it with fuel and oxygen. In fact, a heart muscle uses about half its own weight in blood every minute when you are resting and up to twice its weight in blood per minute during heavy exercise. As a result, the heart retains about $\frac{1}{20}$ of its total blood output to meet its own needs. The primary fuel of the heart is fatty acid molecules, but it can use other fuels instead if fatty acids are in short supply. The heart "burns" these fuels (through cellular respiration) to produce ATP, which powers its rhythmic contractions.

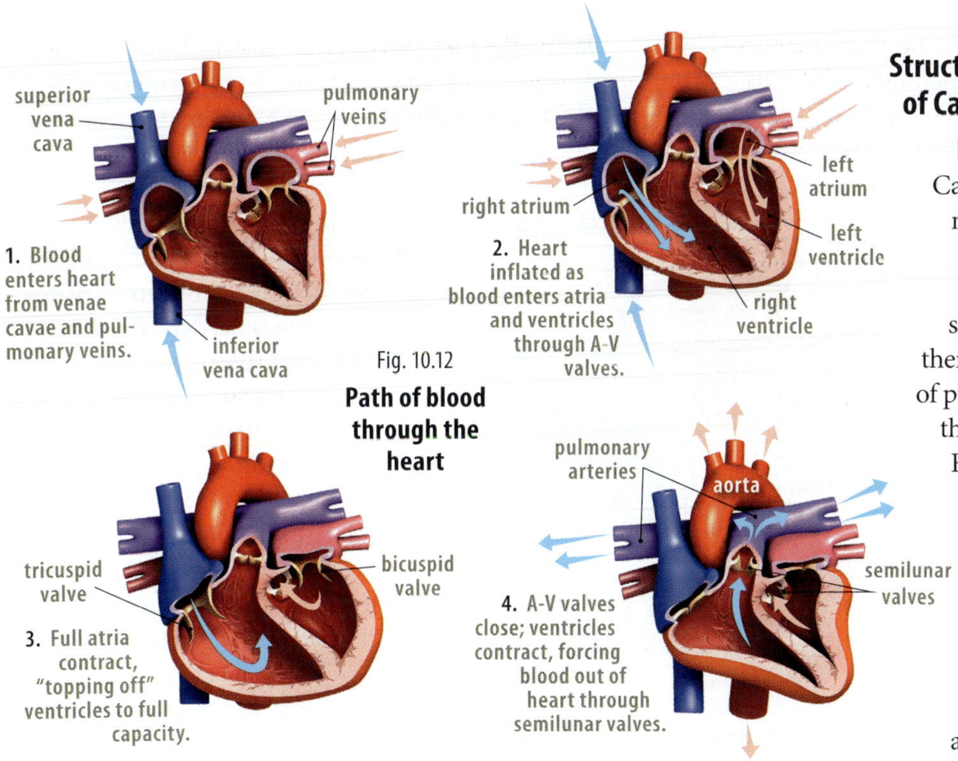


Fig. 10.12
**Path of blood through the heart**

### Structure and Function of Cardiac Muscle

**Design of cardiac muscle.** Cardiac muscle is similar in many ways to skeletal muscle. However, God designed cardiac muscle cells with special features that make them uniquely suited to the job of pumping blood throughout the body, day in and day out. For example, unlike skeletal muscle cells, which take the form of long, thin cylinders with many nuclei, *cardiac muscle cells are short, branched, and contain only one nucleus per cell* (Fig. 10.13). The cells are connected end to end by special connectors called *intercalated* [ĭn·tûr′kə·lāt′ĕd] *disks,* which allow them to contract together as a unit.

*Cardiac muscle cells lack the fuel and oxygen storage systems of skeletal muscle cells to make room for more mitochondria "power plants."* Therefore, cardiac muscle cells can generate large amounts of power for long periods without becoming tired but are absolutely dependent on a steady fuel supply from the bloodstream.

Blood is distributed to the cardiac muscle cells by an intricate system of blood vessels that pass over the surface of the heart and into its interior. Blood enters this network through the two **coronary arteries,** which branch from the base of the aorta. The coronary arteries, roughly the diameter of drinking straws, divide into smaller branches as they pass downward and around the sides of the heart. Each branch fans out into still smaller branches that penetrate deep into the heart, supplying the cardiac muscle cells with blood.

After supplying the cardiac muscle tissues with fuel and oxygen, blood leaves these tissues through the **cardiac veins.** These veins collect the blood from the tissues of the heart and carry it to a larger vein called the *coronary sinus,* which empties into the right atrium (along with the oxygen-depleted blood from the rest of the body). From there, the heart pumps the deoxygenated blood to the lungs to pick up a fresh supply of oxygen.

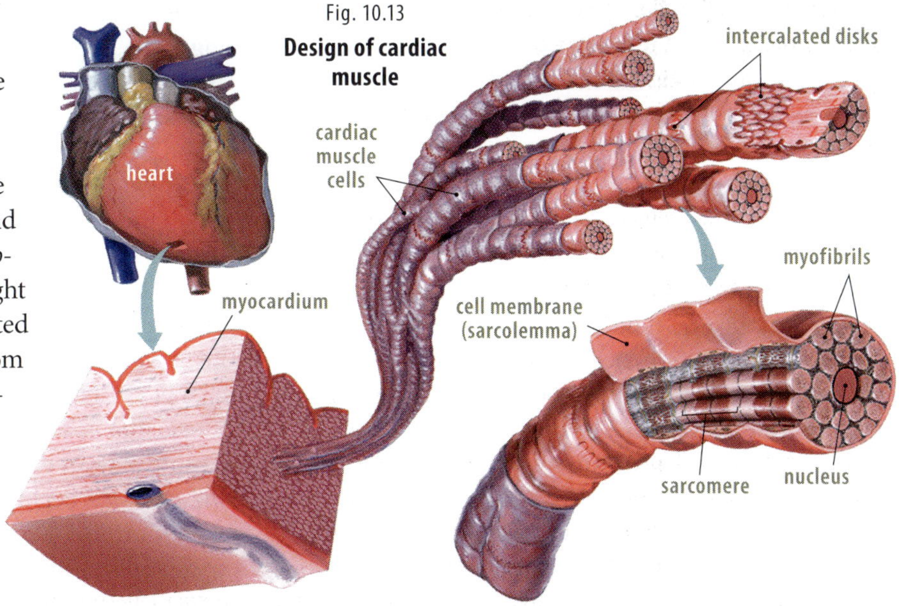


Fig. 10.13
**Design of cardiac muscle**

Finally, cardiac muscle differs from skeletal muscle in that it is *self-exciting.* Although cardiac muscles are stimulated to beat by the heart's built-in pacemaker (the *sinoatrial node*), cardiac muscle cells will beat even if separated from the heart. When several cells are connected by their intercalated disks, they automatically beat together. This unique property serves as a fail-safe mechanism for the heart; even if the heart's electrical system is damaged, the heart can usually continue to beat.

**Electrical system of the heart.** The beating of the heart is regulated by a system of special heart cells known as the **cardiac conduction system.** These highly specialized cardiac muscle fibers generate electrical impulses and distribute them to all parts of the heart.

Fig. 10.14
**Cardiac conduction system**

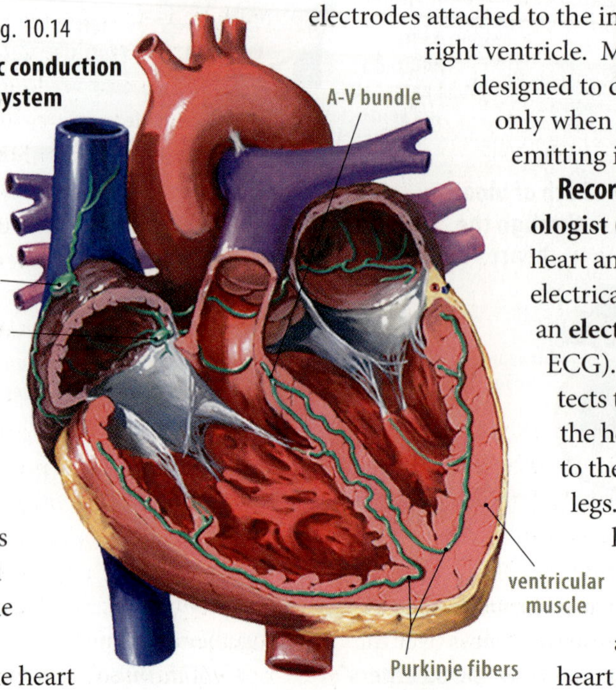


The master control unit of the heart is the **sinoatrial node** (also called the *S-A node*). Situated within the wall of the right atrium, the S-A node is the source of the electrical impulses that stimulate the heart to beat. (For this reason, the S-A node is commonly referred to as the pacemaker of the heart.) These impulses are carried to the rest of the heart by special cells known as *Purkinje* [pûr·kĭn′jē] *fibers,* which spread through the heart like electrical cables, carrying electrical impulses from the S-A node to all parts of the heart.

When the S-A node generates an electrical impulse, the electric charge begins to race through the heart along the Purkinje fibers. The signal reaches the atria first, stimulating them to contract and force additional blood into the ventricles. An electrical barrier between the atria and ventricles prevents the signal from reaching the ventricles before they are filled; the only electrical pathway to the ventricles is a "bridge" called the *atrioventricular node* (A-V node), which delays the impulse about $\frac{1}{10}$ second, until the ventricles are fully inflated with blood. After this, the impulses cross the A-V node and reach the ventricles, triggering them to contract and pump the blood through the body.

**Artificial pacemakers.** When the electrical system of the heart fails, an artificial *pacemaker* can be surgically implanted in the heart. A pacemaker consists of a battery and a microprocessor buried beneath the skin, wires inserted through a small hole in a vein, and electrodes attached to the inside of the right atrium and right ventricle. Most modern pacemakers are designed to deliver a shock to the heart only when one is needed rather than emitting impulses continually.

**Recording heart impulses.** A **cardiologist** (a doctor specializing in the heart and its disorders) can study the electrical activity of the heart using an **electrocardiogram** (EKG or ECG). The electrocardiogram detects the tiny electrical currents in the heart through electrodes taped to the patient's chest, arms, and legs. The electrocardiogram allows the cardiologist to watch for unusual heart rhythms and can reveal the location and extent of damage to the heart as a result of a heart attack.

## The Heart at Work

**Systole and diastole.** The heart pumps about 2000 gallons of blood each day and about 680,000 gallons each year. To accomplish this task, the atria and ventricles contract and relax rhythmically in a highly coordinated series of actions known as the *cardiac cycle.* The contracting and pumping phase of heart action is called **systole** [sis′tə·lē], while the relaxing and filling phase is known as **diastole** [dī·ăs′tə·lē]. When the atria contract and force blood through the A-V valves, the ventricles relax and fill with blood. When the ventricles contract and squeeze blood through the semilunar valves, the atria relax and fill with blood.

**Heartbeats and heart sounds.** The cardiac cycle is perceived as a **heartbeat.** When a physician listens to your heart with a stethoscope, he hears two distinct sounds—a long, low-pitched "lub" and a shorter, higher-pitched "dup." The "lub-dup" sound is produced primarily by the closure of heart valves. The first sound is produced when the ventricles begin to contract, causing the A-V valves to slam shut. The second sound is produced when the ventricles finish contracting and the semilunar valves shut.

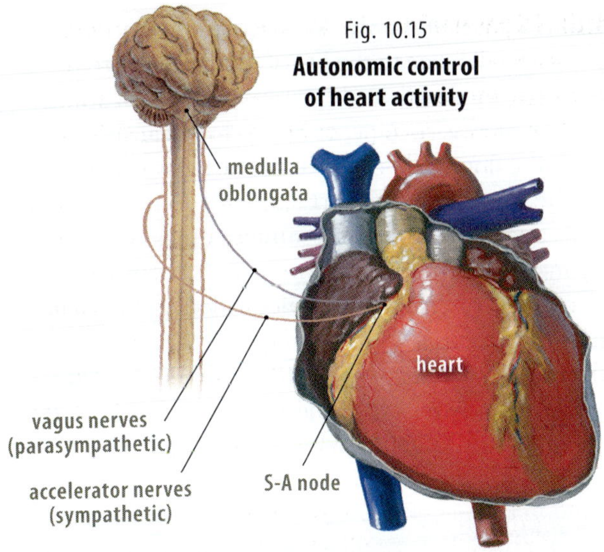


Fig. 10.15
**Autonomic control of heart activity**

Doctors often listen carefully to heart sounds because valve problems can often be detected by the abnormal heart sounds they produce. An abnormal rushing or blowing sound produced by valve damage is called a **heart murmur.** A heart murmur can be caused by a valve that does not close properly, allowing blood to leak back into the chamber, or by a valve that does not open properly, forcing the blood to rush noisily through a small opening. Valve damage can be caused by a bacterial infection in the heart or may result from genetic malformations of the valves.

**Regulating heart output.** The blood requirements of your body's tissues and organs are not constant but depend on your activity level. The brain increases the heart rate through the *autonomic nervous system,* which transmits signals directly to the sinoatrial node. Signals that increase the heart rate are transmitted to the S-A node by a pair of *accelerator nerves* that link the heart to the spinal cord. Signals that slow the heart rate are transmitted by branches of the *vagus nerves* (the tenth pair of cranial nerves). The accelerator and vagus nerves are part of the autonomic nervous system; the accelerator nerves belong to the sympathetic division, while the vagus nerves belong to the parasympathetic division. During highly vigorous activity or when you are extremely afraid or excited, the brain may increase your heart rate to 120–200 beats per minute via the accelerator nerves. When you are asleep, the brain may reduce your heart rate to 50–60 beats per minute via branches of the vagus nerves.

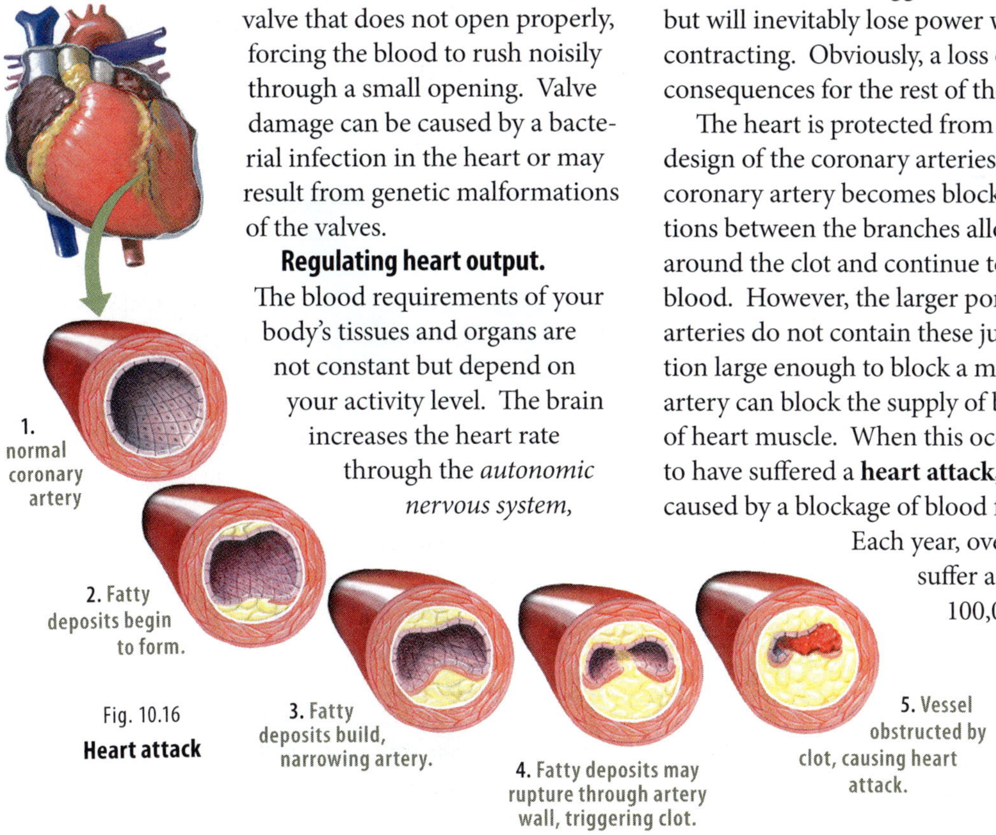


Fig. 10.16
**Heart attack**

## Heart Failure

**Heart attack.** Unlike most muscles in the body, the muscles of the heart must continually contract without ceasing. Although the heart muscles are marvelously designed for this rigorous duty, they require a constant supply of fuel and oxygen. If deprived of fuel and oxygen by a blockage in a blood vessel, the heart muscle cells will struggle valiantly to continue beating but will inevitably lose power within minutes and cease contracting. Obviously, a loss of heart function has dire consequences for the rest of the body.

The heart is protected from minor blockages by the design of the coronary arteries. If a minor branch of a coronary artery becomes blocked by a blood clot, junctions between the branches allow the blood to detour around the clot and continue to supply the heart with blood. However, the larger portions of the coronary arteries do not contain these junctions, so an obstruction large enough to block a major branch of a coronary artery can block the supply of blood to a large portion of heart muscle. When this occurs, the person is said to have suffered a **heart attack,** a type of heart failure caused by a blockage of blood flow to the heart muscles. Each year, over 700,000 Americans suffer a heart attack, and over 100,000 die as a result.

When a heart attack occurs, the part of the heart that loses its blood supply may die unless the blood flow is quickly restored.

If the affected part is small, the heart can often continue beating despite the damage. In a short time, the dead muscle cells will be replaced by scar tissue, and the heart is said to have healed. The heart may not be quite as strong as it was before the heart attack, but it can still pump enough blood to meet the needs of the body. However, if a large portion of the heart loses its blood supply, the heart may lose its ability to beat, causing death within minutes unless immediate medical treatment can restore the flow of blood through the body.

**Ventricular fibrillation.** A heart attack is not the only type of heart failure. **Ventricular fibrillation** [fĭb′rə·lā′shən] occurs when the heart stops beating and instead twitches randomly. Ventricular fibrillation can be caused by damage to the heart's electrical system (such as from a heart attack), by an overdose of stimulant drugs such as cocaine, or by an electrical shock. Because the flow of blood through the body suddenly stops, the victim will lose consciousness in ten to fifteen seconds and die in four to six minutes unless CPR is performed or his heart can be restarted. Generally, the heart must be restarted by a device called a *defibrillator,* which applies a brief electrical shock to the heart to jolt its electrical system back to normal. Although CPR cannot ordinarily restart the heart, it can often keep the person alive until a defibrillator is available.

## Developments in Heart Surgery

Understanding the LIVING CREATION

Many amazing advances have been made in the field of heart surgery. Generally, the simplest types of heart surgery involve repairing wounds in the heart by stitching them closed; similar techniques can be used to repair complex birth defects of the heart. Many heart repairs require the aid of a *heart-lung machine* because the heart must be stopped and, often, opened. This complex machine pumps blood from the body, removes carbon dioxide, adds oxygen, and pumps it back to the body, allowing the heart to be stopped by suitable drugs and operated upon.

Most heart surgeries are performed to open or bypass coronary arteries that have been clogged by fatty deposits. The simplest type of coronary artery repair is *angioplasty,* which does not involve opening the chest. One type of angioplasty involves threading a very thin, flexible tube called a *catheter* into an artery in the patient's leg. The catheter is guided through the artery to the heart and threaded into the blocked coronary artery. A balloon at the tip of the catheter is inflated and deflated several times to open the blockage and stretch the artery. Sometimes a catheter is used to insert a tiny metal brace called a *stent* into the coronary artery to keep the artery propped open.

Sometimes the clogs in the coronary arteries are too extensive to be helped by angioplasty and a *coronary artery bypass* must be performed. In the most common form of this procedure, a large vein is removed from the patient's leg and cut into pieces. One end is sewn into the aorta above the heart, and the other end is sewn into the lower portion of a blocked coronary artery, bypassing the blockage. This allows blood to reach oxygen-starved heart tissue.

In *valve-replacement surgery,* the surgeon removes a faulty heart valve and replaces it with a mechanical valve made of stainless steel or plastic, a valve from a pig's heart, or a valve made of cattle pericardium tissue. Replacement valves from animals or made from animal tissue are chemically treated to remove antigens so the patient's immune system will not reject them.

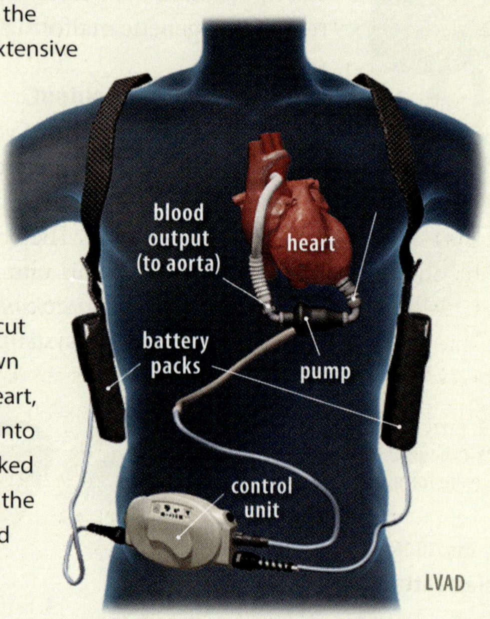

LVAD

Sometimes the heart is damaged beyond repair and a *heart transplant* becomes necessary. The damaged heart is replaced with a donor heart from a fatal-accident victim. Because the body's immune system will often attack a donor heart and reject it, numerous drugs must be used to suppress the patient's immune system, leaving the patient susceptible to infectious diseases. And there are only a few donor hearts available, so patients with badly damaged hearts may die before receiving a transplant.

To avoid these problems, doctors sometimes use mechanical artificial hearts in patients awaiting transplants. The patient's heart may be removed and replaced with a mechanical pump until a transplant heart is available. If the patient's heart still has partial function, a *left-ventricular assist device* (LVAD) may be used instead; an LVAD pumps blood from the left ventricle into the aorta, decreasing the load on the heart. LVAD's are typically used temporarily until a transplant heart is available or the patient's heart recovers, but they can also be implanted permanently.

## Section Review 10.2

1. What is the purpose of the atria? Of the ventricles?
2. Briefly describe the path of blood as it circulates through the heart.
3. Describe the structure of cardiac muscle cells.
4. Describe the path of a cardiac impulse through the myocardium.
5. Describe the regulation of heart activity by the autonomic nervous system.
6. What is the cause of a heart attack?

**Identify**

pericardium, septum, myocardium, endocardium, epicardium, valves, venae cavae, pulmonary veins, aorta, cardiac veins, electrocardiogram, heart murmur, ventricular fibrillation

# 10.3 Circulation: To Every Cell

## The Blood Vessels

The heart and the blood vessels form a completely closed cardiovascular system. There are three main kinds of blood vessels: *arteries,* which carry blood away from the heart; *veins,* which bring blood to the heart; and *capillaries,* the tiny blood vessels that link arteries and veins and are the functional units of the circulatory system. As arteries branch deep into the body, they form smaller blood vessels known as *arterioles,* which branch into myriads of microscopic capillaries. From the blood flowing through capillaries, oxygen and nutrients diffuse into hungry body cells. The capillaries absorb metabolic wastes back into the blood to be excreted through the lungs or kidneys. The capillaries merge to form small blood vessels known as *venules,* which, in turn, become veins. The system of arteries, veins, and capillaries, no less than the heart itself, bears witness to the infinite wisdom of its Creator and Designer.

**Arteries. Arteries** are thick-walled blood vessels that transport blood away from the heart. These vessels must be tough and somewhat elastic because of the pressure placed on them by the pumping action of the heart. Arterial walls consist of three layers of tissue. The outer layer is made up of tough, fibrous tissue. The middle layer, which is the thickest, is made up of smooth muscle fibers arranged in a circular manner around the innermost layer. These muscles allow the brain to control the diameter of the blood vessels to raise or lower blood pressure or adjust blood flow between different parts of the body. The inner layer is a thin, smooth sheet of epithelial cells formed into a tube through which blood can stream.

**Veins.** The walls of **veins** are just like those of the arteries except they are not as thick or as elastic as arteries. Since the blood in the veins is returning to the heart, it is under less pressure than the blood pumped into the arteries from the heart. Periodically, veins have valves to prevent blood from flowing in the wrong direction. These valves also allow your skeletal muscles to assist in returning blood to the heart. Every time you move, your skeletal muscles squeeze the veins, forcing blood through them; the valves ensure that the blood flows only toward the heart. Thus, physical activity helps your blood return to the heart.

**Capillaries.** **Capillaries** are microscopic structures; the smallest are so tiny that red blood cells must "squeeze" through them, traveling in single file. The wall of a capillary is only one cell layer thick; this layer is the same type of epithelial tissue that forms the inner layer of the arteries and veins. All of the exchanges between the cells of the body and the blood take place through the walls of these blood vessels, and because every cell of the body is very close to a capillary, the exchanges take place quite easily. The amount of blood that reaches the capillaries is based on demand. When you are running, the capillaries in your legs require more blood, while other capillaries may close down temporarily. After you eat a large meal, the capillaries of the intestines, which normally require very little blood, expand and make great demands on the circulatory system.

**Atherosclerosis.** Most heart attacks result from a blockage of the arteries caused by a slowly developing disease called **atherosclerosis,** a form of *arteriosclerosis* [är·tēr′ē·ō·sklə·rō′sĭs: "hardening of the arteries"]. Blood vessels are normally very smooth inside to allow blood to slip by. Deposits of fatty material and calcium build up to form **plaque,** which makes the linings rough. Scar tissue forms around plaque deposits because of ruptured cells and irritated artery linings. Over time, the plaque deposits and scar tissue build up, narrowing the inside diameter of the artery. This means that the heart must work harder to pump the same amount of blood through smaller channels. To compound the problem, the disease is narrowing the heart arteries just when the heart needs more blood due to the extra workload. The progress of the disease accelerates in later years until a heart artery becomes completely blocked.

Fig. 10.17

**Artery, vein, and capillaries**

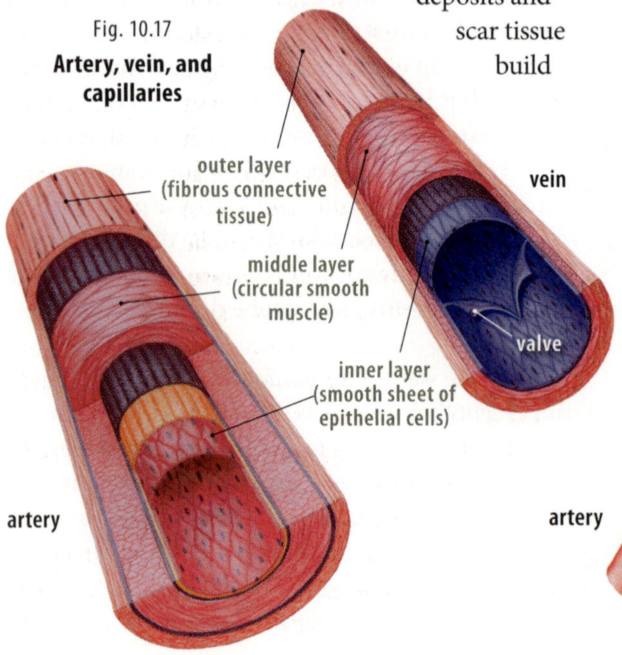


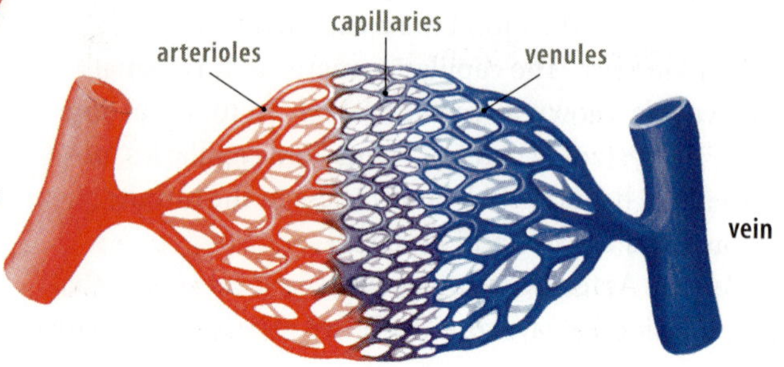


Atherosclerosis can also lead to *hypertension* (high blood pressure) and to kidney failure. As arteries in the kidneys become clogged and hard, they cannot properly filter the wastes out of the blood. Waste materials remain in the blood, putting more stress on the body. The kidneys sense their failure and secrete a substance that makes the heart work harder and the blood pressure go up.

## Blood Circulation

**Systemic circulation.** The blood follows several pathways through the body. The movement of the blood through all parts of the body except the lungs is **systemic circulation.** Arteries branching from the *aorta,* which receives blood from the left ventricle of the heart, enter every organ of the body. In the lower back, the aorta itself divides into two arteries, which supply the legs with blood. After entering the organs, the arteries branch until they form capillaries. The capillaries carry the blood from the smallest arteries to the smallest veins. The small veins unite to become larger veins, much as a river is built up by its tributaries, until they form the *venae cavae,* which empty into the right atrium of the heart.

Systemic circulation includes three special branches: coronary circulation, portal circulation, and renal circulation. *Coronary circulation* is the flow of blood to the heart muscle through the coronary arteries and back to the right atrium. Portal circulation and renal circulation allow the blood to be filtered and processed so that proper body function can be maintained.

Deoxygenated blood from the digestive organs and spleen does not flow directly back to the heart but first travels to the liver through the *hepatic* [hĭ·păt′ĭk] *portal vein;* the movement of blood from these organs to the liver is **portal circulation.** In the liver, carbohydrates absorbed by the digestive system are converted to *glycogen* for storage, toxins and bacteria in the blood are destroyed, and old red blood cells are dismantled. The blood is then returned to the inferior vena cava through the *hepatic veins.*

The movement of blood through the kidneys is **renal circulation.** A branch of the aorta (the *renal artery*) carries blood to the kidneys, where the blood is filtered and various wastes are removed. The *renal vein* returns blood from the kidneys to the inferior vena cava.

**Pulmonary circulation.** The movement of blood between the heart and the lungs is known as **pulmonary circulation.** (The word *pulmonary* refers to the lungs.) Pulmonary circulation begins when blood leaves the heart and travels to the lungs through the pulmonary artery. This artery branches into two arteries; one branch carries blood to each lung. After the blood is oxygenated in the capillaries of the lungs, it returns to the left atrium through the pulmonary veins.

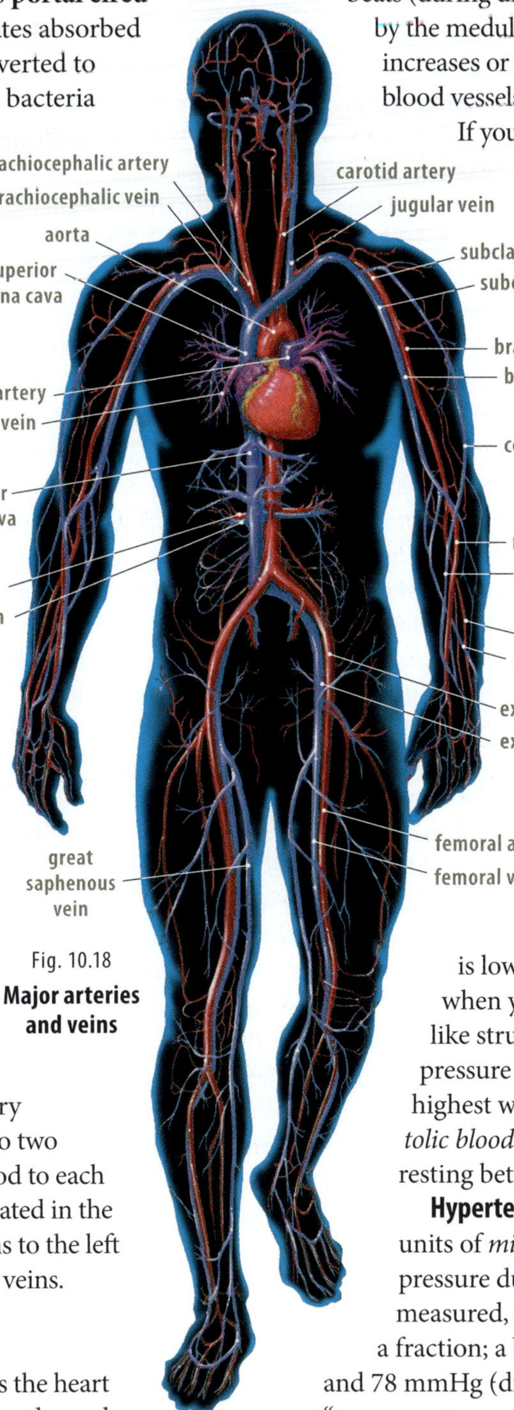


Fig. 10.18
**Major arteries and veins**

## Blood Pressure and Pulse

**Purpose of blood pressure.** As the heart beats, it forces blood into the blood vessels under pressure. The force that blood exerts upon the walls of the blood vessels is known as **blood pressure.** Blood pressure forces blood through your arteries, veins, and capillaries; helps keep blood vessels inflated; and prevents fluids from pooling in the tissues. Blood pressure is also responsible for filling the heart with blood between beats (during diastole). Blood pressure is adjusted by the medulla oblongata in the brain stem, which increases or decreases the tension in the walls of blood vessels to raise or lower the blood pressure.

If your blood pressure drops too low, your heart will not fill properly during diastole and will not pump sufficient blood throughout the body, causing circulatory shock.

**Blood pressure variation.** Blood pressure is not constant throughout the body. It is highest in the arteries that carry blood from the pumping heart to the rest of the body and lowest in the veins, which return blood to the heart after it has passed through all the body's tissues. When you are standing, your blood pressure is somewhat greater in your feet than in your head because of gravity. It also varies with your level of activity; it is lowest when you are sleeping and highest when you are performing a strenuous action like struggling to lift a heavy weight. Blood pressure also varies as your heart beats; it is highest when the ventricles are contracting *(systolic blood pressure)* and lowest when the heart is resting between beats *(diastolic blood pressure).*

**Hypertension.** Blood pressure is measured in units of *millimeters of mercury* (mmHg). The pressure during systole and during diastole are measured, and the two numbers are expressed as a fraction; a blood pressure of 117 mmHg (systolic) and 78 mmHg (diastolic) is recorded as 117/78 (read "one-seventeen over seventy-eight").

The normal upper limit of systolic/diastolic blood pressure is 120/80 ("one-twenty over eighty"), but if you are not overweight and engage in athletic activities, your resting blood pressure is probably lower (such as 115/65).

# The Liver: Chemical Factory of the Body

Understanding the LIVING CREATION

The **liver** is one of the largest and most versatile organs in the body. It is difficult to classify the liver into a single system because it serves vital functions in the digestive system, the circulatory system, the excretory system, and even the lymphatic and immune systems.

Your liver is located in the upper right side of your abdomen just beneath your diaphragm. It consists of two *lobes,* one of which (the right lobe) is considerably larger than the other. Each lobe is composed of tens of thousands of *lobules,* which are the functional units of the liver.

About 1.5 liters of blood flows through the liver every minute. This blood is filtered by the liver cells, which remove substances from the bloodstream that the kidneys cannot excrete, such as used hormones, excess calcium, protein fragments, and toxic substances. Another important function of the liver involves intestinal bacteria that escape the intestine and enter the bloodstream through the intestinal walls. God designed the body so that the blood vessels from the intestines pass through the liver before returning to the heart. Special cells *(Kupffer cells)* in the walls of liver blood vessels snatch bacteria out of the bloodstream and destroy them. So efficient is this system that 99% of the bacteria that escape the intestine are captured and destroyed by the liver on the first pass.

The liver is also an important storage organ for the body. When blood sugar is high, liver cells remove glucose from the blood and store it as *glycogen;* when blood sugar is low, the liver cells convert the glycogen to glucose and release it back into the bloodstream. The liver also stores vitamins (particularly A, D, and $B_{12}$) and minerals (especially iron). About a half-quart of reserve blood is also stored in the liver.

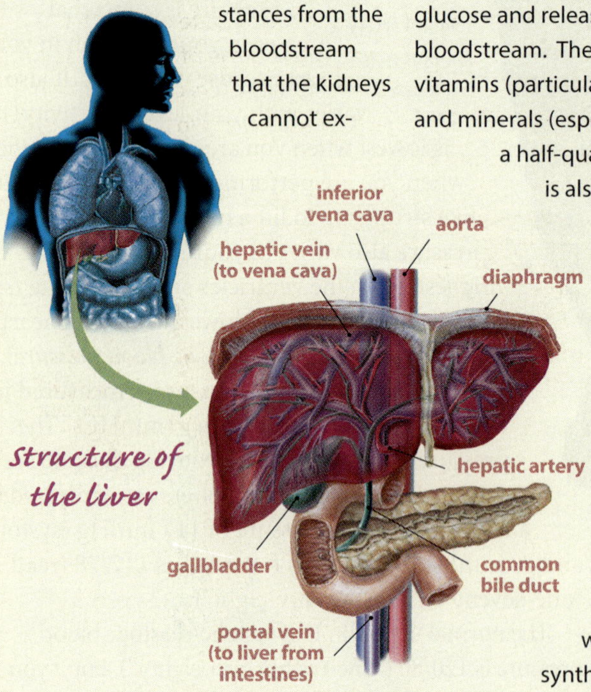


Structure of the liver

The liver is known as the body's chemical factory, and with good reason. Cells in the liver produce all of the blood-plasma proteins except antibodies (which are produced by white blood cells); synthesize lipoproteins (used for fat transport throughout the body) and bile salts (used for digestion of fats); convert amino acids into glucose (when the body is low on carbohydrates) and carbohydrates into fats (when the body is low on fats); and produce about half of the body's lymph fluid.

Because the liver performs so many jobs, liver damage is extremely serious; the body cannot survive for more than a day or two without at least some liver function. Among the most common causes of liver failure are viral hepatitis and alcohol abuse. Liver infections known as **viral hepatitis** can be acquired by several means, including skin contact, contaminated food, and blood transfusions; it is especially common among intravenous drug users. *Alcohol abuse* can cause **cirrhosis** of the liver, the buildup of fatty deposits and fibrous tissue in the liver. Liver failure can also be caused by cancer, genetic diseases, and overdoses of certain drugs.

Because of the liver's great importance to the body, God has designed the liver with a tremendous ability to heal. Unlike most other organs, the liver can regenerate (grow back) even if three-quarters of it is removed. This remarkable regeneration allows liver patients to receive a transplant from a living donor. Part of a donor's liver can be removed and transplanted into the patient. If all goes well, both the donor and the recipient will have normal-sized livers in a few months.

Systolic pressures above 120 mmHg and diastolic pressures above 80 mmHg are considered abnormally high. An individual whose resting blood pressure exceeds either 140 mmHg systolic or 90 mmHg diastolic is suffering from *high blood pressure,* known medically as **hypertension.** Hypertension is harmful because it makes the heart work much harder than normal around the clock, preventing the heart from resting properly and often leading to heart failure. If a person's systolic blood pressure rises too high, there is increased danger of an artery rupturing and causing death. Physicians are also concerned about the diastolic reading because high diastolic pressure prevents the heart from getting sufficient rest and nourishment between beats. Although physical exercise causes your blood pressure to rise temporarily during the exercise (a normal response that helps supply your muscles with extra blood), exercise actually lowers your resting blood pressure (both systolic and diastolic), helping to prevent heart disease.

**The pulse.** As your heart beats, the blood pressure in your arteries rises as the ventricles contract and falls as the heart rests between beats. This change of pressure in the arteries causes their walls to bulge outward and return to normal in time with your heartbeat. This rhythmic pressure change is your **pulse.** You can feel your pulse wherever an artery is located near the surface of your skin; some of the easiest places to feel your pulse are the *carotid* [kə·rŏt′ĭd] *artery* (just forward of the sternocleidomastoid muscle and an inch or two beneath your jawbone), the *radial artery* (just inside the radius, about 1–2 inches below the base of the thumb), and the *temple.* Because your pulse occurs in time with your heartbeat, you can measure your heart rate by measuring your pulse.

## Care of the Cardiovascular System

Most people are born with a strong heart and an efficient system of blood vessels. But with age, many people develop

### Checking Your Blood Pressure

Understanding the LIVING CREATION

If you have ever had your blood pressure checked at a doctor's office, the doctor probably used an apparatus called a sphygmomanometer [sfĭg′mō·mə·nŏm′ĭ·tər]. Essentially, a sphygmomanometer is composed of an inflatable arm cuff, a device (either a rubber bulb or an electrical pump) to inflate the cuff with air, and a gauge for measuring the air pressure in the cuff. The arm cuff is fastened around the patient's arm above the elbow; the cuff is then inflated with air until the pressure squeezes the brachial artery shut. The doctor then slowly lets the air out of the cuff while listening to the lower arm with a stethoscope. When the pressure in the cuff drops to the same as the peak (systolic) pressure in the artery, some blood is then able to force its way through the artery as the heart beats, making a distinctive sound that can be heard through the stethoscope. The air pressure in the cuff when this sound is heard is equal to the systolic blood pressure. When the pressure in the cuff drops to the level of the diastolic pressure, the cuff can no longer hold the artery shut at all, and the sounds suddenly fade. The air pressure in the cuff when the sounds fade is recorded as the diastolic blood pressure. Instead of "listening" for blood flow, digital sphygmomanometers calculate blood pressure based on pressure fluctuations in the cuff caused by the patient's pulse.

Sphygmomanometer

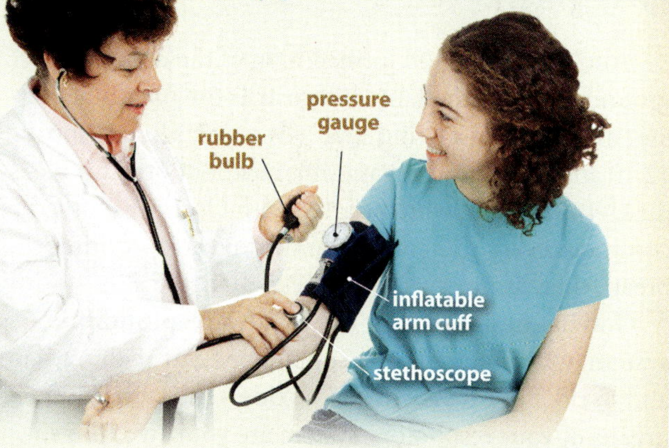

circulatory problems. *The leading cause of death among Americans is cardiovascular disease (CVD).* The term *cardiovascular disease* actually refers to several conditions of the heart, veins, and arteries, including hypertension, atherosclerosis, coronary heart disease, and stroke. After years of extensive research, scientists have identified several factors that contribute to CVD, including diabetes, obesity, a sedentary (inactive) lifestyle, and harmful habits such as smoking.

*Obesity* is a problem because excess weight puts a great stress on the heart. Fats circulating in the blood become deposited in the arteries leading to the heart muscle, clogging them. Most experts recommend that people limit their consumption of animal fats, eat more fresh fruits and vegetables and whole grain products, and maintain a healthy weight. Exercise is also important because it promotes the growth of new blood vessels feeding into the heart, lowers heart rate, prevents obesity, and helps keep blood vessels free of deposits.

Another factor that contributes to CVD is *smoking*. Cigarette smoke contains *nicotine*, which causes the arteries to constrict and elevate the blood pressure, making the heart work harder. A person who smokes is up to four times more likely to die from a heart attack than a person who does not smoke.

A healthy lifestyle is not the only solution, however. A person's spiritual condition also plays an important role in cardiovascular health; adverse effects from stress, bitterness, and guilt degenerate the cardiovascular system. Thus, one of the most important measures for our health is to know and serve the God of creation.

## Section Review 10.3

1. What type of blood vessels carries blood away from the heart?
2. What type of blood vessels carries blood toward the heart?
3. What is the purpose of the capillaries?
4. Name three important functions of the liver.
5. What is the leading cause of death in America?

### Application

6. Name three factors that commonly contribute to cardiovascular disease.

### Critical Thinking

7. Defend this statement: "The way a person lives often determines how long he will live."

### Identify

arterioles, venules, atherosclerosis, viral hepatitis, cirrhosis, hypertension, pulse

### Observe

- Obtain a sphygmomanometer and check your blood pressure.
- Find your pulse and count the beats for fifteen seconds. Multiply your count by four to get the number of heartbeats per minute.

## 10.4 Respiration: The Breath of Life

"And the Lord God formed man of the dust of the ground, and breathed into his nostrils the breath of life; and man became a living soul" (Gen. 2:7; see also Job 12:10 and Isa. 42:5). What is man without breath? He is dust. That breath, which started with God's own breath, continues every minute of your life on earth. Without breath there is no physical life (Psa. 104:29).

God has designed your body with a **respiratory system,** which functions to supply the body with oxygen and to rid it of excess carbon dioxide. The primary organs of your respiratory system are your two **lungs,** large, spongy organs located inside the thoracic cavity on opposite sides of your heart.

In the lungs, the two respiratory gases (oxygen and carbon dioxide) are exchanged between the blood and the air. These gases pass from the air sacs of the lungs to the capillaries, and the blood carries the oxygen to the cells of all the tissues. The process of breathing—inhaling and exhaling air—and gas exchange is called **external respiration.**

The exchange of oxygen and carbon dioxide between the blood cells and the tissue cells is called **internal respiration.** Within the cells themselves, *cellular respiration* takes place. Cellular respiration is the process of releasing energy from food. This process requires oxygen and releases carbon dioxide.

## The Organs of Respiration

**From nasal cavity to bronchi.** When we breathe, air enters the *nose* and passes into the pharynx. This air is warmed, moistened, and filtered as it passes through the **nasal cavity** and circulates through the **nasal meatuses** [mē·ā′təs·ēz], the irregularly shaped pockets, or air spaces, in the nasal cavity. On its passage to the lungs, air passes through the glottis and the voice box (larynx), which are located at the upper end of the **trachea** (windpipe). The **glottis** is the opening of the trachea; the flap of tissue that covers the opening of the trachea is called the **epiglottis.** The epiglottis is open when we are breathing but is closed when we are swallowing. The *larynx,* which is made of cartilage, can be felt under the chin as the Adam's apple. Inside the boxlike larynx are two folds of elastic tissue called *vocal cords* or *vocal folds.* The vocal cords produce sound for speech. The vocal cords can also be closed so tightly together that they prevent air from flowing in or out of the lungs, allowing you to hold your breath. For instance, this action is used when straining to lift a heavy object.

The trachea divides into two branches, the **bronchi** [brŏng′kī; sing. bronchus], when it enters the thoracic cavity. One bronchus enters each lung. The trachea remains open for the free passage of air because horseshoe-shaped pieces of cartilage are spaced regularly along its entire length.

**Lungs.** As each bronchus enters a lung, it divides into several branches called **secondary bronchi.** The secondary bronchi branch many times to form still smaller tubes called **bronchioles** [brŏng′kē·ōlz′]. (The bronchi and their branches, including the bronchioles, are *bronchial tubes.*) The smallest bronchioles end in small saclike structures called **alveoli.** Alveoli are the numerous microscopic, thin-walled structures that make up the tissue of the lungs; each alveolus is surrounded by a network of capillaries. The average adult has between 300 and 400 million alveoli per lung. The design of the alveoli allows molecules of gases to pass freely through the membranes by *diffusion,* the movement of molecules from an area of high concentration to an area of low concentration. Oxygen molecules diffuse easily into the bloodstream through the thin walls of both the alveoli and the capillaries; carbon dioxide and water vapor diffuse just as easily from the bloodstream into the air sacs. These sacs combined can hold about as much air as a basketball.

The lining of the bronchial tubes can be irritated by infection or by smoking; the inflammation of this lining is called **bronchitis.** Severe bronchitis can lead to **pneumonia,** a bacterial or viral infection of the linings in the lungs that causes the lungs to secrete large quantities of fluid. If untreated, pneumonia can be fatal, causing the person to drown in his own body fluids.

The bronchial tubes may also be affected by allergies. **Bronchial asthma** is an allergic condition characterized by spasms of the bronchial tube muscles, which cause the tubes to suddenly constrict, making breathing difficult. Most forms of asthma are caused by immune system malfunctions. When irritants such as dust, pollen, or smoke are inhaled, cells in the lining of the bronchial tubes release chemicals that tighten the tube muscles and cause the linings to swell. The victim can still inhale, but exhaling becomes difficult. Asthma attacks can be treated by inhaling medications that cause the bronchial tubes to relax; regular exercise is also helpful.

Fig. 10.19
**The respiratory system**

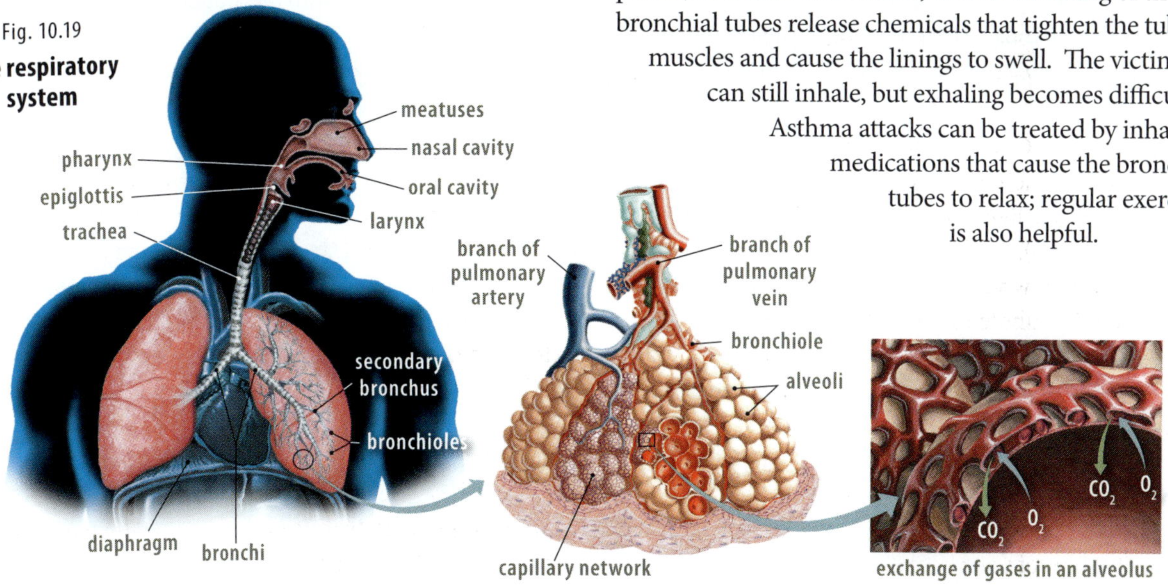

**Pleura.** All of this complex system is housed in an expandable cage of bone called the *rib cage.* God ingeniously designed the ribs with just the right amount of flexibility to permit your lungs to expand and just the right amount of rigidity to shield them from impact. A double membrane called the **pleura** [plo͝or′ə] lines the inside of the rib cage, providing lubrication to prevent friction between the ribs and the delicate surface of the lungs. Sometimes the pleura becomes inflamed, a painful condition known as **pleurisy.**

**Diaphragm.** The thoracic cavity is an airtight chamber, but it has a movable floor called the **diaphragm,** which is a thick sheet of muscle that separates the thoracic cavity from the abdominal cavity. In its relaxed state, the diaphragm bulges upward into the thoracic cavity; when contracted, it moves downward and thus enlarges the thoracic cavity.

## Respiration and Speech

Understanding the LIVING CREATION

The ability to speak in an articulate language and the ability to sing are two wonderful gifts of God given to mankind. We not only have the mental ability for such action but the body parts to make it possible. Speech originates in the throat, inside a delicately designed organ called the **larynx,** or *voice box,* a special enlargement of the trachea. The larynx is located in the front of the neck at the top of the trachea, between the base of the jaw and the sternum. It is constructed of strong cartilage, similar to the cartilage of the external ear. Several **laryngeal** [lə · rĭn′jē · əl] **muscles** act together to move the **vocal cords.** Other laryngeal muscles adjust the tension of the cords.

Normally, your vocal cords are held out of the way, allowing a large opening in the trachea for air to pass through when breathing. When you wish to speak or sing, the vocal cords are swiveled into position by the laryngeal muscles until there is only a small slit between the cords. Air passing through this slit causes the cords to vibrate, producing sound waves in the air. The pitch (frequency) of these sound waves can be adjusted by varying the tension in your vocal cords, just as tightening a string on a guitar or stretching a rubber band changes the pitch of the sounds they produce. The tension of the cords is adjusted by laryngeal muscles that rotate the lower portion of the larynx toward the rear, stretching the vocal cords. When these muscles are relaxed (as in normal speaking), the vocal cords are not pulled tight and the sound is relatively low in pitch. When these muscles are tightened, the vocal cords are pulled tight and the pitch of the sound is raised.

Speech is produced when air is forced out of the lungs and passes over the vocal cords, producing steady sounds. These tones are modified into articulate speech by your mouth, throat, nasal cavity, and tongue.

The ability to speak and sing with quality requires great coordination and training of the muscles associated with respiration. For that reason, learning to speak and sing well takes much practice and work.

Organs of speech

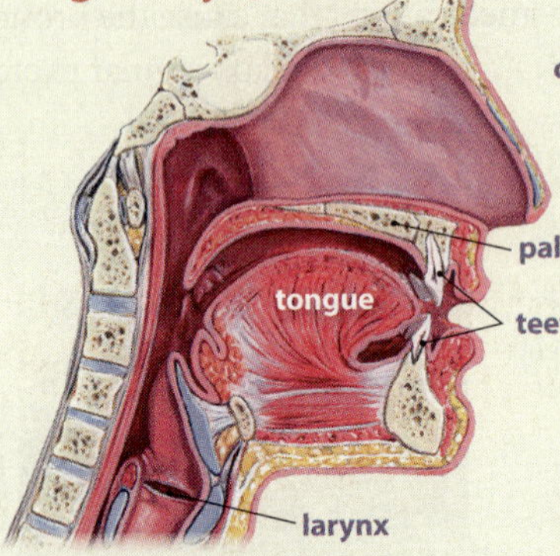


breathing
(cords retracted; muscles not shown)

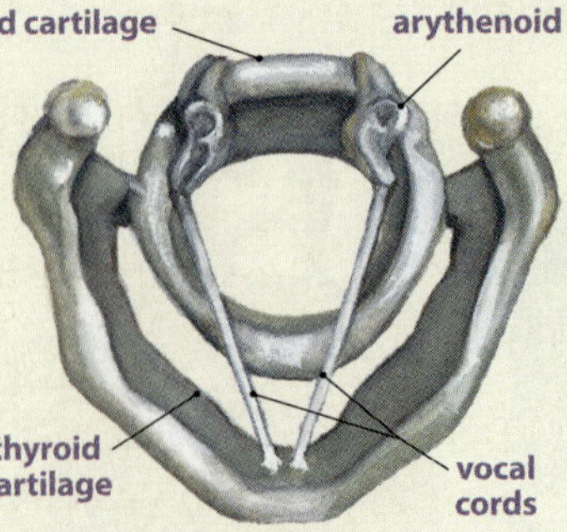


speaking or singing
(cords rotated into position)

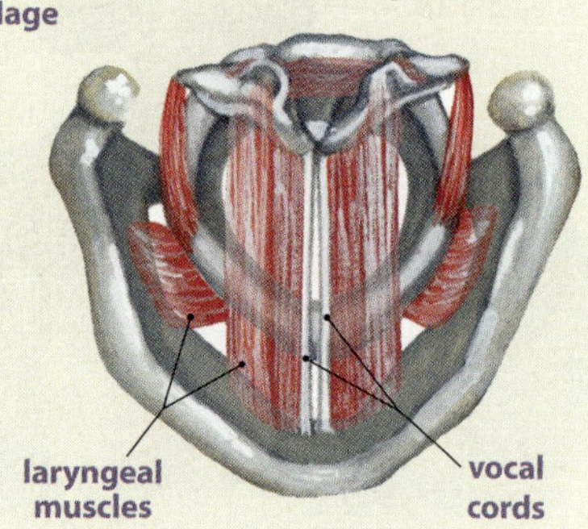

# Trans-Vision™: The Human Body

The following Trans-Vision™ plates reveal the internal anatomy and miraculous design of the human body. Even-numbered plates show the body from the front; odd-numbered plates show the body from the back. Because the plates are transparent, you can easily see how different parts of the body fit together. Each organ is numbered; the names of the organs are in the opaque margins of the plates. You will find it helpful to refer back to these plates to see how the parts of each system relate to other systems and the body as a whole.

## Description of the Plates

**Plate I** – front walls of the thoracic and abdominal cavities

**Plate II** – overview of musculature and internal organs

**Plate III** – organs of the thoracic and abdominal cavities; cross section of the lungs; bones and blood vessels of the arm

**Plate IV** – deeper organs of the thoracic and abdominal cavities

**Plate V** – thoracic and abdominal organs; brain; major blood vessels

**Plate VI** – rear walls of the thoracic and abdominal cavities; skeleton; major nerves

PLATE I
Back view

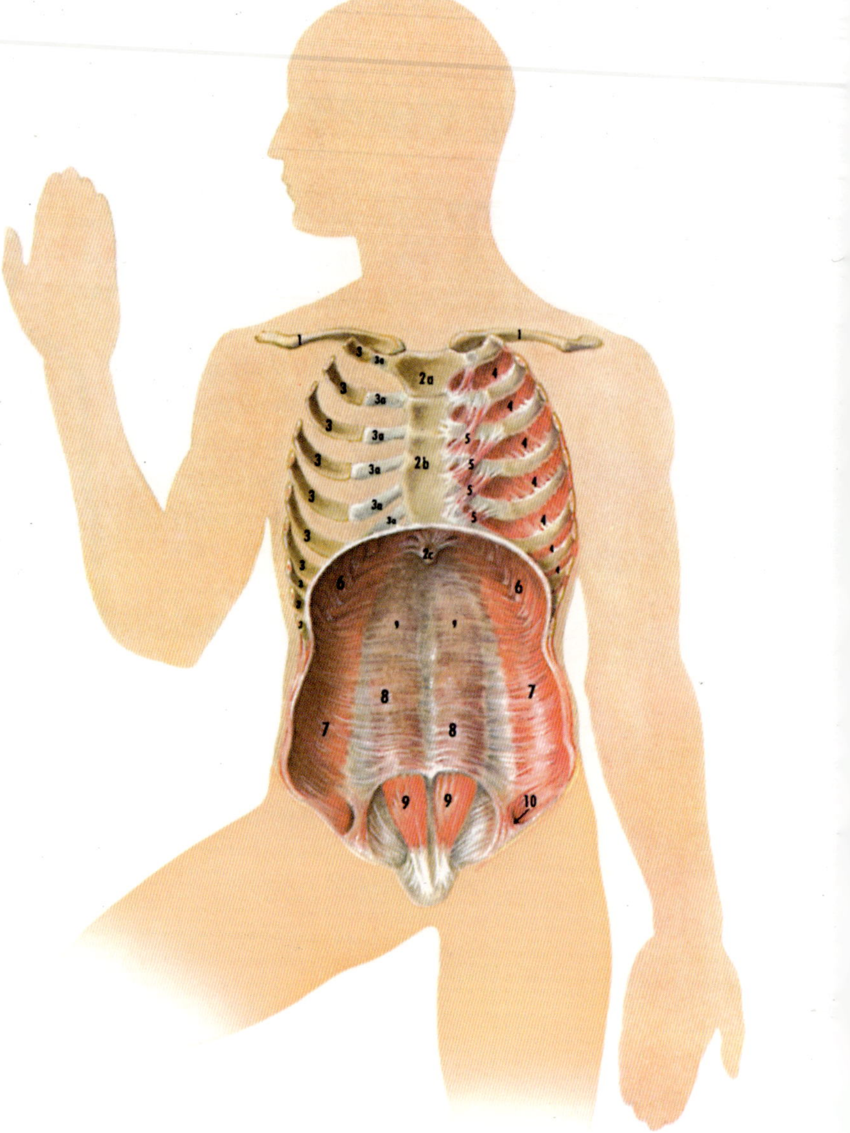


This plate shows the inside surface of the front wall of the chest cavity and the abdominal cavity.

1. Clavice
2a. Manubrium
2b. Body of the sternum
2c. Xiphoid process
3. Ribs
3a. Costal cartilage
4. Intercostal muscles
5. Transverse thoracic muscles
6. Diaphragm
7. Transverse abdominal muscles
8. Sheath of rectus abdominis muscles
9. Rectus abdominis muscles
10. Deep inguinal ring

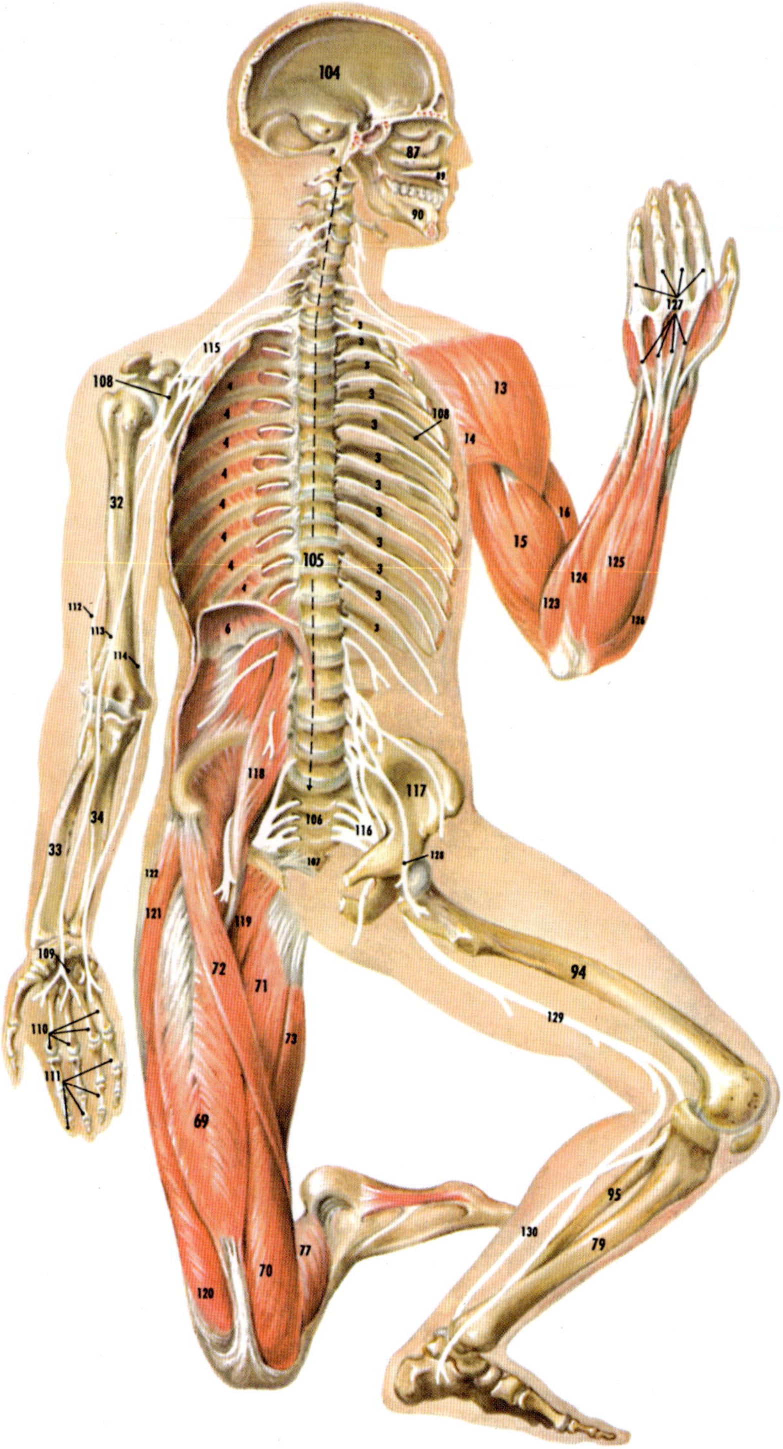


PLATE VI
Front view

104. Skull bones
87. Nasal conchae
89. Maxillary bone
90. Mandible
105. Vertebral column
106. Sacrum
107. Coccyx
108. Scapula
32. Humerus
33. Radius
34. Ulna
109. Carpal bones (wrist)
110. Metacarapal bones
111. Phalanges (finger bones)
112. Radial nerve
113. Median nerve
114. Ulnar nerve
4. Intercostal muscles
3. Ribs
115. Brachial plexus
116. Lumbo-sacral plexus
6. Diaphragm
117. Ilium (pelvic bone)
118. Psoas muscle
119. Pectineus muscle
71. Adductor longus muscle
72. Sartorius muscle
73. Gracilis muscle
77. Gastrocnemius muscle
69. Rectus femoris muscle
70. Vastus medialis muscle
120. Vastus lateralis muscle
121. Tensor fasciae latae muscle
122. Gluteus medius muscle
13. Deltoid muscle
14. Pectoralis major muscle (cut)
15. Biceps muscle (of arm)
16. Triceps muscle
123. Flexor carpi ulnaris m.
124. Extensor carpi ulnaris m.
125. Extensor digitorum communis m.
126. Extensor carpi radialis longus m.
127. Extensor tendons & sheaths
128. Femoral nerve
94. Femur
129. Peronaeus communis nerve
130. Tibial nerve
79. Tibia
95. Fibula

TRANSVISION® CURWOOD, INC.
FREMONT, OHIO

## Table 10.2 Gases Inhaled and Exhaled

| | Inhaled air | Exhaled air |
|---|---|---|
| Oxygen | 20.96% | 16.02% |
| Carbon dioxide | 0.04% | 4.29% |
| Nitrogen | 79% | 79% |

### The Breathing Process

**Changing pressure.** Breathing is based on the same physical principle as a medicine dropper. If the bulb of the dropper is squeezed while the tip is immersed in water, air bubbles will show that air is being squeezed from the bulb. If the bulb is released while the tip is still immersed, the partial vacuum (reduced pressure) in the bulb will cause the water to rise in the glass tube. The water is pushed up into the dropper because the pressure of the air pushing down on the surface of the water is greater than the pressure of the air in the evacuated bulb. A similar process occurs when we inhale and exhale.

The lungs are elastic and can expand or contract according to the pressure surrounding them. When you make your chest cavity bigger, air rushes into the lungs to fill the partial vacuum, and they expand. During the times when you are neither breathing in (inhaling) nor breathing out (exhaling), the pressure in the alveoli and outside the body is equal, and air moves neither into nor out of the lungs. When you make the chest cavity smaller, pressure from the lungs is increased and air rushes out of the lungs until the pressure inside is equal to the pressure outside the body.

**Equalization of pressure.** When a person inhales, muscles attached to the ribs and breastbone (the **intercostal muscles**) contract and lift the ribs and breastbone up and out. At the same time, the diaphragm contracts and moves down. This combined activity results in a larger chest cavity and a reduced pressure in and around the lungs. Air rushes in to equalize the pressure; this process is called **inhalation.**

When a person exhales, the rib and chest muscles and the diaphragm relax. The ribs fall back into place and the diaphragm bulges up into the chest cavity, its relaxed position. The thoracic cavity is thus reduced in size, and the pressure surrounding the surface of the lungs and in the individual air sacs is increased. Air rushes from the lungs to equalize the pressure; this process is called **exhalation.**

**Control by medulla oblongata.** Breathing is controlled by the *medulla oblongata* region of the brain. Impulses are sent from this breathing center through the upper spinal cord to the diaphragm and chest muscles when it is time to inhale and exhale. Special sensory structures in the medulla monitor the carbon dioxide levels in the blood; when the blood is carrying a high concentration of carbon dioxide, the breathing center increases the rate of breathing. When the concentration of carbon dioxide returns to normal, the breathing center adjusts the breathing rate back down to normal. This mechanism explains why you cannot hold your breath for long periods.

Fig. 10.20
**The process of breathing**

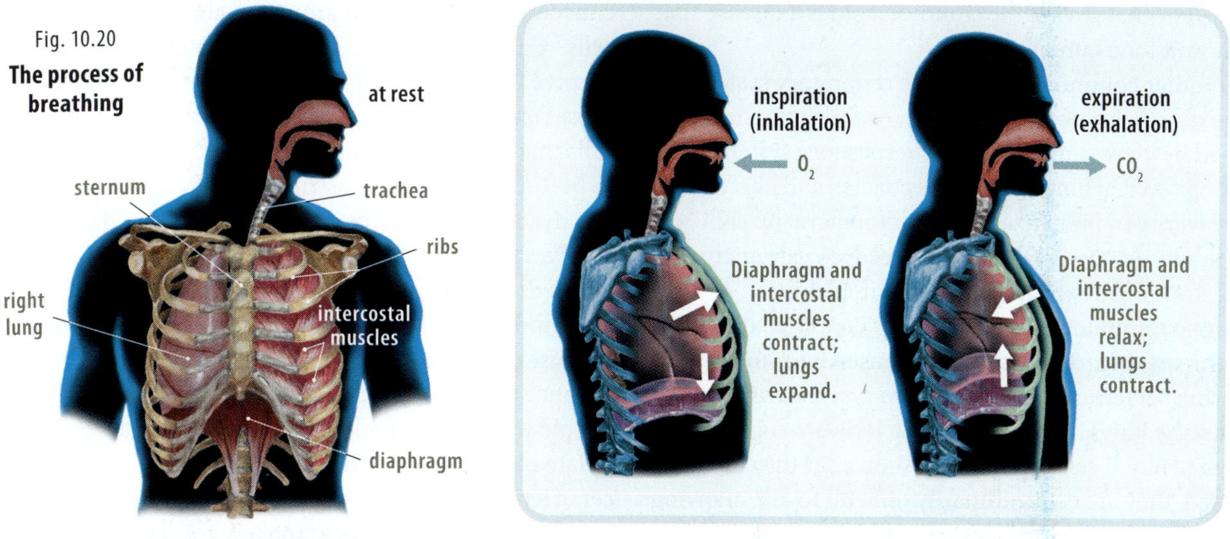

(The average person cannot hold his breath for much more than a minute.) After you voluntarily hold your breath for a while, you are involuntarily forced to breathe deeply and gasp for air. The carbon dioxide that builds up in your blood while you are holding your breath signals the breathing centers of the brain to force you to breathe. Carbon dioxide is not the only thing in the blood that affects breathing rate; the sympathetic nervous system and certain hormones also influence how fast you breathe.

**Lung capacity.** The total lung capacity of an average adult is about 350 cubic inches, or slightly less than six liters. Not all of this lung capacity is usable, because some air will always remain in the lungs no matter how forcefully you exhale. The maximum volume you can inhale and exhale is your **vital capacity.** You would use your vital capacity when running a race. It can be measured by filling your lungs and then blowing all of the air possible into a *spirometer* [spī·rŏm′ĭ·tər], an instrument designed to measure air volumes. The vital capacity of an average-sized adult is about 280 cubic inches (4.6 liters). When we breathe normally, we inhale and exhale approximately 30 cubic inches (½ liter) of air. This normal amount is the **tidal volume.**

## Respiratory Diseases

You have already learned about several disorders and conditions that may affect the respiratory system. Some respiratory diseases, like bronchitis and pleurisy, are more common than others. The most common respiratory disease is undoubtedly the **common cold.** More than a hundred different types of viruses can cause colds. For that reason the common cold is a disease that must wait for a cure.

Another common threat to the respiratory system is the disease known as **influenza,** or flu. Influenza is caused by viruses that infect the respiratory system, causing fever, aching muscles, headache, and chest pain. The severity of influenza varies; although many children and adults get the flu and quickly recover, extremely dangerous forms of influenza sometimes appear.

Some respiratory diseases are less common and more life-threatening. **Tuberculosis** is caused by a slow-growing bacterium *(Mycobacterium tuberculosis)* that infects the lungs. Although the bacteria are engulfed by white blood cells in the lungs soon after they are inhaled, they are extraordinarily difficult to kill, allowing them to survive and reproduce inside the white blood cells. Because of the continuing infection, the lung tissue forms fibrous lumps around the infected sites. The fact that the bacteria can evade the immune system makes tuberculosis difficult to treat even with modern medicine, especially since some strains of the bacteria are resistant to antibiotics.

**Emphysema** [ĕm′fĭ·sē′mə] is a progressive deterioration of the alveoli caused by malfunctioning enzymes; emphysema can result from genetic causes, chronic bronchitis, and continuous exposure to smoke or other irritants. As the alveoli deteriorate, the alveolar walls lose their elasticity and break apart. The tiny chambers form larger chambers, which are less efficient in exchanging gases between the air and the blood. As a result, the person must struggle to get enough oxygen, causing a characteristic shortness of breath. Severe emphysema is especially prevalent among smokers.

**Cystic fibrosis** *(CF)* is one of the most common hereditary diseases in the United States. CF is caused by a genetic defect that affects the chlorine gates of lung cell membranes, hindering the lungs' defenses against bacterial infection and causing the lungs to fill with a thick, gummy mucus. The mucus makes breathing difficult, and the constant lung infections eventually result in lung failure and death. Fortunately, some of the symptoms of CF can be treated by therapy and medications, and considerable progress has been made in understanding the causes and nature of CF. Thus, there is hope that in the relatively near future, CF patients may be able to lead normal lives.

The most devastating respiratory disease is **lung cancer,** the uncontrolled growth of abnormal lung cells. Cancer cells in the lung crowd out normal cells, deprive them of nutrition, and cause their death. The most common form of lung cancer originates in the cells that line the bronchial tubes. The cancer grows to become a mass that obstructs the passage of air to the alveoli. Lung cancer is difficult to treat and usually involves removing the diseased portions of the lung affected. Lung cancer is the most deadly of the more common cancers. The survival rate is extremely low because it is rarely diagnosed until the cancer is well advanced. Fortunately, lung cancer primarily affects people who smoke, making it a disease easy to prevent. Relatively few people who do not smoke get lung cancer. Here is a perfect example of the adage that the best cure is prevention. Do not smoke!

## Section Review 10.4

1. What is the basic purpose of the respiratory system?
2. What is asthma?
3. Briefly describe the process of inhalation.
4. Briefly describe the process of exhalation.
5. What structure of the brain regulates the rate of breathing?

### Critical Thinking

6. Using what you know about inhalation and exhalation, explain why a puncture wound to the chest cavity that allows air to pass through can cause a lung to collapse.
7. In a paragraph, describe how the design and function of the lungs is evidence of a wise Creator.

### Identify

meatuses, bronchus, bronchioles, bronchial tubes, bronchitis, pneumonia, pleurisy, diaphragm, vital capacity, tidal volume, tuberculosis

## Chapter 10 Review

### Define

1. plasma
2. hemoglobin
3. anemia
4. Rh factor
5. pericardium
6. septum
7. coronary arteries
8. systole, diastole
9. heart attack
10. arteries, veins, capillaries, arterioles, venules
11. systemic circulation, coronary circulation, portal circulation, renal circulation, pulmonary circulation
12. cirrhosis
13. blood pressure
14. pulse
15. trachea, bronchi, bronchioles
16. bronchial asthma
17. diaphragm
18. intercostal muscles
19. vital capacity, tidal volume

### Identify

1. the primary organs of the cardiovascular system
2. the three categories of blood proteins and their functions
3. the functions of red blood cells, white blood cells, and platelets
4. the four basic blood types in the ABO group
5. the function of the atria
6. the function of the ventricles
7. the four valves of the heart and their locations
8. three functions of the liver
9. another name for high blood pressure
10. the leading cause of death in the United States
11. three habits that can lead to heart disease later in life
12. the structure that prevents food from entering the lungs
13. five common respiratory diseases and their causes

### Explain

1. How can excessive blood loss cause shock and heart failure?
2. Why are mismatched blood transfusions dangerous?
3. Describe the path of blood through the heart and lungs, from leaving the venae cavae to entering the aorta.
4. How do the differences between cardiac muscle cells and skeletal muscle cells make cardiac muscle more suitable for pumping blood?
5. What is the difference between external, internal, and cellular respiration?
6. Describe the structure and function of the lungs.
7. How is speech produced?
8. How does the brain regulate your heart rate and breathing rate?

### Apply

1. Why are red blood cells dismantled by the spleen and liver instead of continually repairing themselves as most other cells do?
2. During a heart transplant procedure, the nerves to the heart are not always reattached. How can the donated heart beat without being connected to the brain?
3. Normal inhalation at rest is considered an active process, but normal exhalation at rest is considered a passive process. Why is this?

CHAPTER 11

# Integumentary, Excretory, Endocrine & Reproductive Systems

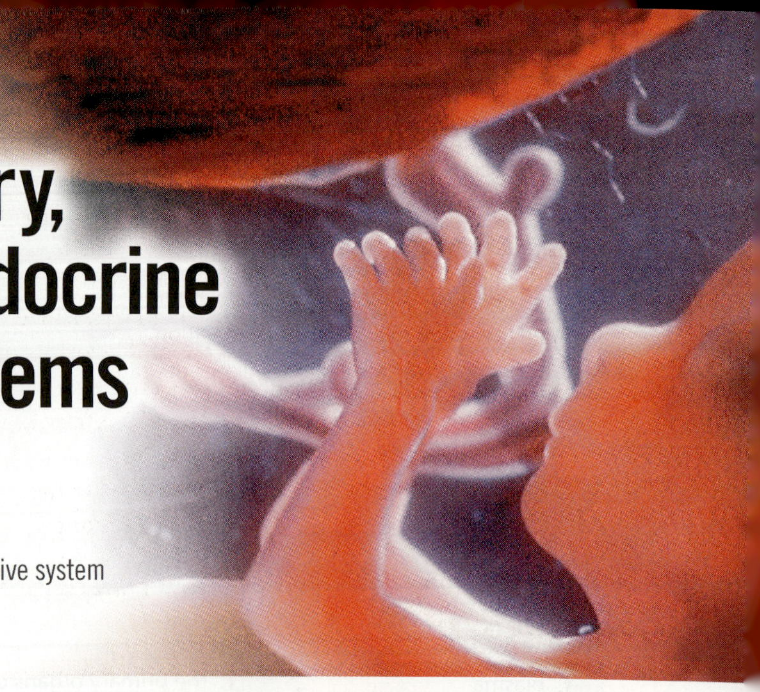

**Key Concepts**

- Design for maintaining homeostasis
- Skin, hair, oil, and sweat
- The marvelous design of the kidneys
- Hormones
- Reproductive system

God designed your body to maintain a stable internal environment so that life may continue. This built-in maintenance of a stable internal environment is **homeostasis.** Three of your body's systems that are heavily involved in maintaining homeostasis are the integumentary, excretory, and endocrine systems. The *integumentary system,* which consists of the skin and various glands associated with it, functions as a means of communication between your body and the outside world, as an organ of secretion, and as a shield to prevent your body from drying out and to protect you from bacteria and viruses. The *excretory system,* composed of the kidneys, ureters, urinary bladder, and urethra (the *urinary system*), as well as the lungs, sweat glands, and rectum, rids the body of metabolic wastes. The *kidney* is one of the most important organs of homeostasis, maintaining the proper balance of water and electrolytes, filtering wastes out of the blood, and conserving certain important substances while eliminating others. The *endocrine system* includes all the glands that secrete chemical messengers known as *hormones* directly into the bloodstream. Hormones help maintain homeostasis by influencing the activity of particular tissues or organs. Closely related to the endocrine system is the *reproductive system,* which God designed to allow the continued existence of human life.

## 11.1 The Integumentary System

The system of the body designed specifically for homeostasis is the **integumentary system.** This system, which most of us know as the **skin,** covers the outside of the body. Scattered throughout the skin are several types of organs: hair follicles [fŏl′ĭ·kəlz], sweat glands, and sebaceous [sĭ·bā′shəs] glands.

### The Skin—A Protective Shield

With a total surface area of 20–25 square feet (in adults), the skin is the largest and most visible organ of the human body. *This amazing organ provides three critical services to the body:*

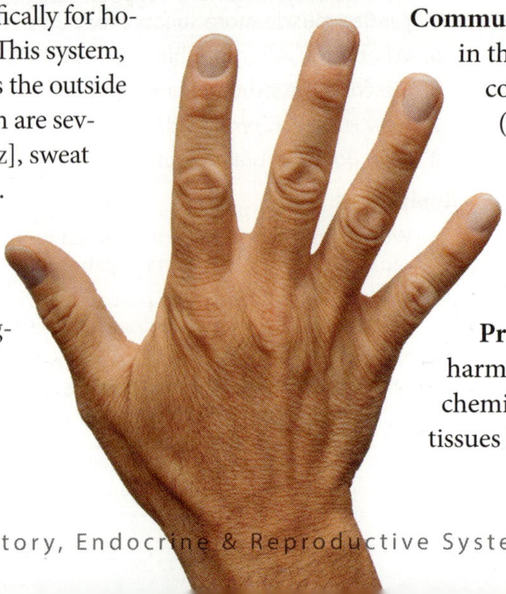

**Communication**—Sense receptors in the skin notify the brain of conditions affecting the body (temperature, pressure, pain).

**Secretion**—The skin provides moisture (sweat) to cool the body and oil to keep it from drying out.

**Protection**—The skin prevents harmful microorganisms and chemicals from reaching delicate tissues beneath it.

## Layers of the Skin

In the 17th century, an Italian scientist, *Marcello Malpighi* [măl·pē′gē: 1628–1694], learned of the design of skin through his microscopic studies. Since then, anatomists have identified all four tissue types (epithelial, muscle, nerve, and connective) in the skin and have organized the skin into distinctive organs and layers. The layers of the skin, from top to bottom, are the *epidermis, dermis,* and *subcutaneous* [sŭb′kyo͞o·tā′nē·əs] *layer* (Fig. 11.1).

**Epidermis.** The part of the skin that is exposed to the environment is the **epidermis** (*epi-*, "above" + *dermis,* "skin"). You can see how thick the epidermis is when you have a *blister.* The transparent skin on a blister is the epidermis. The outermost cells of the epidermis, the *squamous* [skwā′məs] *epithelial cells,* are flattened and dead. Beneath the squamous epithelia are the *germinative* [jûr′mə·nā′tĭv] *cells,* living cells that rapidly reproduce. Your skin is constantly losing cells on the surface from rubbing against objects and clothing. The germinative cells replace those that are lost by the millions each day. The epidermis has no blood vessels, yet the germinative cells are nourished by the blood vessels in the underlying layer. As new cells are produced, older ones are pushed closer to the surface and away from the life-giving blood vessels. This is one reason that the upper epidermal cells die. The cell division in the germinative layer is stimulated by rubbing. Excessive rubbing causes a *callus,* a buildup of epidermal cells. About every 27 days, all epidermal cells have been replaced once. A condition called **psoriasis** [sə·rī′ə·sĭs] results when the immune system attacks epidermal cells, causing them to be formed and discarded too rapidly. When the epidermal cells of the scalp replenish themselves too rapidly, the dead skin cells begin to flake, resulting in *dandruff.*

As epidermal cells move from the lower epidermis to the upper epidermis, they undergo a hardening process called *keratinization.* **Keratin** [kĕr′ə·tĭn], a tough, waterproof protein fiber, is manufactured in the cytoplasm, strengthening and hardening the cell. This keratin is the source of the protective qualities of the upper epidermis. About 90% of the epidermal cells undergo keratinization, and they are called *keratinocytes* for this reason. Fingernails and toenails are epithelial cells that have undergone more keratinization than other epidermal cells. Notice the light-colored, crescent-shaped area at the base of your fingernails (most visible at the base of your thumbnails). Beneath this area lies the region where new epithelial cells are produced.

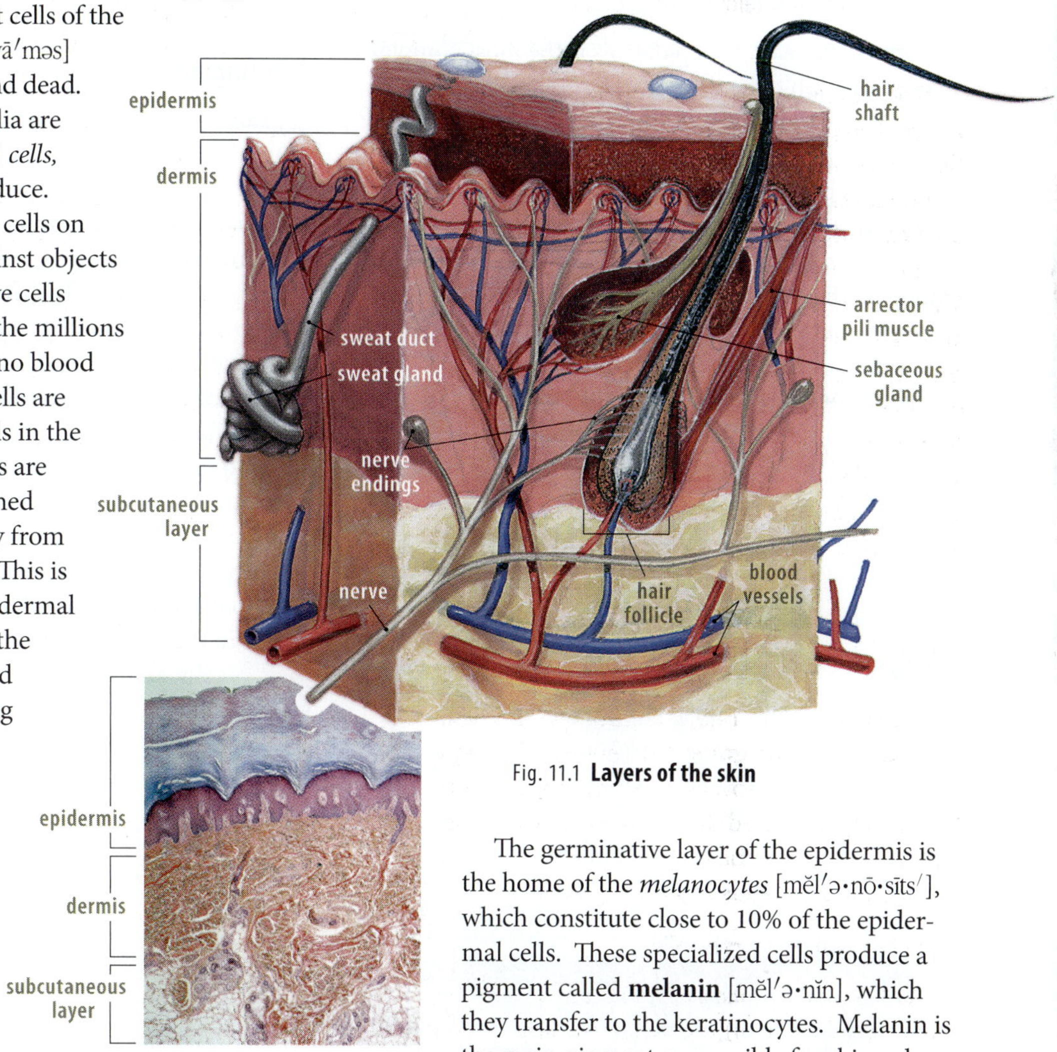


Fig. 11.1 **Layers of the skin**

The germinative layer of the epidermis is the home of the *melanocytes* [mĕl′ə·nō·sīts′], which constitute close to 10% of the epidermal cells. These specialized cells produce a pigment called **melanin** [mĕl′ə·nĭn], which they transfer to the keratinocytes. Melanin is the main pigment responsible for skin color. Everyone has the same number of melanocytes, but the amount of melanin in the skin varies among individuals. People with dark skin have more melanin; those with light skin have less. People with freckles have an uneven distribution of melanin in the skin. Skin color is also affected by *carotenoids* in the skin, which can give the skin an olive tint, and by *hemoglobin* in the blood vessels of the dermis, which gives the skin a reddish or pinkish tint.

Skin color is a hereditary trait involving a number of genes (the units of heredity within the cell). Because of the many genes involved, skin color and hair color are variable traits. There are also certain genes that enable melanin to be produced in the first place. If one of these genes is defective, the person's skin cannot produce melanin and is thus extremely pale. An individual that cannot produce melanin is known as an *albino.*

**Dermis.** Beneath the epidermis is the most complex layer of the skin—the **dermis.** *Elastin* and *collagen* are two protein fibers in the dermis that give skin its ability to stretch and its elasticity (the ability to return to its original shape). As a person ages, the elastin begins to break down, causing the skin to sag. Ultraviolet rays from the sun hasten the aging process of the skin by damaging the elastin fibers. In addition to connective tissue, the dermis houses nerve endings, blood vessels, hair follicles, and sebaceous glands.

**Subcutaneous layer.** Underneath the dermis is a layer of tissue called the **subcutaneous layer.** It is a fatty layer of loose connective tissue that binds the skin to the underlying organs. It acts as an insulator and shock absorber.

Fig. 11.3 **Variety in skin color**

## Tanning and Sunburn

**Tanning** is a natural process that protects the body from the harmful ultraviolet rays of the sun. When exposed to sunlight, the melanocytes in the skin are stimulated to produce more melanin temporarily. This increased amount of melanin shields the skin from overexposure. Too much sun, however, leads to *sunburn,* a painful inflammation resulting from toxic substances released by dead or damaged skin cells. The damage that the sun does to the skin is cumulative and irreversible. Over time, excessive exposure to the sun leads to premature wrinkles, sags, and discoloration.

A more serious problem for sunbathers is **skin cancer.** As in all cancers, cells reproduce more rapidly than normally. Scientists believe that the ultraviolet radiation not absorbed by melanin causes cells deep in the epidermis to mutate, interfering with the cells' ability to pass on the proper genetic information to their replacements.

The most common types of skin cancer, *basal-cell carcinoma* and *squamous-cell*

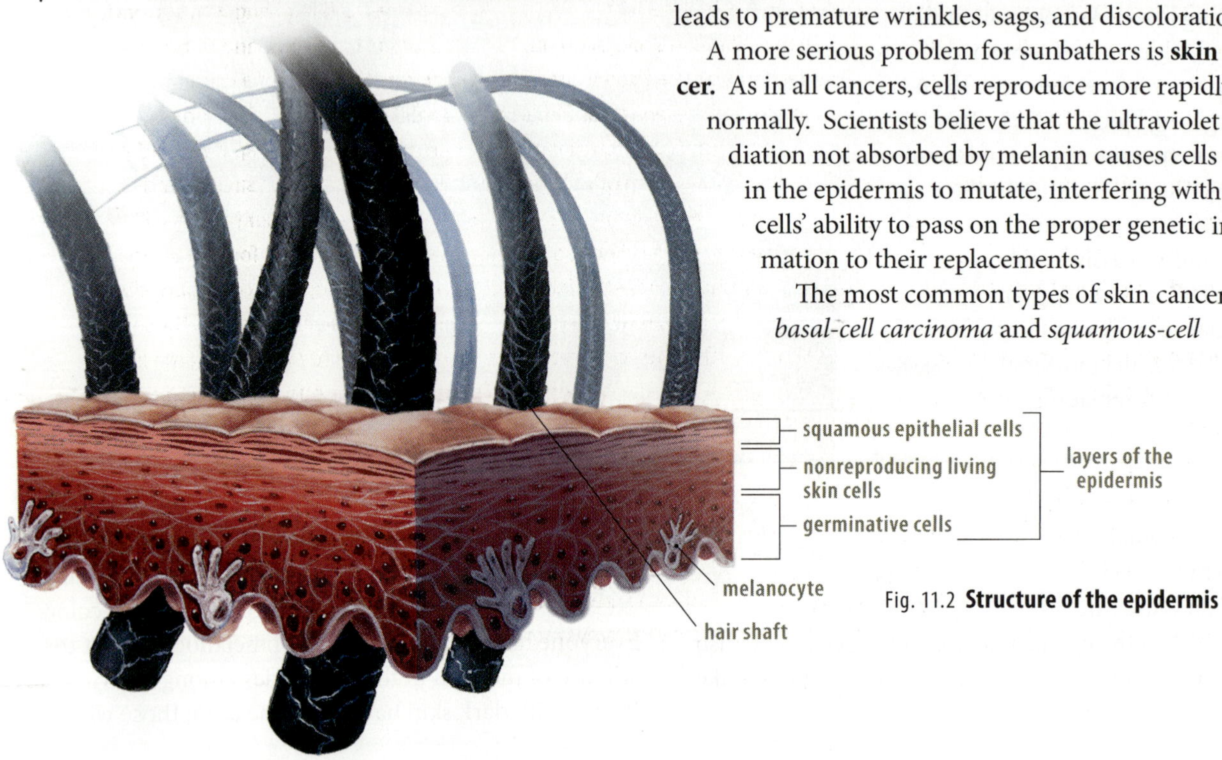


Fig. 11.2 **Structure of the epidermis**

*carcinoma,* appear as pale, waxy nodules or rough scaly patches. In a rarer but much more serious form of skin cancer, *malignant melanoma,* the melanocytes become cancerous, causing the skin to appear dark brown or black in patches.

Before the discovery of cancer drugs and effective surgical techniques, skin cancer was deadly. Today, doctors can treat most forms with a 95% cure rate. The best medicine for most forms of skin cancer is prevention. Do not overexpose your body to the radiation of the sun, and use a sunscreen formula with an SPF (sun-protection factor) of 15 or higher.

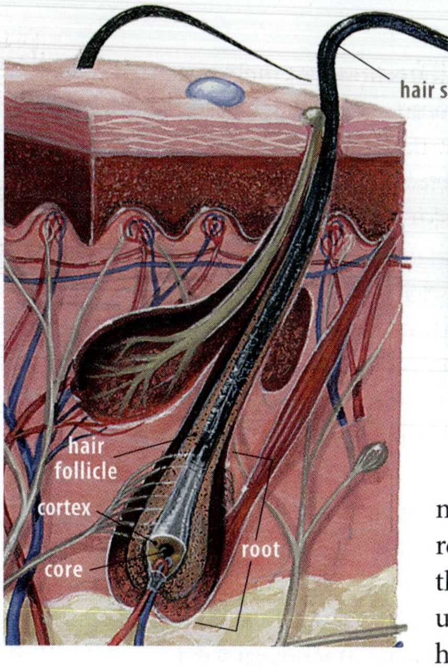

Fig. 11.4 **Anatomy of hair follicle**

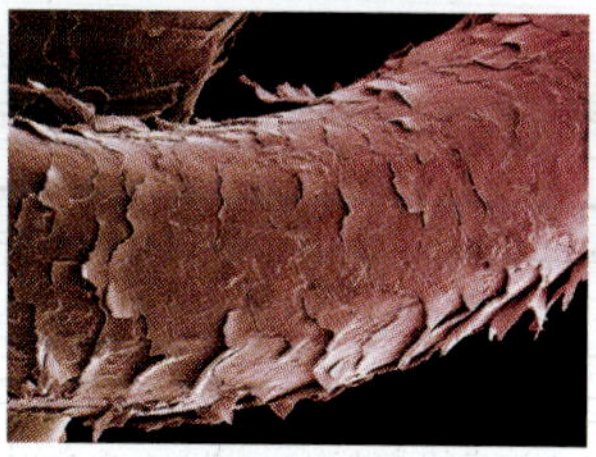
Fig. 11.5 **Close-up of hair shaft**

## Hair

**Structure.** Another characteristic of human skin is of **hair**. Hair grows from a **hair follicle,** a tube lying vertically in the dermis and designed to manufacture hair. The **hair shaft** itself is built from epidermal cells that fully keratinize. Epidermal cells lining the hair root reproduce rapidly, mature, and then die, pushing dead cells up through the follicle. The hair lengthens as more cells are added. Even though the hair shaft is composed of dead cells, it looks nothing like the cells that made it but rather looks like shingles on a roof (Fig. 11.5).

### The Compliant Skin *by Dr. Paul Brand and Philip Yancey*

Design in the LIVING CREATION

I grasp a hammer in the palm of my hand. Each cluster of fat cells changes its shape in response to the pressure. It yields but cannot be pushed aside because of the firm collagen fibers around it. The resulting tissue, constantly shifting and quivering, becomes compliant, fitting its shape and its stress points to the precise shape of the handle of the hammer. Engineers nearly shout when they analyze this amazing property, for they cannot design a material that so perfectly balances elasticity with viscosity.

If my skin tissue had been made harder, I might insensitively crush a goblet of fine crystal as I hold it in my hand; if softer, it would not allow a firm grip. When my hand surrounds an object—a ripe tomato, a ski pole, a kitten, another hand—the fat and collagen redistribute themselves and assume a shape to comply with the shape of the object being grasped. This response spreads the area of contact, preventing localized spots of high pressure, limiting stress while giving firm support.

Take a clattery hand of bone—such as the hand of a human skeleton on display in a biology classroom—and wrap it around a hammer. Against such a hard surface, the hammer handle will contact only about four pressure points. Without my compliant skin and its supporting tissues, those four pressure points would inflame and ulcerate if I pounded the hammer just a few times. But because of compliancy, my entire skin-covered hand absorbs the impact.

Hair color is determined by the amount of melanin and other pigments produced by a few melanocytes at the bottom of each hair follicle. The amount of pigments is controlled by genes inherited from one's parents. If pigments are missing, hair appears white or gray.

**Purpose.** Some ask why God designed the body to include hair. Hair aids homeostasis in several ways. Eyebrows keep sweat from the forehead out of the eyes. Eyelashes help protect the eyes from insects, dust, and other particles. Tiny hairs in the ear and nose filter out unwanted particles before they enter the body. Even hair of the armpits has the purpose of reducing friction between the arms and the body during walking or running. Hair also helps in the sense of touch. Each hair follicle is wrapped with nerve endings. Whenever the hair is moved by being touched or blown, your brain is alerted. The hair follicle is also the attachment for a group of smooth muscle cells called the **arrector pili** [ə·rĕk′tər pī′lī′] muscles. When you are cold, these muscles contract, giving your skin *goose bumps.* This muscle action helps warm your skin and alerts you to further protect yourself. Besides these functional purposes, the Bible seems to indicate that hair was also given for decoration and distinction (1 Cor. 11:14–15).

## Sebaceous Glands

**Sebaceous glands** are usually associated with each hair follicle. These glands produce **sebum** [sē′bəm], an oil that waterproofs your skin and keeps it soft and flexible. When there is an imbalance of hormones during adolescence, the sebaceous glands may produce too much oil. This excess oil tends to clog the hair follicle opening, resulting in *blackheads.* Bacteria are always present on the skin, and if they infect a clogged pore, a blemish may form.

## Regulation of Body Temperature

**Sweat glands.** The primary function of **sweat glands** is to help regulate body temperature: the body is cooled as sweat evaporates. Sweat glands also help remove nitrogenous wastes and mineral salts from the blood as part of the excretory system.

A person normally loses from 100 to 600 mL (3 to 20 oz) of water and dissolved wastes each day through the sweat glands; but on hot days or during periods of much exercise, 1 to 2 liters of water can be lost in just one hour. *Dehydration* induced by exercise, especially in hot weather, can be fatal if proper precautions are not taken to remain well hydrated. Experts recommend drinking 400 to 500 mL (14 to 16 oz) of water or a sports drink prior to exercise and 200 to 400 mL (7 to 14 oz) of water every 15 to 30 minutes of activity to replace water lost through sweating. Symptoms of dehydration may include feeling lightheaded or disoriented and experiencing muscular cramps.

**Pores,** or openings, in the skin are connected to the sweat glands by **sweat ducts.** Each sweat gland is surrounded by a network of capillaries, which allows water, sodium chloride (table salt), and other salts to diffuse out of the blood into the sweat glands. This mixture is excreted by the glands to the surface through the sweat ducts and collects on the skin's surface as beads of sweat.

The skin is vital to good health because of its ability to regulate body temperature. The *hypothalamus* signals the sweat glands to begin producing perspiration when the body temperature rises above normal (approximately 98.6°F). Evaporation of perspiration from skin causes cooling. The blood vessels of the dermis also aid in heat regulation. They cool the body when they dilate because an increased flow of blood near the surface allows heat

Fig. 11.6
**Clogged pore**

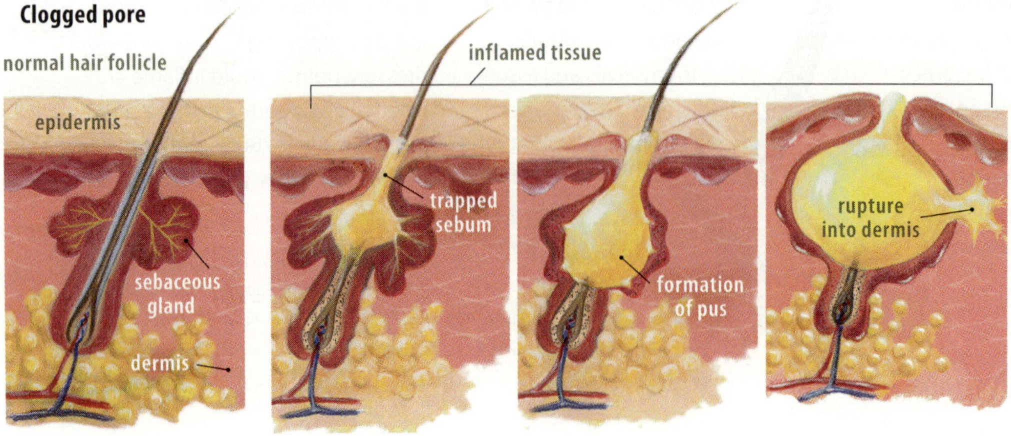


Fig. 11.7
**Close-up of skin pore**

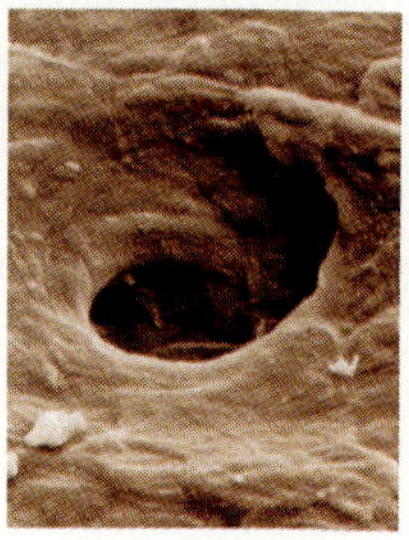

from the blood to be radiated into the atmosphere. When the body must conserve heat, the blood vessels in the dermis constrict, reducing the blood flow near the surface and preventing excessive loss of heat from the blood.

As a person ages, some oil glands are replaced by connective tissue, causing older people to generally have drier skin than young people. Some sweat glands cease to function with age. Perhaps you have heard of instances where many elderly individuals have died of heatstroke during a particularly hot summer. A loss of sweat glands results in a less efficient cooling mechanism for the body.

## Effects of a Burn

Understanding the LIVING CREATION

What happens to the body when the skin is removed is vividly illustrated by the tragedy of a severe **burn.** *Third-degree burns* destroy the entire thickness of skin, leaving the body without its organ of homeostasis. (*Second-degree burns* reach into the dermis, and *first-degree burns,* such as scalds, damage only the epidermis.) The most immediate and severe effect of a serious burn is a loss of water or body fluids. When the skin is not present to stop this loss, fluid literally seeps out as fast as it can be replaced by drinking. A badly burned person will be driven by a sense of thirst. As the water escapes, the blood volume drops dramatically, resulting in a serious drop in blood pressure. A person with a severe burn over a large area may suffer from shock because of this fluid loss; the immediate treatment for shock is replacement of the blood fluids in the form of intravenous plasma. If a severe burn patient survives the loss of fluids, his next battle is with infection. Without the keratin layer of protein, bacteria invade and grow rapidly. Within hours of a burn, doctors may notice signs of infection; therefore, immediate isolation in a sterile room and covering the burns with sterile bandages helps. Skin grafts (transplantation of skin to the burned area) or temporary artificial skin may also be applied. A third-degree burn over a large area can be one of the most serious life-threatening injuries to the body.

| Burn | Symptoms | Treatment |
|---|---|---|
| **First degree** | *pain, redness* | *Put under cold water and apply first-aid cream.* |
| **Second degree** | *blisters and sores* | *Immerse in cold water, cover with cold compress, bandage, and see doctor. Do not break blisters.* |
| **Third degree** | *skin turns black or white; tissue is exposed; damaged nerves may numb pain* | *Do not remove clothing stuck to the burn. Bandage if possible. Get medical attention immediately.* |

## Section Review 11.1

1. What are three functions of the skin?
2. Describe the three layers of the skin.
3. What function does keratin serve in the skin?
4. What are the two most common types of skin cancer? Which type of skin cancer is the most dangerous?
5. How do sweat glands help your body maintain a stable internal environment?
6. How does aging affect the skin?

### Identify

psoriasis, melanin, albino, elastin, collagen, hair follicle, arrector pili, sebaceous glands, sebum, pore, sweat ducts

### Observe

- View your skin through a magnifying glass. Observe the difference between the palm, the back of the hand, and the fingertips.

## 11.2 The Excretory System

The removal of waste substances from the body is **excretion.** Excretion is vital to the health of the body because the wastes are poisonous. If the wastes build up and are not eliminated, they can cause serious problems. As you know, carbon dioxide and water vapor are removed by the lungs, and waste products from food digestion are removed by the large intestine. Other wastes, including urea, uric acid, various salts, and assorted nitrogenous wastes, are removed by the kidneys and sweat glands.

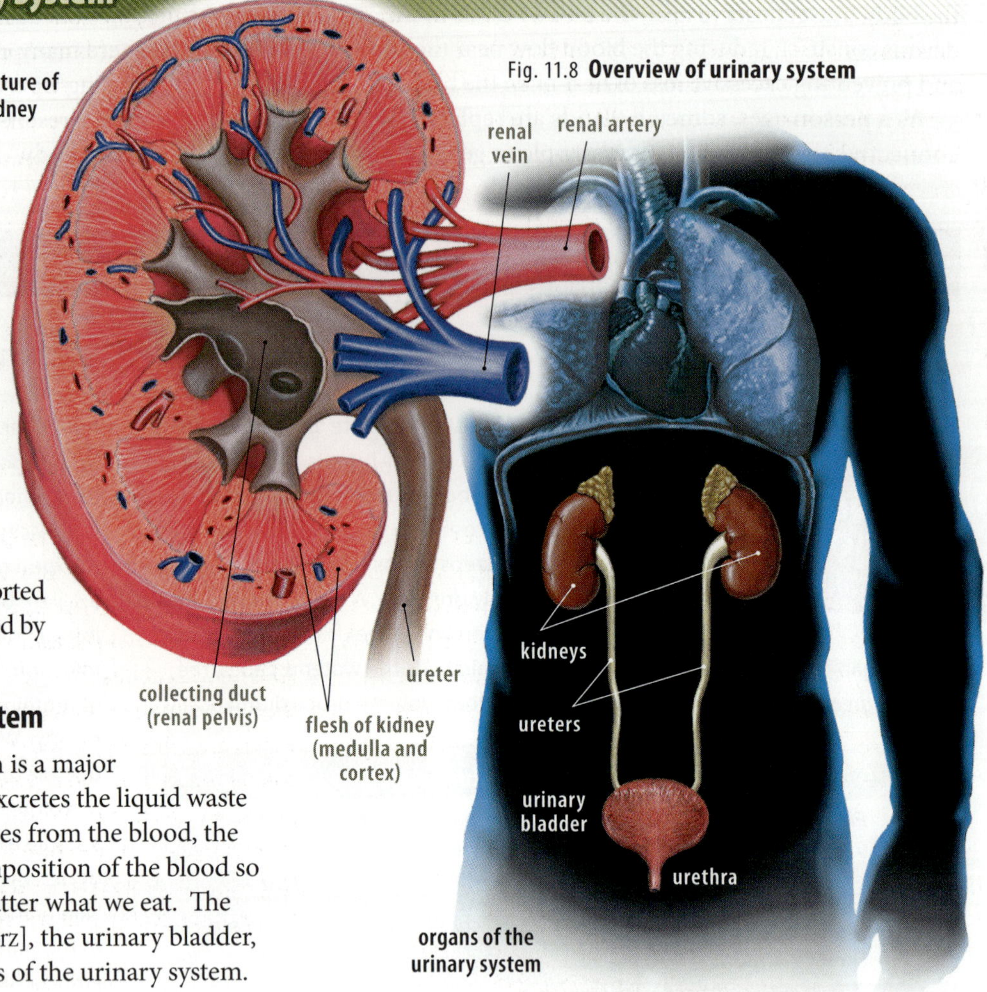


Fig. 11.8 **Overview of urinary system**

### Structure of the Urinary System

The **urinary system,** which is a major part of the *excretory system,* excretes the liquid waste *urine.* Besides removing wastes from the blood, the kidneys help regulate the composition of the blood so that it is fairly constant no matter what we eat. The kidneys, the ureters [yo͞o·rē′tərz], the urinary bladder, and the urethra are the organs of the urinary system.

### The Kidneys—Divinely Designed Filters

The **kidneys** are the most important excretory organs; their primary function is to cleanse the blood of wastes and remove those wastes from the body. Our pair of bean-shaped kidneys are attached to the back wall of the abdominal cavity, one on either side of the spinal column. The wastes removed from the blood by the kidneys are mixed with water to form urine. Urine passes through tubes called *ureters* to the bladder. The **urinary bladder** stores urine until it is discharged to the outside through the **urethra,** the tube that carries urine from the bladder to the outside of the body.

### How the Kidneys Work

If you were to take a kidney from an animal and slice it in half with a sharp knife, you could see the flesh of the kidney surrounding a collecting duct. It is in the flesh of the kidney that the work is done. Each kidney is made up of about a million working units called **nephrons** [nĕf′rŏnz]. The nephron is a microscopic network of blood vessels and renal tubules (small tubes) where blood is filtered and the purified liquids are absorbed.

When blood enters a nephron, it first passes through a sievelike structure called a **glomerulus** [glō·mĕr′yə·ləs], which consists of a cluster of capillaries with perforated walls. The holes in the capillaries allow some of the water and small molecules in the blood to escape from the blood vessel; these holes are too small to allow blood cells and large blood proteins to escape. The glomerulus is surrounded by a spherical container called a **Bowman's capsule,** which collects the water and dissolved substances that escape the blood through the holes.

The fluid in the Bowman's capsule then flows through a long, thin tube called the **renal tubule,** which is surrounded by hundreds of blood capillaries. In this tube, glucose molecules, hormones, and other useful substances are removed from the filtered fluid and returned to the blood. The return of these molecules is called **reabsorption.** About 99% of the water is also reabsorbed by the capillaries. Waste products such as urea are not reabsorbed but are left behind in the small amount of water remaining in the renal tubule. This fluid, now known as **urine,** drains out of the nephron and is pumped down to the bladder for later expulsion from the body.

This process of filtration and conservation is remarkably efficient. Every day, the equivalent of 48 gallons of blood passes through your kidneys to be filtered. (Since you have only five or six quarts of blood in your body, a given unit of blood passes through your kidneys many times—about 38 times per day for the average person.) Of this 48 gallons, only a small fraction is discarded as waste; most is returned to the blood. The remaining one to two quarts is retained by the kidneys to carry away wastes as urine. This ingenious process of filtration allows toxic substances to be removed from the blood very efficiently while conserving useful substances.

Fig. 11.9 **How kidneys filter the blood**

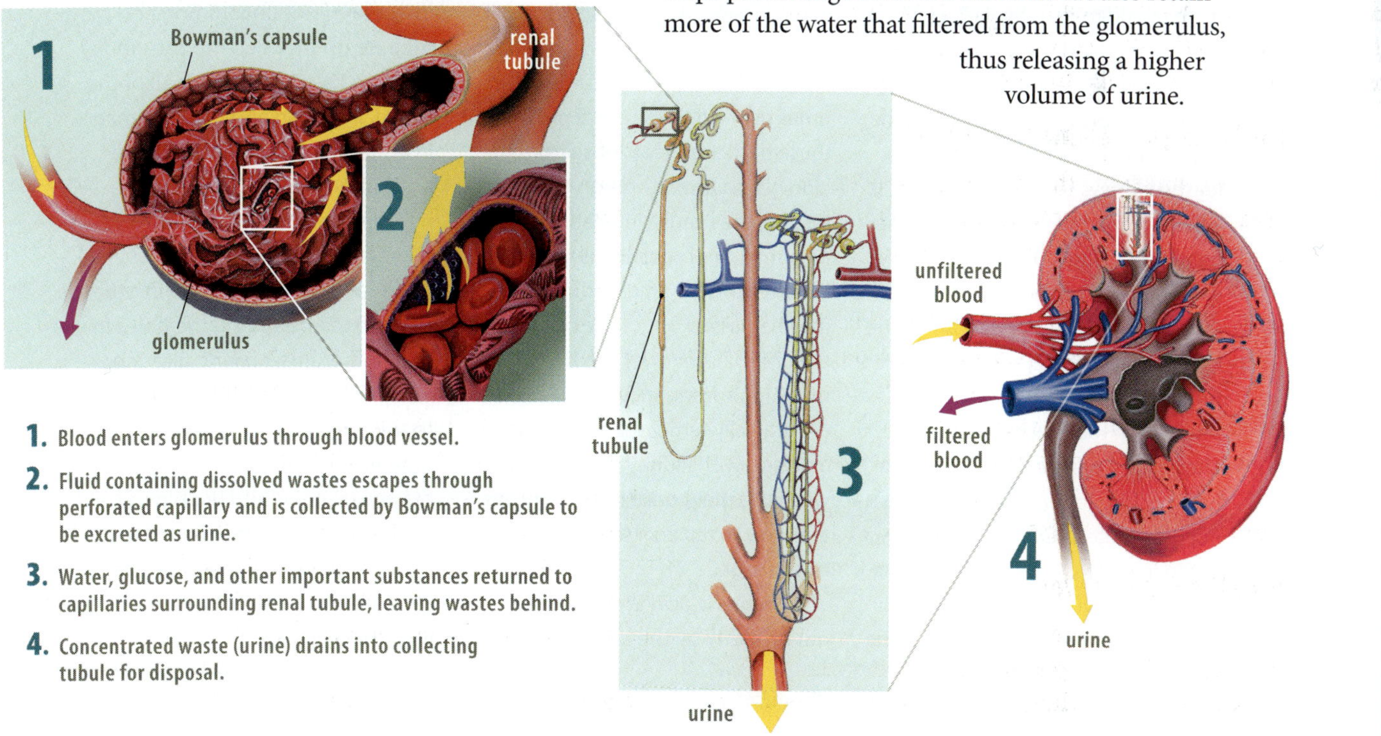


1. Blood enters glomerulus through blood vessel.
2. Fluid containing dissolved wastes escapes through perforated capillary and is collected by Bowman's capsule to be excreted as urine.
3. Water, glucose, and other important substances returned to capillaries surrounding renal tubule, leaving wastes behind.
4. Concentrated waste (urine) drains into collecting tubule for disposal.

## Regulation of the Kidneys

The filtering action of the glomerulus requires a certain amount of blood pressure to push the molecules through the holes in the capillary walls. If the blood pressure drops too low, the kidneys may have difficulty functioning. If this situation occurs, the kidneys release an enzyme called **renin** [rē′nĭn], which acts on the circulatory system to increase blood pressure. This helps the kidneys properly filter the blood. If the kidneys are failing, however, they may trigger high blood pressure to compensate for their failure, leaving the person with constant high blood pressure. This is why many people suffering from kidney failure have other symptoms like high blood pressure.

The regulation of water reabsorption is controlled by the brain using **ADH,** or *antidiuretic* [ăn′tē•dī′ə•rĕt′ĭk] *hormone*. The hypothalamus monitors the concentration of water in the blood and body fluids; when these fluids get too thick, the hypothalamus signals the pituitary gland to release ADH into the bloodstream. ADH that reaches the capillaries in the kidney changes the ability of the renal-tubule membranes to hold water, allowing more water to be reabsorbed and conserved by the body. Consequently, the kidney produces less urine. When the hypothalamus detects enough water (or too much water) in body fluids, it stops producing ADH and the renal tubules retain more of the water that filtered from the glomerulus, thus releasing a higher volume of urine.

### Kidney Failure

The kidney is one of the most important organs of homeostasis. A person cannot survive for very long without at least one functioning kidney. The design of the kidney, although understood by scientists, is difficult to duplicate. One of man's inventions that attempts to take over the job of the kidney is the *artificial kidney* or *dialyzer.*

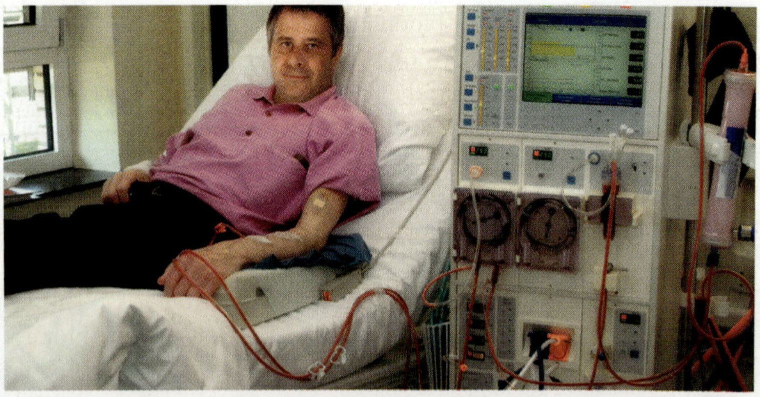

Fig. 11.10

**A man using kidney dialysis machine**

## The Kidneys: Use and Abuse

The kidneys are vital organs of the human body. To keep them as healthy as possible, you should know what conditions affect them. Normally, they do their job without any indication of stress, so you may not feel that you are harming one. One of the easiest and most sensible things to do to help the kidneys is to drink plenty of clean, clear water. Water is a filtering medium for the wastes and poisons in the blood; the more water present, the easier it is for the kidneys to clean the blood. If you drink too little water or become dehydrated, the kidneys must work extra hard to concentrate the urine.

Certain drugs adversely affect the kidneys. *Alcohol* depresses the ADH production of the hypothalamus, resulting in too much urine output. Therefore, a person who drinks alcohol may lose too much water, becoming dehydrated. *Caffeine,* a substance found in coffee, tea, colas, and some foods, can have similar effects if excessive amounts are ingested. The nicotine in *tobacco* has an opposite effect by stimulating ADH production and helping the kidneys concentrate urine. When using tobacco, a person accumulates too much water in the blood, causing increased blood pressure.

If the kidneys are abused or if they malfunction, concentrating the urine too much, some substances that normally dissolve in urine crystallize in the kidneys, forming solid, rocklike particles of calcium and other substances; these particles can sometimes grow to the size of a pea. When these particles break loose from the kidney and travel through the urinary system, they cause intense pain. A person in this condition is said to have **kidney stones.** Small kidney stones may be passed out of the body with the urine; larger ones can be broken up by high-frequency sound waves or surgically removed. In any case, a person with kidney stones should watch his diet because the particles are likely to return if the conditions do not change.

When uric acid is not filtered out properly by the kidneys, it may accumulate in the joints of the body, causing pain. This condition is known as **gout.** The joints of the toes are the most commonly affected.

Of all the problems associated with the kidneys, one of the worst is **urinary tract infection** *(UTI).* Because the urethra empties to the outside of the body, bacteria have a path by which to enter. A bladder infection is not uncommon and may migrate up the ureter to the kidney. Because the kidney filters all of the blood, infection and tissue damage in any area of the body have the potential of accumulating in the nephrons of the kidney. Here excess complexes of molecules left over from battles with infection in other parts of the body clog the holes in the glomerulus, making filtering less efficient. White blood cells rush to the affected kidney, making the glomerulus walls so permeable that plasma proteins and red blood cells are lost into the urine. This sequence of nephron inflammation can be initiated by streptococcal infections like strep throat.

An artificial kidney uses the process of **dialysis,** which refers to separating molecules of different sizes using a membrane. In one form of dialysis, blood from a patient with failed kidneys is pumped into a dialysis machine. Once in the machine, the blood is passed through tubes constructed of an artificial material possessing a multitude of tiny holes like the holes of the glomerulus. Waste molecules are pushed out of the tubes by blood pressure into the wash solution of the tub. The wash solution contains the exact concentration of water and other molecules as blood; therefore, there is no net gain or loss of these important substances. The blood, after leaving the machine, has been ridded of the urea and other wastes and is returned to the person. Dialysis makes life possible without kidneys, but the treatments are extremely time-consuming; thus a kidney patient may remain healthy but dependent on the kidney machine.

Dialysis is not the only option if a person's kidneys fail. It is possible to transplant a kidney from one person to another. Because a person can live a normal life with only one kidney, a single-kidney transplant can allow a person with damaged kidneys to recover fully. However, to prevent the recipient's immune system from attacking the transplanted kidney, special immune-suppressing drugs must be taken for the rest of the recipient's life.

### Section Review 11.2

1. What is the function of the excretory system?
2. List the organs of the excretory system.
3. Describe how the kidneys filter the blood.
4. How is the activity of the kidneys regulated?
5. What are some ways to take proper care of your kidneys?
6. How can the process of dialysis allow a machine to serve as an artificial kidney?

**Identify**

excretion, ureters, urinary bladder, urethra, nephron, glomerulus, Bowman's capsule, renal tubule, reabsorption, renin, kidney stones, gout

## 11.3 The Endocrine System

### Nerves and Glands

In chapter 8, we discussed the marvelous communication system of the body, the nervous system. Controlling our bodies to accomplish what God wills for our lives is possible because of this specialized and powerful system. However, the results of a nerve impulse typically last only a short time and affect only a small portion of the body; controlling homeostasis through the nervous system alone would require continual impulses in many nerves.

To allow efficient control of the entire body over longer periods, God designed the **endocrine system.** This system communicates, not using electrochemical impulses, but using chemicals released into the bloodstream. Because these chemicals are released into the bloodstream, they are carried to all parts of the body and can continue their effects for long periods.

Certain organs of the *reproductive system* (especially the *ovaries* in women and *testes* in men) also function as part of the endocrine system; they stimulate the body to develop the characteristics that distinguish the sexes and make reproduction possible. We will study these reproductive glands in the next section.

### Types of Glands

All the glands of the body, whether they be only one cell or large masses, can be divided into two types, endocrine and exocrine. The **endocrine glands** manufacture and secrete their products into the blood. *Endo-* means "into" and *crine* means "secrete"; therefore, endocrine glands secrete into the body. **Exocrine glands** manufacture and secrete their products to the outside of the body. The glands that secrete digestive juices into the digestive tract are exocrine glands because the contents of the digestive tract are outside your body (remember that the alimentary canal is a continuous tube through the body). Glands in your skin secrete products to the surface of the skin, so they are also exocrine glands.

### Hormones

The endocrine glands are the organs of the chemical messenger system. The messengers (mostly protein compounds) produced by these glands are called **hormones,** a term meaning "stimulators," due to their powerful effects at low concentrations. Your body is under a constant influence of hormones that help regulate almost every function. Hormones help you to grow, to use your food, and to react to dangerous situations and help many organs to operate. A healthy body requires a proper balance of hormones.

Hormones are specific in their target organs, tissues, and cells because of special proteins in the membrane or cytoplasm of the target cells. These proteins, called *receptors,* bind specifically to one kind of hormone: insulin binds to insulin receptors, adrenaline to adrenaline receptors, etc. Receptors for a specific hormone are located only in the organs and tissues where the hormone acts; organs and tissues without a receptor are insensitive to the corresponding hormone.

## The Pituitary Gland

The **pituitary** [pĭ·to͞o′ĭ·tĕr′ē] **gland,** which is sometimes called the master gland because it affects the other endocrine glands, is embedded in a bony compartment just above the roof of the nasal cavity. A nerve stalk attaches the pituitary gland directly to the *hypothalamus* of the brain. You may recall that the hypothalamus regulates many automatic body activities. Nerve impulses from the hypothalamus signal the pituitary gland to secrete hormones that control other parts of the body. Although the pituitary is a very small gland, it is divided into anterior and posterior lobes; each lobe secretes several hormones.

The anterior lobe of the pituitary gland produces hormones that regulate the body's growth and development. The *growth hormone,* **somatotropin** [sə·măt′ə·trō′pĭn], regulates skeletal growth. When somatotropin is present, cells of the epiphyseal disk and periosteum reproduce more rapidly and are more capable of making new bone tissue. Somatotropin also helps other body organs to reproduce and grow. Too much of this hormone produced during the growing years causes **pituitary gigantism,** in which a person may grow seven to nine feet tall. If too little somatotropin is produced during childhood, **pituitary dwarfism** may result. In this situation, an individual may develop into a well-proportioned little person. Excessive somatotropin production after normal childhood development can cause **acromegaly** [ăk′rō·mĕg′ə·lē], a condition in which the bones of the head, hands, and feet become enlarged.

The pituitary gland also secretes **gonadotropins** [gō·năd′ə·trō′pĭnz], hormones that affect the development of the reproductive organs. Gonadotropins also influence the secretions of the mature reproductive organs. Other hormones secreted by the pituitary gland stimulate the production of milk in the mammary glands; affect the activities of the thyroid, parathyroid, and adrenal glands; and exercise some control over blood pressure and the body's water balance.

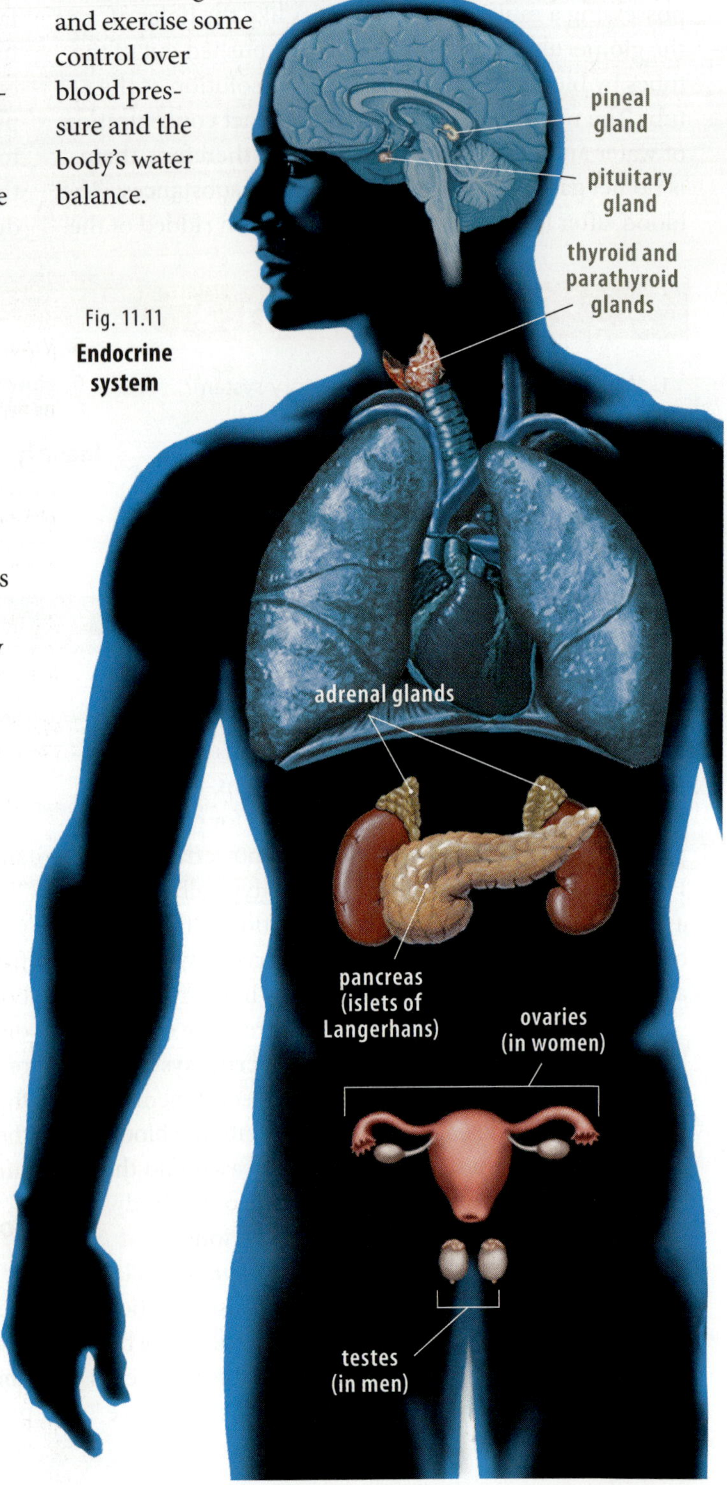


Fig. 11.11 **Endocrine system**

## The Thyroid Gland

**Regulates metabolism.** The **thyroid gland** is located in the neck, slightly below the larynx. It regulates the rate of chemical activity in the body's cells. The twin lobes of the thyroid are located on opposite sides of the windpipe and are connected by a thin layer of tissue. This gland secretes **thyroxine** [thī·rŏk′sēn′], a hormone that regulates general activities related to metabolism and oxidation. *The thyroid gland determines the rate at which food is changed into energy.* Thyroxine indirectly affects the heart and blood vessels and the nervous system; it also affects growth. The thyroid gland also functions with the *parathyroid glands,* which are located behind the thyroid, to regulate calcium in the blood. When blood calcium is too high, the thyroid gland releases *calcitonin,* which increases the storage of calcium in the bone tissue. The *parathyroid hormone,* produced in the parathyroid glands, does the opposite by stimulating the bone tissue to give up calcium to the blood. The proper amount of calcium is vital for proper nerve and muscle function.

### Table 11.1 The Endocrine System

| | Major hormone(s) produced | Hormone functions |
|---|---|---|
| **Pineal gland** | melatonin | triggers sleepiness; functions as antioxidant |
| **Pituitary gland** | somatotropin (growth hormone) | stimulates growth of bones and other organs |
| | gonadotropins | stimulate hormone production by the gonads |
| | prolactin | stimulates mammary glands to produce milk after childbirth |
| | thyroid-stimulating hormone (TSH) | stimulates production of hormones by the thyroid gland |
| | adrenocorticotropic hormone (ACTH) | stimulates production of some hormones by adrenal glands |
| | antidiuretic hormone (ADH or vasopressin) | stimulates kidneys to conserve body water; helps regulate blood pressure |
| | oxytocin | stimulates mammary glands to release milk to nursing infant; stimulates uterine muscle contraction during childbirth |
| **Thyroid gland** | thyroxine and triiodothyronine | regulate metabolism; help regulate growth |
| | calcitonin | decreases blood calcium |
| **Parathyroid glands** | parathyroid hormone | increases blood calcium |
| **Adrenal glands** | epinephrine (adrenaline) and norepinephrine | prepare body as a whole for emergencies by increasing heart and breathing rates, diverting blood flow to muscles, and otherwise keeping body in "high alert" status |
| | cortisol | stimulates body repair and reduction of swelling |
| | aldosterone | regulates sodium and potassium concentrations |
| **Islets of Langerhans** | insulin | decreases blood sugar by stimulating body cells to absorb glucose |
| | glucagon | increases blood sugar by stimulating liver to produce glucose from glycogen |
| **Ovaries** (in women) | estrogens | stimulate development of feminine secondary sex characteristics; perform various functions during pregnancy |
| | progesterone | performs various functions during pregnancy |
| **Testes** (in men) | testosterone | stimulates development of masculine secondary sex characteristics; stimulates production of sperm; increases muscular strength and endurance |

**Thyroid disorders.** A deficiency of thyroxine during infancy may prevent children from growing and developing as they should. Children affected by this condition are said to be suffering from **cretinism** [krēt′n·ĭz′əm]. People suffering from cretinism develop stunted bodies with short limbs and enlarged heads. Their mental growth may also be retarded. Fortunately, cretinism can be corrected if the condition is treated early enough with thyroxine.

Sometimes the thyroid gland of adults becomes less active than normal and many body functions slow down. The heart rate is reduced, and sometimes the heart becomes enlarged. This condition is **hypothyroidism.** Extreme cases of thyroxine deficiency may bring about *myxedema* [mĭk′sĭ·dē′mə]. A person with myxedema becomes weak, often gains weight, develops thickened skin and baggy eyes, and occasionally becomes mentally sluggish. Treatment with thyroxine reverses these conditions and often causes rapid recovery.

**Hyperthyroidism** results from an overactive thyroid gland. People who have this condition are generally very thin, restless, and nervous. The heart beats faster than normal, and sometimes a shaking of the hands develops. Hyperthyroidism may also result in a bulging of the eyes. Even though the person's appetite is greatly increased, he loses weight.

Since thyroxine is a protein-iodine compound, it cannot be produced without a supply of iodine. A lack of iodine causes the thyroid gland to enlarge, producing a **simple goiter,** a swelling in the neck. The condition is easily prevented by eating iodized salt or otherwise obtaining sufficient iodine in the diet.

## The Pancreas

We have seen already that the *pancreas* contributes pancreatic juice to the digestive system. This organ also contains clumps of endocrine glands called the **islets of Langerhans** [lăng′ər·häns′]. These glands produce the pancreatic hormones *insulin* and *glucagon,* which regulate the body's use of blood glucose. High levels of glucose in the blood cause the pancreas to secrete **insulin.** Insulin signals the liver and muscles to remove glucose from the blood and store it as glycogen, thereby reducing the sugar in the bloodstream. It also influences the use of fats and proteins by the body's cells. Whereas insulin reduces the level of sugar in the blood, **glucagon** increases it. Low levels of glucose in the blood cause the pancreas to secrete glucagon. Glucagon signals the liver to convert glycogen to glucose and release the glucose into the bloodstream, making additional glucose available to the body's cells. In this manner, insulin and glucagon regulate the amount of glucose in the blood.

A deficiency of glucagon, resulting in low blood sugar *(hypoglycemia),* can be compensated for by limiting the time between meals and by limiting the output of insulin by the pancreas. A deficiency of insulin is a more serious matter, however. As we digest a meal, large amounts of sugar enters our blood quickly. Insulin helps channel this sugar into the body's cells and into the liver. If too little insulin is produced by the pancreas, high blood sugar results with no means of quickly lowering it.

A disease called **diabetes mellitus** results when the pancreas does not produce adequate amounts of insulin *(type 1 diabetes),* or when the body does not respond properly to the insulin that is produced *(type 2 diabetes).* Diabetes causes sugar to accumulate in the bloodstream instead of being stored as glycogen in the muscles and liver. Diabetes was named after the Greek word meaning "siphon" because a person with the disease produces excessive amounts of urine. People with type 1 diabetes can regulate their blood sugar by taking insulin (usually by injection), which is extracted from genetically engineered bacteria, and by carefully monitoring their diet. Type 2 diabetes, which is often associated with obesity, can often be treated by dietary changes, regular exercise, and medications.

## The Adrenal Glands

The **adrenal** [ə·drē′nəl] **glands** sit like small caps on top of the kidneys. One of the hormones produced by the central part of these twin glands is known as the "fight or flight" hormone because it prepares the body for emergency situations. This hormone is **epinephrine** [ĕp′ə·nĕf′rĭn; also called *adrenaline:* ə·drĕn′ə·lĭn]. Epinephrine enters your blood in increased amounts when you are excited or alarmed and has a marked influence on your body. For example, it speeds up the heartbeat and thus increases the supply of blood to the skeletal muscles; it stimulates the liver to add extra sugar from its store of glycogen into the bloodstream. These activities are important during times of emergency, when strength and endurance are required. Because of its stimulating effect on the heart, epinephrine is sometimes injected directly into a heart that has stopped beating.

The outer shell of the adrenal glands produces as many as thirty other hormones. These hormones are special lipid molecules called **steroids** [stĕr′oidz′]. One steroid, **cortisol** [kôr′tĭ·sôl′], stimulates the body to repair itself following an injury or stress. A shortage of cortisol in the body results in *Addison's disease,* which is characterized by the skin turning brown and the muscles becoming weak. Another steroid hormone produced by the adrenal glands is **aldosterone** [ăl·dŏs′tə·rōn′], which stimulates the kidneys to conserve sodium and excrete potassium. Aldosterone is very important to the body's balance of minerals. While conserving sodium, it also promotes the conservation of water, preventing excess urine production.

### The Pineal Gland

Between the cerebral hemispheres of the brain lies the small **pineal** [pĭn′ē·əl] **gland.** The primary function of the pineal gland seems to be the regulation of the body's waking and sleeping cycles. The pineal gland is connected by nerve fibers to the optic nerve, allowing this gland to measure patterns of light and darkness through the eyes. When the pineal gland detects that it is getting dark, it begins to secrete a hormone called **melatonin** [mĕl′ə·tō′nĭn], which triggers the autonomic nervous system to relax, slows the activity of the reticular formation, and causes other changes that make you feel sleepy. The production of melatonin is inhibited by bright light, allowing you to feel awake and alert during the day. The pineal gland is a remarkable timing mechanism. You may tend to wake up each morning at about the same time even without an alarm. The pineal gland is measuring the nighttime hours and can alert your brain when it is time to rise. God has designed a magnificent order into His creation, as evidenced by the cycles of nature. The pineal gland helps our bodies to synchronize with the rhythms of God's creation.

## Section Review 11.3

1. What is the difference between endocrine and exocrine glands?
2. Which gland is called the master gland of the body? Why?
3. What is another name for the hormone somatotropin?
4. What is the purpose of the thyroid gland?
5. Which endocrine gland regulates blood sugar levels?
6. What hormone is responsible for preparing your body for emergency action?
7. Where are the pituitary, thyroid, adrenal, and pineal glands located?
8. List five diseases or conditions caused by malfunctions of the endocrine system.

### Critical Thinking

9. How do the endocrine system and the nervous system complement each other in regulating the body's functions?

### Identify

hormone, gonadotropins, islets of Langerhans, diabetes mellitus, steroid

## 11.4 The Reproductive System

In Genesis 1:28, God commanded Adam and Eve to reproduce and populate the earth. After the Flood, this command was repeated to Noah and his sons (Gen. 9:1, 7). Like most other living things, man is designed to reproduce by *sexual reproduction,* the union of **gametes** *(sperm cells* and *egg cells)* through the process of *fertilization* to form a new living thing. The **reproductive system** is the system designed to carry out this essential process.

### Anatomy and Physiology of the Reproductive Organs

The main reproductive organs are the **gonads** [gō′nădz′], the organs that produce the gametes. Gametes are produced through the process of *meiosis* [mī·ō′sĭs], in which a normal cell (which contains two copies of each gene) divides to form gametes that each contains only one copy of each gene. (We will study the details of meiosis in chapter 23.) *The gonads also function as part of the endocrine system;* most of their endocrine function is controlled by the pituitary gland. These hormones produce the body changes that occur during **adolescence,** the time of physical transition from child to adult. Adolescence begins with **puberty,** the stage in which the reproductive organs mature and the person becomes physically able to reproduce. Other physical changes during adolescence include the formation of the **secondary sex characteristics,** such as the growth of hair on the face and body, a deepening of the voice in males, and an enlargement of the breasts in females.

The reproductive system includes several other organs. Although these organs are not involved in the production of gametes, they play important roles in the process of human sexual reproduction.

**Male reproductive system.** The male gonads are the **testes** (sing. testis). Because sperm cells cannot efficiently form and grow at humans' normal internal body temperature, God placed the testes in the *scrotum,* a small pouch outside the body cavities. Inside each testis are several hundred *seminiferous* [sĕm′ə·nĭf′ər·əs] *tubules;* **testosterone** [tĕs·tŏs′tə·rōn′], the main hormone produced by the testes, triggers special cells in the seminiferous tubules to continually divide and undergo meiosis, producing sperm cells. Attached to each testis is a small organ called the **epididymis** [ĕp′ĭ·dĭd′ə·mĭs: plural epididymides], in which sperm mature and are stored.

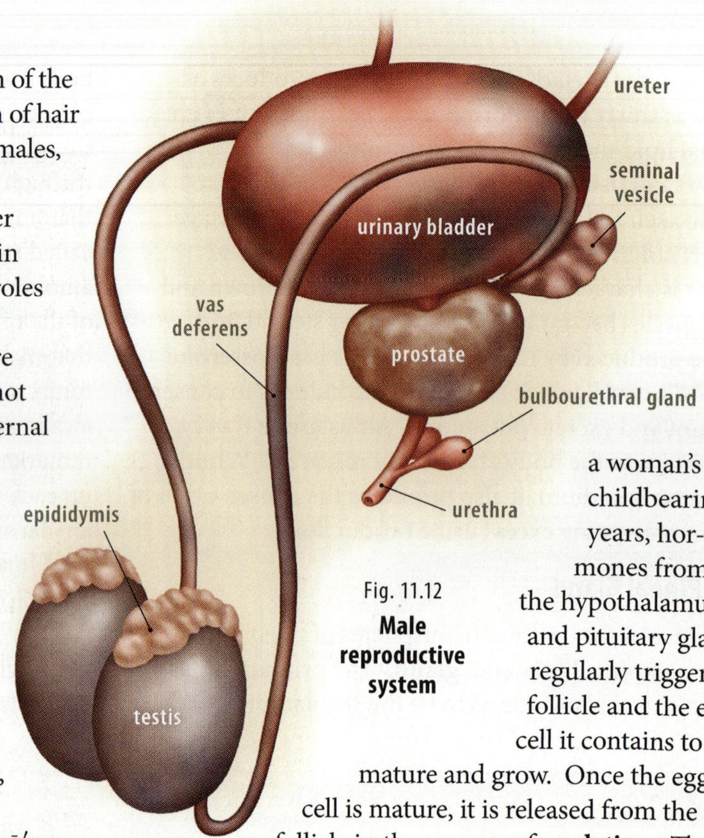


Fig. 11.12
**Male reproductive system**

Mature sperm cells enter the **vasa deferentia** [vā′zə dĕf′ə·rĕn′shē·ə; sing. vas deferens: văs dĕf′ər·ənz], two ducts that transfer sperm from the epididymides to the urethra. These ducts can also store sperm cells for some time; old, unused sperm cells are reabsorbed by the body. As sperm travel along the ducts of the male reproductive system, the *seminal vesicles* [vĕs′ĭ·kəlz], the *prostate,* and the *bulbourethral* [bŭl′bō·yōō·rē′thrəl] *glands* release secretions containing nutrients and other factors the sperm need to survive until fertilization; the fluid consisting of sperm and these secretions is **semen.** During mating, semen leaves the body through the *urethra,* which also functions as part of the urinary system.

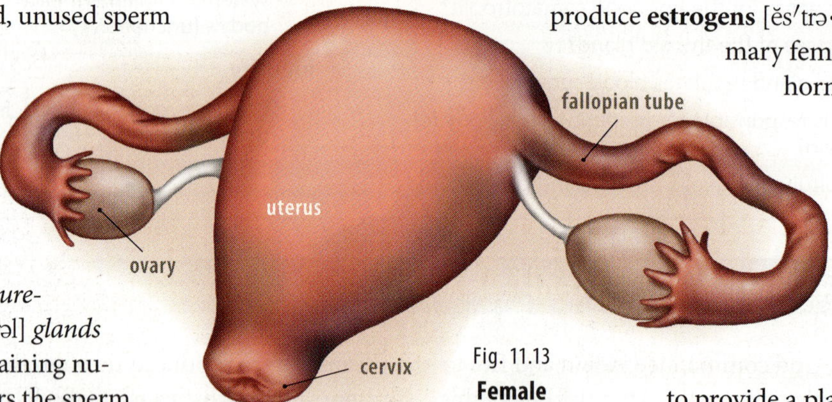


Fig. 11.13
**Female reproductive system**

**Female reproductive system.** The female gonads are the **ovaries.** Each ovary contains several thousand structures called *follicles,* and each follicle contains a developing egg cell. The ovaries produce egg cells only before birth; all of the egg cells a girl will ever have are present (but immature) when she is born. During a woman's childbearing years, hormones from the hypothalamus and pituitary gland regularly trigger a follicle and the egg cell it contains to mature and grow. Once the egg cell is mature, it is released from the follicle in the process of **ovulation.** The egg then enters one of the **fallopian** [fə·lō′pē·ən] **tubes,** or *oviducts,* which connect the ovaries to the uterus (womb). Besides producing egg cells, the ovaries also produce **estrogens** [ĕs′trə·jənz], the primary female reproductive hormones.

The **uterus,** or *womb,* is a hollow muscular organ in the lower abdomen. The primary purpose of the uterus is to provide a place for an unborn child to develop. Normally, the uterus is about the size and shape of a pear; but during pregnancy, it stretches to fill the entire abdominal cavity. *God designed the uterus to regularly prepare itself for pregnancy:* estrogens produced by a developing ovarian follicle stimulate the uterus to produce a thick inner lining with blood vessels that can nourish an unborn child. If fertilization does not occur, the uterus sheds this lining through the process of *menstruation.*

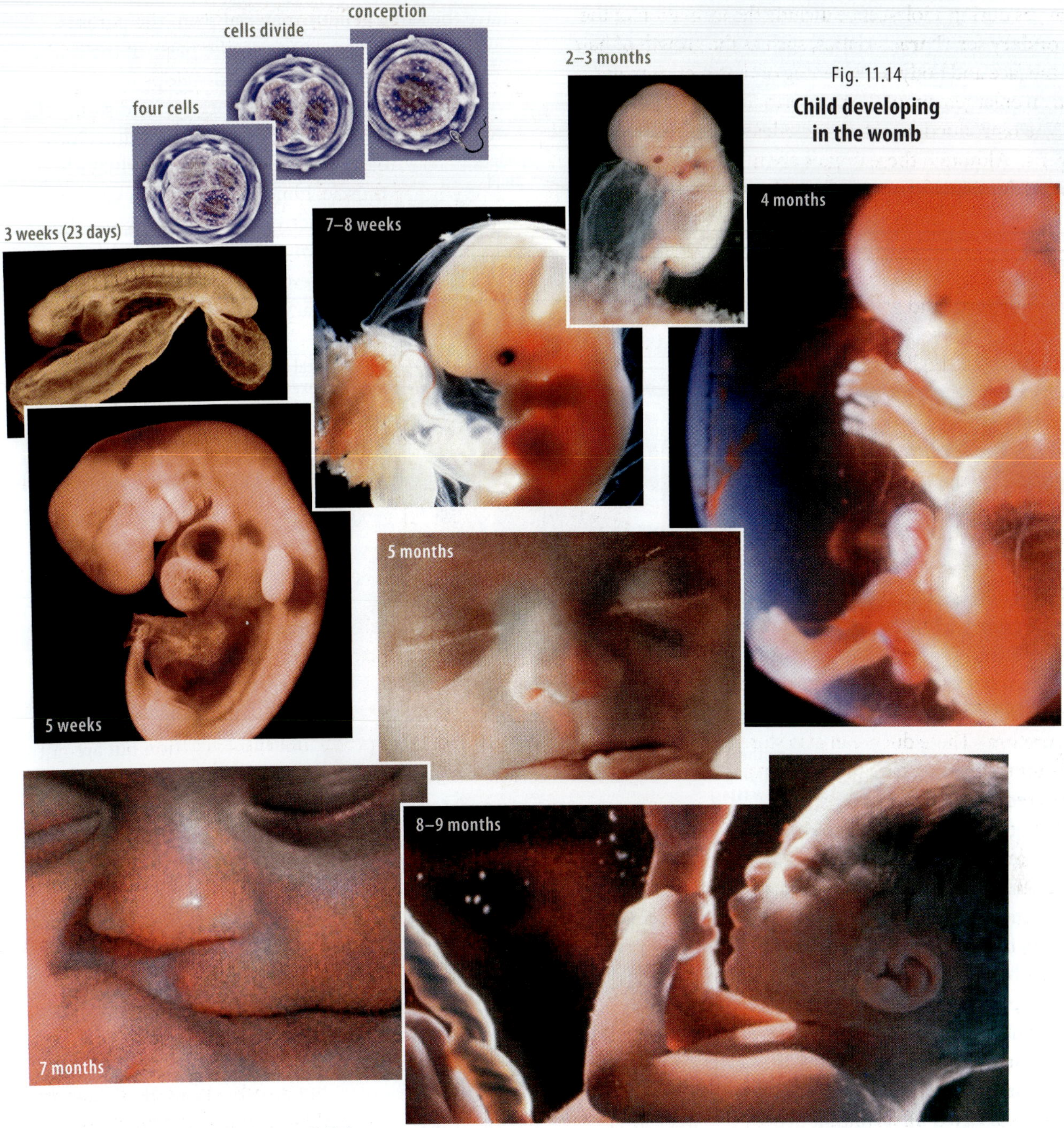


Fig. 11.14
**Child developing in the womb**

## Development of the Unborn Child

As a human being is growing in his mother's womb, the systems, organs, and tissues develop and grow in a specific sequence. From the moment of **conception** *(fertilization)*, the new life begins to develop with unique physical characteristics according to the genetic blueprint from the sperm and egg. Not only are a person's physical traits initiated at conception, but his spiritual nature is also present:

For thou hast possessed my reins: thou hast covered me in my mother's womb. I will praise thee; for I am fearfully and wonderfully made: marvellous are thy works; and that my soul knoweth right well. My substance was not hid from thee, when I was made in secret, and curiously wrought in the lowest parts of the earth. Thine eyes did see my substance, yet being unperfect; and in thy book all my members were written, which in continuance were fashioned, when as yet there was none of them. —Psa. 139:13–16

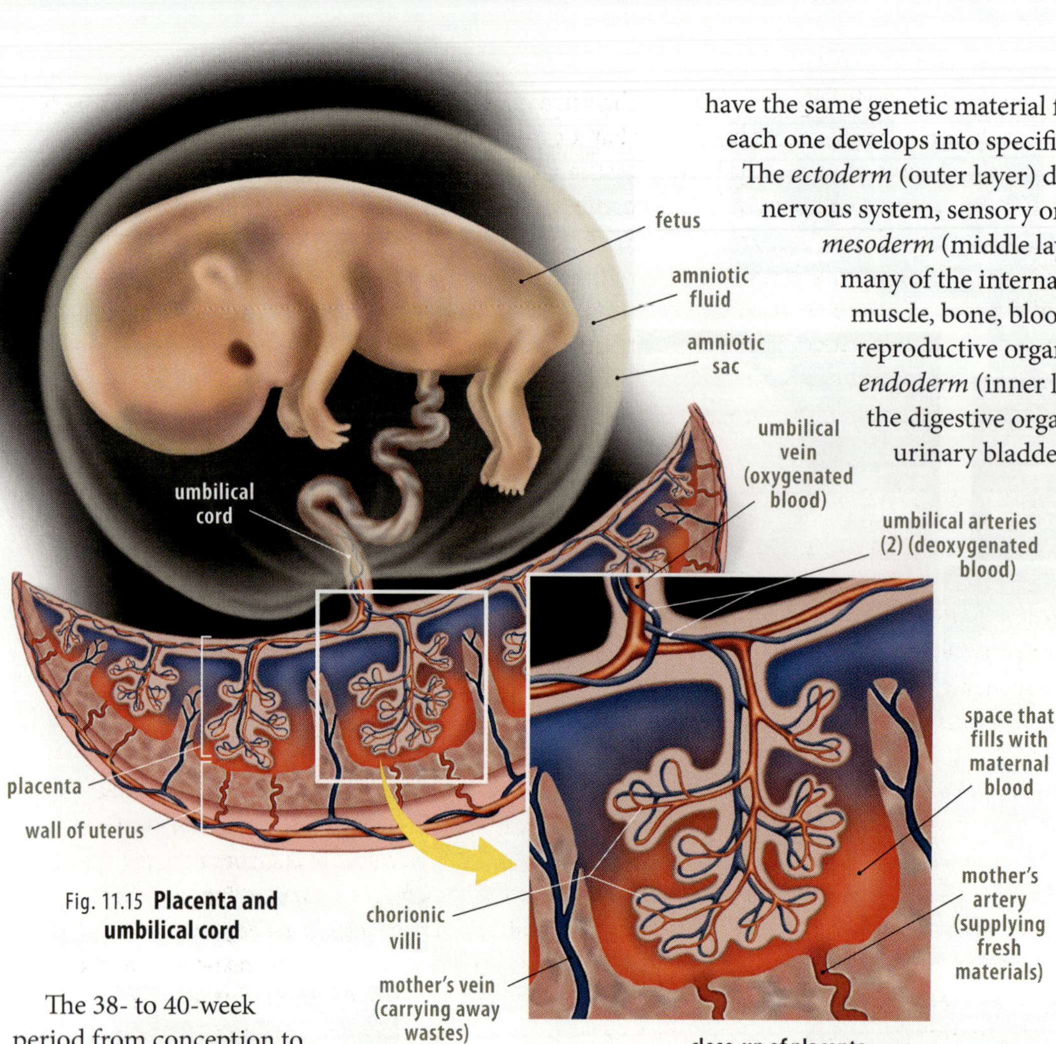


Fig. 11.15 **Placenta and umbilical cord**

The 38- to 40-week period from conception to birth is **gestation** [jĕ·stā′shən]. Gestation is divided into three *trimesters* of approximately three months each. During the first eight weeks after conception, when the basic body structure and the major organs and systems are forming, the unborn child is considered an **embryo;** from the ninth week of development until birth, he is considered a **fetus.**

**Embryonic development.** The development of a child begins in his mother's fallopian tubes. The **zygote,** or fertilized egg, repeatedly divides to form a hollow ball of cells called a **blastocyst.** About one week after conception, the blastocyst attaches to the uterine wall; this process is called **implantation.** Shortly after implantation, the **amniotic** [ăm′nē·ŏt′ĭk] **sac** begins to form around the child. This membranous sac is filled with *amniotic fluid* and protects the child as he develops.

About two weeks after conception, the blastocyst forms three distinct layers of cells called **primary germ layers.** Although these layers appear identical and have the same genetic material from the parents, each one develops into specific parts of the body. The *ectoderm* (outer layer) develops into the nervous system, sensory organs, and skin. The *mesoderm* (middle layer) develops into many of the internal organs, including muscle, bone, blood, lymph vessels, reproductive organs, and skin. The *endoderm* (inner layer) develops into the digestive organs, respiratory tract, urinary bladder, and urethra.

God designed the *placenta* [plə·sĕn′tə] and *umbilical* [ŭm·bĭl′ĭ·kəl] *cord* to allow the growing child to receive needed nourishment from his mother. These structures begin forming shortly after implantation but are not fully developed until about the twelfth week of gestation; they remain functional until birth. The **placenta** is a sac of blood vessels connecting the mother and child. *The placenta is unique because it is a single organ that develops from the tissues of two different people:* half forms from the *chorion* [kôr′ē·ŏn′], a membrane produced by the child, and half forms from the wall of the mother's uterus. The chorion develops fingerlike *chorionic villi* containing the child's capillaries. The space around the chorionic villi fills with the mother's blood, allowing the exchange of oxygen, nutrients, and wastes between the mother and child but keeping their blood from mixing. The placenta connects to the baby by the **umbilical cord.** This cord consists of three blood vessels and supporting connective tissue: the *umbilical vein* carries oxygen and nutrients from the placenta to the baby, and two *umbilical arteries* carry wastes from the baby to the placenta.

The third through eighth weeks of an unborn child's development are marked by rapid formation of his organs and systems. In the *third week*, the *digestive tube*, which will develop into the digestive organs, and the *neural plate*, which will develop into the nervous system, both begin to form. The blood vessels also begin to form, and the heart develops and starts beating. In the *fourth week*, the child grows four bumps that will become his limbs, and the organs of the respiratory system begin to take shape. The kidneys form in the *fifth week*. By the end of the *sixth week*, a cartilage skeleton has formed, and the stomach begins to produce digestive juices. Also, the brain has developed enough that it can send impulses to control body function; by the *seventh week*, the nervous system is developed enough that the child responds to touch. By the *eighth week*, the heart has developed to be a miniature replica of the adult heart. During this week, the process of *ossification* begins to replace the cartilage skeleton with bone; ossification does not complete until adulthood. *By the end of week 8, the unborn child is distinctly recognizable as a tiny human being.*

## The Miracle of New Life *by Alfred M. Rehwinkel*

Design in the LIVING CREATION

Man begins his existence in the merger of two microscopic cells, the male spermatozoon and the female ovum or egg. These cells are derived by cell division in the testicles of the male and the ovaries of the female. Under the stimulus of hormones from the pituitary gland, the ovum ripens and separates itself in regular intervals from the ovary to begin its passage through the Fallopian tubes into the cavity of the uterus. The lining in the uterus is prepared by hormones from the ovaries to receive and nourish the newly arrived ovum. If the ovum has not been fertilized, it will be removed together with the lining of the uterus in the menstrual flow. If the ovum is fertilized it develops rapidly, and by the end of the first month this microscopic formless cell will already begin to assume a recognizable human form.

By a process of incredible precision, the fertilized egg is endowed by means of the chromosomes and genes with a chemical blueprint derived half from the mother and half from the father, which directs all the details of its further development. This blueprint . . . determines how the cells will divide and multiply, how they will form into tissues and bones, limbs and organs, toes, teeth, and fingers, and into all the other parts of the body. These invisible molecules will regulate the harmonious development of the body, will determine whether these cells will develop into a male or a female, have a white, black, or brown skin, be tall or small; or, in short, will select and combine those genes which make every human being different from every other human being in the world. The mother's body is the mold in which this miracle is performed. Her bloodstream provides the material from which this new body is being formed.

This stage of development is terminated at a fixed period of time, and then a miniature human being will leave its original habitat and a new child is born into the world. A new human personality who has not existed before has come into being and is endowed with an immortal soul and all those characteristics which distinguish it as a human being. . . .

And while all this is happening, another remarkable miracle is performed in the body of the mother, giving her the power to provide the proper kind of nourishment for her newborn child from her own body. During the period of gestation, a miraculous change occurs in the body of the mother. Presently her mammary glands begin to develop, and at the precise moment when the child needs the food, these glands produce a nourishing milk that was not there before (which dries out when the child is no longer in need of it); and this milk is exactly the type of food that the newborn child requires for its normal development. All this is so marvelous that it is beyond understanding; neither an Aristotle nor an Einstein can explain or comprehend it. Thereafter, the growth and development of the human child continues for about 20 years, until he has reached the stage that we call "maturity." When maturity has been reached, this process comes to a stop.

*Alfred M. Rehwinkel,* The Wonders of Creation, *Bethany House Publishers, a division of Baker Publishing Group, Copyright ©1974, Used by permission.*

**Fetal development and birth.** In the fetal period of development, from week 9 to birth, the baby grows to the size and proportions it will have at birth; during this time, his weight will increase about 1400-fold. All the organs continue growing and developing to prepare the child for life after birth. By the *tenth week,* the urinary system is fully functional and begins operating. By the end of the *first trimester* (the twelfth week), the child's muscles are developed enough that he can move around; but his mother usually will not feel his movements until partway through the second trimester. By the end of the *second trimester* (the twenty-fourth week), the respiratory system, although not completely developed, is able to function should the child be born early. The baby's hair and nails form during the *third trimester.*

## The Tragedy of Abortion

Thinking about the LIVING CREATION

Since early times, sinful men and women have devised methods of killing unwanted children as they are developing in the womb, and those who recognize the sanctity of human life have denounced this practice. Although the ancient Hebrews learned the value of life from God's Word, in most pagan cultures killing unborn children was commonly practiced. With the coming of Christianity, the sanctity of human life became a widespread belief in the West, and for centuries the killing of unborn children was outlawed in civilized nations. Medically, **abortion** is the death of a developing child by either natural means (a miscarriage) or artificial means; in common use, however, *abortion typically refers to any action taken to kill a child before birth.* Today, the practice of abortion is common and even legal in most parts of the world. Since 1973, when the U.S. Supreme Court overturned state laws prohibiting abortion, between 40 and 50 million unborn children have been killed in the United States alone—about one abortion for every three live births. Worldwide, it is estimated that between 1 and 2 billion abortions have been performed since 1970, and abortions continue at a rate of between 40 and 50 million per year.

Many people who support abortion do not even know what it is. Since it is a vital issue, responsible citizens must understand it. Several methods of abortion are commonly used. (Some of these techniques are also used for purposes other than abortion.) Learning what abortion is and how it is performed has caused many people to feel that it should again be abolished by civilized nations.

Most surgical abortions performed in the United States employ the **suction method** (also called *suction curettage* [kyo͝or′ĭ-täzh] or *vacuum aspiration*), used during the first three months of the child's development. In this method, a tube connected to a suction device is inserted into the mother's womb; the suction draws the unborn child into a container for disposal. A growing number of early-term abortions are **medical abortions,** in which the mother is given drugs that kill the developing child.

After the first three months of pregnancy, the most common method of abortion is **dilation and evacuation** (D&E; also called *sharp curettage*). In this method, a pair of forceps inserted into the mother's womb is used to pull the child out of the womb, often dismembering him in the process. The tiny parts may be reassembled to ensure that none of the child's remains are left in the womb.

Although most abortions are performed during the first three months of pregnancy, 10,000–15,000 per year in the U.S. are performed during the final three months of development, when the baby could survive outside the womb. In these later months of pregnancy, the unborn child is often injected with poison to kill him before D&E is performed.

After 20 weeks, **salt poisoning** (medically called *intrauterine instillation* [ĭn′trə·yo͞o′tər·ĭn ĭn′stə·lā′shən]) is occasionally used. A solution of salt or urea is injected into the amniotic sac to poison the child. The child thrashes about as he slowly dies, usually within an hour and a half, and his body is delivered through the birth canal within 2 to 3 days.

A child's fetal development ends with **labor,** the process of birth. Throughout labor, the powerful muscles in the mother's uterus contract strongly; as labor progresses, the contractions become stronger, more frequent, and more painful. *Labor is divided into three main phases.* In the first phase of labor, the *cervix* (opening of the uterus) opens, or *dilates,* allowing room for the baby to exit; the contractions in this phase typically break the amniotic sac, releasing the amniotic fluid. In the second phase, the strong contractions of labor push the baby out of the birth canal, usually headfirst. In the third phase of labor, the *placenta* detaches from the uterine wall and is delivered through the birth canal; the delivery of the placenta marks the end of labor.

*Several changes occur in the baby's body at birth.* Since he no longer depends on the placenta for oxygen, nourishment, and waste removal, the umbilical arteries and umbilical vein close almost immediately after birth; over the next several months, the portions of these vessels that are inside his body will be replaced with ligaments made of connective tissue. A buildup of carbon dioxide in his tissues causes the respiratory center in the medulla oblongata to activate the respiratory system and initiate breathing. Shortly after birth, the umbilical cord is cut; after about two weeks, the dried remains of the cord fall off, leaving the **umbilicus** [ŭm·bĭl′ĭ·kəs: *navel*] as a scar.

Fig. 11.16 **Marriage is God's way**

## Sexual Morality

God ordained sex to be an expression of love between a husband and wife as part of the marriage relationship but condemns sexual relations outside of marriage (Exo. 20:14; 1 Cor. 6:18; Eph. 5:3). Throughout history, Satan has glamorized the sins of *fornication* (sexual activity before marriage) and *adultery* (sexual activity outside of marriage by a married person), leading many people into sin (see, e.g., Prov. 9:13–18). Many of the personal and societal problems in the world today, such as sexually transmitted diseases, abortion, divorce, and teenage pregnancy, are partly caused by people ignoring God's command to be sexually pure. *The only way to avoid the serious problems caused by sexual impurity is God's way: abstain from sexual relations before marriage and remain faithful in marriage* (Prov. 5:1–23).

## Section Review 11.4

1. What are the two types of gametes?
2. What hormone triggers the production of sperm cells?
3. What are the main organs of the female reproductive system? Which ones produce estrogens?
4. List the three primary germ layers and the structures into which each develops.
5. Which organ is made of tissues from both the child and his mother? What is its purpose?
6. Name the process of birth.

### Application

7. How can you avoid the problems caused by sexual impurity?

### Identify

gonads, adolescence, puberty, secondary sex characteristics, testes, epididymis, vas deferens, embryo, fetus, zygote, amniotic sac, umbilicus

## Chapter 11 Review

### Define

1. homeostasis
2. keratin
3. melanin
4. sebaceous glands, sweat glands
5. kidneys
6. urinary bladder, urethra
7. nephron, glomerulus
8. reabsorption
9. urine
10. dialysis
11. hormone
12. thyroxine
13. islets of Langerhans, diabetes mellitus
14. adrenal glands
15. steroids
16. gonads
17. adolescence, puberty
18. ovaries, fallopian tubes, uterus
19. gestation
20. zygote, blastocyst, implantation
21. umbilical cord

### Identify

1. three functions of the skin
2. the three layers of the skin
3. the organs of the urinary system
4. five organs of the endocrine system
5. the growth hormone produced by the pituitary gland
6. two hormones produced by the islets of Langerhans in the pancreas that help regulate blood sugar
7. the hormone produced by the pineal gland that makes you feel sleepy
8. two hormones produced by the gonads that are responsible for secondary sex characteristics
9. the three primary germ layers
10. two changes that occur in a baby's body after birth

### Explain

1. How does the skin help regulate body temperature?
2. How do the different skin pigments contribute to skin color?
3. How do nephrons filter molecules of waste from the blood?
4. Explain why your body cannot survive for very long without kidney function.
5. How does the endocrine system complement the nervous system?
6. What is the difference between endocrine and exocrine glands?
7. How does epinephrine affect your body's systems?
8. How do the gonads function as part of both the reproductive and endocrine systems?
9. List the main organs of the male reproductive system and describe the purpose of each.
10. Medically, what is the difference between an embryo and a fetus? How does this distinction relate to the child's development?
11. Describe the purpose of the amniotic sac.

### Apply

1. How does aging affect the skin? How can overexposure to the sun accelerate skin damage?
2. Why is it important to drink more water on hot days?
3. How can malfunctions of the endocrine system cause diseases or other disorders?
4. A common condition in the elderly is osteoporosis, in which bone mass decreases, increasing the risk of fracture. Some forms of osteoporosis can be prevented or mitigated by taking calcium supplements. Changes in which two endocrine glands are responsible for osteoporosis? Explain your answer.
5. Why is it essential that the placenta does not allow the child's and mother's blood to mix?

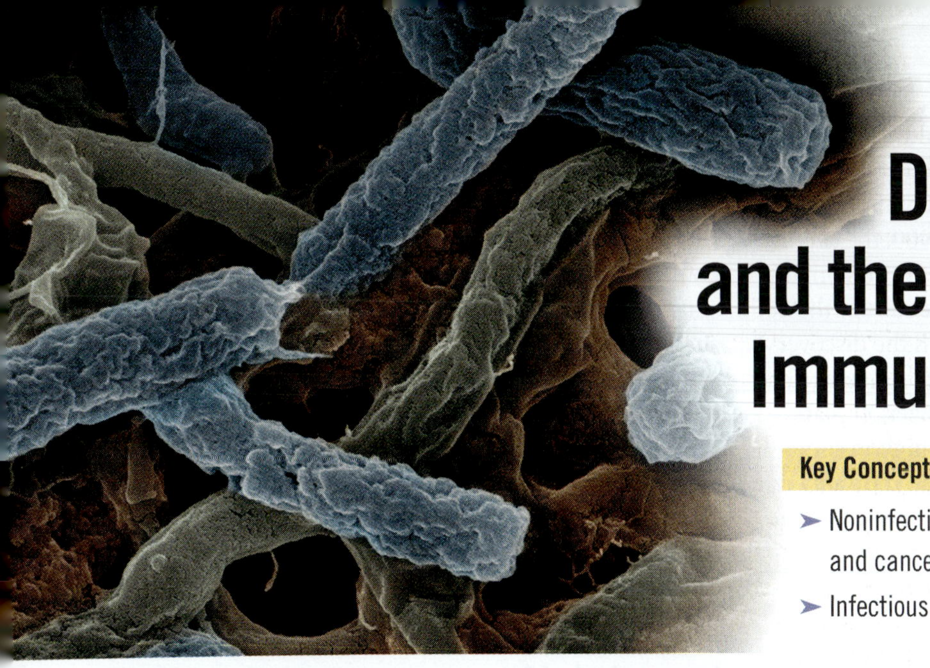

# Disease and the Body's Immune System

CHAPTER 12

**Key Concepts**

- Noninfectious diseases and cancer
- Infectious diseases
- Pathogens
- The immune system
- Defenses against disease

## 12.1 Noninfectious Diseases

### Introduction to Disease

A **disease** is any condition that interferes with the normal functioning of the body. Since the fall, man has struggled to conquer disease, one of the curses brought by sin; but only in the past two centuries has God allowed man to make great strides toward curing and preventing disease. Knowledge of the anatomy and physiology of the human body grew rapidly during the 17th century, but less achievement was made in therapy or in preventing disease. In the 18th and 19th centuries, however, biologists began to recognize the link between microscopic living creatures (called **microbes** or *microorganisms*) and disease. Microbes that cause disease are commonly called *germs.* Any organism that causes disease is a **pathogen;** pathogens include microscopic creatures such as bacteria, fungi, protozoans, and viruses and some larger creatures such as parasitic worms.

Diseases caused by invading pathogens are **infectious;** those caused by other factors, such as aging, hormones, genetics, or nutrition, are **noninfectious.** Both types of diseases may be further classified as **acute,** being severe but lasting only a short time, or **chronic,** being long lasting or recurring.

*Table 12.1* ***Classification of Diseases***

| *Infectious diseases* | *Noninfectious diseases* |
|---|---|
| **communicable diseases**<br>colds, influenza, tuberculosis, leprosy, sexually transmitted diseases | **degenerative diseases**<br>heart disease, arteriosclerosis, osteoarthritis, Alzheimer's disease, dementia, macular degeneration |
| **noncommunicable diseases**<br>tetanus, gangrene, botulism | **immunological diseases**<br>allergies (hay fever, hives, asthma), autoimmune diseases (lupus, rheumatoid arthritis, type 1 diabetes) |
| | **congenital and genetic diseases**<br>cystic fibrosis, Down syndrome, sickle cell anemia |
| | **hormonal diseases**<br>diabetes, Addison's disease, cretinism |
| | **nutritional diseases**<br>deficiency diseases (scurvy, pellagra, rickets, etc.), diseases caused by overnutrition (obesity, etc.) |
| | **diseases caused by harmful substances**<br>emphysema, cirrhosis, lead poisoning, asbestosis |
| | **cancer** (most forms are noninfectious)<br>carcinomas, sarcomas, lymphomas, leukemias |

Until about 1870, over one third of the children in the world died during early childhood from infectious diseases such as measles, mumps, whooping cough, diphtheria, typhoid fever, and scarlet fever. In the year 1900, the average life expectancy at birth was only 47.3 years because so many people died in childhood. However, since the start of the 20th century, God has blessed medical science with the discovery of numerous defenses against disease, so that the average life expectancy in the United States has risen by over 30 years. Although the number of infant deaths is still high in some parts of the world, by the early 1990s the infant mortality rate in the United States had dropped to less than one out of one hundred.

Because most of the childhood diseases that once concerned society have been conquered, a major emphasis of medical science today is on diseases that affect the adult population. As more people live to be old, there are more cases of diseases associated with aging. With fewer body resources and less resistance to disease as a result of aging, the elderly are less able to cope with infections and other illnesses; they also tend to have more than one disabling condition at a time.

## Degenerative Diseases

Diseases in which tissues and organs gradually lose their ability to function properly are called **degenerative diseases.** Although degenerative diseases are more common in middle-aged and older adults, a person may suffer unnoticed symptoms of a degenerative disease, such as cardiovascular disease, for years.

Common degenerative diseases include many forms of heart disease; arteriosclerosis, or hardening of the arteries; osteoarthritis, degeneration of the articular cartilage in the joints; Alzheimer's disease, Parkinson's disease, and dementia, which affect the brain; macular degeneration and cataracts, which affect the eyes and can cause blindness; and osteoporosis, which causes the bones to become weak and brittle.

## Immunological Diseases

Diseases caused by malfunctions of the immune system are broadly termed **immunological diseases.** The most common immunological diseases are **allergies,** which occur when the immune system

### Table 12.2 Selected Autoimmune Diseases

| | Description | Cause |
|---|---|---|
| **diabetes mellitus** | inability to produce insulin; patient's body is unable to regulate blood sugar | immune system kills insulin-producing cells of pancreas (insulin-dependent form), or immune-system antibodies deactivate insulin receptors of body cells (non-insulin-dependent form) |
| **Graves' disease** | thyroid gland swells and produces too much thyroid hormone, leading to nervousness, weakness and excessive fatigue, insomnia, susceptibility to heat, diarrhea, loss of weight, and often bulging of the eyes (exophthalmia) | antibodies of the immune system stimulate the thyroid gland to produce too much hormone |
| **lupus** (systemic lupus erythematosus) | skin, joints, or internal organs become inflamed, causing severe pain; can affect nearly every body system; severe cases can cause death; 90% of victims are women | immune system launches widespread attack on cells and tissues throughout the body |
| **multiple sclerosis** | gradual deterioration in muscle strength and coordination and sensory perception | immune system damages the myelin sheaths of motor neurons, leaving nerves vulnerable to "short circuits" |
| **myasthenia gravis** | muscle cells gradually lose the ability to receive impulses from motor neurons, causing steady deterioration in muscle strength and endurance | immune system attacks acetylcholine receptors of muscles, destroying their ability to receive nerve impulses |
| **psoriasis** | red, patchy areas of skin; overproduction of skin cells causes skin to peel off in large flakes | immune system becomes activated against a particular skin protein, triggering inflammation and rapid skin-cell production |
| **rheumatoid arthritis** | joints become inflamed, painful, and stiff; joints may eventually fuse together | immune system attacks articular cartilage of joint(s) |

overreacts to foreign substances such as pollen, dust, or certain foods. When an allergic person comes in contact with a particular substance, the immune system springs into action as if that substance were life-threatening. *Asthma* is a type of allergy in which the bronchial tubes constrict tightly when an irritating substance is inhaled, making exhalation difficult.

Other immunological diseases result when the immune system attacks the body it is supposed to protect; such diseases are called **autoimmune diseases** (from the prefix *auto-*, "self"). One well-known autoimmune disease is *rheumatoid arthritis*, in which the cells of the immune system attack the lining of the joints, causing them to become inflamed and painful. Some other autoimmune diseases are listed in Table 12.2.

## Hormonal Diseases

Disorders that occur because of malfunctions of the endocrine system are sometimes called **hormonal diseases.** The most common hormonal disorder, *diabetes mellitus,* results from defects in the hormone mechanisms that regulate blood glucose levels. Diabetes mellitus can result from a shortage of the hormone insulin or from problems with cellular insulin receptor proteins. Other hormonal diseases include *Addison's disease,* in which the adrenal glands cease to produce sufficient hormones, and *cretinism,* which is caused by inadequate secretion of hormone by the thyroid gland in newborn babies. Many hormonal diseases, such as insulin-dependent diabetes mellitus and many thyroid disorders, are also autoimmune diseases.

### Table 12.3 Selected Genetic Diseases and Their Causes

| | Description | Defective enzyme or protein | Normal function of defective enzyme or protein |
|---|---|---|---|
| **cystic fibrosis** | lungs become clogged with a thick, gummy mucus; disease also affects pancreas | CFTR (chlorine gate found in cell membranes; protein) | regulates salt content of extracellular fluids |
| **Down syndrome** (trisomy 21) | child is born with 3 copies of chromosome 21 instead of the normal 2 copies, causing widespread impairments in cell and body functions | NA | NA |
| **Duchenne's muscular dystrophy** | muscle cells die, causing the muscles to weaken, atrophy, and become replaced by fatty tissue; often causes death by age 20 | dystrophin (protein) | part of framework of muscle cell; without dystrophin, muscle cell membranes tear under stress |
| **hemophilia** | patient's blood does not clot or clots very slowly | clotting factor VIII, IX, or XI | carries out important steps in clotting process |
| **Lou Gehrig's disease** (amyotrophic lateral sclerosis) | motor neurons deteriorate, causing difficulty in moving, speaking, swallowing, and breathing; usually leads to paralysis and death within 5 years after symptoms appear | unknown; probably ubiquilin 2 and others | ubiquilin 2 "recycles" proteins in neurons |
| **phenylketonuria** (PKU) | patient cannot break down the amino acid phenylalanine; molecules can build up in body, causing brain damage | phenylalanine hydroxylase (enzyme) | helps break down phenylalanine |
| **sickle cell anemia** | hemoglobin molecules in red blood cells crystallize under low-oxygen conditions, causing the red blood cells to become brittle and triggering severe internal clotting | hemoglobin (oxygen-carrying protein) | stores oxygen in red blood cells, allowing them to transport oxygen throughout the body |

## Congenital and Genetic Diseases

Diseases that occur during development in the mother's womb and are present at birth are **congenital diseases** (*con-*, "with" + *genitus*, "birth"); some congenital diseases are commonly called *birth defects.* Many congenital diseases, such as hemophilia, sickle cell anemia, and Tay-Sachs disease, result from genetic or inherited characteristics. Such diseases are **genetic diseases.** Several types of genetic diseases and their causes are listed in Table 12.3. Congenital diseases may also occur if the mother suffers certain infections or nutritional deficiencies or abuses harmful substances such as tobacco, alcohol, or drugs.

Not all congenital diseases are noninfectious. The virus associated with *AIDS* can be transmitted from an infected mother to child during pregnancy or birth, resulting in a congenital infectious disease.

## Nutritional Diseases

Diseases caused by an improper diet *(malnutrition)* are broadly classified as **nutritional diseases.** A diet low in vitamins, minerals, or certain other nutrients can cause a **deficiency disease** (Table 12.4). Deficiency diseases are the most common nutritional diseases in underdeveloped countries due to limited food supplies and an insufficient variety of available foods. In contrast, *nutritional diseases in developed nations are usually caused by an excess of nutrients* due both to overeating and to eating unbalanced diets high in fat, sugar, and salt. *Obesity* due to improper diet increases people's risk of heart disease, cancer, diabetes, and many other diseases.

## Diseases Caused by Harmful Substances

Some diseases can be caused by excessive exposure to harmful substances. For example, heavy smoking can cause emphysema and chronic bronchitis; excessive consumption of alcohol can cause nutritional diseases and cirrhosis of the liver; and use of certain illicit drugs can cause liver failure, sterility, and Parkinson's disease. Some forms of the mineral asbestos can cause a lung disease called asbestosis if excessive amounts are inhaled as dust, especially in people who smoke.

## Cancer: Cells Gone Awry

The second-leading cause of death in the United States is **cancer,** a condition that results from the runaway growth of the body's own cells. Cancer is not really a

### Table 12.4 Selected Deficiency Diseases

| | Description | Deficient nutrient |
|---|---|---|
| deficiency anemia | ability of red blood cells to carry oxygen diminishes, resulting in weakness and fatigue | iron, copper, or vitamin $B_6$ |
| goiter | abnormal swelling of the thyroid gland | iodine |
| kwashiorkor | stunted growth, ulcerated skin, weakness, susceptibility to disease; can cause mental retardation | protein (in early childhood) |
| night blindness | inability to see in dim light | vitamin A (mild shortage) |
| osteomalacia | gradual loss of bone mass | vitamin D or calcium |
| pellagra | abdominal discomfort and diarrhea, soreness of the mouth and tongue, headache, irritability, and skin disorders; can result in depression and other mental disturbances | vitamin $B_6$ |
| pernicious anemia | numbness in the fingers and toes, sore tongue, loss of weight, weakness, and mental and nervous disorders; red blood cells are fewer than normal but abnormally large | vitamin $B_{12}$ |
| poor blood coagulation | blood clots slowly, if at all, because of the inability to properly produce the protein prothrombin | vitamin K |
| rickets | skeletal disorder characterized by fragile bones, malformed skeleton, enlarged bones at the joints, and an underdeveloped thoracic cage | vitamin D (in childhood) |
| scurvy | swollen, painful joints, spongy gums, loose teeth, delayed wound healing, bleeding of the mucous membranes, and general fatigue | vitamin C |
| xerophthalmia | cornea of eye dries out and becomes opaque, leading to blindness | vitamin A (severe shortage) |

single disease but is a group of hundreds of diseases that all result from unregulated cell growth. Although cancer can occur in people of any age, most cancers occur in people over the age of 50.

Fig. 12.1 **Progression of cancer**

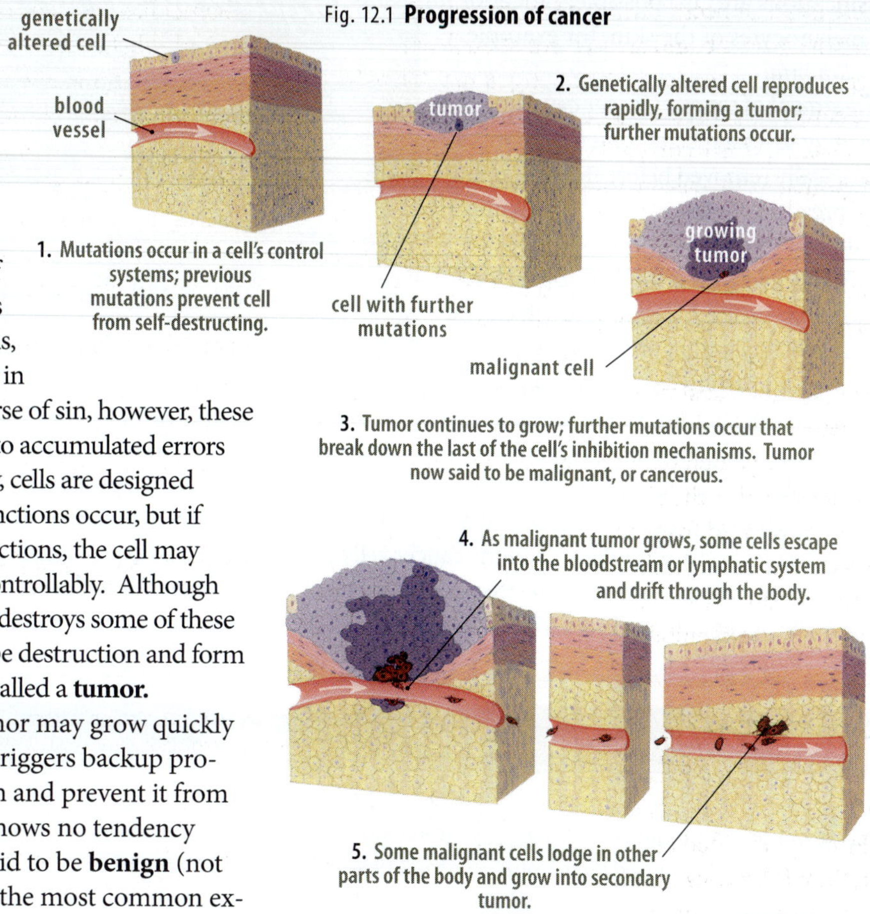


**Beginning of a cancer.** God has programmed each cell of your body to reproduce only when new cells of that type are needed. This process is regulated by the cell's control systems, which follow the instructions found in the cell's nucleus. Because of the curse of sin, however, these control systems sometimes fail due to accumulated errors *(mutations)* in the DNA. Ordinarily, cells are designed to self-destruct when serious malfunctions occur, but if the self-destruct system also malfunctions, the cell may continue to live and reproduce uncontrollably. Although your immune system seeks out and destroys some of these out-of-control cells, a few may escape destruction and form a mass of rapidly reproducing cells called a **tumor.**

**Benign tumors.** Although a tumor may grow quickly at first, its rapid growth normally triggers backup protective systems that slow its growth and prevent it from spreading. As long as the tumor shows no tendency to spread through the body, it is said to be **benign** (not cancerous). Moles and polyps are the most common examples of benign tumors, but benign tumors may occur in any part of the body.

**Malignant tumors.** Occasionally, the rapidly dividing cells of a benign tumor undergo additional malfunctions that allow them to leave the tumor and spread through the body. When this occurs, the tumor is said to have become **malignant.** *A person with a malignant tumor is said to have cancer.* A malignant tumor is dangerous because the malfunctioning cells can take up residence elsewhere in the body, producing additional malignant tumors in vital organs. If the cancer is not stopped, malignant tumors in the vital organs will eventually damage the organs beyond repair, causing death.

The size of a malignant tumor is generally limited only by its blood supply; as the tumor grows, the cells toward the inside of the tumor may begin to die from lack of nourishment. The most dangerous types of malignant tumors trigger the production of new blood vessels that "feed" the tumor, allowing it to grow uncontrollably.

**Causes of cancer.** The development of a cancer is a complex process and may be triggered by a wide variety of causes. It was once thought that harmful substances in our air, food, and water caused most cancers. However, more recent discoveries have shown this idea to be false. *Scientists now think that four major factors—hereditary defects, nutrition, smoking or drug abuse, and viruses—together account for as many as 98% of all cancer deaths.* The remaining cancer deaths are ascribed to other factors, such as excessive exposure to sunlight or other radiation (natural and artificial), excessive exposure to harmful substances in the environment (natural and artificial), and chronic tissue irritation.

**Types of cancer.** Most cancers can be divided into four broad groups, based on the type of tissue that becomes cancerous. Cancers that develop from mutant epithelial cells are called *carcinomas,* while cancers that develop from mutant connective-tissue cells are *sarcomas.* Cancers that develop from mutant lymphocytes (a particular type of white blood cell) or from lymphatic tissue are called *lymphomas,* while cancers that develop from other blood cells are called *leukemias.* More specific classifications are also possible; cancer of the melanocytes of the skin, for example, is *melanoma.*

**Treatment of cancer.** Cancer can be treated by several means. Often, tumors can be surgically removed before the malignant cells spread, curing the cancer. If the cancer has already spread or the tumor cannot be removed because of its location, other techniques must be used. Tumor growth can be inhibited by drugs that prevent new blood vessels from growing into the tumor to supply nutrients to tumor cells. Because rapidly dividing tumor cells are usually more susceptible to damage than normal cells, it is often possible to destroy these cancer cells (through high doses of drugs or radiation) while leaving normal body cells intact. If all the cancer cells are killed, the person can recover and be free of the cancer. Drugs that trigger damaged cells to self-destruct can also be very effective against cancer. Progress is also being made in learning how to stimulate the immune system to destroy cancerous cells.

At present, *one of the most important factors in treating cancer is to treat the cancer early, before it has spread.* Once the cancer has spread throughout the body, treatment is much more difficult (although still possible). Thus, it is vital that cancer be detected as early as possible. The American Cancer Society lists specific danger signals that warn a person to see a doctor.

### Danger signals of cancer

- A persistent sore that does not heal
- A lump or mass that appears where none was previously
- An increase in size or change in color or appearance of any wart, mole, or birthmark
- An unexplained or abnormal bleeding or discharge
- Persistent indigestion or change in bowel habits
- Persistent hoarseness or cough
- An unexplained loss of weight

**Prevention of cancer.** Although many cancers result from circumstances beyond human control (such as inherited

## Carcinogens and Cancer

Understanding the LIVING CREATION

It was once thought that nearly all cancers were caused by harmful substances in food, air, or water. Although more recent discoveries have refuted this notion, it is true that excessive exposure to certain substances can cause specific forms of cancer. A substance that is thought to be capable of causing cancer is a **carcinogen** (*karkinos,* "cancer" + *-gen,* "to produce"). Carcinogens contribute to the development of a cancer by causing repeated mutations in the DNA of body cells.

One of the most common and most deadly cancers, *lung cancer,* can be caused by carcinogens in cigarette smoke. Many years of constant exposure to high levels of these carcinogens can cause mutations in the smoker's lung cells, which can eventually lead to lung cancer.

Whether or not a substance might be a carcinogen is often determined by *high-dose rodent tests,* in which a group of special tumor-prone laboratory rats are fed extraordinary doses of the chemical being tested. Often, the dose is so high that the rats must be force-fed in order to hold it all. Unfortunately, many substances have been erroneously labeled "carcinogens" and banned based on high-dose rodent tests; because of how these tests are performed, about half of all chemicals tested (both natural and artificial) prove to be "carcinogenic."

In reality, any substance can be toxic if the dosage is high enough. Scientific research has shown that for most carcinogens, there is a threshold of exposure that must be exceeded for the substance to cause cancer. That is, a lifetime of exposure to extremely low levels of a particular carcinogen may not increase your risk of cancer at all, whereas a single instance of high exposure can cause cancer in a significant percentage of cases.

A more reliable test known as the *Ames test* is sometimes used instead of the high-dose rodent test. This test involves applying small amounts of the chemical to a culture of cells in a laboratory dish and testing the cells for genetic mutations. If the chemical is *mutagenic* (causing mutations), it is probably a carcinogen; if the chemical is not mutagenic, it is probably not a carcinogen at normal doses, even if extraordinary doses of the chemical can cause cancer in laboratory rats.

genetic defects), it is possible to significantly reduce your risk of cancer. *One of the greatest ways to reduce your risk of cancer is to avoid smoking;* as many as a third of all cancer deaths result from cigarette smoking. *Another major factor in reducing your risk of cancer is to eat a healthy, balanced diet with plenty of vegetables.* People who eat five or more servings of fresh fruits and vegetables each day are only half as likely to develop several forms of cancer as people who eat fewer vegetables. Regular *exercise* may help prevent cancer by keeping your body's systems in proper operating condition. A healthy diet and regular exercise also help to prevent *obesity,* a major risk factor for cancer.

## Section Review 12.1

1. Describe the two main types of disease.
2. What distinguishes an acute disease from a chronic disease?
3. Why has the emphasis on the prevention and cure of diseases largely changed from childhood diseases to adult diseases?
4. List, describe, and give examples of five categories of noninfectious diseases.
5. What is an autoimmune disease? List three examples of autoimmune diseases.
6. What is the most common cause of nutritional diseases in underdeveloped nations? In developed nations?
7. What four causative factors are thought to account for most cancer deaths? What are some ways to reduce your risk of cancer?

### Critical Thinking

8. Describe how the process of aging relates to the fall of man and man's present sinful condition.

### Identify

microbe, pathogen, degenerative disease, immunological disease, allergy, hormonal disease, congenital disease, malnutrition, cancer, carcinogen

## 12.2 Infectious Diseases

Infectious diseases, caused by *pathogens,* are the most common form of disease. Bacteria and viruses, the most common pathogens, reproduce rapidly under the favorable conditions of warmth, moisture, and nourishment that are present within the body of a human host. Other pathogens include protozoa (tiny, single-celled animals), fungi, some algae, and certain worms and other creatures.

### Types of Infectious Diseases

There are at least two ways of classifying infectious diseases. One way is to group them into the categories of communicable and noncommunicable diseases. **Communicable diseases** are those in which the invading microbe can be transferred from person to person (directly or indirectly). The common cold is a very prevalent communicable disease; other examples are influenza, AIDS, leprosy, and plague. **Noncommunicable diseases** are caused by microorganisms that are not transferred from one person to another. Tetanus is noncommunicable; you cannot get tetanus from another person.

A second way to classify infectious diseases is by the organism responsible for the disease. Diseases caused by bacteria, such as plague, pneumonia, diphtheria, tetanus, tuberculosis, and leprosy, are *bacterial diseases.* Diseases caused by viruses, such as the common cold, influenza, rabies, smallpox, yellow fever, poliomyelitis, and viral pneumonia, are *viral diseases.* Diseases caused by protozoa, such as African sleeping sickness, malaria, and amoebic dysentery [ə·mē′bĭk dĭs′ən·tĕr′ē], are *protozoal diseases.* Diseases, such as athlete's foot and ringworm, that are caused by fungi are *mycotic* [mī·kŏt′ĭk] *diseases;* diseases, such as trichinosis [trĭk′ə·nō′sĭs] and hookworm, that are caused by roundworms, tapeworms, or flukes are *helminthic* [hĕl·mĭn′thĭk] *diseases.*

### Bacteria

The microscopic, unicellular (one-celled) creatures known as **bacteria** are some of the most widespread organisms in all of God's creation. Originally, bacteria were classified as nonvascular plants, but more recent discoveries about the structure of bacteria have shown that they are quite different from cells of plants (and animals) in many fundamental ways.

**Structure of bacterial cells.** Algae, fungi, protozoa, and all plants and animals are *eukaryotes,* having cells with membrane-bound nuclei and membrane-enclosed organelles. In contrast, bacteria are *prokaryotes;* instead of being in a nucleus, the genetic material is clumped in a dense region of the cell called the *nucleoid.* The DNA of bacterial cells is unique, for bacteria have only one chromosome, which is circular. They may also have small DNA circles called **plasmids.** Plasmids can be exchanged among bacteria and very often contain the genes that make bacteria resistant to antibiotics.

**Structural variations in bacteria.** For additional protection, some bacteria are equipped with a "slime layer" outside the cell wall; the slime layer *(glycocalyx)* of the bacterium *Streptococcus mutans* allows it to adhere to your teeth, where it can cause tooth decay. Some other bacteria have tough *capsules* that protect the bacteria from drying out and help shield them from attack by viruses; these capsules also help protect some bacteria from your immune system. Some bacteria cannot move on their own, while other types are *motile* [mōt′l], or capable of motion. Many of those that are capable of motion have flagella for movement, anchored at the basal body and driven by reversible rotary motors.

**Basic bacterial shapes.** Although bacteria show great diversity in shapes, biologists recognize three basic shapes of bacterial cells (Fig. 12.3). **Cocci** [kŏk′sī: sing. coccus] are spherical bacteria that usually grow in clusters or chains.

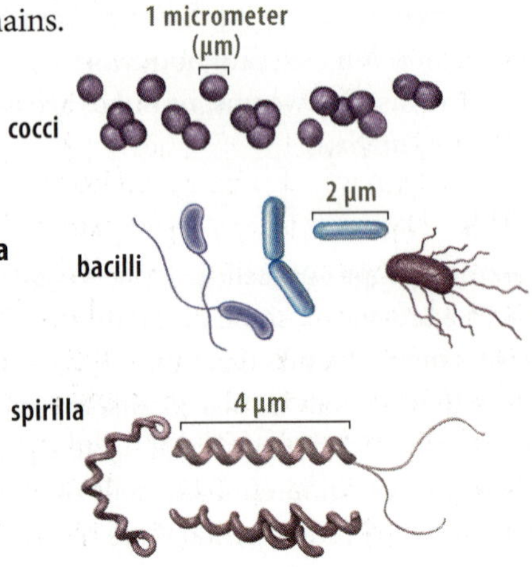


Fig. 12.3
**Basic bacteria shapes**

**Bacilli** [bə·sĭl′ī′: sing. bacillus] are rod-shaped bacteria, and **spirilla** [spī·rĭl′ə: sing. spirillum] are spiral-shaped bacteria. Bacilli and spirilla often possess flagella.

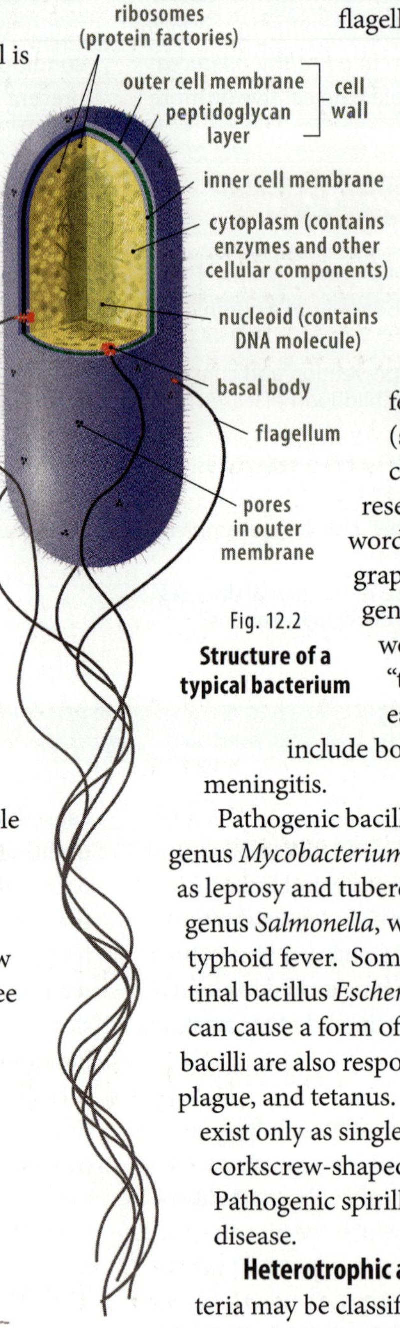


Fig. 12.2
**Structure of a typical bacterium**

Common pathogenic cocci include *Staphylococcus aureus,* which causes staph food poisoning and a variety of skin disorders; *Streptococcus pneumoniae,* the chief microbe responsible for bacterial pneumonia; and *Streptococcus pyogenes,* which causes strep throat, scarlet fever, and several other diseases (see Table 12.5). Staphylococci generally occur in clumps resembling bunches of grapes; the word *staphylo* means "a bunch of grapes" in Greek. Streptococci generally occur in chains; the word *strepto* means "turned" or "twisted" in Greek. Other diseases caused by pathogenic cocci include boils, carbuncles, gonorrhea, and meningitis.

Pathogenic bacilli include the bacteria of the genus *Mycobacterium,* which cause such scourges as leprosy and tuberculosis, and bacteria of the genus *Salmonella,* which cause food poisoning and typhoid fever. Some forms of the common intestinal bacillus *Escherichia coli* are pathogenic and can cause a form of food poisoning. Pathogenic bacilli are also responsible for botulism, diphtheria, plague, and tetanus. The spirilla, which generally exist only as single cells, include several large, corkscrew-shaped, decay-producing bacteria. Pathogenic spirilla cause syphilis and Lyme disease.

**Heterotrophic and autotrophic bacteria.** Bacteria may be classified as either *autotrophic* ("self feeder": able to make their own food) or *heterotrophic* ("other feeder": unable to make their own food). Bacteria that are heterotrophic must live as parasites or saprophytes. Fortunately, not all types of heterotrophic bacteria are harmful; most, in fact, are beneficial. We could not live without some of the bacteria that live in our intestines and help us digest our food.

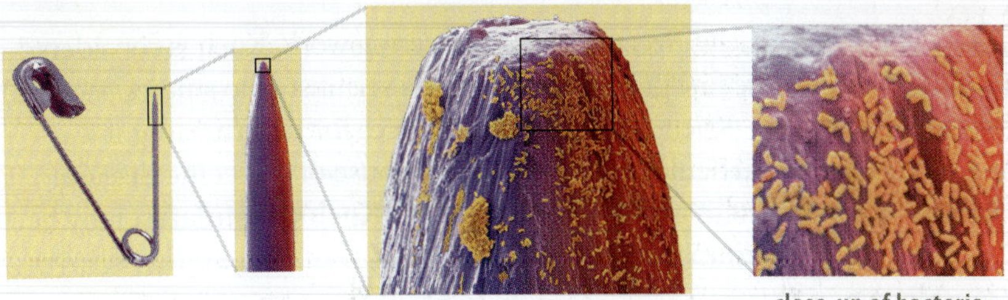


Fig. 12.4 **Bacteria on point of pin**

Some bacteria are autotrophic. Many of these bacteria contain chlorophyll or other light-harvesting compounds and carry on *photosynthesis.* However, they lack chloroplasts and have different photosynthetic machinery than plant cells. In fact, some bacteria can carry on photosynthesis using hydrogen sulfide or other toxic chemicals in place of water. Another group of autotrophic bacteria produce food by a process called *chemosynthesis,* in which iron, sulfur, or nitrogen compounds take the place of sunlight, water, or carbon dioxide in the food-making process.

**Reproduction of bacteria.** Bacteria reproduce rapidly by an asexual process called **binary fission.** One cell splits into two cells, each one being a new bacterium. Although bacteria can grow under widely varying conditions, each kind tends to grow best in a rather narrow range of heat, light, and moisture; when these conditions are optimum, some bacteria can reproduce every 20 minutes. In just two days of reproduction at this rate, the descendants of just one small bacterium could form a colony 86 times the volume of the earth. Fortunately, bacteria do not grow unchecked; when bacteria become crowded and begin to compete for food or space, their growth rate levels off.

Some small bacteria known as **rickettsiae** [rĭ·kĕt′sē·ē′: sing. rickettsia] are identical to other bacteria in structure but reproduce only inside other living cells. They are found in nature as parasites, especially in arthropods, including insects, spiders, and crustaceans. Rickettsiae cause several human diseases, including typhus (not to be confused with typhoid fever) and Rocky Mountain spotted fever. Most rickettsial diseases are transmitted by ticks, mites, and biting insects.

**Conditions favorable for bacterial growth.** God has designed bacteria to be able to grow in an amazing variety of environments. Some bacteria can grow at temperatures well below freezing, while some others thrive in deep-sea hot springs with water temperatures exceeding 260°C (500°F) and pressures exceeding 4000 pounds per square inch. Some bacteria thrive in brine (concentrated salt water), while others can live in distilled water; some thrive in strong acids, while some prefer neutral or basic environments; some require oxygen from the air, while others live only where there is no oxygen. Bacteria are found at the top of Mount Everest, at the bottom of the Challenger Deep, and almost everywhere in between. Some chemosynthetic bacteria are even found several miles deep within the earth's crust, where they live in microscopic pores in the rocks and produce food by chemosynthesis.

**Reaction to unfavorable conditions.** Although different types of bacteria can tolerate vastly different environments, no single type of bacterium can thrive in all of them. When unfavorable conditions occur, such as extremes of heat, cold, or dryness, many types of bacteria protect themselves by becoming dormant. Under unfavorable conditions, the bacterium forms a thick-walled, protective capsule around its nucleoid and other key machinery and becomes inactive until conditions are again favorable for growth. This internal "life-support" capsule is called an **endospore** (Fig. 12.5). Endospores

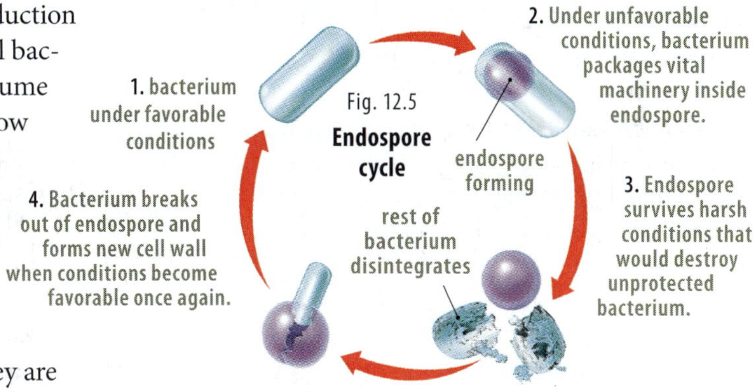


Fig. 12.5 **Endospore cycle**

are extremely resistant to harsh environmental conditions and allow bacteria to survive extremely long periods of unfavorable conditions, even if the portion of the bacterium remaining outside the endospore is destroyed. When the environment again becomes favorable for growth, the endospore germinates and develops into a bacterial cell. Endospores are extraordinarily difficult to kill; spores of the bacterium *Clostridium botulinum,* which causes botulism food poisoning, cannot be destroyed even if they are boiled for five hours. Endospores of some bacteria can survive for thousands of years; live endospores have been found in ancient villages devastated by anthrax, in ancient tombs, and reportedly even in the gut of insects trapped in amber (presumably from before or during the Flood).

**Aerobic and anaerobic bacteria.** Although most living things must have oxygen to live, some types of bacteria can live where there is no air. In fact, many of these bacteria cannot survive in the presence of oxygen. Bacteria that live where air is present and obtain oxygen from the air are **aerobic bacteria;** those that live in the absence of oxygen are called **anaerobic bacteria.** Botulism, a very dangerous type of food poisoning, is produced by anaerobic bacteria that can grow in tightly sealed cans. Tetanus, sometimes referred to as lockjaw, is another disease caused by anaerobic bacteria. Tetanus endospores enter the body through puncture wounds and become active deep in the flesh, where little free oxygen is present.

## Bacteria and Disease

**Pasteur's germ concept of disease.** In the mid-1800s, **Louis Pasteur** [lo͞o′ē păs·tûr′], the famous French scientist, observed that bacteria sometimes spoil wine by turning it to vinegar. This observation led Pasteur to wonder if bacteria might not also have a detrimental effect upon the body. At that time no one knew what caused diseases, and Pasteur wondered if bacteria might not be a cause. Eventually, he formulated the **germ concept of disease:** every infectious disease is caused by a specific germ.

**Koch's postulates.** About the same time that Pasteur was doing his experiments, a German physician named **Robert Koch** [kôk: 1843–1910] was experimenting with the bacteria he suspected of causing anthrax, which was a great killer of cattle and sheep. While experimenting with anthrax bacteria, Koch developed the techniques scientists needed to study bacteria safely. Because of his pioneering work, Koch is considered the first great *bacteriologist* (a scientist who studies bacteria).

Koch formulated guidelines for determining if a particular bacterium is the cause of a specific disease. These guidelines, now called **Koch's postulates,** may be summarized as follows:

1. The suspected pathogen must be identified in each of the diseased animals.
2. The suspected pathogen must be grown in a pure culture.
3. Organisms from the culture should cause disease when injected into healthy animals.
4. The suspected pathogen must be identified in these diseased animals.

Koch proved Pasteur's germ concept of disease, and medical scientists today use Koch's postulates in their own studies of disease and disease pathogens.

Fig. 12.6 **Applying Koch's postulates**

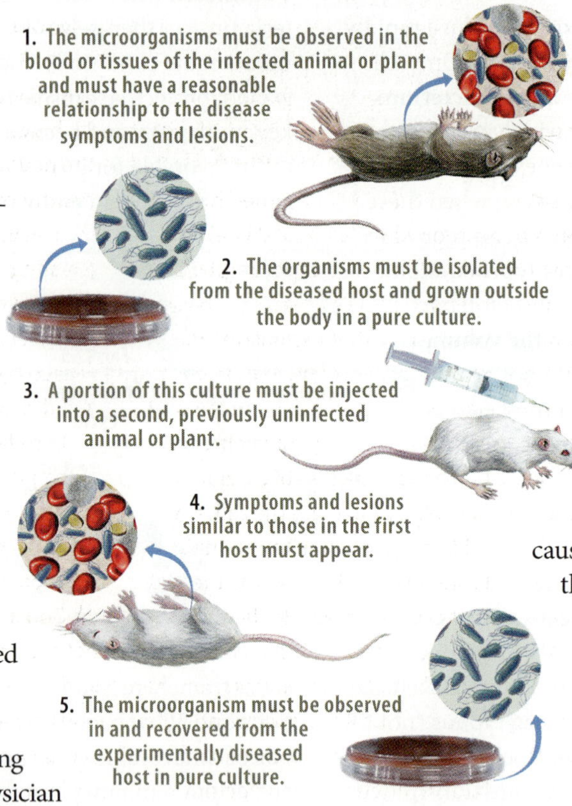


**Pathogenic activities of bacteria.** As pathogens, bacteria cause problems in several ways. *The simplest way bacteria harm the cells of the infected individual is by stealing the cells' nutrients. Many bacteria also cause harm by producing toxins* that damage or kill cells of the infected individual. Tetanus bacteria, for example, produce a toxin that damages key components of nerve cells. The bacterium that causes strep throat and scarlet fever can produce a variety of harmful enzymes, some of which cause red blood cells to burst so that the bacteria can scavenge the iron they contain, and some of which loosen the bonds between body cells so that the bacteria can spread more easily through the tissues. The bacterium that causes gas gangrene produces a toxin that breaks open body cells so that the bacterium can feed upon their contents.

*The third way that bacteria cause harm is by turning body tissues into a battleground.* During the fighting between your immune system and invading bacteria, body cells near the pathogens are often caught in the crossfire and killed. Swelling and formation of scar tissue can block blood vessels and other important ducts, causing additional harm. Also, some bacteria cause your immune system to malfunction and attack your own body tissues. For instance, rheumatic fever, which is not an infection but an inflammatory reaction following an infection, can harm the heart valves.

## Louis Pasteur

Explorers of the LIVING CREATION

*Louis Pasteur* (1822–1895), a French chemist and microbiologist, applied scientific theory to the practical problems of industry and medicine. One of Pasteur's early discoveries was that fermentation is caused by microorganisms. This discovery led to one of his most practical developments, the process named *"pasteurization"* in his honor. Before pasteurization was discovered, the only known way to kill microorganisms in milk and to delay spoilage was to boil it for several minutes, but this process altered the taste of the milk. Pasteur realized that it is not necessary to kill every bacterium in the milk, however—just those that are hazardous to humans. Because these bacteria are rather sensitive to heat, the milk can be heated to a lower temperature for a shorter time and still be safe to drink. In modern practice, pasteurization involves heating the milk to 160°F or more for a few seconds and then cooling it quickly to a low temperature. The brief application of heat kills the pathogenic bacteria, and the low temperature hinders the reproduction of the nonpathogenic bacteria that remain. This process makes the milk safe to drink and keeps the milk from quickly spoiling but does not change the milk's flavor.

Louis Pasteur in his laboratory

In 1882, Pasteur turned his attention to rabies, a fatal viral disease that attacks the hippocampus and cerebrum of the brain, causing delirium and other mental disturbances, convulsions, and death. Rabies affects many mammals and is most commonly transmitted to humans by the bite of a deranged animal. By injecting the saliva of rabid dogs into healthy animals, Pasteur was able to isolate the virus that causes rabies. He then produced a weakened form of the virus by carefully drying the spinal cord of an infected rabbit in a special jar. After two weeks of drying, the viruses had degenerated to the point that they could not cause an infection. When this weakened *(attenuated)* virus was injected into healthy animals, it did not cause disease but merely stimulated their immune systems to produce antibodies against the virus. The immunized animals were henceforth immune to the effects of rabies. A substance that stimulates the body to develop an acquired immunity to a disease is a *vaccine*.

In 1885, Pasteur was given the opportunity to test his treatment for rabies on a human. A nine-year-old boy, Joseph Meister, was badly bitten by a rabid dog. The boy's mother pleaded that Pasteur try his cure on her son. Because the boy was certain to die without treatment, Pasteur consented, and the treatment was successful. It is interesting to note that viruses had not yet been discovered in Pasteur's day. He developed a cure for a disease caused by an agent he could not see and did not know existed.

In today's world, science is often presented as being in opposition to faith. The trend in science today is that as men learn more about creation, they tend to depend less on God and lose their faith. This was not the case with Pasteur. He once said, "The more I know, the more does my faith approach that of a Breton peasant. Could I but know all, I would have the faith of a Breton peasant woman." In other words, the more he knew, the more his faith in the Creator was increased.

## Table 12.5 Selected Bacterial Diseases

| Disease | Bacterium |
|---|---|
| anthrax | *Bacillus anthracis* |
| botulism | *Clostridium botulinum* |
| cholera | *Vibrio cholerae* |
| diphtheria | *Corynebacterium diphtheriae* |
| food poisoning (gastroenteritis) | *Staphylococcus aureus*, *Escherichia coli* (some strains) |
| gas gangrene | *Clostridium perfringens* |
| gonorrhea | *Neisseria gonorrhoeae* |
| leprosy | *Mycobacterium leprae* |
| Lyme disease | *Borrelia burgdorferi* |
| necrotizing fascitis, rheumatic fever, scarlet fever, strep throat, toxic shock syndrome | *Streptococcus pyogenes* |
| peptic ulcers | *Helicobacter pylori* |
| pinkeye (bacterial conjunctivitis) | *Haemophilus aegyptius* |
| plague (both bubonic and pneumonic forms) | *Yersinia pestis* |
| Rocky Mountain spotted fever | *Rickettsia rickettsii* |
| syphilis | *Treponema pallidum* |
| tetanus | *Clostridium tetani* |
| tooth decay (dental caries) | *Streptococcus mutans* |
| tuberculosis | *Mycobacterium tuberculosis* |
| typhoid fever | *Salmonella* Typhi |
| typhus | *Rickettsia typhi*, *Rickettsia prowazekii* |

### Viruses

Viral diseases are probably the most common of all diseases. Most of us contract at least one viral infection each year—the flu, an intestinal "bug," or a cold. Unlike bacteria, protozoa, and other microscopic creatures that are clearly alive, viruses are difficult to classify as living things

Fig. 12.7 **Relative sizes of bacteria and viruses**

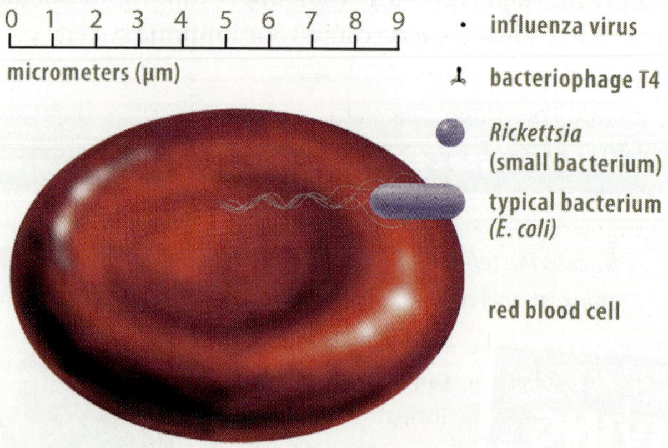


at all. Rather, a **virus** is a molecule of genetic material (DNA or RNA) enclosed in a geometric protein structure called a *capsid*. In some viruses, the capsid is surrounded by additional layers of lipids, proteins, and carbohydrates. Viruses are classified into families primarily based on their structure and how their genetic information is stored.

**Replication of viruses.** Viruses do not really reproduce in the ordinary sense of the word. Rather, *new viruses are manufactured and assembled by infected cells.* Ordinarily, a cell's *ribosomes* constantly make proteins to replace worn-out cellular components or for export to the rest of the body. These "factories" are programmed by RNA (ribonucleic acid) "blueprints" that are produced by the

Fig. 12.8 **Structure of selected viruses**

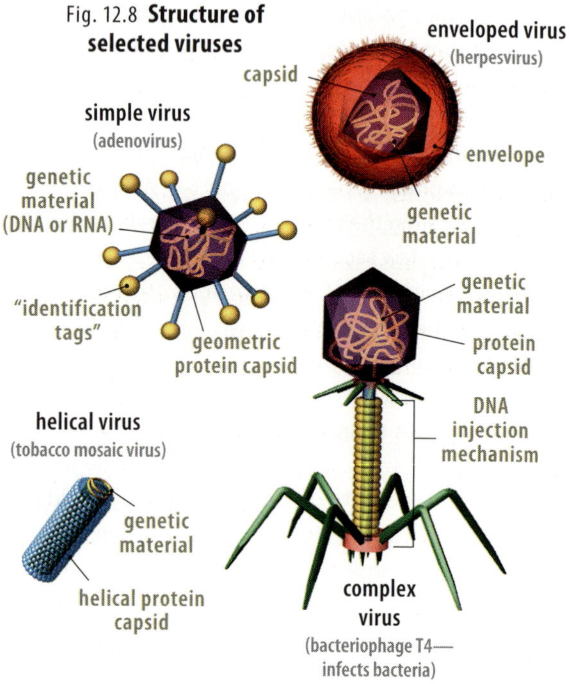

nucleus. When the cell needs a particular protein, the nucleus produces RNA for that protein and ships it to the ribosomes. When the ribosomes receive the blueprints, they automatically start producing the protein from the available amino acids.

*A typical virus is essentially a capsule of genetic material for new viruses.* When virus particles invade a cell, its genetic material takes control of the cell. The cell's nucleus is reprogrammed to make copies of the virus's genetic material, and the cell's ribosomes are instructed to produce virus proteins. The cell assembles these components into new viruses, which are ejected from the cell by the cell's export mechanisms. (Sometimes the viruses are not ejected but trigger the host cell to self-destruct and burst, spreading the viruses all at once.)

We can see from this process that a virus cannot reproduce independently; it is merely a set of instructions encoded in DNA or RNA that are copied and spread by living cells. Nor does a virus carry on metabolism or perform any other life activities. For these reasons, *virologists* (scientists who study viruses) are uncertain whether to classify viruses as living things.

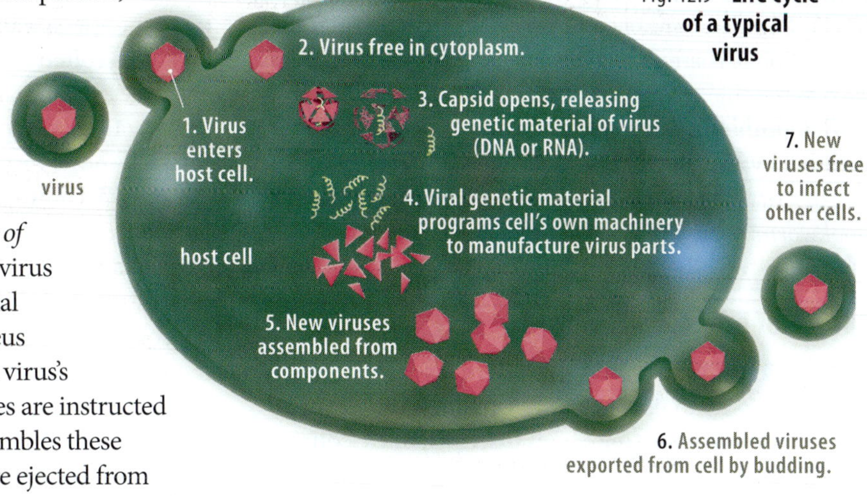


Fig. 12.9 **"Life cycle" of a typical virus**

**Viral diseases.** Viruses cause disease in bacteria, plants, animals, and humans. Cold sores, warts, rabies, smallpox, chickenpox, yellow fever, poliomyelitis, mumps, measles, influenza, mononucleosis, and the common cold are all viral diseases. Some viruses contribute to certain forms of cancer.

**Table 12.6 Selected Viral Diseases**

| Disease | Virus family | Virus genetic material |
|---|---|---|
| AIDS | retrovirus (HIV) | RNA |
| chickenpox (varicella) | herpesvirus | DNA |
| cold sores | herpesvirus (herpes simplex virus-1) | DNA |
| common cold | picornavirus *(Rhinovirus)* | RNA |
| Ebola | filovirus (Ebola virus) | RNA |
| influenza (flu) | orthomyxovirus | RNA |
| measles | paramyxovirus | RNA |
| mononucleosis | herpesvirus (Epstein-Barr virus) | DNA |
| poliomyelitis | picornavirus (poliovirus) | RNA |
| rabies | rhabdovirus | RNA |
| warts | papovavirus *(Papillomavirus)* | DNA |
| West Nile fever | flavivirus (West Nile virus) | RNA |

## Protozoa, Fungi, and Parasites

**Protozoal diseases.** Although most types of infectious disease are caused by bacteria and viruses, other types of organisms can also cause disease. **Protozoa** (single-celled animals) are responsible for several well-known diseases common in tropical regions. **Malaria,** for example, is caused by protozoa of the genus *Plasmodium,* which is very common in tropical regions; it is estimated that as many as 150 million new cases of malaria occur each year. *Other diseases caused by protozoa include amoebic dysentery, African sleeping sickness, leishmaniasis, and toxoplasmosis.*

**Mycotic diseases.** Several diseases are caused by fungi. *Athlete's foot* and *ringworm* are both skin conditions caused by fungal infections; they are characterized by red, inflamed, and flaking skin and can cause great discomfort unless treated. *Thrush* is a fungal infection of the mouth that sometimes affects infants. Most cases of *dandruff* seem to result from fungal infection of the scalp by the yeast *Malassezia furfur* (also called *Pityrosporum ovale*), which irritates scalp cells and causes them to reproduce too rapidly. This type of dandruff can be treated by regular application of shampoos containing antifungal compounds such as pyrithione zinc, sulfur, or selenium sulfide.

**Helminthic diseases.** Occasionally, the body may be invaded by parasitic worms such as roundworms, tapeworms, or flukes, causing such diseases as *trichinosis* and *hookworm.* You will learn about some of these diseases when we study these worms in chapter 21.

## Spread of Infectious Diseases

Infectious diseases are spread in many ways. Pathogens may come to us through the air, on food, in water, by contact with a person who is diseased, or by contact with a person who is not ill but harbors pathogens. Insect and animal bites may also transfer pathogens from one person to another.

**Modes of transmission.** Most infectious diseases are spread by human or animal sources. Coughs and sneezes are responsible for the spread of many infectious diseases. A **droplet infection** is carried by water droplets, such as those discharged during a sneeze or a cough; microscopic droplets are also created by the vocal cords during talking and singing. The common cold and many other diseases are spread by this method. Airborne pathogens are not the only problem; when pathogens from a sneeze or cough get on the hand, they are easily transferred to other surfaces and picked up by other people. For this reason, it is important to sneeze or cough into facial tissues and dispose of them properly and to wash your hands frequently.

Some diseases can be transmitted by infected individuals even after they have recovered from the sickness. A person who displays no symptoms of a disease but harbors the pathogens in his body and can pass them on to infect others is a **carrier.** Perhaps you have heard of a famous carrier nicknamed Typhoid Mary (Mary Mallon: 1869–1938). Mary had a mild case of typhoid fever and recovered, but the bacteria in her body were not all killed; although she was immune to these pathogens, they could infect other people. Typhoid Mary moved from household to household working as a cook. Pathogens were spread to people in each home (and in one case, in a New York maternity hospital) through contaminated food she handled, killing at least three people and causing scores of others to become seriously ill. She was finally arrested and kept in quarantine in a health facility for the last 25 years of her life.

Some diseases are not transmitted directly from person to person but are transmitted indirectly by other creatures (often insects or other small arthropods). For example, malaria and yellow fever are not transmitted directly from person to person but are spread from infected individuals to noninfected individuals by mosquitoes. An animal that transmits infection is a **vector.** Other vectors include the flea (the vector for bubonic plague) and the louse (the vector for typhus).

Diseases may also be transferred by substances that are contaminated with pathogens. Many forms of food poisoning, such as salmonellosis, are transmitted through contaminated food. The bacteria that cause ulcers can be spread by contaminated water, and a variety of pathogens can be spread by contaminated blood.

**STD's.** Some diseases, known as **sexually transmitted diseases** (STD's) or *venereal diseases,* are almost always spread by sexual relations, especially those outside of God's institution of marriage. The Centers for Disease Control estimates that 20 million new STD infections occur each year; about half are contracted by young people ages 15 to 24. Three million new STD cases occur among teenagers every year. The reason for this terrible scourge is not a lack of scientific information or of sex education but a lack of morality and righteousness. STD's are almost never acquired by people living according to God's standard of waiting until marriage to have sexual relations (see Exo. 15:26). God's command of *abstinence* is the best protection for every teenager.

**Incubation and infection.** When pathogens enter the body of a human and grow there, that person has an **infection.** Pathogens require a period of development before their presence produces signs or symptoms of disease. This is called the **incubation period.** Incubation periods range from a few hours to several years, depending upon the pathogen and other factors.

## Preventing the Spread of Disease

When a disease spreads rapidly and extensively among a population, an **epidemic** [ĕp′ĭ·dĕm′ĭk] is said to occur. Some notable worldwide epidemics include the plague epidemics of the Middle Ages, the flu epidemic of 1918–1919, and the AIDS epidemic of today.

Historically, several means have been used to prevent or control epidemics. *One of the simplest and most effective ways to stop the spread of many diseases is frequent hand washing.* Cleanliness is an important good health factor that is specifically addressed in the Bible.

Not until the late 1800s did medical science discover that God's instructions to the Israelites on washing prevent the spread of diseases. For example, Leviticus 15 instructs the Israelites to wash after they had touched anything a person with an infection had touched. Numbers 19 gives further instructions to the Israelites concerning washing after handling a dead body. Until the late 1800s, doctors did not wash their hands before examining patients. It was common for medical students to go straight from the morgue to hospital wards to examine their living patients. Of course, the diseases that had killed those in the morgue were spread to the patients. Hospitals became a breeding ground for disease. When doctors began to follow the simple practice of washing their hands, mortality rates in hospitals were drastically reduced.

*Another method of disease prevention mentioned in the Scriptures is temporarily avoiding unnecessary contact* between contagious individuals and healthy people (such as in Lev. 13:46) if such contact could spread the disease. Staying home from school when you have the flu, for example, helps prevent you from spreading the disease to others.

*Diseases spread by contaminated objects or substances can be prevented by avoiding the contamination or by disinfecting the contaminated objects* with chemicals, heat, or radiation. Modern sewage systems (from septic tanks to municipal water-treatment plants) and the chlorination of drinking water have possibly saved more lives from disease than any other modern technological advances by preventing the spread of cholera, typhoid fever, and other diseases through the water supply. Cooking food thoroughly to prevent food poisoning and sterilizing surgical instruments before they are used are also examples of this type of disease control.

*Diseases spread by vectors can be controlled by avoiding or eliminating the vectors.* For example, the incidence of malaria in America was drastically reduced through the use of insecticides and other mosquito-control methods. When these mosquito-control methods were cut back for environmental reasons, malaria became

## Ignaz Semmelweis

Explorers of the LIVING CREATION

*Ignaz Semmelweis* [ĭg′näts′ zĕm′əl·vīs′: 1818–1865], a Hungarian doctor, was the first physician to realize that handwashing prevented the spread of diseases among patients in hospitals. In the Vienna hospital where he worked, 25 to 30 percent of the maternity patients died after childbirth. All this suffering and pain drove him to relentlessly seek the cause of this astronomical number of deaths. Childbirth at home did not have anywhere near the mortality rate of childbirth in a hospital. Careful observation led him to discover that most deaths occurred in the ward that was examined by medical students. He theorized that they were spreading disease to living patients from dead bodies on which they were performing autopsies. To test his theory, he required that the students and doctors wash their hands with an antiseptic solution before examining patients. The mortality rate in his ward dropped to about 1%. Because of personality conflicts with his superiors, Semmelweis was relieved of his position. His handwashing procedures were dropped, and mortality rates again soared. He instituted his procedures in other hospitals and experienced the same results—a drastic reduction in the spread of disease. Though his methods worked, they were met with hostility and ridicule by traditional medicine. It was not until after his death in 1865 that his procedures were fully accepted by mainstream medical science. Similar instructions for cleanliness were written in Numbers 19:11–17 about 1400 years before the birth of Christ.

widespread once again. Killing mosquitoes, lice, ticks, and rats can also prevent diseases such as yellow fever, typhus, Lyme disease, and plague.

**Vaccination** can also be used to prevent the spread of diseases. A **vaccine** is a substance that stimulates your body to develop an acquired immunity to a disease. The memory cells and circulating antibodies produced in this way remain in the bloodstream for months or years, giving the body the same acquired immunity that it would have developed from an infection. If enough people in an area are vaccinated against a particular disease, outbreaks of the disease will quickly subside because the disease will not infect enough new victims to sustain itself.

## Table 12.7
### Some Diseases Fought with Vaccines

| Disease | Duration of immunity |
|---|---|
| chickenpox | life |
| cholera | 6 months |
| diphtheria | 5 to 10 years |
| hepatitis A | unknown |
| hepatitis B | 20 years to life |
| influenza | 2 to 12 months |
| measles | life |
| meningococcal meningitis | unknown |
| mumps | life |
| pneumococcal pneumonia | unknown |
| poliomyelitis | life |
| rabies | unknown |
| rotavirus | unknown |
| rubella | life |
| tetanus | 5 to 10 years |
| tuberculosis | unknown |
| typhoid fever | 2 to 5 years |
| whooping cough (pertussis) | several years to life |
| yellow fever | 10 years |

## Joseph Lister

Explorers of the LIVING CREATION

*Joseph Lister* (1827–1912) was the British doctor who established the use of *antiseptics,* chemicals that kill bacteria and help prevent infections. Lister grew up in a strongly religious family in an atmosphere that promoted science. His father was elected to the Royal Society because his work led to the development of an achromatic microscope (one that does not distort color).

After completing his studies and practicing medicine for nine years, Lister began his experiments with antiseptics in 1861. He suspected that pollenlike dust caused infections. In 1865, he became familiar with the work of Louis Pasteur and accepted Pasteur's view that microorganisms cause disease. Lister first incorrectly thought that pathogens were carried only by the air, but his work taught him that they could also be carried by the surgeon's hands and instruments.

When Lister instituted his antiseptic procedures, which involved using carbolic acid (phenol) as a surgical antiseptic, the mortality rate among his patients dropped from 45 percent to only 15 percent. Like so many other discoveries of his time, it took many years for doctors to generally accept Lister's innovative practices.

Lister's family life and early education were centered on the truths of the Scriptures. Though he was widely acclaimed during his lifetime, his accomplishments did not give him a false sense of his own worth. He was a humble, unassuming man who often declared that he believed the fundamentals of the Word of God.

## Section Review 12.2

1. Explain the difference between communicable and noncommunicable infectious diseases.
2. What are the basic bacterial shapes?
3. List three major ways that bacteria can cause harm during an infection.
4. How are new viruses produced? How might this process cause harm to the infected cell?
5. List three diseases caused by protozoa, three diseases caused by fungi, and two diseases caused by parasitic worms.

### Application

6. How can Koch's postulates help researchers determine whether a particular bacterium causes a particular disease?
7. What are some ways to control the spread of infectious diseases?

### Critical Thinking

8. *Clostridium tetani,* which causes the potentially fatal disease tetanus, is an anaerobic bacterium frequently found in soil. Why is tetanus primarily associated with puncture wounds and not with wounds that affect only the surface of the skin?

### Identify

plasmid, endospore, rickettsiae, germ concept of disease, virus, STD, incubation period, epidemic

# 12.3 Protection against Disease

The body is constantly under attack by disease. Every day, the average person inhales from 1 to 10 million pathogens and is exposed to countless more through his digestive system, mucous membranes, and microscopic skin abrasions. But the immune system is so effective that the average person gets sick only a few times a year. Even during these bouts of infection, the body's marvelously designed defense system almost always wins—which is fortunate, because losing even once means death from infection.

The most important part of the immune system is the *white blood cells,* special cells programmed to search out and destroy any organism or substance that they do not recognize as part of the body. The white blood cells are assisted by the *lymphatic system,* a complex network of fluid-filled vessels and chambers that helps transport white blood cells throughout the body and aids the white blood cells in cleansing internal body fluids of pathogens.

The immune system is assisted by special features of other systems. The skin serves as a barrier to keep bacteria and viruses from entering the body. The respiratory passages are lined with sticky hairs and other features that trap airborne bacteria before they can enter the lungs; the lungs themselves secrete special antibiotics that quickly destroy most bacteria that reach the alveoli. (Special white blood cells also roam the alveoli to guard against airborne pathogens.) The brain aids the immune system by adjusting hormone levels, by raising the body's temperature *(fever)* to help fight certain kinds of pathogens, and by signaling the body to rest when necessary.

Fig. 12.10 **White blood cells**

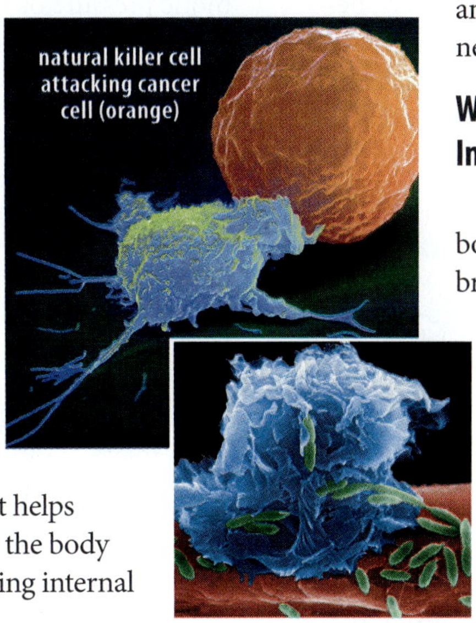


macrophage engulfing bacteria

## White Blood Cells: Agents of the Immune System

The key components of most of the body's systems are organs, such as the brain and spinal cord, the heart and blood vessels, the stomach and intestines, and the endocrine glands. The immune system is unique in that its key components are not organs but individual cells. These cells, called **white blood cells** or *leukocytes,* do the work of protecting your body against invaders.

*God has designed the immune system with several types of white blood cells, each of which has a specific function.* Some white blood cells called **phagocytes** [făg′ə·sīts′], roam through your tissues and body cavities, engulfing and disposing of any bacteria, viruses, and cellular debris they encounter.

When faced with tougher pathogens, some phagocytes can field an impressive array of cellular weaponry—secretions of acids, hydrogen peroxide, poisonous free radicals, nitric oxide (a toxic gas), cell-wall-bursting enzymes, and other poisons. Other white blood cells are sentinels that warn the body of invaders. Still others are designed to patrol your body's tissues and kill any body cells that become infected or cancerous. New white blood cells are constantly produced in the bone marrow to replace those that wear out or are destroyed in combat. Together, the white blood cells of your body are a formidable defense against invading pathogens.

**Communication between cells.** God has designed white blood cells so that they can communicate with body cells and with each other. They communicate over short distances by docking and exchanging information through receptor proteins on the cell surfaces. Over longer distances, white blood cells communicate using chemical signals called *cytokines.* Some white blood cells also produce a chemical messenger called **histamine** when invading microorganisms or foreign substances are detected in the body. Histamine causes the blood vessels to dilate and become more permeable. This allows blood fluids to enter into the surrounding tissues, causing swelling or *inflammation.*

Interestingly, body cells also communicate among themselves to fight pathogens. When a body cell becomes infected with a virus, the infected cells produce a chemical messenger called **interferon,** which limits the spread of the virus. This chemical tells nearby uninfected cells to produce antiviral compounds that protect them from becoming infected by the virus. Interferon also warns white blood cells that a viral infection has been detected.

**Antibodies.** White blood cells called *lymphocytes* produce one of the most important weapons that your immune system uses to fight pathogens: **antibodies.** Antibodies are Y-shaped protein molecules that act like molecular homing missiles. Both arms of the Y are tipped with special receptors that bind tightly to a particular surface molecule *(antigen)* of the pathogen. *Antibodies defend against pathogens in two main ways:* by inactivating the pathogens until they can be destroyed (either by causing the pathogens to "stick" together or by covering all their docking sites) and by marking them for destruction by other defense mechanisms, such as white blood cells. Because each type of antibody binds to only a single type of pathogen, antibodies and the lymphocytes involved in their production are considered the body's *specific defenses.* In contrast, the body's *nonspecific defenses* are those, such as the action of phagocytes and the effects of histamine, that are the same for any pathogen.

Antibodies are especially effective against bacteria, most of which live and reproduce outside of body cells, making them easily reached by antibodies. Viruses traveling though the body can be deactivated by antibodies as well, although antibodies are less effective against viruses already inside a cell. Antibodies can also help protect you from protein poisons (such as tetanus toxin or snake venom) by attaching to the poison molecules, thereby deactivating them and marking them for destruction as if they were bacteria or viruses.

## The Lymphatic System: Transportation and Sanitation

**Tissue fluid.** The body's cells, such as epithelial cells, muscle cells, and connective cells, are surrounded by **tissue fluid,** which originates in the plasma of the blood. Tissue fluid provides the cells with an environment that helps maintain a proper balance of water and electrolyte concentration. Without this balance, the cells would take in too much water and explode because of the osmotic pressure or would lose too much water and shrivel like prunes. The tissue fluid also acts as a battlefield in the fight against invading microorganisms.

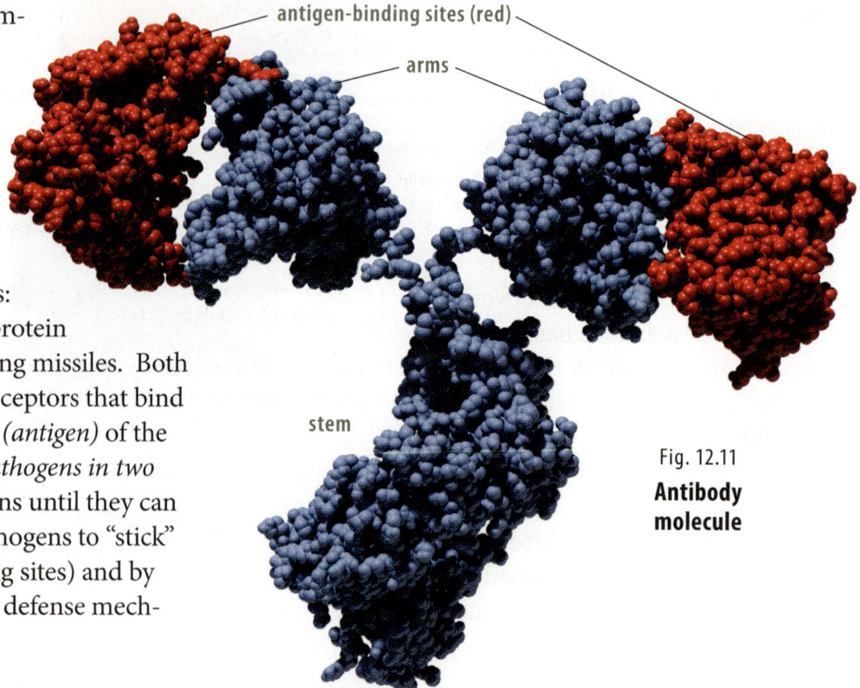


Fig. 12.11
**Antibody molecule**

**A one-way system.** The **lymphatic system** works as a drainage and sanitation system for tissue fluid. Tiny *lymph capillaries*, which originate in the tissues, collect excess fluid with its cellular debris and microbes and carry it through a closed system of **lymph vessels.** The lymph system is a one-way transportation system and does not circulate like the blood. The lymph vessels drain the tissue fluid, which is called **lymph** once it enters lymph vessels, and eventually empty it into the blood. Because the lymph system has no pump to move it along, it relies on body movements to propel it. The movement of skeletal muscles is the major force that moves lymph. As you move your body, your skeletal

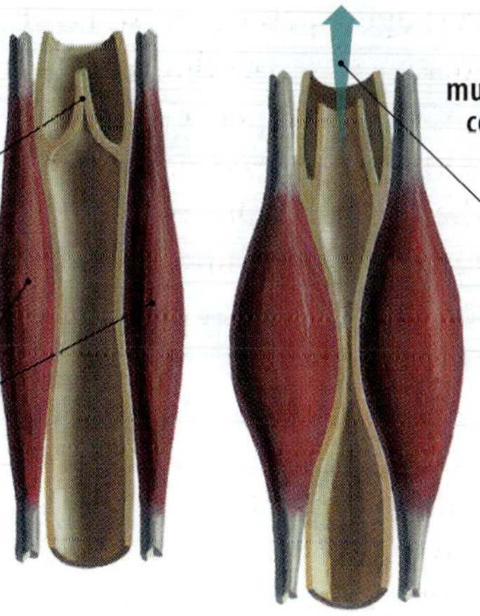


Fig. 12.12
**Movement of lymph**

## Table 12.8 Types of White Blood Cells and Their Functions

| Category | Cell type | Function(s) |
|---|---|---|
| **phagocytes** | neutrophils | – short-lived phagocytes that respond very quickly to invading pathogens, holding them at bay until slower white blood cells can join the battle<br>– most numerous white blood cells |
| | macrophages (monocytes) | – long-lived and largest phagocytes; clean up cellular debris, engulf and destroy pathogens, and report infections to T cells<br>– When fully activated, mobile macrophages bring an impressive array of cellular weaponry against pathogens, including acids, hydrogen peroxide, poisonous free radicals, nitric oxide (a toxic gas), special antibiotics called defensins, cell-wall-bursting enzymes, and other poisons. They can ingest large microorganisms like protozoa.<br>– Stationary macrophages are found in the spleen, liver, brain, and other organs. |
| | eosinophils | – long-lived, heavily armed phagocytes specialized for attacking large, multicellular invaders such as parasitic worm larvae by secreting destructive enzymes; have limited phagocytic ability |
| **lymphocytes** | B cells (plasma cells) | – when activated, produce antibodies that target specific invaders<br>– convert to memory cells to prepare for future attacks by the same pathogen, helping to prevent recurrence of the disease |
| | helper T cells | – activate B cells; supervise the immune system's response to infection |
| | killer T cells (cytotoxic T cells) | – body's main defense against pathogens found inside of the body's cells (viruses, mycobacteria, rickettsiae)<br>– when activated, roam through the body killing all body cells that show signs of infection<br>– If a killer T cell notices that a cell is infected, the T cell docks with the infected cell and blasts it at close range, riddling its cell membrane with holes and injecting DNA-shredding enzymes into its core. |
| **other white blood cells** | dendritic cells | – remain stationary in the tissues and watch for pathogens (particularly for viruses)<br>– travel to lymph nodes to present suspicious particles to T cells |
| | mast cells | – stationary white blood cells located in various body tissues<br>– sound the alarm by releasing histamine when they detect tissue damage or invading pathogens |
| | basophils | – similar to mast cells but stationed in the bloodstream<br>– help sound the alarm by releasing chemical messengers when they detect tissue damage or invading pathogens |
| | natural killer cells | – roam through the body killing any body cells that appear cancerous, infected by intracellular pathogens, or appear abnormal; do not attack microorganisms directly |

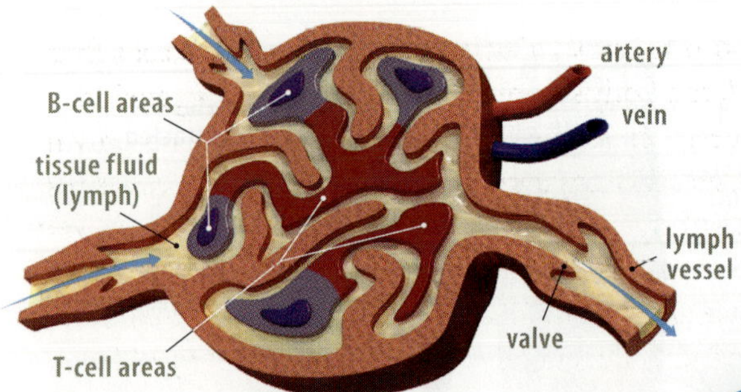


Fig. 12.13
**Structure of a lymph node**

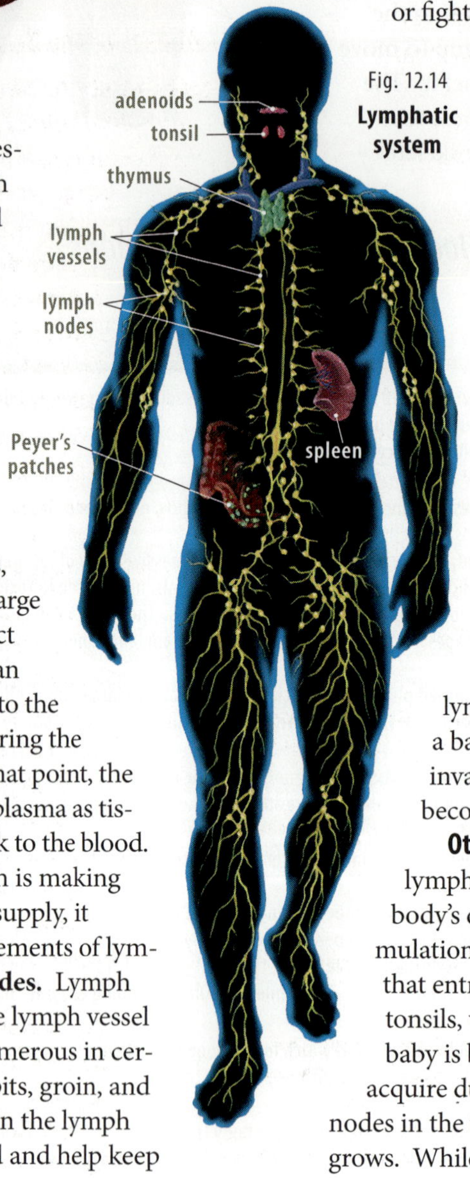


Fig. 12.14
**Lymphatic system**

muscles squeeze the lymph vessels, pushing the fluid through them; the vessels are designed with valves to prevent the fluid from moving backward. The movements involved in the constant act of breathing also help move lymph up from the lower body.

After lymph vessels collect lymph from several lymph capillaries, they converge into larger lymph vessels, which empty into one of two large **lymph ducts.** Each lymph duct empties lymph into a subclavian vein; the contents then flow into the superior vena cava before entering the right atrium of the heart. At that point, the fluid that originated in blood plasma as tissue fluid has made its way back to the blood.

**Lymph nodes.** As the lymph is making its journey toward the blood supply, it passes through special enlargements of lymphatic tissue called **lymph nodes.** Lymph nodes are located all along the lymph vessel network but are especially numerous in certain areas, including the armpits, groin, and neck. White blood cells within the lymph nodes monitor the tissue fluid and help keep it healthy and free of pathogens.

Lymph nodes come in a variety of shapes and sizes, but most are less than one inch in length and somewhat bean-shaped. Inside the lymph node are small compartments made of connective tissue. The compartments near the rim of the lymph node are used by resting *B cells,* while the compartments farther from the rim are used by resting *T cells.* The center of the lymph node is used as a meeting place where *macrophages* can rendezvous with T cells; it also serves as a staging area for *plasma cells* (activated B cells). The core of a lymph node contains a special segment of blood vessel that allows white blood cells to enter or exit the bloodstream at will.

If white blood cells are analogous to soldiers or fighting machines in a war, then lymph nodes can be compared to the military bases from which the attacks are launched. *Lymph nodes serve as bases and staging areas for white blood cells.* Whenever a macrophage or *dendritic cell* picks up a fragment of an invading pathogen, it moves to the nearest lymph node, where it presents the fragments to any B and T cells that pass through the node. Any cells that recognize the fragment become activated and enter the compartments of the lymph node, where they reproduce in preparation for battle. Once the army of activated cells is assembled, they exit the node to do battle.

Because the lymphatic system drains tissue fluid from all parts of the body, pathogens will often enter lymph nodes near an infected site directly, through the lymph vessels. The lymph node then becomes a battleground between white blood cells and invading pathogens, causing the lymph node to become swollen, inflamed, and painful.

**Other lymph organs.** Some special areas of lymph node tissue play an important part in the body's defense. **Tonsils** and **adenoids** are accumulations of lymph nodes in the throat that protect that entrance to the body. Children have large tonsils, which shrink as they grow older. When a baby is born he lacks much of the immunity he will acquire during life; therefore, the exaggerated lymph nodes in the throat give him extra protection as he grows. While filtering bacteria, the tonsils often become infected, resulting in *tonsillitis.* If they are consistently inflamed, doctors may remove them because they are causing more harm than good. **Peyer's** [pīʹərz] **patches** are lymph node groups located in the intestinal wall of the *ileum* (the lower part of the small intestine). They work to protect that digestive area from pathogens.

The **spleen,** the body's largest lymph organ, is about the size of your fist and is located beneath the diaphragm just behind the stomach; it resembles a normal lymph node except that it also has chambers containing blood capillaries. Stationary macrophages within the spleen clean the blood of dead cells and any microbes that may have entered it. Also, the spleen is soft and elastic, holding a large volume of blood that can be placed into circulation when there is an emergency.

The **thymus,** one of the most inconspicuous lymphatic organs, is located between the lungs, above the heart. The thymus is the center where immature T cells are matured and "trained" after being produced in the bone marrow. During maturation, any T cells that attack the body's noninfected tissues are triggered to self-destruct, so the T cells that remain respond only to invaders. The thymus is very large at birth because the child's body must produce large numbers of T cells very quickly so his immune system can protect him from disease after birth. The thymus gradually decreases in size as a person ages, becoming quite small by old age.

**Table 12.9 The Body's Defenses**

| |
|---|
| The **bone marrow** produces white blood cells. |
| The **Peyer's patches** are lymph tissue in the intestines that help protect the intestines from pathogens. |
| The **brain** controls many aspects of the immune system and triggers fever when necessary; special macrophages (microglia) stationed permanently in the brain engulf and destroy any pathogens that manage to cross the blood-brain barrier. |
| The **large intestine** contains harmless bacteria that prevent harmful bacteria from colonizing. |
| The **lacrimal glands** secrete tears, which wash dust and dirt from the eyes; tears contain lysozymes, which destroy bacteria. |
| The **liver** filters toxins from the blood and breaks them down for elimination; stationary macrophages (Kupffer cells) in the liver engulf and destroy any intestinal bacteria that escape into the bloodstream. |
| The **lymph vessels** carry lymph around the body; lymph contains white blood cells and carries bacteria to the lymph nodes. |
| The **lymph nodes** serve as "bases" for white blood cells. |
| The **mucous membrane** in the nose, mouth, and trachea contains mucus and cilia, which prevent pathogens from entering the body. |
| The **skin** has a shield of keratin, which protects the body from invasion of organisms; most organisms can enter through the skin only when it is damaged. Perspiration contains lysozymes, which destroy bacteria; sebum contains antibacterial substances. |
| The **spleen** is composed of lymphatic tissue that removes unwanted debris from the blood. |
| The **stomach** produces hydrochloric acid and enzymes that destroy bacteria; stomach acid converts nitrites into nitric oxide, which destroys bacteria. |
| The **thymus** serves as the "boot camp" for T cells, which fight intracellular pathogens. |
| The **tonsils** and **adenoids** are lymph node tissue; they act as barriers to bacteria and viruses. |

## Other Defenses of the Body

**Barriers to infection.** *Most bacteria and viruses cannot penetrate the tightly knitted cells of the skin;* as long as the skin is unbroken, it is one of the body's most effective defenses against infection. In addition, harmless bacteria that live on the skin protect the body by hindering the growth of any harmful bacteria that try to take up residence. The *sebum* (oil) produced by the sebaceous glands contains substances that hinder the growth of many pathogenic bacteria.

The nose and mouth have a sticky coating of *mucus,* which is secreted from the *mucous membrane;* the cells lining the mucous membrane are equipped with *cilia* that sweep pathogens caught in the mucus to the throat, where they can be expelled by a cough or swallowed and destroyed. Sneezes clear pathogens from the nasal passages. Most pathogens that reach the stomach, including those that enter via the foods we eat, are killed by the stomach's hydrochloric acid.

**Lysozymes.** Tears and perspiration contain **lysozymes,** enzymes that kill bacteria by destroying their cell walls. Tears also wash pathogens from the eyes into the tear ducts; tear ducts carry the germs to the nasal passages for disposal.

**Microorganisms of the digestive system.** Certain "friendly" bacteria that reside in the digestive system interfere with the growth of invading pathogens. The best known of these are the bacteria in the intestines. These bacteria break down some of the sugars and cellulose in the diet to form various other substances that inhibit the growth of most pathogens. These beneficial bacteria may also crowd out harmful bacteria, preventing them from taking up residence in the intestines.

**Fever.** An elevated body temperature, or **fever,** is another effective weapon against many bacteria and some viruses. A fever speeds up the body's cellular activities, helping the body fight disease. Fevers also help fight infections by hindering the growth of some pathogens that reproduce more slowly at higher temperatures. Interestingly, fevers are also used by coldblooded animals to fight infections. Experiments with lizards have shown that if they are injected with bacteria, they will lie in the sun until their body temperature reaches a level sufficient to fight the infection. If they are not allowed exposure to the sun, they often die.

Fevers are beneficial unless they last too long or become too high. Anyone with an elevated body temperature needs additional liquids to prevent dehydration. Infants and preschoolers tend to run higher fevers than older children and adults. Generally, children above two years of age with fevers under 103°F do not need an antifever drug unless they become quite uncomfortable. Most adults become uncomfortable at 102°F. Regardless of the person's age, if the temperature remains above 103°F after a few hours of medication and sponge bathing, a doctor should be consulted. A prolonged high fever can inactivate the enzymes that are essential to the normal functioning of the brain, leading to delirium and possibly brain damage. Fevers may also accompany noninfectious disorders such as heat exhaustion, sunstroke, and heart attacks.

## Immunity

The first time you encounter a particular pathogen, your immune system's response may take several days to fully develop. Often, when the infection is over, many antibodies remain in circulation, and some lymphocytes become *memory cells* to guard against that particular pathogen for the rest of your life. If you encounter that pathogen again, your immune system will be in full attack mode within hours. As a result, these pathogens will be destroyed before you develop symptoms of that infection again. Such a condition of resistance to a pathogen is **immunity.**

**Types of immunity.** There are two main types of immunity: acquired immunity and innate immunity. Immunity that results from circulating antibodies or memory cells is **acquired immunity.** Acquired immunity may be either active or passive. *Active acquired immunity* is developed through an immune response to a pathogen; this may be caused by being infected with the pathogen or by receiving a *vaccine* for the pathogen. In active immunity, the body's specific defenses manufacture antibodies and memory cells that will help fight the pathogen in the future. *Passive acquired immunity* develops when someone receives an infusion of antibodies from someone else. A newborn has antibodies from his mother's blood that give him natural passive immunity to many diseases until his immune system is completely functioning; he may also receive more antibodies from her milk. Passive immunity may be artificially caused by injection with a mixture containing antibodies. Such mixtures are used to give temporary immunity to people whose immune systems do not function properly. They are also used to treat some infections before a dangerous disease can develop; for example, the tetanus shot given to a person who has stepped on a rusty nail is typically not a vaccine but rather a concentrated solution of antibodies that will give him passive immunity to the tetanus toxin.

**Innate immunity** results from factors other than the presence of antibodies and memory cells; some of these factors include heredity, health of the immune system, diet, and exercise. Innate immunity includes the body's nonspecific defenses, those natural defenses (such as the skin barrier, fever, and the action of phagocytes) that do not involve antibodies or lymphocytes. *Species immunity* is a type of innate immunity that occurs because many pathogens infect only particular types of organisms. For example, most people are immune to the hoof-and-mouth disease virus because that particular virus primarily affects livestock. Likewise, cattle have species immunity to measles, which infects only humans.

**Immune deficiencies.** Certain conditions can cause a person's immune system to break down, causing an **immune deficiency** and leaving him susceptible to infections to which he normally would be immune. For example, adults may lose their immunity as a result of excessive exposure to radiation or certain chemicals; some cancer treatments and drugs used to prevent rejection of transplanted organs can also cause immune deficiencies.

In addition to these factors, certain genetic defects can prevent a child's white blood cells from working properly, leaving the child defenseless against certain types of diseases. Examples of congenital immune deficiency diseases include *XLA* (*X-l*inked *a*gammaglobulinemia), commonly known as *Bruton's disease,* and *SCID* (*s*evere *c*ombined *i*mmuno*d*eficiency). By far the most prevalent immune deficiency disease is *AIDS* (acquired immune deficiency syndrome), which is caused by a virus known as *HIV* (human immunodeficiency virus).

## AIDS

Understanding the LIVING CREATION

**AIDS,** or *acquired immune deficiency syndrome,* is a very serious disease that affects the immune system, destroying the body's ability to fight diseases. People with AIDS will typically get tuberculosis and various fungal, protozoal, and viral infections (including forms of cancer caused by viruses) that are rare in people with healthy immune systems.

AIDS is caused by **HIV** *(human immunodeficiency virus).* HIV is a *retrovirus,* a type of RNA virus that makes DNA copies of its genetic material and inserts these copies into a host cell's DNA. In this way, a retrovirus can remain latent (inactive) in a host cell for many years while its genetic material is replicated along with the host cell's DNA. The main host cells of HIV are T cells, a type of lymphocyte.

*HIV is transmitted by direct contact with body fluids of an infected person.* This contact most often takes place through promiscuous sexual and homosexual activity and intravenous drug abuse. However, many people have become infected through accidents involving bleeding, marital relations with a spouse that did not know he or she was infected, contaminated blood transfusions, birth from an infected mother, and other factors. Because of the long latency period, one promiscuous person can spread the virus to many other people.

When a person becomes infected with HIV, the virus reproduces rapidly and attacks T cells. In response, the body produces antibodies against HIV and destroys most of the viruses present in the blood, but many viruses remain latent in T cells. For many years, the virus silently replicates in the body, producing few if any outward symptoms yet steadily destroying T cells faster than the body can replace them. Eventually, the HIV-infected person's immune system begins to fail; only when the infection reaches this stage is the person considered to have AIDS. He may eventually succumb to fatal fungal infections of the lungs, bones, and nervous system; protozoal infections of the brain; antibiotic-resistant tuberculosis; or other diseases.

As of 2012, the World Health Organization estimated that there are about 35 million people worldwide infected with HIV. They also estimated that an additional 36 million people have died because of AIDS since the first cases were identified in 1981.

Currently, there is no cure for HIV infection or AIDS, nor is there a vaccine that can prevent the disease. *Antiretroviral drugs,* which block certain phases in the reproduction of HIV, can slow the development of AIDS and reduce the transmission of HIV but cannot cure the disease completely. Various treatments and mandatory screening of blood donors have greatly reduced the possibility of HIV transmission through accidents and blood transfusions; as a result, most new cases of HIV infection in the U.S. result from disobedience to God's standards of morality. *Thus, following God's standards—waiting until marriage to have sexual relations, remaining faithful to your spouse after marriage, and taking care of your body by avoiding drug abuse—dramatically reduces your risk of acquiring the disease.*

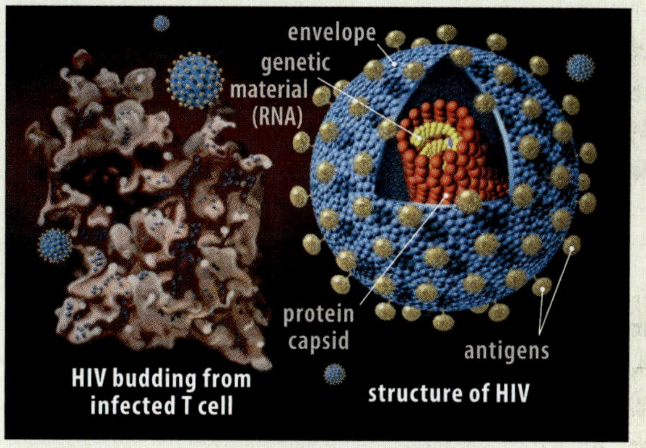

HIV (Human Immunodeficiency Virus)

## Medical Defenses

Medical scientists have made amazing advances in the war against infectious diseases; many diseases that once struck fear into the heart are rarely heard of today. Some communicable diseases (such as smallpox) have even been completely eradicated from the earth; however, new diseases seem to be discovered continually, and old diseases keep flaring up. Physicians use two general defenses to protect against infectious diseases—immune therapy and chemotherapy.

**Immune therapy.** The main tools of **immune therapy** are vaccines and antiserums. Immune therapy using *vaccines* is simply the process of exposing the body's immune system to a weakened form of a pathogen in order to produce memory cells and antibodies against it. This method of defense provides the same kind of acquired immunity that you would develop from an infection. *Antiserums* (blood extracts that contain antibodies) provide passive immunity.

Unfortunately, not all pathogens can be fought with vaccines or serums. Some pathogens (such as HIV or the protozoan that causes African sleeping sickness) have antigens that can vary widely from generation to generation. A vaccine against such a highly variable pathogen is often useless because the pathogen's changing external features prevent it from being recognized by the body's circulating antibodies or memory cells.

**Chemotherapy.** The term *chemotherapy* is most commonly used to refer to the treatment of cancer by strong chemical compounds that kill cancer cells. In a much broader sense, however, **chemotherapy** can refer to any procedure in which chemical compounds are used to treat a disease. Ideally, chemotherapy assists the body's natural defenses and causes the patient as little harm as possible.

The most familiar chemicals used in chemotherapy are **antibiotics,** substances produced by bacteria, molds, and certain other organisms that are effective in stopping the growth of microorganisms. *Penicillin,* for example, is a powerful antibiotic that is extracted from a yellowish excretion of the *Penicillium* mold, a bluish-green fungus that is sometimes found growing on bread, cheese, and fruits.

*Alexander Fleming* (1881–1955), a bacteriologist working in London, discovered penicillin by accident in 1928. Fleming saw *Penicillium notatum* growing in some culture dishes he was using to grow bacteria for an experiment. The *Penicillium* spores had somehow fallen into his bacteria cultures and had begun to grow. Fleming did not toss out his ruined cultures in exasperation, as you might expect, because he noticed that none of the bacteria he was culturing were growing near the fungus. It occurred to him that *Penicillium notatum* produced a substance that was harmful to the bacteria. Fleming's discovery of penicillin opened a new era for medicine. Today, most of the "antibiotics" we use are produced synthetically.

Most antibiotics work by interfering with specific chemical processes in microorganisms. For example, penicillin blocks an enzyme that a bacterium normally

### Careers *in* BIOLOGY

### Human Anatomy & Physiology

**Forensic science** – using medical and scientific knowledge to help solve crimes

**Physical Therapy** – treating patients with injuries or disabilities to increase mobility

**Medicine** – using scientific knowledge to minister to others by helping heal injuries, cure diseases, correct physical problems, alleviate pain, and prolong life

- general practitioner (physician)
- medical specialist
- nurse
- dentist
- laboratory technician
- x-ray/MRI technician

**Microbiology** – understanding the causes of diseases and helping discover cures and treatments by studying microorganisms, particularly pathogens

uses to build its peptidoglycan cell wall. When penicillin is present, bacteria cannot build cell walls and soon burst from osmotic pressure. Penicillin does not harm humans or animals because only the specific bacterial enzyme is affected.

Although antibiotics are potent weapons against bacteria, they are useless against viruses, which have no internal metabolic processes to interfere with because they are manufactured by the host's own cells. Since the mid-20th century, several *antiviral drugs* have been developed; these drugs work by interfering with the virus's ability to reproduce.

## Prevention: The Best Cure

*The best way to fight diseases is to take measures to prevent them.* Eating well-balanced meals is a sure way to prevent many diseases. Certain diseases are caused by a lack of essential nutrients in the diet. When the body is weakened from poor nutrition, a person is more susceptible to any disease. Fatigue can also make the body's systems more susceptible to disease. A regular schedule (including eight hours of sleep) will help the body maintain a high resistance to disease. Because of rapid growth, teenagers and children often need more than eight hours of sleep. Exercise is one of the most important ways to prevent disease. The body was designed to work and move around. Movement aids circulation and transportation of fluids that include the body's most powerful weapons against disease. A body without movement is like a nation at war that cannot move its army to the battlefield.

The Bible teaches that the body of a Christian is the temple of the Holy Spirit. Therefore, we should be careful to take care of our bodies according to the teachings of Scripture, avoiding harmful habits (such as drinking, smoking, and overeating) and being careful to get enough exercise.

### Health Practices in the Scriptures

Understanding the LIVING CREATION

Modern science knows enough about the physical causes of disease to prevent or treat a number of ailments that were fatal in the past. However, man has not even come close to conquering disease. The Bible has something to say about the root causes of many diseases.

Many Scripture verses tell us that *attitude is an important factor in good health.* For example, Proverbs 14:30 says that envy is "the rottenness of the bones." Proverbs 17:22 advises the medicine of a merry heart and warns that a broken spirit is unhealthful for the bones. Studies have shown that emotional stress, the effects of which are governed by attitude, is a contributing factor in many degenerative diseases, particularly heart problems.

Other Scripture verses indicate that *the fear of the Lord is essential to good health.*

> *Be not wise in thine own eyes: fear the LORD, and depart from evil. It shall be health to thy navel, and marrow to thy bones.*
> —*Prov. 3:7–8*

Proverbs 4:20–22 tells us that taking heed to God's instructions results in "health to all their flesh." By simply observing the lives of those who do not obey God's commands we can see the truth of these verses. For example, God commands purity in our relationship with those of the opposite sex. Some of the most widespread communicable diseases are sexually transmitted diseases. Another example that shows that disobedience to God's Word leads to poor health is the drinking of alcoholic beverages. Proverbs 23 describes the poor physical condition of a drunken man.

*Another area in which following God's instructions leads to good health is in the foods that we eat.* Several passages of Scripture condemn gluttony and overeating; obesity, which results from consuming more Calories than you burn, is a major cause of degenerative diseases in developed nations.

## Section Review 12.3

1. What is considered the key component of the immune system?
2. How do antibodies help fight infections?
3. What is the primary function of lymph nodes?
4. Describe the purposes of the tonsils and adenoids, Peyer's patches, spleen, and thymus.
5. How do your skin, mucous membranes, stomach acid, and tears help prevent harmful bacteria from entering the body?
6. How do the "friendly" bacteria of your digestive tract help protect you from pathogens?
7. Name two ways that a fever helps your immune system fight pathogens.
8. How do you become immune to a disease?
9. Describe the two main types of immunity and their causes.
10. What are three immune-deficiency diseases mentioned in the text? Which one is a communicable infectious disease?
11. What is HIV? What disease does this virus cause?
12. List some ways mentioned in the text to greatly reduce a person's risk of becoming infected with HIV.
13. How do vaccines and antibiotics help your body fight diseases?
14. Explain the importance of proper diet, rest, and exercise in preventing disease.

### Identify

white blood cells, phagocytes, interferon, lymphocytes, antibodies, tissue fluid, lymphatic system, lymph vessels, lymph, lymph ducts, lysozymes, chemotherapy

## Chapter 12 Review

### Define

1. disease
2. microbe, pathogen
3. autoimmune disease, hormonal disease, congenital disease
4. nutritional disease, deficiency disease
5. cancer
6. benign tumor, malignant tumor
7. carcinogen
8. virus
9. carrier, vector
10. STD
11. infection
12. incubation period
13. epidemic
14. vaccine
15. tissue fluid, lymph
16. tonsils, adenoids, Peyer's patches, thymus
17. immunity
18. AIDS, HIV
19. retrovirus
20. antiserum
21. antibiotic

### Identify

1. four factors thought to account for the vast majority of cancer deaths
2. the five main types of pathogens that cause infectious disease
3. three ways that bacteria can harm the body
4. four ways to reduce the spread of infectious diseases
5. the group of cells that includes phagocytes and lymphocytes
6. two main types of immunity

### Explain

1. What is the difference between an infectious disease and a noninfectious disease?
2. What is the difference between a noncommunicable disease and noninfectious disease?
3. How do Koch's postulates allow a scientist to determine whether a bacterium causes a particular disease?
4. Describe the "life cycle" of a virus.
5. How do antibodies fight infections?
6. What is the primary function of lymph nodes?
7. Describe the lymphatic function of the spleen.
8. How do your skin, mucous membranes, stomach acid, and tears assist your immune system?
9. How does fever contribute to the defense of the body against pathogens?
10. Why does your body typically fight off a pathogen much more quickly the second time you are exposed to it than the first time?

### Apply

1. Why are many degenerative diseases associated with aging?
2. Are most noninfectious diseases acute or chronic? Explain your answer.
3. Use the germ concept of disease to explain why washing your hands with an antibacterial soap can help prevent the spread of disease.
4. Considering the functions of the lymphatic system, explain why this system is designed to have one-way flow (instead of circulating flow, like blood).
5. How can proper nutrition, rest, exercise, and following God's standards of morality and self-control help prevent disease?

# Natural History and Scientific Investigation

CHAPTER 13

**Key Concepts**

- Man's understanding of nature
- Philosophy and science
- The rise of modern science
- Presuppositions of science
- The scientific method
- Limitations of science

## 13.1 Natural History through the Ages

Throughout all of recorded history, there have been people who have sought to understand the natural world around them. *Man, who was created with an inquisitive nature, was commanded by God in Genesis 1:28 to subdue the earth to the glory of God and for the good of humanity.* The history of science demonstrates man's desire to know what is in the world and why it was put there so he can use it for the benefit of mankind; this is the study of the naturalist. Biology, as we know it today, had its beginnings in man's endeavor to understand nature, especially the living things of nature.

### Biblical Beginnings

The Bible, the oldest record of man's past, reveals to us the close working association between created man (Adam) and other living things. Adam gave names to the animals created by God to inhabit the earth (Gen. 2:19). Adam and Eve, living in perfect harmony with all of the earth's living creatures, exhibited God's perfection in creation. *After the fall of man because of sin, God cursed the earth and altered man's association with its creatures.* Because of the curse of God, man was driven from the Garden of Eden to work among other living creatures for his survival. Henceforth, finding answers for his inquisitive mind would be much more difficult and would require laborious study. Being separated from God, many men sought for their answers in nature alone and totally ignored their Creator. Some, however, remaining true to God, realized that they were created in the image of their Maker and that by studying nature they were "thinking God's thoughts after Him."

In this unit, we will take a look at one of man's many attempts to explain the natural world apart from its Creator, the theory of evolution. To fully understand the flaws of this theory, however, we will first consider briefly the history of natural science and the basis of scientific investigation.

### Early Civilizations

After the Flood of Noah's day and the dispersion of mankind from the Tower of Babel, cultures arose with only a vague memory of God's truth. *As civilization continued to develop, men moved further and further away from the truth of God's Creation.* To satisfy their need to know, men devised ideas or myths of how the universe came into being and how nature operates.

Many of these people, having rejected God, worshiped the things of nature, for all men must have something to worship. Religions of superstition arose, holding people captive to fear and ignorance and often hindering their pursuit of true science. In spite of their superstitions, the ancients did make some significant discoveries.

### The Hebrews

Of all ancient peoples, the **Hebrews** or *Jews* had a distinct advantage in their study of the natural world, for they possessed God's truth as preserved in the Scriptures. The most famous early Hebrew naturalist was **Solomon,** a king of Israel who ruled around 1000 B.C. The Bible says that God made him wiser than "all the children of the east country, and all the wisdom of Egypt" and that people came from all over the known world to hear him tell the things he had learned about trees, beasts, fowl, creeping things, and fishes (see 1 Kings 4:29–34). Other men in the Old Testament who displayed a great deal of understanding of the natural world included *Job, Jacob, Moses, David,* and *Daniel.* Most notable was the Jews' progress in the study of human anatomy, physiology, and medicine. By the Middle Ages, Jewish physicians had become famous for their skill and could be found in many of the royal courts in Europe.

## The Greeks

The **Greeks** borrowed the wisdom of the ancients and built upon it. But in their interpretation of the natural world, Greek naturalists made the same unfortunate mistake that most ancient civilizations made. They failed to acknowledge the true God who created them and attemped to offer alternative explanations for the origin of life.

**Anaximander: spontaneous generation.** The early Greek naturalist *Anaximander* [ə·năk′sə·măn′dər: c. 610–c. 547 B.C.] advocated an explanation for the origin of life based solely on natural principles. Anaximander observed the presence of fossil shells in places far from the sea and concluded that the whole earth was once covered by water. He assumed that, as the earth dried by action of the sun's heat, life appeared in the warm slime. He taught that all land animals and man developed from fishlike creatures produced in the warm mud or slime. The idea that living things can arise from nonliving things is **spontaneous generation.** Observing that some large fish protect their helpless young until they mature, much as humans do, he also assumed that humans originated from lower forms of life. The ideas of creation without a Creator were perhaps man's first attempt at developing a philosophy of origins called **evolution,** the teaching that all living things developed from a common ancestor.

**Plato: only ideas are real.** Another early Greek philosopher whose thoughts influenced Western science was **Plato** (c. 427–c. 347 B.C.). He put together a system known as the **doctrine of ideas.** Plato viewed the physical world as not genuinely real and therefore changeable and relative. Thus, according to Plato, man cannot trust his senses for an understanding of physical reality. He taught that we learn by "remembering" knowledge gained through countless cycles of reincarnation. Plato founded a school in Athens called the *Academy,* where he taught this philosophy.

Fig. 13.1 **Plato (left) and Aristotle (right)**

**Aristotle: order and design in nature.** Plato's most famous disciple, the Greek naturalist and philosopher **Aristotle** (384–322 B.C.), made some great advances in the knowledge of the structure, habits, distribution, and development of animals. Included with the works of Aristotle were the first anatomical drawings. Aristotle rejected much of Plato's system and developed his own **doctrine of intellect.** According to this philosophy, intellectual speculation is the highest form of reality, and a personal God is left out of the picture. Aristotle is elevated by biologists today for his achievements in defining interrelationships among living things. Captive to Greek philosophy, however, he was searching for the meaning of the world in nature alone and fell into scientific error.

Besides their confusion of the Creator with the creation, many Greek scientists also based their hypotheses upon nothing more than speculation; relatively little effort was made to test these ideas by experimentation. As a result, although some Greek scientists stumbled upon important truths about the universe, these ideas were often obscured by a multitude of errors.

*Despite the Greek misunderstandings of the universe, their studies led them to a demonstration of the unity and consistency in nature.* The gentile world, although not acknowledging God, was beginning now to see the intricate design of the Creator. Because of their rejection of God, however, fact continued to be mingled with fancy, and true science could not arise.

## The Romans

The rise of the Roman Empire shifted world power to a people more practical and less intellectual than the Greeks, although they still held to the naturalism of the Greeks. The **Romans** were excellent compilers of knowledge, and many original advances in natural science resulted from their efforts. *One important benefit of Roman civilization was the spread of knowledge.*

**Pliny: fact and fancy.** *Platonism* (the philosophy derived from Plato's ideas) continued in one form or another throughout the Roman world, partly because of its influence on **Pliny the Elder** (Gaius Plinius Secundus: A.D. 23–79), a Roman general who wrote prolifically in his spare time. Pliny's major scientific work, ***Natural History*** *(Historia Naturalis),* touched on many scientific subjects, including the nature of the heavens, technology, zoology, botany, mineralogy, and medicine. Unfortunately, Pliny mixed fact, speculation, and superstition so thoroughly that *Natural History* was of little scientific value.

**Galen: long-lived authority.** **Galen** (Claudius Galenus: A.D. 129–c. 199), a Greek physician who lived in Rome about one hundred years after Pliny, was especially interested in the human body. Galen stressed direct observation as the key to science. He dissected animals to better understand how the human body is put together and works, and he experimented with animals to disprove many of the superstitious teachings of his contemporaries. (Dissection of human bodies was illegal in Galen's day.) Through his experiments and observations, Galen discovered that the brain controls the muscles by means of nerves, that the kidneys produce urine, and that damage to the spinal cord can cause paralysis. *Galen's work became the absolute authority in human anatomy for over a thousand years.*

Fig. 13.2 **Galen**

## Some Scientific Truths Revealed in the Bible

*Many important scientific truths were revealed in the Bible thousands of years before they were discovered by modern scientists.*

- Earth is suspended in space. **Job 26:7**
- The water cycle keeps the land watered. **Job 36:27–28; Eccles. 1:7; Amos 5:8**
- The universe is running down. **Isa. 51:6; Psa. 102:26**
- Ocean currents flow through the sea. **Psa. 8:8**
- Blood sustains life. **Lev. 17:11**
- The stars are incredibly distant from the earth and cannot be numbered. **Gen. 15:5, 22:17; Job 22:12; Jer. 33:22**
- The winds form a circulating system. **Eccles. 1:6**
- Man's body is composed of the same materials as the earth. **Gen. 2:7, 3:19; Psa. 103:14**

### Ancient Science vs. the Bible

Someone has said that *although the Bible is not a science text, whenever it speaks of scientific matters it speaks truly and accurately.* Many items of "modern discovery" were written in Scripture thousands of years ago. God tells us of the heavenly bodies and about the climate and weather; He reveals truths about chemical and biochemical processes, the ocean with all its mystery, the earth, animals and plants, the composition of the human body and its characteristics, and diseases and their prevention.

Within the Bible lie the truths and inspiration man needed to learn the facts about the world in which he lives. Only when men finally turned from pagan philosophies, superstitions, and magic to the inspired truth of Scripture did God open their eyes to see truth—not only spiritual truth but also physical truth.

God in His grace chose to give a number of scientific truths to man in Scripture through the nation of Israel. In addition to scientific truths, Scripture also teaches us about the relationship of the Creator and all of creation (nature) to man. Many naturalists have used the ideas of Scripture in their own work in natural history.

The truths of Scripture that are foremost in importance for the study of nature are these:

1. Man and all of nature are the creation of one rational, loving, orderly God.
2. God is the Creator of matter, but He is not matter Himself; He is a spirit. There is more to the universe than simply matter and energy.
3. Man was created in part to search out and discover the ways of God in creation. He is to subdue the earth and have dominion over it.
4. As man seeks to subdue the earth, he can gain an understanding of many of the secrets God has placed in creation and can use this understanding for the benefit of mankind and the glory of God.

### The Middle Ages

*During the Middle Ages, though a form of Christianity was the official religion in the Western world, pagan philosophers like Plato, Aristotle, and Pliny continued to be the primary scientific authority.* Few people were educated except those who were destined for the priesthood, and ignorance prevailed across all ranks of society. In most cases, even the priests were largely ignorant of the Scriptures, and they considered the study of nature to be a rejection of religion and therefore not a legitimate work.

Much of the learning that did survive this era of suppression and superstition was preserved by a Greek sect of Christianity commonly known as the **Nestorians.** The Nestorians (named after *Nestorius,* a 5th-century bishop of Constantinople) were driven from Constantinople because of doctrinal disagreements with the religious leaders there. Like most religious groups of the Middle Ages, they had some beliefs that were not in line with the Scriptures, but they did have enough interest in truth to preserve learning. They moved east, carrying with them the accumulated knowledge of that time, and were gladly received by the Persian Sassanid rulers, who were eager to increase their store of knowledge.

By establishing schools based on Greek learning, the Nestorians passed Greek science on to the Arabs. Syrians, Persians, Jews, and Nestorian Christians were very active in science, and they translated Greek manuscripts into Arabic and conducted new investigations, especially in the field of medicine. The city of Baghdad became the intellectual center of the world for a time.

When the Moors (Muslims from North Africa) conquered Spain, they established universities there, carrying knowledge back to Europe. Many western Europeans attended those schools and kept the spark of scientific knowledge alive. According to science historian William Locy,

> Jews and Nestorian Christians played an important part as dispensers of intellectual treasures, and to a certain extent the keen, bright mind of the Arabians shone as by light reflected from them.[1]

[1] *The Story of Biology* (Garden City, NY: Garden City Publishing Co., 1925), 87.

## Section Review 13.1

1. What is the basic purpose of science according to Genesis 1:28?
2. How did the fall of mankind into sin affect man's scientific endeavors?
3. What hindered many early civilizations in their pursuit of true science?
4. List some of the ideas and contributions of the Hebrews, Greeks, and Romans to science.
5. Why did medieval Europe make such little progress in science?
6. What were the contributions of Nestorians, Muslims, and Jews to medieval science?

### Application

7. In a paragraph, explain why the Bible is completely true and accurate when it speaks of scientific matters, although it is not a scientific textbook or treatise. Give examples of truths recorded in the Scriptures many years before they were recognized by scientists.

### Identify

spontaneous generation, evolution

# 13.2 Modern Science

## The Reformation Era

The Reformation in Europe loosened the grip of superstition and made possible a new era of progress in science. *The return to the authority of the Scriptures, which was the hallmark of the Protestant movement, made people very interested in the natural world that the God of the Scriptures had created.* They wanted to discover the secrets that God has concealed in the universe (see Prov. 25:2) and, when possible, to use their discoveries for the benefit of others. As science turned to the principles of the Bible, a great new era began—the era of modern science. *Modern science was built on biblical principles.* Today, if man rejects God and His Word, man's science risks deteriorating into pagan superstitions as it did for Pliny and many other ancients. Only by holding to the truth of Scripture and a biblical worldview can man hope to know more of God's creation.

The great German Reformer *Martin Luther* (1483–1546) set a good example to his followers for using the Scriptures to arrive at an understanding of reality. Luther ridiculed the medieval notion of spontaneous generation because of his belief in the Genesis account of a finished Creation. He also refuted the ancient pseudoscience of astrology, which was still believed in and practiced by many of his contemporaries (and by superstitious people even today). Luther had a lifelong interest in science and true learning, and his son Paul became a noted physician who helped transform the superstition of alchemy into the science of pharmacology.

Fig. 13.3 **Konrad Gesner**

## Advances in Botany and Zoology

Some of the earliest signs of the renewal of interest in nature came in the field of botany, which at that time was closely connected to medicine. **Otto Brunfels** [brŏŏn·fĕls′: c. 1488–1534], a Protestant pastor and schoolmaster who lived in Strasbourg (now in eastern France), wrote *Herbarum Vivae Eicones (Living Pictures of Herbs)* for the use of physicians and apothecaries, who used plants for medicinal purposes. *Herbarum* is valuable for the more than 250 plant illustrations made from actual observations of the flora of Strasbourg and vicinities.

Fig. 13.5 **William Harvey**

At the same time that Brunfels worked, **Leonhard Fuchs** [fyŏŏks: 1501–1566], who taught at the Protestant university in Tübingen [tōō′bĭng·ən] in southwestern Germany, wrote *The Natural History of Plants,* in which he accurately illustrated and described about 500 medicinal plants. He encouraged medical students at the university to spend their free time in the summers roaming the fields and mountains collecting and studying plants.

Swiss naturalist **Konrad Gesner** (1516–1565) was probably the best-educated naturalist of his day. Gesner served as a professor of natural history and medicine at the Protestant university in Zurich, Switzerland, where the Swiss Reformation had begun in 1518. Gesner was a staunch Protestant, as were many of his scientific contemporaries. His primary contributions to science were a five-volume work called *Historiae Animalium (Histories of Animals)* and his botanical work called the *Opera Botanica.*

## Advances in Anatomy and Physiology

**Vesalius.** The year 1543 marks a milestone in the history of biology. That was the year that **Andreas Vesalius** [vĭ·sā′lē·əs: 1514–1564] presented to the world his book on the structure of the human body. Vesalius, a devout Roman Catholic from Brussels (now part of Belgium), studied the work of Galen, who was considered by most scientists as being without error. However, Vesalius saw many contradictions between Galen's books and the bodies that he was dissecting. Vesalius wrote a book about his observations called *De Humani Corporis Fabrica (The Structure of the Human Body),* usually called the ***Fabrica.*** This book, published when Vesalius was only 28, became a classic for biological study and laid the foundation for generations of study. Because of his contributions, Vesalius is considered *the Father of Anatomy.*

**Harvey.** Anatomy, the structure of living things, was the first effort in understanding natural history. Man had to be able to know how living things were put together before he could know the function of the parts (physiology). **William Harvey** (1578–1657), an English physician and a great physiologist, was known for his classic work on the circulation of blood through the body. Living just one generation after Vesalius, Harvey used the more advanced knowledge of anatomy to study how the heart worked. His prolonged and searching work was broadly comparative. He mentioned some 40 kinds of animals upon which he worked, including fish, amphibians, reptiles, birds, and mammals. He also studied human cadavers. Harvey's extensive experimentation revived biological investigation. He wrote,

Fig. 13.4 **Andreas Vesalius**

> I have also observed, that almost all animals have truly a heart, not the larger creatures only, and those that have red blood, but the smaller, and pale-blooded ones also, such as slugs, snails, scallops, shrimps, crabs, crayfish, and many others; nay, even in wasps, hornets, and flies, I have, with the aid of a magnifying glass, and at the upper part of what is called the tail, both seen the heart pulsating myself, and shown it to many others.[2]

Harvey was finding the similarities placed in nature by nature's one Designer, and he used these similarities to better understand man, the apex of God's design.

[2] *On the Motion of the Heart and Blood in Animals,* trans. Robert Willis, ed. A. Bowie (London: George Bell and Sons, 1889), 31.

## *Bible-Believing Scientists of the Past* by Henry M. Morris, Ph.D.

Explorers of the LIVING CREATION

Many of the greatest scientists of the past were creationists and, for that matter, were also Bible-believing Christians. . . . They believed that God had supernaturally created all things, each with its own complex structure for its own unique purpose. They believed that, as scientists, they were "thinking God's thoughts after Him," learning to understand and control the laws and processes of nature for God's glory and man's good. They believed and practiced science in exactly the same way that modern creationist scientists do. . . .

This attitude did not hinder them in their commitment to the "scientific method." In fact, one of them, Sir Francis Bacon, is credited with formulating and establishing the scientific method! They seem also to have been able to maintain a proper "scientific attitude," for it was these men (Newton, Pasteur, Linnaeus, Faraday, Pascal, Lord Kelvin, Maxwell, Kepler, etc.) whose researches and analyses led to the very laws and concepts of science which brought about our modern scientific age. The mechanistic scientists of the present are dwarfed in comparison to these intellectual giants of the past. Even the achievements of an Einstein

### Table I *Scientific Disciplines Established by Creationist Scientists*

| Discipline | Scientist | Discipline | Scientist |
|---|---|---|---|
| *Antiseptic surgery* | *Joseph Lister* | *Hydraulics* | *Leonardo da Vinci* |
| *Bacteriology* | *Louis Pasteur* | *Hydrography* | *Matthew Maury* |
| *Calculus* | *Isaac Newton* | *Hydrostatics* | *Blaise Pascal* |
| *Celestial mechanics* | *Johannes Kepler* | *Ichthyology* | *Louis Agassiz* |
| *Chemistry* | *Robert Boyle* | *Isotopic chemistry* | *William Ramsay* |
| *Computer science* | *Charles Babbage* | *Model analysis* | *Lord Rayleigh* |
| *Dimensional analysis* | *Lord Rayleigh* | *Natural history* | *John Ray* |
| *Dynamics* | *Isaac Newton* | *Non-Euclidean geometry* | *Bernhard Riemann* |
| *Electronics* | *John A. Fleming* | *Oceanography* | *Matthew Maury* |
| *Electrodynamics* | *James C. Maxwell* | *Optical mineralogy* | *David Brewster* |
| *Electromagnetics* | *Michael Faraday* | *Paleontology* | *John Woodward* |
| *Energetics* | *Lord Kelvin* | *Pathology* | *Rudolph Virchow* |
| *Entomology of living insects* | *Henri Fabre* | *Physical astronomy* | *Johannes Kepler* |
| *Field theory* | *Michael Faraday* | *Reversible thermodynamics* | *James Joule* |
| *Fluid mechanics* | *George Stokes* | *Statistical thermodynamics* | *James C. Maxwell* |
| *Galactic astronomy* | *William Herschel* | *Stratigraphy* | *Nicholas Steno* |
| *Gas dynamics* | *Robert Boyle* | *Systematic biology* | *Carolus Linnaeus* |
| *Genetics* | *Gregor Mendel* | *Thermodynamics* | *Lord Kelvin* |
| *Glacial geology* | *Louis Agassiz* | *Thermokinetics* | *Humphrey Davy* |
| *Gynecology* | *James Simpson* | | |

(not to mention Darwin!) are trivial in comparison. The real breakthroughs, the new fields, the most beneficial discoveries of science were certainly not delayed (in fact probably were hastened) by the creationist motivations of these great founders of modern science.

Nor should anyone suppose that their commitment to theism and creationism was only because they were not yet acquainted with modern philosophies. Many (Agassiz, Pasteur, Lord Kelvin, Maxwell, Dawson, Virchow, Fabre, Fleming, etc.) were strong opponents of Darwinism. Even those who lived before Darwin were strong opponents of earlier evolutionary systems, not to mention pantheism, atheism, and other such antisupernaturalist philosophies, which were every bit as prevalent then as now.

To illustrate the calibre and significance of these great scientists of the past, Tables I and II have been prepared. These tabulations are not complete lists, of course, but at least are representative, and they do point up the absurdity of modern assertions that no true scientist can be a creationist and Bible-believing Christian.

Table I lists the creationist "fathers" of many significant branches of modern science. Table II lists the creationist scientists responsible for various vital inventions, discoveries, and other contributions to mankind. These identifications are to some degree oversimplified, of course, for even in the early days of science every new development involved a number of other scientists, before and after. Nevertheless, in each instance a strong case can be made for attributing the chief responsibility to the creationist scientist indicated. At the very least, his contribution was critically important and thus supports our contention that belief in creation and the Bible helps, rather than hinders, scientific discovery.

In each case, the scientists listed were strict creationists, unreservedly believing in the Bible and the God of the Bible. Some were "progressive creationists," but none were theistic evolutionists, so far as can be determined. They came from a variety of denominational backgrounds and doctrinal persuasions, but all were at least professing Christians, committed to the basic doctrines of Christianity.

*From Impact #103 in Acts & Facts, Vol. 11, No. 1, January 1982. Used by permission of the Institute for Creation Research.*

## Table II Scientific Contributions by Creationist Scientists

| Contribution | Scientist | Contibution | Scientist |
|---|---|---|---|
| *Absolute temperature scale* | *Lord Kelvin* | *Global star catalog* | *John Herschel* |
| *Actuarial tables* | *Charles Babbage* | *Inert gases* | *William Ramsay* |
| *Barometer* | *Blaise Pascal* | *Kaleidoscope* | *David Brewster* |
| *Biogenesis law* | *Louis Pasteur* | *Law of gravity* | *Isaac Newton* |
| *Calculating machine* | *Charles Babbage* | *Mine safety lamp* | *Humphrey Davy* |
| *Chloroform* | *James Simpson* | *Pasteurization* | *Louis Pasteur* |
| *Classification system* | *Carolus Linnaeus* | *Reflecting telescope* | *Isaac Newton* |
| *Double stars* | *William Herschel* | *Scientific method* | *Francis Bacon* |
| *Electric generator* | *Michael Faraday* | *Self induction* | *Joseph Henry* |
| *Electric motor* | *Joseph Henry* | *Telegraph* | *Samuel F. B. Morse* |
| *Ephemeris tables* | *Johannes Kepler* | *Thermionic valve* | *John A. Fleming* |
| *Fermentation control* | *Louis Pasteur* | *Transatlantic cable* | *Lord Kelvin* |
| *Galvanometer* | *Joseph Henry* | *Vaccination and immunization* | *Louis Pasteur* |

## The Spread of Science

By the time of Vesalius and Harvey, the work of naturalists was becoming more accepted and popular. But the advancement of science was being threatened by false philosophies such as **materialism,** the idea that the universe consists of nothing but matter and energy and has no spiritual or supernatural aspects. Fortunately, men like *Robert Boyle* (1627–1691), the "Father of Chemistry," and *Sir Isaac Newton,* the "Father of Physics," based their work on the truth of the Scriptures, laboring to find the design of creation and ultimately formulating the theories that enabled science to explode with discoveries.

The people who took the greatest interest in science during the 1600s were the English Puritans. In 1645, *John Wilkins* (1614–1672), a Puritan clergyman, led in the formation of the *Philosophical College,* which met regularly in London to conduct experiments and discuss scientific theories. These men did much to spread the work of naturalists. Later, Wilkins and many other Puritans were leading members in the **Royal Society of London for Improving Natural Knowledge.** This group, whose membership also included Robert Boyle and Sir Isaac Newton, was created to repudiate the unbelieving ideas of materialistic philosophers who said that man could arrive at all knowledge through science. The founders of the Royal Society believed that science can teach us about physical reality but not about the mind, the soul, the spirit, and the supernatural.

The **French Academy of Sciences** (Académie des Sciences) was founded in Paris in 1666. It was supported largely by *Huguenots* [hyōō′gə·nŏts′: French Protestants] and by *Jansenists,* a group of Calvinist French Catholics who emphasized salvation through God's love and grace. (The most outstanding Jansenist was the great mathematician and scientist *Blaise Pascal* [blĕz pă·skăl′: 1623–1662].) By 1800 there were scientific societies, where the great scientists and naturalists of the day could share their findings, in many parts of Europe and in America.

## Microscopy

**Early work with magnification.** There has perhaps been nothing more energizing to the study of biology than the development of the *microscope.* Trying to view nature with the unaided eye limited man's knowledge of the structure of living things and therefore limited his understanding as to the function God has assigned to each part of each living thing. When naturalists began using magnification, a whole new world opened to them. The concept of using glass for magnification can be traced back to the medieval English monk *Roger Bacon* (c. 1214–c. 1292). By the time of William Harvey, magnifying glasses were no novelty and were used by most naturalists. In 1665, **Robert Hooke** (1635–1703) published his work *Micrographia,* in which he described the cells of cork. He was the first to devote an entire book exclusively to microscopic observations.

**Leeuwenhoek's animalcules.** The Dutch naturalist **Anton van Leeuwenhoek** [vän lā′vən·hŏŏk′: 1632–1723] was the first person to devote his whole life to studies with the microscope. Although he lacked extensive schooling, he was financially independent and became exceedingly interested in microscopy. He made his own microscopes (over two hundred) from homemade parts. Leeuwenhoek had no system or method of study; he simply observed, wrote, and illustrated whatever was of interest to him. By 1675, his expertise was sufficient for him to describe "animalcules," which today we call protozoa. Remarkably, in 1683 he described bacteria from material scraped from his teeth. The Royal Society published Leeuwenhoek's work for naturalists in other localities to see and attempt to duplicate. As a result of scientific publications, microscopy became a widely studied field by naturalists, and its study brought about a new era of biology.

## The Cell Theory

One of the most fundamental principles of biology resulted from the work of the early microscopists. The **cell theory** maintains that *all living things are composed of living units called cells and of cell products* and that *all cells come from preexisting cells.* This principle was not widely accepted until the mid-19th century, at which time the study of microbiology became a recognized field of science. The theory of the cell was a direct result of man's searching for order in living things. On the premise that all creatures were created at one time by one Creator and according to one design, biologists suggested that all life forms followed this same pattern. In the years that followed, thousands of investigations were conducted that verified the theory. Man had found one more aspect of God's living design.

## *The Discovery of Vorticella* by Anton van Leeuwenhoek

Explorers of the LIVING CREATION

In the year 1675, I discover'd living creatures in Rain water, which had stood but few days in a new earthen pot, glazed blew within. This invited me to view the water with great attention, especially those little animals appearing to me ten thousand times less than those represented by Mons. *Swamerdam,* and by him called *Water-fleas* or *Water-lice,* which may be perceived in the water with the naked eye.

Leeuwenhoek using his microscope

The first sort by me discover'd in the said water, I divers times observed to consist of 5, 6, 7, or 8 clear globuls, without being able to discern any film that held them together, or contained them. When these *animalcula* or living Atoms did move, they put forth two little horns, continually moving themselves: The place between these two horns was flat, though the rest of the body was roundish, sharpening a little towards the end, where they had a tayl, near four times the length of the whole body, of the thickness (by my Microscope) of a Spiders-web; at the end of which appear'd a globul, of the bigness of one of those which made up the body; which tayl I could not perceive, even in very clear water, to be mov'd by them. These little creatures, if they chanced to light upon the least filament or string, or other such particle, of which there are many in water, especially after it has stood some days, they stook intangled therein, extending their body in a long round, and striving to disintangle their tayl; whereby it came to pass, that their whole body lept back towards the globul of the tayl, which then rolled together Serpent-like, and after the manner of Copper- or Iron wire that having been wound about a stick, and unwound again, retains those windings and turnings. This motion of extension and contraction continued a while; and I have seen several hundreds of these poor little creatures, within the space of a grain of gross sand, lye fast cluster'd together in a few filaments.

*From* Philosophical Transactions *12 (1677): 821. Italics in original.*

## Section Review 13.2

1. How did the Protestant Reformation assist the rise of modern science?
2. What contributions did Andreas Vesalius and William Harvey make in the fields of human anatomy and physiology?
3. Why can the researches of Robert Hooke and Anton van Leeuwenhoek be considered the beginnings of microscopy and microbiology?
4. State the cell theory.

### Critical Thinking

5. What might be some reasons that English Puritans and French Huguenots and Jansenists were so prominent in the advancement of 17th-century science?

### Identify

*Fabrica,* Royal Society, French Academy of Sciences, Huguenots, Jansenists

# 13.3 Biology and Scientific Investigation

## The Role of Scripture in Scientific Investigation

*The Scriptures not only provide us with a true account of the origin of all things but also furnish us with an interpretive framework by which to view life and the world.* Basic to this framework are the two biblical presuppositions without which science would be impossible:

1. There is a cause, or reasonable explanation, for every effect, or event, in creation.
2. There is something real to be discovered and understood about creation.

All scientists must accept these postulates in order to do scientific work of any kind. They must also assume that nature can be truly known and understood by human minds. However, these presuppositions are difficult to justify apart from the biblical worldview. Without the restoration of the authority of God's Word brought about by the Protestant Reformation, modern science and technology may not have arisen. Without the open book of Scripture, there is no open book of nature. *Sir Francis Bacon* (1561–1626), a devout Protestant philosopher, essayist, and statesman and formulator of the scientific method, declared:

> Let no man . . . think or maintain, that a man can search too far or be too well studied in the book of God's word, or in the book of God's works, divinity or philosophy; but rather let men endeavour an endless progress or proficience in both.[3]

## Solving Scientific Problems

The biologist faces many problems. What causes cancer? What can be done to prevent heart attacks? What can be done to make the body accept an organ transplant? How can better crops be grown? How can insect pests best be handled? Scientists employ a special process known as the **scientific method** to solve scientific problems. Scientific research and investigation, when not conducted in areas forbidden by God's law or hostile to the doctrinal or ethical content of God's Word, are a valid part of man's task to subdue the creation and exercise dominion over it under God.

Based upon their observations and intuition, scientists propose an idea about a particular phenomenon. Then, using the scientific method, they attempt to verify that theory to see if it remains credible under many different conditions. The great theories in science are the result of man's creative mind building upon what he already believes to be true about the real world. Scientific theories such as gravitation in physics, atomic theory in chemistry, and cell theory in biology were not products of method but of intellectual power. The scientific method serves to verify or nullify a theory by providing evidence for it or against it.

The scientific method is a thought process developed by Francis Bacon and other natural philosophers over the past centuries. The scientific method is a form of **inductive reasoning** (reasoning from the specific events in nature to their general causes) that involves the activities of theorizing, observing, and experimenting. The purpose of the scientific method is to collect data and then assemble that data to confirm or discard a hypothesis.

## Steps in the Scientific Method

**State the problem clearly.** One of the first steps to solving any scientific problem is to state the problem clearly.

**Think of possible solutions.** When the problem is clearly defined, the scientist attempts to think of as many possible solutions to his problem as he can. It is often necessary for the biologist to obtain information about the problem by consulting scientific journals, books, and scientific papers and by gathering facts from direct observations. After consulting these sources of information, the biologist is ready to formulate his possible solutions. These tentative solutions to a scientific problem are called **hypotheses** (sing. hypothesis).

**Test the hypotheses.** *Hypotheses are not the answer to the problem,* but after they are tested, one or more of the hypotheses may be the solution. Each of the hypotheses must be tested, generally by a series of experiments. The biologist uses **controlled experiments,** which are experiments in which all factors are identical except the one being tested; the one factor being tested is called the *variable factor,* or **variable.**

Scientists must conduct their experiments in a carefully monitored situation. For example, suppose you are a biologist seeking to find medication to kill a certain pathogenic bacteria. To determine if a particular substance is effective for killing the bacteria, you would have

[3] *The Advancement of Learning,* ed. William Aldis Wright, 2nd ed. (Oxford: Clarendon Press, 1876), 10.

to run a controlled experiment using two identical groups of the bacteria. One group would be exposed to the substance being tested while the other group would not. The presence of the substance being tested would be the variable in your experiment because it would be the only difference between the two groups. The group of bacteria that received the chemical would be the *experimental group;* the group of bacteria that were not treated would form the **control group.** *Every good experiment must have a control group to prove that the variable produced the results of the experiment.* If the experimental group of bacteria died, but the control group continued to live, you could assume that your substance was the killer of the bacteria. By repeating the experiment, you could be even more certain of your results being correct if they were the same. This repeating of experimental work is called *replication.* The more replications you do the more confident you would be in your results.

**Reach a conclusion.** Experiments such as the one just described do not always support a hypothesis. Sometimes the experiment shows the hypothesis to be partially or totally wrong. When this occurs, the biologist has to reexamine his facts and alter his hypothesis or form an entirely new hypothesis. Before the solution to a scientific problem is found, the whole process may have to be repeated many times.

When a hypothesis has passed the test of many well-designed experiments and has the support of other scientists, it is referred to as a **theory.** Although a theory is considered more reliable than a hypothesis, it is possible that the theory will still be proved wrong by more precise experiments or that it will have to be modified in light of new knowledge.

Fig. 13.6
**The scientific method**

If a theory stands the test of time and is verified by experiment after experiment, it may eventually be referred to as a **scientific law.** Scientific laws represent man's best understanding of how God's universe works. However, it should still be kept in mind that man's understanding is often imperfect, so that even scientific laws may sometimes have to be revised in light of unexpected observations or experimental results.

**Summary of the scientific method.** The major steps to be followed when using the scientific method may be summarized as follows:

1. State the problem clearly and completely.
2. Examine the available facts and suggest as many probable solutions (hypotheses) as possible.
3. Test every hypothesis; modify or reject faulty ones.
4. Form a conclusion that is based upon all known facts, causes, and effects.
5. If the facts are not sufficient to justify forming a conclusion, keep an open mind toward the problem until enough is known to justify a conclusion.
6. Test the conclusion with additional controlled experiments.

**Science and the laws of nature.** Conclusions that cannot be duplicated in controlled experiments by other scientists regress to become unverified hypotheses. Conclusions that hold up in test after test will strengthen in time to become laws. *The goal of science is to determine God's laws of nature and to use them for man's benefit and God's glory.*

God created all of nature to follow certain regular patterns. Organisms live in a world of organization and design. The scientific method applied in this context allows man to discover what God has ordained in nature and how God providentially governs His creation. *Scientific facts exist. Answers to scientific questions exist.* The job of the biologist (and other natural scientists) is to find these truths and use them for the benefit of mankind.

## *Nature to the Eye of a Christian* Anonymous

Thinking about the LIVING CREATION

Whatever leads our minds habitually to the Author of the Universe;—whatever mingles the voice of nature with the revelation of the gospel;—whatever teaches us to see, in all the changes of the world, the varied goodness of him, in whom "we live, and move, and have our being,"—brings us nearer to the spirit of the Saviour of mankind.... To all indeed the same objects appear—the same sun shines—the same heavens are open; but to the Christian alone it is permitted to know the Author of these things; to see his spirit "move in the breeze and blossom in the spring;" and to read, in the changes which occur in the material world, the varied expression of eternal love. It is from the influence of Christianity, accordingly, that the key has been given to the signs of nature. It was only when the Spirit of God moved on the face of the deep that order and beauty were seen in the world.

...The *beauty of nature,* as felt in modern times, seems to have been almost unknown to the writers of antiquity. They described occasionally the scenes in which they dwelt, but, if we except Virgil, ... never with any deep feeling of their beauty.... The Alps themselves, ... the most glorious objects which the eye of man can behold, were regarded by the ancients with sentiments only of dismay or horror; as a barrier from hostile nations, or as the dwelling of barbarous tribes. The torch of religion had not then lightened the face of nature; they knew not the language which she spoke, nor felt that holy spirit which to the Christian gives the sublimity of these scenes.

*The heavens declare the glory of God; and the firmament sheweth his handiwork. Psa. 19:1*

*From "On the Pulpit Eloquence of Scotland, No II—Alison,"* Blackwood's Edinburgh Magazine *2 (December 1817): 325–326. Italics in original.*

### Repudiation of Spontaneous Generation

**Francesco Redi.** A good example of the use of the scientific method for scientific investigation was the repudiation of the ancient idea of spontaneous generation. The Italian physician **Francesco Redi** [frän·chās′kō rĕd′ē: 1626–1697] was one of the first scientists to present evidence against spontaneous generation. In 1668 he published a series of experiments disproving the belief that maggots (the larvae of flies) are generated by decaying meat. With his carefully conducted experiments, Redi demonstrated that maggots are produced when living flies lay eggs on meat. His experiments involved three jars into which he placed pieces of meat. One jar was left open, and of course, it attracted flies. The flies laid their eggs, and soon maggots appeared. The second jar was covered with gauze. It, too, attracted flies. But the flies could not get to the meat because of the gauze. They laid their eggs on the gauze, and soon the maggots appeared on the gauze but not on the meat. The third jar was sealed with thick parchment. The flies were not attracted to that jar, and no maggots appeared either on the meat or on the parchment. The popular theory of spontaneous generation began to be seriously questioned.

**Pasteur.** In the 19th century, **Louis Pasteur** of France conducted experiments that marked the end of spontaneous generation as a scientific possibility. *His experiments showed that not even the simplest of organisms can develop from nonliving matter.*

Pasteur prepared a broth and let it stand uncovered until bacteria began to grow in it. He killed the bacteria by boiling the broth. Then he sealed some of the broth in flasks and left some in open flasks. Soon he found bacteria in the open flasks but none in the sealed flasks. He broke open some of the sealed flasks, and within a few days bacteria appeared. It seemed clear that bacteria had gotten into the broth from dust carried by the air. To test this latter idea, Pasteur took some sealed flasks to the top of a mountain, where there was little dust. He opened some of the flasks and found that few became contaminated with bacteria.

Some scientists, however, were not convinced. They explained that the lack of air (air is thin on a mountaintop) in the sealed flasks kept the bacteria from forming in the broth. Pasteur designed one last, convincing experiment. He boiled the broth in flasks but did not seal the flasks. Instead, he stretched the necks of the flasks into a zigzag shape. Air was free to enter the flasks, but bacteria were trapped by the zigzag neck of the flasks and could not reach the broth. Pasteur boiled the broth in the flasks a final time and let it stand. Even after four years, no bacteria were found in the flasks. Several of Pasteur's flasks are on display at the Pasteur Institute in Paris.

**The law of biogenesis.** Scientists were finally convinced. They agreed that *living things can come only from other living things.* It is impossible for nonliving substances to change into even simple living things. This is such an important scientific concept that it is known as the **law of biogenesis.**

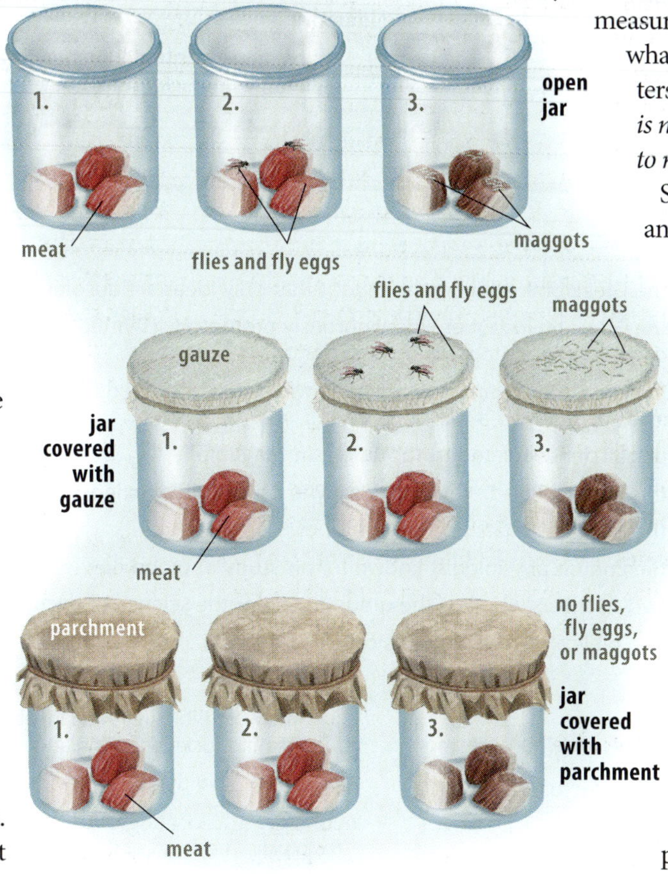


Fig. 13.7
**Redi's experiment**

## The Limitations of Science

**Science and scientism.** When God began to be pushed out of academic life by some scholars in the 19th century, materialistic science filled the void. Some people began to think that everything could be explained through scientific investigation. This thinking became so prevalent among intellectuals by the 20th century that it was thought that nothing could be known apart from scientific investigation. The idea that science can find answers for all the problems in life is called **scientism.** The religion of scientism eliminated several important forms of human knowledge, namely the Scriptures, the Christian heritage of Western civilization, and plain common sense.

**The "science" of man.** Modern science was created as a way to explain and control nature. Today, many people are attempting to use what they call science to explain and control human nature. If man were only a product of nature, then he could be measured and predicted, but he is not. Man is a spiritual being created in God's image and reaches beyond the ability of science to observe and measure. An understanding of who and what man is lies outside the parameters of science. *The scientific method is not a valid instrument with which to measure that which is spiritual.*

Sociologists attempt to explain and measure man's interpersonal relationships, but they often fail to see man's relationship to God. Psychologists attempt to explain man's behavior, but many of them fail to see man's sinful nature as a result of his rebellion against God. Economists attempt to explain and predict man's worth, but they usually fail to see man's value in God's eyes. All of these fields of study and many others claim to be science and to use the scientific method to verify their theories, but by rejecting God's design, purpose, and revelation, they often overstep the bounds of true science.

**The limits of science.** True science is limited to what is observable and repeatable in the natural world. Science cannot tell us who God is, what man is, or how everything got here. Biologists have a legitimate science as long as they observe and measure living things, including man's physical body; but because many of them are not Christians, they are blind to spiritual reality as taught in the Bible. Many biologists take what they know about life and speculate on matters outside the realm of science, such as human behavior and the origin of life.

Science is limited to discovering and formulating God's laws of nature and using them to benefit mankind. *Biology, then, is limited to finding God's design in the physical part of the living creation and applying that knowledge to help man dominate the earth.*

## Section Review 13.3

1. What is the role of Scripture in scientific work?
2. Name two biblical presuppositions upon which modern science is based.
3. Explain the steps in the scientific method.
4. How did Redi's and Pasteur's experiments repudiate the idea of spontaneous generation?
5. State the law of biogenesis.

**Application**

6. Explain why modern "social scientists" such as anthropologists, economists, psychologists, and sociologists often err in their conclusions about the nature of man and his society.
7. Why is science unable to answer questions about spiritual realities or ultimate origins?

**Identify**

hypothesis, controlled experiment, variable, theory, scientific law, scientism

## Chapter 13 Review

**Define**

1. spontaneous generation
2. evolution
3. materialism
4. scientific method
5. inductive reasoning
6. controlled experiment, variable, control group
7. scientism

**Identify**

1. the Greek philosopher who taught that only ideas are absolute
2. the Greek philosopher who vigorously promoted scientific work and made advances in comparative anatomy and classification
3. the Greek physician and anatomist who was considered the absolute authority in human anatomy for over a thousand years
4. three science-related truths taught in the Bible
5. the group of medieval Christians who carried Greek science to the Arabs
6. the Father of Anatomy and author of *Fabrica*
7. the English physiologist known for his study of blood flow
8. two scientists who laid the groundwork for the science of microbiology
9. two biblical presuppositions of science

**Explain**

1. Explain how Genesis 1:28 relates to the purpose and goal of science.
2. How did the fall of man and the dispersion of man at the Tower of Babel affect natural knowledge?
3. Why was Pliny's work of little scientific value?
4. Why was Galen wrong in many of his statements about the human body, even though he was a careful observer?
5. Describe the scientific contributions of Otto Brunfels, Leonhard Fuchs, and Konrad Gesner.
6. How were Bible believers involved in the founding of scientific societies in England and France?
7. Explain how the Bible provides an interpretive framework by which to view life and the world.
8. Explain the steps of the scientific method.
9. Distinguish between hypotheses, theories, and scientific laws.
10. How did Redi's and Pasteur's experiments repudiate the theory of spontaneous generation?
11. State the cell theory and explain how it relates to the law of biogenesis.

**Apply**

1. What were some of the successes and failures of Greek science? Of Roman science?
2. Contrast the accuracy of the Bible with that of the ancient philosophers.
3. How did the Protestant Reformation help spur the development of modern science?
4. A biologist hypothesizes that a certain plant is triggered to bloom by the rising temperature in the morning. Describe a controlled experiment to test this hypothesis; identify the variable and the control group.
5. Why is the scientific method not a valid instrument with which to measure that which is spiritual?

# Chapter 14

# Evolution: A Retreat from Science

**Key Concepts**

- The faith of the great scientists
- Rejecting the truth
- The faith of Darwin and his disciples
- Why many scientists opposed Darwin
- Fossil evidence against evolution
- Biological evidence against evolution
- Why evolution cannot be properly called a science

## 14.1 Science and Faith

All scientific work is ultimately based on certain assumptions that are accepted by faith. These presuppositions are the basis for one's philosophy, or worldview. For many scientists today, that faith is in materialism, and everything they observe in the natural world is filtered through the theory of evolution. However, many of the world's greatest scientists—including those who made the discoveries upon which modern science was built—based their scientific thinking on faith in the fact that God created the world and all that is within it and that therefore the physical part of the universe operates by orderly laws, which science seeks to discover. Sir Isaac Newton, Lord Kelvin, Michael Faraday, James Maxwell, and Louis Pasteur were just a few of the great scientists who believed in divine creation. Others include Joseph Lister, Gregor Mendel, Johannes Kepler, Samuel Morse, Carolus Linnaeus, Wernher von Braun, and many more.

### The Scientist Needs the Bible

The scientist uses his powers of reasoning to find out the truth about the universe; in a sense, he reads the physical universe as if it were a book written by the hand of God. However, many vital facts about the origin of the universe, the origin of life, and the ultimate destiny of the cosmos cannot be discovered through science alone. To answer these questions, one needs the Book of books, the Bible, which contains information one cannot get from the study of nature. Thus, the Bible is very important to the scientist because it records important facts about the history and future of the physical universe that he would not otherwise know.

The question God asked Job is a pertinent one, not only for the scientist but also for us:

> Where wast thou when I laid the foundations of the earth? . . . —Job 38:4

Only God was present in the beginning, when the universe was created. The facts about the manner and order of Creation that God has chosen to reveal to us in the Bible are all that we (including scientists) can know with certainty about the beginning.

### Rejecting the Truth

During the 1700s and the centuries that followed, some scientists and philosophers chose to reject the truth of the Scriptures. In its place, they substituted various false philosophies. Some rejected the Bible completely, and others tried to mix scriptural truth with the false teachings of ancient Greek philosophers. The result was a disastrous mix of truth and error that laid the foundation for a philosophical shift in the scientific community.

Having abandoned faith in the Scriptures, many turned to the false philosophy of *materialism,* the idea that the universe consists of nothing but matter and energy and has no spiritual or supernatural aspects. Another name for materialism is *naturalism,* the idea that nature is all that exists. To the materialist, there is no higher authority than the forces of nature; therefore, he believes that all things are relative, having no value but that given them by nature or by man himself. *The philosophy of materialism is actually a faith—a faith not in the Creator but in nature and matter itself.* Richard Lewontin, a world-famous evolutionist from Harvard University, has written:

> We take the side of science *in spite* of the patent absurdity of some of its constructs, . . . because we have a prior commitment, a commitment to materialism. It is not that the methods and institutions of science somehow compel us to accept a material explanation of the phenomenal world, but, on the contrary, that we are forced by our *a priori* adherence to material causes to create an apparatus of investigation and a set of concepts that produce material explanations, no matter how counterintuitive, no matter how mystifying to the uninitiated. . . . for we cannot allow a Divine Foot in the door.[1]

Many modern scientists place their faith in materialism and in modern scientific methodology based on materialistic assumptions. Science therefore is no longer the unbiased observation of facts and the impartial evaluation of theories but rather is the application of materialism to the interpretation of nature. One of the many fruits of materialism is the prevalent dedication among modern scientists to the theory of *evolution.*

## The Faith of Charles Darwin

The acceptance of evolution by the scientific community can be credited largely to the British naturalist **Charles Darwin** (1809–1882). Darwin's naturalistic ideas have had an enormous impact upon science and upon society at large.

**The voyage of the *Beagle*.** In December 1831, Charles Darwin set sail with a surveying expedition on the H.M.S. *Beagle* around South America and to islands in the Pacific Ocean. The voyage, which lasted five years, offered many opportunities for observation. "The voyage of the *Beagle* has been by far the most important event in my life, and has determined my whole career," he wrote.[2] He spent his time on the voyage observing rain forests, unusual land formations, and other natural wonders that were new to him; collecting strange animals from oceans, shores, and rivers; and taking painstaking notes on all his observations.

Fig. 14.1 **Charles Darwin (c. 1855)**

Darwin took with him a copy of *Principles of Geology* by **Charles Lyell** (1797–1875) and was thus introduced to Lyell's false doctrine of **uniformitarianism,** the idea that the present is the only key to the past and that all things continue by natural processes at the same rates as they always have. Darwin applied Lyell's ideas to try to explain the origin of the landscapes he observed and extended the idea to explain the origin of the many varieties of plants and animals that he saw on his voyage. He concluded that "all the organic beings which have ever lived on this earth may be descended from some one primordial form."[3]

***The Origin of Species.*** Darwin made many observations during his voyage aboard the *Beagle* that he thought supported his hypothesis of evolution. He took copious notes of his observations, and when he returned to England in 1836, he began to assemble his ideas and observations into coherent form. In 1859, more than 20 years after the *Beagle* voyage, Darwin finally published his ideas in a book entitled *On the Origin of Species by Means of Natural Selection, or the Preservation of Favoured Races in the*

---

[1]"Billions and Billions of Demons," *The New York Review of Books* 44 no. 1 (Jan 9, 1997).
[2]*The Life and Letters of Charles Darwin,* ed. Francis Darwin (New York: D. Appleton, 1887), 1:51.
[3]Darwin, *The Origin of Species,* 6th ed. (London: John Murray, 1873), 425.

*Struggle for Life.* (Because its title is quite long, this book is often referred to as ***The Origin of Species,*** or simply as *Origin.*) In this lengthy book, Darwin presented his speculations about the origin and development of living things, based on the many observations he had made in the course of the years.

In summary, Darwin proposed the following theory of evolution:

1. Species exist but are not immutable. They have changed over time and are evolving constantly into new species.
2. The diversity that we observe in nature is not the result of a special creation by God but rather the result of a natural and random evolutionary process.
3. The main mechanism that drives this evolutionary process is natural selection or the survival of the fittest.

**"One long argument."** *Origin* was not a scientific treatise but rather a series of thousands of wild speculations strung together—in Darwin's words, "one long argument."[4] Its easygoing, conversational style and its intricately woven arguments were dangerously disarming, and many people who read *Origin* found Darwin's philosophy quite plausible. By the end of the 19th century, *The Origin of Species* was regarded by many as the greatest intellectual discovery of the century and the greatest thought to enter the mind of man.

**Natural selection.** Various hypotheses of evolution had existed before Darwin wrote *Origin,* but Darwin was the first to propose a plausible means by which evolution might occur. The cornerstone of Darwin's hypothesis was **natural selection,** the idea that the fittest and strongest of each species were more likely to survive and reproduce than weaker, poorly adapted animals. This concept is sometimes referred to as *survival of the fittest.* Eventually, according to Darwin, whole new types of creatures would appear. Of course, natural selection is observed in nature as a conservative force. An animal well suited to its environment is certainly more likely to thrive than an animal poorly suited to its environment. However, Darwin believed that natural selection would act upon the variety that naturally occurs within kinds to gradually produce new kinds of organisms.

[4]Ibid., 404.

**The failures of Charles Darwin.** Darwin's reasoning is faulty because *variety within kinds has definite boundaries*—a fact that Darwin would not accept. Because natural selection itself produces no new characteristics, natural selection cannot create new kinds of organisms. Rather, it keeps a kind strong and healthy by suppressing harmful changes. In other words, *natural selection acts to preserve existing kinds, not to create new kinds.* Natural selection as a creative force has never been observed in nature, nor is it documented in the fossil record.

## Reactions Abroad

**The faith of Darwin's disciples.** Some people who embraced Darwinism did so for philosophical reasons; they wished to remove God from their thinking. Many people had a simpler reason for accepting evolution: they simply believed that it was scientific. Science had brought wonderful changes to the world of the 19th century, and some people would believe anything if they thought it had the support of science.

Charles Darwin's philosophy of evolution really had little to do with science, however. In fact, one of Darwin's first supporters was a liberal theologian named *Charles Kingsley,* who worked hard to integrate the ideas of evolution into Victorian society. Most scientists were initially skeptical of Darwin's argument because they realized that it was based largely upon speculation.

**Scientists who rejected Darwin.** One of the great writers on scientific thought in that day was *William Whewell* [hyo͞o′əl], professor of science and college master of Trinity College, a part of Cambridge, and author of *History of Inductive Sciences.* Whewell realized that merely imagining how something might have occurred is not scientific proof that such a thing did occur. Because Darwin's work was based entirely on speculation instead of on scientific fact, Whewell would not even allow the book in the college library. *Sir John Herschel,* the great English astronomer, chemist, mathematician, and physicist (son of Sir William Herschel, the discoverer of the planet Uranus), called Darwin's ideas the "law of higgledy-piggledy." *Philip Gosse,* a noted biologist, was not convinced by Darwin's arguments, and neither were other esteemed scientists such as *Adam Sedgwick,* a noted mathematician and geologist; *Sir Richard Owen,* an anatomist; and *Andrew Murray,* an entomologist. Each of these men firmly declined to accept the hypothesis.

Adam Sedgwick, who was one of Darwin's mentors and an evolutionist of sorts himself, denounced Darwin's hypothesis of evolution by natural selection as "a dish of rank materialism cleverly cooked and served up . . . to make us independent of a Creator."[5] Two of the world's greatest physicists, *James Clerk Maxwell* and *Lord Kelvin,* strongly opposed Darwinism and developed mathematical and scientific refutations of evolution. French scientists were generally no more enthusiastic about *The Origin of Species* than were English scientists.

In America, the scientific community, unlike liberal theologians and socialists, largely avoided the philosophy of evolution at first. One of the most influential American naturalists of the day, *Louis Agassiz* [ăg′ə·sē] of Harvard, remained unmoved by Darwin's arguments. *Matthew F. Maury,* the "Pathfinder of the Sea" and founder of the science of oceanography, also strongly opposed evolution and insisted that the Bible be accepted as true in matters of science. However, Charles Darwin did find a few scientists willing to support his views, one of whom was *Asa Gray,* a noted American botanist. Gray worked hard to try to convince the scientific community that Darwin's ideas were not inconsistent with a belief in God. This effort gave Darwinism a big boost in America.

## The New Faith's Effects

**Acceptance of naturalism.** Despite the fact that several notable scientists rejected Darwin's notions, one of the astonishing facts about Darwinism is how fast and uncritically it was accepted by so much of the scientific and theological communities in Europe and America. Within just 20 years of the publication of *The Origin of Species*, nearly every naturalist in the Western world had embraced the theory of evolution in one way or another. As the hypothesis of evolution was debated, many scientists became detoured from their true calling of mastering nature for the benefit of mankind and devoted their energies instead to the task of trying to prove Darwin's ideas. More surprisingly, within the major Evangelical denominations, a great percentage of educated ministers also accepted Darwin's idea.

**Effects upon society.** The acceptance of Darwin's hypothesis in society caused a dramatic shift away from the traditional Judeo-Christian worldview toward a naturalistic worldview. Instead of recognizing man as the special creation of God, people began to regard mankind as a mere animal, with no more worth than any other part of nature. Right and wrong came to be thought of as relative, defined either by the whims of the individual or by the will of the majority. Some Darwinists twisted the biblical concepts of labor and reward into a ruthless "kill-or-be-killed" distortion of capitalism, while others promoted various forms of socialism. *Karl Marx,* the "Father of Communism," was thrilled with Darwin's speculations and wanted to dedicate his own book *Das Kapital* to Darwin (Darwin declined).

Some of Darwin's followers founded the "science" of **eugenics** [yo͞o·jĕn′ĭks], which sought to improve the human species by selectively breeding humans to produce a "master race." Eugenics laws were passed by many nations, under which thousands of "genetically inferior" individuals were forcibly sterilized to prevent them from having children. Years later, the philosophy of eugenics would culminate in the agenda of the National Socialists (Nazis) in Germany, who used abortion, euthanasia, and mass murder to eliminate millions of people they deemed "genetically inferior" in order to "improve" the German race.

**God versus chance.** The debate over Darwin's *The Origin of Species* was no mere battle over evolution or Creation. French-American scholar Jacques Barzun calls it a "major incident . . . in the dispute between the believers in consciousness and the believers in mechanical action; the believers in purpose and the believers in pure chance. The so-called warfare between science and religion thus comes to be seen as the warfare between two philosophies and perhaps two faiths."[6] The great novelist Aleksander Solzhenitsyn, who came face-to-face with the materialistic faith in Communist Russia and rejected it, eloquently expressed the opposite faith in these words:

> Our life consists not in the pursuit of material success but in the quest for worthy spiritual growth. Our entire earthly existence is but a transitional stage in the movement toward something higher, one rung of the ladder. *Material laws alone do not explain our life or give it direction. The laws of physics and physiology will never reveal the indisputable manner in which the Creator constantly, day in and day out, participates in the life of each of us, unfailingly granting us the energy of existence; when this assistance leaves us, we die.* And in the life of our entire planet the Divine Spirit surely moves with no less force: this we must grasp in our dark and terrible hour.[7]

---

[5] Sedgwick to Miss Gerard, Norwich, January 2, 1860, in John Willis Clark and Thomas McKenney Hughes, *The Life and Letters of the Reverend Adam Sedgwick* (Cambridge: Cambridge University Press, 1890), 2:360.

[6] Jacques Barzun, *Darwin, Marx, Wagner: Critique of a Heritage,* ed. 2 (New York: Doubleday, 1958), 37.

[7] Aleksander Solzhenitsyn, "Men Have Forgotten God," trans. by A. Klimoff, *National Review,* 22 July 1983, 876. Emphasis added.

## An American Scholar Speaks for Creation by William Jennings Bryan

Thinking about the LIVING CREATION

The world is now learning—most of the world for the first time—that evolution, as the scientists teach it, is an imaginary process, wholly unproved, that begins with life but does not attempt to explain life, and represents man as the climax of a series of changes coming up from a simple cell through millions of forms of life different from man.

This hypothesis makes every living thing known in animal life a blood relative of every other living thing in animal life, and makes man a blood relative of them all—either an ancestor or a cousin. If this hypothesis were true, we would all be murderers if we swatted a fly or killed a bedbug, for we would be killing our kin, and we would be cannibals whenever we ate any of the mammals. . . .

But that is not all. If the evolutionary hypothesis is true, man has come up through the animals below him by a cruel law under which the strong kill off the weak. Darwin argues that the race was necessarily impaired by the suspension of this cruel law. He commended by implication the savages who are eliminating the weak, saying that it left the survivors strong.

He even suggested that vaccination had saved the lives of thousands who would otherwise have succumbed because of weak constitutions—the implication being that the race would have been benefited by allowing them to die instead of prolonging their lives and permitting them to propagate. He complained that civilized society and medical men attempt to prolong life every last moment.

No more cruel doctrine was ever promulgated. Those who believe it are robbed of the pity and the mercy that comes of civilization.

To show that Darwin's heartless doctrine has not been abandoned one has only to read a book that came out about three years ago. I will not give the name of the author, for I do not care to advertise his name.

In his preface, he says that he is indebted to some twenty eminent scientists, "professors and doctors," and he singled out for special gratitude a young man recently elected president of a great state university, a man whose career the author predicts will "be one of the world's events of the coming generation." This eminent educator read the manuscript over twice and "made many invaluable suggestions."

On page 34 of this book we are told that "evolution is a bloody business, but civilization tries to make it a pink tea." Then he adds:

"Barbarism is the only process by which man has organically progressed and civilization is the only process by which he has organically declined. Civilization is the most dangerous enterprise on which man ever set out. For when you take man out of the bloody, brutal but beneficent hand of natural selection, you place him at once in the soft, perfumed, daintily gloved but far more dangerous hand of artificial selection." Here we have evolution unmasked.

The evolutionists have not been honest with the public. Even ministers who believe in evolution have assured their congregations that there is no inconsistency between Darwinism and Christianity. Do they know its effect on Darwin, or, knowing its effect, do they dare conceal it from their congregations?

The ministers should tell their congregations that evolution leads logically to agnosticism; they should tell them of the wail of Romanes, sometimes called the successor of Darwin, who said in his book, written to prove that there is no God:

> I am not ashamed to confess that with this virtual negation of God, the universe to me has lost its soul of loveliness, yet when at times I think, as think at times I must, of the appalling contrast between the hallowed glory of that creed that once was mine and the lonely mystery of existence as now I find it—at such times, I shall ever feel it impossible to avoid the sharpest pang of which my nature is susceptible.

. . . The Christian world is not going to give up its belief in God or its belief in the Bible as our only standard of morals or in Christ as our only Savior and wisest guide. The Christian world will not give up these sacred things at the demand of these intolerant champions—not of science but of an unproven guess—the logical tendency of which is to rob man of his moral standards in this world and of hope of immortal life in the world to come.

*Quoted in Leslie H. Allen, ed.,* Bryan and Darrow at Dayton *(New York: Russell and Russell, 1967), 103–106.*

## Section Review 14.1

1. Explain the relationship between the modern concept of science and the philosophy of materialism.
2. What are some of the factors that prevent natural selection from creating new kinds?
3. Give some of the reasons why many well-known 19th-century scientists rejected Darwin's hypothesis of evolution.

**Application**

4. What were some of the effects of the naturalistic worldview upon science? Upon society at large?

**Identify**

materialism, naturalism, uniformitarianism, *The Origin of Species,* natural selection, eugenics

# 14.2 Paleontology: Evidence against Evolution

## Important Concepts in Evolution

What is evolution? *Evolution has been defined as changes in the hereditary characteristics of groups of organisms over the course of generations.* However, this is a very general definition and it is important to be more specific to avoid confusion. For this reason, we need to distinguish between the concepts of speciation and macroevolution. **Speciation** is the changes within a particular kind of organism, referring to variations in the gene pool within a population. For instance, dogs, wolves, and coyotes are obviously members of the same kind even though they represent different species. They probably proceed from an original created pair that over time generated these modern animals. The human race is an excellent example of speciation. All the diversity that we see in the human race today comes from an original couple. Evolutionists often refer to speciation as *microevolution.* Creationists believe in variations within kinds because it is a fact of nature and because it agrees with biblical teachings.

On the other hand is the concept of **macroevolution,** the hypothetical process by which new kinds of creatures emerge from existing kinds over time. This is the concept referred to simply as "evolution" in this textbook. According to this concept, amphibians came from fish, reptiles from amphibians, and so on. *Macroevolution has never been observed in nature.* It is not happening among living organisms, and there is no indisputable record of it among the fossils.

Fig. 14.2 **Selected extinct animals**

## Fossils: Record of the Past

**Fossils** are the remains or impressions of plants, animals, and humans preserved in sedimentary rock. Countless billions of fossils are found in the earth's crust, most of which were probably buried during the worldwide Flood of Noah. The study of fossils is called **paleontology** [pā′lē·ŏn·tŏl′ə·jē].

**Fossil variety.** Although many fossils represent plants and animals we see all around us (such as pines, ferns, insects, horses, snakes, etc.), some fossils represent plants and animals that have become extinct. It is possible that some of these organisms had difficulty surviving the post-Flood environment and gradually died out. Some of the interesting animals that are known only from fossils include mammoths, dinosaurs, 40-foot-long crocodiles, 2000-pound turtles, giant birds, and eagle-sized dragonflies.

**Fossils: evidence against evolution.** *The fossil record, when viewed from a biblical perspective, is actually one of the most powerful evidences against evolution.* This is true because if evolution had occurred, it would have left traces of the process in the fossil record. Yet, after a century and a half of fossil excavation, evolutionists have failed to discover the proof they seek. When kinds of animals appear in the fossil record, they appear fully formed, showing no traces of having evolved from something else or of evolving into something else. In addition, the fossil record reveals a record of sudden death and destruction that is consistent with the biblical teaching concerning a worldwide Flood.

**The necessity of transitional forms.** Darwin's hypothesis of evolution proposed that one kind of organism gradually changes into another kind over many generations by extremely slight changes in each generation. Geologists of Darwin's day expected the fossil record to prove evolution by providing fossils of **transitional forms,** or *"missing links."* (Transitional forms would be fossils that connect one kind of organism with another kind by a series of tiny steps.) Darwin insisted that transitional forms would connect every kind of organism now living with a primeval single-celled ancestor "by differences not greater than we see between the natural and domestic varieties of the same species at the present day."[8]

*If evolution were true*—if organisms have gradually changed into other organisms over time—*there would be countless fossils of transitional forms,* connecting every kind on earth with their common ancestors by tiny steps. However, this is not the case; *there are actually large gaps between different kinds.* That these transitional forms have not been found is perhaps the greatest evidence against evolution.

The most ardent opponents of Darwin's ideas during his lifetime were not theologians but paleontologists. *Louis Agassiz* from Harvard, *John William Dawson* from McGill, and *Arnold Guyot* from Princeton, although not biblical literalists, rejected Darwinism because they knew the fossil record very well. Darwin himself acknowledged that the absence of transitional forms in the fossil and living records was a mortal weakness to his theory. He wrote:

> But, as by this theory innumerable transitional forms must have existed, why do we not find them embedded in countless numbers in the crust of the earth? . . . Why then is not every geological formation and every stratum full of such intermediate links? Geology assuredly does not reveal any such finely-graduated organic chain; and this, perhaps, is the most obvious and serious objection which can be used against the theory.[9]

The fossil evidence was sketchy and incomplete when Darwin wrote *The Origin of Species* because relatively few fossils of any sort had been excavated. Darwin appealed to the lack of fossil evidence, hoping that future excavations would uncover transitional forms:

> The explanation lies, as I believe, in the extreme imperfection of the geological record. . . . That our collections are imperfect is admitted by every one. . . . Those who believe that the geological record is in any degree perfect, will undoubtedly at once reject the theory. For my part, following out Lyell's metaphor, I look at the geological record as a history of the world imperfectly kept, and written in a changing dialect; of this history we possess the last volume alone, relating only to two or three countries.[10]

**Lack of transitional forms.** Darwin freely admitted that no transitional forms had been found in his time, but he attributed this lack of evidence to the small number of fossils that had been excavated. Since Darwin wrote those words, over 100 million fossils, representing a quarter of a million species, have been excavated, cataloged, and placed in museums. The fossil record cannot be dismissed as incomplete anymore. Michael Denton, in his book *Evolution: A Theory in Crisis,* reports that out of the 43 living orders of terrestrial vertebrates, 42 of them (98%) are represented in the fossil record. At the family level, 261 (79%) of the 329 living families of terrestrial vertebrates are represented in the fossil record. If we exclude birds, which are poorly fossilized, 88% of the families are represented.[11] Yet, despite 150 years of intense digging all over the earth, the fossil record still does not provide the evidence Darwin was hoping for.

The fact is, *no true, unquestionable transitional forms or "missing links" have ever been found to bridge the gaps between different kinds of organisms.* Thousands of extinct kinds of animals have been revealed, but all are distinct kinds; none can be regarded as truly transitional forms. Rather, the fossil record vividly illustrates the biblical truth that kinds of living things do not change into other kinds of living things; every creature reproduces "after its kind" (Gen. 1:11–12, 21, 25). The glaring contradiction between

[8] *Origin,* 266.
[9] Ibid., 134, 264–265.
[10] Ibid., 265, 270, 289.
[11] *Evolution: A Theory in Crisis* (Bethesda, MD: Adler & Adler, 1968), 189–190.

Darwin's predictions and the facts of the fossil record make it clear that evolution has *not* occurred.

> This regular absence of transitional forms . . . is an almost universal phenomenon. . . . It is true of almost all orders of all classes of animals, both vertebrate and invertebrate.
>
> —*George Gaylord Simpson, vertebrate paleontologist*[12]

**Punctuated equilibrium.** To attempt to explain the lack of transitional forms in the fossil record, paleontologists *Stephen J. Gould* (1941–2002) and *Niles Eldredge* proposed the **punctuated equilibrium hypothesis** in 1972. Whereas Darwin taught that new organisms came about through the constant, gradual accumulation of small changes, Gould and Eldredge suggested that evolution occurs in sudden spurts, followed by long periods without noticeable change. According to this idea, new kinds of organisms arise because of drastic environmental changes that cause rapid genetic changes (over several thousand years instead of several million years) in small groups of animals and plants isolated from their main populations. Since genetic change and natural selection worked too quickly for fossil formation to be likely, no transitional forms were left in the fossil record. But there is no evidence that punctuated equilibrium could or did occur. Rather, punctuated equilibrium is simply an attempt to explain away the fact that the fossil record does not support the idea of evolution.

## The Fossil Record versus the Geologic Column

The Bible teaches that the earth and living things did not develop gradually but were divinely created. Evolutionists, however, commonly present the fossil record as a straightforward evolutionary progression from "primitive" organisms deep in the earth's crust to more "modern" organisms nearer the surface. This simple-to-complex sequence of fossils is known as the **geologic column** or *geologic time chart.* The implication is that if you were to take a vertical slice through the earth's crust, you would see a record of evolution from the simplest invertebrates to the living things we see around us today, supposedly representing some 4.6 billion years of earth's history. The hypothetical column is divided into four major time divisions called *eons,* which are subdivided into *eras, periods,* and *epochs* [ĕp′əks] (see Table 14.1 on the next page).

[12]*Tempo and Mode in Evolution* (New York: Columbia University Press, 1944), 107.

**An imaginary arrangement.** Although it is presented as conclusive evidence for evolution, the geologic column is not really a description of the order of rocks and fossils in the earth's crust. This is because *the geologic column is a hypothetical arrangement of fossils and rocks* collated from many different locations and habitats around the world, arranged according to evolutionary assumptions; even in the few locations where a large portion of the column may be found, individual layers or systems are incomplete. (In fact, if all the rock strata in the hypothetical geologic column were present at one location, it would be about 100 miles thick.) The most of the geologic column that you can see anywhere on the earth are a few rocks containing "simple" fossils overlain by a few strata containing "complex" fossils, generally representing only two or three periods and often widely separated in "age." In many places, the fossils are in their "proper" order, but in some places the order is actually reversed.

**Arrangement by assumption.** *The succession of fossils indicated by the geologic column occurs nowhere in the world.* The actual fossils in the earth's crust are not

Fig. 14.3 **Construction of the geologic column**

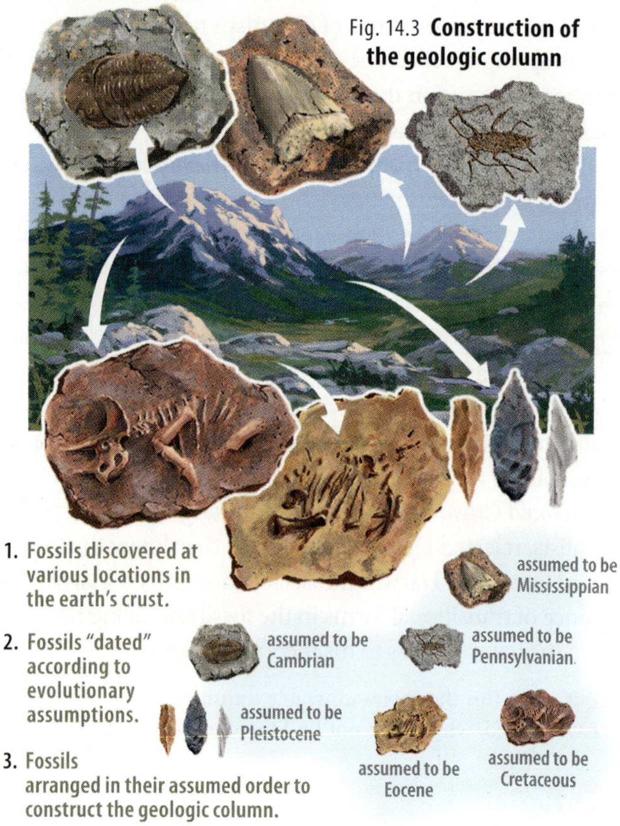


4. Simple-to-complex progression in the geologic column is claimed to be evidence for evolution (circular reasoning).

arranged in a strict evolutionary progression but rather are sorted mainly by habitat and mobility. That trilobites (a type of small, extinct marine arthropod) lived before dinosaurs, and dinosaurs lived before mammals, is an assumption based upon the hypothesis of evolution; the fossil record merely indicates that trilobites, dinosaurs, and mammals were usually buried in different places (perhaps because they lived in different habitats). Layers in the geologic column are identified by the presence of characteristic **index fossils.**

## Table 14.1 The Geologic Column

| Eon | Era | Period | Assumed events | Approximate assumed beginning date (years ago) |
|---|---|---|---|---|
| **Phanerozoic** | **Cenozoic** | **Quaternary** | | |
| | | Holocene Epoch | end of last Ice Age; rise of human civilization | 11,500 |
| | | Pleistocene Epoch | Ice Age(s); mass extinctions; rise of man | 1,800,000 |
| | | **Neogene** | | |
| | | Pliocene Epoch | peak of mammals; modern invertebrates | 5,300,000 |
| | | Miocene Epoch | rise of grazing mammals | 23,000,000 |
| | | **Paleogene** | | |
| | | Oligocene Epoch | primitive apes; whales; first modern mammals | 33,900,000 |
| | | Eocene Epoch | first horses | 56,000,000 |
| | | Paleocene Epoch | early mammals become dominant; rise of modern birds | 66,000,000 |
| | **Mesozoic** | **Cretaceous** | mass extinctions (including dinosaurs); flowering plants appear | 145,500,000 |
| | | **Jurassic** | giant dinosaurs; early birds; first mammals | 201,000,000 |
| | | **Triassic** | first dinosaurs; rise of mammal-like reptiles | 252,000,000 |
| | **Paleozoic** | **Permian** | modern insects; reptiles spread; evergreens; extinction of trilobites | 299,000,000 |
| | | **Pennsylvanian*** | giant insects; first reptiles | 323,000,000 |
| | | **Mississippian*** | amphibians spread; sharks and bony fish; winged insects | 359,000,000 |
| | | **Devonian** | first amphibians; freshwater fish; wingless insects | 419,000,000 |
| | | **Silurian** | land colonized by arthropods and plants | 444,000,000 |
| | | **Ordovician** | trilobites abundant; vertebrates increase | 485,000,000 |
| | | **Cambrian** | sudden explosion of life; trilobites dominant; rise of other marine invertebrates; some vertebrates | 541,000,000 |
| **Proterozoic**†‡ | | | simple multicellular sea creatures develop; algae and plankton develop; eukaryotic cells develop | 2,500,000,000 |
| **Archean**†‡ | | | earth becomes habitable; spontaneous generation of first cells; rise of bacteria and cyanobacteria (blue-green algae) | 4,000,000,000 |
| **Hadean**† | | | formation of earth (uninhabitable) | 4,800,000,000 |

* The Pennsylvanian and Mississippian periods are often together referred to as the Carboniferous Period.

†The Proterozoic, Archean, and Hadean eons are also informally referred to as the Precambrian.

‡Although not shown, the Proterozoic and Archean eons are also divided into eras; the Proterozoic is further subdivided into periods.

**Circular reasoning.** Using the geologic column as "evidence" to support the evolutionary hypothesis is an example of *circular reasoning*—an argument that is based on the very assumption it attempts to prove. There is no objective way to look at a sample of sedimentary rock and determine its age. Rather, index fossils considered characteristic of a specific period are used to identify rock layers in the field.

The geologist looks at the rock, determines what types of fossils it contains, and dates the rock according to the presumed age of those fossils (based on the estimation of when the organism evolved). If the fossils are of organisms that are assumed to have evolved recently, the geologist automatically assumes that the rock layers are young. On the other hand, if the fossils represent organisms thought to have evolved many millions of years ago, the rock strata are automatically assumed to be very ancient. For example, a rock layer containing a certain type of trilobite would be classified as Cambrian.

Fossils from around the world are dated in this manner and then arranged in their assumed order—a simple-to-complex progression—to compose the geologic column. This simple-to-complex progression is then said to prove the hypothesis of evolution. In other words, the major "evidence" for evolution is based upon the assumption of evolution: the evolutionary hypothesis determines the "age" of the fossil-bearing rocks, the "age" of the rocks determines the "sequence" of fossils, and the "sequence" of fossils is said to support the hypothesis.

**Radiometric dating.** Evolutionists also use a technique known as **radiometric dating** to lend credence to the ancient dates used in the geologic column. Radiometric dating is based on the fact that atoms of certain elements break down into atoms of other elements (known as their daughter elements) at relatively constant rates. The decay of these naturally occurring radioactive elements can be used to calculate the age of a rock or fossil. The carbon-14 method is commonly used for samples of biological origin, and the potassium-argon method is the most popular for dating rocks.

There are several problems with these techniques. The greatest problem is that the decay of an element cannot be used to calculate the age of a rock or fossil unless both the original and final amounts of radioactive element in the sample are known. Although the present composition of the sample is easily measured, *there is no way to measure how much of the "parent" and "daughter" elements were originally in the sample.* Thus, the scientist must estimate these amounts using methods that are based on assumptions (usually evolutionary assumptions) and are prone to error.

## The Cambrian Explosion: Evidence against Evolution

One of the serious contradictions between the facts of the fossil record and the hypothesis of evolution is known as the **Cambrian explosion.** If what evolutionists believe were true, there should be a gradual progression of extremely simple cells to more "advanced" creatures, with a gradual increase in variety and diversity. *But the fossil record contains no such progression.* Cambrian rocks show a sudden outburst of complex organisms of great variety, with no evidence of ancestors in the practically empty Precambrian rocks. Cambrian rocks contain representatives of every major animal phylum in existence today in addition to extinct animals. Evolutionists call this the Cambrian explosion because life seems to have "exploded" onto the scene. However, *this arrangement is just what we would expect if life were divinely created.*

> If there has been evolution of life,
> the absence of the requisite fossils in the rocks
> older than the Cambrian is puzzling.
>
> —*Marshall Kay and Edwin H. Colbert*[13]

> To the question why we do not find rich
> fossiliferous deposits belonging to these
> assumed earliest periods prior to the Cambrian
> system, I can give no satisfactory answer.
>
> —*Charles Darwin,* The Origin of Species[14]

## Missing "Missing Links": Evidence against Evolution

The fossil record contains a wonderful variety of living creatures, some of which have become extinct and are no longer alive today. Unfortunately, in an attempt to get around the embarrassing lack of transitional forms in the fossil record, unusual extinct animals are often pressed into service as "transitional forms." Invariably, however, a closer look at these creatures reveals that they are either creative "reconstructions" based on tiny fragments of bone and large quantities of imagination, varieties of known kinds, or new kinds of animals altogether.

[13]*Stratigraphy and Life History* (New York: John Wiley & Sons, 1965), 102, quoted in Henry Morris, ed., *Scientific Creationism*, ed. 2 (El Cajon, CA: Master Books, 1985), 81.

[14]P. 286.

Fig. 14.4 **Coelacanth** ***(Latimeria chalumnae)***

It would merely indicate that God created dinosaurs and birds with very similar design characteristics.

**The horse series.** In the past, many evolutionists regarded the supposed evolution of the horse as the best example of an evolutionary transition found in the fossil record. However, the **horse series** actually provides no evidence for evolution.

In the late 1800s, fossils of a three-toed, dog-sized mammal were discovered in North America. This fossil was named *Eohippus* [ē′ō·hĭp′əs: "dawn horse"] because it had several characteristics similar to modern horses. Several decades later, these fossils were reclassified as part of the genus *Hyracotherium* [hī′rə·kō·thĕr′ē·əm] (the original members of which were discovered in the 1830s) based on presumed evolutionary relationships; other horselike fossils were later added to this genus. Because the original *Hyracotherium* fossils resemble modern tapirs more than they resemble horses, some scientists again classify *Eohippus* and other horselike fossils in their own genera.

Speculating that *Eohippus/Hyracotherium* evolved through a series of stages into the modern-day horse, *Equus* [ē′kwəs], evolutionists constructed an elaborate "family tree" for the horse, emphasizing an increase in size and a reduction in number of toes over time. Because many horselike fossils do not fit this simple progression, the tree is sometimes shown with branches that end in extinction.

**Coelacanth.** Rocks of the Devonian Period contain fossils of an unusual 6-foot-long fish called the **coelacanth** [sē′lə·kănth′]. (*Coelacanth* is a general term used to describe any fish of the order Crossopterygii.) In coelacanths, unlike most fish, the fins are attached to the body by thick, fleshy lobes that allow the fins to be more freely rotated. Because of these unusual fins, evolutionists taught for many years that coelacanths were shallow-water fish and the ancestors of the first amphibians. The fish were often depicted crawling onto land from shallow water, using their lobed fins as feet. Scientists often speculated about their "amphibianlike" anatomy and how a few minor changes could have produced a genuine amphibian. In 1938, however, a live coelacanth was caught in the Indian Ocean; it was the first of many live coelacanths that would eventually be found. Evolutionists were surprised to find that coelacanths live very deep in the ocean and only rarely ascend to within 500 feet of the surface—making them highly unlikely to ever crawl out on land. In addition, their internal organs are completely fishlike, bearing no resemblance to those of amphibians, and the bones of their fins are not connected to the spine, preventing them from being used as legs. Using deep sea submarines, scientists tried to see whether coelacanths walk on the bottom of the ocean. Much to their disappointment, they do not.

***Archaeopteryx.*** The fossil bird ***Archaeopteryx*** [är′kē·ŏp′tər·ĭks] is often presented as an evolutionary link between dinosaurs and modern birds. Although it is true that the skeletal structure of certain dinosaurs is surprisingly similar to that of *Archaeopteryx* (except for the feathers), it does not necessarily follow that these animals were similar in other ways. However, even if fossil discoveries were to prove that *Archaeopteryx* and theropod dinosaurs were anatomically very similar, *that would still not imply that birds evolved.*

Fig. 14.5 ***Eohippus/ Hyracotherium***

One problem with the horse series is that fossils from all over the world had to be lumped together to produce it. One of the "links" was found in India, some were found in Europe, and most were found in the United States. Also, many of the species are actually found at the same "time" in the geologic column as their "ancestors" or "descendants," indicating that they could not have evolved from each other.

A third problem with the horse series is that the currently living members of the horse kind vary greatly in size and other characteristics. The "horse ancestors" found in the fossil record are within the size range of modern horses, and some modern horses even contain the genetic information needed to have three toes. Because of these similarities, several creationist researchers have hypothesized that the "horse series" represents variation within a single created kind. Another creationist hypothesis divides these animals into three kinds based on differences in tooth structure and other characteristics.

## Hominid Fossils: Evidence against Evolution

**Darwin and *The Descent of Man*.** In *Origin,* Darwin did not directly address the origin of mankind. He discussed this subject 12 years after the publication of *Origin,* in a second book entitled ***The Descent of Man*** (1871). In this work, Darwin pointed out similarities between man and "other animals" and came to the false conclusion that man is simply the most evolved form of animal:

> . . . man is descended from a hairy, tailed quadruped, probably arboreal [living in trees] in its habits, and an inhabitant of the Old World.[15]

This directly contradicts the biblical teaching that God created man:

> And the LORD God formed man of the dust of the ground, and breathed into his nostrils the breath of life; and man became a living soul. —Gen. 2:7

Fig. 14.6

**Differences between skulls of human and ape**

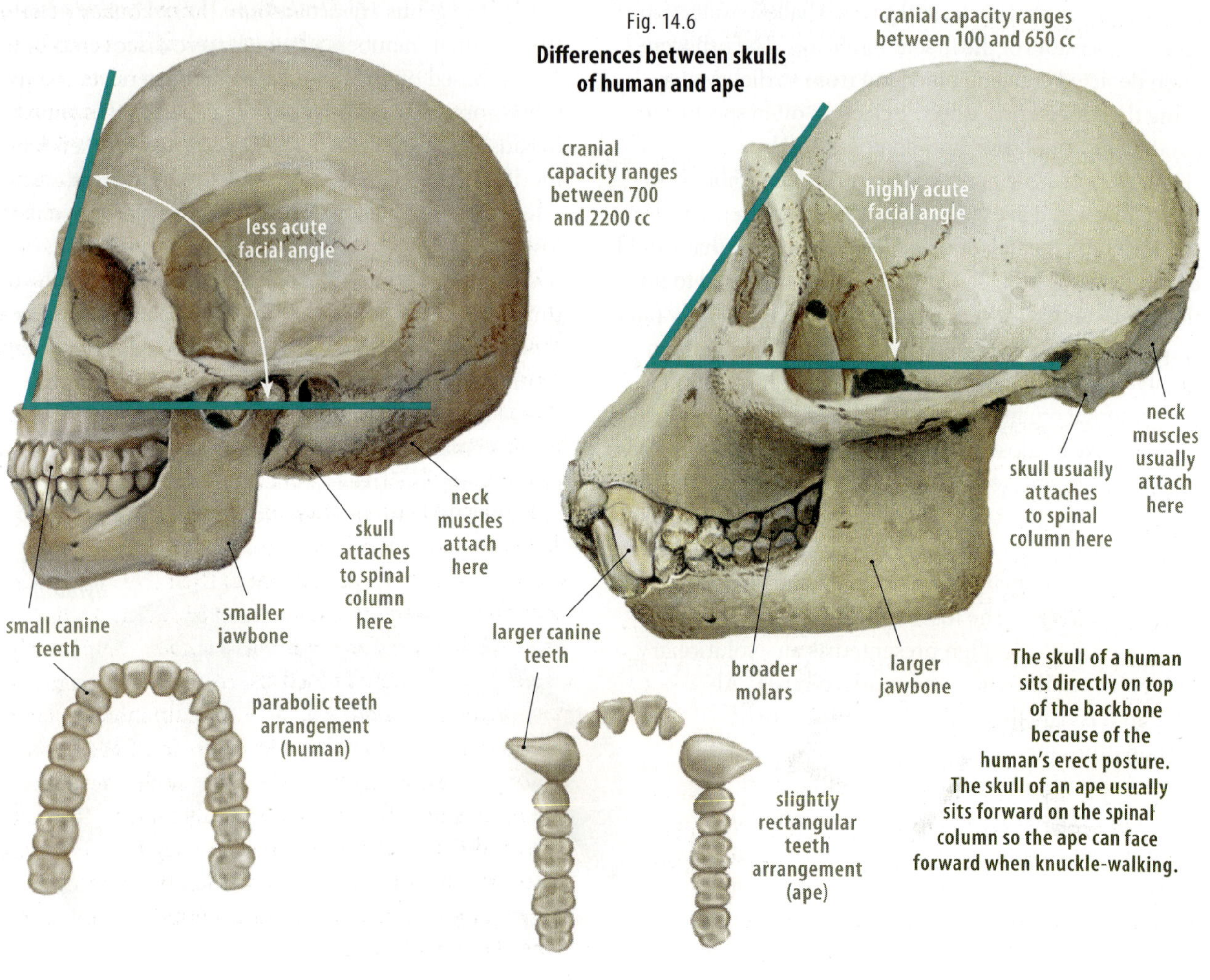


**The skull of a human sits directly on top of the backbone because of the human's erect posture. The skull of an ape usually sits forward on the spinal column so the ape can face forward when knuckle-walking.**

Many evolutionists claim that Darwin never taught that man evolved from monkeys, but that is exactly what *The Descent of Man* claims:

> The Simiadae then branched off into two great stems, the New World and Old World Monkeys; and from the latter, at a remote period, Man, the wonder and glory of the Universe, proceeded.[16]

Modern evolutionists have modified Darwin's ideas somewhat and now claim that man evolved from apes or apelike creatures that are now extinct.

**Differences between man and ape.** It is true that God made man and apes with many body similarities. For example, both apes and humans have a similar overall body plan, having five-fingered hands with opposable thumbs, suitable for grasping; five-toed feet; and facial muscles that are superficially similar. However, *there are also many important differences between man and ape,* not only physical differences but also moral and spiritual differences. Some of the differences between the skulls and teeth of a typical human and a typical ape are shown in Fig. 14.6 (p. 260). The arms of an ape are longer in relation to the legs than those of a man. In addition, man's feet are arched, have relatively short toes, and are designed primarily for upright walking. An ape's feet are typically flat, have longer toes, and are more suitable for grasping. The differences between man and ape go far beyond differences in body structure, however. *The most important difference between man and apes is that man is created in the image of God* (Gen. 1:27).

Fig. 14.7 **Piltdown man: a famous hoax**

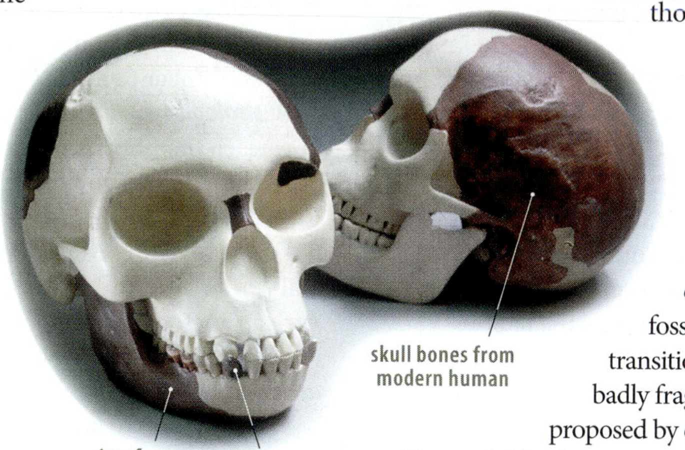


**An imaginary "family tree."** Since Darwin published *The Descent of Man* in 1871, paleontologists have found many fossils from around the world of extinct apes and ancient humans. Like the horse series, man's evolutionary "family tree" is constructed of skeletons taken from many different parts of the world, including East Africa, Indonesia, northern China, and northern Europe. When arranged in a particular order, they appear to show how man gradually became taller, more erect, more intelligent appearing, and more humanlike; evolutionists particularly emphasize that, in this sequence, the brain becomes larger and the jaw becomes smaller. The order in which these skeletons are arranged is, however, determined by the beliefs of those who assemble them.

In reality, scientists are no closer to proving Darwin's hypothesis of human evolution than they were in Darwin's day. Despite over a century of searching and the discovery of thousands of relevant fossils, only a handful of alleged transitional forms (many of them badly fragmented) have even been proposed by evolutionists over the years.

From a biblical perspective, *all of these "transitional forms" can be considered either 100% ape or 100% human.*

***Eoanthropus* (Piltdown man).** One of the most famous "early human" fossils of the 20th century was *Eoanthropus* [ē′ō·ăn′thrō·pəs: "dawn man"], commonly known as **Piltdown man** because the fossil was discovered in a gravel pit near the village of Piltdown, England. Discovered between 1908 and 1911 by an amateur paleontologist named *Charles Dawson,* the fossil consisted of a fragmented but clearly humanlike skull, along with an apelike jaw. Further investigations at Piltdown found crude tools carved from rocks and from bones of extinct creatures.

*For over 40 years, Piltdown man was used as "proof" that man had evolved from apes.* As a result, some scientists who had spent much of their lives studying *Eoanthropus* were devastated in 1953, when it was revealed that *"Piltdown man" was a hoax.* Apparently, someone had carefully filed the teeth of an orangutan's jaw fragment to make the jaw look humanlike, treated it with chemicals to make it look "ancient," and buried it alongside similarly treated human skull fragments in a place where they would be easily found. The fact that Piltdown man was accepted as a legitimate "human ancestor" for nearly four decades demonstrates that paleontology is often based on assumptions and faith instead of scientific fact.

[15] Charles Darwin, *The Descent of Man,* 2nd ed. (London: John Murray, 1890), 609.
[16] Ibid, 165.

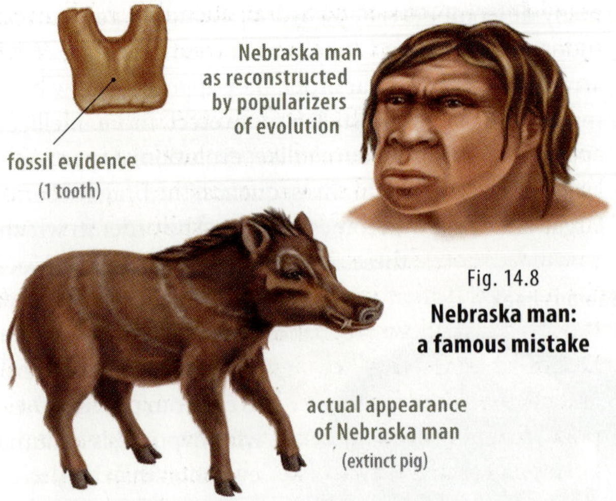

Fig. 14.8 **Nebraska man: a famous mistake**

***Hesperopithecus* (Nebraska man).** In many cases, the reconstructions of man's "subhuman" ancestors that appear in books and magazines are based on surprisingly little evidence. One example is the famous "early human" known as *Hesperopithecus* [hĕs·pĕr′ō·pĭth′ē·kəs], or **Nebraska man.** This missing link was "reconstructed" from a single tooth found in Nebraska in 1922. Popularizers of evolution produced imaginative illustrations that portrayed Nebraska man as a brutish creature, half ape and half man, who lived in caves and walked with a shuffling, stooped posture. "Nebraska man" was accepted as a legitimate "ape-man" for a few years, until it was revealed that the tooth did not come from a man at all, but from an extinct pig.

Fig. 14.9 ***Ramapithecus:* a discredited "missing link"**

Imaginary "reconstruction" of *Ramapithecus* by evolutionists (attempts to make the animal look humanlike)

Probable appearance of *Ramapithecus* (very similar to modern orangutans)

Interestingly, humorous mistakes such as these still occur. Some fossils apparently mistaken for "early human" remains in more recent years include a dolphin's rib, an extinct horse's toe, and an alligator's leg bone.

***Ramapithecus* ("Rama's ape").** Another "missing link" between apes and humans that has been discarded is ***Ramapithecus*** [rä′mə·pĭth′ē·kəs: "Rama's ape"]. *Ramapithecus* was a medium-sized ape originally "reconstructed" from a few teeth, a heavy jawbone, and fragments of facial bones. *Ramapithecus* was quickly assumed to be an ancestor of man because it possessed certain "humanlike" features, such as smaller front teeth than other apes and a humanlike jaw. (It was later determined that the fossil skull fragments had been assembled incorrectly to give them a humanlike shape.) Many evolutionists further speculated that *Ramapithecus* walked upright, even though no fossil hip or leg bones were found. However, the discovery of a complete *Ramapithecus* skull in 1979 revealed that the creature was not humanlike at all but was very similar to modern orangutans.

***Australopithecus* ("southern ape").** Evolutionists now teach that the first "links" in the chain of supposed human descent were the *australopithecines* [ô·strā′lō·pĭth′ĭ·sēnz: "southern apes"]. The first australopithecine fossil to be discovered was the "Taung [tông] child," a small ape skull discovered in 1924 by a South African anatomy professor named Raymond Dart. Dart named the animal ***Australopithecus*** [ôs′trə·lō·pĭth′ĭ·kəs] ***africanus.*** The Taung skull was that of a juvenile ape, but fossils of adults were later found. Evolutionists regarded *A. africanus* as a human ancestor until the early 1970s, when it was dismissed as merely an extinct ape. However, some evolutionists still teach that *A. africanus* was an early ancestor of humans.

Another variety of *Australopithecus* that was once regarded as a "human ancestor" is ***Australopithecus boisei.*** This fossil was discovered by Dr. *Louis Leakey* (1903–1972) and his wife *Mary* (1913–1996) in 1959 in the Olduvai [ōl′də·vī′] Gorge in Kenya. The Leakeys found a skull broken into 400 different pieces but were convinced even before they completely unearthed the fossil that it represented a human ancestor. The Leakeys originally named the fragments *Zinjanthropus* [zĭn·jăn′thrō·pəs: "East Africa man"], and the fossils were declared "almost human" by many evolutionists. However, it soon became clear that *Zinjanthropus* was not humanlike at all but was fully ape. As a result, the fossil was quickly reclassified as *A. boisei* and is now considered to have been merely an extinct ape.

The most famous australopithecine is probably ***Australopithecus afarensis*** [ăf′är·ĕn′sĭs], originally based on a chimpanzee-sized fossil nicknamed "Lucy." *A. afarensis* females such as "Lucy" were about 3½ to 4 feet tall, while the males may have been as much as 5 feet tall; all had ape-sized brains. The long, curved toes of *A. afarensis* were presumably suitable for grasping, like those of most other apes, and the skulls were chimpanzee-like. Thus, the animals can undoubtedly be considered extinct apes; calling them "human ancestors" is nothing more than speculation.

Some evolutionists teach that *A. afarensis* evolved from another extinct ape named *A. anamensis,* based on fossils found in 1995. Several other varieties of australopithecines have been found, including *A. robustus* and *A. aethiopithecus,* but these are uniformly acknowledged to have been extinct apes.

***Homo habilis* ("handy man").** Evolutionists teach that an australopithecine ape evolved into ***Homo habilis*** [hō′mō hăb′ə·ləs: "handy man"], a "missing link" announced in 1964. *Homo habilis* appears to have been a small, chimpanzeelike creature with a small brain. Although these shattered fossils were originally classified in the genus *Homo* ("man"), many scientists argue that *Homo habilis* was actually a type of australopithecine and should be classified as such. This conclusion is supported by the fact that more complete skeletons (such as O.H. 62) have shown that some of these animals were only 3½ feet tall as adults (even smaller than *A. afarensis*) and not humanlike at all.

Further problems with *H. habilis* were revealed when much larger skulls and leg bones (KNM-ER 1470, 1481, and 1590) were found in Kenya in 1972. Skull 1470 was incomplete and so badly shattered that its reconstruction is very subjective, but some creationists believe it could have been a human skull. Other creationists believe that it is merely the skull of an ordinary australopithecine ape. But there is no such doubt about the leg bones (KNM-ER-1481) that were found some distance away; they are in relatively good condition and unmistakably human.

Many creationists believe that the smaller fossils, such as O.H. 62, should be considered apes, while the larger fossils, such as KNM-ER 1481, should be considered fully human. In this view, "Homo habilis" *is actually a mixture of human fossils and ape fossils* and should not be considered a "transitional form" between apes and humans. A number of evolutionists agree with this position.

***Homo erectus* ("upright man").** In 1893, a Dutch scientist named *Eugene Dubois* [dü·bwä′: 1858–1940] discovered a fossilized skull fragment and leg bone on the Indonesian island of Java. The skullcap had some unusual characteristics, such as heavy brow ridges, but the femur was identical to that of a modern human. Although the skullcap and femur were nearly 50 feet apart, Dubois assumed they belonged to the same individual and claimed they represented an "ape-man"—a transitional form between apes and humans. Dubois may have exaggerated the "apelike" characteristics of the skullcap in order to call it an "ape-man." Dubois called the fossil *Pithecanthropus erectus* [pĭth′ĭ·kăn′thrə·pəs ĭ·rək′təs: "upright ape-man"], but it was popularly referred to as **Java man.** Although Dubois claimed the bones of Java man were 500,000 years old, some scientists have speculated that the rock in which these bones were found could not be more than 500 years old because of the frequent floods and volcanic eruptions that occur in that part of Java. Dubois also discovered, reportedly in the same strata, a "normal" human skull (later dubbed *Wadjak man*), although he kept this fossil more or less a secret for 30 years. Obviously, if Java man was buried in the same rock layer with a "modern" man, he could not be man's ancestor.

Three decades later, similar fossils were unearthed in China near the city of Beijing (Peking). These fossils were dubbed *Sinanthropus pekinensis* [sī·năn′thrə·pəs pĕ′kĭn·ĕn′sĭs: "Chinese man from Peking"], commonly known as **Peking man.** It was not long before scientists realized that "Java man" and "Peking man" were very similar and that the "apelike" characteristics of Java man may have been exaggerated. Thus, both Java man and Peking man were reclassified as ***Homo erectus*** ("upright man"). However, because of Dubois's haphazard methods, and because the Peking man fossils were lost during World War II, both of these fossils are somewhat controversial. Some creationists consider these fossils to represent some sort of giant, extinct gibbon—a type of ape—while others believe they were fully human.

*Other fossils classified as* Homo erectus *include more than a hundred fossils found in Africa and Asia that are undoubtedly fully human,* although they do possess some unusual skeletal features. They are characterized by thick skull bones; heavy ridges above the eyes; a low, sloping forehead and relatively acute facial angle (for humans); large teeth; and a receding chin. The brain size of these fossils, 700 cc (in an infant) to 1200 cc, is well within the range of "modern" humans. The rest of the skeletons were practically identical to those of "modern" humans, and fossil evidence indicates that they were just as tall; a careful analysis of one *H. erectus* boy's skeleton (KNM-WT 15000) indicated that he would have been about 6′1″ tall had he reached adulthood. The fossil evidence indicates that these individuals were intelligent; used tools, paint, and fire; and may have buried or cremated their dead. Thus, it is likely that these fossils represent an extinct race of humans (or possibly humans with vitamin deficiencies that caused unusual skeletal features). For these reasons, *H. erectus* should probably be classified as a variety of *Homo sapiens* (true man).

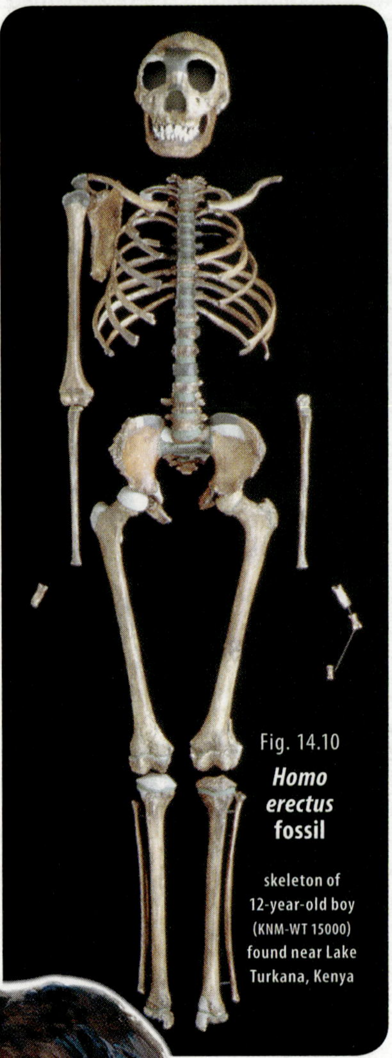

Fig. 14.10 ***Homo erectus* fossil**

skeleton of 12-year-old boy (KNM-WT 15000) found near Lake Turkana, Kenya

In addition, Dr. Marvin Lubenow (of the Institute for Creation Research) points out in his book *Bones of Contention* that according to the fossil record, *H. erectus* was contemporary with both *"H. habilis"* and *H. sapiens* throughout most of the hominid fossil record. Thus, *H. erectus* should not be considered an evolutionary link between the australopithecine apes and "modern" humans.

***Homo neanderthalensis* (Neanderthal man).** In 1856, a limestone cave perched on a cliff in Germany's Neander Valley yielded the first skeletal remains of what would become known as **Neanderthal** [nē·ăn′dər·thôl′] **man** (originally classified as *Homo neanderthalensis*). These fossils were said to possess "apelike" features such as heavy brows and sloping foreheads. The skeletons were also assembled in an artificially stooped posture to make them seem "subhuman." As other fossilized remains of Neanderthal skeletons began to turn up across much of western Europe, imaginative paleontologists in Europe and America reconstructed an entire race of brutish "cave men" with a slouched over, gorillalike posture. Models of Neanderthal families, naked, brutish, and savage, began to fill museum exhibits as further "evidence" of a "missing link" in man's supposed evolution from an apelike animal ancestry.

Fig. 14.11 **Neanderthal man**

Although evolutionists characterized the Neanderthals as apelike brutes for more than a century, closer examination of Neanderthal skeletal remains eventually burst this evolutionary bubble. Scientists now know that Neanderthal man was not a shuffling brute but rather a true human who stood upright with the posture, gait, and intellect of a person living today. Neanderthal skeletons and tools suggest that Neanderthals were physically superior to humans today, with larger, more powerful muscles and stronger bones. In addition, the brain of the average Neanderthal was nearly 11% larger than that of modern man. Neanderthal man was also a moral and spiritual being as man, made in the image of God, has always been: he buried his dead and cared for the old and crippled among the living. Evolutionists now admit that *the Neanderthals were fully human;* they are now classified as *Homo sapiens neanderthalensis,* designating them as a variety of true humans.

One final problem with classifying Neanderthals as an evolutionary ancestor of modern man is the fact that *Neanderthal man and modern man lived as contemporaries.* Bones of Neanderthal and bones of "modern" man, along with some simple tools, were found in two caves only yards apart on the slopes of Mt. Carmel in Palestine. The tools found in both caves apparently date from about the same period. This suggests that the Neanderthals either were members of an extinct race that lived alongside "modern" man or (possibly) were modern humans afflicted with certain deforming and crippling diseases. The Bible confirms these scientific discoveries; man has always been man, and Neanderthal man is no exception.

**Cro-Magnon man.** The last "link" in the supposed evolution of man is **Cro-Magnon** [krō·măg′nən] **man,** who was discovered in 1868 in southwest France. (Cro-Magnon is the local French name for the cave in which the first remains were found.) At the time, the Cro-Magnon "cave men" were described as subhuman. However, evolutionists now admit that the *Cro-Magnons were identical to modern humans* although, like the Neanderthals, they often had larger brains than the average person today. Far from being "subhuman," they were merely a tribe of people that often dwelled in caves and hunted bison. Today, Cro-Magnon man is classified as *Homo sapiens sapiens,* the same as humans today.

Fig. 14.12 **Cro-Magnon man**

**Confusion and turmoil.** Although a majority of evolutionists speculate that *A. afarensis* evolved into *H. habilis,* which evolved into *H. erectus,* which evolved into *H. sapiens,* there is actually much confusion and turmoil within evolutionary ranks concerning the supposed evolution of man. Evolutionists sometimes refer to ancient humans by names other than *H. erectus* or *H. sapiens.* Fossils such as "Heidelberg man" and "Rhodesian man," fully human fossils with unusual skeletal features, are called "archaic *Homo sapiens*" by some paleontologists and *Homo heidelbergensis* by others. Some evolutionists call some of the *H. erectus* skeletons *H. ergaster,* while others use the term *H. ergaster* to refer to apelike *H. habilis* fossils. Still others call some of the *H. habilis* fossils *H. rudolfensis* or *H. microcranous!* However, from a creationist perspective, *all of the fossil humans can be regarded as* Homo sapiens, while the apelike fossils included in *H. habilis* can be regarded as australopithecines or other extinct apes.

**Man has always been man.** The confusion among evolutionists over man's ancestry has existed for well over a century. Yet there is seemingly as much controversy among evolutionists over the course and procedure followed by the evolution of man as there is controversy between creationists and evolutionists. In contrast, God's account of how man was created is clear, concise, and complete. There is nothing in His record that would lead us to believe that man arose from the animals by evolutionary processes. On the contrary, we are told that God Himself shaped and fashioned man from the "dust of the ground" and then "breathed into his nostrils" to give him life (Gen. 2:7).

## Section Review 14.2

1. Explain the difference between the terms *microevolution* and *macroevolution.* Which term is referred to simply as "evolution" in this textbook?
2. Why does a lack of transitional forms pose problems for the evolutionary hypothesis?
3. Name the fossil fish that was supposed to be a "missing link" between fish and amphibians until a live specimen was caught in the 1930s.
4. What details suggest that the well-known "horse series" is not a series at all?
5. Which famous "early human" fossil turned out to be a hoax crafted from the jaw fragment of an orangutan?
6. Which supposed "missing link" in the evolution of man was constructed from what proved to be the tooth of an extinct pig?
7. What name has been given to the human fossils discovered in Germany's Neander Valley?

### Application

8. What are some of the differences between man and ape? Which is the most important?
9. Name three supposed "missing links" between apes and humans that were concluded to be extinct species of apes.

### Critical Thinking

10. From a biblical perspective, why might we say that fossils such as KNM-WT 15000 should be classified as a variety of *Homo sapiens* instead of *Homo erectus?*

### Identify

fossils, paleontology, transitional forms, punctuated equilibrium hypothesis, geologic column, radiometric dating, Cambrian explosion, *The Descent of Man,* Cro-Magnon man

# 14.3 Biology: Evidence against Evolution

The lack of transitional forms in the fossil record provides strong evidence that evolution has not occurred. However, modern biological knowledge provides additional evidence against evolution. Many fascinating examples, such as the differences between the reptile and bird respiratory systems, the complexity of the cell, the laws of heredity, and the occurrence of natural selection, reveal the faults of evolution and the reality of God's design in the living creation. Indeed, the more we learn about living things, the more insurmountable the difficulties of evolution become.

## Impossibility of Transitional Forms

**Full function at every step.** In *The Origin of Species,* Darwin makes clear that evolution requires that all structures had to develop one small step at a time. For example, evolutionists believe that the four-chambered heart of birds and mammals evolved from the one-chambered heart of invertebrates. Because Darwin realized that gradual change requires that the organism be fully functional and capable of survival at every step, he knew that the discovery of any structure that could not have developed gradually would destroy the hypothesis of evolution:

> If it could be demonstrated that any complex organ existed, which could not possibly have been formed by numerous, successive, slight modifications, *my theory would absolutely break down.*[17]

Although materialistic "reason" and a firm belief in evolution convinced Darwin that precisely designed structures could have evolved gradually, even he admitted that the evolution of complex structures like the eye seems absurd to common sense.

**Bat wings: evidence against evolution.** Zoology and botany reveal many other features of animals and plants that could not have developed by numerous slight modifications. A good example of an organ that could not have developed gradually is the wing of a bat. Evolutionists teach that bats evolved from small, four-legged, rodentlike mammals similar to modern shrews. However, a bat's wings are composed of extraordinarily long finger bones connected by a thin web of skin. For a shrew's forepaws to gradually become wings, the forepaws would become useless for grasping or running long before they would enable the creature to fly. Thus, instead of being an "improved" creature favored by natural selection, the long-fingered transitional creature would be a deformed monstrosity unable to walk, run, hold food to its mouth, or fly; it would be eaten or would starve to death before it could evolve further. This fact is also consistent with the fossil record, which shows no transitional forms between small insectivores and bats. The "earliest" known fossil bat is the Eocene bat *Icaronycteris,* which was 100% bat.

Fig. 14.13
**Bat and shrew**

**Amphibian egg to reptile egg.** Another example in which transitional forms could not have survived is the transition between the jelly-and-embryo egg of an amphibian and the complex amniotic egg of a reptile. Although evolutionists teach that amphibians evolved into reptiles, there is no way to gradually change an amphibian egg into a reptile egg that would allow the developing embryo to survive the gestation period. Fig. 14.14 illustrates several differences between amphibian and reptile eggs, many of which would have to be overcome simultaneously for an amphibian to evolve into a reptile. Not only would the structure of the egg have to be changed, but the structure of the animal's entire reproductive system would have to be altered. From these facts, it is abundantly clear that the reptile egg could not have evolved by a series of successive, slight modifications from an amphibian egg.

Fig. 14.14
**Amphibian eggs and reptile eggs**

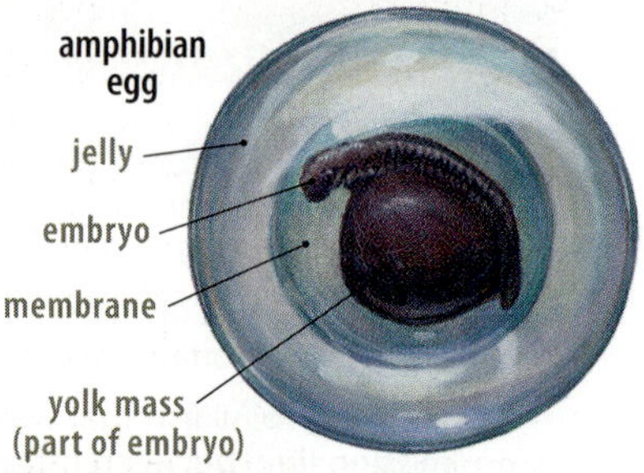


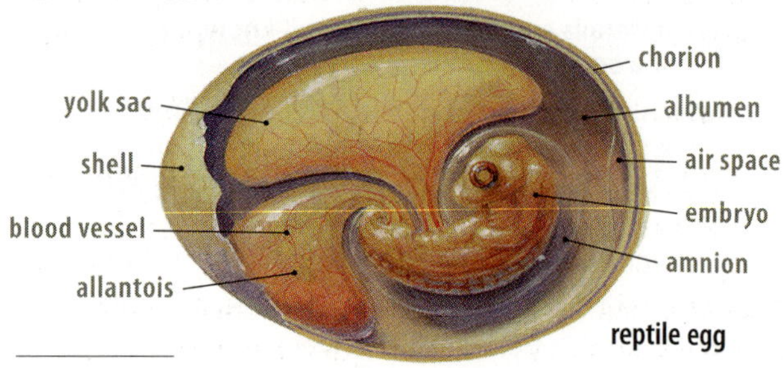


[17] *Origin,* 146. Emphasis added.

Fig. 14.15

**Reptile and bird respiratory systems**

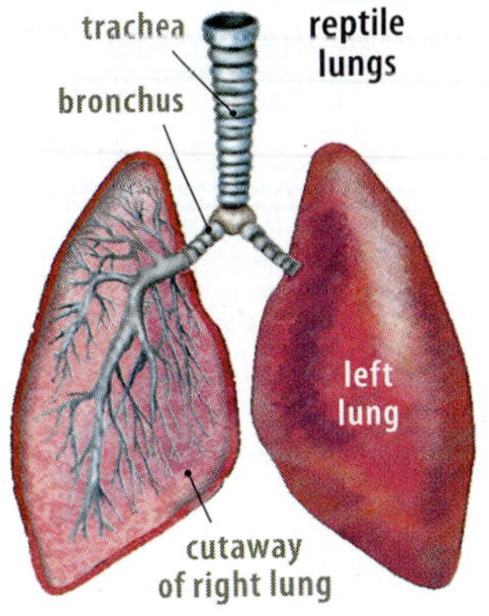


Air flows into the lungs during inhalation and out of the lungs during exhalation.

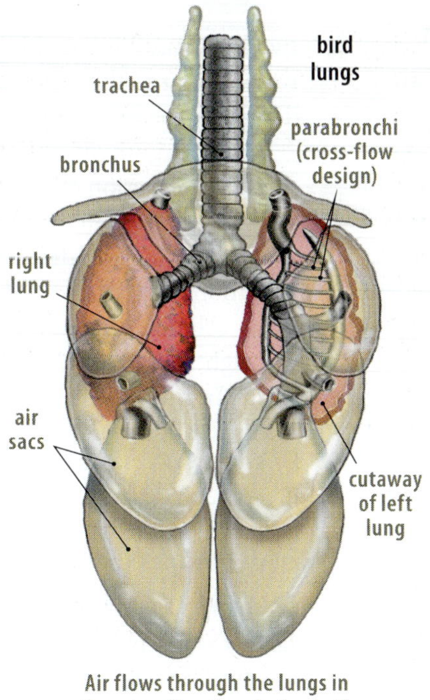


Air flows through the lungs in the same direction during both inhalation and exhalation.

**The bird respiratory system.** The respiratory system of birds is completely different from that of any other vertebrate. Although evolutionists teach that birds evolved from reptiles, there are fundamental differences between the reptile and bird respiratory systems that would have prevented transitional forms from surviving (Fig. 14.15). In addition to the miraculous changes in the lungs that would be required to change a reptile into a bird, air sacs would have to be created in the body cavity, the muscles of the chest and abdomen would have to be redesigned, and the manner in which the lungs are attached to the chest wall would have to be changed. In fact, even the manner in which the lungs are prepared in the egg to take their first breath would have to be completely redesigned. Most of these changes would have had to take place simultaneously for the creature to survive. Clearly the bird respiratory system could not have been produced by a series of "numerous, successive, slight modifications."

## Homology: Evidence of Common Design

Around 1800, the science of **comparative anatomy,** which is the study of similarities and differences in the body structure of organisms, was founded by **Georges Cuvier** [kyo͞o′vē·ā′: 1762–1832]. Cuvier, a French scientist, showed that *the bodies of man and animals have certain basic similarities in their overall designs.* Two concepts are important in this respect: analogy and homology. **Analogy** refers to similarities among organisms designed for the same environment but different in internal structure. For instance, whales share many common features with fish although whales are mammals. Butterflies and birds have wings for flying but the two animals are very different. On the other hand, **homology** refers to similar organs that are also similar in internal structure. The most popular example of homology is the pentadactyl (five-digit) design in the limbs of terrestrial vertebrates, which share an underlying common pattern.

For Darwin, homologies among animals exist because they all descended from a common ancestor. Since Darwin, evolutionists have presented homology as one of the most clear "proofs" of evolution. This is, however, a hypothesis that needs to be proved. The observation that vertebrates have similar patterns in their limbs does not necessarily represent proof that they are related by common ancestry. A very powerful alternate hypothesis is the one presented by creationists, who believe that it is only logical that God would use the same basic plans for many different animals when He created them. Man uses this same technique when designing his own creations. For example, the vast majority of passenger vehicles have four wheels, with an engine in the front and a passenger cabin located near the center of the vehicle. Just as engineers modify this same basic design to make freight trucks or buses, God modified His basic design when necessary.

There are other inconsistencies in the evolutionary interpretation of homology. If homologous organs are due to common ancestry, then it follows that the embryological development of those organs should be similar. In other words, they should derive from the same embryonic tissues and follow the same steps in development. They also should be homologous at the genetic level if indeed there was a common ancestor. They should be formed under the action of the same genes or at least genes that are very similar. An examination of the many cases of homology in the animal world fails to provide this evidence. *Many homologous structures look similar only as the final product and are strikingly different at the embryonic and genetic level.*

## Molecular Biology: Evidence against Evolution

Much of the newest evidence against Darwinism comes from the field of **molecular biology,** the science that seeks to discover how the mechanisms of living cells work. In the 1950s and 1960s, evolutionists expected that as scientists learned more about living cells, they would discover how cellular mechanisms could have evolved from nonliving matter. However, the explosion of knowledge about cells in recent decades has only caused more problems for evolution.

**A naive view of living things.** In Darwin's time, living cells were regarded as simple bags of an amorphous gel or slime called *protoplasm,* which was thought to be a slush of amino acids, protein, fat, and carbohydrates. This naive view of the cell allowed Darwin's followers to speculate that a living cell could easily have evolved from some sort of primordial ooze. They reasoned that it would not be all that difficult for a piece of warm mud in some primordial pond to acquire life and begin to reproduce.

**Design and complexity.** Today, we understand the structure of living cells much more clearly than Darwin's followers ever could have dreamed. But instead of finding simple "goo" at the heart of cells, molecular biologists have stumbled on a miniature world of fantastic complexity—a world of microscopic pumps, motors, sensors, turbogenerators, information storage and retrieval systems, and other complex mechanisms. It is now very obvious that these intricately designed mechanisms could not have been constructed by chance but were formed by the purposeful design of a wise Creator.

Fig. 14.16 **Gregor Mendel**

**Design demands a Designer.** The 18th-century British theologian **William Paley** (1743–1805) remarked that if we were to spot a rock lying on the ground in a deserted place, we might assume that the rock had come to be there by the processes of nature. However, if we were to find a pocket watch lying on the ground, we would naturally conclude that the watch resulted, not from a chance collection of minerals, but from the hand of a watchmaker. Despite the fact that the watch might be made out of the same minerals as the rock, it exhibits clear evidence of purposeful design by an intelligent watchmaker. Paley argued that the intricate mechanisms of living things, such as muscles and eyes, were similar to the watch in that they exhibit evidence of design by their Creator.

Paley's argument, although powerful and based on the observation of nature, has been dismissed by evolutionists, on the basis that nature has only the "appearance" of design caused by the random evolutionary processes. However, the more we learn about cellular biology, the more we see that Paley was right: the tiny molecular machines in the cells of living things make man's most complex creations look primitive. *Living systems, which all show evidence of careful design, precision engineering, and deliberate purpose, did not occur by chance.*

## Evidence from Genetics

**Unlimited variety: a false view.** Additional evidence against Darwin's hypothesis of evolution comes from *genetics,* the scientific study of heredity. In Darwin's time, very little was known about heredity and genetics. Biologists knew that traits were passed from parents to offspring, but no one knew how traits were passed down or if new traits could arise.

Ignorant of the true nature of heredity, *Darwin assumed that the characteristics of a given plant or animal change randomly from generation to generation, and that there is no limit to this variability.* Darwin assumed that there was no fundamental genetic barrier that would prevent a sea sponge from gradually changing into a fish, or a fish into an amphibian, or an amphibian into a reptile, or a reptile into a bird or a mammal.

**Limited variety.** At the same time that Darwin published *Origin,* an obscure Austrian monk named *Gregor Mendel* [mĕn′dl: 1822–1884] was uncovering the true nature of genetics and heredity. Mendel carried out exhaustive experiments on plants to discover how hereditary traits are passed down from generation to generation. Mendel and later geneticists showed that most random variation occurs by rearrangement of genes that are already present. In other words, *variety within kinds generally results from preexisting genetic variety* (genes that were present from the beginning), and *there are fixed limits to biological change.*

Fig. 14.17 **Variety within kinds versus evolution**

Many varieties of canines have developed from a single pair of canines that left the Ark.

Dogs, seals, raccoons, bears, cats, and hyenas supposedly evolved from a single weasel-like animal.

Many varieties of sparrows have developed from seven sparrows that left the Ark.

All birds supposedly evolved from a dinosaur ancestor.

Much variety within the human race has developed from the eight people who left the Ark.

Man supposedly evolved from an apelike creature to his present stature.

Although Mendel's work was ignored until after Darwin's death, later scientists realized that the laws of heredity contradicted Darwin's concept of gradual, unlimited change. By about 1900 it was clear to most scientists that Darwin's natural selection process alone could not be the mechanism for evolution, and in 1901 *Hugo de Vries* [də vrēs′: 1848–1935] published his mutation theory. De Vries was convinced that **mutations** (random errors in an organism's genetic material) and natural selection working together provided the mechanism of evolution. Mutations would provide the source of new variation, and natural selection would preserve the advantageous ones.

**Mutations: harmful, not helpful.** This line of reasoning has serious flaws, however. After decades of experimentation, geneticists have proved that the overwhelming majority of mutations are detrimental for the organism. Mutations in body cells do not result in new, useful organs; they typically result in cellular death, impaired cell function, or runaway cell growth (cancer). Mutations in reproductive cells do not result in new, more capable organisms; they result in less capable organisms. If we add these facts to the requirement by Darwinism of an infinite number of positive mutations needed for evolution, we clearly see the impossibility of such a process.

## Natural Selection and Genetic Variety

Although it does not produce new characteristics, *natural selection does help reveal latent genetic variety within a kind.* For example, Darwin observed that finches on the Galápagos Islands had bills that were quite different from those of European species and were much better suited to eating the seeds and fruit found on the islands. The changes actually occurred, however, because *the finches had possessed the genes for those bill shapes from the very beginning.* **Inbreeding** (breeding within a small, isolated group) revealed these latent characteristics, and natural selection caused them to become predominant among the finches on the islands.

It is important to remember that although God created genes for great variety within each kind, one kind does not change into another kind (Fig. 14.17). Rather, God ordained that living things should always reproduce "after their kind." Finches may develop larger bills, or change color, or grow longer tail feathers, but they still belong to the same kind; they never become ducks, ostriches, or eagles.

Fig. 14.18 **Peppered moths**

**The peppered moth.** A commonly cited example of natural selection in action is that of the **peppered moth** *(Biston betularia),* found in England. The moths range in color from black to light gray with a few dark spots. In the 1850s, many trees in Britain were covered with a type of mottled gray lichen. The light-colored moths blended in well with this background, but the darker moths stood out clearly. As a result, birds spotted and ate the dark moths more quickly. The result was that 98% of the peppered moth population consisted of the light-colored variety. Over the next hundred years, air pollution killed most of the lichen and darkened the tree bark. Then it was the light-colored moths that stood out and were eaten first. After the introduction of pollution controls in the mid-20th century, the lichen regrew and the light moths again became predominant.

The peppered moth provides an excellent example of how natural selection can act on the created variety within a kind. Changes in the environment changed which color of moth was better able to survive, and the moths of that color became predominant. But the peppered moth does not demonstrate evolution because no new kinds, or even new species within a kind, were produced: all of the moths, light and dark, are still *Biston betularia.*

**Pesticide resistance in insects.** The increasing percentages of houseflies, cockroaches, and other insects that are found to be resistant to certain insecticides is often cited as an evidence of evolution in action. Experiments have shown, however, that the resistant insects did not evolve resistance as a result of insecticide applications but rather already possessed the resistance to the insecticide before the insecticide was used. The use of insecticides simply made the nonresistant insects less likely to survive, so the resistant insects became predominant.

Furthermore, observations on DDT-resistant mosquitoes show that once the applications of DDT ceased in the 1960s, the percentage of insecticide-resistant mosquitoes dropped to levels similar to those present before the applications started. No further "evolution" took place.

**Antibiotic resistance in bacteria.** A similar phenomenon occurs when bacteria are exposed to antibiotics such as penicillin. Some bacteria have genes for a group of enzymes called *beta-lactamases,* which break down penicillin and similar antibiotics into harmless substances. When bacteria are exposed to penicillin, the bacteria lacking these enzymes are killed while the resistant ones survive. Importantly, no new kinds of bacteria are formed in response to penicillin applications. Some bacteria already have the penicillin-resistance gene before the antibiotic is applied, and except for that, they are still the same as the nonresistant bacteria. The resistant bacteria contain genes for resistance to penicillin whether or not they have been exposed to the substance.

**Bacterial proliferation: evidence against evolution.** In fact, the proliferation of bacteria actually provides evidence that kinds do not change into other kinds. Under favorable conditions, many bacteria can reproduce every 20 minutes or so. In just three years of reproduction at this rate, a colony of bacteria would go through more generations than humans could in one million years. Yet when today's bacteria are compared with dormant bacteria from ancient tombs, fossilized amber, or archaeological remains, their DNA shows little change—in spite of the fact that some of these bacteria may be separated from each other by thousands of years and millions of generations.

## Embryonic Recapitulation: A False Hypothesis

Evolutionists once presented the hypothesis of **embryonic recapitulation** as "proof" of evolution. This hypothesis states that an unborn baby, while developing in its mother's womb, goes through various stages of development that resemble different animals, providing a "replay" of man's evolution from a simple invertebrate to his present form. Although the hypothesis of recapitulation has been disproven, it still is presented in many biology texts as a strong evidence for evolution.

The hypothesis of embryonic recapitulation was developed by a German evolutionist named **Ernst Haeckel** [hĕk′əl: 1834–1919]. It is now known that *Haeckel, in order to support his views, deliberately falsified the pictures he published of certain embryos.* Haeckel first published his "findings" in 1868. Within 6 years the truth was discovered, but over 100 years later this hypothesis is still being taught in many schools.

Fig. 14.19 **Pharyngeal arches of human embryo**

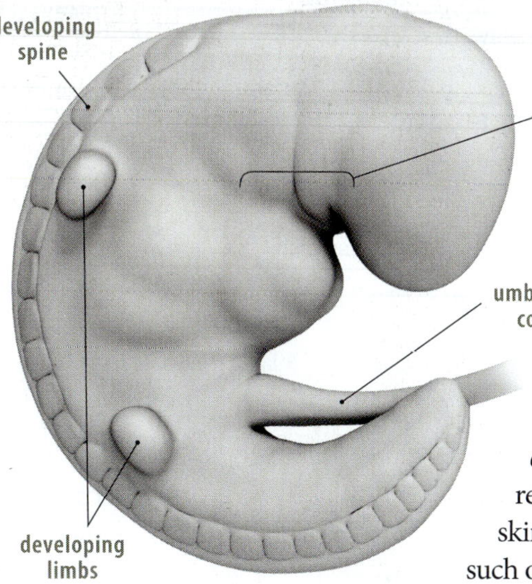


The function of certain parts of the embryo is also sometimes misrepresented by evolutionists. For example, they sometimes say that the human embryo has gill slits. *A mammal embryo never develops gill slits, and neither does a human embryo.* The so-called gill slits are in reality merely pouches in the embryo's skin. These pouches later develop into such organs as the middle ear canals, the thymus, and the parathyroid glands. They have no ability whatsoever to extract oxygen from water at any stage in the embryo's development, and they do not even develop into respiratory-related structures. Surely, if these were true gill slits, they would function as gill slits or at least become respiratory structures.

## Section Review 14.3

1. How does comparative anatomy furnish evidence of a single Creator?
2. How was Darwin's view of living cells oversimplified?
3. What discoveries by Gregor Mendel and later geneticists contradicted Darwin's ideas of gradual, unlimited change?
4. Why do mutations tend to be harmful instead of helpful?
5. How do natural selection and inbreeding help to reveal latent variety within animal and plant kinds?
6. Describe the hypothesis of embryonic recapitulation and explain why it is faulty.

### Application

7. Explain why examples such as the peppered moth, pesticide-resistant insects, and antibiotic-resistant bacteria do not demonstrate evolution.

### Identify

analogy, homology, molecular biology, genetics

### Observe

- To illustrate some of the difficulties of the evolutionary hypothesis, take one of Shakespeare's sonnets (sonnet 18, for example) and try to change it into sonnet 29 by changing one letter at a time (analogous to mutations). If the sonnet ceases to make perfect sense or deviates from the proper sonnet form in rhyme or meter, it has "died," and you must go back to the previous step. (Remember that in nature, natural selection weeds out organisms with harmful mutations.) Then try changing letters randomly instead of by intelligent direction and see if you make less or more progress than before.

## 14.4 Evolution Is Not Science

*The concept of biological evolution is faith in the idea that all presently existing kinds of plants and animals developed by natural causes from earlier forms, which originated by chance from nonliving matter in the distant past.* One evolutionist, G. A. Kerkut, says that this idea involves seven basic assumptions:

1. Non-living things gave rise to living material; i.e., spontaneous generation occurred.
2. Spontaneous generation occurred only once. The other assumptions all follow from the second one.
3. Viruses, bacteria, plants and animals are all interrelated.
4. The Protozoa [one-celled animals] gave rise to the Metazoa [multicellular animals].
5. The various invertebrate phyla are interrelated.
6. The invertebrates gave rise to the vertebrates.
7. Within the vertebrates the fish gave rise to the amphibia, the amphibia to the reptiles, and the reptiles to the birds and mammals.[18]

Kerkut calls these seven assumptions the *general theory of evolution,* and he points out that all evolutionists accept them as true and valid. He then makes a very interesting and revealing admission about this philosophy of evolution and its assumptions.

> The first point that I should like to make is that *these seven assumptions by their nature are not capable of experimental verification.* They assume that a certain series of events has occurred in the past.[19]

Professor Kerkut, although an evolutionist himself, candidly tells us that the philosophy of evolution is based upon assumptions (or a faith) that cannot be scientifically verified. He then admits that *whatever evidence can be assembled for evolution is both limited and circumstantial in nature.*[20]

Fig. 14.20
**General theory of evolution**

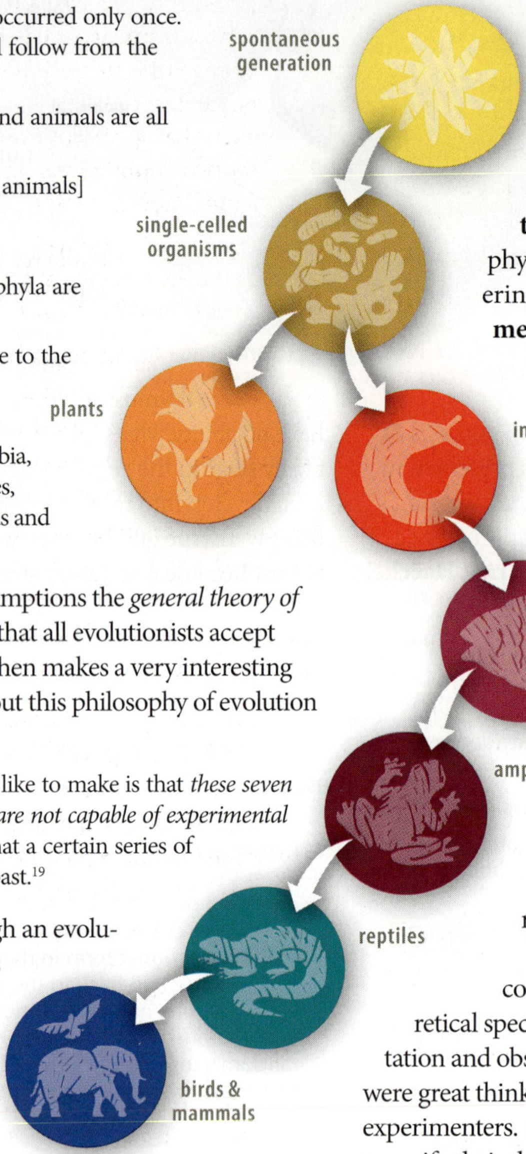


### Evolution versus Science

**What makes science?** The great scientific work of men like Johannes Kepler, Galileo, Sir Isaac Newton, and Albert Einstein is of inestimable value to science. Three things characterized their approach to solving scientific problems and answering puzzling questions about the physical universe: theoretical speculation, accurate observation, and precise experimentation.

**Theoretical speculation** involves thinking seriously and rigorously about the phenomena of the physical universe and forming general principles to explain them. **Observation** involves looking very closely at the physical universe for the purpose of gathering scientific knowledge about it. **Experimentation** is a special kind of observation that involves testing theoretical speculations in a controlled and systematic way. Experimentation seeks to determine the truth or falsity of theoretical speculations.

In modern science, these three characteristics (theoretical speculation, observation, and experimentation) are closely linked to each other, and they are often difficult to separate. Regardless of the order in which they occur, these three scientific activities are closely interrelated, and one always leads to the others.

Those great scientists whose major contributions have been in the realm of theoretical speculation appreciated the role of experimentation and observation. Sir Isaac Newton and Galileo were great thinkers, but they were also great observers and experimenters. Both specifically designed experiments to verify their theories, and much of what they speculated

[18]Adapted from G. A. Kerkut, *Implications of Evolution* (Oxford: Pergamon Press, 1960), 6.
[19]Ibid., 7. Emphasis added.
[20]Ibid.

about can be traced to some observation that intrigued them. These men and scientists like them did not want to lose touch with reality in their thinking because they knew that their hypotheses would have to be tested by observation and experimentation. Since the real world reflects the rationality of its Creator, they expected that good reasoning would produce theories that correspond to the real world, and, in that sense, scientists are able to "think God's thoughts after Him." *Scientists expect rigorous testing in the real world to confirm their theories;* if rigorous testing fails to confirm the theory, then the theory must be discarded or modified.

Theorists also expect to make occasional mistakes in their theoretical speculations. Experimental testing reveals these mistakes and helps eliminate erroneous ideas in the scientific world. *The testing process ensures that, over the long run, scientific theories that contain errors will be eliminated.* This in turn ensures that most scientific theories will be in touch with reality, which means they can be used by mankind to master nature. Experimentation either confirms or disproves scientific theories.

> . . . but I still think that to the unprejudiced, the fossil record of plants is in favor of special creation.
>
> —*E. J. H. Corner, botanist*[21]

**Is evolution a good science?** The hypothesis that life developed by spontaneous generation and evolved to its present form and variety is not testable by experimentation. Thus, *it must always remain in the realm of philosophy, for no one can go back to the time of the beginning.* Only God was present at the Creation, and no man was there to observe what took place.

All science begins with some sort of observation. The scientist must observe carefully and accurately if he is to gain useful scientific knowledge. Scientific observation may be of two types: *direct observation* and *indirect observation.* The latter type employs a variety of scientific instruments, whereas the former involves only the scientist and his own sensory abilities. All good scientific observations, whether direct or indirect, share the following characteristics:

1. An observation, to be scientific, must be repeatable. *Any event that is not subject to confirming observation (the origin of life, for example), is beyond the realm of science.*
2. An observation, to be scientific, must be as free from *bias* (prejudice) as possible. In other words, the scientist must learn to faithfully report what he actually observes, no more and no less. (Scientists universally agree that this is a very difficult task.)

Evolution does not measure up to the tests required of scientific observations.

1. Evolution, according to evolutionists themselves, is not repeatable. Furthermore, if change of one kind of creature into another is occurring today, it is occurring at such a slow pace that it cannot possibly be observed. Observations reveal variety within kinds but cannot reveal evolution.
2. Evolution begins with a bias (the prior assumption of philosophical naturalism), and the theory is kept alive by reliance upon this assumption.

One of the scientist's goals is to determine the cause-and-effect relationships that exist in the physical world. Cause-and-effect relationships are best determined by experimentation, and no one has yet devised experiments that show a cause-and-effect relationship for evolution. Many scientists have attempted to conduct controlled experiments to prove that life arose by chance or that one kind of creature can change into another kind, but they have all failed. A brief summary of some of these experiments and their results follows.

1. *Synthesis of amino acids.* Various experimenters, beginning with *Stanley Miller* (1930–2007), have produced certain amino acids with specialized apparatus and conditions that were supposed to correspond to the imagined conditions on the primitive earth. Miller's apparatus included a trap to separate the amino acids as soon as they were formed; otherwise, they would have quickly been broken down by the same "atmospheric" conditions that produced them. In addition, many geologists today doubt that the primitive atmosphere was anything like the one used in Miller's experiment.
2. *Synthesis of sugars, nucleotides, and other simple molecules.* Several researchers have devised elaborate chemical "simulations" that use

[21]In Ann M. MacLeod and Leslie Samuel Cobley, eds. *Contemporary Botanical Thought* (Chicago: Quadrangle Books, 1961), 97.

electricity or ultraviolet light to produce simple sugars and nucleotides from other chemicals. However, just as with amino acids, these simple molecules are no closer to being living things than a tiny blob of aluminum is to being a Boeing 747, or a blob of wood pulp is to being the *Encyclopædia Brittanica.* The difference is not in the basic constituents, but in the design (information content) of the structure.

3. *Linking of amino acids. Sidney Fox* (1912–1998) and others have been able, by very special heating techniques and certain conditions that could never have existed on the hypothetical primeval earth, to bond amino acids together to form what he called proteinoids. These are not in any sense the incredibly complex molecular machines found in living cells, however. They were mere "blobs," with no order and no utility, and would quickly have been destroyed if they had ever been actually produced on the primeval earth.

4. *RNA world.* The theory that the molecule of RNA was the primeval molecule from which all macromolecules and eventually the cell evolved is well established in current evolutionary thought. The main facts supporting this contention are that RNA can carry biological information and that some RNA molecules possess enzymatic activity. These characteristics made them the hypothetical potential self-replicating primeval molecules. However, this scenario is seriously challenged by the facts that RNA is a very difficult molecule to synthesize and that it does not replicate easily even under optimal laboratory conditions. In addition, the nucleotides, sugars, and phosphates needed are neither present nor can be generated in a stable form in the assumed Earth's primitive atmosphere.

5. *Synthesizing cells.* In 1970, *James Danielli* (1911–1984) was reported actually to have synthesized a living cell. (Actually, he supposedly started with living cells, then disassembled them, then refabricated a cell from parts of the dismantled cells.) However, it has since been shown that Danielli did not succeed at reassembling a living cell from its component parts. Indeed, given today's knowledge of cell complexity, many scientists are now doubtful that a completely disassembled cell could be assembled into a workable cell by human instruments, even if we knew how everything in the cell actually worked.

A common fault with these experiments is that even if they were successful (which they have not been), they do not address whether or not life actually arose by chance and evolved into its present form. The origin of living things is a historical event, not a day-to-day process of nature. *Experiments demonstrating the possibility that a historical event occurred in a certain way cannot prove whether the historical event actually occurred that way.* (For example, it can be scientifically demonstrated that certain kinds of bacteria can survive the vacuum of space for long periods of time while in a dormant state, but this does not mean that life arrived on earth from outer space.) Nevertheless, *experiments have failed to show that spontaneous generation and biological evolution are even possible.*

We can see that evolution is a retreat from science because it is not observable and it cannot be proven by experiment. *Evolution is not science. It is merely a philosophy based on faith in nature rather than faith in God.*

## Evolution Is a Threat to Modern Science

**The decline of scientific integrity.** The kind of thinking that leads one to accept evolution as an established fact goes against the kind of thinking that is important to science. *Historically, modern science was characterized by its emphasis on provable facts.* Hypotheses were regarded with skepticism until fact after fact had supported the hypothesis and disproved others. In *The Origin of Species,* however, Darwin pioneered a new technique of "science": presenting vague, improvised speculations one after another, supported only by imagination and conjecture, as "proof" of a hypothesis. If facts are used, they are used vaguely and selectively to give the appearance that the hypothesis is supported and never contradicted.

Such jumping to conclusions is characteristic of the writings of many evolutionists today, whose works abound in such statements as "it might have been—it is conceivable—apparently—not difficult to conceive—it is probable—perhaps—maybe." *William R. Thompson* (1887–1972) stated in his introduction to a 1963 reprint of *The Origin of Species* that the widespread adoption of the methodology of Darwinism has harmed science in general:

> *The success of Darwinism was accompanied by a decline in scientific integrity.* This is already evident in the reckless statements of Haeckel and in the shifting, devious and histrionic argumentation of T. H. Huxley. . . .
>
> To establish the continuity required by [evolutionary] theory, historical arguments are invoked even though historical evidence is lacking. Thus are engendered those fragile towers of hypotheses based on hypotheses, where *fact and fiction intermingle in an inextricable confusion.*[22]

In an article on the "Cambrian explosion," *Time* magazine noted that in order to reconcile this sudden appearance of complex forms with the hypothesis of evolution, "scientists delicately slide across data-thin ice, suggesting scenarios that are *based on intuition rather than solid evidence.*"[23] Unfortunately, hypotheses supported only by intuition are accepted as fact by many people and are taught as fact in many science textbooks. True science, however, regards intuition with great suspicion and always subjects ideas and speculations to experimental test.

**A return to superstition.** The widespread acceptance of this "new scientific method," in which imagination, speculation, and consensus take the place of rigorous experimental proof, threatens to pull science back toward the days of superstition. For example, mainstream scientists and science journals now regularly discuss the **Gaia** [gā′ə] **hypothesis,** a quasi-religious idea that states that the earth and all the living things that inhabit it constitute a single living thing. Some proponents of the hypothesis (named after Gaia, the Greek goddess of the earth) even imply that the earth purposely directs the processes of evolution for "her" own benefit. The fact that this philosophy is accepted in many circles as a sound scientific theory demonstrates the direction in which science is sliding.

**The politicization of science.** Although many fields of science are still characterized by an emphasis on proven facts, the widespread acceptance of the Darwinian method of science (speculation and argumentation) has allowed some branches of science to become motivated by politics or ideology instead of by absolute truth. For example, Dr. Stephen Schneider, a former climate researcher at the National Center for Atmospheric Research and the recipient of several scientific awards, stated that for scientists to influence environmental policy:

> We have to offer up scary scenarios, make simplified, dramatic statements, and make little mention of any doubts we might have. . . . Each of us [scientists] has to decide what the right balance is between being [politically] effective and being honest.[24]

As this attitude toward science spreads, it threatens to undermine the very existence of science as a search for truth and order in the universe.

**Conclusion.** Earlier in this chapter we read about the seven assumptions that form the general theory of evolution as listed in *Implications of Evolution* by G. A. Kerkut. At the end of his book, Kerkut lists those assumptions again and summarizes the scientific evidences pertaining to them. It is interesting to read what he says:

1. The first assumption was that non-living things gave rise to living material. *This is still just an assumption.* . . . There is, however, little evidence in favour of biogenesis [spontaneous generation] and as yet we have no indication that it can be performed. . . .
2. The second assumption was that biogenesis occurred only once. *This again is a matter for belief rather than proof.* . . .
3. The third assumption was that Viruses, Bacteria, Protozoa, and the higher animals were all interrelated. . . . *We have as yet no definite evidence about the way in which the Viruses, Bacteria or Protozoa are interrelated.*
4. The fourth assumption was that the Protozoa gave rise to the Metazoa. *This is an interesting assumption* and various schemes have been proposed to show just how the change could have taken place. . . . *Here again nothing definite is known.* . . .
5. The fifth assumption was that the various invertebrate phyla are interrelated. . . . *The evidence . . . is tenuous and circumstantial;* not the type of evidence that would allow one to form a verdict of definite relationships.
6. The sixth assumption, that the invertebrates gave rise to vertebrates, has not been discussed in this book. There are several good reviews on this subject. . . . *Here again it is a matter of belief which way the evidence happens to point.* As Berrill states, "*in a sense this account is science fiction.*"

---

[22]William R. Thompson, introduction to *The Origin of Species,* by Charles Darwin (London: J. M. Dent, 1963), xxi, xxiv. Emphasis added.

[23]J. Madeleine Nash, "When Life Exploded," *Time,* 4 December 1995, 73. Emphasis added.

[24]Stephen Schneider, quoted in Jonathan Schell, "Our Fragile Earth," *Discover,* October 1989, 47.

Thinking about the LIVING CREATION

## Creation, Evolution, and Their Correspondence to Reality

| | A scientist who accepts the Bible would predict | A scientist who rejects the Bible might predict | Modern science has uncovered these facts |
|---|---|---|---|
| **Concerning the universe and the earth** | The universe is orderly, law abiding, and rational. | The universe is disordered, without rhyme or reason. | Order is the hallmark of the universe. Evolutionists cannot explain its lawfulness, but neither can they deny it. |
| | The universe and its matter had a definite beginning. | Either matter is eternal, or the universe sprang into being by solely natural processes. | The universe is "running down"; if it is "running down," there must have been a time when it was "wound up." We have no way of knowing, through science, whether matter is eternal or not. |
| | The earth is relatively young. | The earth is extremely old. | The earth's magnetic field, the absence of meteoritic dust accumulations on the earth, the scarcity of helium in the earth's atmosphere, the lack of certain chemical concentrations in the oceans, etc. suggest that the earth is relatively young. |
| | Various catastrophies have affected the earth (catastrophism). | There has been a uniform continuation of past processes into the present (uniformitarianism). | Evolution must violate uniformitarian doctrine to have occurred. Some evidence of change in the earth's crust seems best explained by catastrophic conditions. |
| **Concerning living things** | Living things are distinct from nonliving things. | At some point, the differences between living and nonliving things are obliterated. | The distinction between living things and nonliving things is clear. The definitions for what constitutes a living thing will have to be rewritten before differences can be obliterated. |
| | There is much common design in organisms (because of a common Designer), with modification in the basic design where needed; however, organisms are always of distinct types (kinds) and thus can be easily classified. | There is similarity in body structure resulting from the evolution of one into another, but a continuum of diverging organisms makes classification impossible. | There is much similarity in basic structure and modification for a particular function. Continuum of intermediate forms is not found. Classification is very possible and is based upon basic body structure involving both differences and similarities in the structures of organisms. |
| | Living things have a limited ability for variation. | The possibilities for variation are unlimited. | The distinct groups of living things observed today are the same ones we find in the fossil record. Breeding experiments (especially in certain plants) have nearly exhausted the possible variations. |
| | Man is a special creation, unique from other organisms. | Man is nothing more than a highly evolved animal. | Man's unique characteristics cannot be denied, and evolution cannot account for them. |
| **Concerning the fossil record** | Distinct types of organisms will be found in the fossil record with no transitional forms. | Many transitional forms connecting one type of organism with others would be expected, forming a smooth continuum with no major gaps. | There are many gaps in the fossil record, transitional forms are absent, and many evolutionists advocate forgetting about the "missing links." |
| | Fossils should be arranged mainly by mobility and habitat. | Fossils should be arranged in a simple-to-complex sequence. | There is no straightforward, simple-to-complex gradation in the fossil record; habitat arrangement has been verified. |
| | Fossils appear abruptly with varying levels of complexity. | Living things should appear as a "trickle" in the fossil record with only "simple" forms in older rock strata. | There is no uniform gradation of fossils from simple to complex in the fossil record. |

7. We are on somewhat stronger ground with the seventh assumption that the fish, amphibia, reptiles, birds, and mammals are interrelated. There is the fossil evidence to help us here, though many of the key transitions are not well documented and we have as yet to obtain a satisfactory objective method of dating the fossils. . . . In effect, much of *the evolution of the major groups of animals has to be taken on trust. There is a certain amount of circumstantial evidence but much of it can be argued either way. . . .* [25]

**Acceptance by faith.** The weight of evidence is clearly not in favor of evolution; therefore, an evolutionist cannot honestly claim that the hypothesis of evolution has been scientifically "proved." A person can accept evolution and believe that it is true if he chooses, but his decision must be based upon something other than the facts of science—*it must be based on faith.* A belief in special creation is no less "scientific" than a belief in evolution, for both are based on faith. It is clear, however, that *the facts of modern science agree marvelously with the biblical account of Creation.*

**Direction in science.** Science, to be of any practical use, is not a backward activity but is an attempt to understand how the world operates in the present and how to use that understanding to benefit man now and in the future. Science deals with the here and now—the real world in which we live. The direction of science is to be forward, making progress in man's subduing of the earth as commanded by God in Genesis 1:28. Evolution threatens to destroy science both by fostering a "new scientific method" based on ideology and speculation and by insisting that science spend its energies trying to prove evolution. As scientific "truth" becomes more dependent on superstition, politics, and ideology, man's ability to fulfill the mandate of Genesis 1:28 can only decline—leaving the cures for cancer and other diseases undiscovered, laws of the universe unknown, and inventions uninvented.

Some people believe that many fields of science, as they are known today, are on a course of gradual decline. Only a revival of spirit among the great minds of the scientific community can turn it around. If this does not come about, the fabric of man's knowledge about the world in which he lives may unravel, leaving mankind in an inescapable tangle of ignorance of God's living creation.

> The more one studies paleontology, the more certain one becomes that *evolution is based on faith alone.*
>
> —*Louis T. More, physicist*[26]

[25]Kerkut, 150–154. Emphasis added.

[26]*The Dogma of Evolution* (Princeton: Princeton University Press, 1925), 160. Emphasis added.

## Section Review 14.4

1. Contrast how the predictions of Creation and evolution correspond to reality.
2. Why have some suggested that the success of Darwinism contributed to a decline in scientific integrity?

### Application

3. Explain why acceptance of either creation or evolution is ultimately based on faith and on one's presuppositions.

### Critical Thinking

4. Why is the evolutionary hypothesis not experimentally verifiable?

### Identify

Gaia hypothesis

## Chapter 14 Review

### Define

1. materialism, naturalism
2. uniformitarianism
3. natural selection
4. eugenics
5. speciation, macroevolution
6. fossil
7. paleontology
8. transitional form
9. punctuated equilibrium hypothesis
10. geologic column, index fossil
11. circular reasoning
12. radiometric dating
13. Cambrian explosion
14. comparative anatomy
15. molecular biology
16. mutation
17. inbreeding
18. embryonic recapitulation
19. Gaia hypothesis

### Identify

1. the scientist whose doctrine of uniformitarianism influenced Charles Darwin
2. the book in which Charles Darwin first published his evolutionary hypothesis
3. the book in which Charles Darwin published his hypothesis concerning the origin of mankind
4. the French scientist who showed that the bodies of man and animals have certain design similarities
5. the 19th-century Austrian monk who discovered laws of genetics that contradicted Darwin's hypothesis
6. the 19th-century German evolutionist who proposed the false idea of embryonic recapitulation

### Explain

1. Explain how natural selection helps suppress harmful changes in a species.
2. What are some of the factors that prevent natural selection from creating new kinds?
3. How do fossils such as *Australopithecus, Homo habilis, Homo erectus,* Neanderthal man, and Cro-Magnon man provide evidence against evolution?
4. Why is it important to the theory of evolution that great numbers of transitional forms, separated by only tiny, gradual steps, be found?
5. Why would an organ have to remain functional at every step in order for it to "evolve"?
6. How do the discoveries of modern molecular biology pose problems for the hypotheses of spontaneous generation and evolution?
7. Why are mutations generally harmful (instead of helpful) to the overall health of an organism?
8. Why can we say that the examples of the peppered moth, pesticide-resistant insects, and antibiotic-resistant bacteria do not demonstrate evolution?
9. Explain why evolution cannot be considered true science.

### Apply

1. Why is the Bible particularly helpful to the scientist when studying the history and future of the physical universe?
2. How did the rejection of biblical truths by Darwin and others contribute to their acceptance of evolution?
3. Why might the 40-year acceptance of *Eoanthropus* (Piltdown man) as a "human ancestor" suggest that fossil interpretations are sometimes based on faith?
4. What are some features or organs mentioned in the text that could not have developed by numerous slight modifications while remaining fully functional at every step? Can you think of any others?
5. Briefly explain Paley's "argument from design." Do you think this argument is any stronger today than it was in 1802?

CHAPTER 15

# Introduction to Ecology

**Key Concepts**

- Levels of ecology
- Ecosystem factors
- Food chains and niches
- Symbiotic relationships
- Nutrient cycles
- Major biomes
- Ecological succession
- Man's role in ecology

You are learning about a fascinating variety of plants, animals, and other creatures in your study of God's living creation. Through this study, you have seen abundant evidence of the work of a wise Designer in the structure and function of the "simplest" bacteria and protozoa. You are also learning to appreciate the brilliant engineering of the more familiar creatures you see every day. The wings of a bird, the gills of a fish, the skeleton of a snake, the silk of a spider—all point to the existence of God the Creator.

God's handiwork is also clearly demonstrated in the complex interrelationships that exist among the millions of living organisms that He created. Everywhere on earth—on the highest snowy mountain peaks, in the depths of the ocean, on desert sands, in tropical rain forests, and even in boiling hot springs—you can find living things existing side by side and interacting with one another. No organism—plant, animal, or microbe—exists alone. Each depends in some way on other living or nonliving things to stay alive. For example, animals must have plants for food and oxygen. Plants, in turn, rely on animals for nutrients and carbon dioxide. God designed plants and animals to benefit each other in this and other ways. The amazing relationships you are about to study in this chapter did not just "happen" by chance. They are yet another irrefutable evidence of a master plan.

The study of how living things interact with one another and with their physical environment is **ecology.** An *ecologist* is a scientist who specializes in studying these interactions and discovering how organisms survive in their environments. The word *ecology,* from the Greek *oikos,* means "the study of houses"; ecology is the study of where organisms live. Ecologists refer to an organism's environment or home as its **habitat.** A trout's habitat is a lake or pond; an earthworm's habitat is the soil. Every habitat includes both living and nonliving things. The living things—animals, plants, fungi, and microbes—are the **biotic factors** of the habitat. The nonliving things are **abiotic factors.** No organism can be completely studied apart from its environment. This is why the study of ecology overlaps with geography, including the distribution and movement of plants and animals around the world.

Ecology is an important area of study for the Christian because it pleases God when we seek to understand His ways and His plans for the world. Whenever you enjoy the outdoors or travel to new places, take advantage of the opportunity to observe God's living creation and to learn more about it. This chapter on ecology will help you to better understand and describe the relationships among living things in your community and in the places you visit.

## 15.1 Levels of Ecology

God designed the earth to support the many different forms of life He created. The relatively thin layer of the earth's surface in which life exists is called the **biosphere.** This layer includes a portion of the *atmosphere* (air), the *lithosphere* (land), and the *hydrosphere* (water). The biosphere is the broadest or most general level of ecology.

Geography, climate, and other conditions vary throughout the biosphere; so if you were to take a

Fig. 15.1

**Levels of ecology**

worldwide tour, you would encounter deserts, grasslands, forests, and many other regions, each with its own characteristic vegetation and animal life. These large geographic regions are called **biomes.** A biome is identified by its **climax vegetation,** the species of plants that dominate the region, and also by the distinctive animal species that dwell there.

Within each biome you would find many individual lakes, marshes, forests, grasslands, and other smaller natural areas, each with its own specific environmental conditions and living organisms. Ecologists call these smaller areas **ecosystems.** The ecosystem is the basic unit of ecology and may be defined as the interrelated network of all organisms and their environment within a limited area. All ecosystems fall into one of two general categories, **terrestrial** (land) and **aquatic** (water), with aquatic ecosystems comprising more than 70% of the earth's surface.

The diverse ecosystems of the world make a fascinating study because God designed specific creatures to thrive in each one. A **community** is all of the living things in an ecosystem. A forest community, for example, includes not only trees and plants but also the animals living there.

Within a community, living things are grouped by individual species. A **population** is a group of organisms of the same species living in the same ecosystem. All the white-tailed deer in a forest or all the sugar maple trees in the forest are examples of populations. The populations in a community are *interdependent:* they depend on each other in some way for nutrients, food, oxygen or carbon dioxide, or protection. Therefore, a change in one population may cause a change in other populations. If the population of hawks in an area increases, the population of rabbits will decrease. If the population of grasses in an area decreases, the deer population will also decrease.

Each population may be further divided into individual living things, or **organisms.** An organism is the smallest living unit of the biosphere. A single white-tailed deer or sugar maple tree is an organism.

Ecologists study the relationships among living things and their environment at all of these levels: biome, ecosystem, community, population, and organism. From their research, we can learn what makes an ecosystem thrive and what harms it. *Ecosystems are constantly changing* due to changes in abiotic (nonliving) factors such as water or nutrients and changes in populations; an ecosystem's **stability** is its ability to withstand and recover from changes. Population size increases as a result of births and *immigration,* or the movement of animals into an ecosystem. On the other hand, population size decreases as a result of deaths and *emigration,* or the movement of animals out of an ecosystem. Often these changes are natural: lightning may cause a grass fire, the arrival of a new species may cause another species to diminish, or a drought may kill much of the vegetation in an area and force many animals to leave. Man also brings change to ecosystems as he uses and develops the land and its resources. Some of these changes are beneficial, and others are not. But we must always remember that God has given man *dominion*, or authority, over creation (Gen. 1:26–28). We have a God-given command to subdue the earth and manage its resources, and with that right comes the responsibility to use our natural resources wisely.

## John Ray: Father of Natural History

Explorers of the LIVING CREATION

There is for a free man no occupation more worthy and delightful than to contemplate the beauteous works of nature and honour the infinite wisdom and goodness of God.
—John Ray, 1660

Mankind has studied nature since the beginning, when Adam named the creatures in the Garden of Eden. But during the Middle Ages, natural science, like many areas of learning, was neglected. In fact, as late as the 1500s, some leaders of the Church of England still taught that attention to the natural world was harmful because it distracted people from religious devotion. But at least one theologian in England thought differently. Anglican minister *John Ray* (1627–1705) believed that nature ought to be studied; in fact, he said that God is pleased when man studies His creation.

Like many pioneers of modern science, Ray was a Christian who recognized nature as the handiwork of the Creator. While recuperating from an illness in 1650, he took up walking about the grounds of Cambridge University. As he walked, he admired the trees, flowers, and other plants he saw. But when he looked for a book describing the local vegetation, he could not find one. Over the next ten years, Ray studied the surrounding area and compiled notes on all of the plants he observed. In 1660, he published *Catalogus plantarum circa Cantbrigiam nascentium* (known today as *Ray's Flora of Cambridgeshire*), describing more than 600 species. It was the first book to catalog the plants of a county so thoroughly, and it set John Ray on his way to fame as the "Father of Natural History" in Britain.

Ray went on to write several more systematic catalogs of plant life as well as works on fish, birds, mammals, and insects. His greatest work, *Historia Plantarum,* described more than 18,600 species of plants. Unlike other naturalists of his day, who classified organisms on the basis of just one feature or use, Ray devised a more natural system of classification that reflected the Divine order he observed in creation. He based this system on the overall structure and function of organisms; for example, when classifying plants, he considered flowers, seeds, fruits, and roots. His approach revolutionized the science of classification and brought order to the study of nature.

John Ray's belief in natural theology also had a profound impact on natural science. *Natural theology* is simply the biblical doctrine that man can learn of the wisdom and power of God by studying His creation. The idea was not a new one—God revealed it to man in His Word long ago—but Ray helped to establish the concept in England and throughout Europe. *The Wisdom of God Manifested in the Works of the Creation* (1691), which he based on a series of sermons he preached at Cambridge, became so popular that several foreign-language editions were produced, and the book still enjoyed a widespread audience fifty years after its publication. It was in this monumental work that Ray advised his readers to open their eyes and study the natural world for themselves:

> Let it not suffice us to be Book-learned, to read what others have written and to take upon Trust more Falsehood than truth: But let us ourselves examine things as we have opportunity, and converse with Nature as well as Books.*

The student of biology can still benefit from this bit of wisdom today. Your textbook is important; it gives you the basic framework of knowledge you need to understand biology. But without observation, the subject will be dry and lifeless. To truly comprehend and appreciate the living creation as God intended, you must get outside and study it with all of your senses. Only then will you begin to grasp the power and wisdom of our amazing Creator and His master plan.

**The Wisdom of God Manifested in the Works of the Creation,* ed. 7 (London: William Innys, 1717), 172.

## The Amateur Ecologist

Observing the LIVING CREATION

You don't need a college degree to start studying the environment around you. You simply need an inquisitive mind, your five senses, and a little basic equipment. Perhaps the most basic piece of equipment is a *notebook*. By making brief notes and simple sketches, you can keep a record of what you have seen and heard for future reference and enjoyment. Write about the creatures you observe—not only their appearances but also their behavior and interaction with each other. Also jot down questions that come to your mind. Often, you will find the answers in your own notes.

You can aid your eyes with a pair of *binoculars* or a *camera* with a zoom lens. Most animals won't let you get very close, but with binoculars or a camera you can zoom in on your subjects without creating a disturbance and can watch them interact with their habitat. You may also wish to build a blind with which to camouflage yourself. You can construct a simple blind with brush and fallen branches. Sometimes a natural blind will be available, such as a shrub or tree; but beware of poisonous plants, such as poison ivy or poison oak. Another helpful tool for observation is a *hand lens* or *magnifying glass* for examining smaller organisms.

Pocket-size *field guides* are an indispensable aid for the amateur ecologist. Use these guides to identify the vegetation and wildlife you observe. *Maps* can be helpful if you are observing in an unfamiliar area or want to locate nearby sources of water, where animals might be found. You can also use a map to mark places where you have observed different plants or animals.

If you plan to collect any specimens such as insects or plant cuttings, you will need an assortment of clear *plastic bags and containers*. Baby food jars, small jam jars, and pill bottles make good containers for insect specimens. Simply punch holes in the lid or cover the opening with a piece of cheesecloth and a rubber band. For collecting and handling specimens, you will want a *butterfly net, tweezers,* a sharp *pocketknife*, and a pair of *gloves*.

Finally, for safety, it is wise to bring along a *compass* and, if possible, a *cell phone*. (Make sure you keep the cell phone muted to avoid disturbing any wildlife you observe.) It is also vital that you tell your parents exactly where you will be observing and when you plan to return. If possible, bring a friend along.

With this basic equipment and the appropriate clothing for crawling around in the wild, you are ready to explore and study the marvels of God's living creation.

## Section Review 15.1

1. How does the meaning of the word *ecology* relate to this branch of science?
2. Under what general category of ecosystem would a freshwater lake fall?
3. What are the six levels of ecology? List them from the broadest to the most specific.
4. Would a hive of bees in a tree be best described as a community, a population, or an organism?

### Critical Thinking

5. At what level of ecology would you place yourself and your classmates if you were describing your school as an ecosystem?

### Identify

ecology, habitat, biotic factor, abiotic factor, biosphere, climax vegetation, ecosystem, stability, John Ray

### Observe

- Notice the different species of plants that grow in your local area. What do you think would be considered the climax vegetation?

## 15.2 Influences in an Ecosystem

In the previous section, we defined an ecosystem as the interrelated network of all the living things and nonliving things within a limited area. The number and variety of species living within an ecosystem is known as the **biodiversity** of the ecosystem, while the ability of the ecosystem to support these organisms is called its **carrying capacity.** For example, a desert has less biodiversity than a forest because it has a smaller carrying capacity due to the heat, scarcity of water, and fewer varieties of plants for food.

*Both biodiversity and carrying capacity depend on several factors.* One important factor is the energy supplied from one organism to the next. The source of energy for every organism is its food. Energy is transferred between organisms as one feeds upon another. When a rabbit eats clover, energy is supplied to the rabbit from the clover. When a hawk eats the rabbit, some of the energy is in turn transferred to the hawk. Another important factor is the recycling of nutrients from the environment to the organisms that populate it and back to the environment again. In the above example, nutrients from the soil pass through the clover, the rabbit, and the hawk; and when the hawk dies and decays, they will return to the soil once again. No ecosystem can sustain itself without this renewing of energy and recycling of nutrients.

Fig. 15.2 **The amount of sunlight that reaches a forest floor affects the growth of plants there.**

### Abiotic Factors

*Every ecosystem consists of both living, or biotic, factors and nonliving, or abiotic, factors.* Plants and animals depend directly on these factors to grow and reproduce successfully. The following are the main *abiotic factors* considered when scientists study an ecosystem:

**Radiation**—*The sun provides all the basic energy of an ecosystem.* This is true because sunlight powers photosynthesis in all green plants, which are the ultimate source of food for almost every creature on the earth. The amount of sunlight an ecosystem receives depends on its geographic location. Because the sun's radiation is more direct at the equator and less direct at the poles, daylight hours, temperatures, and energy levels vary around the globe. Other physical factors also limit radiation. For example, the distance sunlight penetrates into a pond affects where algae live, and the thickness of a forest canopy affects the growth of plants on the forest floor.

**Temperature**—Radiation that penetrates the atmosphere warms the earth's surface. Temperature is very important, for it influences all plant activities: the absorption of water, photosynthesis, transpiration, respiration, germination, growth, and reproduction. Temperature also affects the activity and metabolism of animals. Animals expend much of their metabolic energy regulating their body temperature.

**Water**—Water is required for all life. In fact, living organisms are composed largely of water, and virtually all physiological processes take place in the presence of water, from photosynthesis in plants to digestion in animals. As rainwater drains into streams and streams drain into rivers, nutrients from the soil are carried by the water to other ecosystems, usually moving from higher to lower elevations.

**Atmosphere**—The availability of gases such as oxygen, nitrogen, carbon dioxide, and water vapor affects the survival of plants and animals in several ways. The oxygen levels in an ecosystem, for example, determine the variety and growth of animal populations. Fewer plants and animals live at high elevations because of the thin atmosphere and the lower oxygen levels found there.

**Wind**—Global wind patterns give rise to differences in temperature and rainfall at different latitudes of our planet. Wind also disperses

Fig. 15.3 **Rocky, sandy soil limits the growth of plants in an ecosystem.**

plant pollen and seeds, affecting the distribution of plant species, and produces currents in lakes, streams, and oceans, bringing up nutrients from the bottom. The earth's annual revolution around the sun introduces seasonal changes in wind patterns, temperature, and rainfall.

**Soil**—The type and quality of soil in an area influences the kinds of plants that grow there. The plants upon which man depends for food and oxygen are dependent in part upon the soil. Good-quality soil supplies the minerals needed by food crops. Soil is also a critical factor in an aquatic environment. For example, underlying rock may inhibit the growth of aquatic plants. If the environment does not favor the growth of plants, there will be little food and few hiding places for animal life as well.

Other physical factors that influence ecosystems include weather conditions, climate, the topography of the land, the depth and mineral content of aquatic environments, and natural disturbances such as fire, floods, and earthquakes.

Organisms live within a **tolerance range** for each one of these factors. A tolerance range is a range of values that is needed for the organism's survival. For example, although some bacteria thrive in hot springs and in deep-sea vents at temperatures of 80 °C (176 °F) or higher, most living things have a narrow tolerance range for temperature. It is usually between 0 °C to 50 °C (32 °F to 122 °F). Within an organism's tolerance range for temperature or for another factor is an ideal range called the **optimum range.** In this range, the creature will not only survive—it will thrive. In ecosystems where temperatures are mild year-round, a great variety of living things flourish. Any factor outside an organism's tolerance range becomes a **limiting factor** and may threaten survival. Low food supply, decreasing oxygen supply, lack of space, predators, lack of water, and disease are examples of limiting factors in a population.

## Biotic Factors

*Biotic factors* vary according to the ecosystem. In a mature forest of oak and maple trees, the living factors would include ferns and mosses on the forest floor, along with a variety of trees and shrubs, animals, insects, birds, and organisms such as millipedes, worms, protozoa, decay bacteria, and fungi, which feed on forest litter and humus. Several thousand different species may live together in a single ecosystem.

Fig. 15.4 **Biotic factors in a forest include a variety of plants and animals.**

## Section Review 15.2

1. Why does geographic location affect the amount of sunlight an ecosystem receives?
2. How does altitude limit plant and animal life at high elevations?
3. Describe the influence of global wind patterns on different ecosystems.
4. How does soil affect the animal life in an ecosystem?
5. Contrast the terms *tolerance range* and *optimum range.*

### Application

6. List the biotic and abiotic factors in your home.

### Identify

biodiversity, carrying capacity, limiting factor

# 15.3 Nutritional Relationships in an Ecosystem

## Producers and Consumers

The welfare of an ecosystem depends on the flow of energy and recycling of nutrients. God has built these important natural processes into His creation. The Bible tells us that

> He sendeth the springs into the valleys, which run among the hills. They give drink to every beast of the field.... He watereth the hills from his chambers: the earth is satisfied with the fruit of thy works. He causeth the grass to grow for the cattle, and herb for the service of man.... The young lions roar after their prey, and seek their meat from God. —Psa. 104:10–14, 21

All living things require a continuous supply of energy to sustain life processes. Organisms that make their own food using an inorganic (nonliving) energy source are called **autotrophs** or *producers. Photoautotrophs,* like green plants, algae, and cyanobacteria, are autotrophs that obtain energy from sunlight. *Chemoautotrophs* are autotrophs that obtain energy from chemical substances rather than from sunlight. For instance, sulfur and nitrogen are used by some bacteria to get energy.

All other organisms in an ecosystem are called **heterotrophs** or *consumers.* They cannot produce their own food and must obtain energy from organic sources (living things), usually by eating other organisms. **Herbivores** are consumers that feed on plants. Herbivores include deer, rabbits, and songbirds. **Carnivores** are consumers that feed on other animals. Hawks, lions, and snakes belong to this category. Some carnivores, called *scavengers,* do not kill their own food but feed on dead animals instead. Consumers that feed on both plants and animals are **omnivores.** For example, bears eat both berries and fish. Opossum, vultures, buzzards, and raccoons are other examples of omnivores.

When producers and consumers die, their remains provide food and energy for two special groups of organisms, decomposers and detritivores [dĭ·trī′tə·vôrz′]. These organisms recycle nutrients back into the soil by breaking down dead organic matter, called *detritus* [dĭ·trī′təs]. **Decomposers** are bacteria and fungi that feed on detritus by secreting enzymes that break it down into simpler molecules. The simple molecules not absorbed by the decomposers become part of the soil and are available for uptake by plant roots. Like decomposers, **detritivores** feed on detritus. They cannot actually convert the detritus into a soil component as decomposers do, but they help to break it down into smaller pieces and speed up the process of decay. Detritivores include crabs, earthworms, and soil insects like ants. All ecosystems include detritivores and decomposers. They play a vital role in the process of recycling nutrients back to the soil to be used by plants.

## Trophic Level

Energy flows through an ecosystem as consumers eat producers and other consumers. We know that in nature hawks eat weasels, weasels eat rabbits, and rabbits eat plants. Each organism in this process represents a **trophic** [trŏf′ĭk] **level,** a classification that describes its feeding relationship to other organisms in its ecosystem. Plants, being the producers, represent the *first trophic level.* Herbivores, like rabbits, represent the *second trophic level* because they feed on plants. Carnivores like weasels feed on herbivores, rabbits in this case, and therefore represent a *third trophic level.*

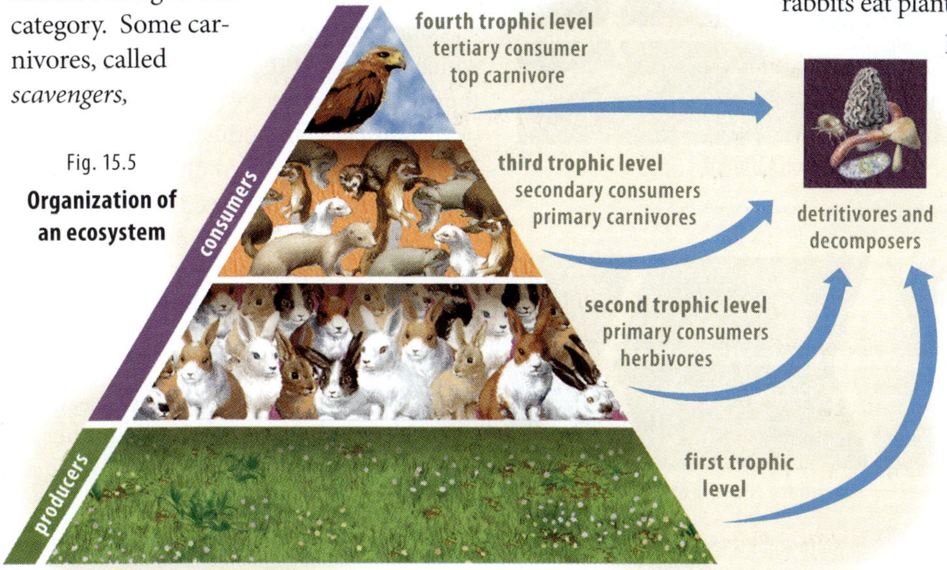


Fig. 15.5 **Organization of an ecosystem**

In a *fourth trophic level,* another carnivore, in this case a hawk, eats the weasel. Often the carnivore at the top of the food chain is called the **top carnivore.** In our example, the hawk is the top carnivore.

## Food Chains and Webs

A **food chain** is a model used by ecologists to show the nutritional relationships among organisms in an ecosystem. It is a one-way series of steps showing the transfer of food and nutrients from one trophic level to another. *Living things are classified into trophic levels by their position in a food chain;* each trophic level is simply a step in the food chain. At the bottom of the food chain, the first trophic level, we find the producers.

Fig. 15.6 **Food chains**

Producers are typically green plants, which supply the whole ecosystem with food through photosynthesis. At the second trophic level, we find herbivores. Because they eat plants directly, they are called **primary consumers.** Then, at the third trophic level, we find animals that eat herbivores. They are called **secondary consumers.** We can also find tertiary or quaternary consumers if the chain has higher trophic levels.

Nutritional relationships are usually more complex than what the simple food chain shows. Food chains often overlap with one another because most animals feed on two or more species. A **food web** is a model used by ecologists to show all possible feeding relationships at each trophic level. A food web consists of several interconnected food chains involving many different species. In most cases, food webs involve tens or even hundreds of species. Figure 15.7 shows how food chains interconnect to form a food web. Some consumers may feed at different trophic levels and be involved in more than one food chain.

## Ecological Niches

Food chains and webs help us understand much about the relationships of organisms in an ecosystem, but living things do more than just feed or provide food for others. The function or "occupation" of a living thing is its **niche.** A trophic level refers specifically to the position of an organism in the food chain, but a niche includes the organism's entire life: its habitat, its relationship to other living things, and its role in the ecosystem.

For example, a live oak is a producer, which places it at the first trophic level. It provides food for a variety of insects, birds, squirrels, and other consumers. But the live oak serves multiple purposes within its ecosystem. Besides providing a source of food for animals, the tree

- serves as a home for other organisms. Birds and squirrels nest in its branches, various insects live in its bark, and other insects and microbes live in the soil among its roots.

- provides support for creeping vines and plants like Spanish moss.
- absorbs water from the soil and releases water vapor into the air through transpiration, contributing to the water cycle.
- uses the energy from sunlight to produce new wood, leaves, and acorns through photosynthesis.
- provides shelter from the sun for shade-loving plants and fungi.
- contributes dead leaves, twigs, and acorns to the forest floor as detritus, enriching the soil with nutrients.

All of these functions compose the niche of the oak tree and help us to better understand its importance to the ecosystem.

Fig. 15.7
**Food webs**

## Energy Flow and the Ecosystem

You can see now how producers and consumers depend on each other for energy and nutrients. Plants use energy from the sun to produce food through photosynthesis. Primary consumers obtain energy by eating plants, secondary consumers obtain energy by eating those animals, and so on up the food chain.

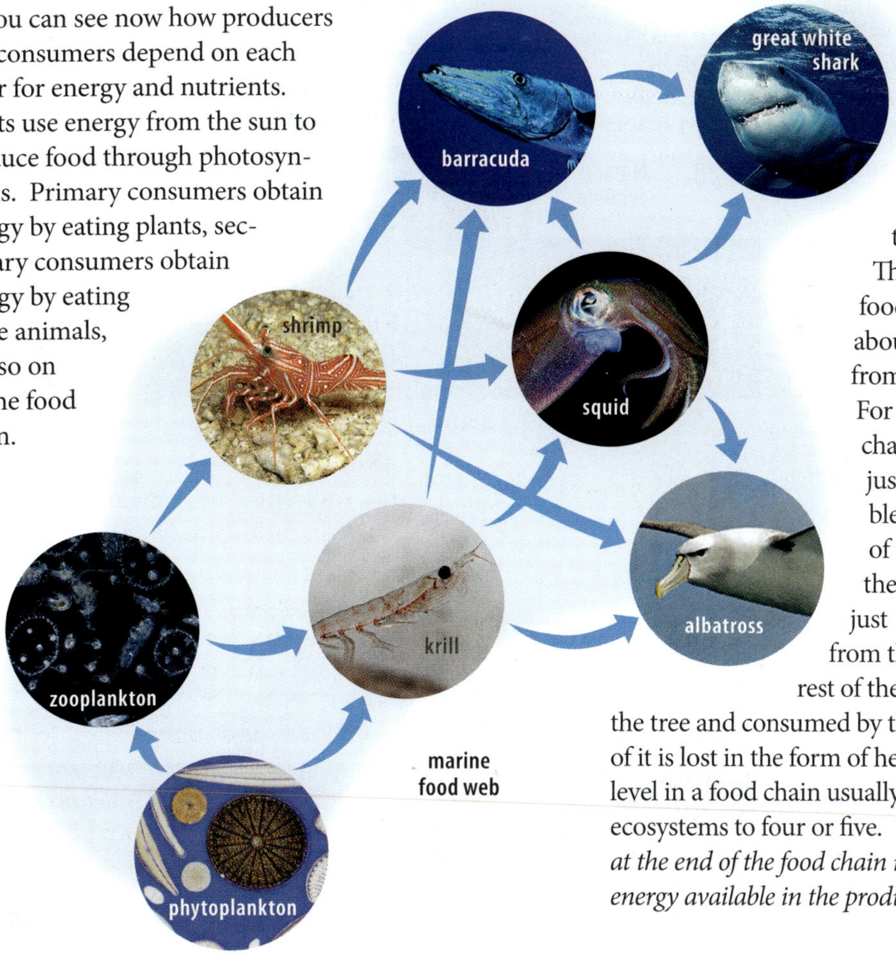


*Not all of the solar energy that reaches the earth's surface enters the food chain,* however. Plants can capture and store only about 1% of the solar energy that falls on their green leaves and stems. The successive members of a food chain, in turn, process only about 10% of the available energy from the previous trophic level. For example, in the grazing food chain in Fig. 15.6, the tree stores just 1% of the solar energy available to it; the deer stores just 10% of the energy it consumes from the tree leaves; and the wolf stores just 10% of the energy it consumes from the deer. What happens to the rest of the energy that was captured by the tree and consumed by the deer and the wolf? Most of it is lost in the form of heat. The loss of energy at each level in a food chain usually limits the number of levels in ecosystems to four or five. *The amount of energy available at the end of the food chain is only a small percentage of the energy available in the producer.*

## Pyramid Diagrams

One helpful way to represent information about an ecosystem is to construct a *pyramid diagram.* Ecologists use three types of pyramid diagrams to illustrate the transfer of energy and the distribution of organisms at different trophic levels in an ecosystem. In each pyramid, producers form the base and consumers form successive levels above it.

An **energy pyramid** shows the energy transferred from one trophic level to the next. Remember, producers capture and store only about 1% of the solar energy they receive and, on average, only 10% of the energy at one trophic level is used by the next trophic level. Therefore, energy pyramids are always "right-side up." They have a large energy base at the bottom, and each successive layer is smaller than the one below it.

A **biomass pyramid** represents the total mass of living matter per unit area, or **biomass,** at each trophic level in an ecosystem. Producers make up most of the biomass, with the pyramid narrowing at each successive trophic level, with fewer carnivores at the top. *Some biomass pyramids are exceptions because they are "inverted"* or top-heavy with the smallest biomass on the bottom. For example, in ocean or lake ecosystems the primary producers *(phytoplankton)* have less biomass than the primary consumers *(zooplankton).* This is possible because the phytoplankton grow and reproduce very rapidly, providing food for the much greater biomass of zooplankton.

A **number pyramid** represents the number of organisms at each trophic level and shows that population size usually decreases at each higher trophic level. Most are "right-side up," with the greatest numbers at the base and smaller successive levels similar to those of energy pyramids. There are more producers than primary consumers, more primary consumers than secondary consumers, and so on. For instance, thousands of plants are needed to feed tens of rabbits that will feed just one red fox. Some number pyramids are inverted, however, such as a single acacia tree that supplies food for many primary consumers (mainly ants).

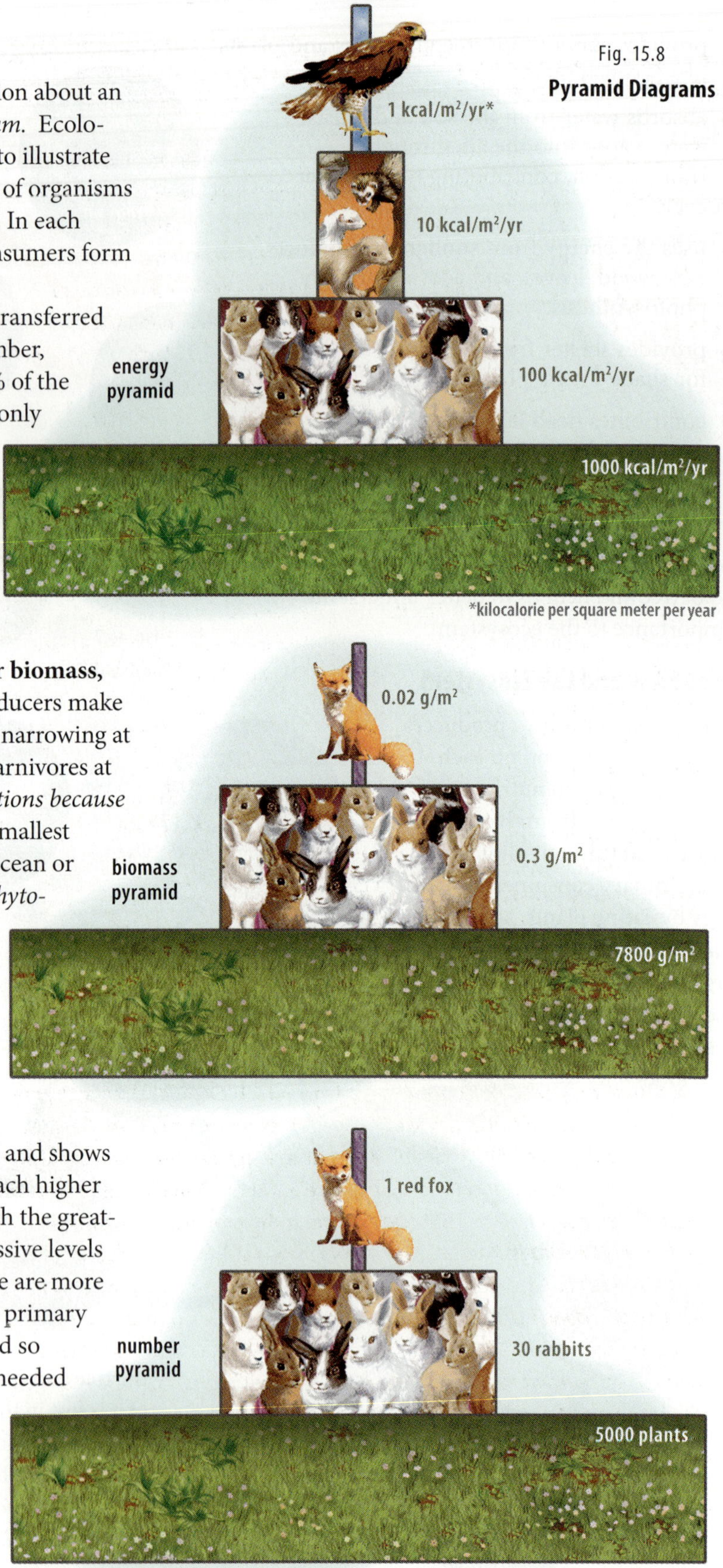


Fig. 15.8 **Pyramid Diagrams**

# Explore a Miniature Ecosystem

Observing the LIVING CREATION

A flower garden is a convenient ecosystem in which to practice your observation skills. Flowers and other blooming plants attract a wonderful variety of birds and insects. As you observe these creatures flitting, fluttering, and crawling about the garden, you will begin to see that they provide an important service to this miniature ecosystem. Nectar feeders such as butterflies, bees, and hummingbirds, for example, help to pollinate plants while they feed. If you watch closely, you can see the golden pollen collect on the legs of a bumblebee or the wings of a butterfly.

These beautiful creatures are not the only residents of this miniature ecosystem. Step up to a shrub or a flower bed and you will see a host of other garden dwellers, ranging from beetles and snails to ants tending herds of tiny aphids. The closer you get, the more organisms you will see. The plants and soil of a flower garden are teeming with a variety of life. It is an ecologist's job to study these different species and to learn how they all live together.

To lure nectar feeders into your garden, plant honeysuckle, hibiscus, aster, cosmos, and zinnia, or try wildflowers typical of the area where you live. Some trees and shrubs also attract nectar feeders; examples include lilac, tulip tree, citrus trees, and azalea. Since pesticides will not encourage insect life in your garden, use pesticides sparingly and try to use substances that deter insects rather than kill them.

Another thing you can do to create a desirable habitat is to provide a source of water. This can be achieved with a birdbath, a fountain or pond, or even a rain barrel.

Now that you have created an inviting garden ecosystem, it is time to observe and learn. Using a field guide and a notebook, keep a log of the different species you see. Note which plants they seem to prefer, what time of day they are most active, and how they interact with one another. Don't forget to write about the predators you discover, such as praying mantises and garden spiders, stalking their prey or waiting patiently at the edge of a web. Flying predators, such as yellow jackets and birds, may also make an appearance.

Besides keeping a journal, you may want to collect some specimens—insects and other garden creatures—for closer observation. Butterflies and other flying insects can be captured with a butterfly net. One good way to collect crawling specimens is with a simple pitfall trap. Simply bury a large glass jar (at least 6 inches deep) in the soil so that its rim is at ground level. If you erect a short section of garden-edging material leading to your pitfall trap, many wall-hugging insects will be directed right into the trap. Creatures that fall into the jar will be unable to climb the smooth glass walls to escape. Place four stones around the mouth of the jar and lay a piece of wood or tile over them to keep predators out and to protect your specimens from the elements. You may want to bait the traps with fruit, cheese, or meat. By varying the location of the traps and the type of bait, you can collect a variety of specimens.

Even the smallest rose garden is a living ecosystem, complete with its own habitats and niches. The same is true of vegetable gardens, orchards, and front lawns. It isn't hard for the ecologist to find a place to study: all he has to do is step outside his front door. Make it your goal to start an ecosystem study this week somewhere close to home. You will be surprised by the variety of living things that you discover.

## Section Review 15.3

1. Explain the relationship between producers and consumers.
2. How do the feeding habits of herbivores, carnivores, and omnivores differ?
3. What purpose do decomposers serve in the food chain?
4. How does a food chain differ from a food web?
5. Approximately how much of the sun's energy that strikes a plant is actually captured and stored by the plant?
6. Why are energy pyramids always "right-side up"?
7. How does a biomass pyramid differ from a number pyramid?
8. Could a number pyramid be inverted? Explain.

### Application

9. Describe man in terms of his nutritional relationship to the ecosystem. Is he a producer or a consumer? A herbivore, carnivore, or omnivore?
10. Make a number pyramid representing a small, family-operated farm with the following populations: a family of four, five thousand plants providing feed grain for livestock, and six head of cattle.

### Critical Thinking

11. At what trophic level in man's food chain are plants that produce grain, such as wheat, barley, and oats? At what level is man in relationship to these plants?
12. What group of organisms would make up the largest level of a biomass pyramid for a grassland ecosystem? What kind of organisms might make up the second-largest level?

### Identify

autotroph, heterotroph, detritivore, food chain, niche, energy pyramid, biomass

## 15.4 Special Nutritional Relationships and Nutrient Cycles

### Special Nutritional Relationships

**Symbiosis.** If you carefully observe an ecosystem, like a forest or a pond, you will find that God designed organisms to interact in a variety of ways. Many of them live in close association with one another in a type of relationship called **symbiosis** ("living together"). Symbiosis is defined as a close relationship between two different species over a period of time. The effects of this relationship may be positive, negative, or neutral. In many cases, species depend on symbiotic relationships for survival. Three forms of symbiosis often seen in nature are *mutualism, parasitism,* and *commensalism.*

**Mutualism** is a relationship in which both organisms benefit. You studied an example of mutualism in chapter 5 when you learned about lichens. Each lichen is a combination of a fungus and an alga. Through photosynthesis, the alga provides food for the fungus, while the fungus provides water and nutrients for the alga. The following are just a few of many more examples of mutualism in nature:

Fig. 15.9 **Lichen**

- The rhinoceros is rarely seen without tickbirds (oxpeckers) perched on its head and back. The ticks that infest the rhino's hide provide food for these birds; not only do the birds clean the rhino's hide but they also provide warning of danger.
- A termite can digest the woody cellulose in plants because of a tiny protozoan that lives inside the insect's intestine. The termite provides a habitat for the protozoan, and the protozoan secretes enzymes that enable the termite to digest its food.
- A fascinating case of mutualism has been described between certain species of ants and acacia trees. The ants live in the large hollow

Fig. 15.10 **Mutualism: rhinoceros with tickbirds**

thorns of the tree and feed on special modified leaves. The acacia tree produces new leaves year-round to provide food for the ants. The ants, in turn, continually patrol the leaves and branches of the acacia and immediately attack any herbivore that tries to feed on the tree. Not only that, but the ants also destroy any plant that touches the tree, clearing up the area around it and providing more space for tree growth. Scientists have observed that when they removed the ants from the trees, the acacias grew less and very often died.

Which is the more logical explanation for these amazing examples of symbiosis—the design of an intelligent Creator or evolution? Could chance produce such complex yet mutually beneficial relationships? Consider the termite. Which came first, the termite or the protozoan within its intestine? Such questions cannot be satisfied by any answer other than the existence of a purposeful Designer. God designed these creatures to live together and benefit from one another.

Fig. 15.11 **Commensalism: remora on the underside of a shark**

**Parasitism** is a relationship between two organisms in which one, called the *parasite,* benefits while the other, called the *host,* is harmed. The parasite lives on or in a living host, drawing energy and nutrients from it. The host is harmed or weakened but usually not killed. Mosquitoes, for instance, draw blood from the bodies of animals and humans without killing them in the process. Other examples include ticks, tapeworms, roundworms, fungi, and some bacteria.

**Commensalism** is a relationship in which one organism is benefited and the other is neither harmed nor helped. Some birds use tree branches for nesting sites. The trees benefit little, but they are not harmed. The remora is a small fish that hides on the belly of a shark, where it is protected from predators and eats scraps from the shark's meal. Orchids, mosses, and ferns grow on the limbs of large plants without harming the plants. Spanish moss, a flowering plant that hangs from the branches of oak trees, is benefited by the support provided by the tree, but it neither harms nor helps the tree.

**Other special relationships.** Several other special nutritional relationships can be observed among organisms in an ecosystem. Some of these are predation, competition, amensalism, neutralism, and herbivory. **Predation** involves a predator-prey relationship. One organism, the *predator,* hunts, kills, and eats another, the *prey.* A hawk capturing a rabbit, a great blue heron diving for a fish, and a cat hunting mice are examples of predation.

Fig. 15.12 **Predation: blue heron catching a fish**

**Competition** is a relationship in which two organisms compete for the same limited resources such as food, water, and shelter. The more limited the resource, the greater the competition. Hawks and eagles are both birds of prey, and they compete with each other for food such as small rodents. Weeds and carrots compete for water and nutrients from the soil.

Fig. 15.13 **Herbivory: cattle and sheep grazing**

**Amensalism** is a relationship in which one organism inhibits another. Many molds produce antibiotics that inhibit bacterial growth, but the mold is neither helped nor harmed by the presence of the bacteria.

In **neutralism,** organisms share only an indirect relationship. For example, the bobcat and the plants that share its habitat do not affect each other directly. However, an indirect relationship between bobcats and plants has been observed. The bobcat feeds on wild rabbits. Plants are the main food of rabbits; therefore, the bobcat depends indirectly on the plants that fatten its prey. Most species have neutral relationships with most other species in their habitat.

**Herbivory** is a relationship in which grazing animals feed on plants by cropping portions of the plant without killing it. Animals such as deer, elk, sheep, and cattle are all examples of herbivores.

## Nutrient Cycles

Food chains and energy pyramids demonstrate that *energy moves in only one direction in an ecosystem:* from the producers, the first trophic level, upward through successive trophic levels (primary consumers, secondary consumers, and so on). However, you will also notice that God provided a means of replenishing that energy: *photosynthesis.* In addition to designing this constant supply of energy for His creation, God set in motion self-perpetuating systems to recycle matter in the earth's ecosystems. He devised, for example, *nutrient cycles,* or **biogeochemical cycles,** to recycle minerals and other nutrients in the environment. Organisms depend on these nutrient cycles for survival. For example, green plants require carbon, hydrogen, and oxygen to manufacture carbohydrates. They also require nitrogen, phosphorus, and other minerals for growth. Three types of nutrient cycles are generally recognized: the hydrologic cycle, atmospheric cycles, and sedimentary cycles.

In the **hydrologic cycle,** water moves from the soil and from the water surfaces of the earth, through the atmosphere, and then back to the earth again. Three major processes—**evaporation, condensation,** and **precipitation**—continuously cycle water between the earth and the atmosphere (Fig. 15.14). Liquid water from the earth's surface evaporates and enters the atmosphere as water vapor. The water vapor condenses in the air to form clouds made of small droplets of water. These small droplets of water then come together to form larger droplets of water that return to the earth as precipitation in the form of rain, snow, or ice. Precipitation accumulates in lakes, rivers, and oceans, where the water eventually evaporates and returns to the atmosphere once again.

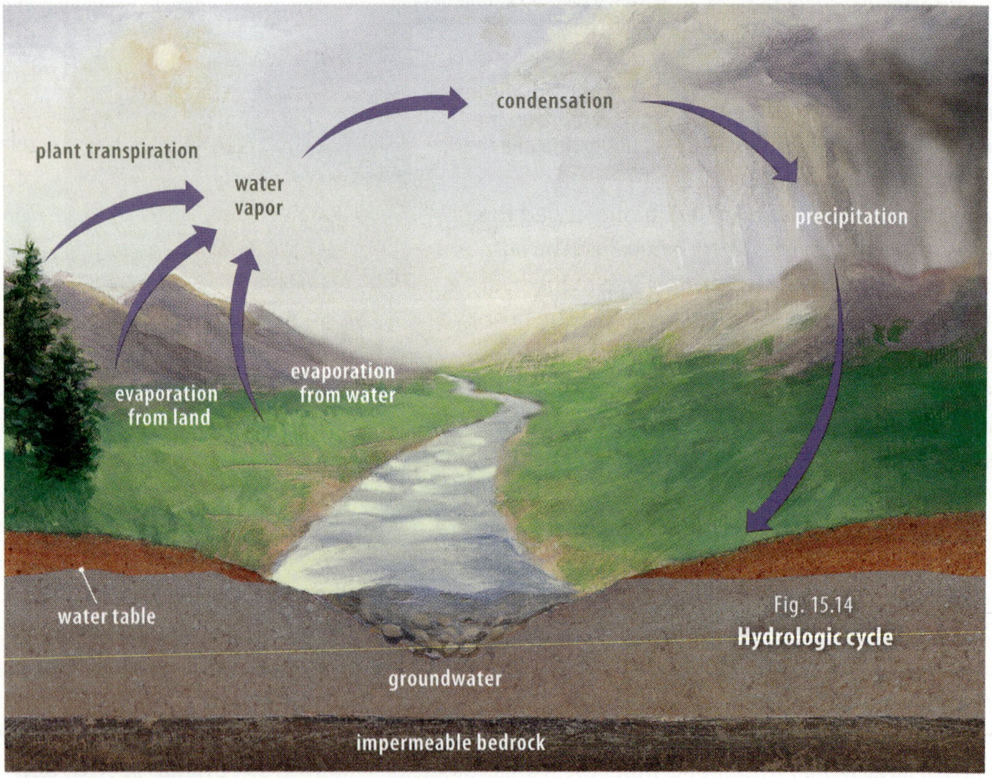

Fig. 15.14
**Hydrologic cycle**

**Atmospheric cycles** include the carbon-oxygen cycle and the nitrogen cycle. Atmospheric cycles involve nutrients found mainly in gaseous form in the atmosphere. The *nitrogen cycle* is described in detail in chapter 1. In the **carbon-oxygen cycle** (Fig. 15.15), the elements carbon and oxygen are recycled through people, animals, and plants. Like oxygen, carbon is essential for all life; it is a necessary component of carbohydrates, lipids, proteins, and many other important molecules. Plants use carbon dioxide ($CO_2$) and water ($H_2O$) to form carbohydrates during photosynthesis and, in the process, release oxygen ($O_2$) to the atmosphere. People and animals take in oxygen from the air and from the carbohydrates produced by green plants and release carbon dioxide and water as waste products of respiration. Carbon and oxygen continually cycle between plant and animal life.

**Sedimentary cycles** involve nutrients found in the soil and rocks of the earth's surface. The **phosphorus cycle** is one of the main sedimentary cycles. Phosphorus moves from the soil to producers and then to consumers. It returns to the soil when dead plants and animals are decomposed by the action of detritivores and decomposers. Phosphorus is required for the growth and development of living things and is needed for the formation of nucleic acids. *The sedimentary and hydrologic cycles together transport most minerals through all terrestrial and aquatic environments.*

In all of these cycles, we once again find evidence of the master plan devised by our all-wise Creator. Everything that He created and every natural process or cycle that He set in motion was carefully designed to reveal His majesty and power (Rom. 1:20).

Fig. 15.15
**Carbon-oxygen cycle**

## Section Review 15.4

1. Contrast mutualism and parasitism.
2. Heartworms are organisms that live in a dog's circulatory system. The worms benefit from the relationship, but the dog is harmed. What term would best describe this symbiotic relationship?
3. What type of nutritional relationship does a lion have with an antelope?
4. What is a nutrient cycle?
5. Name some of the resources and nutrients that are recycled by the various nutrient cycles.
6. What three major processes are involved in the hydrologic cycle?

### Application

7. Why would the relationship between a predator like a wolf and a plant like clover be considered neutral? Could these organisms be indirectly related, and if so, how?
8. How is a farmer's relationship with his crops mutually beneficial?
9. List some of the biotic factors involved in the carbon-oxygen cycle and explain their particular roles.

### Critical Thinking

10. Think back to chapter 1, where you learned about the nitrogen cycle (p. 8). What are some biotic factors involved in the nitrogen cycle? Are there any abiotic factors to be considered?

### Identify

symbiosis, commensalism, competition, amensalism, herbivory, biogeochemical cycle

## 15.5 Exploring Earth's Biomes

You learned earlier that a *biome* is a large geographical region of the world with a characteristic *climax vegetation* and distinctive animal populations. The term **climax community** is often used to refer to the combination of a region's climax vegetation and its animal populations. Each biome has its own unique climax community.

Figure 15.16 shows the distribution of the major biomes in the world. Ecologists use mainly plant life to classify biomes because plants are stationary and do not move like animals do. *Each biome is unique because of its characteristic climate and geography.* Different biomes have different climates because they are located at different

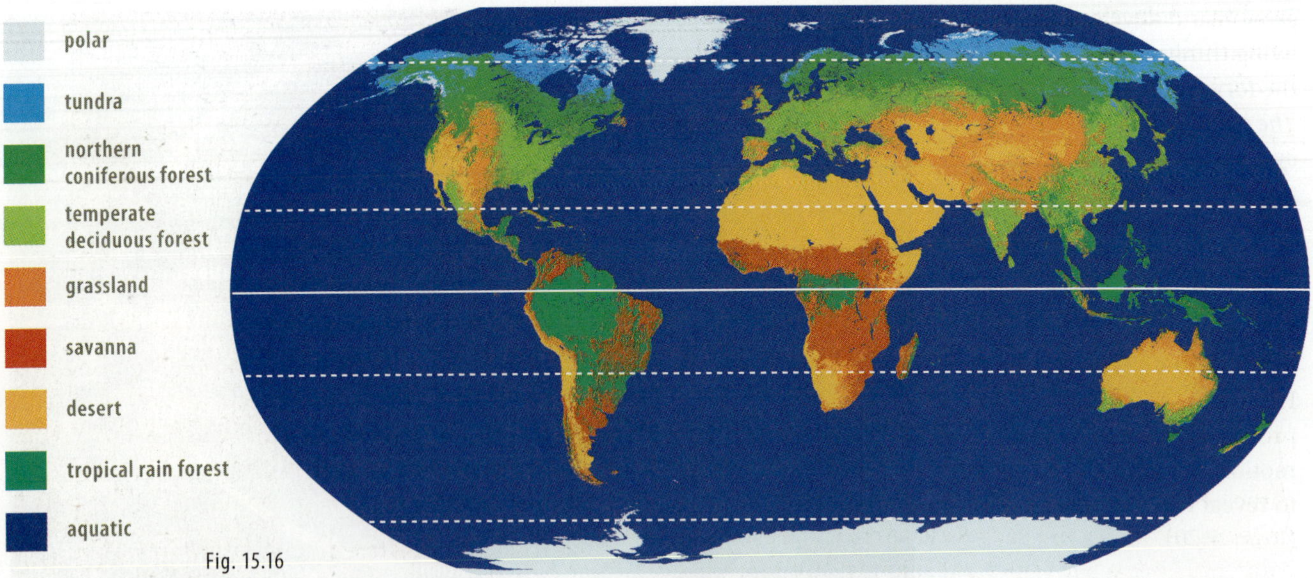


Fig. 15.16
**Distribution of major biomes**

altitudes and latitudes and also because worldwide wind circulation patterns produce regions with differences in humidity, temperature, and precipitation. Geographical differences include topography, the presence of large bodies of water, and soil variations. Together, these geographic and climatic factors limit the kinds of plants and animals that live in a region. The relative influence of these factors helps explain the different climax communities found in each biome.

As we study each major biome, notice how it compares and contrasts with the others. Also note the climate, geography, vegetation, and other characteristics that make it unique.

Fig. 15.17 **Arctic tundra**

## Tundra

If we start our journey at the North Pole, the first major biome we encounter outside the polar ice cap is the **Arctic tundra,** which extends across the northernmost parts of North America, Europe, and Asia. This biome has long, harsh winters, with freezing temperatures, snow, ice, and high winds. Winter days are short, and in some places the sun sets almost as soon as it rises. Summers are cool, short, and dry. Summer days, however, are long and sunny because the North Pole leans toward the sun during this season. A variety of colorful mosses, lichens, grasses, and dwarf shrubs grow and cover the ground during the summer. There are no trees on the tundra because the growing season is very short and the **permafrost,** a permanently frozen layer of soil 2 to 3 feet below the surface, does not allow roots and water to penetrate deeply into the soil.

Reindeer, muskoxen, lemmings, arctic hares, and caribou feed on the tundra vegetation, and predators like wolves and arctic foxes prey upon them. You can find a great number of insects on the tundra during the summer. It is common to see clouds of mosquitoes and black flies.

Tundra biomes also exist far away from the North Pole. At high mountain altitudes, we find the **alpine** tundra, which has vegetation and climatic conditions similar to those of the Arctic tundra, although the soil has better drainage and lacks permafrost.

## Northern Coniferous Forest

South of the tundra, we find the **northern coniferous forest,** also called *boreal forest* or *taiga,* which extends throughout much of Canada and the northern United States and through the northern parts of Europe and Russia. As in the tundra, the climate of the northern coniferous forest is harsh, with long winters and heavy snowfalls. However, trees prosper in this biome because the summer is longer and permafrost is rare. In fact, conifers, especially white and black spruce, are the dominant vegetation in this biome. Other conifers like pine, hemlock, larch, and fir are also common. Fire is one important abiotic factor in the taiga. Natural forest fires are common during the summer because of the dry air and the flammability of the conifer trees. Unfortunately, forest fires are often caused by irresponsible human activity. As good stewards of God's creation, we should always be cautious and aware of the dangers of fire when visiting natural areas.

Trees provide shelter for more diverse and abundant animal life. Great flocks of migratory birds spend their summers here, at the end of which they migrate further south for the winter. Many taiga animals are particularly large compared to similar animals in warmer climates. Here we find the *moose,* which is the largest of all deer; the *wolverine,* the largest member of the weasel family; and the *capercaillie* or *cock-of-the-wood,* the largest of all grouse. Their larger size allows them to conserve heat because the area of their skin is small in relation to their whole mass. Therefore, the animals lose heat more slowly.

Fig. 15.19 **Northern coniferous forest**

The *temperate rain forests* found along the West Coast in Washington, Oregon, and northern California are also part of this biome, though the climate is warmer and wetter there. The *southern pine forests,* which dominate the southeastern states along the Atlantic and Gulf Coast regions, are another variation of this biome.

Fig. 15.18 **Altitude / latitude biome comparison**

## Temperate Deciduous Forest

Further south, we come to the **temperate deciduous forest,** which extends from the East Coast of the United States to the border of the grasslands in the Midwestern states. This biome also covers much of Europe and eastern Asia. Seasons are well-defined here, with cold winters and hot summers. The generous amount of precipitation in this biome is sufficient to support large trees and diverse animal species.

This region is dominated by a great variety of deciduous broadleaf trees, which lose their leaves completely each year. Maples, oaks, hickories, birches, beeches, basswoods, ashes, black locusts, and poplars produce a rainbow of color as their leaves die each fall, providing the scenic landscapes that characterize the season. Interspersed among the deciduous trees, you will find conifers such as hemlock, pine, spruce, and fir.

Spring flowers are a particularly beautiful characteristic of temperate forests. These flowers come from herbs that grow and bloom on the forest floor before the leaves of the trees grow to block the sunlight.

Yet another important characteristic of this biome is the great amount of decaying deadwood it contains. In some cases, almost half of the biomass of an undisturbed forest is deadwood. A rich diversity of microorganisms, fungi, and insects thrive in the decomposing leaves, twigs, and trunks that lie on the forest floor.

Fig. 15.20 **Temperate deciduous forest**

Some of the herbivores you might see in a temperate forest are white-tailed deer, squirrels, mice, rabbits, and many birds, including blue jays and woodpeckers. You can also find carnivores like the owl, weasel, and fox and omnivores like the bear and opossum.

## Grassland

The rich and beautiful *prairies* and *plains* of North America, whose fertile soils are almost without rival in the world, are part of the **grassland** biome. The *savannas* of Africa, South America, and Australia; the *steppes* of Russia; and the *pampas* of Argentina also belong to this category. As you probably guessed, grasses are the dominant plants in this biome. When the leaves and stems of grasses die each year, they add rich organic matter, which accumulates quickly in the topsoil. The grass roots survive the winter, allowing the plants to grow back in the spring. Fire is important in the grasslands because it promotes the recycling of nutrients by burning the dead plant material that has accumulated on the surface of the soil. Most plants in this biome are designed to survive fire and contain underground parts that enable them to return after a fire. The low amount of rainfall and high rate of evaporation hinders the growth of forests here, but trees are found near rivers, lakes, and other sources of water.

Not long ago the North American prairies were the home of millions of bison and pronghorn. The *bison* were a main source of food and clothing for Native Americans, but unfortunately they were almost exterminated when settlers arrived and started trading their meat and hides. The *pronghorn,* the fastest animal in the Western Hemisphere, can run up to 60 miles per hour. Like the bison, it was almost exterminated. Careful management of these animals has helped to increase their numbers dramatically, and they are no longer in danger of extinction. Another curious animal of the prairies is a rodent that barks like a dog: the *prairie dog.* Prairie dogs live together in large, elaborate communities or "towns" consisting of underground tunnels and chambers that sometimes extend for miles. Wolves, coyotes, eagles, and hawks are common predators that feed on prairie animals.

Fig. 15.21 **Temperate grassland**

Fig. 15.22 **Savanna (tropical grassland)**

A very special form of grasslands found not in temperate regions but in the tropics is the **savanna.** Savannas contain some of the fastest and fiercest predators in the world—lions, leopards, wild dogs, and cheetahs. The largest savanna in the world is found in Africa. Trees are widely distributed here, and grasses grow rapidly, providing food for zebras, wildebeests, giraffes, Cape buffaloes, and antelopes, among many other animals.

## Deserts

If you live in the western United States, particularly the Southwest, you are probably familiar with the desert biome. **Deserts** are areas where more water is lost through evaporation than is gained from rain. These are the hottest and driest biomes of the world. The world's largest desert is the *Sahara,* with 3.5 million square miles of mostly barren land covering much of North Africa. God designed perennial desert plants with the ability to minimize their loss of water in order to survive harsh, dry conditions. The cactus, for example, does not have noticeable leaves but rather has green stems designed to store water and carry out photosynthesis. Rain does fall occasionally in a desert, providing water for annual plants, which germinate quickly and grow rapidly, adorning the desert with a spectacular display of colorful flowers. Animal life includes numerous insects, birds, rodents,

Fig. 15.23 **Desert**

reptiles, and arthropods such as the scorpion. Tortoises, diamondback rattlesnakes, coyotes, hawks, roadrunners, and the kangaroo rat are common desert animals.

## Tropical Rain Forest

As we continue south, we come to the **tropical rain forests,** found near the equator in America, Africa, and Asia. These hot and humid regions receive abundant rain and have a year-round growing season. Biodiversity is at its maximum in this biome; more than half of the world's species of plants and animals are found in the rain forests. The creative power of God is evident here in the astounding and beautiful variety of plants and animals. Evergreen broadleaf trees create a dense canopy of leaves that blocks sunlight from the forest floor. Vines climb hundreds of feet to reach the sunlight at the treetops. In the shadow of the forest canopy, orchids, ferns, and shrubs thrive. Elephants, tapirs, gorillas, and other large mammals feed on the leaves close to the floor of the forest. Jaguars in South America and clouded leopards in Southeast Asia are constantly hunting for prey. Colorful birds of incredible beauty fill the trees of the rain forest, including toucans, quetzals, macaws, and birds of paradise. In fact, many of the creatures that inhabit the rain forest are *arboreal*

Fig. 15.24 **Tropical rain forest**

(living in trees), including the jaguars and leopards mentioned earlier, as well as a variety of monkeys and other primates, lizards, and snakes.

## Aquatic Biomes

All of the biomes we have explored so far are terrestrial. However, two thirds of the earth's surface is covered with water. To continue our journey, we would have to don scuba gear or climb into submersibles, small research submarines. **Aquatic biomes** are larger than the terrestrial biomes and show great variations in temperature, dissolved nutrients, gases, and sunlight penetration. Marine ecosystems and freshwater ecosystems are the two major divisions of aquatic biomes. *High salinity distinguishes marine ecosystems from freshwater ecosystems.* The water in marine ecosystems has a salt content of approximately 3.5%. Freshwater ecosystems have no salt or extremely low amounts of it.

Fig. 15.25 **Aquatic biomes**

**Freshwater ecosystems** are classified according to whether the water is flowing or standing. Rivers, streams, and brooks have water in motion. Because the waters are flowing, they contain high levels of dissolved oxygen, beneficial for freshwater fish and other aquatic organisms. Carp and trout are typical freshwater fish. Ecosystems with standing water include ponds and lakes. Aquatic plants and animals are usually found in the upper layers of pond water, where dissolved oxygen is most abundant. Wetland areas such as marshes, swamps, and peat bogs are typical of areas where the underground water table is near the ground surface. Wetlands are important hatcheries and rearing grounds for wildlife.

**Marine ecosystems** are ecosystems associated with the ocean, starting from the shore and moving toward the open ocean. Two special types of marine ecosystems are the estuary and the coral reef. An **estuary** occurs where nutrient-rich fresh water from rivers and streams meets and mixes with seawater. Because the salinity of the seawater is lowered, estuaries are very productive ecosystems in which phytoplankton, grasses, crabs, fish, and many migratory birds thrive. Examples include the Hudson River estuary in New York Harbor, the Susquehanna River estuary in Chesapeake Bay, the Mobile River estuary in Mobile Bay, and the salt marshes found along the New England coast. **Coral reefs** are formed from the remains of stony corals and coralline algae. Coral reefs can be found in warm, clear, relatively shallow water parallel with the shoreline of land and near volcanic islands. These ecosystems provide a variety of habitats for plant and animal life because they are resistant to waves and are rich in nutrients. The longest coral reef, the *Great Barrier Reef,* extends over 1500 miles along the east coast of Australia.

If you stand at the shore of the ocean and imagine sailing toward the open ocean, you will observe the intertidal, neritic, and oceanic ecological zones. **Intertidal zones,** also called *littoral zones,* are the areas located between the highest and lowest tides along the coast. Waves and the alternation of the tides create a variety of habitats for plants and animals. Algae, marine worms, small fish, and a variety of crabs, sea stars, and sea cucumbers live here. Because there is easy access to

Fig. 15.26
**Ocean**

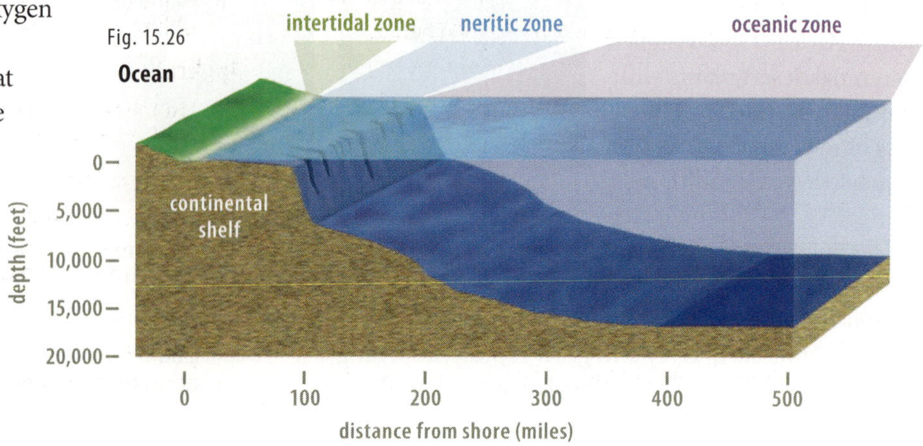

## Rules for the "Eco-explorer"

Observing the LIVING CREATION

You might enjoy exploring one of the ecosystems in your local area. Depending on where you live, you may have several from which to choose. "Eco-adventure" is a great way to get some exercise and enjoy creation at the same time. Before you begin your investigation, be sure to review the information in this chapter, especially The Amateur Ecologist on page 282. Your local library may also contain some good references on the subject of ecology. Here are some principles to consider before you get started:

- Remember, "The earth is the LORD's, and the fulness thereof" (Psa. 24:1). We are stewards of God's creation; it belongs to Him. Keep in mind whose property you are exploring!
- Practice personal safety. The Amateur Ecologist lists several safety precautions you can take.
- Make sure you know the local laws governing wildlife and their natural habitats. Some species of animals and plants are protected by law. In many places, picking flowers from roadsides and public parks is illegal. Your local chapter of the Audubon Society is a good source for this kind of information.
- Never start a fire in the wild. Campfires are for campgrounds.
- Always get permission before exploring private property.
- Learn about the dangerous insects, spiders, animals, and plants in your area before you start collecting specimens.
- When you are collecting specimens, don't take more than you need and try not to disturb their habitats.
- Cut plant samples with a knife or pair of clippers—never break or tear a sample from a plant.
- If you capture live specimens, be certain that you provide the necessary food, water, and living conditions to sustain them. When you have finished observing the live specimens, return them to the exact place you found them.
- Try to leave an ecosystem as you found it. Dismantle any blinds you built, remove your traps, and above all, don't leave litter behind.

these areas, intertidal communities are the best known and most studied of all the marine communities.

**Neritic zones,** also called *coastal zones,* are the regions that extend past the intertidal zone to the edge of the *continental shelf.* This zone is home to the coral reef ecosystem and supports most of the world's commercial fishing industry. The shallow waters also allow for the harvest of many bottom-dwelling creatures, such as crabs and lobsters.

**Oceanic zones** extend from the continental shelves into the open ocean. Whales, squid, octopuses, and other large sea creatures are found here, including several migratory species of tuna that comprise a large part of the ocean fishing industry.

## Section Review 15.5

1. Why do ecologists classify biomes according to plant life rather than animal life?
2. What are some abiotic factors that influence the development of different climax vegetation in different biomes? You may need to refer to section 15.2.
3. Where is the Arctic tundra biome found? Alpine tundra?
4. How does the vegetation of a northern coniferous forest differ from the vegetation of a temperate deciduous forest? Which type of forest is known for its brilliant display of color in the fall season?
5. Why are there fewer varieties of wildlife in tundra biomes than in forest biomes?
6. What makes the soil in a grassland biome very fertile? What abiotic factor inhibits the growth of trees in grasslands?
7. Both deserts and tropical rain forests are found near the equator, yet these biomes contain very different ecosystems. Explain.
8. Name the two major divisions of aquatic biomes. What is the main difference between the two?

### Application

9. List the seven major biomes discussed in the text and briefly describe the climax vegetation in each.

### Critical Thinking

10. Australia's Great Barrier Reef abounds with a tremendous variety of aquatic life. Why do you suppose this marine ecosystem is well populated with living things?

### Identify

permafrost, estuary, coral reef, intertidal zone, neritic zone, oceanic zone

## 15.6 Ecological Succession and Man's Role

Every day you can see in the news how floods, storms, fires, droughts, and volcanic eruptions change communities around the world in drastic ways. Although most of these events are destructive, some of them do contribute to the overall well-being of the ecosystem. Natural forest fires, for instance, destroy vegetation and kill animals but at the same time provide the extreme heat needed by the seeds of some plants in order to sprout.

But what happens in the ecosystem after the disturbance? Does the ecosystem recover, and if so, how? Consider two dramatic ecological events: a volcanic eruption and a forest fire. In 1980, the violent volcanic eruption of Mount Saint Helens transformed over 100,000 acres of hemlock and Douglas fir forest in the state of Washington into barren land. Several years later, in 1988, great fires destroyed thousands of acres of forests in Yellowstone National Park. The recovery of these ecosystems helps us to understand how God designed the earth to "heal" itself.

What we observe in both cases is that after the fire is gone or the volcanic activity ceases, a variety of plant and animal species colonize the disturbed area and begin to bring it back to life. The first organisms that colonize a disturbed ecosystem are called **pioneer species.** These plants and animals enrich the soil and prepare the way for other species to come in and repopulate the area. A series of changes that may take months, years, decades, or even centuries begins. These changes are called an **ecological succession,** which is defined as the replacement of early pioneer species by later species in an orderly progression until the climax community is established. The type of ecological succession, primary or secondary, depends on the starting material.

### Primary Succession

**Primary succession** starts with barren ground, similar to what you observe after a volcanic eruption or in sand dunes. A dramatic case of primary succession was observed after the eruption of a volcano on the island of Krakatau, Indonesia, in 1883. More than 35,000 people died from the great tsunami generated by the eruption, and the dust thrown into the atmosphere clouded 10% of the sunlight in Europe for three years.

No trace of plant or animal life remained on the island. Two years later, pioneering blue-green algae, scattered mosses, and some ferns appeared in the interior of the island. Thirteen years later, there were some scattered trees, wild sugarcane, and a few orchids. Twenty-three years later, grass and a few scattered trees covered the island. Ants, mosquitoes, birds, bats, and lizards were also present by then. Thirty-seven years after the eruption, trees occupied half of the island, and ten years later a tropical forest had reclaimed the island.

### Secondary Succession

While primary succession starts with barren ground, **secondary succession** starts with ready soil and pioneer species, as in an abandoned field or a burned forest. This type of succession occurs more rapidly than primary succession because some organisms already inhabit the area. What happens, for example, when farmland is abandoned and no longer cultivated? First, annual weeds, such as crabgrass, horseweed, and dandelions invade the field. Then, perennial plants and grasses displace the annual plants and take over. If you look at a field three or four years after it was abandoned, it is common to see broomsedge, a perennial grass, as the dominating species, and also numerous scattered pine seedlings. In a few more years, the broomsedge will be displaced by pine, and in a short time the field will have become a pine forest. If you look carefully then, you will see oak, hickory, and maple seedlings beginning to grow in this pine forest. Eventually, the pines will be replaced by these deciduous trees, which will constitute the climax forest of this ecosystem.

### Man's Dominion

Although the Bible does not use the word *ecology,* it is certainly not silent on the subject. The following are a few of its many references to living things and the environment.

- The use of animals and plants as food for man is mentioned in Genesis 9:2–3.
- The habitats of birds and other animals are described in Psalm 104:16–18.
- Wind currents and the water cycle are mentioned in Ecclesiastes 1:6–7.
- Man's use of trees as fuel is described in Isaiah 44:14–15.

Scripture also declares that man has **dominion** over nature. From the very beginning, God placed the earth and all living things under man's authority and commanded him to subdue and have dominion over them:

> And God said, Let us make man in our image, after our likeness: and let them have dominion . . . over all the earth. . . . —Gen. 1:26

> And God blessed them, and God said unto them, Be fruitful, and multiply, and replenish the earth, and subdue it: and have dominion . . . over every living thing. . . . —Gen. 1:28

This means that mankind is to use God's living creation to meet his physical needs. Human life depends on the use of creation's resources, both living and nonliving.

## Good Stewardship

With man's right of dominion comes the responsibility to exercise good **stewardship.** Mankind has a responsibility to manage creation in a way that balances the wise use of our natural resources and their **conservation.** This can be accomplished in many different ways:

- Conservation and management of natural resources. Managing resources effectively requires distinguishing **renewable resources** such as soil, water, and wood, which regenerate very quickly, from **nonrenewable resources** such as fossil fuels and minerals, which regenerate slowly, if at all.
- Replanting of trees and other vegetation to continue the carbon-oxygen cycle and supply the crucial first link in all food chains.
- Proper management of natural ecosystems to provide food for man as well as for plants and animals.

Fig. 15.28 **Forestry worker planting trees**

Fig. 15.27 **Secondary succession**

## Life Science

**Ecology** – using the relationship between organisms and the biotic and abiotic factors in their environment to conserve living resources and benefit man

**Forestry** – planting and managing forests or harvesting forest products

**Scientific journalism** – writing about science for magazines and newspapers and writing technical publications accompanying scientific supplies, instruments, and machinery

**Wildlife management / game management** – using scientific knowledge to manage wildlife populations

- Wise use of biotechnology to produce higher-yielding crop varieties and medicines such as insulin.
- Reducing unnecessary consumption of resources and also reusing and recycling resources whenever feasible.

To carry out this stewardship responsibility in a practical way, we must understand how our actions affect all natural life and accept our responsibility to manage it wisely.

### Unbiblical Viewpoints

Unfortunately, some people take conservation to such an extreme that they stress the welfare of nature above that of man. In their zeal to protect the environment, they dismiss man's needs as unimportant. For example, some environmental extremists object to the harvesting of trees for any reason. Yet one of the reasons God created trees was to provide people with lumber for homes, furniture, paper, household heating, and cooking. Forestry not only benefits mankind but also benefits the ecosystem by thinning a forest of diseased and dying trees to make room for new growth. Of course, if too many trees are cut down in an area the results can be detrimental, but cutting down a few trees where a highway or building is planned is reasonable. Most logging companies replant in forests where they harvest, and trees will reproduce on their own unless too many are harvested.

*The most dangerous flaw of the modern environmental movement is that it promotes a reverence for nature that belongs to God alone.* We are to worship the Creator, not the creation. God controls the world's ecosystems, but He expects man to study creation, to share His concern for it, and to bring honor and glory to Him through the practice of good stewardship.

## Section Review 15.6

1. Give three examples of natural disturbances that could change a community.
2. Which type of ecological succession begins with a totally barren, lifeless area? What kind of natural disturbance might prepare a community for this process?
3. Which type of succession occurs more rapidly? What kind of disturbances could lead to this process?
4. For what purpose did God give man authority over His living creation?
5. Define the term *stewardship* in the context of man's God-given dominion over creation.

### Application

6. Why are lichens and mosses usually the first plants to colonize a community in primary succession? How do these organisms help prepare the ground for other plant species?
7. Give some examples of biblical stewardship that you have observed. Refer to the list on pp. 301–302 for ideas.

### Critical Thinking

8. Some environmentalists oppose the development of natural resources for man's benefit. What biblical principle does this contradict? Support your answer with Scripture.

### Identify

pioneer species, ecological succession, renewable resources, nonrenewable resources

## Chapter 15 Review

### Define

1. ecology
2. habitat
3. biosphere
4. climax vegetation
5. stability
6. biodiversity
7. carrying capacity
8. limiting factor
9. autotroph, heterotroph
10. scavenger
11. decomposer, detritivore
12. trophic level
13. food web
14. niche
15. symbiosis, mutualism, parasitism
16. biogeochemical cycle
17. permafrost
18. pioneer species
19. ecological succession
20. renewable resource, nonrenewable resource

### Identify

1. six levels of ecology
2. two general kinds of ecosystems
3. three processes that make up the hydrologic cycle
4. two main atmospheric cycles
5. the main sedimentary cycle
6. two main types of aquatic ecosystems

### Explain

1. What is the difference between biotic and abiotic factors in an ecosystem?
2. How is an organism's optimum range related to its tolerance range for an abiotic factor such as temperature?
3. How is a producer different from a consumer?
4. How do herbivores, carnivores, and omnivores differ in their feeding habits, and how are they all related to one another in the food chain?
5. Describe the first four trophic levels of a food chain.
6. How do the three types of pyramid diagrams illustrate the flow of energy and the distribution of organisms at different trophic levels in an ecosystem?
7. Describe the three types of symbiotic relationships described in this chapter and explain one example of each type.
8. List the five types of marine ecosystems and where each type is found.
9. What is the difference between primary and secondary ecological succession?

### Apply

1. Draw a diagram (similar to Fig. 15.6) of a food chain in a local ecosystem. Identify the trophic level of each organism and classify the organisms by their form of nutrition (producer, herbivore, etc.).
2. When the European gypsy moth was accidentally released in the United States in the 1860s, the gypsy moth larvae damaged and destroyed many oak trees by eating their leaves. Because the oak trees damaged by the moths did not produce acorns, the population of squirrels, which depended on the acorns for food, shrank. What type of special nutritional relationship existed between the moths and the squirrels? Explain your answer.
3. What are some things that man can do to fulfill his biblical role as God's steward in nature?

# UNIT 4 | Zoology

## Variety & Design in the Animal World

# CHAPTER 16 Mammals

**Key Concepts**

- Characteristics of animals, vertebrates, and mammals
- Mammal reproduction
- Variety of mammals

## 16.1 Animals—Vertebrates—Mammals

On the fifth and sixth days of Creation, after God had created matter, energy, light, the earth, the heavenly bodies, land, sea, and plants, He made a new kind of being: "the moving creature that hath life" (Gen. 1:20). As you read Genesis 1:20–25, you will become aware of two of the main attributes of the animal kingdom: mobility and diversity. As you become more familiar with kingdom Animalia, you will appreciate more fully the biblical description of the animal kingdom as "every living creature that moveth" (Gen. 1:21).

### Mobility

Biologists characterize **animals** as living organisms that move from place to place in their environment, mainly for the purpose of obtaining food. Such mobility is not necessary in the plant world because most plant life is autotrophic (makes its own food); therefore, plant movement is limited to tropisms, to growth, and to drifting with water and air currents. Animals, on the other hand, have been created with the ability of locomotion in order to obtain their food. Most animals are **motile;** that is, they can move their bodies from place to place. A few animals are attached to one location but can move the environment—air or water—toward themselves for the purpose of trapping food; these are **sessile.**

### Diversity

The diversity found in the animal world is enormous. *There are over one million known and classified species of animals alive today.* Some biologists estimate that there may be as many as 8 million living species, most of which have not yet been classified. Size ranges from microscopic creatures to the 100-foot-long blue whale. Some have definite body shapes, but others seem to have no shape at all. Some have rigid parts on the inside of the body, others have rigid coverings, and still others are quite flexible. Some live absolutely alone, and others live in colonies of thousands of individuals. As you look closely at the animals of the earth, you will again appreciate the complex design and order of the living creation.

> And God said, Let the waters bring forth abundantly the moving creature that hath life, and fowl that may fly above the earth in the open firmament of heaven. And God created great whales, and every living creature that moveth, which the waters brought forth abundantly, after their kind, and every winged fowl after his kind: and God saw that it was good. And God blessed them, saying, Be fruitful, and multiply, and fill the waters in the seas, and let fowl multiply in the earth. And the evening and the morning were the fifth day. And God said, Let the earth bring forth the living creature after his kind, cattle, and creeping thing, and beast of the earth after his kind: and it was so. And God made the beast of the earth after his kind, and cattle after their kind, and every thing that creepeth upon the earth after his kind: and God saw that it was good. —Gen. 1:20–25

## Symmetry in the Animal Kingdom

As biologists study the living creation, they observe that many living things possess the quality of **symmetry.** In general, living things can be grouped into three basic kinds of symmetry. Animals are particularly interesting in respect to symmetry because of their internal parts and the way that these parts develop from the embryo.

**Bilateral symmetry.** Humans and most animals are constructed with a design called **bilateral symmetry.** An organism with bilateral symmetry can be cut in half in only one plane to make each half like the other. Your own body is an example of bilateral symmetry. You have many body parts in pairs: ears, eyes, arms, legs, etc. For your body to be divided into equal parts, the division would need to be in a plane between your eyes. As you studied the internal organs of the human body, you may have noticed a number of pairs: the brain has two hemispheres, there are two lungs and two kidneys, and the muscles, bones, blood vessels, and nerves usually come in pairs. Even the single heart is designed as a two-sided organ. *This design is also the basic blueprint from which God made most animals.*

**Radial symmetry.** A few animals possess **radial symmetry.** The best way to illustrate radial symmetry is to picture a pie. Cut the pie in half any way you wish; you will always have the same shape on one side as on the other. The starfish is a good example of radial symmetry. Its internal design is such that it can be cut in half in several different ways through the center and both halves will be the same.

**Asymmetry.** A few animals, like the amoeba, display great variation in form, changing shape almost constantly. Biologists call these animals **asymmetric** [ā′sĭ·mĕt′rĭk], which literally means "without symmetry." The beauty of these creatures is like the constantly changing beauty of the clouds.

## Vertebrates: Animals with Backbones

Zoology is such a broad study that historically it has been divided into smaller parts, beginning with

Fig. 16.1 **Types of symmetry**

**vertebrates** (animals with a backbone) and **invertebrates** (animals without a backbone). These two groups are not at all the same in size, either in number of kinds or number of individuals. The largest and most diverse group of animals is the invertebrates. Invertebrates include such varied creatures as insects, clams, starfish, centipedes, worms, sponges, and amoebas. *Vertebrates, which make up only about 3% of all animals, include the more familiar animals:* mammals, birds, reptiles, amphibians, and fish.

**Characteristics of vertebrates.** Every vertebrate has a *spinal cord* with a *brain* at its anterior (front) end. The brain is protected by a *skull,* and the body is supported or stiffened by either a *backbone* or a *notochord* [nō′tə·kôrd′: a flexible, unsegmented rod].

All vertebrates have an internal framework, or skeleton, called an **endoskeleton;** the endoskeleton is made of bone, cartilage, or a combination of the two. As in man, the skeleton is divided into two parts, the *axial skeleton* and the *appendicular skeleton.* The axial skeleton consists of the skull and backbone. Most vertebrates are **tetrapods,** animals with four *appendages* or *limbs* attached by special bones to the body at the hips and shoulders. Whales and some other water mammals have only one pair of appendages; snakes, which have received special treatment by the Creator, have no appendages. The bones of the appendages and the bones that attach them to the body make up the appendicular skeleton.

Vertebrates are noted for their complex body systems. They are characterized by the same 11 body systems as man: skeletal, muscular, integumentary, nervous, endocrine, digestive, circulatory, respiratory, excretory, reproductive, and immune. Each of these systems contains many highly developed organs.

*All vertebrates exhibit bilateral symmetry.* Each has a body consisting of a head and a trunk. Many vertebrates also have a neck and a tail. The body parts, such as the appendages, eyes, ears, and nostrils, are usually paired, each a mirror image of the other.

**Phylum Chordata.** Vertebrates are typically classified as subphylum **Vertebrata** [vĕr·tĕ′brā′tə] within phylum **Chordata** [kôr·dă′tə]. Phylum Chordata also includes *amphioxus* [ăm′fē·ŏk′səs: a fishlike marine invertebrate] and *tunicates* (sea squirts and similar creatures). These invertebrates have notochords during some parts of their lives but lack some of the other distinguishing features of true vertebrates.

## Mammals: Vertebrates with Warm Blood, Hair, and Milk

Animals of the class Mammalia, better known as **mammals,** are the most familiar and the most dominant group of vertebrates on the earth today. These animals are warm-blooded, have hair, and are provided with mammary glands for producing milk. All

### Table 16.1 Classification of Vertebrates

(Kingdom Animalia; phylum Chordata; subphylum Vertebrata)

| Class[1] | Description | Number of living species (approximate) | Examples |
|---|---|---|---|
| **Mammals** (class Mammalia) | warm-blooded vertebrates that breathe with lungs and have hair and mammary glands | 5000 | dogs, cats, cattle, rodents, primates |
| **Birds** (class Aves) | warm-blooded vertebrates that breathe with lungs and have wings, feathers, and two feet | 9600 | eagles, songbirds, penguins |
| **Reptiles** (class Reptilia) | cold-blooded vertebrates that breathe with lungs and have dry, scaly skin | 8700 | snakes, lizards, turtles |
| **Amphibians** (class Amphibia) | cold-blooded vertebrates with soft, moist skin; larval forms have gills; most adults breathe with lungs | 6100 | frogs, salamanders, caecilians |
| **Fish** (classes Osteichthyes, Chondrichthyes, and Cyclostomata[2]) | cold-blooded vertebrates with gills for breathing, fins, and a skeleton of bone or cartilage | 28,000 | trout, perch, sharks, eels, rays |

[1]All living vertebrates can be classified as either mammals, birds, reptiles, amphibians, or fish. Some authorities also place the extinct dinosaurs in the class Reptilia, while others place them in a separate class (Dinosauria) based on differences in body structure.

[2]Or Agnatha.

mammals breathe air by means of lungs, and all have a four-chambered heart. Most have two pairs of limbs and seven neck vertebrae, and most are born alive and nourished by the mother.

Some of the smallest mammals are the mouselike shrews, which can be less than 2 inches long (excluding tail) and weigh less than a penny. The largest are the blue whales, the largest animals that have ever lived, which can be up to 110 feet long.

Mammals have many characteristics in common, but there are also many structural differences in their design. For example, there is much variation in the teeth of mammals. The "incisor teeth" of an elephant take the form of tusks; these same teeth in a squirrel are sharp and chisel-like; rabbits are designed with chewing incisors. Some mammals have few teeth; others have no teeth at all. The design of each is tailored to fit its needs.

Fig. 16.2 **Mammal limb variation**

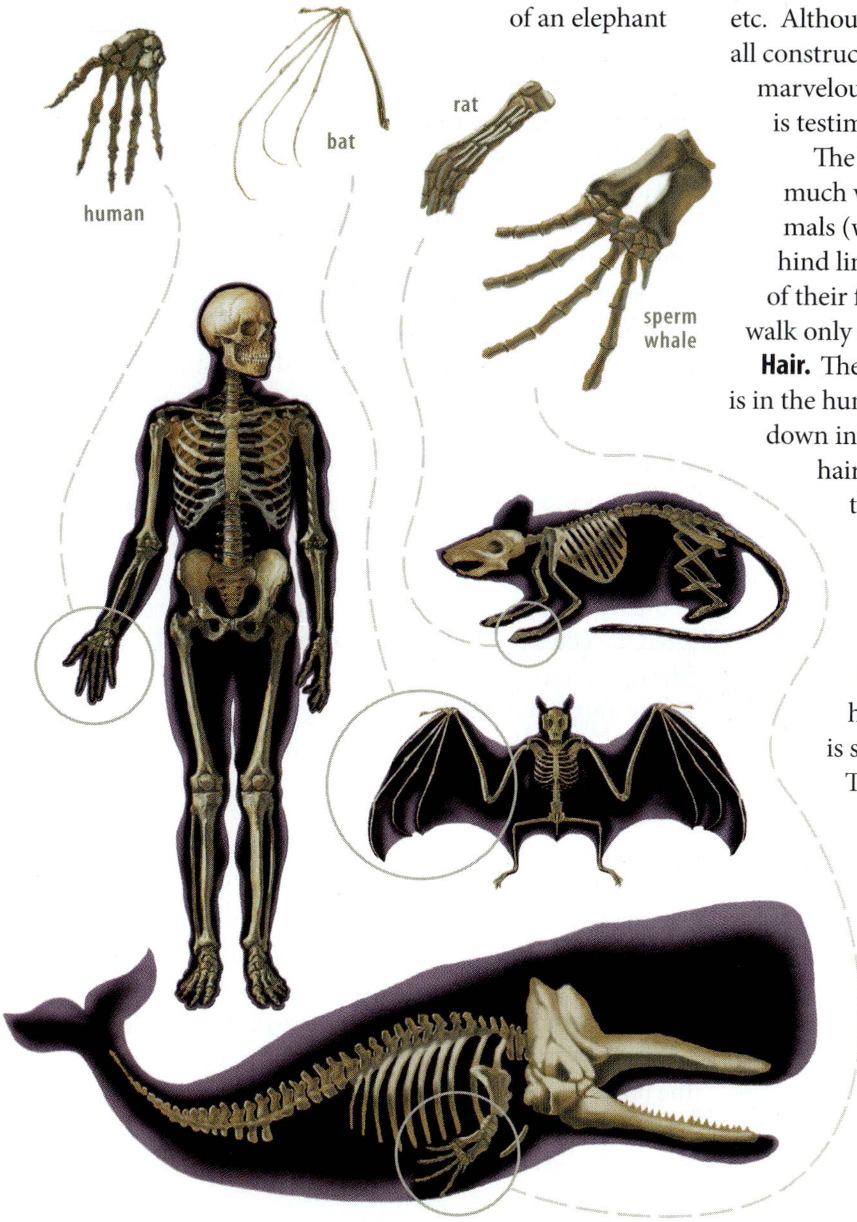


**Skeleton.** The *skeleton* of all mammals is built on a similar pattern. For example, the number of separate vertebrae in the neck region of the dog, bat, man, horse, giraffe, etc. is the same, although the neck regions are not the same length. The vertebrae differ greatly in size, but not usually in number.

**Limbs.** Most mammals have two pairs of *limbs*—forelimbs and hind limbs. The forelimbs may be designed for special uses such as running on hard ground, climbing trees, burrowing, flying, swimming, etc. Although the forelimbs may differ in use, they are all constructed along the same general pattern. The marvelous variation in the ways these limbs are used is testimony to the ingenuity of the Creator.

The hind limbs of mammals do not show as much variety as do the forelimbs, but some mammals (whales and other marine mammals) have no hind limbs at all. Some mammals walk on the soles of their feet (metatarsal and calcaneus bones); some walk only on their toes (phalanges).

**Hair.** The *hair* in all mammals is produced just as it is in the human. Epithelial cells of the epidermis reach down into the dermis to form a closed-end tube, or hair follicle. The cells multiplying at the root of the hair follicle fill with the protein *keratin* and force the hair shaft up. *Horns, claws, nails,* and *hoofs* are special structures in the skin of some mammals that are produced the same way as hair and are composed of keratin. Many mammals have two coats of hair to protect their bodies. The underhair is soft and fine to provide a thick, warm coat. The outer hair is longer and stiffer to give shape to the animal's coat. Some mammals have thick, long hairs coming from their mouths or heads, which are used for the sense of touch. We commonly call them *whiskers.* Hair not only keeps the animals warm and helps them with the sense of touch; it is also important in hiding them from enemies. The coloration and pattern of hair often blend in well with the surroundings, protecting the mammals from predators.

**Milk.** The production of *milk* is one of God's most marvelous provisions for mammals. Mammals are born rather immature compared to many other animals. Their bodies need special nourishment that is found perfectly in the mother's milk. **Mammary glands,** like all glands of a mammal's body, are specialized epithelial cells. When the female is pregnant, the mammary cells of the mammary gland multiply. The gland may nearly double in size as it manufactures milk to provide nourishment for the newborn. Parts of the cells pinch off to form milk-filled droplets that accumulate in ducts. In response to the release of hormones in the mother, the milk is released and available at exactly the time it is needed by the young. God designed this system to work as long as the young require the nourishment; as soon as they stop taking the milk, the mammary gland shuts down production, reduces in size, and awaits the arrival of more young.

**Temperature regulation.** Animals are often classified as either warm-blooded or cold-blooded. A "warm-blooded" animal does not necessarily have warmer blood than a cold-blooded animal; the difference is that a **warm-blooded** animal regulates its temperature by internal mechanisms (altering blood flow, shivering, sweating, panting, etc.) whereas a *cold-blooded* animal must regulate its temperature by external factors (such as lying in the sun when its body temperature is too low or cooling off in the shade when its temperature is too high). Animals classified as warm-blooded are mammals and birds; animals classified as cold-blooded are reptiles, amphibians, fish, and invertebrates. *All mammals are considered to be warm-blooded.*

The ability to regulate their temperature internally allows warm-blooded animals to maintain a stable body temperature regardless of their environment. Thus, warm-blooded animals such as mammals are also said to be **homeothermic** [hō′mē·ə·thûr′mĭk: "maintaining the same temperature"]. Cold-blooded animals, on the other hand, have a body temperature that fluctuates with the environment. A cold-blooded animal, like a snake, is subject to its environment in that when the weather grows cold, the animal slows down; a warm-blooded animal, on the other hand, adjusts quickly to cold weather and remains active.

Because mammals regulate their temperature internally, a significant percentage of the food they eat is used to produce heat to keep the body warm. This is particularly true of mammals in cold climates and of very small mammals. During winter, some small mammals like ground squirrels and woodchucks would find it difficult to gather enough food to keep their body temperature high, so God has given them a special ability to **hibernate.** Their body temperature falls to within a degree or two of the freezing point, and their respiration and heartbeat slow as well. If you find a woodchuck in hibernation, it will be stiff and cold to the touch, breathing only about once every five minutes. Although it appears dead, it will awaken in the spring. In this state, the creature uses far less food and can survive long, harsh winters. *A few other mammals, such as bears, raccoons, and opossums, sleep during the winter but do not hibernate.* Their body activities slow down but their temperatures remain high. A hibernating animal will not wake up if disturbed, but a sleeping one will, as one ambitious biologist found out when he entered a cave to take the temperature of a sleeping bear.

**Lungs.** The *lungs* of mammals are larger in relation to body size than those of reptiles and amphibians, primarily because mammals need more oxygen to maintain their high body temperature. Only mammals have a muscular *diaphragm* that separates the body cavity into two main areas—abdominal and thoracic. The diaphragm, when contracted, expands the lungs and causes air to enter. When the diaphragm relaxes, the lungs contract and air is exhaled. Most mammals breathe through their noses or mouths, but the whales and dolphins have special blowholes on top of their heads through which they can breathe.

**Heart.** Warm-blooded animals such as mammals need the efficiency provided by a *four-chambered heart.* The *atria,* muscular pouches on top of the heart, act as a priming pump to fill the strong *ventricles* with blood. This design ensures that the heart will always be full. Thus, when the heart exerts effort to pump the blood throughout the body, little energy is wasted. The four-chambered heart is like having two hearts in one. The right side pumps blood to the lungs. The separate action is required because blood flows through capillaries in the lungs, providing much resistance; as a result, *the blood pressure falls greatly after the blood gets through the lungs.* The blood pressure is so low at this time, in fact, that it could not reach the other parts of the body with fresh oxygen if the blood did not return to the heart to receive a powerful push from the left side of the heart.

## Mammals and Man

From your study of human anatomy and physiology, you no doubt recognize the mammalian characteristics just given to be true also of man. Is man, then, a mammal? Physically speaking, man shares many characteristics with mammals. The Bible teaches that he was created on the same day as the land animals (Gen. 1:24–26), after the creation of the mammals. The Creation account clearly distinguishes man from the animals, however:

> And God said, Let us make man in our image, after our likeness: and let them have dominion over the fish of the sea, and over the fowl of the air, and over the cattle, and over all the earth, and over every creeping thing that creepeth upon the earth. —Gen. 1:26

What is the difference? *Man was created in the image of God and was created to rule over the animals.* Man is more than a physical being: he is a spiritual being created to know and serve his Creator. Unlike the animals, mammals included, man is an immortal being who will live somewhere forever. Man also differs from the animals in that he is a moral being; that is, he has the ability to make decisions about right and wrong and is accountable for his decisions.

Mammals were created to be a helper to man. Adam named all the animals, and an affection between man and beast must have resulted. But since the animals could not meet the deepest needs of the man, God created woman. It was not until after the fall of man into sin that man's relationship with nature, including the animal kingdom, changed. Man then found it necessary to kill animals for clothing, sacrifice, or in self-defense. Scripture tells us that after the Flood, God put a fear of man into the animals and allowed man to eat animals for food. One day God will restore the original tranquility of nature (Isa. 11:6–9).

Through the ages, mankind has used mammals for companionship, protection, and service. Some mammals, such as dogs and cats, are popular as pets today. Others are used for food; many meats come from the flesh of slaughtered mammals, including beef from cattle, pork from hogs, and mutton from sheep. Horses, oxen, donkeys, camels, and elephants are mammals that have been used as a source of power to do work. Leather is made from the skin of cattle and other mammals. Because of their physical similarities to man, mammals are also useful in medical research when used in a humane way for testing drugs, manufacturing vaccines, and studying diseases that affect humans.

## Mammals and Nature

*Mammals are important in nature beyond their benefit directly to man.* They help plants to grow by fertilizing the ground with their nitrogen-containing wastes and by adding carbon dioxide to the atmosphere. Without these raw materials, plants could not manufacture their food by photosynthesis. Mammals help keep the balance in the populations of other animals and plants. For instance, anteaters eat thousands of ants at each meal, keeping the ant population in check. Bats eat flying insects, shrews eat ground insects, the mongoose eats snakes, cats eat birds, seals eat fish, coyotes eat rabbits, and lions eat antelope. Plant populations would dwindle if herbivores were not kept in check by carnivorous mammals. Since the curse of sin fell upon creation, God has maintained a delicate balance among these creatures in order that His world might remain alive. God even provides scavengers among mammals, such as hyenas, jackals, skunks, and opossums, to eat the bodies of dead animals and scraps left by others. Mammals, once created to be the closest of man's earthly friends, remain a most important source of raw materials for humans and nature.

## Mammal Reproduction

**Placental mammals.** Like humans, plants, and most other animals, mammals reproduce after their kind by *sexual reproduction,* the union of a sperm cell from a male with an egg cell from a female to form a new organism. Most mammals are considered **placental mammals** because their developing young receive nourishment from a *placenta* until birth. Reproduction in the placental mammals is very similar to reproduction in humans. Fertilization takes place in an *oviduct,* which corresponds to a fallopian tube in humans. Some mammals, such as dogs, cats, rabbits, and pigs, release several eggs at once, resulting in multiple births. Others, such as horses, cows, and elephants, release only one egg at a time, resulting in single births. The fertilized egg divides to form a hollow sphere of cells called a *blastula,* which attaches itself to the wall of the *uterus.*

Like an unborn human, the *embryo* of a placental mammal is nourished by the **placenta** formed from the tissues of both the young and the mother. During its growth, the embryo is protected by the **amniotic fluid,** which absorbs the shocks of its mother running or being bumped by an object. Once it is recognizable as a young individual, the animal is called a *fetus.* The entire period of growth from fertilization to birth is the **gestation period;** its length ranges from 16 days in a hamster to almost 2 years in an elephant.

When gestation is complete, the fetus moves out of the uterus through the birth canal and is born alive; it is then ready to breathe air for the first time and to receive food through its digestive system. The **umbilical cord** will be cut by the mother, and the placenta will be expelled by the uterus so the mother can heal and prepare for another pregnancy. Because their young are born alive, placental mammals are **viviparous** [vī·vĭp′ər·əs: from Latin *vivus,* "alive" + *-parus,* "giving birth"]. Some mammals, such as mice, cats, and rabbits, are born blind and very helpless. Some mammals with a longer gestation period, such as cattle, horses, and elephants, are able to stand, walk, and even run immediately after birth.

**Nonplacental mammals.** In two unusual orders of mammals, the young are not nourished inside the mother's body through a placenta. In the *platypus* and the *echidna* [ĭ·kĭd′nə: spiny anteater], the members of the order *Monotremata,* the egg is covered with a leathery shell after fertilization and expelled from the mother's body. These **oviparous** [ō·vĭp′ər·əs: egg-laying] mammals differ from viviparous mammals in that they do not bear their young alive. The female echidna carries her eggs for 7 to 10 days in a temporary mammary pouch on her abdomen until they hatch. The egg is nourished by the yolk. When the baby echidna hatches, it remains in the pouch for about two months, lapping milk that is secreted by the mother's mammary glands. The female platypus does not develop a temporary pouch. Instead, she curls around her young after they hatch, and they lap milk that is secreted through pores in her abdomen. Most young mammals are born with a sucking instinct; the little platypus, because of the shape of its bill, cannot suck, and so it has been given a lapping instinct instead.

*Marsupials* (kangaroos, opossums, etc.) do not lay eggs; they bear their young alive. But unlike most other mammals, marsupials spend the majority of their gestation outside of the womb. The tiny, helpless young are usually nourished inside a pouch, where they attach themselves to the mother's nipples to be fed milk from the mother's mammary glands. Although marsupials do have a simple placenta, they are generally considered nonplacental mammals because the function of the placenta in marsupials is limited and brief (just a few days in opossums, for example).

**Parental care.** Parental care is an important factor for mammals. Dependent upon the mother's milk, the young must have constant attention. They need to learn the many activities of life, such as hunting, home-building, and cleanliness, in order to survive. Some mammalian behaviors are instinctive, while others are taught by the parents.

## Section Review 16.1

1. What primary characteristics are shared by almost all animals?
2. What are three characteristics of all vertebrates?
3. What are five characteristics common to all mammals? What are three additional characteristics common to most mammals?
4. How does a warm-blooded animal regulate its temperature? How does a cold-blooded animal regulate its temperature?
5. What purpose does hibernation serve?
6. What are three major categories of mammals, based on the way the developing young are nourished?

### Application

7. What is the purpose of a mammal's mammary glands? Why is this necessary?
8. Explain why we do not consider man a mammal even though he shares many mammalian characteristics.

### Identify

motile, sessile, symmetry, oviduct, placenta, amniotic fluid, gestation period, viviparous, oviparous

## 16.2 The Mammalian Orders I

In recent decades, evolutionists have begun dividing mammals into about 30 orders based on presumed evolutionary relationships; in this text, however, we will use a more traditional system that divides the roughly 5000 species of mammals into 18 orders based on physical characteristics. (Recall that an *order* is a category in the Linnaean system of classification.) One order, *Tubulidentata* [tū′byo͞o·lĭ·dĕn·tā′tə], contains only one species, the aardvark. The largest order, *Rodentia,* contains over 2000 species of beavers, gophers, mice, porcupines, rats, squirrels, and other gnawing mammals. Because they are warm-blooded, mammals are not restricted to warm regions and can live in almost all habitats. They are extremely diversified and enjoy many different modes of life.

### Table 16.2 Classification of Mammals

(Kingdom Animalia; phylum Chordata; subphylum Vertebrata; Class Mammalia)

| Order | Description | Number of living species | Examples |
|---|---|---|---|
| Monotremata | egg-laying mammals | 5 | platypus, echidna |
| Marsupialia | pouched mammals | 276 | koala, kangaroo |
| Cetacea | marine mammals | 81 | bottle-nose dolphin, blue whale |
| Pholidota | mammals with horny scales | 7 | pangolin |
| Sirenia | aquatic mammals called "sea cows" | 4 | dugong, manatee |
| Perissodactyla | odd-toed hoofed mammals | 15 | horse, rhinoceros |
| Proboscidea | mammals with trunks | 3 | African elephant, Asian elephant |
| Artiodactyla | even-toed hoofed mammals | 213 | giraffe, hippopotamus |
| Chiroptera | flying mammals | 1166 | gray bat, fruit-eating bat |
| Insectivora | insect-eating mammals | 482 | mole, shrew |
| Lagomorpha | mammals with four upper incisor teeth | 87 | hare, rabbit, pika |
| Primates | tree-dwelling mammals | 300 | ring-tailed lemur, gibbon |
| Rodentia | gnawing mammals | 2055 | beaver, rat, red squirrel |
| Hyracoidea | defenseless mammals with padded feet | 6 | hyrax |
| Tubulidentata | ant-eating mammals with tubed teeth | 1 | aardvark |
| Xenarthra | mammals with extra joints between vertebrae | 29 | sloth, armadillo, anteater |
| Dermoptera | arboreal gliding mammals | 2 | flying lemur |
| Carnivora | flesh-eating mammals | 274 | seal, lion, bear |

### The Solitary Aardvark (Order Tubulidentata)

The **aardvark** is custom-designed for two things: digging holes and eating ants and termites. It lives in the savannas of Africa south of the Sahara and was named by the Afrikaners, Dutch people who settled in South Africa in the 1600s. *Aardvark* is an Afrikaans name meaning "earth pig." The "earth pig" has a long, piglike snout; long, tough ears; short, stocky legs; and a long, thick, muscular tail that tapers to a point. An adult aardvark weighs about 140 pounds and is 4 to 5 feet long, excluding its tail, which may be 18 to 25 inches long. Its tough skin has only a few scattered hairs.

The aardvark has strong, blunt claws for digging burrows and tearing into ant and termite nests; it also has nostrils that can close to protect the animal from flying dust while

Fig. 16.3 **Aardvark**

it digs. With its slender, sticky tongue (up to 18 inches long), the animal scoops up its meal of insects, which it finds using its keen sense of smell. The aardvark's thick skin protects it from insect bites as it eats.

Aardvarks make a good meal for lions and are also hunted by the African people. Finding them, however, is not easy, because they are entirely nocturnal: they stay buried in their underground homes during the day and come out only at night to feed. Even at night, the aardvark can be difficult to surprise because it has a sharp sense of hearing for detecting predators.

Fig. 16.4 **Elephants**

Aardvarks usually live and hunt alone, and they are very timid. The female usually bears one young in the midsummer and keeps it close to her for about six months, until it can dig its own tunnels.

Aardvarks have only a few teeth, all far back in the jaw. They differ from all other mammals in that their permanent teeth possess little tubes that radiate from the central pulp cavity; hence, scientists have named the order Tubulidentata ("tube teeth"). Aardvark teeth have no roots or enamel.

## Three Species of Elephants (Order Proboscidea)

**Elephants,** the largest living land animals, are the mammals with trunks (order Proboscidea: [prō′bə·sĭd′ē·ə]). There are only three living species, the *African bush elephant,* the *African forest elephant,* and the *Asian elephant* (also known as the *Indian elephant*), but nearly 300 species of elephants are known from fossils. The African bush elephant is larger and has bigger ears than the Asian elephant; the African forest elephant is the smallest species but has larger ears than the Asian elephant. The largest elephant ever recorded measured 13 feet 4 inches in height, was 35 feet long (including trunk and tail), and weighed over 13 tons. An elephant's massive weight is supported by a bulky body and columnlike legs, and its large head is designed with unusually lightweight bone. The trunk is about six feet long, weighs about 300 pounds, and may contain more than 30,000 different muscles.

With its trunk the elephant can smell, drink, and feed itself. Its sense of smell is very keen, as is its hearing. The trunk can hold about a gallon and a half of water, which can be squirted into the elephant's mouth or into the air for a shower.

The elephant is herbivorous, eating grass, leaves, and small branches in great quantities. A wild bull (male) elephant may consume more than 300 pounds of food a day. The animal often uses its tusks to uproot small trees in order to eat the leaves. **Tusks,** made of ivory, are actually enlarged incisor teeth. Each of an elephant's molars is a foot long and weighs eight pounds. Chewing huge quantities of food wears the teeth down, but fortunately they are replaced five times until the last set arrives when the elephant is about 40 years old. When those teeth wear out (usually around age 60), the elephant may die from starvation because it cannot chew.

## Sea Cows (Order Sirenia)

Aquatic dugongs [do͞o′gŏngz′] and manatees are the only living families in the order Sirenia [sī·rē′nē·ə].

Fig. 16.5 **Manatee**

They are often called **sea cows** because they graze like cows; they can be seen in shallow waters of tropical rivers, lagoons, and coasts eating algae and submerged grasses and coming to the surface every minute or two for a breath of air. Manatees and dugongs have no hind limbs, and their forelimbs act as flippers. They live entirely in the water, requiring the support of the water to help them move and breathe. If one were to leave the water, its rib cage would collapse under the mass of its body, and the sun would burn its dried skin. Ancient sailors called these animals sirens, which is Greek for "sea nymph," or mermaid; therefore, scientists named the order Sirenia.

**Dugongs,** which have forked tails, reach 10 feet in length and weigh about 400 pounds. One kind lives in the Red Sea and is believed by some to be the source of the hide that covered the Hebrew Tabernacle (see Exo. 26:14; 36:19; the Hebrew word for "badgers" can be translated "porpoises" or "sea cows"). Others live in the Indian Ocean. The **manatee** is larger than the dugong; the largest ever caught measured over 15 feet long and weighed over 1000 pounds. Manatees live in the coastal waters and rivers of Florida, the Caribbean Sea, the northeastern coastal waters of South America, the Amazon and Orinoco rivers, and the rivers and coastal waters of western Africa. The manatee's tail flipper is not forked, and its upper lip is divided into two halves, each of which moves independently of the other. In the past, manatees have been extensively hunted for their abundance of fat, their tough skin, and their meat. However, most manatees are now protected species because of their rarity.

## Echidna and Platypus
(Order Monotremata)

The echidna and the platypus live exclusively in Australia, New Guinea, and Tasmania in areas ranging from hot, dry deserts to humid rain forests and alpine mountain heights. *They are the only oviparous (egg-laying) mammals in existence.*

**Echidnas.** The **echidna** *(spiny anteater)* is between 12 and 18 inches long and weighs about 10 pounds. It has short spines (about two inches long) over its body. Interspersed among the spines are coarse brown hairs, and on its underside there is hair only. As you may suspect, the spiny anteater has a long snout; a long, sticky tongue that it uses to lap up ants and termites; and a strong forefoot with flattened claws for digging. It has a long, narrow beak with no teeth and a very small tail. Although the echidna has weak vision, it has a good sense of hearing and a keen sense of smell. The creature can dig its way very suddenly into the earth when threatened, leaving only its sharp spines showing. If the ground is too hard for this disappearing act, it rolls itself into a spiny ball. For a home, the echidna burrows into the earth or finds a crevice in the rocks. It usually lives and hunts alone.

**Platypuses.** The **platypus,** which weighs only about 4½ pounds and is 1½ feet long, is designed to live both on land and in water. Its velvety fur is waterproof, and its long, flat body and paddlelike tail make it a graceful swimmer. The strong claws are webbed for swimming, but when the creature comes ashore, the webs on its forefeet pull up under its paws, exposing the claws for walking and digging. The distinctive bill is one of the platypus's most useful pieces of equipment. It is not hard like a duck's bill but is soft and rubbery and is covered with soft, sensitive skin. On land, the platypus uses the bill for digging. In water, the bill is the animal's chief sensory organ, because the eyes and ears close under water. The sensitive touch organs of the bill enable the platypus to find its way around in the water even though the animal is practically blind and deaf in that environment. In addition to its sense of touch, the platypus has special organs in its bill that detect electric fields, allowing the platypus to locate a frog or crayfish at a distance by sensing the tiny electrical currents generated by the prey's muscles. This fascinating creature lives in tunnels that it digs on the banks of streams, where it preys on worms, frogs, crayfish, and water insects. The platypus scoops up dirt and sand along with the food and keeps everything in its cheek pouches until it surfaces for air.

Fig. 16.6 **Animals of the order Monotremata**

Then the toothless animal apparently uses the minerals to help grind up the food. The platypus is similar to the echidna in that its stomach is lined with a hornlike epithelium and lacks digestive glands (digestive enzymes are produced in the intestine, however). On each hind foot, the male platypus has a sharp spur connected to a venom gland, which he uses for defense against predators. The platypus, strange as its appearance may be, has a body that has been well planned for its way of life.

## Armadillos, Sloths, and Anteaters (Order Xenarthra)

Order Xenarthra [zĕn·ărth′rə] was given a name meaning "strange joints" because the lower vertebrae of its members possess extra joints that are not found in any other group of mammals. This order consists of three families: sloths, armadillos, and anteaters. These families are all found only in the New World, mainly in South and Central America, but are very different from each other in size, shape, body covering, feeding habits, metabolism, and habitat.

Fig. 16.7 **Animals of the order Xenarthra**

**Armadillos.** Sixteenth-century Spanish explorers gave the **armadillo** its name, which means "little armored one." The armadillo is covered with neat rows of tough plates that are jointed across the back, allowing some species to curl up in a ball and tuck their head and feet in for protection when they are threatened. Armadillos live in South and Central America, ranging as far north as the southern United States. The kind that lives in the United States has nine bands of armor on its body; others living farther south have three to six bands. Some kinds are only about 6 inches long, but the *giant armadillo* of South America is 4 feet long and weighs over 100 pounds. The armadillo is a shy nocturnal animal that burrows into the ground to rest and hide. It uses its claws to scrape the ground in search of insects, spiders, earthworms, and snails. The armadillo always bears identical twins or quadruplets, causing all offspring in one litter to be of the same sex.

**Sloths.** The **sloths** spend the majority of their lives hanging upside down with their long, curved claws hooked around tree branches; they eat, move from branch to branch, and even sleep in this position. These arboreal animals have almost no tail or ears, a blunt nose, and peglike teeth. The trees in which they live provide their home, food, and protection. They feed on leaves, buds, and young twigs. They are camouflaged among the leaves and branches by the greenish-gray color of their algae-covered bodies. Because the hair of a sloth lies from belly to back—the opposite direction to that of other mammals—rainwater runs off easily while the animals are in a hanging position.

The two types of sloths, *two-toed* and *three-toed,* are among the slowest-moving animals. One reason they move so slowly is that their body temperature is relatively low (75°F–91°F). On the ground, they can stand on their feet but cannot walk. To move, they sprawl on their bellies and drag their bodies forward with their forelimbs.

The *ai* [ī], a three-toed sloth that lives in the tropical forests of South America, is about two feet long and gives the distress call "ai-ai" that inspired its name. *The ai is the slowest land mammal,* with a usual speed of 6 to 8 feet per minute. On one occasion a mother sloth was clocked at the rapid pace of 14 feet per minute ($\frac{1}{6}$ mi/hr) while responding to distress calls from her offspring. In the trees, this speed is slightly accelerated.

**Anteaters.** The largest **anteater** is the *giant anteater* or *ant bear.* It grows up to 6 feet long and weighs 75 pounds. With its long claws and powerful forelimbs it tears down large anthills and consumes as many as 30,000 ants and termites a day. Its long tongue is almost two feet in length and is covered with a sticky secretion. Giant anteaters live in the tropical forests of Central and South America as far south as Paraguay. For homes, they use the burrows of other animals or hollow logs. Giant anteaters are **diurnal** (active during the daytime) in uninhabited regions and **nocturnal** (active at night) in densely populated areas. The giant anteater is covered with a coat of shaggy, bristly hair and has a long, bushy tail that is about 2½ feet long. Although anteaters have no teeth and have relatively poor eyesight and hearing, they do have an acute sense of smell—about 40 times keener than that of humans.

Fig. 16.9 **Marsupials**

## **Pangolins** (Order Pholidota)

A **pangolin** [păng′gə·lĭn], or *scaly anteater,* has a protective hide like an armadillo and eating habits like the anteater. The characteristic that makes it different is the composition of its skin. The order name, Pholidota, means "horny scale" and describes the overlapping plates that cover the body of the pangolin. The plates are made from fused bundles of hair in the skin rather than from horn as in armadillos. Pangolins live in Southeast Asia, Indonesia, and parts of Africa south of the Sahara. They grow up to five feet long, but much of that length is a long, thick tail. Their sticky tongue is up to 10 inches long. They are sometimes hunted for their meat, but because they are nocturnal and shy, they are difficult to find. When threatened, they can roll up into tight balls. Like anteaters, pangolins have poor vision and hearing but a keen sense of smell.

Fig. 16.8 **Pangolin**

## **The Pouched Mammals** (Order Marsupialia)

All of the more than 270 species of pouched mammals **(marsupials)** are identified by an abdominal pouch, or **marsupium,** in which the young are raised. It is characteristic of marsupials to give birth to very tiny young (the young of the *mouse opossum* of Central and South America is no larger than a grain of rice). The young are kept within the protection of the marsupium and nourished by the mother's milk until they are large enough to move around on their own. Nearly all of the pouched animals live in Australia and nearby islands; the only exception is the opossum, found throughout North and South America.

**Kangaroos.** The largest living marsupial is the **kangaroo.** The *red kangaroo* weighs around 160 pounds and measures about 8 feet long. Males continue to grow throughout their entire lives, but as they get bigger they are less efficient at moving around. Kangaroos will hop on their powerful hind legs when pursued, achieving speeds of 25–30 mph for several miles and 40 mph for short distances. Biologists have found that a kangaroo, bouncing along on two legs, is actually more energy-efficient than a similarly sized four-legged animal running at the same speed. The kangaroo's design may seem clumsy at first, but the animal is built to carry a large body long distances in a hurry.

Some kangaroos can go for long periods, perhaps two to three months, without water, making them able to live in arid land. The animal conserves water by concentrating urine efficiently and by resting during the heat of the day.

**Opossums.** The **opossum** (commonly known as the *possum*) is the only marsupial outside Australia and the surrounding islands. It is about the size of a cat, with hairless ears and tail and a narrow, tapered snout. After young opossums mature in their mother's pouch, they cling to her back. One of the most unusual characteristics of the opossum is its acting ability. If attacked by a dog, an opossum will play dead by going limp, rolling over with its eyes shut and tongue lolling out. After the attacker loses interest, the opossum will get up and run away. Opossums are seldom seen by hikers because they are nocturnal. They eat just about anything, especially insects and carrion.

Fig. 16.9 (cont.) **Marsupials**

numbat

koala

wombat

**Wombats.** The **wombat** [wŏm′băt′], an animal 27 to 47 inches long and weighing up to 80 pounds, is one of the best diggers among mammals. Its body is thick, with little or no tail, and its legs are short and strong. The wombat's toes are armed with stout claws that help it dig through just about anything. Resembling a woodchuck (groundhog), the wombat has been described as a living tank as it digs through rough terrain littered with fallen trees and cut with ditches and streams. If the wombat did not have a pouch in which its young developed, it would fit nicely into the order of rodents, since its incisor teeth are very sharp and grow continually.

**Koalas.** The animal with the most confining diet may be the **koala.** It eats only the buds and leaves of the eucalyptus, an evergreen tree native to Australia. It spends most of its time in the trees, seldom coming to the ground. When it does descend, the koala will lick the soil to add grit and bacteria to its digestive tract, helping it digest the leaves. It also has a special stomach that helps to digest the bulky, fibrous food.

A newborn koala is only the size of a bumblebee and weighs only a fifth of an ounce; its mother is about 24 inches long. The young stays in the pouch for six months and then rides its mother's back until it can get around on its own.

**Numbats.** The **numbat,** or *banded anteater,* is a squirrel-sized marsupial that inhabits the eucalyptus forests and sandy deserts of southern and western Australia. Unlike most marsupials, the numbat is diurnal and retires for the night into a hollow log or a bed of leaves and grass it makes for itself. It has six or seven white stripes between the mid-back and the base of the tail. All numbats have a dark cheek stripe that runs through the eye, with a white line above and below. The body fur is short and coarse, and the tail is long and somewhat bushy. Numbats have long, tapering, pointed snouts and small mouths. The long, slender, cylindrical tongue can be extended for at least four inches—just right for extracting termites and ants from crevices in rotten logs and the subsurface soil. Termites are the preferred diet of numbats, and a numbat in captivity consumes up to 20,000 termites a day. The numbat uses its strong claws to scratch into decaying wood and soil, and its long snout sometimes functions as a lever in the extraction of termites, ants, and other small invertebrates. It swallows smaller termites whole, but the larger ones it chews up thoroughly with its 52 teeth—more than any other land mammal.

Numbats are just one of the many kinds of ant-eating mammals that we have observed. Fortunately for these mammals, there are as many as 8000 species of ants (and probably many more undiscovered and unclassified species). Besides ants, between 1600 and 2000 species of termites, or "white ants," exist—predominately in Africa, Asia, and South America, where termite mounds may measure 20 feet high. The total estimate of scientists is that there may be as many as 100,000,000,000,000,000 (100 quadrillion) ants and termites in the world. *These insects are God's provision for ant-eating mammals and for many reptiles, amphibians, and birds.* These creatures, in turn, help keep the ant population under control.

## Section Review 16.2

1. What is the largest living land animal?
2. What animals are commonly called sea cows?
3. What characteristic is most unusual about the monotremes (echidna and platypus)? What other unusual features does the platypus have?
4. What animal is the slowest land mammal?
5. What two animals mentioned in this section are covered with protective armor?
6. What physical characteristic do all marsupials have in common?
7. What is the largest living marsupial?
8. What is the only marsupial common outside Australia and the surrounding islands?
9. What mammals mentioned in this section feed primarily on ants and termites? Why might it be said that there is an abundance of food for these animals?

**Identify**

tusk, diurnal, nocturnal, marsupium

# 16.3 The Mammalian Orders II

## Insect-Eating Mammals (Order Insectivora)

As we have seen, many mammals are **insectivorous** (insect eating). One order of small mammals, the **insectivores** [ĭn·sĕk′tə·vôrz′], gets its name from the insect-eating habits of its members. The nearly 500 species of insectivores include many kinds of shrews, moles, and hedgehogs. These animals feed mostly on insects and worms but they will also eat seeds, nuts, and other plant materials. Because of their eating habits, insectivores are of great benefit to man, especially in agricultural areas, where they help to keep the vast numbers of insects in check. It is also fortunate for these creatures that insects are so plentiful, because many insectivores have such a high metabolism that they must eat almost constantly. Even several hours without food is enough to cause some insectivores to starve to death. These animals live such a rapid life, working furiously to collect enough food to survive, that many live little more than one year. The rapid life is evident in the heartbeat, which is normally about 600 beats per minute in the smallest insectivores.

Fig. 16.10 **Insectivores**

**Shrews.** **Shrews** are common insectivores that inhabit moist areas. They are small, short-legged, mouselike animals with long, pointed noses. The lightest mammal (although not the one with the smallest body) is believed to be the *Etruscan shrew* found in the Mediterranean region, South Africa, Pakistan, and India. An adult may measure only 1½ inches in length and weigh less than a penny. Its body is so small that it can crawl through a tunnel left by an earthworm. Biologists believe that this shrew has the smallest possible design for a warm-blooded animal because of the tremendous amount of food that it must eat to remain alive. A mammal much smaller than the shrew would not be able to eat fast enough to avoid starvation.

Shrews are busy 24 hours a day with only brief periods of rest. They spend most of their time underground, searching for insects, snails, and worms. A shrew may occasionally kill a mouse several times its size. Shrews will eat the equivalent of their own body weight each day. The salivary glands of some shrews secrete a poisonous substance that quickly immobilizes prey. Shrews have been known to bite people and inflict painful wounds.

**Hedgehogs.** The **hedgehog** is an interesting insectivore covered with spines that are shorter and softer than a porcupine's spines and always lack barbs. When threatened, the hedgehog rolls into a ball, exposing its spines for protection. A hedgehog is about 10 inches long and weighs 2 pounds. With its pointed snout, it searches for snails, slugs, insects, and worms. Hedgehogs are the longest-living insectivores (they may live up to 15 years) and are immune to the poisons of most insects. They are extremely durable animals, unlike the nervous little shrew, which may drop dead if startled by a loud noise. Perhaps some of their endurance is gained by their long periods of rest and hibernation. From October to April, these animals have a greatly lowered body activity and rely on stored fat supplies for energy.

**Moles.** Living almost exclusively underground, **moles** burrow and tunnel close to the surface, searching for their favorite food, earthworms. They will also eat insects and other small invertebrates. Although a mole's eyes are hidden within its pelt, and its eyesight is poor, it has no trouble finding its way around in the dark soil. In place of sight it uses its keen senses of smell and hearing. Able to detect vibrations at a distance, a mole can sense its surroundings by getting a picture of vibrations rather than of light.

Fig. 16.10 (cont.) **Insectivores**

Moles are excellent diggers. They can bury themselves, even in hard soil, within seconds. The snout is used to push and shove up the earth within the tunnel while the animal digs with its short, clawed limbs. Another design feature that enables the mole to move quickly through its maze of tunnels is its flexible coat of hair. Unlike the hair of other mammals, the hair of a mole bends in any direction. This provision allows the mole to move forward and backward in the tunnel with little resistance.

Moles eat as furiously as shrews and cannot survive more than a few hours without food. To provide for times of short supply, the mole stores up earthworms. Biting off the head end, the mole twists the worm into a knot and pushes it into a cavity in the soil. At times, a mole may have hundreds or even thousands of stored earthworms. The earthworms are not dead, though, but are in the process of growing new heads. When the process is complete, the worms will be able to escape.

## Flying Mammals (Order Chiroptera)

**Bats,** the flying mammals, differ in structure from other mammals in that their fingers are greatly elongated. Attached to these fingers and to each side of the bat's body, to its hind legs, and to its tail is a thin web, or membrane, of skin. The web consists of two layers of skin with very little flesh between them, only a little connective tissue containing blood vessels, nerves, and a few tiny muscles. This special structural design allows bats to fly. Most bats fly and feed at night. Because their wings have such great evaporative surfaces, bats demand more water or moist food than other mammals of comparable weight. The *Kitti's hog-nosed bat,* which is the size of a large bee, is believed to be the smallest mammal by overall body size.

Bats are among the most numerous of all mammals. Over 1000 species are known, and they live in all parts of the world except the Polar Regions. Probably the largest concentration of wild mammals found anywhere in the world is the bat colony in Bracken Cave near San Antonio, Texas. In the summer during a peak population time, as many as 20 million bats are found in that one place, eating an estimated 250 tons of insects every night.

**Echolocation system.** Contrary to popular belief, bats can see well, and some species have excellent vision. *Insectivorous bats, however, are amazingly well designed for catching flying insects in the dark by sound alone.* These small bats home in on flying insects by emitting

Fig. 16.11 **Bat using sonar**

high-pitched sounds and using their extra-large ears to detect the echoes that bounce from the insects. This amazing sonar system surpasses the precision, compactness, and flexibility of the best manmade sonar. With it, a bat can not only detect flying insects as much as 100 feet away but can also determine their size, texture, direction, and speed. Biologists believe that the bat's ability to use its echolocation system is so accurate that it can actually build a mental image of its surroundings, much as other animals do with their eyes.

The sounds that bats emit are in short blasts, lasting only 5 to 10 milliseconds and directed in a broad beam from the mouth; each pulse is accurately timed so that a new pulse is not sent until the previous one has been received. The frequency (pitch) of bat sonar pulses is generally between 12,000 and 150,000 hertz (cycles per second). (Humans cannot hear sounds with frequencies higher than about 20,000 hertz.) In "search mode," when the bat is looking for insects or obstacles, the pulses are produced at a rate of 30 to 50 per second. In "target mode," when the bat has located an insect, the pulse rate may increase to as much as 200 per second for maximum precision, enabling the bat to snatch its prey out of the air. Amazingly, even when tens of thousands of bats of the same species are flying in close quarters (such as in a cave) and generating sonar pulses, each bat can recognize and analyze its own faint echoes.

**Behavioral characteristics.** As a rule, bats are timid creatures and do not attack people. Although they are generally harmless to man, they should not be handled because they can carry rabies. Since bats eat large numbers of insects, they are actually valuable to man. Tropical bats eat fruit, drop the seeds, and help spread the plants. The *long-nosed bat* of Mexico and the southwestern United States has a long snout and tongue, which it uses to lap nectar from flowers, pollinating the plants upon which it feeds.

All bats are nocturnal creatures, even the fruit-eating bats that use vision and smell to locate food. After flying at night, bats seek cool, dark shelter during the day. This can be found in caves, crevices, tree cavities, and buildings, although some bats sleep in exposed locations on trees. Many bats like the undisturbed regions of dark caves because, with their sonar, they can safely maneuver in total darkness.

In colder areas, *bats spend longer periods of uninterrupted hibernation than any other mammal.* Most other hibernating mammals wake up from time to time throughout the winter, but bats stay in a hibernating state for as long as 64 days. They withstand cold so well that bats have been known to survive though ice crystals formed in their wing membranes. The tree-roosting *red bat* can even tolerate parts of its body freezing solid.

Perhaps the most feared bat is the common *vampire bat* of Central and South America. With razor-sharp incisor teeth, it makes a small incision in the skin of its victim (often livestock) and proceeds to lap up the blood that flows out. An anticoagulant in the bat's saliva prevents clotting of the blood so the bat can obtain a full meal. Although this causes little direct harm to the victim, vampire bats can carry rabies, making them a threat to livestock.

## Marine Mammals (Order Cetacea)

**Whales** (including dolphins and porpoises) are marine mammals that make up the order Cetacea. While closely resembling fish, cetaceans are true mammals in that they are warm-blooded, breathe with lungs, bear live young (called *calves*), and nurse their young with milk. Cetaceans are most easily distinguished from fish by their horizontal tail fins; fish have vertical tail fins.

Fig. 16.12 **Blue whale** (baleen whale)

**Baleen whales.** Whales are by far the largest of the mammals, and are probably the largest animals that have ever lived. **Baleen whales** are whales that have two rows of comblike plates (known as **baleen** or *whalebone*) that hang like curtains from each side of the upper jaw. A baleen whale uses its baleen as a gigantic strainer to filter its food from the water. To feed, the whale swims through a school of its prey (such as *krill,* a small marine crustacean) with its mouth open, trapping thousands of the animals in its mouth; it then closes its mouth, strains out the excess water through the baleen plates, and swallows the prey whole. In the days before plastics, whalebone was prized as just the right material for corsets, bustles, hoopskirts, buggy whips, umbrellas, fishing rods, and other articles that had to be elastic, lightweight, and strong. The giant **blue whale,** the world's largest animal, may attain a length of 110 feet and a weight of more than 150 tons—even more massive than the largest dinosaurs that ever lived. Blue whales spend the summers in the polar regions and migrate to equatorial waters in winter, navigating by the earth's magnetic field and by long-distance sonar.

**Toothed whales.** The **toothed whales** include such marine mammals as sperm whales, killer whales, dolphins, and porpoises. The giant **sperm whale,** which may be as long as 65 feet and weigh as much as 60 tons, has between 36 and 56 teeth in its jaws. Its enormous head is about a third of its body length. Sperm whales prey on large marine animals (most notably giant squid) and have been known to dive more than two miles beneath the ocean's surface in search of prey. During the heyday of American whaling, from 1815 to 1860, the sperm whale was the most sought-after whale, and each yielded huge quantities of whale oil for fuel. In addition, the sperm whale's huge head contains a large amount of a waxy substance known as *spermaceti* [spûr′mə·sē′tē], which was once used as a lamp oil, as a lubricant, and as a base for cosmetics. The sperm whale's intestinal tract produces a grayish, waxy material known as *ambergris* [ăm′bər·grĭs′], which is often found floating in tropical seas and was once used to make perfumes. The American whaling industry gradually declined after the beginning of the petroleum industry in 1859, and there has been no American commercial whaling since 1971 because of concern for the survival of certain species of whales.

Many scientists think that **dolphins** may be among the most intelligent of the mammals, ranking second only to the chimpanzee in intelligence. Two of the most common types of dolphins are the *bottle-nose dolphin* and the *common dolphin,* both of which live in all of the oceans of the world. Like bats, dolphins have a sophisticated sonar system for detecting underwater objects. They make sounds and then listen for the echoes that are produced when the sound is reflected from an object. The sounds are produced by blowing air through the air passages used in breathing. These passages lead to the blowhole, which has flaps of muscle that move to help produce different sounds. The sound is then sent out through the *melon,* a fatty organ lying in the bulge of the head. Like bats, dolphins can sense the shape and size of small objects at a distance by sound alone. Dolphins also have keen eyesight and hearing but are thought to lack a sense of smell. Some dolphins can grow up to 12 feet long and weigh 800 pounds, but most are smaller. They like the company of other dolphins, congregating in herds often numbering into the hundreds. Dolphins can apparently swim at more than 30 miles per hour and are noted for their habit of leaping out of the water. Dolphins feed on fish and squid. Some people confuse the **porpoise** with the dolphin. A dolphin has a beaked snout; the porpoise does not. The porpoise is not as playful as the dolphin and rarely leaps out of the water.

**Orcas** *(killer whales)* are another well-known toothed whale. These strikingly patterned black and white whales prey on such creatures as seals, penguins, dolphins, and small baleen whales. Although they commonly hunt in packs and are ferocious hunters in the wild, they are surprisingly gentle in captivity and, like dolphins, are noted for their intelligence.

Fig. 16.13 **Bottle-nose dolphin**

Fig. 16.14 **Orca (killer whale)**

## Tree-Dwelling Mammals (Order Primates)

The order Primates [prī·mā′tēz], which includes about 300 species, includes monkeys, apes (chimpanzees, orangutans, gibbons, and gorillas), aye-ayes, tarsiers, lemurs, galagos, lorises, and pottos. The **primates** [prī′mīts′] are mostly arboreal animals and are noted for their intelligence. Primates all have five fingers and toes, and their hands, which usually have opposable thumbs, are designed for grasping. Most also have toes that are designed for grasping. All have *nails* (flat structures designed to protect the fingers and toes) instead of claws. The eyes of these mammals look forward, and their field of vision overlaps in such a way that primates enjoy excellent depth perception. Most primates live in tropical or subtropical climates.

**Apes.** **Apes** are primates without tails. The largest of the apes is the **mountain gorilla** of the volcanic mountain ranges in East Africa. The average male stands about 5½ feet high when upright, has a chest 59 inches around, and weighs about 430 pounds. These large, powerful creatures have gained an unwarranted reputation as fierce killers when actually they are rather gentle vegetarians. They feed mainly on nuts, fruits, and other plant material, but many will also eat insects, eggs, and small lizards. Although not openly hostile, the gorilla is very powerful and able to fight to defend its family if necessary.

The smallest ape is the **gibbon,** which uses its long arms to swing hand over hand through the trees. When on the ground it walks erect, with a characteristic waving of the arms above the head. These small apes, 16 to 26 inches long, are a popular attraction in zoos because of their funny movements and loud voices.

The **orangutan** is characterized by its long, shaggy, orange-brown hair and large, sad-looking face. Although much smaller than the gorilla, weighing up to 165 pounds, it appears fairly large because of its shaggy coat. In its homeland, the swamp forests of Borneo and Sumatra, it lives in trees and eats mainly fruits. Many zoos have orangutans, which have adapted well to captivity. In 1971, one orangutan, in its boredom at the zoo in Topeka, Kansas, took up painting "abstract" pictures, some of which its keepers entered into an art contest. The ape won first prize and gained international acclaim as an abstract artist before the judges found out that the "artist" was really an animal.

Fig. 16.15 **Apes**

One of the best-known apes is the **chimpanzee,** which is considered to be the most intelligent of the apes and is probably the most intelligent animal. An adult chimpanzee generally stands about 3–5 feet tall and weighs between 50 and 130 pounds.

**Monkeys.** **Monkeys,** the most familiar primates with tails, vary widely in size and behavior. Monkeys are often divided into two groups, New World monkeys and Old World monkeys. **New World monkeys** live in the tropical forests of Central and

South America. They possess a **prehensile** [prē·hĕn′səl] **tail** (a tail designed for grasping objects), and their nostrils are widely spaced. New World monkeys are among the most graceful and beautiful of the primates. They include the marmosets, capuchins [kăp′yə·chĭnz], douroucoulis [do͞or′ə·ko͞o′lēz], howlers, spider monkeys, squirrel monkeys, and woolly monkeys.

Fig. 16.16 **Woolly monkey**

**Old World monkeys,** noted for their close nostrils and nonprehensile tails, live only in the Old World—Africa and Asia—and not in the Western Hemisphere. Old World monkeys include the *mandrill* of the high forest regions of Africa, the *colobus monkey* of the rain forests of central Africa, and the *baboon* of the African savannas. Leaves make up about 80% of the food of many Old World monkeys. These monkeys have a special kind of stomach that is divided into compartments. They also have large salivary glands. These specialized structures help leaf-eating monkeys digest their coarse food.

**Lemurs. Lemurs** are an unusual kind of primate that live only in Madagascar and nearby islands. They vary in size from that of a mouse to the size of a small dog. Lemurs eat a variety of foods but mostly eat leaves, fruits, and bark; some eat a few insects as well. The *mouse lemur* has the fastest development of any primate, being independent of its mother in only four months and fully mature between seven and eight months. All but one kind of lemur have long, bushy tails and live in trees most of the time.

mouse lemur

ring-tailed lemur

Fig. 16.17 **Lemurs**

## Gnawing Mammals (Order Rodentia)

The **rodents** are the small gnawing mammals that have two pairs of sharp, chisel-like incisor teeth. These large teeth do not wear out because they keep growing as long as the animal lives. Rodents are often very destructive animals. Approximately 2000 species of rodents are known to biologists. *There are more rodents than there are all other mammals combined.* They range in size from the tiny mice to the four-foot-long capybaras of South America. The rodents include such familiar animals as the porcupine, gerbil, hamster, squirrel, chipmunk, beaver, and rat. Rodents are the most widespread of all the mammals, and many of them seem to follow man wherever he goes.

**Mice.** The common *house mouse* is found on every continent, including Antarctica, where it lives in the heated dwellings of man. **Mice** make nests of shredded material wherever suitable cover and food are present. They feed on human food and also paste, glue, soap, and other household articles. They often do more damage by gnawing than by eating, and they seem to survive on very little food. In the wild they eat many kinds of vegetables, such as seeds, fleshy roots, leaves, and stems. Mice are among the most prolific creatures; they breed throughout the year and may have five or more litters of 3 to 12 young in one year. Young mice can begin to reproduce when they are only 35 days old. The most effective control on mice are God's created mouse catchers, such as cats, weasels, owls, and snakes.

**Rats. Rats** look like mice but are larger, stronger, and much more aggressive. The smallest kinds of rats grow longer and weigh more than the largest mice. Some rats are as large as small housecats; body lengths of up to 1 foot (excluding tail) have been recorded, but most rats are somewhat smaller. Rats breed rapidly; a single pair can produce more than 100 young per year under favorable conditions. Rats, like mice, seem almost impossible to exterminate. When their population is threatened, the surviving individuals become stronger and more productive. Man has been able to control the numbers of rats but never been able to eliminate them from areas where people live. *The best methods of control are sanitation* *procedures.* Keeping populated areas clean and free of wastes often does more than all the rattraps that can be set. Rats are of grave concern to health officials, because rats carry diseases fatal to man such as *bubonic plague, food poisoning,* and *typhus.* Most of the time these diseases are not the result of attacks of rats upon humans but of attacks by the fleas and bacteria living on the rats. Rats also cause millions of dollars of damage yearly in the United States by eating stored grains and gnawing on furniture, pipes, and electrical wiring.

**Capybaras.** Fortunately, the mice and rats of populated areas do not reach the size of the **capybara** [kăp′ə·bär′ə], the largest rodent. Growing up to 4 feet long and weighing over 100 pounds, the capybara lives in South America east of the Andes; a smaller species lives in the southern part of Central America. The capybara grazes near lakes and rivers and carries no threat to man; in fact, it is easily tamed and makes a good pet. Looking somewhat like a giant guinea pig, it has webbed toes and swims well. It is a favorite food of the jaguar.

**Squirrels.** Another familiar rodent similar in size to the rat is the **squirrel.** There are over 250 species of squirrels around the world comprising several types: ground squirrels, chipmunks, woodchucks, tree squirrels, and flying squirrels. Most people are familiar with the common *tree squirrels.* The *gray squirrel* lives in hardwood forests, where its grayish color blends well with the bark of trees such as oak, hickory, and maple. The *red squirrel* has a reddish color that blends well with the reddish tint of pine trees; it lives well in coniferous forests, particularly in the southeastern United States. Tree squirrels hibernate during the winter but wake up occasionally to eat from the store of nuts and seeds they collected during the autumn months.

Fig. 16.18 **Rodents**

*Ground squirrels,* which have shorter, less bushy tails and larger feet than tree squirrels, dig burrows in the earth or find shelter in hollow logs, under logs, or among rocks. A typical behavior of ground squirrels, when alerted, is to rise on their hind legs to look for danger. Ground squirrels eat seeds, nuts, roots, leaves, bulbs, fungi, insects, birds, and eggs; but they in turn are the prey of the coyotes, bobcats, weasels, lynxes, and hawks. Chipmunks, marmots, prairie dogs, and woodchucks are often called ground squirrels, but they are not in the same genus.

*Flying squirrels* make their home among the trees and inhabit areas where they find their favorite food: hickory nuts. *Flying squirrels have gliding flight instead of powered flight like the bats.* They glide through the air quite efficiently by spreading a large flap of skin out from their limbs and body. At the start of the flight, after deciding where it will go, the squirrel leans forward, pushes with its hind legs, and spreads all four legs at right angles to the body so the flap of skin on the two sides nearly forms a square. The tail is used as a balance and a rudder. A spur on each wrist joined to the flap of skin is used to tighten or slacken the outstretched flap. By dropping the legs of one side to give added lift to the skin on the other side, the squirrel can bank or turn sharply. It can also dive steeply and use the speed to rise sharply at the end of the dive.

**Woodchucks.** A **woodchuck** or *groundhog* is a stocky, burrowing rodent that belongs to the squirrel family; it may grow more than 16 inches long and weigh up to 30 pounds. Woodchucks, which populate the eastern

United States, are very destructive creatures to farm and ranch lands because of their huge appetites and burrowing habits. These animals build tunnel complexes covered by mounds perhaps several feet high. These mounds interfere with farming and create hazardous holes where livestock may break their legs.

**Gophers. Gophers,** about halfway between a squirrel and woodchuck in size, populate the central and western areas of America and belong to a family of their own. These burrowing rodents display a community behavior, often living in long, complex, underground tunnels that form a gopher community, often consisting of hundreds of individuals.

**Beavers. Beavers** are the largest rodents in the United States; adults can be over 40 inches long (including tail) and weigh as much as 60 lbs. These ingenious engineers construct large dams to support their underwater homes, which they build of sticks and mud. The backed-up water, which never freezes solid in winter, keeps a steady environment for rearing their young and for storing their food, which consists of the bark of young saplings of willow, aspen, poplar, birch, and alder trees. Most beaver dams reach less than 75 feet in length, but one spotted in 1937 in Colorado was over one thousand feet long with a number of smaller dams above and below it. The largest beaver dam ever built may have been on the Jefferson River in Montana; it was roughly 2300 feet long and could bear the weight of a horse and rider. Huge dams are maintained by succeeding generations of animals, but much can be done by one or two animals. For example, observers watched two Canadian beavers cut down 266 trees in 15 months and build three 50-foot dams and a lodge with a capacity of 1000 cubic feet—the equivalent size of a small bedroom. It is interesting to note that these engineers are not especially intelligent but use the instinctive skills with which they were created. The large, broad tails of beavers make them excellent swimmers and also serve as a form of communication; when a beaver senses danger, it loudly slaps the water with its tail as it dives, warning other beavers in the area of the threat.

Fig. 16.18 (cont.) **Rodents**

**Porcupines.** One of the most unusual rodents is the **porcupine,** which has long, stiff *quills* on its back, sides, and tail. Quills are long, sharp bristles composed of keratin. Porcupines attack predators with their quills, which come out easily and stick to the attacker. Some porcupines have barbs at the end of their quills, which make the quills almost impossible to remove without cutting them out. Animals that fall victim to these arrows often die of festering wounds because the quills introduce bacteria into the wound and the barbs prevent their removal. Despite the use of these weapons, porcupines are occasionally preyed upon by carnivorous mammals and by the great horned owl.

## Flying Lemurs (Order Dermoptera)

**Flying lemurs** are nocturnal animals of the tropical jungles of Malaysia, Thailand, Borneo, Java, and the Philippines. These animals were once confused with lemurs, a kind of primate, because of their arboreal habits, and that name remains with them. A flying lemur is about the size of the domestic cat, having a slender body, a long neck, and a doglike head. *It is the longest-gliding mammal,* able to glide several hundred yards at a time. This is made possible by a large, leathery membrane stretched from just behind the head to the tip of the tail with a spread of 20 inches. Flying lemurs eat mainly fruits but also feed on leaves and flowers. Young flying lemurs ride on the mother's back until they are able to care for themselves.

Scientists have given the flying lemurs the name Dermoptera, meaning "skin wing." Flying lemurs are not grouped with flying squirrels, which are rodents, based on one characteristic: their teeth. A rodent's incisor teeth grow continually; the flying lemur's teeth do not.

## Harelike Mammals (Order Lagomorpha)

There are almost 90 species of **lagomorphs.** These animals have teeth similar to those of rodents, but they have four incisors in the upper jaw instead of two. Lagomorphs (hares, rabbits, and pikas) are herbivores. The tails of these mammals are very short, and their hind legs are longer than their front legs.

Fig. 16.19 **Harelike mammals (lagomorphs)**

**Rabbits.** The **rabbit** is a well-known lagomorph. The only means of protection that this animal has are its keen senses and its speed. To escape harm, the rabbit can jump 16 feet and run up to 35 miles per hour for short bursts. Rabbits are very agile and will often zigzag to elude a predator. A rabbit may even backtrack its own trail to confuse predators that hunt by scent.

**Hares.** **Hares** are much like rabbits except they are larger, reaching 25 inches in length, and they tend to have longer ears. Their large ears act as radiators to expel excess heat. The hare is a very excitable animal with rapid body activities that create much heat. An earless hare would overheat and die. Hares give birth to young that have eyes open and a full coat of hair, whereas rabbits give birth to naked young with closed eyes. Hares, like rabbits, eat plants and roam at night, returning to their homes in the morning. Hares live in grassy hollows, whereas rabbits burrow into the ground for a home. The common *jackrabbit* of the southwestern United States is really a hare and not a true rabbit. The *snowshoe rabbit* is also a hare.

**Pikas.** The **pika** is a most unusual lagomorph because it is very small, about 7 inches in length; it has short ears and looks more like a guinea pig than a rabbit. Pikas live in Asia, Europe, and western North America. American pikas live on mountainsides beyond the tree line. They spend much of their time collecting food for winter.

## Hyrax (Order Hyracoidea)

The **hyrax** [hī′răks′], also called a *coney*, is described in the Bible (Prov. 30:26) as a defenseless animal that makes its home in the rocks. This group of rabbit-sized herbivorous mammals is limited to Africa and Syria. Hyraxes look somewhat like short-eared rabbits, but their teeth are arranged differently than those of lagomorphs. Hyraxes have hooves on their toes and pads on the soles of their feet. The soft foot pads allow the otherwise defenseless animal to cling to the rock for protection. Hyraxes usually live in colonies of from 5 to 50 individuals.

Fig. 16.20 **Cape hyraxes**

## Section Review 16.3

1. What insectivore is considered the lightest mammal? Name the other two insectivores described in the text.
2. What are the two broad groups of whales?
3. List four characteristics of primates.
4. What is the most obvious way to distinguish between an ape and a monkey?
5. What is the largest group of mammals?
6. What characteristic do all rodents have in common?
7. What are three kinds of lagomorphs mentioned in the text?

### Application

8. How do insectivores benefit mankind?
9. What are three ways in which bats are helpful to man?
10. What are some diseases that can be spread by rats?

### Identify

insectivorous, insectivore, echolocation, baleen, melon, orca, prehensile tail

### Observe

- Make a chart of 10 nondomestic mammals that live in your region. If possible, include the name and a picture of each, its order, habitat, feeding habits, economic importance, what it does during the winter, and whether it is diurnal or nocturnal.

## 16.4 The Mammalian Orders III

### Odd-Toed Hoofed Mammals (Order Perissodactyla)

**Ungulates.** Some of the most valuable mammals to man are those whose toes end in **hooves** (enlarged and thickened toenails). Hoofed animals are called **ungulates** [ŭng′gyə·lĭts], and scientists divide them into two groups according to the number of toes they have. Most of them (cattle, buffalo, antelope, deer, sheep, goats, camels) have an even number of functional toes (two or four), giving them their order name, **Artiodactyla** [är′tē·ō·dăk′tĭ·lə: Greek *artios,* "even," + *dactulos,* "toe"]. Three families (horses, tapirs, and rhinoceroses) have an odd number of functional toes, giving them their order name, **Perissodactyla** [pə·rĭs′ō·dăk′tĭ·lə: *perissos,* "odd"]. *Most of the ungulates have been designed as swift runners because they need a means of escape from predators.* The *pronghorn* (an even-toed ungulate) of the United States plains is capable of sustained speeds in excess of 50 miles per hour and can briefly exceed 60. Although the African cheetah can sprint 100 to 200 yards at nearly 70 miles per hour, it could never keep up with the pronghorn for mile after mile.

*All ungulates, whether odd-toed or even-toed, are primarily herbivorous.* They graze on grass and herbs; browse on bark and leaves from trees and brush; and forage for nuts, roots, fruits, and seeds. Both groups of ungulates have cropping teeth in the front of their mouths, but horses, tapirs, and rhinoceroses (the perissodactyls) do not have the complex stomach found in the even-toed ungulates.

Fig. 16.21 **Perissodactyls**

**Rhinoceros.** The **rhinoceros** is a very large perissodactyl with three functional toes on each foot. Some species weigh as much as 3½ tons. A rhinoceros has one or two horns on the front of its head, which it uses for defense. Often the rhinoceros is pictured wallowing in the mud. Since it has no sweat glands, the water cools it off and the mud creates a protective coating which shields it from the sun. Rhinos are amazingly fast and agile for their size. A large rhinoceros can charge at nearly 35 mph, out-accelerate a horse from a standing start, and turn faster than a horse.

**Tapirs.** The **tapir** is an unusual creature that looks much like a pig with an extra-long snout or short trunk. It is about the size of a pig and lives in the humid tropical areas of Asia and South America. It has four toes on its front feet and three on its hind feet. The tapir's diet includes fruits and grasses, but it prefers the young, tender shoots, leaves, and branches of certain shrubs. The number of tapirs is quite small. Once they were a source of meat for people, but today they are protected by the countries of their habitation.

**Horses.** Horses, zebras, and donkeys are all **equids** [ĕk′wĭdz], members of the *equine* [ē′kwīn′] *family.* The **horse,** a single-toed perissodactyl, has been one of the most useful animals in God's creation. Since earliest history, horses have been used for hunting, farming, travel, and warfare. In several civilizations, horses were the symbol of power and riches, and much of man's economy revolved around their use. These animals are intelligent; they have a good memory and are easily trained for tasks.

Man has selectively bred horses for particular purposes and environments. The large *Shire* and *Belgian* horses, weighing over a ton each, were used in Europe for working and war. The small *Welsh pony,* standing only 4 feet high at the shoulders, was used to haul burdens in Welsh coal mines.

*Arabian* horses, probably the oldest living line of horses, were bred in the Middle East especially for the desert environment because of their great endurance. Through thousands of years of selective breeding, horses have taken on a wide variety of characteristics, yet a horse is always a horse.

Horses are not used as much in Western man's economy as they once were. For example, in 1910, there were

20 million horses in the United States, while today there are just over 8 million. They have been replaced, of course, by machinery. However, in some nonindustrialized parts of the world, horses are still of great economic importance, and horses were used in military campaigns as late as World War II (primarily for hauling supplies).

Scientists have found fossil evidence of horses in all continents except Australia. After some catastrophe, horses disappeared from the Western world, surviving only in the East. The people of the North and South American continents had no horses until the Spanish brought them to Mexico in 1519. From those few animals came the herds of "wild" American horses. The only horses considered truly wild by scientists are the *Przewalski's* [shə·väl′skēz] *horses* in Mongolia.

> Hath thou given the horse strength? hast thou clothed his neck with thunder? . . . He paweth in the valley, and rejoiceth in his strength: he goeth on to meet the armed men. . . . The quiver rattleth against him, the glittering spear and the shield. He swalloweth the ground with fierceness and rage. . . . —Job 39:19–24

zebra

Fig. 16.22
**Equids**

**Zebras and donkeys.** **Zebras,** also perissodactyls, look much like striped horses. Zebras have a much wilder nature than horses and are nearly impossible to tame, although it has been tried many times. Zebras are resistant to many diseases that afflict horses, making a cross between a horse and a zebra an ideal animal for the African rancher. However, horses and zebras are considered separate species because crosses between them are always infertile.

**Donkeys** are perissodactyls that have also helped man since earliest times. The book of Job, one of the oldest books of the Bible, mentions wild donkeys *(onagers)* in a number of places (see Job 6:5; 11:12; 24:5; 39:5–8); and the Sumerians, Job's contemporaries, used onagers as beasts of burden and animals of war to pull their chariots. Although donkeys have been crossed with horses, their offspring—*mules* and *hinnies*—are almost always sterile.

## Even-Toed Hoofed Mammals (Order Artiodactyla)

There are 213 species of **artiodactyls,** and all of them are herbivores. Many of our domesticated animals are members of this order. Most artiodactyls have two toes, which form hooves, although the hippopotamus and a few other ungulates have four toes. Many artiodactyls have either **antlers** (solid hornlike structures that are shed annually) or **horns** (hollow structures that are usually permanent and thus not shed). Cattle, buffalo, sheep, goats, and antelope all have horns; deer have branching antlers.

Artiodactyls are the most diverse and numerous of all the ungulates, with species native throughout the world, except Australia and Antarctica. They vary in height from the 15-inch *royal antelope* to the 18-foot *giraffe,* the tallest living animal. The long skulls of artiodactyls provide a large surface for the nerve receptors used in the sense of smell. Their eyes are placed on the side of the head, limiting their *binocular vision* (use of both eyes together) but increasing their peripheral (side) vision. This design of nose and eyes allows the grazers to consistently feed and at the same time watch for predators.

**Chewing the cud.** *The artiodactyls can be divided into those that do not chew the cud* (pig and hippopotamus) *and those that do chew the cud* (cow, sheep, etc.). The term *"chew the cud"* refers to the habit of hastily chewing and swallowing food to be later regurgitated and chewed more thoroughly. Animals that chew the cud are called **ruminants** [ro͞o′mə·nənts]. Ruminants have a stomach that is divided into four sections. The first division of such a stomach, called the *rumen,* is the section that holds the unchewed food until it can be masticated (chewed) thoroughly.

# Horns and Antlers

Understanding the LIVING CREATION

True **horns,** found in ruminants such as sheep and cattle, are composed of hollow sheaths of keratinized epidermal cells that surround a core of bone arising from the skull. Horns, which can be quite curved and twisted, are used by animals as protection from predators. Although horns are not usually shed, the horns of the North American *pronghorn* are shed each year after the breeding season. The new horns replace the old ones by growing up inside and pushing off the outer sheath. The largest horns belong to the *water buffalo* of India; horn spreads of 6 feet or more are fairly common, and spreads exceeding 13 feet have been reported in exceptional animals. Horns on cattle that are in close confinement may be a problem, because the animals may injure each other or people. For this reason, farmers often will remove the horns; they also have bred animals that do not grow horns or that grow only small ones. In some species both the males and females have horns; however, the male sex hormone greatly influences the development of larger horns.

The horn of a *rhinoceros,* a perissodactyl, is not a true horn. Hair-like keratin fibers projecting from papillae in the dermis are cemented together to form the rhinoceros horn.

**Antlers** are the bony elevations on the skull of deer. Usually only the male deer will grow antlers, but the female reindeer also possesses them. *Unlike horns, which grow from skin cells, antlers are solid bone.* In the spring of the year, the buck (male deer) begins to develop knobs on the top of its skull.

## Horns

Stimulated by growth hormones from the pituitary gland, the bone grows rapidly, pushing into the air and branching. This growth, which needs much nourishment, takes place beneath a layer of soft skin that is rich in blood vessels. This skin covering is called **velvet** because of its appearance. At just the right time of development, the blood vessels in the velvet constrict (close off), and the skin dies. The animal then sheds the velvet just before the breeding season. The new set of antlers is ready for use as a weapon that the buck uses to fight other deer in competition for territory or to protect his doe (female deer) from predators.

The making of the set of antlers takes place in a matter of weeks and puts great stress on the animal's body, as it requires large amounts of minerals. In the fall and winter, the antlers are dropped, and a few months later, when spring is near, the buck begins the process over again. Each year, as the buck grows older, new size is added so that its antlers are larger than those of the younger animals. The *Alaskan bull moose,* the largest of all the deer, has a huge set of antlers, which may spread over five feet across. To grow such a set of antlers, the moose must accumulate an extra 50 pounds of calcium salts from its vegetable diet, which is not very high in calcium.

## Antlers

**Importance of artiodactyls.** *Artiodactyls are among the most important animals on earth.* As you recall from your study of botany, energy from the sun is captured through the process of photosynthesis. Artiodactyls convert this energy into forms that man and many flesh-eating animals can use—meat and milk.

Fig. 16.23 **Cape buffalo** (bovid)

**Bovids.** The artiodactyls of the family Bovidae, known as **bovids,** have been particularly valuable to mankind since ancient times as sources of food and clothing. All bovids are ruminants with unbranched hollow horns and are especially well equipped for grazing. Most inhabit grassland, scrub country, or desert; but some live in forests, swamps, or the arctic tundra. Familiar animals in this family include cattle, sheep, goats, and antelope. *Goats* are noted for their milk, which is higher in fat and protein than cow's milk and is especially good for making cheese. Probably more people in the world use goat's milk than cow's milk.

More exotic varieties of bovids include the *water buffalo,* a type of ox, which eats the buds of aquatic plants and grasses, and the *Cape buffalo,* a fierce creature that is very strong and easily provoked. The *yak,* an ox that lives in the Himalayas at altitudes up to 20,000 feet, has a thick coat of hair to protect it against the cold and broad hooves that enable it to roam nimbly in rugged mountain terrain.

**Deer.** The **deer** belong to a family of ruminants that are distinguished from the bovids by the growth of antlers rather than horns. Most deer are spotted when they are born, but as they grow older their coats change to a solid white or chocolate brown, yellowish brown, or grayish brown, often with distinct white markings. A few retain their spots throughout their lives. There are 43 species of deer, some of which are important to man. The early Egyptian people used the deer in some sort of managed way as a source of food and leather. The people of the Arctic have for centuries depended heavily upon various species of deer for survival. The Laplanders of northern Europe, for example, tend herds of reindeer; the Inuit peoples of northern Canada hunt caribou, moose, and elk to supply their needs. During the early history of the United States, the hunting of deer supplied many Americans with venison and buckskin. Even today, the common white-tailed deer is one of the most widely hunted animals in the United States.

**Camels.** The **camel** is a rather unusual creature designed to withstand extremes of hot, dry lands. For millenia, desert dwellers have used camels as a prime source of meat, milk, and clothing and as a means of transportation. Camels can survive temperatures ranging from −20°F to over 100°F. Thick pads connect the two toes on each of the feet, giving the animal snowshoelike support on soft sand, traction on steep mountain trails, and insulation from hot and cold ground. The camel eats

Fig. 16.24 **Artiodactyls**

a wide variety of vegetation, including highly alkaline plants that are unpalatable to other animals. The camel's stomach has three chambers, and, as in other artiodactyls, fermentation aids its digestion.

Unlike most animals, the camel stores fat in large humps atop the body instead of in a layer under the skin. This allows the camel to shed heat more easily than an animal with a layer of subcutaneous fat. The camel's large fat stores serve two purposes: they allow the camel to go for long periods without food, and they serve as a source of water (water is produced when fat molecules are broken down in the body). Another unusual

## Artiodactyls and Cellulose

Understanding the LIVING CREATION

Many of God's creatures, including all artiodactyls, are herbivorous. Plant-eating animals must eat large amounts of food to receive the nourishment they require. Those herbivores that eat many grasses, called *grazers,* always seem to be eating. Because no mammal has the ability to digest grasses completely enough to survive, every one of the herbivorous mammals relies on help from tiny microorganisms in its alimentary canal to digest the food for it. Most of the carbohydrate food in grass is in the form of *cellulose,* which is composed of long molecular chains of glucose, a simple sugar. Although mammals do not have the ability to produce an enzyme that will break down the cellulose into sugar, some microorganisms do. The process used by microorganisms to break down cellulose is called **fermentation.** Within the digestive tract of many herbivore mammals is a special place where microorganisms ferment the cellulose of the plant food. In horses (perissodactyls) and rabbits (lagomorphs), a special side pocket, called a **cecum,** lies near the stomach to provide for a fermentation chamber. Sometimes this side-pocket style of digestion does not complete the job, and the food is eaten twice to complete the process. This may be done in two ways. Hares, rabbits, rodents, and pigs often eat their own fecal material, giving the food a second pass through the digestive tract. This activity may be one reason God instructed the Israelites not to eat the flesh of these animals, because the practice of eating fecal material spreads disease.

A second means of twice eating the same food is demonstrated by the ruminants, the majority of artiodactyls, including cattle, bison, goats, antelope, sheep, deer, and giraffes. *The ruminants possess a special four-chambered stomach in which fermentation can take place.* When these animals eat, the food passes down the esophagus to the **rumen,** the largest portion of the stomach, where it is acted upon by a rich supply of bacteria and formed into small balls called **cud.** At the animal's leisure, the rumen returns the cud to the mouth, where the food is chewed at length to crush the cellulose fiber. Swallowed for the second time, the cud returns to the rumen, where it is digested by special bacteria that break down the cellulose. The rumen then moves the food to the *reticulum,* then to the *omasum,* and finally to the *abomasum,* which acts more like the human stomach by digesting the proteins in the vegetation. In this fashion, the ruminants are able to eat their food twice, getting the most out of it and still allowing only one final pass as feces. The quality of flesh within these animals in the wild is far greater than that of those that do not chew the cud.

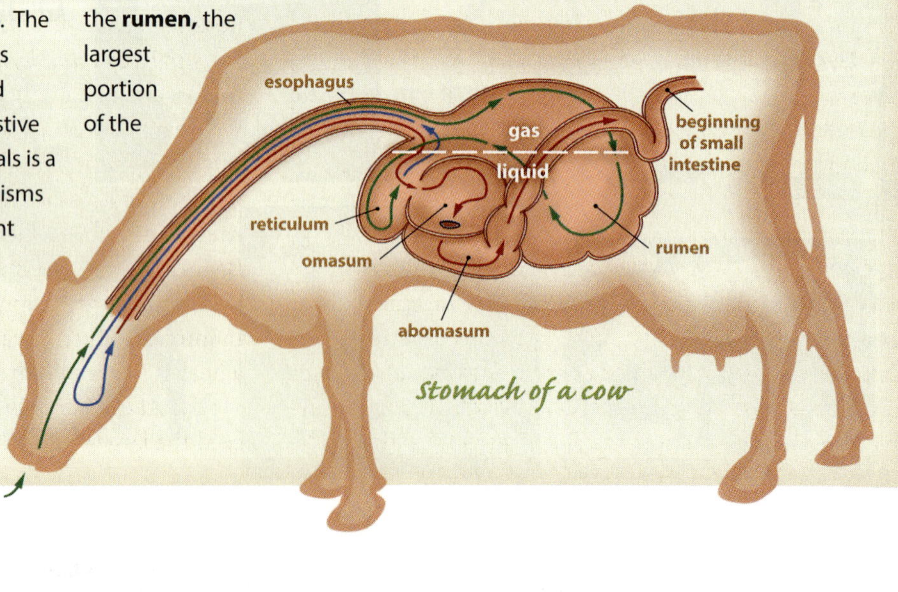

Stomach of a cow

characteristic that helps the camel to live in arid lands is its ability to become dehydrated and still survive. A person who loses 12% of his body fluids is unable to function, but a camel can lose as much as 27%, well over 44 gallons (350 pounds) of water, and still survive. To restore that weight, a camel can drink 200 to 300 pounds of water in a brief time. This ability to handle water comes from the unique way in which the camel processes water. Its water is lost from the tissues of the body rather than from the blood, and thus the heart is not strained by blood that is too thick. An emaciated camel with ribs showing can look fat and swollen within minutes after drinking. Other design advantages given to camels include the ability to excrete concentrated urine, which wastes little water; muscles that can tightly seal the camel's nostrils against blowing sand; and thick eyelashes and ear hairs that keep dust and sand out of the eyes and ear canals. Because of their ability to carry heavy loads under hot, dry conditions, camels have been called "the ships of the desert."

## The Flesh Eaters (Order Carnivora)

Scientists use the Latin words *carnis,* "flesh," and *vore,* "to eat," to name the **carnivores,** or meat eaters. There are over 270 species of carnivores, including many species of bears, cats, dogs, foxes, raccoons, weasels, seals, walruses, sea lions, pandas, skunks, badgers, otters, mongooses, civets, and hyenas. In general, *carnivores lead a more active life than herbivores.* Herbivores must spend much of their lives eating (grazing, browsing, foraging) because their food is not as concentrated with nutrients and is difficult to digest. Carnivores eat the herbivores, which have already done the job of concentrating the nutrients of plants into their flesh. Because flesh is high in protein and minerals and is easily digested, carnivores have a shorter digestive tract, lacking a complex stomach or cecum in which fermentation takes place.

Carnivores eat separate meals and have time in between to play, explore, and develop their hunting skills. They are noted for their cunning abilities of stalking their prey, taking time to be at the right place at the right time to surprise an unattentive herbivore. Using powerful clawed limbs and teeth well designed for biting and piercing, the predator overpowers the prey and easily kills it. Usually, the only hope of survival for prey is a swift exit.

**Bears. Bears,** the least carnivorous of the flesh eaters, take advantage of several kinds of food. *Bears are more accurately described as omnivores* because they live on both plant and animal foods. When they come out of their winter sleep, they probably consume more vegetable material (grasses, leaves, and roots or tubers) than meat. When berries are ripe, they usually eat them. Bears are fond of some insects and ants, and their fondness for honey is proverbial. By turning over stones and moving small logs or masses of vegetation, they catch small mammals such as mice. Those that are near streams will search diligently for food along the shore, and they catch fish, especially salmon where available. Because of their eating habits, bears contain *trichina worms* and need to be cooked thoroughly if they are killed for food. Bears are extremely strong animals, able to kill and drag an animal larger than themselves. Unlike most other mammals, which walk on their toes, the massive body of a bear is supported by a flat-footed stance.

The *Alaskan brown bear* (or *Kodiak bear*) is considered the largest land-dwelling carnivore; some of these

Fig. 16.25 **Bears**

bears have a nose-to-tail length of over 8 feet and weigh over 1600 pounds. These animals live on Kodiak Island and adjacent lands in the Gulf of Alaska. Kodiaks are a type of **grizzly bear,** some of which live in the western United States. Although grizzly bears may be the most dangerous animals of North America, they usually do not attack unless threatened. Their seemingly friendly appearance is often misjudged by people who do not understand that bears are dangerous predators. Their somewhat ungainly appearance is also misleading; a charging grizzly can reach 30 mi/hr on level ground, far faster than the average human can run.

The **polar bear** of the Arctic regions is the only large land carnivore that does not instinctively fear man. The polar bear will often deliberately stalk and kill a human for food. The polar bear is an excellent swimmer and preys primarily on seals.

**Dogs and wolves. Wolves,** which are members of the dog **(canine)** family Canidae, are ruthless and ferocious carnivores with a justified reputation as cunning attack animals. Running in packs of two to three dozen animals, wolves use their intelligence and a high degree of cooperation to select and surround their prey. Wolves are enormous eaters, perhaps consuming one fifth of their body weight at one meal. (That is comparable to a 100-pound person eating forty 8-ounce steaks at one sitting.) After such a meal, a wolf can go for days without eating, or even weeks if necessary. Wolves live in the wilderness areas of northern America, Europe, and Asia. There they feed on caribou, moose, or deer, especially those animals that are sick, injured, or older.

Fig. 16.26 **Gray wolf**

A wolf resembles a malamute or Siberian husky dog but is even larger and more powerful. A large male may be 6 feet long (including tail) and weigh 125 pounds. Wolves usually mate for life, keeping very close-knit family units. Pups are born in a den dug into the earth and are cared for constantly by both parents. Pups are trained in the art of hunting and in social behavior, which is very well-defined among wolves. The wolf pack, usually made up of family groups, travels together in search of food, with the howls of individuals keeping them together.

*Other canines that are apparently members of the same biblical kind as wolves are domestic dogs, dingoes, coyotes, and jackals,* all of which can interbreed with wolves and produce fertile offspring. It is likely that these animals are all descendants of one originally created pair that possessed a great deal of genetic variety. Despite their doglike appearance, *foxes* do not appear to be members of the wolf kind because of fundamental genetic differences.

**Weasels and wolverines.** One of the most relentless hunters among the carnivores is the **weasel.** Stalking its prey mainly by scent, the weasel hunts at night; it pounces swiftly, killing the prey with a bite to the back of the head. Attacking rats, mice, voles, frogs, birds, and occasionally even fish, the weasel is courageous and ferocious. It will attack animals larger than itself, such as rabbits, and has even been known to attack man when he stands between it and its prey. This is amazing considering the typical weasel's size—only about eight and one-half inches long plus a two-inch tail. The male is larger than the female, weighing about four ounces. A weasel marks out a territory about 200 yards across and lives there in a den until practically all of the rodents have been taken as prey before it moves on to a new territory. A male weasel may kill at least 500 small rodents each year, keeping an area fairly free of rats and mice. *The largest member of the weasel family is the wolverine,* which may weigh up to 50 pounds. Wolverines are noted for their strength and ferocity.

Fig. 16.27 **Wolverine**

**Cats.** Another large group of land-dwelling carnivores are the various members of the **cat family.** Among the largest cats are the **tigers;** the largest of the living cats is the *Siberian tiger,* or *Amur tiger*. A male Siberian tiger can weigh as much as 650 pounds and measure 10 feet in length (excluding tail). Found primarily in India, eastern Russia, China, and southeast Asia, tigers prefer more thickly covered terrain than lions and commonly dwell in forests. The striking coloration of the tiger is actually camouflage, allowing the big cat to blend in with vegetation at a distance.

Also among the largest living cats are the **lions;** male lions may attain a length of 7 feet (excluding tail) and a weight of 500 pounds. Lions are unusual among the cats in that they dwell in small groups called *prides* (most cats are rather solitary). Like most cats, lions typically ambush their prey, stalking as close as possible to the victim (often a zebra or wildebeest) before making a short, high-speed charge. Most cats are quite fast for short distances; a charging lion can briefly attain a speed of 50 mi/hr, faster than a horse. Most lions live in sub-Saharan Africa, but a few are found in India. The largest cats found in the New World are the *cougar* (also known as the *mountain lion* or *puma*) and the *jaguar.*

A common resident of homes in many nations is the **housecat,** which is prized in many cultures for its ability to keep homes, warehouses, and ships free of mice; they also make good pets. Housecats are thought to be a domesticated breed of the *European wildcat* or the *African wildcat,* with which it can freely interbreed and produce fertile offspring. These wildcats look just like ordinary household tabbies but have a wild, ferocious nature and are completely untamable.

**Seals.** **Seals** are aquatic carnivores. The seals are well suited to aquatic life. They have webbed flippers instead of paws or feet. Their bodies are streamlined and covered with short, dense fur that serves as insulation. Especially designed to dive deep in pursuit of fish, they have flap valves that keep water out of their ears and nostrils. Their blood carries about five times as much oxygen-carrying hemoglobin as human blood, and their muscles also contain large amounts of the oxygen-storage molecule myoglobin, allowing them to remain submerged for 15 to 20 minutes at a time. During a dive, the seal's heart beats only about 15 times each minute, helping the animal to conserve oxygen. A thick layer of blubber (fat) and the constriction of blood vessels beneath the skin make long, deep dives possible. Some seals routinely dive to 1000 feet below the surface in search of food. The *elephant seal* is said to descend

Fig. 16.29 **Aquatic carnivores**

Fig. 16.28 **Cats**

to 2000 feet to eat bottom-dwelling sharks, squids, and rays. At that depth, the seal withstands pressures of 875 pounds per square inch on its body.

Seals come in a variety of sizes, ranging from the *Baikal seal* of southern Siberia, which is about 4 feet long and weighs about 140 pounds, to the enormous elephant seal of the subantarctic islands, which reaches a length of up to 20 feet and weighs around 3 tons. Most performing seals are *sea lions,* large seals that have ears but lack underfur.

**Walruses.** **Walruses,** which are also aquatic carnivores, look somewhat like seals but are different animals. A walrus has *tusks,* which are elongated upper incisor teeth. The tusks serve as ice axes when the animal hauls itself out of the water or when it chisels into the ice to open an air hole to breathe. Tusks are also used as weapons to repel polar bear attacks. A big bull walrus of the far north may weigh as much as 3000 pounds, and its tusks may measure more than 2 feet long and 10 inches around at the base.

## Extinct Mammals

God created an amazing variety of mammals. Many of these creatures, such as wolves, apes, tigers, and camels, are common around us today. Unfortunately, some very interesting mammals have died out since Creation because of the curse of sin. A type of animal or plant that is no longer found alive on the earth is said to be **extinct.**

Although the most famous extinct animals are probably the dinosaurs (chapter 18), many magnificent mammals are also known only from fossils.

Extinct carnivorous mammals include the *saber-toothed cat,* a lion-sized carnivore characterized by its prominent 8-inch fangs. Analysis of its skeleton reveals that it was even more muscular than most of the big cats still alive today.

Most members of the elephant family have become extinct since the original creation; although only three species survive today, fossils reveal that a much greater variety of elephants lived in the past. Some extinct elephants had broad, shovel-like attachments on their lower jaw that they apparently used to dig with. Another famous type of extinct elephant was the *woolly mammoth,* characterized by its thick fur and huge tusks.

The *glyptodont* was a 3-ton, armadillolike mammal with extremely heavy armor, making the animal a mobile fortress. Shielded by its bony armor and wielding its muscular, spiked tail like a medieval mace, it would likely have been a difficult meal for most predators.

The largest of all extinct land mammals (known by the Latin genus name *Indricotherium*) was nearly as tall as a 3-story building and may have weighed 50,000 pounds. Regarded as a member of the rhinoceros family, this huge animal fed on plants and was larger than many dinosaurs.

Fig. 16.30 **Extinct mammals**

## Endangered Species

Understanding the LIVING CREATION

Because of the curse of sin, death became a part of the living creation. Animals now preyed upon each other, and man was given authority to kill animals to meet his own needs. Many animals that could not survive the sin-caused struggle for life declined in numbers (the process of *natural selection*) and eventually became extinct when the last members of that kind died out.

The great variety of animals in the fossil record reveals that most of the species that were alive at the time of the Flood of Noah's day are now extinct. The vast majority of extinctions in the past probably resulted from the changes in the earth's climate that followed the Flood. However, man has also been responsible for some extinctions. One of the most famous examples is that of the *passenger pigeon,* a type of North American dove that existed in flocks numbered in the billions in the 1800s. Although many people hunted passenger pigeons responsibly, irresponsible and wasteful slaughter by others caused the passenger pigeon to become extinct within just a few decades. Similar irresponsible behavior nearly wiped out the *American bison* (buffalo) during the same period. Other human-related causes of extinction can include destruction of an animal's habitat and introduction of competing species to the area.

Unfortunately, many people today have gone to the other extreme, teaching that man is of no more worth than "other animals" and has no right to use nature for his own benefit. Many of these individuals condemn responsible hunting and oppose the use of animals for medical research; some even decry the use of draft animals as "slavery" and oppose raising animals for food. (This philosophy is in direct contradiction to the biblical teaching that man has been given the authority to subdue nature for his own benefit and God's glory.)

spotted owl (*Strix occidentalis*)

Types of animals and plants that still exist today but are in danger of extinction are said to be **endangered.** Concern for endangered species in the United States led to the *Endangered Species Act,* a Federal law originally enacted in 1973 and later expanded. This law defines the word *species* to include subspecies or even local inbreeding populations. Unfortunately, this allows local varieties of abundant species to be designated as "endangered" or "threatened" even if the species is not actually in danger of extinction. An example of a controversial "endangered species" protected by this law is the **spotted owl** *(Strix occidentalis),* a type of small owl that is fairly abundant in the western United States. In 1990, the U.S. Fish and Wildlife Service classified a population of spotted owls living in Oregon forests (the "northern spotted owl") as a threatened species and declared millions of acres of forest off-limits to logging, putting an estimated 18,000 to 20,000 people out of work. However, some people have questioned the wisdom of classifying the northern spotted owl as a threatened species when it is virtually indistinguishable from relatively abundant owls of the same species *(Strix occidentalis)* living in California.

While concern for and protection of endangered species is a worthy goal, some people have also used these concerns to further a radical social agenda. Many environmentalists oppose practically all mining, logging (even when trees are replanted), drilling for oil and natural gas, construction of hydroelectric dams, and other activities on the basis of these concerns. Some politicians have used concern for endangered species to increase Federal and international control of private property and to promote global authority at the expense of national sovereignty (the right of a nation to run its own affairs as it pleases).

With the authority that God has bestowed upon man to subdue the earth (Gen. 1:28) also comes the responsibility of good stewardship. Although natural resources were made for man's use, man should not use them in a wasteful or careless way, because he is the steward of God's creation. Man should treat the animals that he uses for food, work, or research humanely (Prov. 12:10) and should not waste resources unnecessarily, remembering that "the earth is the LORD's, and the fulness thereof" (Psa. 24:1).

## Section Review 16.4

1. What is the most obvious way to distinguish between an artiodactyl and a perissodactyl?
2. What is the difference between antlers and horns?
3. What family of artiodactyls includes such important animals as cattle, sheep, and goats?
4. What is the largest land-dwelling carnivore?
5. Name two aquatic animals of the order Carnivora.

### Application

6. Describe the digestive system of a ruminant and explain how it is different from that of a carnivore. How does each design suit the animal's diet?

### Critical Thinking

7. How are the tusks of an elephant different from both antlers and horns?

### Identify

hoof, ungulate, ruminant, velvet, fermentation, rumen, extinct, endangered

## Chapter 16 Review

### Define

1. motile, sessile
2. bilateral symmetry, radial symmetry, asymmetric
3. vertebrates, invertebrates
4. endoskeleton
5. tetrapod
6. homeothermic
7. hibernation
8. placenta, gestation period, umbilical cord
9. viviparous, oviparous
10. diurnal, nocturnal
11. marsupial, marsupium
12. insectivorous, insectivore
13. prehensile tail
14. rodent
15. hooves, ungulates
16. ruminant, rumen
17. extinct, endangered

### Identify

1. three characteristics of all vertebrates
2. five characteristics common to all mammals
3. the purpose of mammary glands
4. two egg-laying mammals
5. the largest living land animal
6. the largest living marsupial
7. the lightest mammal
8. two broad groups of whales
9. the world's largest animal
10. three common rodents

### Explain

1. What primary characteristics are shared by all animals?
2. What are three types of symmetry mentioned in the text? Which type of symmetry is exhibited by vertebrates?
3. How does a warm-blooded animal regulate its temperature? How does a cold-blooded animal regulate its temperature?
4. What purpose does hibernation serve?
5. What are the three broad categories of mammals, based on the way the developing young are nourished?
6. How does the reproductive cycle of a marsupial differ from that of a placental mammal?
7. What is the most obvious difference between an ape and a monkey?
8. What characteristic is used to distinguish artiodactyl ungulates from perissodactyl ungulates?
9. What is meant by "chewing the cud"?
10. How is the digestive system of a cow different from that of a wolf? How do these differences relate to the diet of each?

### Apply

1. Why are sessile animals considered capable of movement even though they do not travel from place to place?
2. Why do we distinguish man from the mammals, even though man and mammals share many body similarities?
3. What are some effective ways that you can practice wise conservation and stewardship of the living creation?

# Birds

CHAPTER 17

**Key Concepts**

- Designed for flight
- Observing birds
- Groups of birds
- Migration, courtship, and nesting behavior
- Eggs, incubation, and hatching
- Avian anatomy and physiology

**Birds,** which belong to class *Aves,* are vertebrates with feathers, wings, and scale-covered legs. **Ornithologists,** zoologists who study birds, have identified approximately 9600 species of birds alive today, ranging in size from the tiny *bee hummingbird,* which is about two inches long (including beak and tail) and weighs about $\frac{7}{100}$ of an ounce, to the African *ostrich,* which may stand nearly 8 feet tall and weigh nearly 350 pounds. All birds have wings, but not all birds can fly. No birds alive today have teeth; although God created several kinds of toothed birds, these kinds have since become extinct. *Birds, like mammals, are homeothermic, warm-blooded vertebrates,* and thus they can live in frigid regions of the earth as well as in warm regions.

## 17.1 The Feathered Vertebrates

### Designed for Flight

Most birds are designed for flying. The majority fly at altitudes under 3000 feet, but there are exceptions: migrating flocks of *bar-headed geese* routinely fly over the Himalayas of Asia at a height exceeding 29,000 feet, and one *Rüppell's griffon* (a type of vulture) was killed by an airliner at an altitude of 37,000 feet.

*The following characteristics equip birds for flying:*

- a compact, lightweight, and very strong skeleton
- a nearly rigid backbone that gives a solid support for the strenuous muscular activity required for flying
- hollow spaces in the skeleton that provide room for air sacs, which contribute to the efficiency of the respiratory system and make the bird lighter
- **bipedal** ("two-footed") anatomy, which leaves the wings free for flying
- a prominent ridge, or *keel,* on the breastbone to allow for the attachment of flight muscles by strong tendons
- three bones in the shoulder area to support each wing (clavicle, coracoid, scapula), in contrast to the two bones that support the limbs of other vertebrates (clavicle and scapula)
- lightweight beaks and skulls
- an efficient circulatory system, including a four-chambered heart
- powerful muscles to power the wings during flight

The body characteristics that equip birds for flying will be discussed in more detail in section 17.2.

### Familiar Varieties

**Backyard birds.** One of the most convenient places to observe birds may be your own backyard. Birds are attracted to this environment because it provides protective cover, nesting sites, food, and water. As you observe them, note their bathing and drinking behavior.

# Guide to Familiar Backyard Birds

## Starling

**Feeding habits:**
lawn feeder; feeds on insects and worms; eager, probes constantly

**Nesting habits:**
nests in holes in leaves; lays 4–6 pale blue eggs

**Coloration:**
dark feathers, yellow bill; speckled in winter

## American goldfinch

**Feeding habits:**
feeds on flower seeds and tree buds; is particularly attracted to dandelions and thistles

**Nesting habits:**
lays 4–6 pale blue eggs in a nest in either a bush or a tall tree

**Coloration:**
male is bright yellow with black wings; female is dull yellow with black wings

## Mourning dove

**Feeding habits:**
lawn feeder; eats grass seeds and insects; is often seen eating gravel from driveways

**Nesting habits:**
lays 2 white eggs in a nest on the ground

**Coloration:**
is grayish brown with slight pink tint on breast; tail is long and black spotted with a white border

## Robin

**Feeding habits:**
lawn feeder; hops about and stops as if to listen, then extracts worm

**Nesting habits:**
nests in tall shrubs and small trees and on porches or buildings; lays 3 or 4 blue eggs

**Coloration:**
male is red breasted with a black head and gray back; female is duller than male

## Song sparrow

**Feeding habits:**
picks weed seeds from the ground; eats some insects

**Nesting habits:**
lays 3–5 brown-spotted greenish eggs in a clump of grass or builds a grassy nest in a bush

**Coloration:**
head and back are brown streaked; breast is streaked white

## Baltimore oriole

**Feeding habits:**
eats insects and nectar

**Nesting habits:**
lays 4–6 white eggs with traces of brown; builds nest high in a tree

**Coloration:**
male has black head, wings, and tail, is orange underneath; female has orange-olive upper parts and is yellow-orange underneath

## Brown-headed cowbird

**Feeding habits:**
lawn feeder; waddling gait; often stops foraging to display and sing; eats insects and weed seeds; follows cows to eat insects off them

**Nesting habits:**
lays eggs in nests of other birds

**Coloration:**
male is black with a brown head; female is gray

## Grackle

**Feeding habits:**
lawn feeder; careful prober; eats grubs, worms, mice, and small birds; stops foraging to sing and display feathers

**Nesting habits:**
nests in holes in evergreens

**Coloration:**
male is iridescent black with yellow eyes; female is dull black and smaller than male

## Chipping sparrow

**Feeding habits:**
lawn feeder; hops about eating insects and seeds

**Nesting habits:**
lays 3–5 speckled blue eggs; builds its grassy nest in a shrub, young pine, or porch vine

**Coloration:**
has a gray breast, brown and gray wings, and a rust cap

### Hummingbird

**Feeding habits:**
is attracted to red flowers; uses its long, needlelike beak to extract nectar and insects from flowers

**Nesting habits:**
lays 2 pea-sized white eggs in a nest made of spider webs covered with lichens

**Coloration:**
most are green above and white underneath; male has a red throat; has a characteristic small size and fast, beating wings

### Blue jay

**Feeding habits:**
eats seeds, insects, eggs of other birds, mice, snails, frogs; sometimes buries seeds for storage; is characteristically noisy

**Nesting habits:**
lays 6 olive eggs with brown spots in a nest in a conifer or thicket; male defends nest

**Coloration:**
has gray breast, blue wings and tail with white spots, black neckband

### Mockingbird

**Feeding habits:**
eats fruits, berries, and insects

**Nesting habits:**
nests within 10 feet of the ground in a shrub or tree; lays 4 or 5 bluish-green eggs with brown spots

**Coloration:**
is gray with large white wing and tail patches

### Cardinal

**Feeding habits:**
eats insects and worms

**Nesting habits:**
lays 3 or 4 blue- or green-tinted white eggs with brown spots in a thicket or hedge about 3 feet from the ground

**Coloration:**
male is red with a black face mask; female is brown with red on the tail, wings, and crest

# Guide to Familiar Roadside Birds

### Eastern meadowlark
streaked brown head and back, yellow breast with a black V

### Tree swallow
blue or green back, pure white underneath

### Cedar waxwing
brown plumage, black mask, brown crest, and red "wax" marks on wingtips

### Bobwhite quail
call sounds like its name, "bobwhite"

### Red-winged blackbird
male is black with red shoulders bordered in yellow; travels in colonies

### Brown thrasher
bright reddish brown back, streaked breast; characteristically seen along the ground among the leaves

### Belted kingfisher
mostly seen alone; blue gray back, white belly; ragged crest on top of head; often seen fishing

### Horned lark
light brown body, black head tufts resembling horns, black face mask

### Barn swallow
deeply forked tail; dark blue back

### House finch
brown-streaked plumage; male has bright red headband and throat markings

*Doves* are the only birds that drink with their heads down; other birds lift their heads to swallow. *Blue jays* fly into a birdbath screaming and cause the other birds to scatter, whereas *warblers* cautiously descend a nearby tree and approach the bath vertically.

If you want to attract birds to your backyard, provide an environment that affords protection; a few trees, shrubs, or even a brush pile will attract birds seeking cover. Make sure food is available. Many birds prefer the seeds of wildflowers (weeds). Others are attracted to flowers that produce sweet nectar. Flowers with red blooms will attract *hummingbirds*. If you would like birds to nest in your backyard, provide nesting materials such as pieces of yarn, string, straw, or paper.

**Roadside birds.** Another good opportunity to observe birds is during a drive along a country road. There are certain species that have made their homes along the roadside. Some can be observed on the power lines overhead, while others sit on fence posts. *Crows, jays, eagles,* and *vultures* are often recognized as they feed on dead animals beside the road. Many backyard birds can also be seen along the roadside.

Identifying characteristics to watch for include the following:

- *Unmistakable tails.* The *flycatcher* perches with its tail pointed down. In contrast, the *wren* holds its tail upright at about a 45° angle. The long, pointed tail of the *mourning dove* is easy to distinguish even at a distance. The *starling, nuthatch,* and *meadowlark* have short, square tails. The *barn swallow* has a distinctive forked tail, and the *mockingbird* a neatly rounded tail.
- *Field markings.* The red shoulder patches of the *red-winged blackbird* leave no doubt as to its identity. The *chickadee* wears a distinctive black bib around its neck as if it is preparing for a meal. An *eastern kingbird* can be instantly recognized by the white band on the end of its tail.
- *Identifying calls.* Once you have heard the scream of a *blue jay,* you will recognize it each time you hear it. The *mourning dove* makes a sad, cooing noise. Legend says that it is mourning the loss of its mate. The mourning dove mates for life and does not mate again when its partner dies. Some birds have been named for their call. The *whippoorwill's* call sounds like its name. The *bobwhite's* call is simply "bobwhite, bobwhite."

## Groups of Birds

There is such a great variety of birds that it is sometimes helpful to divide them into common-sense groups for more detailed study and observation. One commonly used system is to place the 27 orders of birds into the following categories: perching birds, birds of prey, water birds, game birds, tropical birds, flightless birds, and extinct birds. This system is very useful for bird watching because the broad groups are easy to remember. Because some birds do not fit in any of these categories, while several orders are split between more than one category, the Linnaean-style classification system is more commonly used by ornithologists.

**Perching birds.** Whenever you think of birds in nature, the first birds to come to mind are often the **perching birds.** Most birds (approximately 60%) fall into this category. These birds have three toes pointing forward and one toe pointing backward. Their feet have a special design that holds the bird securely on its perch even when asleep. Below their toe bones lies a tendon connecting the toe to leg muscles. When the bird's weight is on the tendon and the leg is bent, the toes wrap around the perch and are held taut by hundreds of ratchetlike projections along the tendon that mesh with grooves in the tendon sheath.

Most of the familiar songbirds are perching birds. The beautiful melodies of the *thrushes* enhance any natural area—the *wood thrush* has perhaps the most delightful song of all birds. A familiar thrush is the red-breasted *American robin.* One of North America's favorite songbirds is the *bluebird.* Its bright blue feathers cover all the body except for the chest and belly. This category also includes several songless birds, such as the hummingbird and the woodpecker.

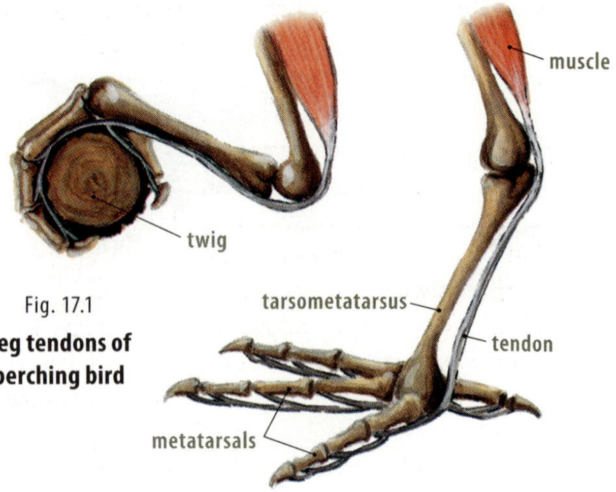


Fig. 17.1
**Leg tendons of perching bird**

**Birds of prey.** Several birds are known for their fierce and precise hunting skills; as a result, these birds are called **birds of prey.** Some birds of prey, such as the eagle and falcon, hunt only during the day, using especially acute vision to spot rodents, rabbits, and snakes from great distances. Others, such as many species of owls, hunt only at night, using eyes especially designed for night vision. All of these birds have sharp, powerful talons, useful in grasping and killing their prey, and strong, hooked beaks, suitable for eating flesh.

*Vultures* are large carnivorous birds with weak feet that cannot carry off large prey; thus they must feed on small animals or carrion (dead animals). *The largest living land bird of North America is the California condor,* a vulture with a wingspan of up to 11 feet and a body nearly 4 feet in length. Condors are almost extinct today, but at one time they ranged over the entire southwestern United States. California condors nest in caves and crevices near the summits of high cliffs. They lay only a single egg each time they breed, and they do not breed every year, which may help to explain why they are almost extinct.

*Eagles* are large, powerful birds of prey. One of the strongest is the *sea eagle,* which lives in coastal regions and has been seen carrying off large geese and fish weighing twice its own weight. There is even one documented case of a sea eagle carrying off a four-year-old girl in Norway (the girl was eventually rescued). The wingspan of an eagle is great, measuring up to 9 feet across. Eagles commonly feed on small rodents and birds, but a few attack rather large animals. Tropical eagles commonly feed on sloths and monkeys.

The *bald eagle* and the *golden eagle* are the only two eagles native to North America. Bald eagles are found in many parts of Canada and the United States, but they are not found in large numbers anywhere except Alaska. These birds are called bald eagles because their white feathers, which stand out in bold contrast to their brown plumage, cause them to appear bald at a distance. Bald eagles prey primarily on fish. The golden eagle is slightly larger than the bald eagle and is identified by the golden brown color of the feathers on the back of the neck. This capable hunter preys on rabbits, ground squirrels, birds, and occasionally young lambs and deer. Similar in appearance to eagles, but somewhat smaller, are the *hawks,* which generally prey on rodents, snakes, rabbits, and other small animals. An exception is the *osprey,* or fish hawk, which lives near the water and feeds on fish.

Fig. 17.2 **Birds of prey**

> Doth the eagle mount up at thy command, and make her nest on high? She dwelleth and abideth on the rock, upon the crag of the rock, and the strong place. From thence she seeketh the prey, and her eyes behold afar off.
> —Job 39:27–29

*Falcons* are smaller birds of prey that are nevertheless remarkable hunters. The *peregrine* [pĕr′ə·grĭn] *falcon* is not only the fastest flying bird; *it is the fastest of any living creature.* Swooping down in a 45-degree dive, this bird can reach speeds of 220 miles per hour or more as it dives to attack its prey (often doves or songbirds) in midair. Falcons achieve these speeds with an exceptionally streamlined body; long, narrow, pointed wings; and long tails.

Most *owls* feed at night and swallow their prey whole. Unlike the eagle and hawk, which tear their food into bits and eat only the acceptable flesh, the owl consumes everything. Owls have no crop for storage, so they process the prey in the stomach and then regurgitate the fur and bones as a pellet.

**Swimming and wading birds.** Birds that commonly live or feed in the water are known as **water birds.** A number of water birds are equipped with water-repellent feathers and webbed or lobed feet for swimming. To maintain their water-repellence, *ducks* and *geese* oil preen their feathers from 2 to 17 times a day using oil secreted by glands at the base of the tail. Oil preening also helps trap air close to the body to keep the animal buoyant.

Fig. 17.3
**Water birds**

Fig. 17.4
**Game birds**

Some water birds have very long legs designed for wading instead of swimming. Storks, herons, flamingos, cranes, and egrets do not swim or dive but walk about in shallow water in search of their favorite food. Flamingos are famous for their pink coloring. It actually takes a young flamingo about three years to develop this color; the birds are born a grayish white. Their distinctive pink coloration comes from a pigment that they obtain from food such as shrimp.

The *white stork* of Europe, the best-known variety of stork, is a large wading bird noted for its beautiful white plumage. These birds, which are reported to mate for life, build nests on roofs or chimneys, where they return to breed year after year. The female lays three to five eggs each breeding season, and both mates provide good care to their young. It is perhaps the care they give their young that originally gave rise to the folklore tale that storks deliver babies to expectant mothers.

Herons, including the *great blue heron,* are known for their graceful lines. Great blue herons stand about four feet tall. They have slender, pointed beaks, long curved necks, stiltlike legs, and large wings. When herons fly, they curl their necks back between their shoulders and stretch their legs straight out behind. A hunting heron stands motionless as a statue until it spies a prey; then with a lightning-fast dart of its beak, the heron spears the prey. The heron advances a step or two while eating before resuming its motionless stance once again.

**Game birds.** Birds that have long been hunted for both food and sport, including quails, pheasants, partridges, grouse, doves, wild turkeys, and many others, are classified as **game birds.** Many of these birds nest on the ground, sheltered in underbrush or tall grass. Their large flight muscles allow them to burst into flight with astonishing speed, making them difficult prey. Most of these birds fly only in short spurts because their flight muscles are designed for strength instead of endurance. (Doves are excellent long-distance fliers, however.) The flight muscles, or breast meat, of game birds provide the tender white meat for which they are prized.

Fig. 17.5
**Tropical birds**

**Tropical birds.** A wide variety of **tropical birds** live in the world's tropical forests. (Some of these birds may be seen in pet shops and aviaries.) Most tropical birds are brightly colored, and many have large, unusual beaks.

*Parakeets* are small parrots with brightly colored feathers and hooked beaks. One of the best-known parakeets is the *budgerigar* [bŭj′ə·rē·gär′] of Australia. This bird is normally yellow and green, but many colors (white, gray, lavender, blue, etc.) have been developed in domesticated birds. Parakeets make good pets, and they can be trained to "talk" by imitating speech.

*Toucans,* which are native to Central and South America, are easily recognized by their large, colorful beaks. The head feathers of the *curl-crested toucan,* which appear to be made of black patent leather, form a crest of curled feathers. The large beak of a toucan, which appears to be quite solid and heavy, is neither solid nor heavy because it is honeycombed with air spaces; but it is quite strong. These birds nest in dense forests and lay their white eggs in treehole nests.

**Flightless birds.** Some birds, although they have wings or winglike structures, cannot fly. There are only a few species of these **flightless birds** alive today, but several more species are known from the fossil record. Some flightless birds are rather large, and a few stand taller than a man. Because these birds are ground dwellers, they are usually equipped with powerful legs for running. This category includes the ostrich, emu, penguin, kiwi, rhea, and cassowary.

*Ostriches* are the largest living birds; some males stand about 8 feet high and weigh nearly 350 pounds. Ostriches can accelerate from a standstill to nearly 50 miles per hour in 2½ seconds and can sustain that speed for nearly half an hour before tiring. These huge birds can even be trained for riding and pulling carts; because they can be used as beasts of burden, they are occasionally called camel birds. They are native to Africa, but similar birds live in Australia, New Zealand, and South America. The ostrich lays its eggs—the largest eggs laid by any nonextinct animal—in crude nests on the ground. Ostriches are the only birds with just two toes on each foot.

One of the most unusual flightless birds is the chicken-sized *kiwi* [kē′wē] of New Zealand. Its useless wings are mere stubs, and its feathers are long and hairlike. Nocturnal kiwis have long beaks and eat worms, which they locate by smell. Foraging kiwis can be heard at night sniffing for worms. For their size, kiwis lay enormous eggs. A kiwi egg weighs almost one third as much as an adult bird and is about one quarter the size of the bird that laid the egg. Sometimes a nest will contain two of these large eggs.

Fig. 17.6
**Flightless birds**

Fig. 17.7
**Extinct birds**

**Extinct birds.** Some birds are extinct because they could not survive pressures brought to bear on them by man and civilization. Other birds have been extinct for thousands of years, and no one knows exactly why they became extinct.

The *dodo,* which was slightly larger than a turkey, has been extinct since the late 1600s. This flightless bird became extinct primarily because it could not compete with the hogs and monkeys that man imported into its habitat. Dodos lived only on small islands in the Indian Ocean. Other birds that have become extinct since the 1600s include the *moa,* a giant flightless bird that was nearly 10 feet tall; the *great auk,* a penguinlike flightless seabird; and the *passenger pigeon,* a type of dove.

Although no living birds have teeth, several toothed birds are known from the fossil record. One unusual toothed bird was the *Archaeopteryx,* whose genus name means "ancient wing." We really do not know precisely what *Archaeopteryx* looked like, although we have some good fossil remains that give us clues to its appearance. Other unusual features of *Archaeopteryx* were small claws on its wings and an extended tail.

Several giant birds are also known only from the fossil record. The largest known flying bird was a condorlike bird of prey *(Argentavis magnificens)* that stood as tall as a man and had a wingspan of more than 25 feet. The largest birds of all, however, were several kinds of giant flightless birds, some of which stood more than 10 feet tall and weighed 1000 pounds or more.

# Variety in Design

*Design in the* LIVING CREATION

One of the most fascinating features of creation is its variety. The God who created all of life displays more imagination than man could ever assemble. Look about you. Are all the buildings that you see exactly alike? Of course not, yet they are all designed with the same basic principles of architecture. Was there no intelligence behind all of those buildings? So it is with our Creator. He designed nature, all of it, following His divine architecture; yet He insisted, as man does, on variety. It is not the nature of God to make every leaf exactly the same shape or every bird exactly the same color. God designed a world of unlimited variety, each living thing precisely formed for a purpose.

Basically, there are two kinds of variety in the living creation: variety within the kind and variety among the kinds. You are an excellent example of the first type of variety. You are different, unlike any other person on the earth, yet you are totally human—not some sort of in-between creature displaying features you received from other kinds of creatures. God's infinite variety is displayed in you for a particular purpose.

Examples of this type of variety are beautifully illustrated in the birds. Consider the sparrows, for example. There are many different varieties of sparrows, each slightly different and yet all similar in many ways. Although they all share certain features characteristic of sparrows, such as small brown bodies with streaked backs, short conical beaks, appetites for seeds near the ground, and a tendency to be found in flocks, they may differ greatly in their individual habits of movement, song, nesting, and feeding. The different varieties may also have strikingly different color patterns. However, all these varieties of sparrows were produced from the few sparrows that left the Ark after the Flood and represent a single kind. Within the genes of these sparrows of biblical times lay the potential for amazing variety, but always within the limits established by God. Although there are many varieties of sparrows, it is always easy to identify a sparrow as a sparrow. No sparrow could ever become an eagle, a reptile, or any other animal.

The second type of variety—variety among kinds—is also well illustrated in the birds. God made many different kinds of birds. Think of the differences between a hummingbird and an eagle, or an ostrich and a toucan, or a dove and a penguin. All of these animals are very different from each other in many aspects of their overall design; although they are all birds, they are all different kinds of birds. The marvelous creativity of God is revealed in the way He constructed such drastically different kinds of birds using the same basic body plan. And in God's provision, all of these birds reproduce after their kind; no eagle will ever become a hummingbird, or a penguin, or an ostrich.

Why did God make all of this variety possible? The answer is the same as why creation exists at all—to glorify God. The Scriptures tell us that "the heavens declare the glory of God; and the firmament sheweth his handiwork" (Psa. 19:1). Yet could it be that the awesome variety in design is to teach man about God? It was rather difficult for mankind to look into the heavens and relate to God, because heavenly bodies seemed so far away. But God made the sparrow and the leaf to live with mankind, declaring unto him the artistry in the mind of God. Consider Job 12:7–9.

> *But ask now the beasts, and they shall teach thee; and the fowls of the air, and they shall tell thee: Or speak to the earth, and it shall teach thee: and the fishes of the sea shall declare unto thee. Who knoweth not in all these that the hand of the Lord hath wrought this?*

Look at the leaves. Look at the sparrows. Each magnificent design declares the creativity and imagination of a supreme Creator.

## Section Review 17.1

1. What are nine characteristics that help equip a bird to fly?
2. Name several birds that can be observed in your area.
3. Into what six categories are birds sometimes classified for purposes of bird watching?
4. Describe the special design of a perching bird's foot.
5. What is the largest living land bird in North America?
6. What bird of prey is noted as the world's fastest animal?

### Application

7. Using such references as *Nave's Topical Bible* and *Davis Dictionary of the Bible,* do a study on birds in the Bible. For example, which birds did God forbid the Israelites to eat (Lev. 11), and what do the poetic books of the Bible (Job, Psalms, Proverbs, Ecclesiastes, Song of Solomon) say about birds? How did Christ use birds in His teaching to illustrate spiritual truth?

### Identify

ornithologist, bipedal

### Observe

- Try to spot and identify at least 12 of the birds mentioned in this section.

# 17.2 Avian Anatomy and Physiology

## Feathers

**Purpose of feathers.** From its lightweight head to its streamlined tail, practically every feature of a flying bird is designed for flight. Even the internal organs of the bird reveal that God designed the bird to fly. One of the most obvious features of a bird that allows it to fly are its **feathers,** a characteristic unique to birds. One purpose of feathers is to provide a lightweight body covering that is an excellent insulator. Without the insulating effect of feathers, birds, which often maintain a body temperature of 110°F, could not survive where the temperature is more than a few degrees below their own. Because of the thermal efficiency of feathers, some birds are able to live actively in temperatures far below zero. In addition, feathers provide the bird with a smoothly tapered, streamlined shape that allows it to move easily through the air. They are also responsible for the broad area of the bird's wings; stripped of its feathers, the bird would be left with narrow, useless appendages unable to lift it off the ground.

**Types of feathers.** There are several different types of feathers on a bird, each with a different purpose. The streamlined shape of a bird is provided by the bird's **contour feathers,** strong feathers found on the bird's body, wings, and tail; all the contour feathers of a bird point toward the bird's tail to aid in streamlining. Special contour feathers on the wings are known as **flight feathers** because they provide the necessary wing shape for flight. Beneath the contour feathers lie the **down feathers,** soft, fluffy feathers close to the bird's body that provide excellent insulation without adding much weight.

Fig. 17.8 **Types of feathers**

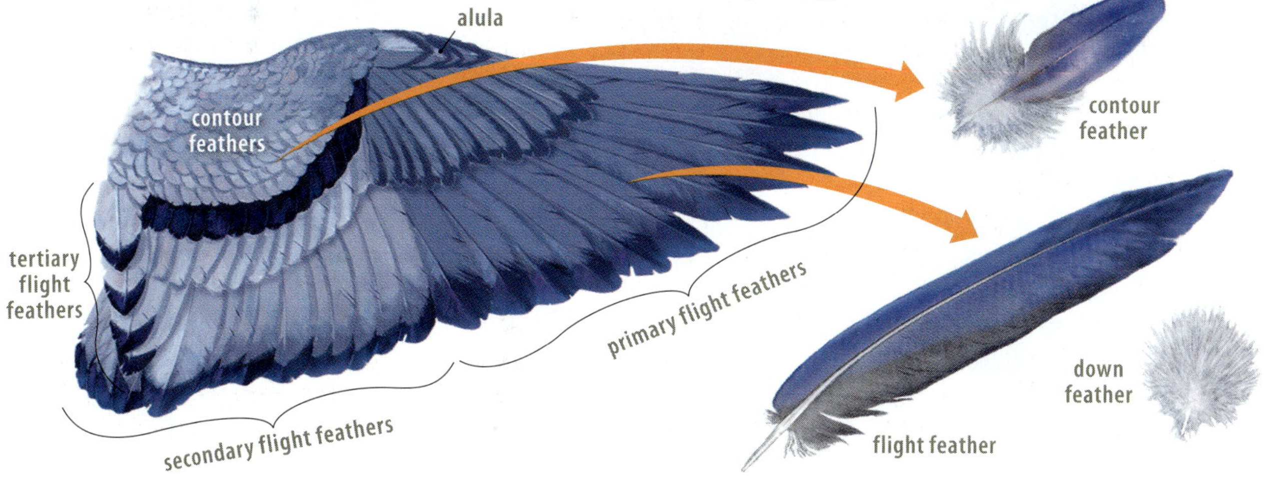

# Designed for Flight *by Asa C. Thoresen*

Understanding the LIVING CREATION

According to an ancient Greek myth, Daedalus, an architect and sculptor of Athens, fell out of favor with Minos, king of Crete. He and his son Icarus were imprisoned in a maze called the Labyrinth, but they determined to be free. Daedalus closely studied the wings of sea birds and, with wax and feathers, made two pairs of wings of similar design. Flapping these wings, father and son soared into the air, intending to escape to the island of Sicily.

Then came disaster. Icarus, excited with his ability to fly, forgot his father's warning and flew too near the sun. The result, according to the poet, was that "with melting wax and loosened strings, sank hapless Icarus on unfaithful wings."

Grieving Daedalus landed safely in Sicily, but Icarus drowned in the Mediterranean.

This old myth illustrates the longing earthbound human beings have always felt to soar into the sky like the birds. Over 2000 years after the days of Minos and the legendary Daedalus, man *did* learn how to fly, but not by attaching wings to his shoulders. Eventually human beings faced the fact that, besides having no wings, their bodies are in no way designed for flight.

seagulls

For example, *the breast muscles which operate the shoulders and arms in man constitute less than 1% of his body weight, while those of some birds weigh as much as 33%!*

From the tip of their beaks to the end of their tail feathers, birds show remarkable aerodynamic design. Birds (except for a few kinds like the ostrich that do not fly) are equipped with intricately constructed feathers; powerful wings; light, hollow bones; rigid skeletal parts; large, strong hearts for circulation of warm blood; a remarkable respiratory system; and a digestive system that quickly absorbs the energy from food.

## The Marvel of Feathers

A bird's feathers make him more efficient in the air than any airplane. Feathers are structured for protection against heat and cold, for fanning the air, and for streamlining the body. *Every slight change of position of a feather during flight is designed to extract energy from the air and use it effectively.* For their weight, feathers are stronger than any manmade substitute. A domestic hen can claim more than 8,300, while a swan sports 25,200 (four fifths of them on the head and neck).

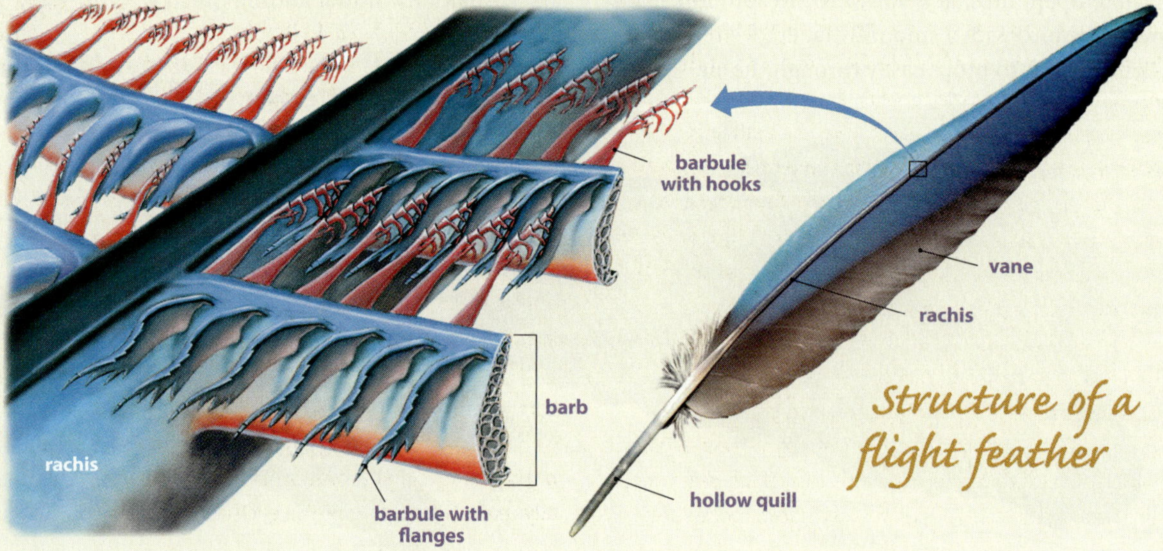


Structure of a flight feather

A ruby-throated hummingbird has only 940, but it boasts more feathers per square inch than the swan.

Look at the construction of a single wing feather under a microscope, and you will see one of the most marvelous sights in all of nature. Compared to the relatively simple scutes of reptiles and scales of other creatures, a bird's feather is superior in design and function. A single wing feather of a pigeon consists of more than a *million* parts!

Along both sides of the stiff quill **(rachis)** are grooves which bear filamentous **barbs,** forming the flat **vanes** of the feather. There are usually several hundred barbs in each vane, held together by tiny **barbules,** which in turn bear flanges and minute hooks called hamuli. These parts work together like a zipper. When these become unzipped, the bird zips them up again by preening. Next time you find a feather, run your fingers over the vanes toward the thick end of the quill. The flanges and hooks will separate and the barbs will look ragged and tattered. Now zip the feather parts together again by slipping the vanes between your fingers toward the tip of the feather. You may have to try several times before you make the feather look neat again.

*A feather's design automatically changes its shape during flight in response to different pressures from the air.* The rachis (quill) may be curved or straight. The feather itself may be thick or thin, round and hollow, square or polyhedral (many-sided), flat or grooved, depending upon air pressure.

Efficient prolonged flight can occur only when every slight change in airflow is automatically compensated by the feathered wings. The flight feathers of the wing tip, like the propeller of an airplane, are mobile and variable in the angle at which they meet the air. The base acts like the wing of a plane to give lift; while the secondary and tertiary wing feathers function as the ailerons and flaps to control direction of flight and give added lift. A slight change in the position or shape of two key primary feathers would ground a bird.

On the leading edge of a bird's wing is the *alula* [ăl′yə·lə], a special aileron consisting of three feathers. This serves to form a wing slot during slow expert maneuvering in a small space. The flight feathers of the wing also slot to prevent a stall—loss of lift, or to lower the stalling speed. *The tail serves as both rudder and elevator* to control direction of flight. Birds such as the puffins, guillemots, and murres have very short tails. To compensate, their webbed feet extend beyond the tail, acting as rudders and elevators by tilting and spreading the extended webs.

*Embedded in the skin near the quill of each flight feather are nerve endings which convert the feathers into sensory receptors.* These record the precise position of each feather, and through intimate connections with reflex arcs in the spinal cord, effect the continuous variations and fine adjustments of more than 12,000 tiny muscles attached to the bases of the feathers.

## Flight Patterns

Different species can be identified by their flight characteristics. A crow does not fly like a blue jay. A titmouse does not fly like a woodpecker. A mallard does not fly like a teal. *Ornithologists recognize four main kinds of flight: powered flying, gliding, soaring, and special flight patterns.*

**Powered flight.** In flapping or powered flight, the down stroking wings develop lift and forward thrust. . . . The upstroke [is] a quick backward flip with partly folded wings. . . . Generally, *the larger the bird the slower it flaps its wings.* A *hummingbird* vibrates its wings about 50 times a second, a *heron* only twice.

**Gliding.** Gliding flight is something like tobogganing downhill. *Giant albatrosses,* wings spanning up to 12 feet, use this type of flight. They glide for hours over the surface of the sea, rarely flapping their wings. A good glider travels the longest distance possible horizontally with the smallest possible loss of height. Flight efficiency depends on the shape of the wings and the angle which the surface of the wings creates with its own direction of motion. The speed of the glide is limited by the size of the wings and the weight of the bird. A light bird with large wings glides slowly, while a heavy bird with small wings develops greater speed.

Sustained gliding flight depends also on the motion of the air. If the air is motionless the glider must eventually reach the surface of the earth or water, the duration

(cont.)

(cont. from previous page)

depending on the height from which the glide began. Upward air currents over the troughs and crests of waves, however, sustain the long flights of *shearwaters* and *albatrosses* at sea.

A shearwater glides down an air slope, losing vertical altitude at 10 feet per second. But an equivalent upward air current will cause the bird to glide in level flight. If the air rises from the water surface faster than the rate of the bird's glide, the bird without effort will gain altitude. When the wind speed equals the forward and downward speed of the bird, its ground speed will be zero and the bird will appear to be suspended absolutely still in the air. Like a man walking down an upgoing escalator at the same speed, he goes nowhere.

Some birds seem to glide about for sheer pleasure on windy days. A few years ago I watched some gulls hover in the breeze above the high point of a rocky island off the west coast of North America. At the highest point stood a lighthouse capped with a slender spire. Several gulls took turns circling, hovering near, and holding the tip of the spire with their bills. When one lost the grip the next in line would move in to hang on. Nearby some guillemots, which have short wings and normally do not glide, circled over the crest of the rocky peaks, laboring slowly against the stiff breezes, then turned to glide down the lee side like playful otters sliding on snow. But the guillemots tobogganed on air.

**Soaring.** Land birds such as buzzards, hawks, and eagles soar in great circles, gradually attaining great heights, then dive swiftly downward to start another spiral. They ride the thermal upcurrents until they reach a downdraft which enables them to reach high speeds without ever flapping their wings.

The bird progresses by expending only a minimum of energy. Everyone has at some time seen birds soaring with their wing tips spread like fingers on a hand. Called *slotting*, this function prevents air turbulence behind the wings when the wings are tilted downward at the trailing edge. A bird slots its wings when it flies at reduced speeds or lands.

Soaring flight

**Special flight patterns.** The extremely complicated patterns of coordination and motion involved in vertical, reverse, and hovering flight—helicopter actions—of swifts, hummingbirds, and skylarks are not fully understood by scientists. These acrobats not only use a powered downstroke but also a powered upstroke.

Hummingbird flight pattern

*Unlike other birds, the hummingbird flies almost exclusively with its hands, the elbow permanently bent.* The shoulder joint is motile, making the hummingbird highly maneuverable. In some ways a hummingbird's wings are like the wings of insects. They operate backward and forward in a rowing motion, but they also tilt like the wings of a dragonfly or the rotors of a helicopter.

No matter what maneuver this bird makes, the wingbeat rate varies no more than 5%. The average wing beat remains at about fifty-two times per second for the female ruby-throat, regardless of her flight motion.

## Too Wonderful for Chance

A bird flying, propelling itself by its own coordinated muscular efforts, is surely one of Nature's masterpieces. To me, the whole phenomenon of bird flight is so fantastically intricate that I can only conclude that its origin must lie, not in the realm of haphazard chance, but in the intelligence of the Designer and Creator.

Thirty centuries ago David, poet-king of the Hebrews, sighed, "Oh, that I had wings like a dove! . . . then would I fly away." Solomon, the wise man, likewise amazed, exclaimed, "The way of an eagle in the air" is "too wonderful for me."

It is "too wonderful" for me also. And I believe it is "too wonderful" for blind chance as well. "Too wonderful," indeed, for anyone or anything except the great Designer, the Creator.

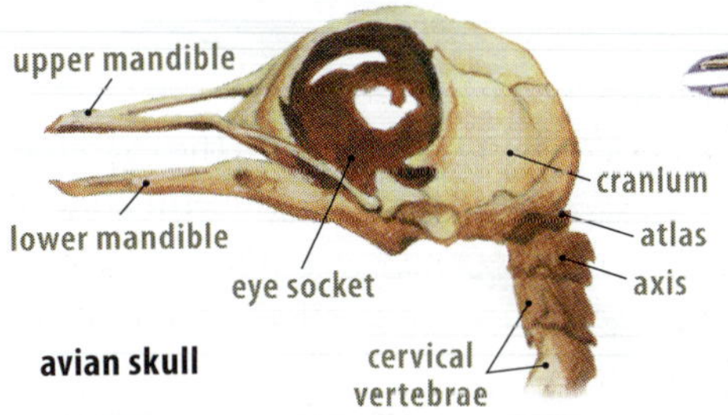


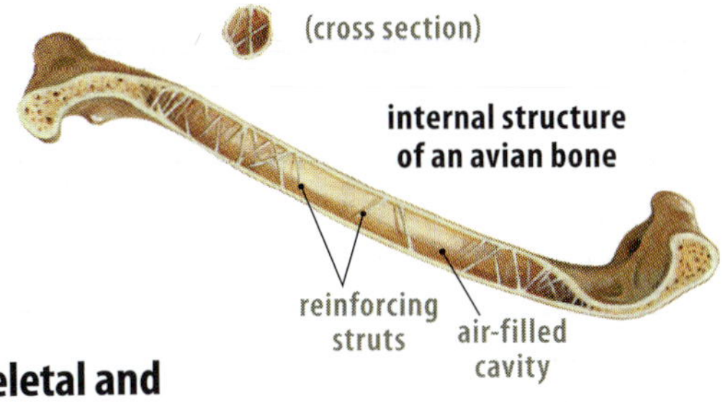


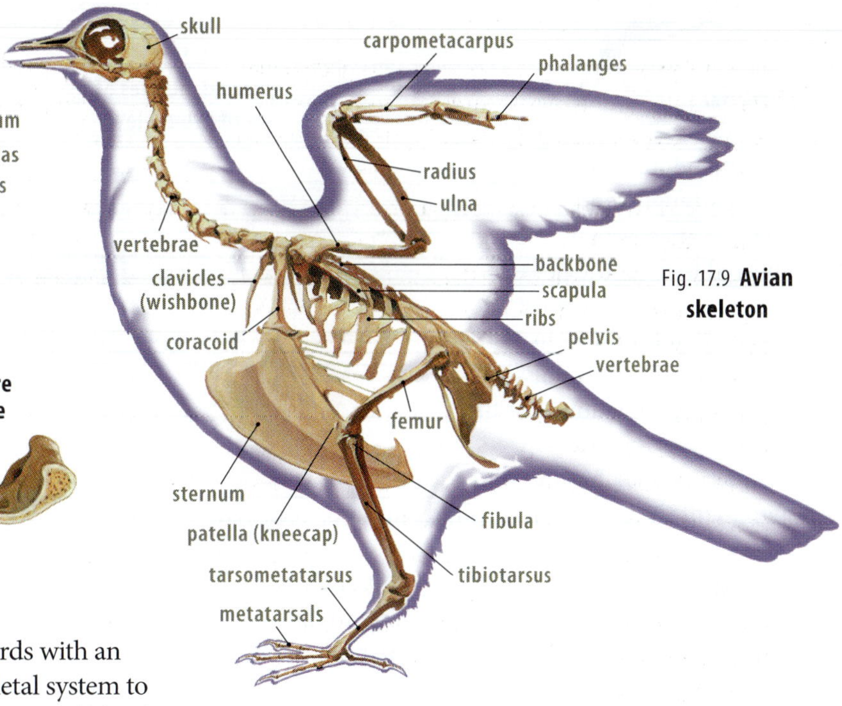


Fig. 17.9 **Avian skeleton**

## Skeletal and Muscular Systems

**Skeletal system.** God designed flying birds with an extraordinarily strong and lightweight skeletal system to meet the demands of flight. Instead of being solid like the bones of most other animals (and of man), *many avian bones are hollow and filled with air instead of marrow.* These hollow bones are strengthened by internal cross bracing similar to that of a steel girder. Because of this special design, a bird's bones are very strong but much lighter than similarly sized bones in other animals (or in man). The lightweight skeleton of a bird accounts for about 5% of its body weight, while the skeleton of a human accounts for about 14% of his body weight.

In addition to these differences, the bladelike *sternum* (breastbone) of a bird is very large and extends outward from the rib cage like a sailboat's keel. This *keel* provides a mounting point for the powerful flight muscles of the bird.

Fig. 17.10 **Bone structure of wing**

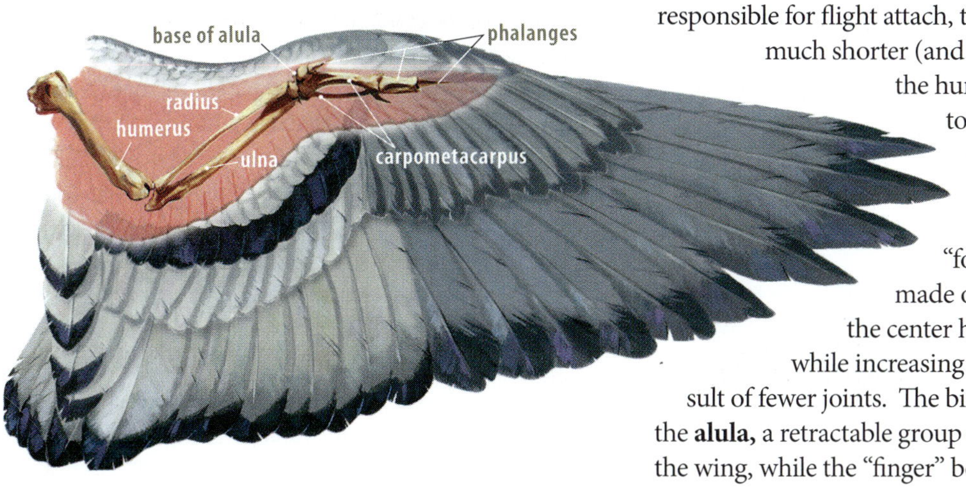


The skull of a bird is also made for lightness and strength. Instead of heavy teeth and jawbones, God designed most birds with toothless bills made from a tough, lightweight protein called *keratin*—the same substance that composes your fingernails—reinforced by lightweight struts of bone. The flat bones of the skull are very thin, and in many places thin girders take the place of the heavier facial bones found in other animals and in humans.

Because it is the bone on which the major muscles responsible for flight attach, the humerus of a bird is much shorter (and therefore stronger) than the humerus of a man in relation to the total arm length. The *radius* and *ulna* of the bird are long in comparison, giving the bird a long "forearm." The avian "palm" is made of one rectangular bone with the center hollowed out, saving weight while increasing strength and rigidity as a result of fewer joints. The bird's "thumb" bones support the **alula,** a retractable group of feathers on the front of the wing, while the "finger" bones support the wingtips.

**Muscular system.** Flying takes tremendous muscular strength, and flying birds are specially engineered to meet the power demands of flight. The most important flight muscles in a bird are the *pectoralis major* muscles, which pull the wing down for the power stroke as the bird flaps its wings. These muscles must be large and strong enough to lift the bird off the ground and accelerate it to flying speed. The wing is raised after each power stroke by the *pectoralis minor* muscles, which connect the sternum to the top of each wing in an arrangement similar to a rope and pulley. This wise design does away with the need for strong back muscles and keeps the heavy flight muscles low in the chest, giving the bird additional stability in flight.

Fig. 17.11
**Pectoral muscles of a bird (front view)**

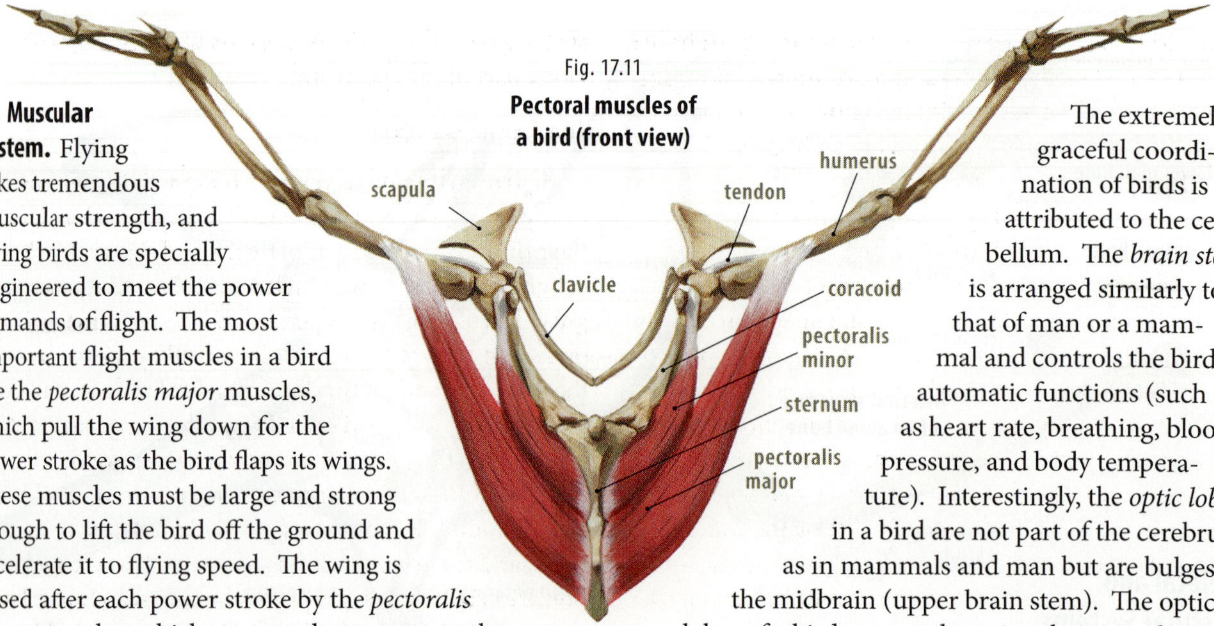


The extremely graceful coordination of birds is attributed to the cerebellum. The *brain stem* is arranged similarly to that of man or a mammal and controls the bird's automatic functions (such as heart rate, breathing, blood pressure, and body temperature). Interestingly, the *optic lobes* in a bird are not part of the cerebrum as in mammals and man but are bulges of the midbrain (upper brain stem). The optic lobes of a bird are very large in relation to the total brain size. The percentage of brain power devoted to vision may help account for birds' amazing eyesight.

## Nervous System

**Brain.** The nervous system of a bird is especially designed to help it in the complex activity of flight. A look at the brain of a bird reveals striking differences from that of a mammal or a human. The brain of a bird can be divided into the cerebrum, cerebellum, and brain stem. Like that of mammals and man, the *cerebrum* of a bird is divided into two hemispheres. The cerebrum is the seat of the bird's intelligence; it controls voluntary behavior and allows the bird to make day-to-day decisions. The *cerebral cortex* of a bird is relatively thin and contains no convolutions like those of man and the mammals. The *cerebellum* of the bird is responsible for muscle coordination. Directed by the cerebrum, the cerebellum processes sensory messages about the bird's body and feather position and transmits impulses to the muscles.

Fig. 17.12 **Brain of a bird**

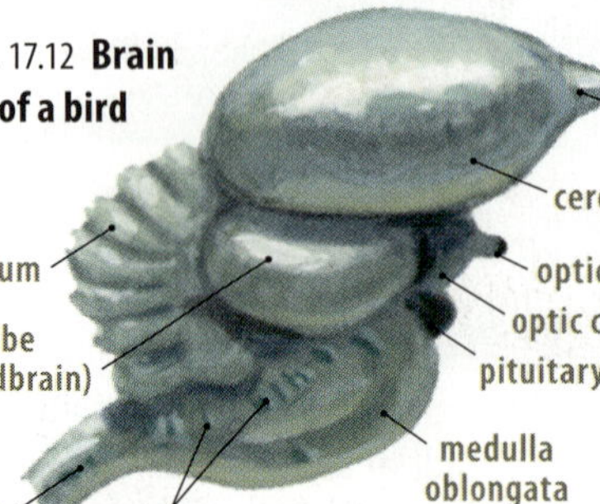


**Senses.** Although the senses of *smell* and *taste* are poor in most birds, their sense of *hearing* is very good. The bird's ear is much the same as our own with a few exceptions. Of course, a bird does not have an external ear projecting from its head; that would make the design less streamlined. The outer ear of a bird consists of a simple tube entering the head; the tube's opening is covered by special feathers. At the end of this tube is the *eardrum,* which vibrates when sound enters the outer ear. A single rodlike bone (instead of three) transmits the vibrations from the eardrum to the *cochlea* of the inner ear, which is much shorter than the coiled cochlea of the human ear. Although the hearing of most birds is comparable to that of a human, *the bird far surpasses the human in the ability to distinguish differences in intensity and respond to rapid fluctuations in pitch.* This feature allows a bird to more easily understand the rapid, intricate calls of other birds. Unlike most birds, owls have hearing that is as much as a hundred times more sensitive to faint sounds as human hearing; it is said that a flying owl can locate a mouse crouching in total darkness by sound alone.

**Eyesight.** Birds have the most remarkable eyesight of any creature—far superior to that of man. A bird's eye resembles our own but is larger in relation to the head, less spherical, and almost immovable. Instead of moving its eyes to look about, the bird turns its head.

In addition to the upper and lower eyelids, birds have a third eyelid called the **nictitating membrane.** This transparent membrane allows a bird to "blink" and moisten its cornea without obstructing its vision, an important ability when flying in close quarters.

The design of a bird's eye allows for much sharper vision than that of a human. This fact is primarily due to the size of the bird's eye, the number of cone or rod cells in the retina, and the design of the image-processing circuits in the retina and the brain. The sharpest focus in our eye is in a depression of the retina called the *fovea.* While we have only one fovea, many birds have two, allowing them to look at two objects at once with the sharpest focus. Birds can distinguish a small object at great distance, an ability made possible by the concentration of cone cells at the fovea. In a hawk's fovea, for example, there are 1.5 million cells, compared to only 0.2 million in a human's fovea. That gives the hawk a visual acuity of almost eight times that of a person. The hawk can see a crouching rabbit more than a mile away. Although most birds that are active in the daytime do not see very well in darkness, nocturnal birds (particularly owls) have extraordinary night vision, far surpassing that of humans.

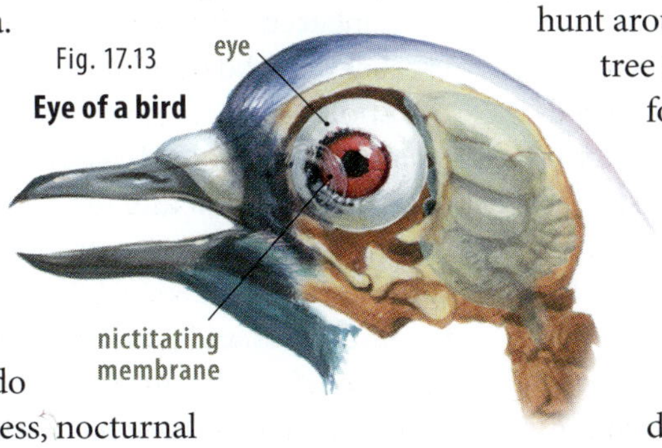

Fig. 17.13 **Eye of a bird**

Most birds also have tremendous *color vision;* again, far better than was given to man. Birds not only see the color spectrum as we do, but they see beyond into the near ultraviolet range of the spectrum. Ornithologists have found that several kinds of hummingbirds can see in the ultraviolet range down to 370 nanometers, a range invisible to humans. This ability helps to attract them to flowers that have **nectar guides,** striking patterns on the flowers visible only in the ultraviolet part of the spectrum.

## Food and Digestion

**High metabolism.** If you watch the activity of birds to any extent, you will be aware that they spend much of their time eating. Because of their small size and furious activity, birds have a very high metabolism that requires a great deal of fuel. The dependence of a bird's metabolism on its size and activity can be illustrated by measuring the intake of food. A tiny hummingbird will eat 100% of its body weight in food each day, while a large, sedentary chicken eats only 3.4% of its body weight in food.

Most birds are voracious eaters, consuming a wide variety of foods. Insects are of particular interest to many birds, and there are few kinds of insects that escape the predation of birds. Birds probe the soil, search tree bark, hunt around every leaf and twig, and even drill into tree trunks in search of insect prey. Other animal foods include worms, mollusks, crustaceans, fish, frogs, reptiles, mammals, and other birds. Many types of birds eat seeds and fruits. A very large group, about one fifth of all birds, eat nectar. All included, birds eat almost anything. God's living creation is delicately balanced, as the thousands of different kinds of birds throughout the world play an active role in checking the populations of many organisms and helping to propagate others.

**Design of bills.** Birds were designed for a particular purpose, as evidenced by their digestive anatomy. Beginning with their specialized bills, each bird fulfills God's plan for nature. Birds like the *toucan* have large bills for seizing

Fig. 17.14 **Variety in bill design**

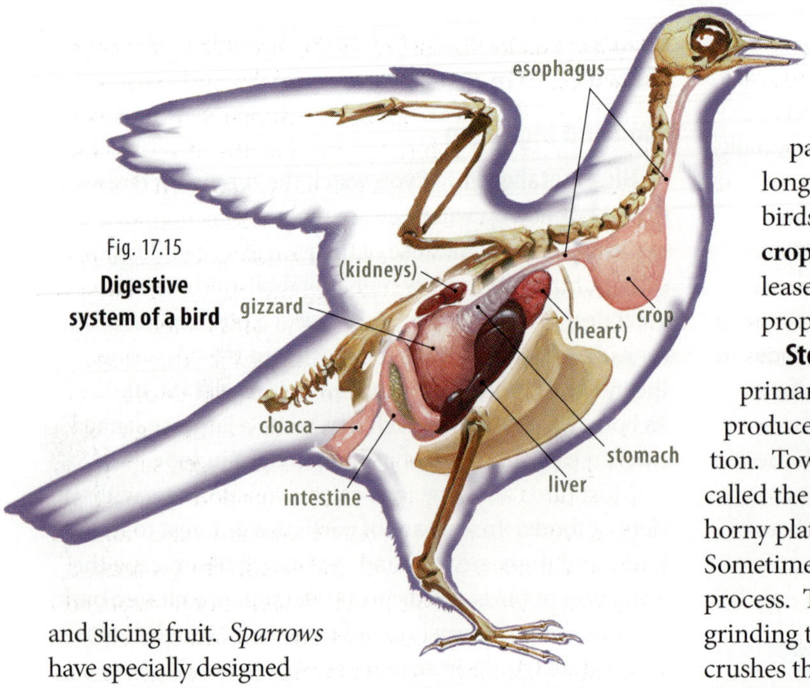


Fig. 17.15
**Digestive system of a bird**

and slicing fruit. *Sparrows* have specially designed seed-eating bills. *Hawks* and other carnivorous birds possess sharp tearing bills. The *woodcock* has a long, probing bill, and the *hummingbirds* have long, slender sucking bills. *Pelicans* have scooping bills to collect fish. *Ducks* and *geese* have harvesting bills to collect various aquatic plants.

**Rapid digestion.** *Birds must rapidly digest whatever food they eat.* Remember that they are designed for flight, and any food remaining in the digestive tract is simply dead weight that pulls the bird down. You have a large stomach and a long intestine that can hold large quantities of food in storage. Because a bird cannot afford the added weight, it eats often and digests food quickly. For example, a *shrike* can digest a mouse and excrete the waste products in just three hours. Your body would take more than two or three times as long to digest a similar amount of food. Even faster is the ability of a *thrush* to digest berries and excrete the waste products in only 30 minutes. Although birds digest food quickly, they use a high percentage of the food they eat, making them very energy-efficient animals. This fact is demonstrated in the grocery store at the meat counter. Chicken is always less expensive than red meat because farmers need not provide as much expensive feed for chickens to gain a pound of meat as they do for hogs or cattle. Some of this efficiency results from the unusually high body temperature of birds (up to 110°F).

**Crop.** The oral cavity of a bird is of little use for digestion. Birds have few taste buds on their tongues and lack teeth for chewing. Salivary glands, however, are present in the pharynx of most birds. Seed-eating birds have abundant salivary glands, which secrete a starch-digesting enzyme. Food passes through a bird's short pharynx into its long, muscular esophagus to its stomach. Many birds have a special sac in the esophagus called a **crop,** which serves as a storage chamber and releases food to the rest of the digestive system at the proper rate.

**Stomach.** The *stomach* of birds consists of two primary chambers. The forward part of the stomach produces and releases gastric juices that aid in digestion. Toward the rear of the stomach is another part, called the **gizzard,** that is very muscular and is lined with horny plates that act like millstones to grind the food. Sometimes birds swallow grit to aid in the grinding process. The gizzard serves the same purpose in birds as grinding teeth (molars) do in humans and mammals—it crushes the food into tiny pieces to prepare it for further digestion.

**Intestine, liver, and cloaca.** Crushed food from the gizzard is emptied into the *intestine,* where digestion continues and nutrients are absorbed. Digestion in the intestine is aided by bile and pancreatic juices from the bird's liver and pancreas. Within the short intestine, digestion and absorption are completed and the remaining waste is readied for elimination in the **cloaca** [klō·ā′kə]. This area of the bird's anatomy is much different than in mammals in that gametes from the reproductive organs (sperm from the testes in the male and eggs from the ovaries in the female) are released into the cloaca rather than to the outside through separate openings. The wastes from the kidneys are also dumped into the cloaca. In young birds, a special organ known as the **bursa of Fabricius** [fə·brĭsh′ē·əs] lies close to the wall of the cloaca. At this site, the lymphocytes that aid in the immune system are produced. The B lymphocytes (B cells) of humans get their name from this bursa, because the chicken was the animal in which scientists first discovered these special cells.

## Excretory System

**Designed for efficiency.** *Birds display a unique ability to concentrate the urine collected by the kidneys.* Unlike the urine of mammals, bird urine contains a high concentration of uric acid rather than urea. Uric acid is not very soluble in water and requires little water to be eliminated; therefore, birds can conserve water, a requirement for an animal that drinks little. A mammal loses 60 times as much water as a bird through the kidneys. The concentration of uric acid in the cloaca, observed as a white paste, is 3000 times more concentrated than it is in the blood. The most effective mammal kidney, found in a desert rat, is able to concentrate urea to only 25 times the urea concentration of the blood.

**Salt gland.** Marine birds have been equipped with a special excretory gland called a **salt gland,** which is located above each eye. These birds consume much salt water in their diet. Seawater is approximately 3% salt, which is three times the concentration of salt in their bodies. Ordinarily, drinking seawater would soon cause them to dehydrate because the salt water would draw fluid out of the cells. As a result, marine birds must eliminate the excess salt. Because their kidneys, like ours, cannot concentrate salt, they have been given the salt gland with an ability to concentrate salt to twice the concentration of seawater. This salty solution runs out through a duct to the nostrils, giving these birds a perpetual runny nose. Salt glands allow marine birds to obtain needed body water by simply drinking seawater and excreting the excess salt.

## Respiration

*The respiratory system of birds is completely different from that of any other living vertebrate.* The lungs in birds follow a different design, from their size and shape to their microscopic structure. Unlike other animals, a bird does not breathe air into the lungs but through the lungs. In the lungs of other creatures, air enters the expanding lungs through the bronchial tubes and flows into closed chambers (alveoli), where the lungs absorb oxygen. After a brief pause, the air is forced back out of the lungs through the bronchial tubes and is exhaled through the trachea.

When a bird inhales, fresh air passes through the trachea into large **air sacs** behind the lungs (step 1 in Fig. 17.16). As the bird then exhales, the fresh air from the posterior air sacs flows forward through the lungs (step 2). Oxygen and carbon dioxide are exchanged before the bird inhales again; this inhalation forces the air out of the lungs into more air sacs in front of the lungs (step 3). Finally, the air leaves the bird during the next exhalation (step 4). Thus, a bird inhales and exhales twice to move each breath of air through its respiratory system; in effect, the bird has two breaths of air moving through its system as it inhales and exhales. In this system, air flows in one direction: from posterior air sacs, through the lungs, and into anterior sacs. In addition, the lungs do not contract and expand within the chest cavity but are rigidly attached to the cavity walls.

As we study the respiratory system of birds, we see how beautifully it is designed to suit the needs of a bird. Because fresh air continually passes through the lungs, providing birds with a constant oxygen supply, this system allows birds to extract considerably more oxygen from the air they breathe than any other land animal. (In mammals, dead-end alveoli always contain a significant amount of stale air that cannot be expelled.) Because of this design, birds can fly vigorously at altitudes at which other creatures would lose consciousness from lack of oxygen.

The air sacs of a bird also cool the bird's body systems during the vigorous muscular activity of flight. The air

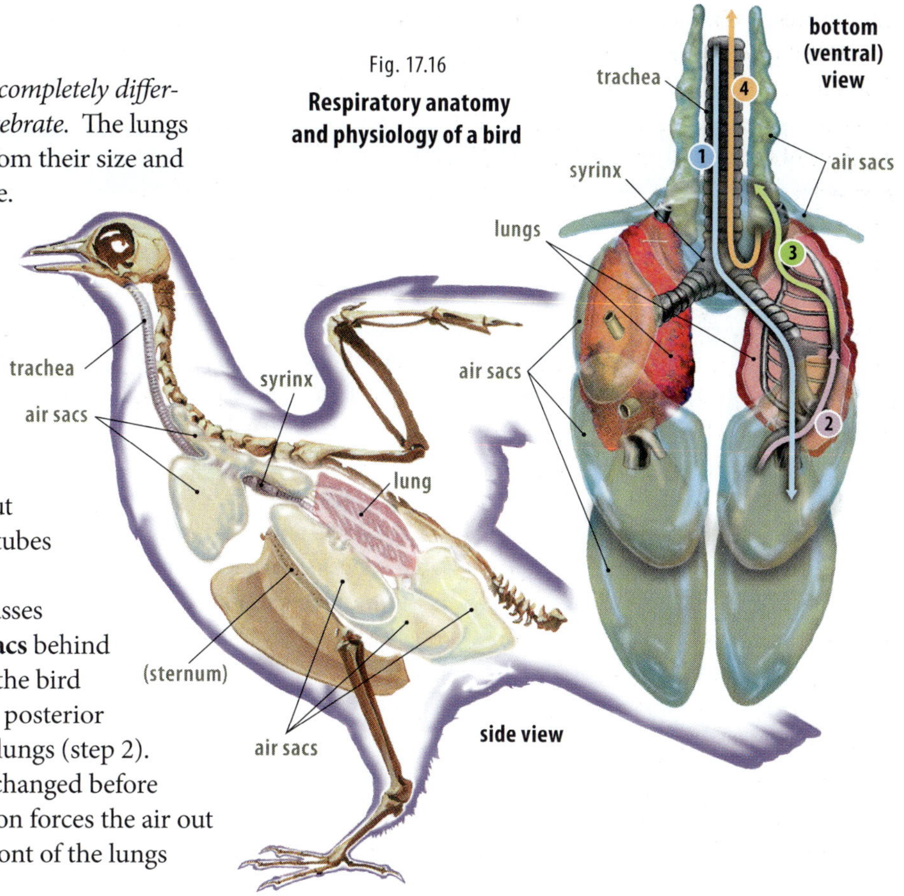


Fig. 17.16

**Respiratory anatomy and physiology of a bird**

sacs are connected by air tubes to the hollow cavities of the major bones, allowing air to flow in and out of the bird's skeleton. With each breath, cool, fresh air moves into the air sacs and hollow bones, absorbing excess heat from the bird's organs. Evaporation of moisture within the air sacs also aids in cooling. Without this special cooling system, birds (which lack sweat glands) would soon overheat.

Another special design in birds is the *voice box* or **syrinx.** Unlike our larynx, the syrinx is at the lower end of the trachea, where it divides into two bronchi. Within the syrinx is a membrane that vibrates when expired air passes over it. Muscle tension and differences in air pressure change the pitch and volume. A few birds especially gifted with intricate song ability, like the wood thrush, have two independent membranes, one in each bronchus. With separate muscles and nerves it is possible for each membrane to produce a different sound. These birds can sing two harmonically unrelated notes simultaneously, making their calls quite complicated.

## Circulation

**Four-chambered heart.** *The general pattern of circulation in birds is not greatly different from that in mammals.* They have an efficient four-chambered heart that completely separates pulmonary and systemic circulation. The heart of a bird is larger and more powerful than that of a mammal of the same size. The two jugular veins in the neck are connected by a cross vein, allowing blood to be shunted from one vein to the other when the bird turns its unusually long, flexible neck. This assures full circulation to the head. As you may suspect, the brachial and pectoral arteries supplying the wings and breast, respectively, are unusually large because these active areas need much blood.

**Heartbeat.** *The principle that we learned in mammals about the heartbeat being proportional to body size holds for the birds as well.* A bird's high metabolism and small body give it a rapid heartbeat, and smaller birds have more rapid heartbeats. A turkey, for example, has a heartbeat of about 93 beats per minute, a chicken's heart beats 250 times per minute, and a chickadee's heart beats 500 times per minute during sleep and increases to 1000 times per minute during exercise.

**Blood.** *Bird blood looks the same as mammalian blood but contains nucleated red cells instead of nonnucleated ones;* the red blood cells are also just a little larger than those of mammals. The white blood cells of birds are more active and efficient than those of most creatures, helping them to repair wounds quickly and prevent disease.

### Prolific Producers by Design

Design in the LIVING CREATION

Have you ever considered why some creatures have just a few offspring while others have many? Have you thought of why plants appear to produce sometimes thousands of seeds or potential offspring or why some insects reproduce in huge numbers? All of these situations are a demonstration of God's provision for other creatures.

God provides for the birds by supplying vast numbers of insects and seeds to be produced. The potential for insect propagation is almost inconceivable. A United States government entomologist has estimated that one pair of Colorado potato beetles, if allowed to go unchecked in one season, can give rise to over 60 million offspring. Another man has said that one female fly, allowed to breed freely starting in May, would by August yield 143,875 bushels of flies. Probably one of the most prolific insects is the plant louse or aphid. Entomologists report that in a single season the aphids reproduce 13 generations, each mating producing over 100 females. If all the aphids survived to reproduce, there would be ten sextillion aphids at the end of this process. That is enough to place them 10 abreast in a string that would reach twenty-five hundred light years into space.

Why would God allow such reproductive potential? The answer is to provide for His creation. Let's consider just one of the beneficiaries of all of these aphids. The *yellow-throated warbler* is a beautiful songbird with a large appetite for aphids. On one occasion, an observer watched one of these warblers on a birch tree eat 89 aphids a minute for 40 minutes. When he seemed to have his fill, the warbler had consumed over 7000 aphids.

Consider how many aphids this fellow and his family and friends would eat during the summer. Is it any wonder why God designed aphids to reproduce in such numbers?

Another bird with a large insect appetite is the *scarlet tanager,* which can eat over 600 gypsy-moth caterpillars in 18 minutes. A *nighthawk* will eat 500 mosquitoes in one feeding. *Chickadees* consume large numbers of the cankerworm. It has been estimated that chickadees in the state of Michigan alone consume over eight billion insects each year.

Birds also consume tremendous numbers of seeds, most of which are weed seeds. From your study of botany you know that most plants produce large numbers of seeds. Have you ever considered why? Of course, an obvious reason is to reproduce their own kind, but are such vast numbers needed? After all, one seed produces one plant. Actually, this is another provision by God for His creation. Birds that eat seeds need large numbers to meet their needs. A *mourning dove* can eat 7500 clover seeds in one feeding. A *bobwhite* has been known to consume over 5000 pigeon-grass seeds or 1000 ragweed seeds or 10,000 pigweed seeds per feeding. It has been estimated that the bobwhite population in Virginia eats 573 tons of weed seeds annually, besides all the insects they eat.

When God created the birds, He designed a means of feeding them. He also provides for His most loved creation, mankind, in much the same way. As people, we eat a very large number of potential living things such as eggs and seeds. You can demonstrate the reproductive capacity of one of our favorite foods—corn. Take an ear of corn and count the grains. Assuming that each grain produced a new corn plant with a new ear, how many grains would you harvest at the end of the second generation? At the end of the third generation? You should have figured a very large number, and all from one ear of corn. Just think of the multiplied ears of corn grown on farms to feed us. Each American consumes about 45 pounds of corn each year, which means that God provides countless corn seeds, most of which are of no use to the species that produces them.

It becomes obvious in our study of nature that God required insects and seeds to feed His larger creatures. Besides creating the beautiful songbird for His glory and for the delight of mankind, God in His wisdom created insects and seeds to feed bird and man alike.

## Section Review 17.2

1. What purposes do feathers serve for a bird?
2. What are the three basic types of feathers on a bird?
3. Describe the structure of a flight feather.
4. What type of flight pattern is characteristic of eagles? Of albatrosses?
5. What is the purpose of a bird's crop? Of the gizzard?
6. How does a bird's excretory system function to conserve water?

**Application**

7. How is the bird respiratory system well suited to meet the demands of flight?

**Critical Thinking**

8. In what ways are birds similar to mammals? In what ways are they different?

**Identify**

nictitating membranes, nectar guides, crop, gizzard, cloaca, bursa of Fabricius, salt gland, syrinx

## 17.3 Family Life of Birds

God ordained that birds, like other creatures, should reproduce after their kind (Gen. 1:21–22). Among the familiar sights and sounds of springtime are those that accompany the return of the songbirds from warmer climates. As the birds prepare to raise their young, they stake out their territory by singing, display beautiful plumage to attract potential mates, and build nests to protect their eggs. The family life of birds is a fascinating example of God's design and provision for His living creation.

## Migration of Birds

**Migration,** or annually moving to a different region, is a characteristic of many kinds of birds. Most songbirds leave the temperate regions at the end of the summer and fly south (north in the Southern Hemisphere) to the tropics, where warm weather and an abundant winter food supply await them. In the spring, when warmer weather returns to temperate areas, the birds return to their summer homes to breed and raise their young.

It is common for birds to **flock,** or gather in large groups, especially just before the breeding season. Flocks usually break up when young birds are being raised, but many birds flock again at the end of the breeding season. These same birds then migrate in flocks. Migrations may be for short distances, or they may be for thousands of miles. The *arctic tern,* the migration champion of the world, travels about 11,000 miles one way from the Arctic to the Antarctic each time it migrates.

As they migrate, birds may move slowly from region to region, or they may make a long, continuous flight. The *red-winged blackbird,* which is often seen in flocks containing hundreds of thousands of migrating birds, moves slowly from region to region. The *ruby-throated hummingbird* flies nonstop across the Gulf of Mexico to Central and South America. The *golden plover* flies nonstop for several days at a time on its migratory flight from Canada to South America.

Much scientific investigation has gone into learning what causes and controls bird migrations, but we do not yet have a complete explanation. We do know the following:

- Migrating birds follow the same route year after year (these routes are called **flyways**).
- A keen sense of direction and a recognition of landmarks is involved.
- The position of the sun and stars guides many migrating birds.
- The earth's magnetic field helps guide certain birds.
- The timing of migration appears to be influenced both by the birds' internal clock (controlled by the pituitary gland) and by environmental factors such as the weather and the length of daylight.
- Migration generally occurs in the spring and fall.
- Most migration is north and south.

It is interesting to note that the flyways of numerous species of birds pass through the country of Israel. This is one reason why birds mentioned in the Bible are familiar to people in many parts of the world.

## Courtship

Birds are noted for their unusual courtship displays, which rival those of any other animal in complexity and beauty. *Finding a mate is usually associated with territoriality and recognition of the proper species.* A male bird will establish his territory in preparation for raising his brood and feeding. Males of most species have unique characteristics of color, plumage, or song; and these characteristics are exaggerated during the breeding season for the purpose of attracting a female of the same species. Different species of the same type of bird in the same general area have distinctively different markings, behaviors, or songs so that the females will not be confused. Some birds mate for only a single season, but many birds mate for life. The "engagement period" of some birds is rather long. For example, *bearded titmice* become "engaged" before their first molt at only 2½ months of age but do not breed until they are one year old. *Wild geese* find their mate in the spring following their birth but do not become sexually mature until a year later. *Albatrosses* pair two years before their first breeding.

*Courtship is ordinarily dominated by the male,* who initiates the courting activities as females wander into his territory. Depending upon the species, he displays some form of behavior that attracts the female of his own species. He may parade, dance, or fly with dazzling acrobatics. The *peacock* approaches the female at an angle so she sees only his drab colors and then, at just the proper distance, suddenly swings himself around to dazzle her with the more than 100 shimmering feather "eyes" vibrating atop his long quills.

Not all birds are so dramatically arrayed with color, and some show no external difference between male and female. For these birds, song is especially critical to the courting process. Males sing loudly, and sometimes they sing special songs that are only sung for courtship. The syrinx changes shape in response to hormones, altering the call of the bird. Special sounds can be produced by a rapid movement of the wings or rattling of the quills. Some *woodpeckers* begin tapping on hollow trees or tin roofs to attract mates. All of these activities can be somewhat costly to the bird; most males lose much weight in the courtship process. At the end of courtship, the *great bustard* male weighs only two thirds of what he weighed at the beginning of courtship.

Mate selection is the primary and most important product of courtship, but not the only one. The activities of the male stimulate the release of hormones in the female that control the production and release of the egg. Courtship also assures the birds of mutual reproductive maturity. Birds cannot afford to waste time and energy in unsuccessful mating; they must reproduce quickly to raise a brood in preparation for migration or a harsh winter.

## Mating and Fertilization

After the elaborate courtship procedures, the birds mate, the male's sperm passing from his testes to the cloaca of the female. In female birds of most species, the left ovary and oviduct develop into functioning organs, but the right side remains small and nonfunctional. Egg cells released by the ovary are collected by the oviduct, which empties into the cloaca. At the proper time, ranging from minutes in the chicken to months in some birds, the sperm migrate up the oviduct and *fertilization* takes place as the egg begins its descent through the oviduct toward the cloaca. The egg expands en route as **albumen** [ăl·byōō′mən: egg white] is added to it, having been manufactured and secreted by special glands. Toward the end of the oviduct, a thin shell membrane is added, and then the shell itself. The egg is laid when it passes from the oviduct into the cloaca and through the opening to the outside.

## The Egg

The avian egg is of amazing design. The egg cell is very large, accounting for the entire center part, the **yolk.** The bulk of the yolk body is a fat-rich food source for the tiny embryo, which develops at a special location called the **germinal spot.**

The shell is designed around a light protein framework heavily deposited with inorganic minerals, mainly calcium carbonate. Most of the calcium used to make the shell comes from the female's daily food intake, but some of it comes from the female's bones. As the chick grows inside the egg, much of the deposited calcium is reabsorbed by the young to produce bones; thus there is a marvelous conservation of minerals. The minerals of the egg shell are arranged as vertical crystals with tiny pores between them. These tiny pores allow the diffusion of water vapor, oxygen, and carbon dioxide while restricting the passage of water and most other liquids. In this respect the shell is a wonderful breathing device. Running through the middle of the albumen layers is a very dense, twisted cord called the **chalaza** [kə·lā′zə]. It is attached to the yolk and acts as a support for the yolk. The flexible chalaza allows the yolk to remain upright as the egg is rolled in the nest, assuring that the germinal spot is always on top (the only position in which the embryo can grow).

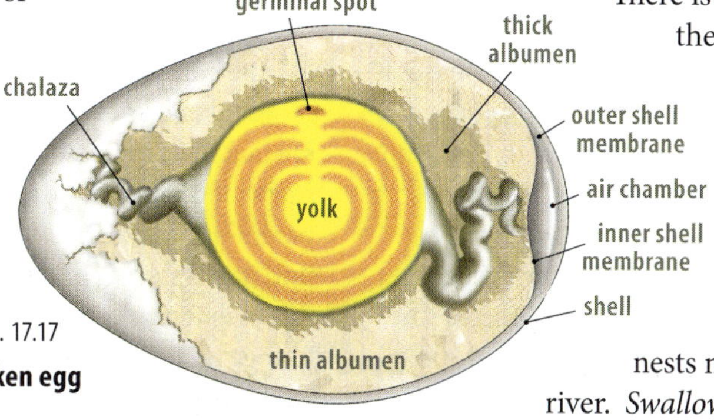

Fig. 17.17
**Chicken egg**

## Nesting

After the male bird has defined its territory and selected its mate, the next major task is building a *nest.* This may require no time and effort at all, as in the case of bare-ground nesters, or it may demand furious activity by both male and female birds for days. Most birds build a new nest for every new brood. Some complete the nest before the eggs are laid; others are still adding to it even after the chicks are hatched. In fact, there are about as many different methods of nest building as there are species of birds.

Nest-building behavior is one of the best examples of animal instinct seen in the living creation. An **instinct** is unlearned knowledge that an animal possesses from birth. The nest-building instinct of birds provides man with a marvelous evidence of the Creator's involvement in nature. Scientists have determined that birds can build nests without any training whatsoever. Hatchlings have been removed from their nest and raised in a solitary environment. When given the right materials, however, these untaught birds immediately began to build a nest just as fast and accurately as the thousands of generations before them. *All of this activity is for a specific purpose: the protection of the eggs.*

There is an amazing variety among the nests of different birds. The *grebe* [grēb], a water bird, builds its nests of aquatic plants in such a way that the nest floats, protecting it from terrestrial predators. *Ospreys* feed on fish, so they build their nests near the shore of a lake or river. *Swallows,* a group of swiftly flying insect catchers, make mud nests. The *cliff swallows* build their nests on the sides of buildings or the peaks of roofs. *Bank swallows* burrow into the face of an earthen bank (such as a riverbank) for nearly a yard to make a small, cavelike nest chamber. The *chimney swift's* nest is made of twigs glued to the corner of a roof or inside a chimney with a sticky salivary mucus. *Orioles* build nests suspended from tree branches. The *tailorbird* actually sews its nest together by piercing holes in leaves and threading strands of plant fiber through, knotting them on the outside. *Weaverbirds* use grasses to weave intricately patterned nests that in some species have two separate chambers.

## Incubation

The purpose of birds' varied nests is one: providing a place for the birds to incubate their eggs. **Incubation** involves keeping the eggs warm, usually by an adult bird sitting on them, so that the young birds will develop properly in the eggs. As in most biological systems, the activity of growth and development is dependent upon an optimum temperature. Left unincubated, the embryo has little or no chance of survival. The efficiency with which birds can provide the heat needed for development is exhibited by the *emperor penguin,* which provides a steady temperature of 93°F to the egg while the air around it is 77°F below zero. The penguin maintains this incubation process day and night for eight to nine weeks.

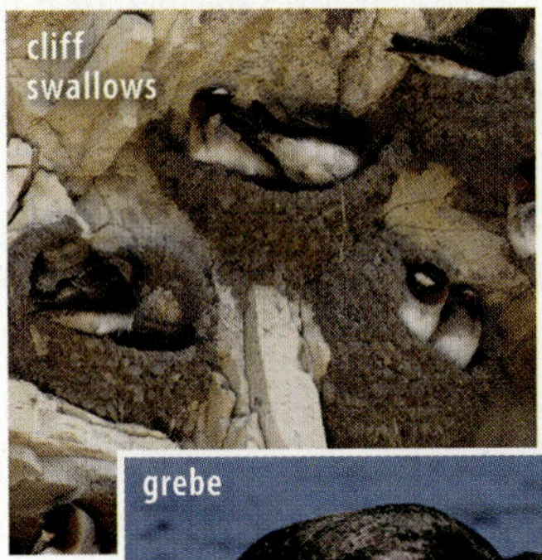

Fig. 17.18
**Nesting birds**

Several changes must occur in the adult bird before it can incubate an egg. While covered with feathers, which are excellent insulators but poor conductors of heat, the bird cannot transfer its body heat to the egg. In response to the presence of the egg, the adult loses feathers on its belly to expose a patch of skin called a **brood patch.** There may be one, two, or several brood patches on the belly of a bird, depending on the species. Also, the skin of this area undergoes changes. The dermis becomes spongy and richly supplied with blood vessels so that it looks inflamed. This exposed patch allows for efficient transfer of heat from the adult to the egg.

In the majority of birds, both male and female incubate the eggs, although in most species the female spends the most time in the nest while the male bird provides food. Many birds, such as ducks, geese, pheasants, and quail, hold off incubation until all of the clutch (eggs in the nest) have been laid. The first egg remains in a state of dormancy until incubation begins. This assures that all of the young will hatch at the same time. To maintain a constant temperature, the adult will occasionally turn the eggs in the nest to eliminate cold spots. (Remember, though, that the embryo will always be in the upward position because of the albumen and chalaza.) The bird instinctively senses the temperature within each egg and adjusts the eggs accordingly.

Most American songbirds lay from three to six eggs in their nests, and they hatch in about 14 days. These birds are **altricial** [ăl·trĭsh′əl], or naked and helpless at hatching. By contrast, the chicks of quail, fowl, ducks, and most water birds are **precocial** [prĭ·kō′shəl], covered with down and able to run or swim almost immediately after hatching.

Because altricial birds cannot maintain a stable body temperature immediately after hatching, the parents must warm them and shelter them from the elements for a while after they are hatched. This may continue for several days, until the young birds have developed enough to maintain their warmblooded temperature control.

Feeding habits are varied and unusual. Precocial birds are taught to eat by the parents. Altricial birds wait, mouths agape, for parents to drop food into their beaks. Many parent birds, like the *little bittern,* swallow the food as they find it and store it undigested in their crop; the parent then regurgitates it at the nest. Young *penguins* and *pelicans* plunge shoulder-deep into the gullets of their parents to feed on shrimp and fish the parents have caught. All *doves* and *pigeons* produce **crop milk** rich in protein and fat; this substance is similar to the milk produced by mammals but is produced by special cells lining the crop instead of by external glands. Other birds, like the *flamingo,* produce oily fluids of the digestive tract that they use to feed the young. In all cases, the chicks eat a protein-rich diet to promote rapid growth. Even when the adults prefer other food, they will gather higher-protein sources for the young.

Young birds stay in the nest for different lengths of time, depending on how rapidly they develop. Often, though, the parents must entice them out or "kick" them out of the nest. Once it leaves the nest, the young bird has much to learn: how to find food, fly, sing, and interact with other birds. Particularly in birds of prey, skills of hunting are practiced over and over until the young are proficient. *Marsh hawks* drop mice in midair for the young to practice catching them. The young *bald eagle* plays with sticks to gain hunting talents. In every case, God has given the birds special abilities that have fascinated man for centuries.

## Birds and Private Property

Understanding the LIVING CREATION

Biologists studying the habits of birds have learned that they are aggressive defenders of individual property. You may have watched the birds in your area, thinking them to fly about helter-skelter, sort of bouncing all over the place like a ball in a pinball machine. Yet upon careful observation you will note that each bird travels about within a tightly guarded space, sometimes overlapping a neighboring bird's area but usually maintaining its home territory very jealously.

Man's understanding of **territoriality** developed during the 19th century when several naturalists noted the seeming impossibility of some species of birds nesting close together. It was evident that these birds remain at precisely fixed distances from each other. The naturalists concluded that this was a necessary design to provide enough feeding area for birds of specialized diets. Simply stated, territoriality is needed to reduce the dangers of overcrowding. The effects of this behavior become quite numerous and usually quite important to a species' survival.

Two major effects of territorial behavior are *isolation* and *dispersion.* When birds establish their territories, they maintain a form of land ownership, or private property. This social arrangement lends itself to a stable family lifestyle, which has many advantages. Ownership of territory proves to be an important factor in reproduction. All of the activities of courting, nesting, and rearing of young go on with minimal interference from other birds. Territoriality ensures ample food and nesting material for the family. As fighting is reduced, energy is saved and life is much more efficient.

Because birds are strong protectors of their territory, newcomers and rivals are dispersed to find and maintain their own territory. This dispersion principle also promotes efficiency in food finding and nest building because it prevents overcrowding. Dispersion also reduces opportunities for disease that may spread through a group of birds living together, and it makes it more difficult for predators to kill them.

A major result of territoriality is that the birds become very familiar with their territory, knowing where to find food and where to seek escape from enemies. In short, community stability is maintained by territory-owning birds. Some biologists have seen such complexity in bird activity that they compare its stability to that of humans who maintain a law-and-order society based on personal property rights. We see by this an example of God's design of order in Creation.

As a rule, a bird will tolerate birds of other species in its territory as long as they are not a threat to its survival.

(cont.)

(cont. from previous page)

Birds of the same species will be driven away because they would compete for the same food. As many as six different species have been found nesting within 30 feet of each other. In Florida, a bald eagle and a great horned owl were once seen nesting in the same tree only three feet apart. Because each of the birds in your area is very specific about its needs of food, protection, and nesting, they may each take advantage of some aspect of the complex nature of that area.

How large an area will a bird defend? That really depends on the kind of bird. Several factors are involved, including the bird's size. Often, *the larger the bird, the bigger its territory.* This is particularly true of predators, because they need a large area in which to hunt. *Predatory species are likely to have larger territories than nonpredatory species.* Birds living in desert areas have larger territories than those living in fertile habitats. Even the volume of a bird's voice may determine its territory size. Birds with a loud call tend to dominate a larger area than birds with a soft call. Birds that live in colonies, such as most waterfowl, will have a very small territory, perhaps an area only a few feet square. The *black-headed gull* in England has a territory of only one square foot, whereas the *meadowlark* has a territory nearly a million times bigger—22 acres. Large predatory birds like the *red-tailed hawk* may defend 325 acres. Perhaps the largest territory is displayed by the *golden eagle,* which may have a territory as large as 22,000 acres (35 square miles).

A bird's territory involves more than distance across the landscape; it involves heights as well. Some birds soar high and nest in the tops of trees and cliffs. Other birds run along the ground and even nest in the ground. This is another reason why you can expect to find several species of birds occupying the same general area.

Birds defend their territory with amazing behavior. When we think of defending property, we usually think of fighting. Birds will fight if necessary, but rarely is that the case. Birds usually defend their property with their song or visual display. Songs are especially effective, because the intruder need not see the defender in the foliage to get the message. A robin was once seen to invade the territory of an established male. The newcomer, not knowing that he was intruding, began to sing. When the property owner heard the newcomer, he gave a warning call. As the newcomer remained still and continued to sing, the owner moved closer, singing more loudly. When about 15 feet away and still hidden from view, the angry robin let out a violent song that drove the intruder away. The result of this activity provided a means of maintaining territory rights without a physical struggle. Both birds benefited from the interaction as they followed the design of their Creator.

Other birds display their colors for defense instead of singing. The *yellow wagtail* puffs out its yellow breast at an intruder as a warning. Experiments with the *red-winged blackbird* showed that when the wing was dyed black, the birds lost their ability to defend their territory because they could not display their colors. Instead of singing or displaying colors, some birds dance around in peculiar motions to ward off rivals. The *American robin* lifts its tail, crouches, and then makes a series of short runs broken by brief pauses. The *mockingbird* performs a dance-display in which two birds face each other across the disputed boundary, making short hops, advancing, and retreating.

As we study the territorial behavior of birds, we are reminded of God's complexity in nature's design. Jesus teaches us in Matthew 6:26 that God is presently, actively involved in caring for the birds. What a marvelous intelligence it is that directs such a magnificent symphony of life!

## Section Review 17.3

1. What are three design features of a bird egg that allow it to protect and nourish the developing bird?
2. How is nest building an example of an instinct?
3. Name three ways that adult birds nourish their developing young.

### Application

4. How does the territoriality of birds help ensure their survival?

### Identify

migration, flock, flyways, albumen, chalaza, instinct, incubation, brood patch, altricial, precocial, crop milk

## Chapter 17 Review

### Define

1. bipedal
2. rachis
3. barbs
4. vanes
5. barbules
6. alula
7. nictitating membrane
8. nectar guides
9. crop
10. cloaca
11. bursa of Fabricius
12. salt gland
13. air sacs
14. syrinx
15. migration
16. flock
17. flyway
18. albumen
19. germinal spot
20. chalaza
21. instinct
22. incubation
23. brood patch
24. altricial, precocial
25. crop milk

### Identify

1. scientists who study birds
2. nine characteristics that help equip birds for flight
3. five common perching birds
4. four birds of prey
5. four water birds
6. three game birds
7. two flightless birds
8. the stiff quill of a feather
9. the organ of a bird that grinds food for digestion

### Explain

1. List the three basic types of feathers and the purpose of each.
2. Which two senses are the best developed in most birds?
3. Why do most birds of temperate regions migrate to the tropics for the winter?
4. What purpose(s) does the yolk of a bird's egg serve? The shell? The chalaza?
5. What is the purpose of a bird's nest?
6. What is the difference between altricial and precocial young?

### Apply

1. Ostriches, kiwis, and other similar species are called ratites (from a Latin word meaning "raft shaped") because their sternums are flat, lacking the keels found on most birds' sternums. How is this design related to ratites' locomotion?
2. One early idea for human flight involved strapping wings made of bird feathers to a person's arms; he would then flap his arms to fly. Describe two differences between the designs of humans and birds that doomed this plan to failure.
3. How is the overall arrangement of a bird's brain similar to that of a human? How is it different?
4. In what ways do the hearing and eyesight of birds surpass that of humans?
5. Why is territoriality important to the survival of birds?

CHAPTER 18

# Reptiles and Amphibians

**Key Concepts**

- Poikilotherms
- Lizards
- Snakes
- Turtles and tortoises
- Crocodilians
- Dinosaurs and similar creatures
- Amphibians

## 18.1 Reptiles: The Creeping Vertebrates

"To creep" is the meaning of the Latin word *repere,* from which the word ***reptile*** is derived. Fewer than 9000 species of these creeping vertebrates are living today. Of the approximately 16 orders of reptiles known from fossils, only 4 have survived to the present. The common reptiles—lizards and snakes, turtles, and crocodilians—represent the three major orders. The fourth order contains only two species, the lizardlike tuatara [to͞o′ə·tär′ə], which are found only on islets off the coast of New Zealand. The largest order, which consists of lizards and snakes, totals about 8400 species.

### Cold-Blooded Animals

Unlike mammals and birds, reptiles are considered **cold-blooded** animals because they regulate their temperature by external factors, such as lying in the sun when their body temperature is too low or cooling off in the shade when their temperature is too high. As a result, *their body temperature fluctuates with the temperature of their surroundings,* cooling greatly at night and rising during the day. Animals whose body temperatures fluctuate in this manner are called **poikilotherms** [poi·kĭl′ə·thûrmz′: "varied temperature"]. The term

**Table 18.1 Groups of Reptiles**

*(Kingdom Animalia; phylum Chordata; subphylum Vertebrata; class Reptilia)*

| Group | Description | Number of living species (approximate) | Examples |
|---|---|---|---|
| **Lizards** (order Squamata; suborder Lacertilia) | elongated, four-legged, scaly reptiles with eyelids, nictitating membranes, ears, legs with claws, and tapered tails | 5500 | Komodo dragon, frilled lizard, Gila monster, anole |
| **Snakes** (order Squamata; suborder Serpentes) | legless, scaly reptiles with long, cylindrical bodies lacking eyelids and outer ears | 2900 | long-nosed tree snake, emerald boa, king cobra |
| **Turtles** (order Chelonia) | reptiles with shells and toothless jaws | 300 | soft-shelled turtle, green sea turtle, box turtle |
| **Crocodilians** (order Crocodilia) | large, thick-skinned, long-tailed, aquatic reptiles | 23 | alligator, caiman, crocodile, gharial |
| **Tuatara** (order Rhynchocephalia) | rare, lizardlike reptile with a crest of enlarged spines along its neck, back, and tail | 2 | tuatara |

*poikilothermic* is the opposite of *homeothermic*, or "same temperature," which describes mammals and birds. A cold-blooded animal does not necessarily have "cold blood"; the ideal body temperature of a lizard may be as high as or higher than that of a mammal.

Because their temperature usually depends on their surroundings, cold-blooded animals must avoid extremely high or extremely low temperatures; in hot weather they overheat, and in cold weather they freeze. Many reptiles that live in hot, tropical climates are nocturnal; most reptiles that are diurnal (active during the day) cool off by moving to shady places. Although reptiles are most abundant in the tropical regions, they live on every continent except Antarctica and in all the oceans except those in the Polar Regions. Reptiles living in climates that have severe winters go into a type of hibernation during this time. Because their body processes nearly cease during cold weather and they cannot become active until the weather is warmer, they are not described as true hibernators. If you unearthed a reptile on a freezing winter day, it would lie there barely able to move. A basic principle in biology states that the lower the body temperature, the slower the actions of the animal.

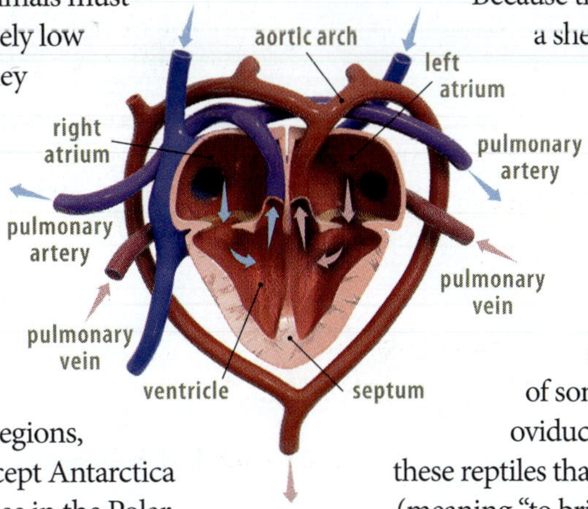


Fig. 18.1
**Reptilian heart**

## Similarities among Reptiles

Although reptiles vary greatly in shape, size, and color, they have several common characteristics. *All true reptiles are vertebrates; are cold-blooded; and have dry, tough skins that are covered by a protective layer of scales.* Many species of reptiles *molt* several times a year when new scales form under the protective layer and loosen it. Since the primary function of their skin is to preserve body moisture, many reptiles can survive without water for long periods. Except snakes and a few lizards, all reptiles are *tetrapods,* having two pairs of appendages; each appendage has *claws* on the toes.

Reptiles breathe with *lungs.* Most species have two lungs, but some snakes have only one. Other snakes and lizards have one small lung and another enlarged, well-developed lung, which extends the length of the body cavity. *Most living reptiles have three-chambered hearts* consisting of two atria and a single partially divided ventricle. Although oxygenated blood and deoxygenated blood are together in the ventricle, muscular ridges almost completely prevent them from mixing. Crocodilians have four-chambered hearts similar to those of mammals and birds.

Most reptiles are *oviparous,* reproducing using eggs. Because the eggs of oviparous reptiles are covered by a shell, they can reproduce on land. Even sea turtles, which are aquatic, must lay their eggs on land. The leathery shell prevents the egg from drying out; inside the shell the young reptile embryo is protected further by a sac of fluid (an *amnion*). The young reptile grows and develops within the amnion and the shell until it can care for itself. The eggs of some snakes and lizards are retained in the oviducts of the female and hatch within its body; these reptiles that bear their young alive are **ovoviviparous** (meaning "to bring forth alive from eggs"). Unlike female mammals, female reptiles of most species do not nourish or care for their young; however, the females of many species do guard the eggs, and some protect their young until they can fend for themselves.

## Lizards: The Most Numerous Reptiles

Lizards and snakes, which form the largest order of reptiles, are alike in that they are covered with scales, as their order name (Squamata: "scaly") indicates; but they differ in several ways. Lizards have eyelids, *nictitating membranes* (transparent inner eyelids), external ears, legs with claws, and an attached lower jaw. Snakes have no eyelids, outer ears, or legs, and instead of the lower jaw being attached directly to the upper jaw, the snake's lower jaw is specially hinged so that it can become disjointed when swallowing an object larger than its head. **Lizards,** which are elongated, four-legged reptiles with a tapered tail, are the most numerous of the reptiles. Iguanas, chameleons, geckos, and skinks are probably the most common.

**Various habitats.** Lizards live mostly in tropical regions and are found in nearly any type of habitat—jungle treetops, grasslands, tidelands, and deserts. Lizards are generally terrestrial or arboreal animals. Some can run erect on their hind feet, although many lizards are fast on all fours. Many species are good swimmers, but only the *marine iguanas* of the Galápagos Islands are actual sea animals. The sea, their source of food, provides the algae and seaweed that they eat. When the marine iguana is not feeding, it usually basks in the sun on a field of lava rock. In the heat of the day, this aquatic reptile may seek shelter under boulders or in crevices.

**Varying sizes.** The majority of lizards today are less than one foot long. They range in size from the tiny *Jaragua Sphaero* [hăr·äg′wă sfē′rō], a 0.6-inch-long gecko from the Dominican Republic, to the 300-pound *Komodo* [kə·mō′dō] *dragon* of Indonesia, which may attain a length of 12 feet. Komodo dragons belong to the family of *monitor lizards,* which shows variation in size but very little variation in form.

Fig. 18.2 **Lizards**

**Poisonous lizards.** Among the thousands of different fierce-looking lizards, only two species, both of which are native to North and Central American countries, are venomous. The stocky, sluggish, belly-dragging *Gila* [hē′lə] *monster,* living in the southwestern United States and much of Mexico, and the *Mexican beaded lizard,* living in Mexico and Central America, are very similar. The Gila monster is predominately pink and yellow with black shading, whereas the beaded lizard is mainly black with pink and yellow patches. Neither of these lizards can strike as a snake does but must hold on with a viselike grip of the teeth. Once a venomous lizard bites its prey, it is difficult for its victim to loosen the tight-gripping jaws. Vigorous chewing causes the venom to flow from glands in the back of the lower jaw.

**Colorful creatures.** The *chameleons,* a common family of about 100 species of lizards, are noted for their ability to change color and blend with their surroundings. The color changes of these arboreal lizards range from pale green to emerald to brown to gray and black. Both their colors and patterns can blend with various backgrounds; this camouflage provides protection from predators. The chameleon's emotions, as well as heat and light, may influence its color changes. Chameleons are most abundant in Madagascar and Africa, and a few are found in southern Asia and Europe. The "American chameleon," or *anole* [ə·nō′lē], is not a true chameleon but belongs to the iguanid family. However, the anole, like many other kinds of lizards, is able to change its color to some extent.

The color markings of some male lizards change during the mating season, turning from a drab brown to vivid blues or scarlets. The brightly striped *racerunner* lizard of North and South America has seasonal color changes. All lizards, regardless of color or size, molt about once a month during the months they are active. The epidermis, which may come off in strips, patches, or one scale at a time, is often eaten by the lizard.

**Detaching tails.** Lizards defend themselves in various ways. The *glass snake,* a legless lizard of the eastern United States, has an extremely brittle tail that is twice the length of its body. If a predator seizes the tail of a glass snake, this lizard pulls its body away from its tail. The long, detached tail continues to wriggle and distracts the predator as the lizard escapes, suffering only the loss of its tail. In a few weeks, a new tail grows to replace the old one. Legless lizards, such as the glass snake and the *blindworm* lizard of Europe and western Asia, are true lizards and not snakes because, except for their lack of legs, they have the structural characteristics that distinguish all lizards from snakes.

Several other kinds of lizards, including about 600 species of *skinks,* readily discard their tails. Most skinks, a common lizard in tropical and temperate regions of the world, have characteristic glossy skin and a cylindrical body. The young of some skinks have a bright blue tail for additional safety. As the tail is more conspicuous than the body, the predator pursues the bounding tail while the young skink scurries to safety.

Some lizards will shed their tails only under great stress. Many of these, such as the *collared lizards* living mainly in the rocky areas of the southwestern United States, may never grow a new tail.

**Other unusual defenses.** Lizards lacking venom, great size, speed, and camouflage were provided by the Creator with other means of defense. When under attack, the *chuckwalla* of the southwestern United States and Mexico hides in a crevice and inflates its body with air to twice its normal size, so that it is nearly impossible for a predator to remove the lizard from the crevice.

Fig. 18.4
**Chameleon**

Although no lizards can fly, the *Draco* [drā′kō], or *flying dragons*, of Southeast Asia can glide from tree to tree by spreading out folds of scaly membrane that are supported by five or six pairs of ribs; this membrane forms a sail between the forelegs and the hind legs. When a flying dragon lands on a tree, it can curl itself into a shape resembling a shrunken leaf.

The *frilled lizard* of Australia rears on its hind legs, unfolds a large frill that encircles its head, and hisses through its open mouth, thus appearing several times larger and more fierce than it actually is.

The *horned lizard* of the southern and southwestern United States has unusual scales that produce a horned appearance. Because the horned lizard has a short tail and a face that resembles a toad, it is sometimes called a *horned toad.* The horns, or sharp spines, on the head of this iguanid are used for protection; the coloring and flat bodies of these lizards provide desert camouflage for further protection. The horned lizard also exhibits an unusual means of protection—the ability to squirt drops of blood from its eyes for a distance of 3 feet. This has an irritating effect if the blood reaches the eyes of its enemy.

Fig. 18.3
**Unusual defenses**

**Eating habits.** A few lizards, such as the chuckwalla, are herbivores. Some lizards eat worms and other small animals; they seize their victims in their mouths and swallow their prey as soon as it stops struggling. The majority of lizards are insectivorous and are therefore important for controlling insect populations. Although the shape of the tongue varies among lizards from forked and snakelike to flattened and fleshy, the tongue serves as an olfactory organ. To catch prey, most lizards require a good sense of smell as well as good eyesight.

The *chameleon* has an extraordinarily long tongue with which it snares insects. When extended, a chameleon's tongue may be as long as its body. When not in use, a chameleon's tongue is kept compressed like an accordion by special muscles that extend the length of the tongue. The center of the tongue is hollow, allowing it to be stored like a coiled spring on the pointed hyoid bone at the rear of the mouth. When the chameleon spies an insect, it aims, opens it mouth, and fires its tongue off the hyoid bone at high speed. As soon as the sticky-tipped tongue slams into the insect, the chameleon contracts its tongue, jerking the insect into its mouth.

A *gecko* is the only lizard that can make a loud noise. This large-eyed little lizard's ability to catch insects makes it a welcome guest in the homes of people living in tropical areas. Using specially designed footpads with microscopic adhesive pads, a gecko can climb slippery walls and scamper across ceilings in its relentless pursuit of flies, mosquitoes, scorpions, and other pests. The *banded gecko* is common in the desert regions of the southwestern United States.

Fig. 18.5
**Gecko**

## Section Review 18.1

1. List the groups of reptiles that are mentioned in the text. Which group has the fewest species?
2. How does a cold-blooded animal regulate its body temperature?
3. List five characteristics common to most reptiles.
4. What group of reptiles has a four-chambered heart?
5. List four differences between lizards and snakes.
6. What is the smallest lizard? The largest lizard?
7. List four lizards that defend themselves in unusual ways.
8. What lizard has an extraordinarily long tongue with which it snares insect prey?

### Identify

poikilotherm, ovoviviparous, lizard

### Observe

- Make a terrarium to house a small lizard and place the terrarium where there is light but not in direct sun. Observe the lizard to determine if it is diurnal or nocturnal and to determine its means of defense. If possible, observe the molting of the epidermis.

## 18.2 Snakes: The Most Widespread Reptiles

**Snakes** are legless vertebrates with long, cylindrical bodies covered with dry scales. These reptiles, which are more widely distributed than lizards, have fascinated mankind for centuries: they are prominent in the Bible, in classical literature, and in mythology. Although snakes live almost everywhere on the earth, they thrive in the tropics, where they may have a life span of over 20 years. Because snakes can survive only until their body temperature reaches the freezing point, they cannot be found in the Polar Regions or at high elevations. In addition, no snakes live in either Ireland or New Zealand. To survive the cold weather of the temperate zones, snakes "hibernate" by burrowing into the soil or some other protected place.

### Varying Sizes

Snakes, like lizards, have great variations in size. One of the smallest snakes, a *blind snake,* may grow from 5 to 6 inches long and weigh only a fraction of an ounce. The giant snakes, which exceed 30 feet in length and weigh over 300 pounds, are the longest land vertebrates alive. The Asian *python* and the South American *anaconda,* or water boa, are the largest known snakes. The longest snake of North America, the *indigo snake,* has been measured at over 9 feet long.

### Methods of Locomotion

**Lateral undulation.** A snake's vertebral column is a long chain of vertebrae designed to allow the animal to bend easily. Most snakes move by **lateral** (sideways) **undulation;** the snake bends into a series of curves as it keeps its body flat on the ground. As the curves of its body push against rocks, twigs, or plants, the snake slithers forward. Each part of the body passes through the same curves, pressing against the same objects to push the body along. Snakes also use this type of movement for swimming. Lateral undulation combined with their ability to hold their breath for a long time makes snakes excellent swimmers.

**Rectilinear movement.** Many thick-bodied snakes, such as African *puff adders* and Asian *pythons,* use **rectilinear** (straight line) **movement** when stalking prey. In this method of movement, the snake alternately stretches and shortens segments of its body to creep forward in a straight line. Some snakes, such as the *rat snake,* use rectilinear movement to climb trees; the scales catch on the bark as the snake creeps up.

lateral undulation (snake slithers forward along same path)

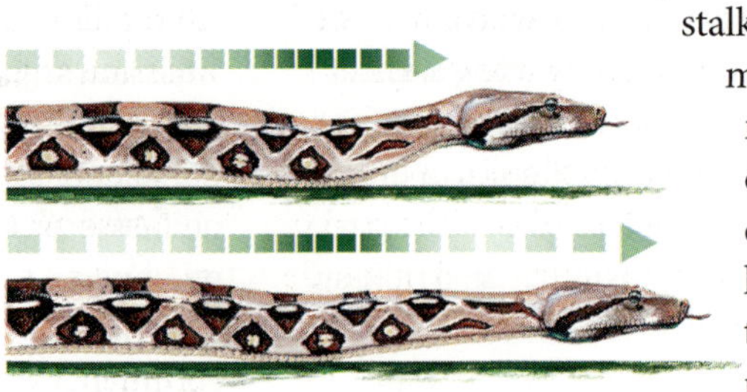

rectilinear movement (snake alternately stretches and shortens body segments)

Fig. 18.6 **Snake locomotion**

**Concertina movement.** Snakes often use **concertina movement** to move across smooth surfaces or to climb through trees. The snake moves the front part of its body forward, coils it slightly to anchor it, and then pulls the back part forward. As the back part is pressed down, the front part moves forward again.

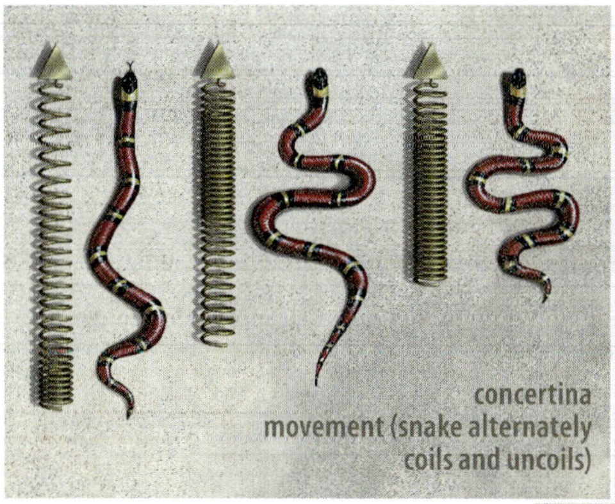


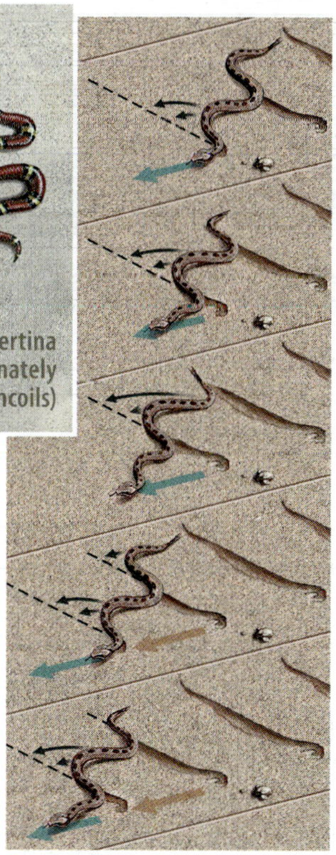

sidewinding movement (snake lifts its body off the ground and throws it sideways)

Fig. 18.6 (cont.)
**Snake locomotion**

**Sidewinding movement.** Desert-dwelling snakes, such as the North American *sidewinder* and the *carpet viper* and *horned viper* of Africa, use a **sidewinding movement** to cross loose sand. As the snake supports its body with its head, tail, and two or three other points of contact, it lifts the remainder of its body off the ground and moves it sideways. The snake then moves its head and tail in line with its body and repeats the sequence.

Although snakes appear to move swiftly across the ground, compared to most animals they are slow-moving; the top speed of most snakes is less than 4 mph. However, a few snakes are considerably faster than this. *The fastest known snake is the black mamba* [mäm′bə], a large (up to 14 ft), extremely venomous snake of Africa; estimates of its top speed over a short distance range from 7 to 20 miles per hour.

## Scaly Skin

The body of a snake is covered with dry scales; these scales, which may be smooth or have ridges, are thickened areas of the epidermis. As the dead skin cells reach the surface, they form tough, keratinized scales. Between the scales are folds of thin, elastic skin that enable the snake to expand to swallow large objects. When the outer layer of skin becomes too small or worn, it is shed. Snakes molt for the first time when only a few days old; in most species, molting occurs more than once a year. Just before molting, the snake loosens the skin around its head by rubbing its nose on a rough surface. The snake then wriggles out of the old skin, turning the skin inside out as it comes off in one piece.

## Sense Organs

**Scale-covered eyes.** A snake has an eye on each side of its head, giving it a wide field of vision. Although snakes can easily see movement, most snakes are rather nearsighted (except for the tree-dwelling snakes, which tend to have excellent vision). Diurnal snakes (snakes that are active during the day) have round pupils, whereas nocturnal snakes have slitlike pupils that can be opened widely at night. A snake's unblinking eyes are different from the eyes of other reptiles because they have no eyelids. Instead, transparent scales fit snugly over the eyes to retain moisture and to provide protection from injury. For a short time before a snake molts, it is less active than usual, and its eyes become clouded, hindering its vision. When the snake molts, the transparent membrane is shed with the skin.

**Hearing without ears.** Snakes lack both outer ears and tympanic membranes (eardrums). Because the inner ear embedded in the skull has no external opening, snakes cannot hear high-pitched sounds; but they are acutely sensitive to low-frequency sounds (100–700 Hz) and to vibrations in the ground. The snake's lower jaws are not attached directly to the skull but are attached to a separate **quadrate bone,** which loosely attaches the skull to the lower jaw on each side. This cartilaginous bone is found in birds, amphibians, and fish as well as in all reptiles. Sound waves traveling through the ground are transmitted from the snake's lower jaws, which are normally touching the ground, through the quadrate bone to the inner ear.

**Sense of smell.** Although the paired nostrils located on the front of a snake's head contain olfactory nerve endings, the snake's forked tongue is the primary instrument of smelling. To smell, the snake flicks out its tongue, which picks up scent particles from the air or ground. When the snake retracts its tongue, these scent particles are carried to the **Jacobson's organ**—two tiny

Fig. 18.7 **Snake's sense of smell**

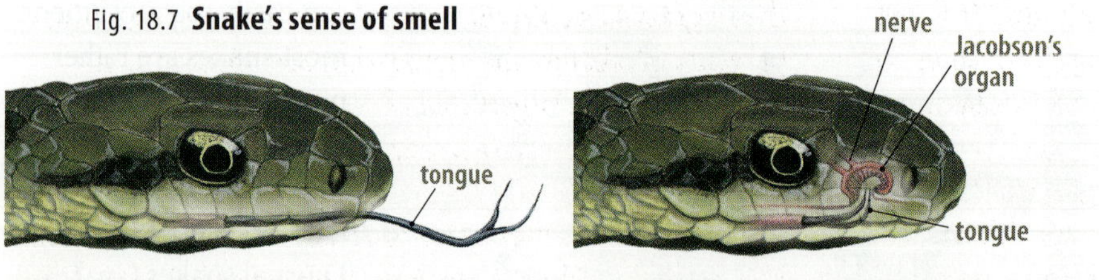


1. Tongue picks up scent particles.

2. Scent particles analyzed by Jacobson's organ.

cavities in the roof of its mouth containing nerve endings that are extremely sensitive to odor. The Jacobson's organ provides a snake with an acute sense of smell, enabling it to follow the trail of its prey.

## Design for Feeding

**Double-hinged jaws.** Many small snakes feed on insects. Larger snakes usually ingest birds, amphibians, fish, other reptiles, and small mammals such as rabbits and rats. Since snakes cannot tear their food into chunks, they swallow everything whole. Because the prey's body is often larger in diameter than the snake's mouth and body, the snake has double-hinged jaws that allow it to open its mouth unusually wide. The quadrate bone loosely attaches the snake's skull to the lower jaw on each side. In the front of its mouth, each side of the lower jaw is attached to the other side by an elastic ligament that allows half of the lower jaw to function independently of the other half.

**Engulfing a meal.** Snakes swallow their prey whole. Although a snake has a double row of teeth on the upper jaw and a single row on the lower jaw, these numerous needlelike teeth, which slant toward the throat, are not used for chewing—they sink into the prey to prevent it from escaping. As the snake alternately pulls each side of its jaw backward, the prey is drawn into the snake's throat. To prevent the snake's trachea (windpipe) from being blocked while its mouth and throat are full, its trachea, which is reinforced by rings of cartilage, is pushed forward over its tongue and out its mouth. Once the prey reaches a certain point, the snake's throat muscles and body movements help to swallow the animal. Although a large amount of saliva is secreted to aid the passage of the animal through the throat, it may take longer than an hour for a snake to engulf some animals.

In most snakes, the stomach can expand considerably, and the ribs, which are not joined together along the belly, can be extended outward as the stomach expands. Enzymes produced in the stomach can digest the entire body of an animal except for hair or feathers; the bones can be digested completely within a few days. Because a snake can consume such large meals, many eat only once or twice

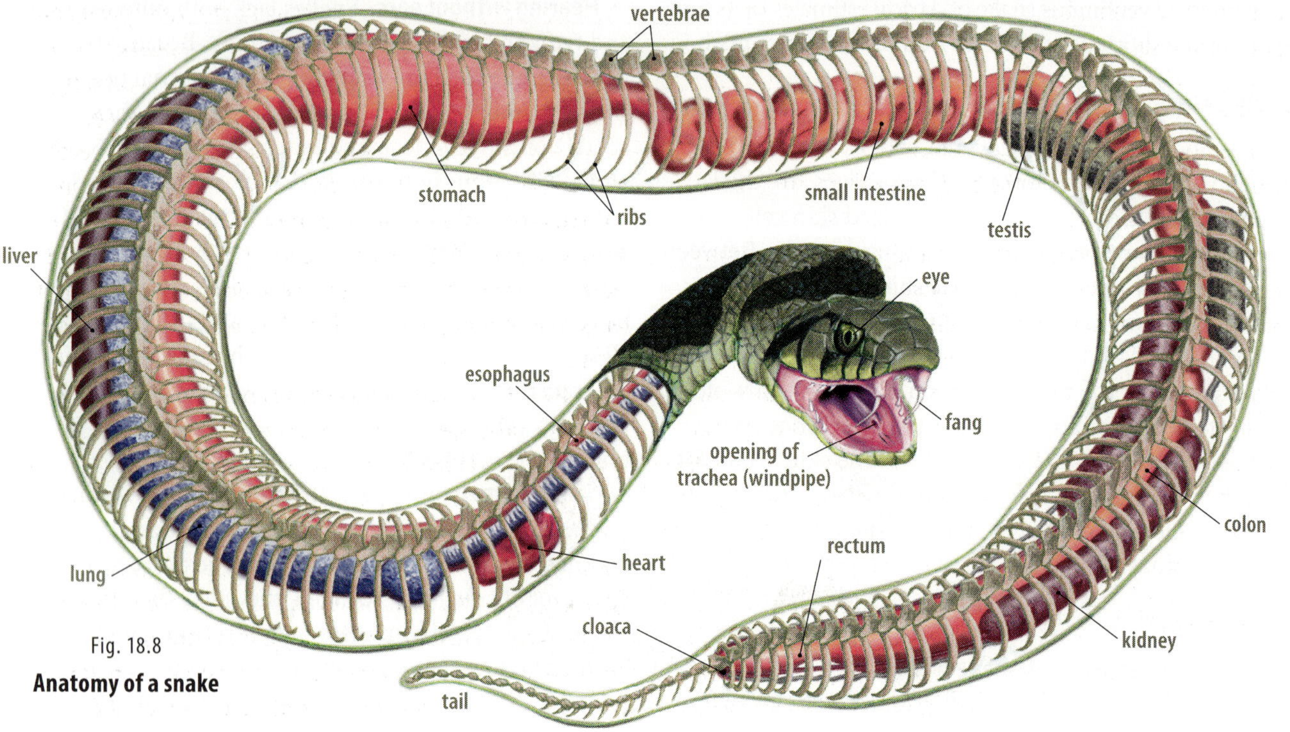


Fig. 18.8
**Anatomy of a snake**

Fig. 18.9 **Snake swallowing an egg**

a month, and it is not unusual for snakes to survive a fast lasting over a year.

Egg-eating snakes have a special row of 6 or 8 sharp spines in the back of the throat, attached to the neck vertebrae. After a snake swallows an egg, the egg is pushed against these spines until it breaks. The liquid portion of the egg is passed along to the stomach, while the bits of shell and membrane are bundled together (with the sharp edges in) and moved forward by the throat muscles to be regurgitated.

## Groups of Snakes

**Blind snakes. Herpetologists,** zoologists who study reptiles and amphibians, have identified several hundred species of *blind snakes,* sometimes called worm snakes, which are tiny, harmless snakes usually 6 or 7 inches long. Because of their size, blind snakes are often confused with earthworms, but whereas worms have no eyes, blind snakes have minute eyes that can distinguish light from dark. Blind snakes live underground or under logs and rocks; they burrow by forcing their strong skull into the ground. A large scale covering the snout of a blind snake helps to protect its head. Its tail, which it drives firmly into the ground for the body to push against, has a sharp spike at the end.

**Constrictors.** A more familiar family of nonvenomous snakes is the **boid** [bō′ĭd: boa] **family.** Because these snakes overcome their prey by constriction, they are commonly called *constrictors.* A constrictor typically wraps its body around its prey and then tightens its coils each time the victim exhales; as this prevents inhalation, the animal dies of suffocation. Although the three largest snakes—*pythons, anacondas,* and *boas*—are all constrictors, many constrictors are moderate or small in size.

Fig. 18.10 **Boids**

**Colubrids.** Over two thirds of all snakes are **colubrids** [kŏl′ə·brĭdz], making it the largest group of living snakes. The snakes belonging to this large family may be terrestrial, aquatic, or arboreal. These small or moderate-sized snakes may be either thick-bodied or as slender as vines. Some familiar colubrids of North America, including *garter snakes, water snakes, hognose snakes,* and *ringneck snakes,* have little economic importance. *Racers, rat snakes, bull snakes,* and *king snakes* are of greater importance to mankind because these colubrids prey on rodents and other pests.

Fig. 18.11 **Colubrids**

Although most colubrids are harmless, a few species are venomous. Scientists group venomous snakes according to their **fangs**—hollow or grooved venom-injecting teeth. Most venomous colubrids have one to three grooved fangs in the rear of the mouth on the upper jaw. Although these rear-fanged snakes, which must chew on their victims to inject the venom, are usually not dangerous to man, the bite of an African *boomslang* can be fatal.

**Elapids.** Only venomous snakes have fangs and venom glands, which they use primarily to kill prey. Their fangs may be shed and replaced several times a year. All **elapids** [ĕl′ə·pĭdz], snakes with immovable fangs in the front of the upper jaw, are venomous. Over half of the species of elapids live in Australia. Although this continent has a greater percentage of venomous snakes than any other continent, not all of the small elapids have a bite that is fatal to man. Many Australian elapids have fangs too small to penetrate clothing and skin. *Cobras* of Africa and Asia, the most renowned elapids, flatten their necks horizontally by movable ribs that stretch the skin, forming a hood. *The 18-foot-long king cobra is the longest venomous snake* (although the eastern diamondback rattlesnake of the United States may be heavier). The *Indian cobra,* which has the largest, most spectacular hood, takes the lives of thousands of people each year in India. The *spitting cobra* of Africa defends itself by squirting venom from its fangs. If the venom reaches the eyes of its predators, it causes great pain and temporary blindness; if it is not immediately washed out of a victim's eyes, it can cause permanent blindness.

Fig. 18.12 **Elapids**

The *mambas* of Africa are known as aggressive elapids with a highly potent venom. Less familiar and less aggressive are the *kraits* [krīts] of Asia. The *coral snake,* an elapid of North and South America, does not strike its prey but approaches slowly, sliding its head over its victim. Because coral snakes are nocturnal, they cause fewer deaths than other elapids. The two North American coral snakes have dark snouts and colored bands around their bodies in the same sequence of black, yellow, red, and yellow—yellow borders each red band. (The *western coral snake,* found in the southwestern states and northern Mexico, may have white bands instead of yellow.) The *scarlet king snake* (p. 369, Fig. 18.11), which is a nonvenomous colubrid, closely resembles the coral snake with bands of black, red, and yellow; but black borders the red bands. The following short poem may help you distinguish the venomous coral snake:

Red on yellow, kill a fellow;
Red on black, venom lack.

Over 50 species of *sea snakes,* all of which are venomous, are aquatic elapids. Most sea snakes are found in the tropical waters of the Indian and Pacific oceans; none live in the Atlantic Ocean. The body of a sea snake is vertically flattened, and its tail narrows to form an oarlike paddle at the end. Although a few species are oviparous and come ashore to lay eggs, most sea snakes are ovoviviparous and give birth to live young in the water.

Fig. 18.13 **Sea snake**

**Vipers.** The **viperid** [vī′pər·ĭd] **family** contains the vipers of the Old World and the pit vipers, which can be found on all continents except Australia in the tropical and temperate regions but are mainly concentrated in the New World. Most venomous snakes of the temperate regions are viperids, which have long, movable fangs attached to the front of the upper jaw. When not in use, their hollow fangs, which are longer than those of other venomous snakes, fold into a sheath on the roof of the mouth. Most viperids have stout bodies with flattened, triangular heads.

A **pit viper** has a sensory depression, or pit, on each side of its head between the eye and the nostril. This sensory organ contains nerves that are highly sensitive to heat, enabling the snake to "see" warm objects in absolute darkness. The sensitivity of pit vipers to temperature differences is amazing; it is thought that a pit viper can "see" an object even if its temperature differs from that of the background

Fig. 18.14 **Pit organ of viperids**

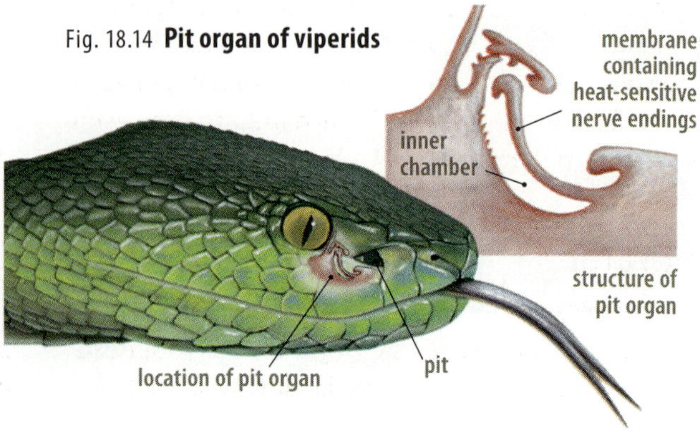

Fig. 18.15
**Viperids**

water moccasin (cottonmouth)

by only $\frac{1}{200}$°F. Although most venomous snakes in the United States are timid and usually attack only when molested, the *cottonmouth,* or *water moccasin,* is more aggressive and may attack with no apparent provocation. Other common pit vipers include the *bushmaster* and *fer-de-lance* [fĕr′dl·ăns′] of tropical America and the North American *copperheads* and *rattlesnakes.*

Remnants of skin moltings form the rattle on a rattlesnake's tail. Each time its skin is shed—usually 3 or 4 times a year—another hollow, horny piece is added. As the snake matures, some of the segments break off, leaving an average of 8 or 9 horny pieces loosely joined together. The rattlesnake shakes its rattle in a spine-chilling warning when it is disturbed; but it shakes no warning before striking a prey or when it is surprised and must defend itself quickly. Because the fangs of a rattlesnake are fragile and easily snapped off, reserve fangs replace them so that the rattlesnake can deliver a fatal bite from the moment it is born until it dies. Even after death, a rattler can inflict a bite by its prolonged reflex action.

## Snake Venom

Of the 2900 species of snakes, fewer than 300 are venomous. After a snake bites its prey, enzymes produced by the venom glands begin to break down the victim's tissues before the snake even swallows the animal. There are two basic types of poisons in snake venom—neurotoxins and hemotoxins. A **neurotoxin** (nerve poison) affects the nervous system, which controls the functioning of the heart, lungs, and muscles. This toxin causes severe stinging, paralysis (making it difficult for the victim to breathe and swallow), and nausea; the victim's pulse becomes weak and rapid. A **hemotoxin** (blood poison) damages blood vessels and destroys red blood cells. At first, hemotoxins cause little pain; later the victim becomes sleepy and his breathing organs are partially paralyzed. The proportions of each type of venom vary according to the family; elapids, such as coral snakes and cobras, have mostly neurotoxins, while the venom of vipers and pit vipers consists mainly of hemotoxins. Both types of venom are deadly; but neurotoxins are faster-acting and therefore more dangerous to man. Fortunately, both types of toxins can be treated by the use of **antitoxins** (also known as *antivenins*), which are antibodies that attack the toxin molecules and neutralize their effects.

## Snakebite Treatment

If the snake is nonvenomous, the bite should be washed thoroughly with soap and warm water and covered with a sterile bandage. A physician should be called, since a booster shot may be needed to prevent tetanus. The victim of a venomous snakebite needs prompt, professional medical attention. In the United States, about 7500 people are bitten each year by venomous snakes. Fortunately only 5 or 6 of these bites are fatal, although amputation of a limb is occasionally necessary. Because activity increases the rate at which the toxin is spread, the person should remain as quiet and motionless as possible; he should never run. Using a splint to immobilize the bitten limb or placing the bitten part of the person's body lower than the level of his heart will reduce the speed at which the toxin spreads. If possible, the dead snake should be taken to the physician along with the victim, as positive identification enables medical personnel to determine the type and amount of antivenin to use. Care should be taken in handling a decapitated snake's head, because reflex actions can occur for 20 to 60 minutes after the snake has been killed.

## Section Review 18.2

1. What are four methods by which snakes can move? Which method is best suited for crossing loose sand?
2. What is the fastest known snake?
3. What organs provide the snake with its excellent sense of smell?
4. What design feature of snakes prevents a snake from suffocating while swallowing an object that completely fills its mouth and throat?
5. What is the largest group of living snakes? Name three familiar snakes from this group.
6. What elapid is the longest venomous snake? What well-known venomous snake of North America is an elapid?
7. What special organ do all pit vipers have in common? What purpose does this organ serve?
8. Name three pit vipers found in North America.
9. Compare the location and length of the fangs of venomous colubrids, elapids, and viperids.

### Identify

quadrate bone, Jacobson's organ, herpetologist, blind snake, boid, constrictors, fangs, neurotoxin, hemotoxin, antitoxin

# 18.3 Turtles, Crocodilians, and Tuatara

## Turtles: Shelled Reptiles

The order name Chelonia [kĭ·lō′nē·ə] is the Greek word for tortoise, or turtle. The **turtle** is a reptile with a shell and toothless jaws. The upper shell of a turtle, called the **carapace** [kăr′ə·pās′], is fused to the turtle's ribs and thoracic vertebrae; the turtle's lower shell, which is fused to the clavicles, is the **plastron.** The dorsal (back) carapace and ventral (belly) plastron are joined along the sides of the body by a bony *bridge.* Because the shell is designed as an integral part of the skeleton, no turtle can be removed from its shell without killing it.

The shell has an inner layer of bone, which is divided into sections called *plates.* The tough, horny outer layer of the shell, made of keratin produced by epidermal skin cells, is divided into sections called *shields,* which are arranged in a symmetrical pattern. As the turtle grows, new layers of keratin are added beneath the old layers.

The shell first develops as keratin epidermal shields during the fetal stage within the egg. When a turtle hatches, the shell is fully formed without the bony plates. As the turtle's ribs are growing, tiny extensions branch out in all directions and penetrate the dermis of the skin until nearly all of the turtle's skin has been invaded with bone-producing cells. Beneath the keratin shields of a baby turtle, the dermis is ossified to bone until the entire bony plate is formed, giving a mature turtle a double-layered shell. The overlapping of the bony plates and epidermal shields contributes to the great strength of the shell.

When disturbed, some turtles (those having hinged plastrons) can pull their head, tail, and legs into their shell and close the plastron tightly against the carapace for protection. Turtles having exposed body parts are protected by scales; as in other reptiles, the skin covering the turtle's head, legs, and tail is completely covered with scales.

**Characteristics of turtles.** Turtles are found in nearly all the warm regions of the world; of the 300 living

Fig. 18.16 **Structure of a turtle**

species, about 90 live in North America or its coastal waters. Each kind of turtle has a specific habitat—desert, grassland, forest, marsh, pond, lake, river, or sea—that fits its special characteristics. Because turtles, like other reptiles, are cold-blooded, most species that live in regions with severe winters "hibernate." Most freshwater species burrow into the muddy bottom of a pond or stream; land turtles bury themselves under rotting vegetation or burrow into the soil. Some species survive hot, dry periods by **estivation,** a state of summer dormancy or limited activity.

Fig. 18.17 **Freshwater turtles**

*Although turtles are toothless and thus cannot grind their food, they can chop their food with their sharp and powerful jaws.* Most species eat both plants and animals; however, some feed mainly on animals while a few are strictly herbivorous. Tough, fibrous plants and the flesh of animals may be torn to bits with the help of the claws on the turtle's forelimbs. Some species that eat shelled animals have ridges inside their mouths to crush the shells.

Although all turtles have *lungs,* their ribcage cannot expand and contract for breathing as in other reptiles because the ribs are attached to the shell and are therefore immovable. Instead, the turtle uses its abdominal and pectoral muscles to squeeze its internal organs, thereby making its body cavity larger and smaller. Some aquatic species can also take in oxygen through the lining of the throat. A turtle's *tympanic membranes* (eardrums) are on the sides of its head. In addition to the upper and lower eyelids, turtles, like most other reptiles and like birds, have a *nictitating membrane,* which helps to lubricate and protect the eye. This transparent membrane, which moves sideways across the eye, also helps turtles to see more clearly underwater.

**Freshwater turtles.** The largest group of turtles, the freshwater turtles, includes a variety of species from small painted turtles to large snapping turtles. Freshwater turtles are usually classified as hard-shelled or soft-shelled. *Hard-shelled turtles* have a flat, hardened carapace made of large keratin scales. They have long toes connected by webbing that can be spread out for swimming in water.

The *snapping turtles* are a group of large turtles found only in North and Central America and northern South America. The heavy-built head and limbs of the snapping turtle cannot be retracted into its small shell, but its strong, sharp-edged jaws give it additional protection. Snapping turtles feed on plants and small water animals, such as fish, frogs, snails, and young waterfowl. Weighing up to 200 pounds, *the alligator snapping turtle is the heaviest freshwater turtle* and lures fish near its mouth by moving its tongue about to simulate a worm.

Some freshwater turtles can swim but spend most of their time on land. These *semiterrestrial turtles* include the *box turtle* and the *red-eared slider. Terrapins* also spend time on land and in water and are the only turtles that live in brackish waters.

*Soft-shelled turtles* have a round, flat carapace covered with flexible, leathery skin. These freshwater turtles have feet that are webbed, and their quick movements propel them rapidly over land and in water. By stretching out their necks and extending their lengthy, tapering snouts above the surface of the water, soft-shelled turtles can stay submerged for long periods of time. Although fleshy lips cover their jaws, these turtles, which feed mainly on animals, can strike with lightning speed and will bite when disturbed.

**Marine turtles.** Found in warm seas throughout the world, *marine turtles are the largest living turtles.* They swim by beating their front flippers up and down, whereas other turtles swim with a back-and-forth paddling motion of their limbs. Because a sea turtle cannot retract into its shell, its defense depends upon its size and swimming speed. Since the seawater would harm or destroy a turtle embryo, sea turtles must lay their eggs on land well above high-tide level. After the eggs are laid, the female turtle covers them with sand and goes back to the sea. Weeks later, when the eggs hatch, the young turtles fight their way to the surface, nearly always emerging at night. The hatchlings have soft shells and are easy prey for many predators as they paddle wildly with their limbs to reach the sea. Because movement on land is laborious, most male sea turtles never return to land, and most females leave the water only to lay eggs. The *leatherback,* a giant measuring up to eight feet long and weighing as much as 1500 pounds, differs from other sea turtles in that it is covered with leathery skin instead of scales.

Fig. 18.18
**Hawksbill (sea turtle)**

Fig. 18.19
**Galápagos tortoise**

**Tortoises.** Although the terms *turtle* and *tortoise* are commonly used interchangeably, the term *turtle* refers to all the members of Chelonia, whether aquatic or terrestrial. The term ***tortoise*** is generally reserved for land turtles with feet and claws suitable for terrestrial life. To carry the heavy, high-domed carapace, a tortoise has large, sturdy limbs. They may also have flattened front limbs to help them dig burrows, like the *gopher tortoise.* The *giant tortoise* of the Galápagos Islands, measuring four feet long and weighing 600 pounds, is one of the world's largest living land turtles. As far as scientists know, it is the longest-lived animal, probably living well over 150 years.

## Crocodilians: Crocodilelike Reptiles

Large, thick-skinned, long-tailed aquatic reptiles, the **crocodilians,** are the largest living reptiles. Crocodiles, alligators and caimans, and gharials [gŭr′ē·əlz], the three groups belonging to the order Crocodilia, all live in tropical and subtropical regions. The term *alligator* is derived from a Spanish phrase *(el lagarto)* meaning "the lizard." Both the alligators of the southern United States and China and the Latin American *caimans* (also spelled *caymans*) have broad, triangular-shaped snouts and concealed lower teeth. *Crocodiles,* found in the

Fig. 18.20
**Crocodilians**

Florida Everglades, South America, Africa, and India, have narrower snouts and some protruding lower teeth. Although crocodiles are more agile and ferocious than alligators, the American crocodiles are not maneaters like some of the Old World crocodiles. The fish-eating *gharials* (also called gavials) of southern Asia have extremely long, narrow snouts containing large numbers of teeth. The male gharial has a hollow hump on the tip of its snout, with its nostrils located at the center of the hump.

**Characteristics of crocodilians.** Crocodilians have partially webbed, clawed feet that facilitate swimming and crawling. They use their short, strong legs for moving on land and use the side-to-side movement of their long, powerful tails to propel themselves through water. Crocodilians often drift along just beneath the surface of the water, submerged except for bulging eyes and nostrils.

All crocodilians have *nictitating membranes,* which clean and moisten their eyes without blocking out the light. They have valves in their ears and nostrils to keep out water. Because their mouths lack lips and thus cannot keep out water, a fleshy flap of tissue in the rear of the mouth prevents water from entering the body through the mouth, even when the crocodilian seizes prey in its jaws and holds the prey underwater to drown. All crocodilians are carnivorous and can replace lost teeth throughout their lifetime. The eyes and nostrils are located high on the skull, enabling them to see and breathe as they stalk turtles, fish, and waterfowl.

Fig. 18.21 **Tuatara**

A few species will attack larger animals such as young giraffes, small hippopotamuses, giant anacondas, and even full-grown wildebeests. Most crocodilians are protected by flexible armor composed of thick, bony plates, especially on the back.

**An unusual reptile.** Like most reptiles, *crocodilians are oviparous;* they come ashore to conceal leathery eggs in nests of vegetation or to bury them in holes in the sand. The female crocodilian, unlike most other reptiles, keeps an almost constant vigil near the nest. As peeping sounds from inside the eggs announce that the young crocodilians are ready to hatch, the female carefully unearths them and transports them, a few in her mouth at a time, to the water's edge. For a few weeks, the young crocodilians swim after their mother in the water, crawl over her on land, and seek refuge in rushes or under rock ledges at night.

Since many adult crocodilians will devour juveniles, young crocodilians are seldom seen near the adults after they leave their mother's protection. The *Nile crocodile,* one of the largest crocodilians, can grow from 9 inches when it is hatched to a normal length of 18–20 feet long. Because crocodilians continue to grow throughout their lives, some types of crocodile can grow to lengths of 25 feet and weights of more than 2000 pounds, given the right conditions. Some extinct crocodilians grew even larger, attaining lengths of 35 to 40 ft and weights approaching 6 tons (12,000 lb).

### Tuatara: Spiny-Crested Reptiles

The two species in *Sphenodon,* the only genus in a family and order of its own, are commonly known as **tuatara,** a word that means "spiny back." These extremely rare lizardlike reptiles, which inhabit a few rocky islands off the coast of New Zealand, have a crest of enlarged spines along their necks, backs, and tails. The tuatara's skull is stronger and firmer than that of lizards. A "third eye," called a **parietal eye,** is situated on top of its brain, with a hole in the skull just above it. The parietal eye has no eye muscles and thus cannot focus. The function of this third eye is to sense radiation; when the eye is covered, the tuatara does not protect itself from the sun as it does when the parietal eye is uncovered. A few lizards also have a parietal eye.

The small, nocturnal tuatara is mainly insectivorous. Male tuatara, which are somewhat larger than females, weigh slightly more than 2 pounds and may exceed 2 feet in length. The growth rate of the tuatara is very slow, and it continues to grow in size for 50 years. The period of incubation for the eggs of this oviparous reptile—12 to 15 months—is longer than for any other reptile. The tuatara's normal rate of breathing while moving about is less than 9 breaths per minute, and they can go for an hour without taking a breath.

## Dinosaurs and Similar Creatures

Many extinct reptiles and reptilelike creatures are known only from the fossil record. Probably the most famous of all extinct creatures are the **dinosaurs,** a large group of extinct creatures that resemble reptiles in some ways. Although the word *dinosaur* (Greek *deinos,* "fearfully great" or "awe-inspiring" + *sauros,* "lizard") implies that dinosaurs were true reptiles, there are many important differences between dinosaurs and modern reptiles. For this reason, dinosaurs are sometimes placed in a separate group (Dinosauria) instead of the class Reptilia. There are many fossil remains of dinosaurs, and the creatures also seem to be referred to in Scripture. Although the Bible does not use the term *dinosaur* (which was first used in 1842), many Bible scholars believe the two creatures God describes in Job 40 and 41 may have been dinosaurs living during the time of Job. Although some dinosaurs were as small as chickens, others were the largest land animals that have ever lived.

**Sauropods.** The largest dinosaurs were the **sauropods,** which were large herbivorous animals characterized by long necks and tails and columnlike legs. The most famous sauropod is undoubtedly *Apatosaurus* [ə·păt′ə·sôr′əs], commonly called *brontosaurus* ("thunder lizard"). This magnificent animal stretched nearly 80 feet from nose to tail and may have weighed more than 40 tons (80,000 pounds). Although huge, *Apatosaurus* was by no means the largest sauropod. The most massive sauropods are thought to have weighed more than 90 tons (180,000 pounds) and been about 100 feet long, while some more slender sauropods may have stretched as much as 130 feet from nose to tail. If such a large sauropod were to rear up on its hind legs (which fossils suggest was possible), it would have been as tall as a five-story building. Many creationists think that *behemoth* [bĭ·hē′məth], a giant land animal described in Job 40 as living near a brook and eating plant matter, may have been a type of sauropod.

**Theropods.** Some of the most fearsome dinosaurs belonged to a group called the **theropods,** which were bipedal dinosaurs characterized by powerful lower legs used for walking and smaller "arms" used for grasping. The most famous theropod is undoubtedly *Tyrannosaurus rex,* one of the largest land carnivores that has ever lived. *T. rex* stretched nearly 50 feet long and may have weighed 5 tons or more; its skull alone was over 5 feet long, and its mouth was equipped with 6-inch teeth that were serrated like steak knives. Despite its size, *T. rex* is thought to have been quite agile; based on analysis of its skeleton, some paleontologists speculate that it could sprint at 30–35 mi/hr. Other well-known theropods include *Allosaurus,* a 40-foot-long predator with stronger arms than *T. rex*, and *Deinonychus* [dī·nŏn′ĭ·kəs], a five-foot-tall predator that must have been extremely fast and agile.

**Ceratopsians.** A third well-known group of dinosaurs were the **ceratopsians,** the most famous of which is *Triceratops* [trī·sĕr′ə·tŏps′]. This elephant-sized herbivore was built much like a rhinoceros, and some paleontologists speculate that it may have been as fast and agile as a rhino in a charge. *Triceratops* was characterized by a massive skull (as much as 8 feet long and 4 feet wide in some animals) topped by two 3½-foot horns above

Fig. 18.22 **Dinosaurs**

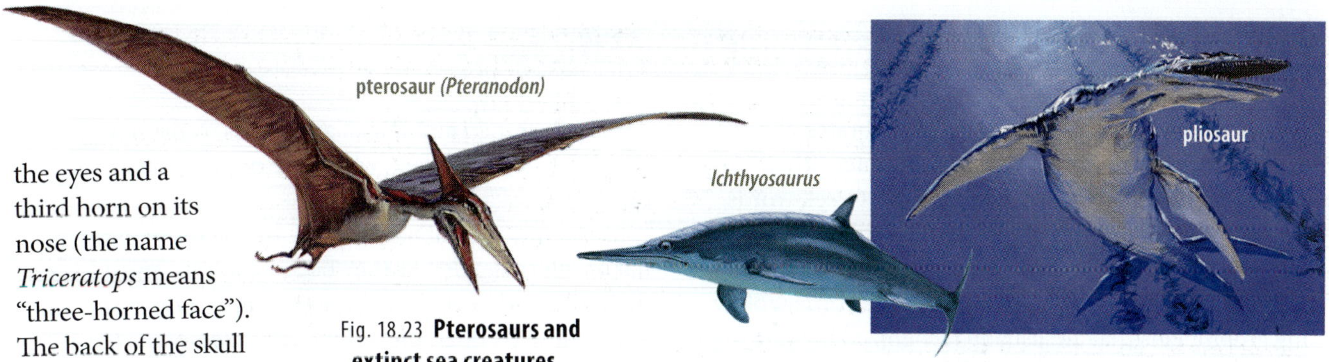

Fig. 18.23 **Pterosaurs and extinct sea creatures**

the eyes and a third horn on its nose (the name *Triceratops* means "three-horned face"). The back of the skull spread out into a massive shield of bone that protected the neck from attack.

**Pterosaurs.** Although the term *dinosaur* refers specifically to land-dwelling creatures, other extinct creatures were very similar to the dinosaurs and are sometimes referred to as "dinosaurs" in common usage. These animals include the **pterosaurs** [tĕr′ə·sôrz], the best-known of which are *pterodactyls*, and various sea creatures.

The pterosaurs were flying creatures characterized by long wings, large heads, and (apparently) fur-covered bodies. Although many were bird-sized, some were as large as some airplanes. One of the largest pterosaurs, known as *Quetzalcoatlus* [kĕt′söl·kō·ät′ləs], is thought to have had a wingspan approaching 40 feet. Although scientists used to picture these creatures as sluggish and inept fliers, they are now thought to have been excellent fliers.

**Sea creatures.** Well-known dinosaurlike sea creatures include the long-necked *plesiosaurs* [plē′sē·ə·sôrz′], which had bodies shaped somewhat like a flattened sauropod dinosaur but with fins instead of legs. They probably swam by "flapping" their fins up and down. *Elasmosaurus* [ĭ·lăz′mə·sôr′əs], a typical plesiosaur, was slightly more than 40 feet long, with a 20-foot neck. Similar to the plesiosaurs were the *pliosaurs* [plī′ə·sôrz], which had larger bodies and shorter necks than the plesiosaurs. *Kronosaurus* [krŏ′nə·sôr′əs], a large carnivorous pliosaur, had jaws 10 feet long and was 40 feet long overall. Many creationists think that *leviathan* [lə·vī′ə·thən], which God describes in Job 41 as a fire-breathing aquatic animal with scaly skin, thick armor, powerful jaws, and terrifying teeth, may have been a type of pliosaur or plesiosaur. *Ichthyosaurus* [ĭk′thē·ə·sôr′əs] looked very much like a dolphin; it was probably a fast swimmer and likely preyed on fish.

**Sudden extinction.** Even though the extinction of dinosaurs and similar creatures has never been fully explained, several theories have been offered. Some scientists think climatic changes may have brought about their extinction; other scientists believe they may have been forced into extinction by mammals that fed on their eggs. In the hypothesis that is presently most popular, scientists think the dinosaurs may have become extinct suddenly as the result of some catastrophe. Many creationists think the dinosaurs probably could not cope with the environmental changes brought about by the Genesis Flood.

## Section Review 18.3

1. Why is it impossible to remove a live turtle from its shell?
2. What is the difference between a turtle and a tortoise?
3. What are the largest living reptiles?
4. What are three types of crocodilians mentioned in the text?
5. What are two obvious differences between alligators and crocodiles?
6. What rare, unusual reptile mentioned in the text is noted for its parietal eye?
7. What group of dinosaurs consisted of large herbivorous animals characterized by long necks and tails and columnlike legs? What group included bipedal carnivores like *Tyrannosaurus rex* and *Deinonychus?*
8. What two creatures mentioned in the book of Job might have been dinosaurs or dinosaurlike creatures?

**Identify**

turtle, carapace, plastron, estivation, terrapin, crocodilian, tuatara, parietal eye, dinosaurs

# 18.4 Amphibians: Vertebrates with a Double Life

Another group of cold-blooded vertebrates, the **amphibians,** live both on land and in water. Their class name, Amphibia, which is derived from Greek words meaning "on both sides of life," implies that amphibians live a double life. Generally, part of their life is spent in water and part is spent on land. The three orders of living amphibians that are usually recognized include *frogs and toads,* the largest order, with 5400 species; *salamanders,* with 550 species; and *caecilians* [sə·sĭl′yənz], with about 170 species.

## Characteristics of Amphibians

*Body forms of amphibians exhibit great diversity of structure,* from a compact body with a fused head and trunk to an elongated trunk having a distinct head, neck, and tail. Amphibians range in size from the tiny species *Paedophryne amanuensis,* measuring only 0.3 inch in length, to the *Chinese giant salamander,* which may measure almost 6 feet long. *Most amphibians are oviparous.* Because the eggs lack a shell, they must be deposited either in water or in a very moist place. After the eggs hatch, young amphibians usually pass through an aquatic, gilled *larval stage;* the transition of the larva into an adult, which usually involves major changes in body structure, is called **metamorphosis.** Although a few species are aquatic their entire lives, most species live part of their life on land close to water or under moist vegetation. The soft, scaleless, glandular *skin* of an amphibian, which is thin and porous and therefore susceptible to drying out, must remain moist for the amphibian to remain alive.

Fig. 18.24
**Anurans**

Larval amphibians and permanently aquatic adult forms live in fresh or brackish water; none live in the sea. Terrestrial amphibians, including many salamanders and toads, emerge from the water to live on land. Although some amphibians are legless, most have two pairs of *limbs;* their feet, which are often webbed, have no claws.

As larvae, amphibians breathe by means of external or internal *gills.* As they become adults, most lose their gills and begin breathing with *lungs;* most species also respire through their skin. *All amphibians have a three-chambered heart,* and their skin is abundantly supplied with blood vessels through which the blood circulates. *Amphibians are poikilotherms* and thus are concentrated in the tropics; however, many live in the temperate regions. Some species estivate by burrowing deep into the soil to avoid extreme heat or drought. Those living in temperate regions "hibernate" during cold weather. Although amphibians are not generally considered of direct value to man, they are beneficial in controlling the population of injurious insects; some amphibians (particularly frogs) are also consumed for food by humans. Amphibians may also feed on young amphibians and other small animals, including earthworms, spiders, and minnows.

## Frogs and Toads: Leaping Amphibians

Frogs and toads (collectively called **anurans** [ə·no͝or′ənz]) are leaping, tailless amphibians belonging to the order Anura (Greek, "without tail") or Salientia (Latin, "leaping"). The term *frog* can refer to all anurans but is especially used for those species that have smooth skin and live in or near water. *Toad* refers to primarily terrestrial anurans with rough, "warty" skin, especially the members of the "true toad" family, Bufonidae. Most members of another family, the *tree frogs,* are usually arboreal. Although anurans inhabit all land areas except the Polar Regions and extreme deserts, the greatest diversity of species occurs in Africa. The seven-pound *Goliath frog* of west-central Africa is the world's largest anuran; its body measures a foot long, and its legs add another two feet to its length.

Because of their powerful hind legs, it is not unusual for a frog to leap 20 times its body length on a level surface. The smaller forelimbs of a frog help absorb the force of landing on firm ground. Since frogs also use their hind legs for swimming, most aquatic species have 5 webbed toes on each of their hind feet; the 4 toes on each forefoot are not webbed. Tree frogs are equipped with tiny suction cups on the ends of their toes to help them cling to the bark. A few burrowing species have short hind legs and thus cannot hop. Toads have shorter legs than frogs and are generally more clumsy.

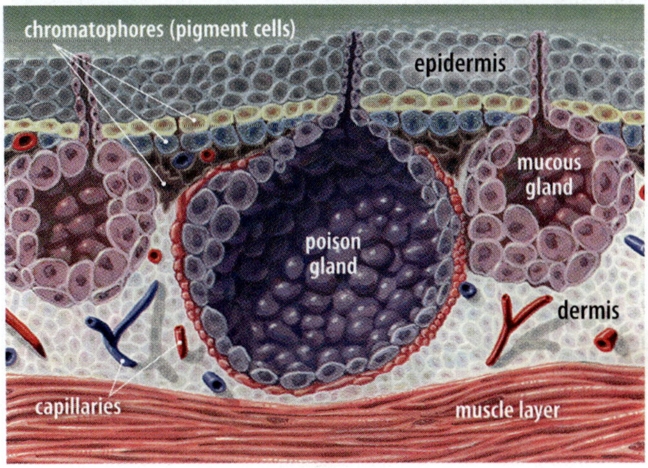

Fig. 18.25 **Frog skin**

**Protective coloring.** Many anurans are protected by camouflage. Three types of branched pigment cells, called **chromatophores** [krō·măt′ə·fôrz′], control the skin coloration of amphibians as well as many reptiles and fish. The chromatophores containing *melanin* are the most important pigment cells for adjusting coloration to blend with a background. When such conditions as light, humidity, and temperature change, a hormone stimulates the dispersion of pigments, allowing some amphibians, reptiles, and fish to alter their skin color.

The outer epidermal layer of anurans (which is shed during molting) contains *keratin,* which provides the skin with some protection against loss of water. The skin of a frog is noticeably moist and slippery, whereas the skin of a toad appears to be drier. Small mucous glands located under a frog's loose skin secrete a protective mucus, which waterproofs its skin.

**Toxic skin secretions.** Anurans and most other amphibians have serous glands that secrete poison through the skin, but the potency of the toxin they produce and its effectiveness varies from species to species and with different predators. The toxin may irritate the predator's mouth, causing the animal to release the amphibian; the toxin may also be nauseating or even cause death. This toxin, however, provides no protection from most snakes.

The *poison-dart frogs* from the forests of South and Central America possess an extremely potent neurotoxin. These tree frogs are collected by local Indians, who skillfully imitate the frog's peeping. The frogs are pierced with a stick and heated over an open fire; the heat contracts the skin, causing the frog to exude the toxin. The tips of 50 arrows can be dipped in the toxin of one tiny frog. Animals, such as jaguars,

## Table 18.2 Groups of Amphibians

*(Kingdom Animalia; phylum Chordata; subphylum Vertebrata; class Amphibia)*

| Group | Description | Number of living species (approximate) | Examples |
|---|---|---|---|
| **Anurans** (frogs and toads; order Anura [or Salientia]) | tailless, leaping amphibians | 5400 | bullfrog, tree frog, American toad |
| **Salamanders** (order Caudata [or Urodela]) | amphibians with elongated bodies; long tails; short legs; and soft, moist skin | 550 | giant salamander, mudpuppy, axolotl, newts |
| **Caecilians** (order Gymnophiona [or Apoda]) | slender, burrowing, wormlike, legless amphibians | 170 | (known by scientific names only) |

Fig. 18.26 **Life cycle of a frog**

deer, monkeys, and birds, that are wounded by a poisonous arrow are paralyzed almost instantly. Its toxin is so potent that one ounce could kill nearly three million people. To be effective, the toxin must enter the bloodstream directly; digestive enzymes destroy the toxin, making it safe to consume poisoned game.

**External fertilization.** The majority of anurans return to the water for reproduction. The male of each species has its own mating call that directs the female of that species to it. Egg laying, which takes about 10 minutes, usually occurs at night or early dawn. Fertilization is external. In most species, the female is clasped by the male in a process called **amplexus** (a Latin word meaning "embrace"), which helps the female expel the eggs. When the eggs pass through the oviducts, they are coated with a jellylike substance. As the eggs are discharged from the female's body, the male discharges sperm, which are spread over the eggs.

After the eggs are released into the water and fertilized, the jellylike substance that coats each egg absorbs water and swells. This process produces a thick *jelly layer* around each egg, which helps stick the eggs together or to objects in the water, making it difficult for fish or other predators to eat the eggs.

True toads lay their eggs in two long strings, which they spread over plants in the water. Most other anurans lay eggs in clusters, which are usually anchored to vegetation. Some tropical species attach their eggs to the underside of leaves growing over water. As the eggs hatch, the tadpoles drop into the water. Even though some species lay thousands of eggs at a time, many eggs are eaten by larger water animals, and therefore only a few reach maturity.

Anurans lay their eggs in a variety of ways. The American *white-lipped frog* lays its eggs in a hole near a pond and whips the jellylike substance into a foam that the tadpoles live in until rain washes them into the pond. The *Surinam toad* of South America lays its eggs and then quickly loops around to catch them on its back; they remain in cavities on the female's back throughout their larval stage. The male *midwife toad* of western Europe carries the eggs wound around his hind legs. When the eggs are ready to hatch, the male toad enters water and the tadpoles swim away. After the female *Darwin's frog* lays eggs, the male frog picks up the eggs with his tongue and carries them around in his vocal sac, where they hatch and the young develop. A few species lay their eggs on land under rocks, logs, or dead leaves. These amphibians, such as the *barking frog* and the *cliff frog* that live on rocky cliffs in Texas, have no tadpole stage; they hatch from the eggs and begin life as a terrestrial animal.

**Development of a tadpole.** Anuran eggs usually hatch in 6 to 9 days, depending on environmental conditions (low temperatures slow the development of the embryo, while higher temperatures speed development). The developing embryo contains a large *yolk sac* that nourishes it. When an egg hatches, the frog larva (called a *tadpole*) is only a fraction of an inch long and is largely helpless. For the first few days, it clings to nearby vegetation while it grows larger, still nourished by the yolk sac. As the tadpole grows, its tail becomes longer and develops a fin, enabling the larva to swim from place to place. When it has developed sufficiently and its yolk sac is used up, the tadpole lets go of the vegetation and swims freely, using its strong jaws to feed on algae and other plant matter in the water. A long, coiled intestine allows the tadpole to digest the plant matter. The tadpole breathes with *gills,* which are external at first but soon become covered by a gill covering. Two deep pits in front of the mouth will eventually develop into nostrils.

**Metamorphosis of a tadpole.** After an extended period of growth (anywhere from a few weeks to several months), other changes begin to take place in the tadpole. Hind legs appear first, followed by forelegs. The tail begins to be reabsorbed, and the digestive system becomes shorter and changes so that the tadpole can eat insects and other small animals. Just before metamorphosis completes, the gills are replaced by lungs, and the gill slits close. When the young anuran emerges from the water, its tail may not be completely reabsorbed. The time required for metamorphosis varies according to water temperature and species. Some species metamorphose in a few weeks; in constrast, the *bullfrog* tadpole may take two or three years to reach maturity.

Interestingly, in their larval stage, anurans can regenerate an injured or lost part of a limb or tail. As tadpoles mature, however, they lose this capability; adult anurans cannot regenerate body parts.

## Anatomy of an Anuran

Anurans have structures similar to those found in other vertebrates. Typical frogs have nine trunk vertebrae. Ribs are absent in most species.

**Features of the head.** The most prominent features of a frog's head are its protruding *eyes,* which are quite large in relation to the head and enable the frog to see in nearly every direction, even when partially submerged. Most frogs have *nictitating membranes* (transparent inner eyelids) attached to the bottom eyelids, which protect and moisten the eyes. A frog closes its eyes by pulling the eyeballs deeper into their sockets. A frog's eyes contain both rod and cone cells, providing the frog with color vision.

A frog's *nostrils* are located on the upper part of the head in front of the eyes. Having the nostrils in this position allows the frog to breathe when most of its body is submerged. Just behind the eyes are the large eardrums, or *tympanic membranes.* The eardrums are connected to the frog's inner ears by a single lightweight bone. Large anurans hear low-frequency sounds well, but their ears are not designed to hear high-frequency sounds (above 4000 Hz or so). Small anurans were created with the ability to produce and hear high frequencies.

**Features of the oral cavity.** A frog has a wide, gaping mouth that is nearly as large as its head. The openings of the *eustachian tubes,* which lead to the tympanic membranes, are located on either side of the roof of the mouth near where the upper and lower jaws hinge. In males, openings to *vocal sacs* (inflatable pouches of skin under the chin that are used to make loud sounds) are located in the floor of the mouth. The long, protrusible *tongue* is one of the most prominent features of the oral cavity. The surface of the tongue produces a sticky secretion that helps the frog snare its prey (usually insects). The base of the tongue is near the front of the mouth, allowing the tongue to be rapidly flicked outward to catch insects.

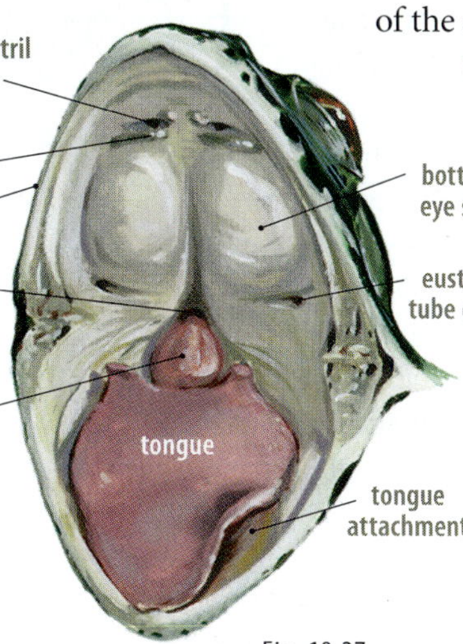


Fig. 18.27

**Oral cavity of a frog**

Tiny *maxillary teeth* project from the upper jaw, and two additional teeth (known as **vomerine teeth** because they are attached to the vomer bones of the skull) extend from the roof of the mouth. A frog's teeth are not used for biting or chewing; rather, they prevent the prey from escaping while the frog is preparing to swallow it. To aid in swallowing, the frog pushes the food into its gullet by closing its mouth and pushing with its eye sockets, which bulge inward from the roof of the mouth.

**The digestive system.** The rear of a frog's mouth opens into a large *gullet,* which leads to an expandable *esophagus.* The lower end of the esophagus joins the *stomach.* The *pyloric sphincter* connects the lower end of the stomach to the coiled *small intestine,* where most nutrients are absorbed. The lower end of the small intestine leads into the *large intestine.* The large intestine opens into the *cloaca,* the cavity that collects and stores wastes from both the large intestine and the kidneys; these wastes are excreted through the *anus.* The cloaca is also the passageway for male and female reproductive cells to the exterior.

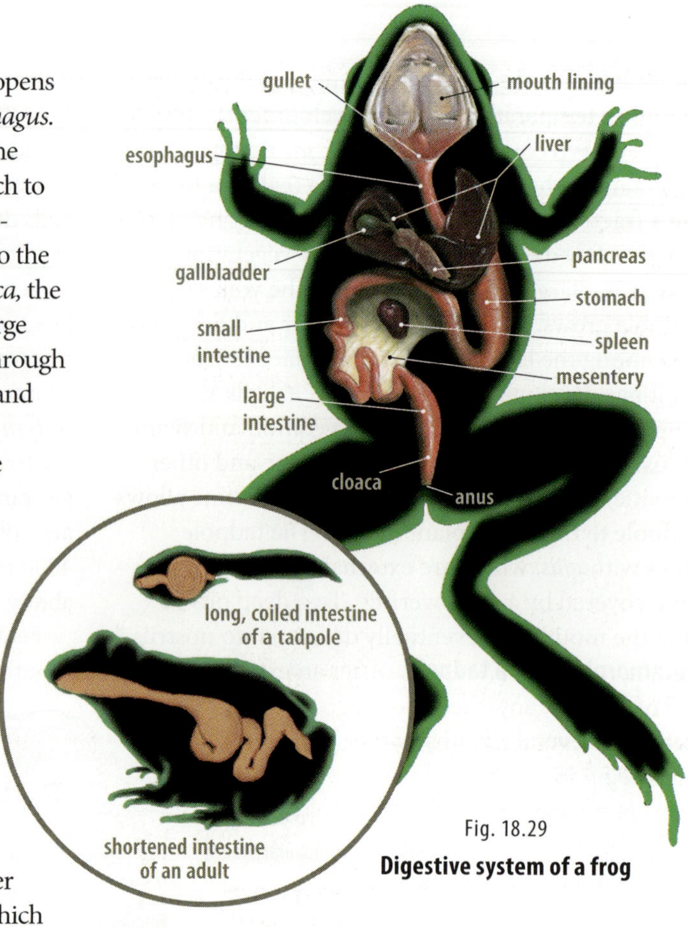

Fig. 18.29 **Digestive system of a frog**

The *liver,* the largest organ in the frog, has three lobes, is dark red in color, and is located in the ventral part of the body cavity. The *pancreas* is a small, light-colored, irregularly shaped organ that is held in place between the stomach and the first fold of the small intestine by a special membrane called the **mesentery.** Both the liver and the pancreas are joined to the small intestine by ducts. The liver secretes bile, which is stored in the *gallbladder.* Pancreatic fluid and bile pass through ducts into the duodenal section of the small intestine. Slow oxidation of glycogen and fat stored in the liver and in special *fat bodies* above the kidneys maintains the frog's life processes during estivation and "hibernation," when it burrows into mud at the bottom of a pond.

Fig. 18.28 **Frog snaring insect**

**The respiratory system.** The respiratory system of an adult frog includes the lungs, the mucous membrane lining the oral cavity, and the skin. Because frogs lack a rib cage and diaphragm, they fill their lungs by forcing air into them. Paired nostrils open into the oral cavity from the exterior. As the floor of the mouth is lowered, air is drawn in through the nostrils. The nostrils are closed by skin flaps as the floor of the mouth is elevated, forcing air through the open glottis in the back of the throat into the lungs. When exhaling, the muscles of the body wall contract, emptying the lungs. Because the thin, moist lining of the oral cavity contains many tiny capillaries, a frog can obtain oxygen through its mouth **(buccal** [bŭk′əl]

Fig. 18.30 **Frog respiration**

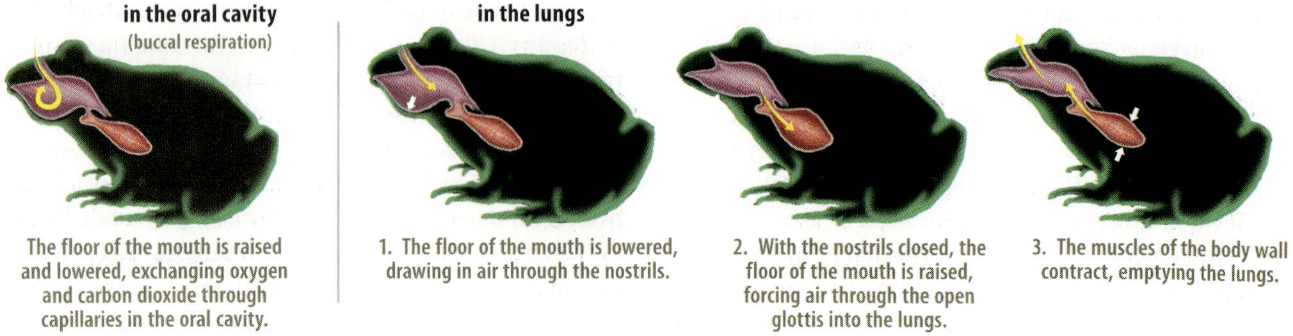

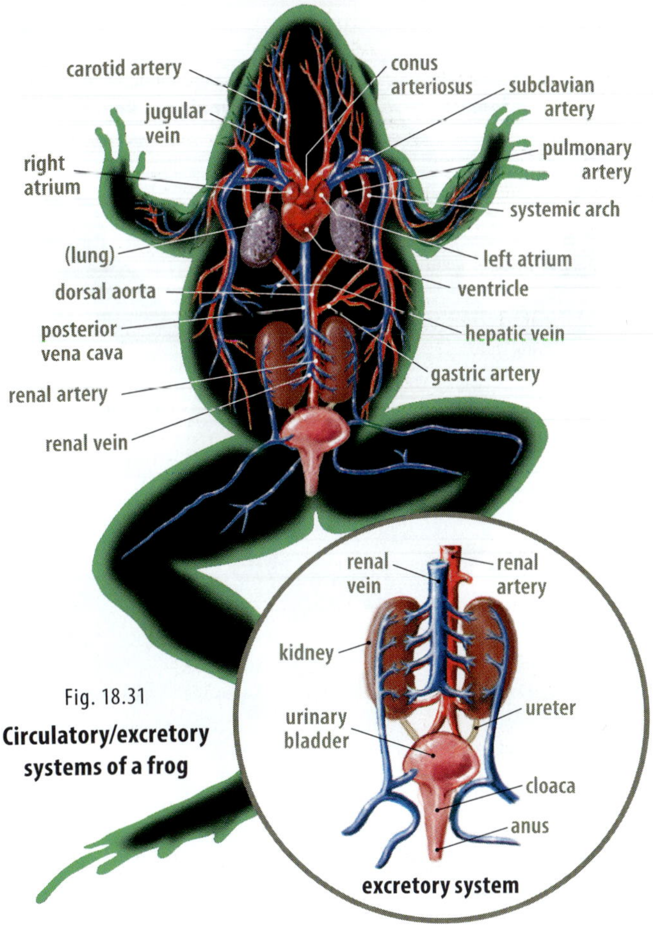


Fig. 18.31
**Circulatory/excretory systems of a frog**

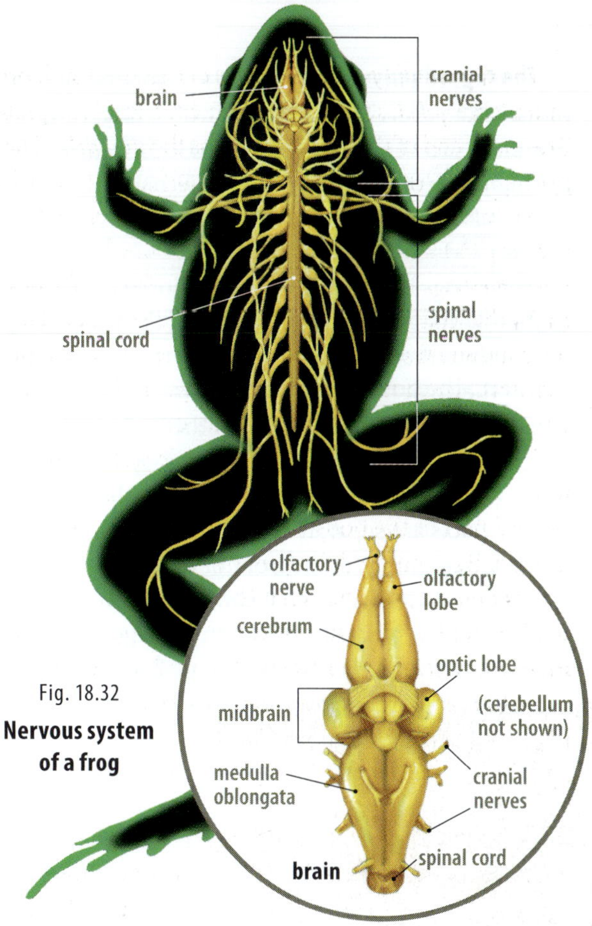


Fig. 18.32
**Nervous system of a frog**

**respiration)** for some time by raising and lowering the floor of the mouth and thus exchanging oxygen and carbon dioxide in its body without the use of lungs. When a frog remains underwater for prolonged periods, its skin absorbs dissolved oxygen from the water and gives off carbon dioxide. Even under normal conditions of lung respiration, most of the carbon dioxide is dissipated through the skin. Since the frog's metabolism is greatly reduced during "hibernation" or estivation, *cutaneous* (skin) *respiration* alone provides all its oxygen.

As air is forced through the glottis into the lungs, it passes through the *larynx* (voice box). A frog croaks when paired vocal cords in the larynx vibrate as air is forced from its lungs. The male uses its voice to attract females during the mating season. If the male has a vocal sac as in some species, the sac, which swells as sound is being made, increases the volume of the sound. In some species, the female also has a voice.

**The circulatory system.** Anurans and other amphibians have a three-chambered heart consisting of an undivided muscular *ventricle,* which pumps blood to all parts of the body, and two thin-walled *atria,* which receive blood from the body. The left atrium receives blood from the lungs, and the right atrium receives blood from the rest of the body. The single ventricle contains both oxygen-rich blood from the lungs and oxygen-poor blood from the systemic circulation; although some mixing occurs, mixing is minimized by a special valve in the blood vessels leaving the heart and by the rhythm of the heart, in which the atria contract at different times.

**The excretory system.** The primary excretory organs of a frog are the *kidneys,* which cleanse the blood of water-soluble wastes and form urine. The kidneys are located close to the vertebral column in the body cavity. A ureter connects each kidney to the cloaca. Attached to the cloaca is the urinary bladder, where liquid wastes are stored until they are eliminated through the cloaca.

**The nervous system.** The frog's *central nervous system* includes the spinal cord, which is encased in vertebrae, and the *brain,* which is divided into two olfactory lobes, two cerebral hemispheres, the midbrain, two optic lobes, the cerebellum, and the medulla oblongata. Ten pairs of cranial nerves and 10 pairs of spinal nerves make up the *peripheral nervous system.*

**The reproductive system.** At the front end of each kidney is a yellowish mass of fat with a fingerlike appearance. The energy stored in these *fat bodies* is used during times of estivation and "hibernation"; the fat bodies also provide a ready source of energy when the frog is producing reproductive cells.

Male frogs have a pair of light-colored testes attached to their kidneys. Sperm produced in the testes are carried into the ureters and through them to the cloaca, where they exit the body. Female frogs have a pair of large ovaries, one on each side of the body cavity, where eggs are produced. Eggs are not released directly into the oviducts but rather are released into the body cavity. The oviducts are separate from the ovaries and have their own openings at the front of the body cavity, just behind the lungs. As eggs accumulate in the body cavity, some enter the oviducts and move through them, where they are given their jellylike coating. Each oviduct ends in a *uterus,* where the coated eggs accumulate until the frog lays them through her cloaca.

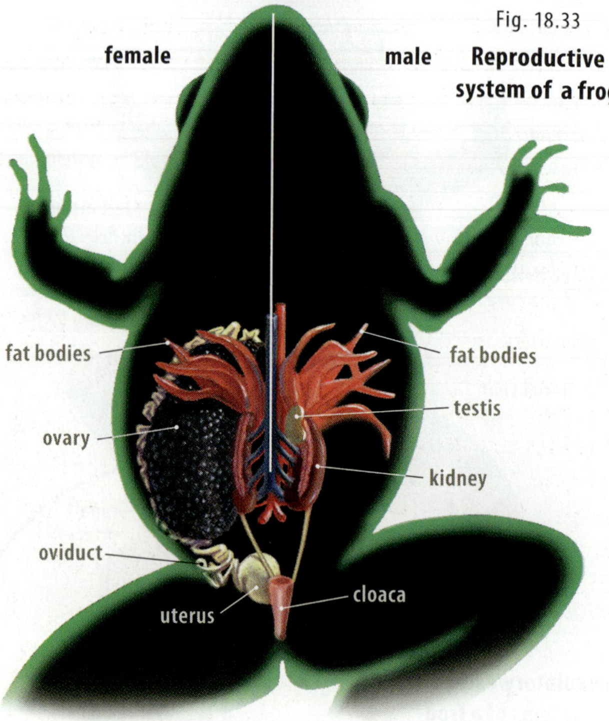


Fig. 18.33 **Reproductive system of a frog**

## Section Review 18.4

1. What are the three major groups of amphibians?
2. What is the transition from a larva to an adult called?
3. What is the collective term for frogs and toads?
4. What is the purpose of a frog's vomerine teeth?
5. What is buccal respiration?

**Critical Thinking**

6. Observe the illustration of a frog's brain on p. 383. Judging from the features of the brain, what two senses would appear to be most acute in a frog? Upon what features of the brain would you base this conjecture?

**Identify**

chromatophores, amplexus, vomerine teeth, mesentery

# 18.5 Other Amphibians

## Salamanders: Amphibians with Tails

The order Caudata [kô·dā′tə: Latin, "having a tail"] or Urodela [yo͝or′ə·dē′lə: Greek, "a visible tail"] includes the **salamanders,** *newts,* and *sirens.* These amphibians are found throughout the Northern Hemisphere and in northern Africa and South America, but most species occur in North America. Although salamanders superficially resemble lizards (and are often confused with lizards), there are actually many differences between the two. For example, a lizard has dry, scaly skin and clawed toes, whereas a salamander has moist, scaleless skin and no claws. Lizards often bask in the sun, but salamanders avoid sunlight and are seldom seen. Salamanders also move with a characteristic undulating gait, bending the body sharply left or right with each step.

**Reproduction and metamorphosis.** Most salamanders are terrestrial and live in moist places under stones and rotten logs, usually not far from water. Although a few species are ovoviviparous, most are oviparous. Some lay their eggs in water; others lay eggs on land in burrows or under debris where they will remain moist. Unlike frogs, salamanders fertilize their eggs internally. The male salamander deposits sperm in a gelatinous structure, called a **spermatophore** [spər·măt′ə·fôr′], on a leaf or stick. The female then settles over the spermatophore and maneuvers the packet of sperm into her cloaca, where the gelatinous material dissolves. *Internal fertilization* occurs as the eggs descend through the female's oviducts.

The embryos of salamanders hatch from their eggs resembling their parents; however, the larvae possess gills and have teeth in both jaws. Most salamander larvae undergo *metamorphosis*, but their structural change is considerably less noticeable than the metamorphosis of frogs and toads. Most salamanders change color, often dramatically, when they emerge from the water. A few species never metamorphose but remain in the water as permanent larvae; they retain gills and other larval characteristics throughout their life, even after becoming sexually mature. This condition is called **pedomorphosis** [pē′də·môr′fə·sĭs: Greek, "child form"]. In most adult amphibians, lungs replace the larval gills; but many adult salamanders have neither lungs nor gills. In lungless salamanders, *cutaneous respiration* is supplemented by *buccal respiration* as air is pumped in and out of the oral cavity. Salamanders lack vocal cords, and they have no outer ear, middle ear cavity, or tympanic membrane. Sound transmission is from the forelimb to a muscle that passes from the shoulder blade to a bone in the *inner ear*. Although most salamanders have sight, only the eyes of terrestrial forms have lids.

Fig. 18.34

**Salamanders**

**Variety in size.** Most salamanders are fairly small, averaging less than 6 inches long. However, some salamanders grow larger than this. The land-dwelling *Pacific giant salamander* of Oregon and Washington may grow more than a foot long, and the aquatic *Chinese giant salamander* can be almost 6 feet in length, making it the largest living amphibian. One of the largest salamanders in North America is the *hellbender*, an aquatic salamander that can exceed 2 feet in length.

**Mudpuppies.** Another aquatic salamander common in the United States is the *mudpuppy* or water dog. The mudpuppy does not undergo metamorphosis but spends its entire life in the larval form, complete with gills. Although the body of the mudpuppy is usually a drab, spotted brown, its gills are red because of their abundant blood supply. The mudpuppy is nocturnal and remains concealed in the mud or beneath underwater debris during the day. At night, the mudpuppy emerges and preys on aquatic insects, crayfish, worms, and other invertebrates. Most mudpuppies are about a foot long.

**American axolotls.** *Axolotls* [ăk′sə·lŏt′lz], found only in North America, are widely distributed over Mexico and the southwestern United States. These aquatic salamanders range from 4 to 7 inches long, and although they are usually black or a dark brown color, albinos are quite common. Axolotls become sexually mature and breed in the larval form, but unlike permanent larvae, they can metamorphose to an adult form in which they lose their gills, develop lungs, and begin a terrestrial life if the water where they live begins to dry up.

**Newts and efts.** The terms *newt* and *eft* are old Anglo-Saxon words meaning "lizardlike." Although the terms are commonly used interchangeably to describe many salamanders, **eft** designates the juvenile terrestrial

Fig. 18.34 (cont.)
**Salamanders**

form of various members of the family Salamandridae; **newt** refers to the aquatic adult form of the same salamanders. Thus, when the larval *red-spotted newt* of the United States loses its gills and its olive-green skin changes to bright red with spots, it becomes the *red eft* and begins its two or three years of terrestrial life. When it matures to adulthood and returns to the water for breeding, it again changes color.

**Sirens.** The North American amphibians of the family Sirenidae [sī·rĕn′ĭ·dē] are sometimes classed as a separate order. *Sirens* (not the large marine mammals known as sirens, or sea cows, which belong to the order Sirenia) have a long, slender body, which has the appearance of an eel. They lack hind limbs and have only minute forelimbs. These aquatic amphibians, which never completely metamorphose, range in size from 7 inches to over 2 feet. External gills provide respiration underwater. They are both herbivorous and carnivorous, consuming plant leaves, algae, worms, snails, shrimp, and crayfish. When water in the ponds begins to dry up, sirens estivate. At the onset of estivation, their skin secretes a slimy mucus that hardens when it comes in contact with air and thus forms a protective coating in the drying mud. During this time, sirens breathe with their small lungs.

## Caecilians: Limbless Amphibians

The smallest of the three orders of amphibians is the order Gymnophiona [jĭm′nə·fī′ə·nə: Greek, "naked snake"] or Apoda [ăp′ə·də: Greek, "without feet"], which contains some 170 species. These slender, burrowing, wormlike tropical amphibians, which range in size from 7 to 53 inches long, are found in the warm, moist climates of South America, Africa, and southeast Asia. Their common name, **caecilians,** comes from the Latin root *caecus,* meaning "blind." Most species of caecilians are strictly underground creatures and thus are seldom seen by man; at least one species is aquatic, having a dorsal fin fold.

The smooth, slimy *skin* of caecilians appears segmented because of circular folds, which are separated by shallow grooves. Unlike other amphibians, most species of caecilians have tiny *scales* embedded in the folds of their skin. Caecilians feed mainly on worms, termites, and other small underground invertebrates. Like other amphibians, caecilians have a three-chambered *heart,* and, although they have *lungs* (one large lung and one small lobe), their breathing is partly by *cutaneous respiration.*

Fig. 18.35 **Caecilian**

## Section Review 18.5

1. What are some of the differences between lizards and salamanders?
2. What is the largest living amphibian?
3. Describe the life cycle of a newt.
4. What eel-like amphibians are characterized by a long, slender body, no hind limbs, and small forelimbs?
5. Why are caecilians rarely seen by man?
6. What feature of the skin distinguishes caecilians from other amphibians?

### Identify

spermatophore, pedomorphosis, axolotl, eft, newt, siren, caecilian

## Chapter 18 Review

### Define

1. poikilotherm
2. ovoviviparous
3. quadrate bone
4. herpetologist
5. fangs
6. neurotoxin, hemotoxin, antitoxin
7. carapace, plastron
8. estivation
9. parietal eye
10. chromatophore
11. amplexus
12. vomerine teeth
13. mesentery
14. spermatophore
15. pedomorphosis

### Identify

1. four groups of living reptiles mentioned in the text
2. five characteristics common to most reptiles
3. the only group of reptiles with a four-chambered heart
4. the largest lizard
5. two species of venomous lizards
6. the three largest snakes
7. the longest venomous snake
8. three pit vipers found in North America
9. the longest-lived animal
10. three groups of crocodilians mentioned in the text
11. five groups of dinosaurs and dinosaurlike creatures mentioned in the text
12. three broad groups of amphibians
13. three ways by which a frog can obtain oxygen from the air

### Explain

1. How do cold-blooded animals regulate their temperature?
2. List four differences between lizards and snakes.
3. Describe the four ways in which snakes move.
4. What organs provide a snake with an excellent sense of smell? How does this sense of smell work?
5. What special organ do all pit vipers have in common? What purpose does this organ serve?
6. What are the differences between the fangs of venomous colubrids, elapids, and viperids?
7. Explain why it is impossible to remove a turtle's shell without killing the turtle.
8. Why must a turtle breathe differently from other reptiles?
9. Describe the life cycle of a typical frog.
10. What is the difference between an eft and a newt?

### Apply

1. The health of anuran populations is often considered an indicator of the health of an ecosystem. Explain why these populations are especially sensitive to disturbances in an ecosystem.
2. What superficial differences are there between lizards and salamanders? What more profound differences are there?

CHAPTER 19

# Fish

## Key Concepts

- ➤ Diversity of species
- ➤ Bony fish
- ➤ Special organs
- ➤ Spawning
- ➤ Parental care
- ➤ Cartilaginous fish

**Fish** are cold-blooded, water-dwelling vertebrates with gills and scales. There is greater variety among fish than among all other land and water vertebrates put together. Scientists have classified about 28,000 species of these cold-blooded vertebrates. This tremendous diversity is the result of God's creative word, uttered on the fifth day of Creation.

The earth's oceans, seas, and freshwater bodies of water abound with many strange and wonderful creatures. When an *archerfish* catches sight of an insect resting above the surface, it takes aim and spits a stream of watery "arrows" (drops of water) at its prey with enough force to knock the insect into the water. Although the *lionfish* has beautiful fins that resemble a bird's feathers, they are as sharp as needles and give off a deadly poison. The *porcupine fish* is covered with protective spines and can swell up into a prickly balloon by filling itself with water. The *sargassum fish* looks just like the seaweed in which it lives and is almost invisible as it lumbers among the algae.

Getting oxygen from stagnant water poses no problem for the *lungfish,* which uses its swim bladder as a lung and can breathe air at the surface. Whenever a dry season occurs, some lungfishes build their own mud shells and stay inside their "cocoons" in a state of estivation until the rainy season arrives. The *stingray,* a broad, flat fish that may be as long as 14 feet, also enjoys mud and makes its home on the sandy to muddy bottoms of shallow, warm waters. Many an unwary swimmer has accidentally stepped on a stingray and received a painful and serious wound from the strong, sharp spine on the creature's powerful, whiplike tail. Another unusual fish that lives in the warm waters of the Pacific Ocean off the

Fig. 19.1 **Variety among fish**

coast of southern California is the *California flying fish.* This creature propels itself up out of the water with the motion of its strong body muscles and powerful tail and then uses its large fins as wings to glide above the surface for 150 to 1000 feet; some flying fish can even turn in flight. The tiny, oddly shaped *seahorse,* with its long snout and prominent eyes, is one of the most unusual-looking of all fish.

The Southeast Asian carp *Paedocypris progenetica,* the world's smallest known fish, measures less than one-half inch in length when fully grown. In contrast, the *whale shark,* the world's largest fish, may attain a length of 60 feet and a weight of 20 tons—more than 2 or 3 African elephants put together. Some commentators speculate that the whale shark may have been the great fish that swallowed Jonah, the reluctant prophet (see Jon. 1:17).

## 19.1 The Bony Fish (Osteichthyes)

Traditionally, **ichthyologists** [ĭk′thē·ŏl′ə·jĭsts: scientists who study fish] have divided fish into two main groups. Those whose skeletons are made either partially or wholly of bone are called **bony fish.** Those whose skeletons are made of cartilage are divided into two classes: the sharks, rays, and chimaeras [kī·mər′əz] and the lampreys [lăm′prēz] and hagfish.

### Abundance and Importance

The bony fish of the class Osteichthyes [ŏs′tē·ĭk′thĭ·ēz: from Greek *osteon,* "a bone"+ *ichthys,* "a fish"] comprise over 95% of all known species of fish. They include such familiar fish as bass, catfish, cod, goldfish, herring, minnows, mullet, perch, pike, red snapper, salmon, trout, tuna, and sturgeon.

Bony fish are one of God's greatest provisions for mankind. He told the Israelites, "Whatsoever hath fins and scales in the waters, in the seas, and in the rivers, them shall ye eat" (Lev. 11:9). In such seafaring nations as Norway and Japan, fish make up the major part of the diet. *Fish are among the most nourishing of all foods,* providing as much protein as red meat as well as necessary vitamins and minerals. Fish-liver oils, such as cod-liver oil, are extremely rich in vitamins A and D; fish are also a good source of vitamin E, certain essential fatty acids, and the mineral iodine (necessary for proper thyroid metabolism). People in Japan, who eat much fish, have a far lower incidence of heart attacks and high blood pressure than Americans, who tend to consume far more red meat than fish. Commercial fishing at sea yields millions of tons of cod, herring, mackerel, tuna, and other ocean food fish each year, while commercial fishermen in inland waters catch such freshwater food fish as perch and trout. American fish farms raise catfish, pompano [pŏm′pə·nō′], and trout.

### Structure of Bony Fish

Because of a common design and Designer, bony fish resemble other vertebrates in certain ways. Like other vertebrates, fish have an internal skeleton, an outer skin, and vital organs such as a heart, intestines, and a brain. Fish, however, have fins instead of legs and gills instead of lungs. These differences show the wisdom of the Creator in making fish ideally suited to their watery environment.

**Table 19.1 Groups of Fish**

*(Kingdom Animalia; phylum Chordata; subphylum Vertebrata)*

| Group | Description | Number of living species (approximate) | Examples |
|---|---|---|---|
| **Bony fish** (class Osteichthyes) | cold-blooded, aquatic vertebrates that have fins and scales; breathe by means of gills; and have skeletons made either partly or wholly of bone | 27,000 | perch, bass, trout, tuna, seahorse, goldfish |
| **Cartilaginous fish** (class Chondrichthyes) | cold-blooded, aquatic vertebrates that have fins and scales; breathe by means of gills; and have skeletons made of cartilage | 940 | sharks, rays, chimaeras |
| (class Cyclostomata[1]) | | 113 | lampreys, hagfish |

[1]Or Agnatha. Some authorities also refer to Cyclostomata / Agnatha as a superclass and classify lampreys and hagfishes in separate classes (Cephalaspidomorphi and Myxini, respectively) within the superclass.

Fig. 19.2
**Structure of a bony fish (yellow perch)**

**Body regions of a fish.** Even at a glance, it is obvious that the body of a fish is marvelously designed for moving easily through the water. A typical fish is highly streamlined, with a tapered head, swept-back fins, and a narrow tail. The body is covered with a slippery coating that further reduces resistance. These attributes allow a fish to move through the water with remarkable grace and agility.

Although the body of a typical fish is smoothly tapered from nose to tail, *ichthyologists speak of all fish as having three main body regions.* The front portion of the fish's body, extending from the mouth to the back of the gill covers, is the *head.* The portion extending from the back of the gill covers to the urogenital opening is the *trunk.* The portion extending from the rear of the trunk to the base of the tail fin (caudal fin) is the *tail.*

**Features of the head.** The *mouth* of a bony fish is usually located at the very front of the fish's body. A complex arrangement of jaw bones gives the fish a very streamlined shape when the mouth is closed while allowing the mouth to open very wide when necessary. Typically, the mouth is studded with many *teeth,* the structure of which depends on the fish's diet. Herbivorous fish typically have flat, grinding teeth, while predatory fish have cutting or spearing teeth. Many predatory fish also have special *throat teeth* that help them grip their prey while swallowing. Although a fish has a stiff, bony *tongue* in its mouth, the tongue contains no taste buds. Instead, a fish's sense of taste is provided by taste buds spread over the outside of its body, especially the head; taste buds may even be found on the fins of some fish.

Above the fish's mouth are two pairs of *nostrils.* These nostrils do not function in breathing (respiration) but allow water to enter special nasal cavities in the head that are lined with sensitive olfactory cells. These cells give the fish an excellent sense of smell.

The actual respiratory structures of a fish are the *gills,* which are located near the back of the fish's head. Ordinarily, the gills are not visible because they are concealed beneath large, flat gill covers known as **opercula** [ō·pûr′kyə·lə: sing. operculum] located at the back of the head on each side. The opercula help protect the delicate gill structures beneath from being damaged.

A fish's *eyes* are located on the sides of its head, giving it a wide field of vision. The eyes are also rather large compared to the head, and the pupils are very wide. This design allows the eyes to receive as much light as possible, helping the fish to see even when swimming in murky water.

**Features of the trunk and tail.** The largest part of a fish's body is the trunk, which narrows into the tail. The most prominent features of the trunk and tail are the fish's **fins.** Most bony fish are said to have *rayed fins* because each fin is composed of a thin web of connective tissue spread over a fan-shaped skeleton of thin rods (known as *rays*) made of bone or cartilage. Most bony fish have two sets of paired fins: **pectoral fins,** located just behind the head on each side of the body, and **pelvic fins,** located to the rear of the pectoral fins and lower on the trunk. The pectoral fins and pelvic fins allow the fish to angle up or down while swimming, help keep the fish from rolling sideways, and serve as brakes when the fish wishes to stop. The fish can also move slowly backward by gently sculling with its paired fins.

A typical bony fish also has two or three unpaired vertical fins; because these fins project from the centerline of the body, they are described as *median fins.* The median fins of a fish include the **dorsal fin,** which projects upward from the fish's back, and the **anal fin,** which projects downward from the rear of the fish's trunk. The dorsal and anal fins work together to stabilize the fish when it is in motion. Some fish have only one dorsal fin, while many have two, located one behind the other. The dorsal fins of some fish, including the perch, contain long, sharp spines made of bone that help protect the fish from predators; a predator attempting to swallow such a fish from behind would likely drive the spines through the roof of its mouth. The *sailfish* and *marlin* are both noted for their large, spiny dorsal fins, which are folded down to reduce drag when the fish is swimming at high speed.

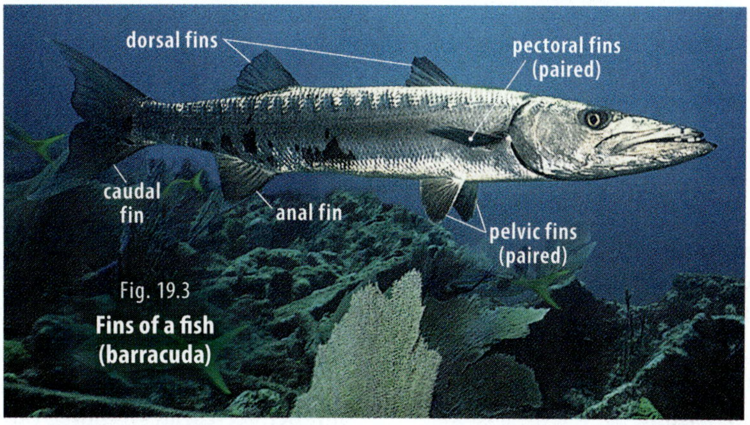

Fig. 19.3
**Fins of a fish (barracuda)**

Projecting from the fish's tail is a third median fin, the large **caudal fin,** which serves as a fish's propeller and rudder. Short, broad caudal fins (such as that of the perch) are designed for rapid acceleration from a standstill, while narrow, sweptback caudal fins (such as those of tuna and sailfish) are designed for cruising at high speed. In some fish, such as eels, the dorsal, anal, and caudal fins merge together at their edges to form a single fin that passes all the way around the tail.

**Skin and scales.** Most bony fish have a fairly tough skin. Like the skin of other animals, it contains blood vessels, nerve fibers, and connective tissues. It also contains *mucous cells,* which produce a slimy mucus that makes fish slippery and gives them their fishy odor; this mucus layer helps waterproof the fish's skin, protects the fish from parasites, and reduces friction as the fish moves through the water. Like many reptiles and amphibians, a fish's skin contains *chromatophores,* or special pigment cells, that give them their richly varied colors. Some species of bony fish have more pigment cells of a certain color; other species have their pigment cells grouped differently, thus accounting for the wonderful variation in color among the bony fish. The whitish or silvery pigments in the skin and scales of many bony fish produce a variety of rainbow colors in the bright sunlight. These colors often protect a fish by enabling it to blend into its surroundings; most fish have the ability to alter their color to match color changes in their environment. A few species can even change the pattern formed by their colors.

Most bony fish have a protective covering of small, bony plates called **scales** located just beneath the epidermis, in the dermis of the skin. Fish scales vary in size and shape among different species of fish. Most fish have rounded or partially rounded scales that overlap considerably. A few fish, most notably *gars,* have thick, diamond-shaped

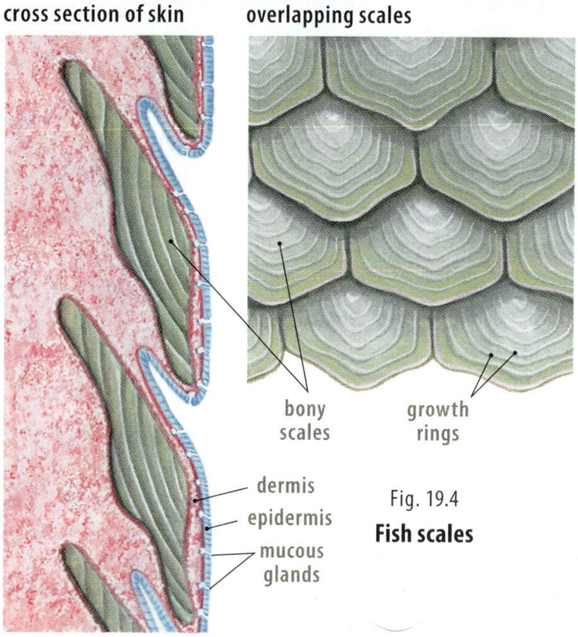

Fig. 19.4
**Fish scales**

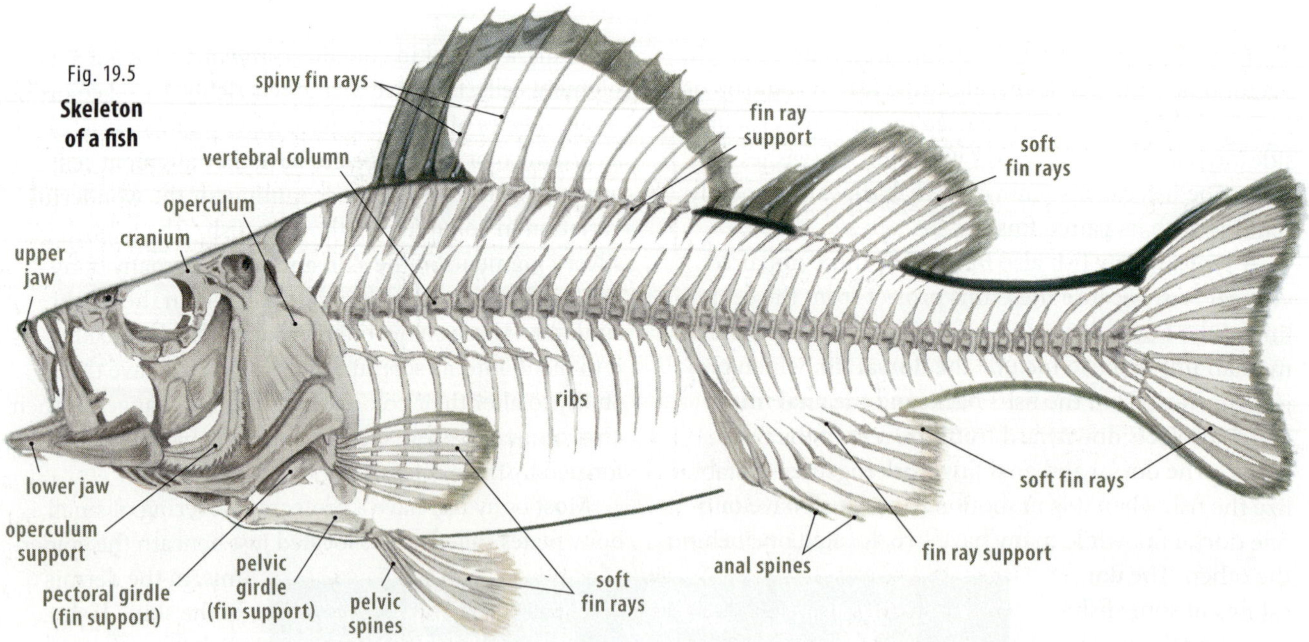


Fig. 19.5
**Skeleton of a fish**

scales that do not overlap. Some fish, such as *catfish* and certain *eels,* lack scales completely.

A fish does not grow additional scales as it increases in size; the existing scales merely become larger. Because most fish grow more rapidly during the summer than during the winter, the scales of many fish show annual growth rings (p. 391, Fig. 19.4). The age of a fish can often be estimated by counting the fish's growth rings with the aid of a microscope.

**Skeleton and muscles.** A fish's *backbone,* which consists of many separate vertebrae of bone or cartilage, provides the central framework and chief means of support for the trunk and tail. Bony fish have a long, spinelike extension at the top of each vertebra. These extensions support the upper part of the fish's body; each tail vertebra has an additional spine at its bottom. The fish's ribs are attached to the vertebrae of the trunk. The *cranium* encloses and protects the brain and provides support for the mouth and gills. The pectoral fins of most bony fish are attached to the back of the skull by a *pectoral girdle.* The pelvic fins are supported either by a *pelvic girdle* or by muscle tissue in the abdomen. Bony or cartilaginous structures that are rooted in muscle tissue above the backbone support the dorsal fins. The tail supports the caudal fin, and the anal fin is supported by structures of bone or cartilage below the backbone.

The muscles of fish, like those of other vertebrates, are of three basic kinds: skeletal muscle, smooth muscle, and cardiac muscle. The *heart* is composed of the specially organized cardiac muscle. *Digestive organs* have smooth muscle as part of their structure. Most of the *flesh* of fish (the part that is eaten) is composed of skeletal muscle. A fish's skeletal muscle, like its skin, has certain unique features that make it different from that of other vertebrates. Unlike other vertebrates, a fish's skeletal muscles are arranged in W-shaped bands called **myomeres** [mī′ō·mērz]. The myomeres can be seen clearly on a skinned fish. *Each myomere is controlled by a separate nerve, allowing the fish to move its body in a wavelike, undulating motion* that propels the fish through the water. The efficiency of this movement is amazing; fish expend far less energy per mile than any walking or flying animal. Powerful muscles and a streamlined shape allow some fish to swim with remarkable speed; it is reported that the *oceangoing sailfish* can exceed 60 mi/hr in short bursts.

Fig. 19.6
**Oceangoing sailfish**

## Section Review 19.1

1. What characteristic distinguishes bony fish from other fish?
2. Name the three regions of a bony fish's body.
3. What purpose do a fish's nostrils serve?
4. List a bony fish's fins and describe their functions.

### Application

5. How does the shape of the caudal fin affect the swimming ability of a fish?
6. How are the rings of fish scales analogous to the growth rings of trees?

### Critical Thinking

7. What are some of the similarities between bony fish and other vertebrates? What are some of the differences?

### Identify

ichthyologist, bony fish, operculum, fin, scale, myomere

### Observe

- Obtain a whole fish or a large fillet with clearly visible myomeres. Using a hand lens, examine the structure of these tissues.

# 19.2 Systems and Internal Organs of Bony Fish

Fig. 19.7
**Gills of a fish**

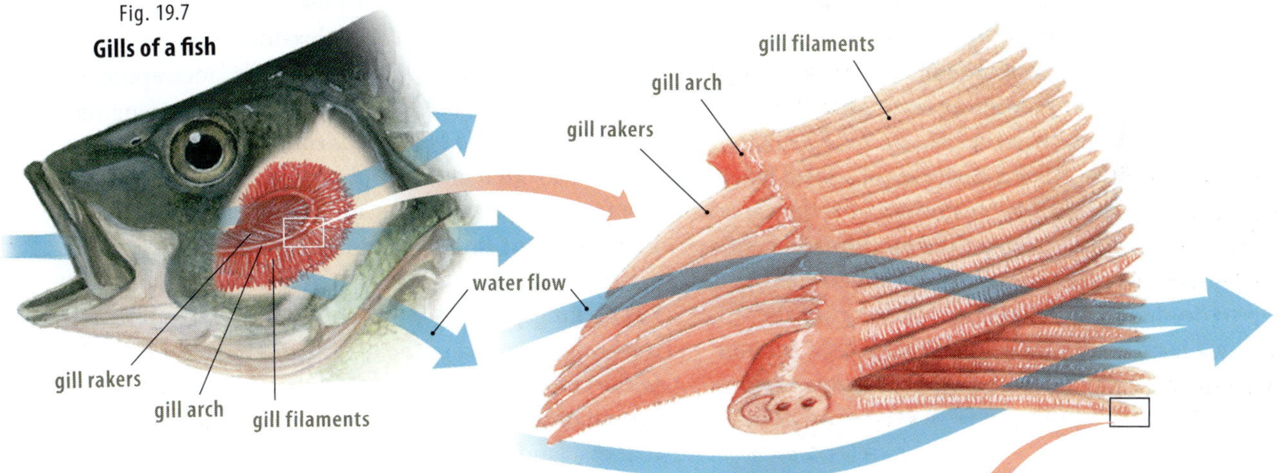


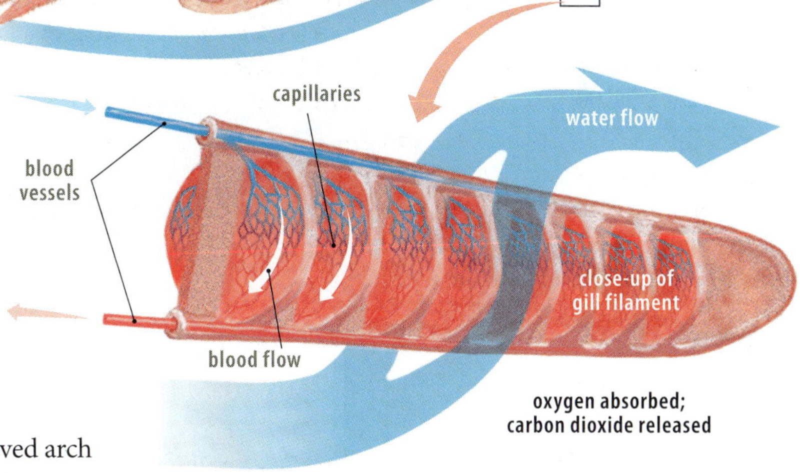


## Respiratory System

Unlike whales and other aquatic mammals, which must periodically come to the water's surface to breathe air, a fish obtains needed oxygen directly from the water. This ability is made possible by **gills,** special organs that allow oxygen and carbon dioxide to pass between the water and the fish's blood. A typical bony fish has eight gills, four on each side of the head.

**Structure of a gill.** Each gill consists of a curved arch of cartilage (the **gill arch**), from which branch two rows of long, narrow **gill filaments.** Under a microscope, each gill filament looks like a stack of thin disks connected by a narrow stalk containing blood vessels. Just beneath the surface of each disk are hundreds or thousands of tiny capillaries. As water flows across the disk, oxygen from the water diffuses into the blood, and carbon dioxide from the blood diffuses into the water. This process is so efficient that a fish can extract as much as 80% of the oxygen from the water that passes across its gills. (By comparison, humans extract only about 25% of the oxygen from the air we breathe under ordinary conditions.) Because water generally contains far less oxygen than air, it is vital for a fish to be able to extract as much oxygen from the water as possible.

**Protecting the gills.** Because the gills are delicate structures, they are covered and protected by a wide plate of bone (the *operculum*) found on each side of the head. The front of each operculum is attached to the fish's head, but the rear is unattached; this arrangement allows the rear of the operculum to swing in and out like a door on its hinges. Beneath each operculum is a large **gill chamber** that opens into the fish's pharynx (throat). The gills are also protected by **gill rakers,** vanelike projections from the front of each gill arch that strain food particles and other debris out of the water, preventing the gills from becoming fouled by these sediments. The gill rakers form a screen that separates the gill chambers from the fish's pharynx, allowing water to flow from the pharynx to the gills but preventing large particles from escaping the mouth through the gills.

**How a fish breathes.** If you have ever observed fish in an aquarium, you have probably noticed how they are constantly opening and closing their mouths and gently flapping their opercula. Stationary or slow-moving fish do this to produce a constant flow of water over the gills, which is necessary for respiration. A fish "inhales" water by opening its mouth and expanding its pharynx; the fish's closed opercula ensure that water enters only through the mouth. The fish then closes its mouth, contracts its pharynx, and opens its opercula slightly. This forces the water in the mouth and pharynx through the screen of gill rakers into the gill chamber, where it passes across the gills and exits the fish by means of the open opercula.

When a fish is swimming, it does not need to open and close its mouth and flap its opercula to breathe. Instead, it simply keeps its mouth and opercula open slightly as it swims. The fish's forward motion causes water to enter the open mouth, flow across the gills, and exit past the open opercula. Sharks commonly use this method of respiration, and a few sharks cannot breathe in any other manner.

## Circulatory System

The *circulatory system* of a fish is a closed system of arteries, veins, and capillaries, broadly similar to the circulatory systems of mammals, birds, and reptiles. Oxygen is transported by hemoglobin-containing red blood cells. White blood cells of the fish's *immune system* also circulate through the blood, protecting the fish from bacteria, viruses, and other pathogens. But there are some important differences between fish circulation and that of other vertebrates. Because a fish has no lungs, for example, there is no need to separate the blood pumped to the lungs from the blood pumped to the rest of the body. As a result, the circulatory system of a fish has a somewhat simpler design than that of a land-dwelling vertebrate.

Fig. 19.8
**Internal organs of a bony fish**

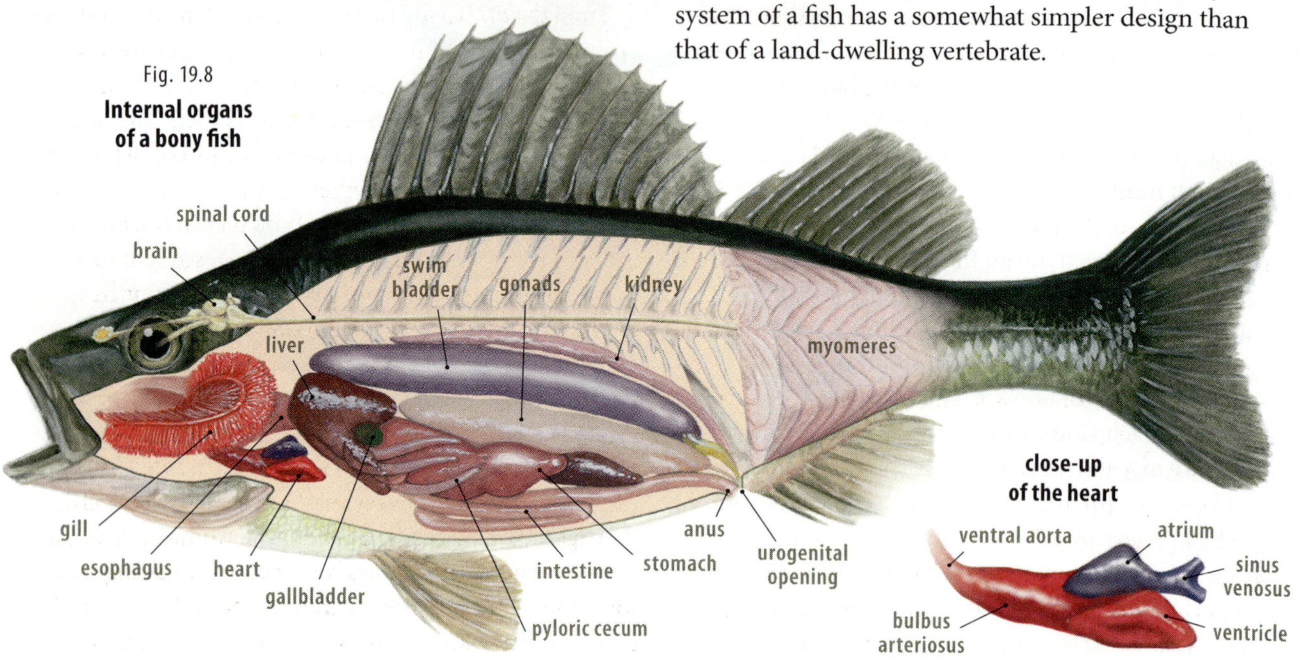

**From the heart to the gills.** *A fish is designed with a two-chambered heart* located in the lower (ventral) portion of the fish, slightly to the rear of the gills. The heart consists of an *atrium* and a *ventricle.* The atrium ensures that the ventricle is fully inflated with blood before the ventricle contracts, and the muscular ventricle provides pressure to force the blood through the body.

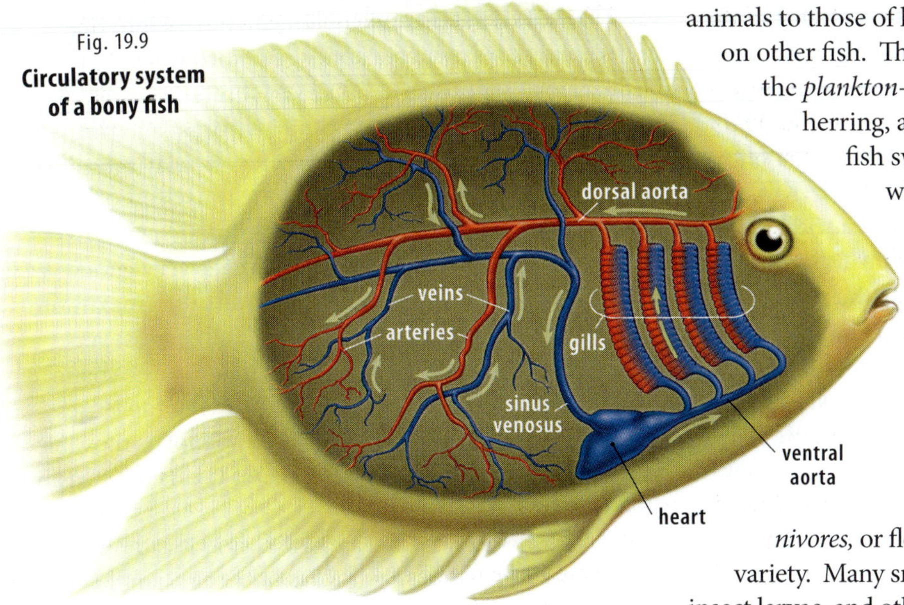


Fig. 19.9
**Circulatory system of a bony fish**

A single major artery, called the *ventral aorta,* exits the front of the heart from the ventricle. As its name suggests, this artery is located on the lower side of the fish. Where the ventral aorta joins the heart, a muscular, funnel-shaped structure called the *bulbus arteriosus* [bŭl′bəs är•tēr′ē•ō′sŭs] helps smooth the blood flow so that the blood pressure through the gills is kept almost constant. After leaving the heart, the ventral aorta divides into eight branches to supply the fish's gills (four on each side). As the blood flows upward through the gills, it absorbs oxygen from the water.

**From the gills to the body.** After absorbing oxygen in the fish's gills, the blood flows out of the gills through eight arteries that rejoin to form the *dorsal aorta.* Branches of this artery supply the fish's head and brain with oxygen and nutrients. The dorsal aorta then extends rearward beneath the fish's spine, where other branches supply the fish's muscles and internal organs with blood.

**From the body to the heart.** After traveling through the capillaries of the fish's muscles and organs, the blood returns to the heart through veins. Veins from all parts of the fish's body converge to form a single large vein called the *sinus venosus* [sī′nəs vē•nō′səs], which flows into the atrium of the heart, beginning another cycle of circulation.

## Digestive and Excretory Systems

Bony fish have a wide variety of diets, from the diets of small fish that feed on microscopic plants and animals to those of large carnivores that prey exclusively on other fish. The most numerous of all fish are the *plankton-eaters,* which include such fish as herring, anchovies, and menhaden. These fish swim near the surface of the water with their mouths open, straining plankton from the water using elongated, closely spaced gill rakers; the plankton accumulates in the throat and is swallowed. Plankton-eating fish are an extremely important part of the marine food chain.

Although they are less numerous than the plankton-eaters, the *carnivores,* or flesh-eaters, are found in the greatest variety. Many small carnivorous fish eat tiny insects, insect larvae, and other small invertebrates, while larger carnivores eat crayfish and other crustaceans, larger insects, and small fish. Large, specialized predators such as tuna and barracuda prey exclusively on other fish.

**From the oral cavity to the stomach.** A fish's mouth, or *oral cavity,* is the beginning of the fish's digestive system. The jaws of a fish are often lined with various types of *teeth* that are designed to suit the fish's diet. The oral cavity opens into the fish's *pharynx,* or throat, which is separated from the gill cavities by a screen of gill rakers on each side of the pharynx. At the rear of the pharynx is the opening of the *esophagus,* a short, straight tube that leads to the stomach. In many carnivorous fish, the mouth, pharynx, esophagus, and stomach form a straight line, allowing large prey to be swallowed whole.

**Digestion, absorption, and excretion.** A fish's *stomach* is located low in the fish, near the center of the body cavity. Food is primarily digested in the stomach, aided by digestive juices and enzymes secreted by the stomach walls.

Digested food exits the rear of the stomach through a pyloric sphincter that leads to the fish's *intestine.* As food enters the intestine, secretions from the liver and pancreas neutralize stomach acids and finish digesting the food, which is quickly absorbed through the intestinal walls. Short extensions of the intestine called the *pyloric ceca* [sē′kə: sing. cecum], found in many fish, produce additional digestive juices and may also aid in absorption.

In most bony fish, the walls of the intestine are convoluted and covered with villi to provide a large area for absorbing nutrients. (In sharks, rays, and a few bony fish, the intestine is equipped with a spiral divider called the *spiral valve,* which serves the same purpose.) Solid wastes and indigestible materials are excreted from the fish through the *anus,* located on the fish's underside near the anal fin.

**Excretion of liquid wastes.** Water-soluble metabolic wastes are filtered from the blood by the *kidneys,* although some are also excreted through the fish's skin and gills. Because they spend their entire lives in water, fish face unique challenges in regulating their water balance. Freshwater fish are constantly gaining water in their bodies because their body fluids have a greater concentration of salts than the surrounding watery environment. Freshwater fish, therefore, ingest little water in proportion to the high volume of urine their kidneys excrete. On the other hand, saltwater fish are constantly losing water because there is greater salt concentration in seawater than in their body fluids. For this reason, saltwater fish ingest large quantities of water and excrete much salt (primarily through the gills) but relatively little urine. In both freshwater and saltwater fish, the principal organ for excreting metabolic wastes is the kidney.

## Nervous System

**Structure of the brain.** Although not as complex as the brains of many other vertebrates, the *brain* of a fish is still a marvelously designed organ consisting of billions of neurons. The two *olfactory lobes,* located at the front of the brain, are responsible for the fish's keen sense of smell. The fish's *cerebrum* consists of two small lobes connected by a *corpus callosum,* as in other vertebrates. However, a fish's cerebrum is smooth and lacks the wrinkles and convolutions that are found in the brains of mammals and humans. The cerebrum is also quite small in relation to total brain size. By contrast, the *optic lobes* of a fish, located just behind the cerebral lobes, are often larger than the cerebrum and cerebellum combined. The optic lobes are connected to the large eyes by *optic nerves.*

Although smaller than the optic lobes, the fish's *cerebellum* is still quite large in relation to total brain size and is often larger than the cerebrum. The cerebellum is responsible for coordinating the fish's voluntary muscles, allowing the fish to move swiftly and gracefully through the water. The rearmost portion of the brain is the *medulla oblongata;* as in other vertebrates, this portion of the brain controls heartbeat, digestive processes, and other activities affected by the autonomic nervous system.

The rear of the medulla oblongata tapers into the fish's *spinal cord,* which extends rearward through the fish's spinal cavity. As in other vertebrates, sensory nerves carry messages from the sense organs to the spinal cord and brain, and motor nerves relay messages from the brain and spinal cord to the muscles.

**Senses of a fish.** *Bony fish have well-designed senses and sense organs.* A typical bony fish has very good eyesight. Fish have no eyelids, thus permitting a maximum amount of light to enter the pupils. Unlike most vertebrates, which focus their eyes by changing the shape of the lens, a fish focuses its eyes by moving the lenses back and forth. The eyes themselves are not as movable as those of most other vertebrates, but the fish can easily turn its body to look about.

Most fish seem to hear fairly well, and it is thought that many fish can hear as well as or better than humans. Unlike human ears, the ears of a fish are completely internal, being concealed inside the skull slightly to the rear of the eyes.

Fish are also acutely sensitive to low-frequency vibrations below the threshold of human hearing. A fish senses faint vibrations and pressure changes in the water with its **lateral line,** a system of special nerve endings that extends over its head and the sides of its body. Protected by a row of pitted scales, these nerve endings give the fish extraordinary sensitivity to water currents, turbulence, changes in pressure, and low-frequency vibrations. The lateral line system allows fish to detect the approach of a predator or prey by vibrations in the water. This system also helps fish avoid unseen obstacles and navigate together when traveling in a *school* (large group).

Fish are gifted with an excellent sense of smell. Many predatory fish use their "noses" to search for food; others use their sense of smell to warn them of predators. At breeding time, several species of fish (most notably the six species of Pacific salmon) use their sense of smell to find their way back to the very stream in which they hatched. Somehow, these fish can recognize "their" stream out of hundreds or thousands of others by the odors of water plants or soil runoff.

Many fish have additional senses with which they navigate over long distances or locate prey. Some fish have special organs that can sense the tiny electric fields produced by the nervous systems of other animals. The *coelacanth* is one type of bony fish that is thought to hunt for food in this manner. Other fish can sense magnetic fields, allowing them to navigate over long distances using the earth's magnetic field as a guide.

## Reproductive System

The reproductive organs, or *gonads,* of a fish are located toward the rear of the fish's body cavity. The *testes* of the male produce sperm cells, while the *ovary* of the female produces large numbers of eggs. In both sexes, gametes (eggs or sperm) are released through the **urogenital** [yo͝or′ō·jĕn′ĭ·tl] **opening,** located slightly to the rear of the anus. Most bony fish reproduce through *external fertilization;* that is, the eggs are fertilized after they are released from the mother's body.

Beginning months before the breeding season, the thousands of egg cells in the female's ovary gradually develop into mature eggs, ready to be fertilized. Often, the egg mass is so large that it produces a noticeable swelling in the sides of the female.

When the fish are ready to **spawn,** or lay their eggs, they may congregate together, or they may migrate to the distant spawning grounds where they themselves hatched. As the eggs (known as **roe**) are laid in the water by the female, the male releases a fluid called **milt** into the water through his urogenital opening; the milt contains large numbers of sperm cells. These sperm cells swim through the water to the eggs, guided by chemical signals that the eggs produce. When a sperm cell reaches an egg, the egg is fertilized and begins to develop into a young fish.

Many kinds of fish make interesting preparations to protect their eggs. For example, when a female trout's eggs are mature, she makes a nest for them by scooping dirt and gravel on the bottom of a stream with her tail. To protect her eggs after they are fertilized, she covers the nest by swimming into the current and swishing her tail to stir up the gravel. The current carries the loosened gravel back over the nest, where it settles on the eggs.

Once an egg is fertilized, it begins to divide and form an *embryo.* Nourished by the *yolk mass* (a nutrient-filled sac attached to the embryo), the developing fish grows larger over a period of several weeks. When the fish is developed enough to move about, it hatches from the egg and begins to swim, although it is still nourished by the yolk mass. By the time the yolk mass is absorbed, the young fish (called a *fry*) is large enough to find its own food, feeding on insect larvae, plankton, or other tiny creatures until it is large enough to feed like an adult.

God in His wisdom provided that female fish would lay huge numbers of eggs so that the species might be preserved and so other animals might be fed. Although a female may lay hundreds, thousands, or even millions of eggs during spawning, only a small fraction ever hatch and produce fry. Many eggs fail to develop because they were not fertilized, and many become food for predatory fish or other creatures. Of the fish that do hatch, many fall prey to predators, so that only a handful of the young fish survive to reach adulthood.

## Special Organs

**Swim bladder.** One of the most interesting organs of a bony fish is the **swim bladder,** a hollow, gas-filled chamber located high in the fish's body cavity. A swim bladder allows a fish to adjust its buoyancy so that it can remain stationary at a fixed depth without having to "tread water." A special *gas gland* in the wall of the swim bladder extracts oxygen or other gases from the fish's blood and uses the gas to inflate the bladder. The fish removes gases from its swim bladder through a small side chamber lined with capillaries. Muscular valves between the swim bladder and the side chamber allow the fish to control how quickly the gas is removed. Some fish that dwell near the surface have a tube that connects the swim bladder with the esophagus, allowing them to simply gulp air at the surface or "burp" to adjust their buoyancy more quickly.

The marvelous design of the swim bladder becomes apparent when we consider deep-sea fish such as the *grenadier* (rattail), which may live at depths of 6000 feet or more. At this depth, the gas in the swim bladder must be pressurized to nearly 2700 pounds per square inch to withstand the crushing water pressure. Because the swim bladder takes time to pressurize and depressurize, the fish must change depth gradually; if a fish whose swim bladder is highly pressurized were to ascend too quickly, the bladder could expand or burst, seriously injuring the fish.

*Flounders* and other bony fish that live near the bottom have no swim bladder. *Lungfish* use their swim bladders as a sort of lung and thus can breathe in the air as well as in water. *Catfish* use their swim bladders to make various sounds—a simple form of communication. Because *coelacanths* feed at depths of 550–650 feet at night but live at much greater depths during the day, their swim bladders are specially designed for large depth changes; instead of being filled with gases, the swim bladder of a coelacanth is filled with a buoyant oil, eliminating the need to adjust its gas pressure.

**Light- and electricity-generating organs.** Many fish have special organs that produce light or generate electricity. Many deep-sea fish, for example, have special *light-producing organs* as a part of their skin or digestive tract, enabling them to pierce the gloom of the murky ocean depths. Other fish, notably the electric eel and electric catfish, have *electricity-generating organs.* The electric current is used to stun or kill an enemy or prey. The largest *electric eel,* a 10-foot freshwater giant native to South America, devotes about 75% of its body to its electricity-generating apparatus. This creature can discharge a flash of 600 volts at just under two amperes—an output of about one kilowatt and enough to shock or electrocute a man.

Fig. 19.10 **Flounder**

Fig. 19.11 **Electric eel**

## Spawning and Parental Care

Understanding the LIVING CREATION

### Pacific Salmon

Spawning behavior varies greatly among the bony fish. Salmon, for example, are famous for their spawning runs from ocean waters to freshwater streams. The most dramatic of these runs are those of the **Pacific salmon,** a group that includes the chinook salmon, sockeye salmon, and coho salmon. Unlike Atlantic salmon, which return to their streams every year to spawn, most species of Pacific salmon make one spawning run when they are several years old and die after they spawn.

The spawning run often begins in the spring of the salmon's fourth or fifth year. Leaving the deep ocean waters where it has spent most of its life, the salmon migrates back to the mouth of the river or stream in which it hatched. Millions of salmon each year migrate up the Columbia River in Washington, the largest river in the Pacific Northwest.

Once a salmon finds its river, it begins to swim upstream, guided by its sense of smell and by memories of its downstream journey several years before. As the river narrows and steepens, the fish fights its way upstream, battling strong currents and struggling up waterfalls without stopping to rest or eat. When the salmon reach the very place where they hatched years before, they spawn; female salmon lay their eggs in a simple nest scooped from the stream bed and cover them with gravel after they are fertilized by male salmon. The exhausted parents then drift downstream and die, leaving their eggs protected in the gravel

sockeye salmon spawning

nest. When the young salmon hatch a few weeks later, they follow the current downstream to the sea, to return several years later when it is their turn to spawn.

## Freshwater Eels

Another interesting migratory fish is the common **freshwater eel** of Europe and North America. Unlike the salmon, which lives in the sea but spawns in fresh water, the eel lives in fresh water but spawns in the ocean. Adult male eels typically live near the mouths of rivers and streams, while female eels live farther inland in the same rivers and streams. When it is time to spawn, both sexes of eels leave their streams for the Atlantic Ocean, navigating deep beneath the surface for hundreds or thousands of miles to their spawning grounds in the *Sargasso Sea* (a region of the Atlantic northeast of the West Indies). There the eels spawn, laying their eggs a thousand feet or more beneath the ocean's surface. After spawning, the parents die and sink to the abyssal plain miles below.

After a short gestation period, the eggs hatch, releasing millions of tiny, transparent eel larvae. Although less than an inch long, the larvae begin a slow migration back to the continents from which their parents came. Larvae traveling to North America complete the journey in 8 to 12 months, while larvae traveling to Europe require 3 years to make the trip. By the time they reach coastal waters, the larvae have developed into toothpick-sized juvenile eels called *elvers.* The male elvers take up residence near the mouths of rivers and streams, while the females journey farther inland. When the elvers mature after several years, they will again take to the sea to repeat the spawning journey of their parents.

stickleback

## Parental Care

Many fish show little concern for their offspring and may even eat them under certain conditions. Certain species, however, do care for their young in unusual ways. The male *stickleback,* a small, scaleless, saltwater and freshwater fish, builds a nest for the female's eggs and guards them carefully with an arsenal of two to eleven sharp spines in front of its dorsal fin. Female *bass,* a spiny-finned food and game fish of fresh or salt water, make nestlike depressions in the sand or gravel for their eggs; the male then stands guard while the fertilized eggs are incubating. Some fish, including certain *ocean catfish,* carry the eggs in their mouths during the hatching period; in some species, the male catfish carries the eggs, while in others the female carries them. Female *seahorses* lay their eggs in a pouch on the male's underside, and the eggs hatch inside the male's pouch. The *Siamese fighting fish,* a popular aquarium fish, has a unique way of caring for its offspring. The male blows bubbles that stick together to make a nest for the female's eggs. As the female lays her eggs, the male collects them in his mouth and blows them into the nest. Any eggs that become dislodged from the nest the male returns to it. He even guards the fry after they hatch until they can swim and hide from predators for themselves.

freshwater eels

## Section Review 19.2

1. Describe the structure of a fish's gills.
2. How does a fish ensure a continuous flow of water across its gills?
3. How does the heart of a fish differ from that of other vertebrates?
4. How do some fish use their gills to trap food?
5. What sensory organ do fish possess that land animals lack? What purpose does this organ serve?
6. What is the purpose of a fish's swim bladder? How is the gas pressure within the swim bladder adjusted?

### Critical Thinking

7. Compared to aquatic amphibians, fish extract very little oxygen from the water through their skin. What differences between fish and amphibians might account for this fact?

### Identify

gill chamber, lateral line, urogenital opening, spawn, roe, milt, swim bladder

## 19.3 The Cartilaginous Fish (Chondrichthyes and Cyclostomata)

Two classes of fish have skeletons made of cartilage rather than bone and thus are called **cartilaginous fish.** These two classes are the sharks, rays, and chimaeras (Chondrichthyes [kŏn′ drĭk′thĭ·ēz]) and the lampreys and hagfish (Cyclostomata [sī′klə·stō′mə·tə]). Although cartilaginous fish are often considered to be "primitive" by evolutionists, they have some features that are characteristic of the "higher" animals. Sharks, for example, like mammals, reproduce by internal fertilization, and some bear live young.

### Sharks, Rays, and Chimaeras

Sharks, rays, and chimaeras constitute only about 3% of all known species of fish. *All have jaws and a cartilaginous skeleton.* As you learned from your study of human anatomy and physiology, cartilage is the tough, sinewy substance that gives shape to your nose and ears and supports the weight of your backbone. Bone is stiff and inflexible; cartilage, however, is stiff enough for support but also highly flexible. The skin of sharks, rays, and chimaeras is covered with toothlike scales known as **denticles,** which resemble tiny, closely spaced teeth. These scales are designed with a flat base and a sharp, hooklike structure that projects outward from the base. Denticles make the skin of a shark or ray feel like rough sandpaper. They do not overlap each other like the scales of most bony fish but simply butt against each other to form a protective covering.

**Sharks.** Of all the creatures that roam the oceans, the **shark** has generally been the most feared by man. According to the International Shark Attack File, at least 35 species of sharks are known to have attacked humans. Man has been concerned about sharks for millennia. The Greek philosopher *Aristotle* (384–322 B.C.) made precise notes on the anatomy and behavior of the shark, and the Roman naturalist *Pliny the Elder* (A.D. 23–79) wrote about sponge divers having "terrible combats with the dogfish [shark]."

Most sharks have a highly streamlined body with fins like those of bony fish. *Sharks can be easily distinguished from most other fish by their asymmetrical caudal fins.* A shark is designed so that its large jaws can retract into the bottom of its head for better streamlining; therefore, it is difficult to appreciate the size of a shark's mouth unless the mouth is open. The highly streamlined body makes sharks swift and efficient swimmers. The broad, wing-shaped head of the *hammerhead shark* also serves as an underwater rudder, making it one of the most maneuverable sharks.

In most shark species, the mouth is set well back in the head, and the jaws are filled with up to 20 rows of jagged, razor-sharp teeth. Unlike other vertebrates, the rows of shark teeth are not anchored to the jaws by roots but are embedded in the gums. The teeth move steadily forward as they increase in size, falling out when they are worn and being replaced by other teeth moving up from behind. A single *tiger shark* may form, use, and lose as many as 24,000 teeth in a 10-year period.

Like some land carnivores, sharks can eat a very large meal and then go for long periods without eating. This is made possible by a widely gaping mouth and an expandable stomach that can stretch to several times its normal size. A shark can also store food or foreign objects within its stomach for weeks if necessary without digesting them. This remarkable feature suggests that Jonah may have been swallowed by a giant shark rather than a whale (see Jon. 1:17; 2:10). *Other features that make a shark different from bony fish include a skeleton of cartilage, the lack of a swim bladder, and a complex reproductive system.* Most sharks bear their young alive, and some are truly viviparous, nourishing their unborn young with a placenta. Without a swim bladder, sharks must keep moving or sink—a fact which means that most sharks never sleep. While a shark is moving, its enlarged pectoral fins and asymmetrical caudal fin provide the lift that keeps the shark from sinking, while its paired pelvic fins provide stability. A notable exception is the *sand tiger shark,* which uses its stomach as a swim bladder and retains enough air in it to keep itself buoyant for hours at a time. Although the *whale shark* lacks a swim bladder, its liver is very large and contains an oil that is lighter than water, helping to make the animal buoyant.

Sharks have an extremely fine sensory system, which they use to locate food. A shark's *lateral line,* running along its body from snout to tail, is extremely sensitive to vibrations and can detect vibrations from an animal in the water for great distances. Special organs located in the shark's head allow it to detect the electric fields produced by the nervous systems of other animals, so that sharks can even locate prey buried in the sand. *The shark also brings into play one of the keenest senses of smell in the animal kingdom.* Some sharks are said to be able to sniff an ounce of fish blood in a billion parts of water; under ideal conditions, these sharks may be able to detect prey by smell more than 2 miles away. The shark's eyesight becomes more important as the shark closes in on its prey; contrary to popular belief, sharks have excellent eyesight and are very sensitive to movement and contrast.

There is great diversity among the over 400 species of sharks, which range in size from 6 inches to 60 feet in length. The most feared of the maneaters is the *great white shark,* which may measure up to 21 feet long and weigh over 3 tons. Great white sharks may attack without warning in warm waters and cool, in the shallows and in the depths; more than a fifth of these attacks are fatal. A great white shark caught off the coast of Australia was discovered to have the remains of an entire horse in its belly. Great white sharks generally

Fig. 19.12 **Sharks**

prey on sea lions, seals, and similar creatures. Although other sharks such as the *hammerhead* and the *tiger shark* can also be dangerous to humans, attacks by any type of shark are (fortunately) relatively rare.

The two largest sharks—the *basking shark* of temperate waters and the *whale shark* of the tropics—are gentle eaters of plankton. The huge whale shark, pushing its enormous mouth through the water in search of plankton, filters about 400,000 gallons of seawater an hour. It also swallows all sorts of strange objects—from discarded crates and trunks to (possibly) the Hebrew prophet Jonah.

**Rays.** The **rays,** or *batoids* [băt′oidz], include around 600 species of true rays, electric rays, skates, and sawfish. These unusual fish have broad, flat bodies and generally a long, whiplike tail. Both eyes are located on the dorsal surface of these cartilaginous fish. Like the sharks, the gills of rays have no gill covers but connect to the outside with gill slits. Unlike the sharks and all other fish, rays breathe through openings just behind their eyes, known as *spiracles.* As in amphibian larvae, water containing dissolved oxygen enters the spiracles, passes over the gills, and then leaves through the gill slits. Some rays spend much of their time lying half buried in the sandy ocean bottom, while other rays spend their time cruising through surface waters feeding on plankton.

The most dangerous ray is the venomous *stingray,* a true ray that abounds in all the world's seas and oceans and is also found in fresh water. Stingrays vary in size from 5 inches to 7 feet across, but each is equipped with a sharp, barbed, poisonous stinger on its whiplike tail. The stingray is a bottom dweller and normally nonaggressive, but when its habitat is disturbed, it stiffens its venom-producing stinger and uses its muscular tail to drive the stinger into the intruder's body, an always painful and sometimes fatal encounter. Interestingly, *stingrays are responsible for more injuries to people than all other types of fish combined.*

The most fearsome-looking ray, the *giant manta* (also known as the *devilfish*), is nonaggressive and rarely attacks anything larger than a shrimp. The giant manta may attain a breadth of 25 feet and a weight of over 3000 pounds. It lives in surface waters of the sea, cruising about by gracefully flapping its "wings." Periodically, mantas make a series of spectacular leaps into the air, jumping as high as 15 feet out of the water and sometimes even performing a series of splashy cartwheels. Like sharks, and unlike most other fish, stingrays, giant mantas, and other *true rays reproduce by internal fertilization and bear live young.*

*Skates* are nonvenomous rays with long noses. They are similar to true rays, and only skilled ichthyologists can distinguish them. *Skates are the only rays that are oviparous.*

A rather strange cartilaginous fish is the *sawfish,* which, as its name implies, has a long, saw-toothed snout that it uses to cut and chop smaller fish to death for food. The sawfish's formidable snout can slash right through a school of fish and rapidly disable its prey. The jagged, serrated teeth pose no real threat to man unless one steps on a sawfish accidentally. If a live sawfish is caught in a fisherman's net or gets loose on the deck of a fishing boat, however, it can be a considerable hazard.

Fig. 19.13 **Rays**

**Chimaeras.** The **chimaeras** include nearly 30 species of unusual cartilaginous fish that live near the ocean bottom. They are medium-sized fish with huge eyes and a long, slender, pointed tail. They have only one gill opening on each side, and unlike other cartilaginous fish, chimaeras have an operculum, or gill cover. There are three families—short-nosed, long-nosed, and plow-nosed chimaeras. Short-nosed chimaeras, with their rounded snouts and large cutting teeth, are commonly known as *ratfish*. Their diet consists mainly of crustaceans and mollusks. The flexible snouts of the plow-nosed chimaeras, which turn down and backward toward the mouth, give these fish their common name of *elephant fish*. They feed mostly on other fish but will also devour mollusks and crustaceans.

Fig. 19.14
**Chimaera (elephant fish)**

## Lampreys and Hagfish

The 43 species of **lampreys** and the 70 species of **hagfish** make up less than 1% of all species of fish. Lampreys live in both salt water and fresh water, while hagfish are at home only in the ocean. Both lampreys and hagfish have slimy, scaleless bodies that resemble eels. Like the other cartilaginous fish, they have skeletons of cartilage; but *they lack jaws*. A lamprey's mouth is mainly a round sucking organ with a toothed tongue. About half of all species of lampreys are parasitic bloodsuckers that use their sucking organ to attach themselves to other fish, including game fish of the Great Lakes. They then cut into their victims with their toothed tongues and feed on their blood; a lamprey may even suck out a fish's vital organs. Hagfish have a slitlike mouth with sharp teeth but no sucking organ. They are scavengers that eat the insides of dead fish.

Fig. 19.15
**Lamprey and hagfish**

## Section Review 19.3

1. What is the primary difference between sharks and bony fish?
2. Why does a shark's skin feel rough? How are shark scales different from those of a bony fish?
3. How are a shark's teeth different from those of other vertebrates?
4. How are a shark's senses well suited for finding prey?
5. What two large sharks are plankton-eaters?
6. What group of unusual cartilaginous fish includes the ratfish and the elephant fish?

### Critical Thinking

7. Compare the fins of the barracuda in Fig. 19.3 (p. 391) with the fins of the blue shark in Fig. 19.12 (p. 401). What similarities and differences in fin structure do you see between the fins of the bony fish and the cartilaginous fish?

### Identify

cartilaginous fish, denticle, shark, ray, skate, chimaera, lamprey, hagfish

## Chapter 19 Review

### Define

1. ichthyologist
2. bony fish
3. fin
4. scale
5. myomere
6. gill, gill arch, gill filament, gill chamber, gill raker
7. lateral line
8. urogenital opening
9. spawning
10. roe, milt
11. cartilaginous fish
12. denticle

### Identify

1. the largest fish
2. three body regions of a bony fish
3. the primary difference between bony fish and other fish
4. the fins of a bony fish
5. five main types of cartilaginous fish
6. the type of fish that causes more injuries to people than all other types of fish combined
7. the only group of cartilaginous fish that have opercula
8. two jawless fish mentioned in the text

### Explain

1. What is an operculum? What purpose does it serve?
2. How does the shape of a fish's caudal fin relate to its lifestyle?
3. What is the purpose of the mucus that covers a fish's skin?
4. How can a fish's scales be used to determine its age?
5. Why are a fish's skeletal muscles arranged in myomeres?
6. What special sensory organ gives a fish great sensitivity to water currents, turbulence, pressure changes, and vibrations?
7. What is the purpose of a fish's swim bladder?
8. Why must most sharks constantly swim to keep from sinking?
9. What scaleless cartilaginous fish is noted as a parasitic bloodsucker?

### Apply

1. How does the heart of a fish differ from that of other vertebrates?
2. According to evolutionists, amphibians evolved from a bony fish similar to the modern lungfish or coelacanth. Describe several differences between fish and amphibians that would make the needed transitional forms unable to survive.
3. Besides skeletal differences, what are some differences between sharks and bony fish?

# CHAPTER 20 Arthropods: Invertebrates with Jointed Feet

## Key Concepts

- Animals without backbones
- Characteristics of arthropods
- Insects
- Metamorphosis
- Anatomy of a typical insect
- Insects and man
- Arachnids, centipedes, and millipedes
- Crustaceans

The largest and most diverse group of animals is the **invertebrates,** animals without backbones. If we could somehow see before us a representative of every species of animal that lives on the earth, 97% of them would be invertebrates. It is apparent that God has chosen the animals without backbones to display His creative powers through the magnificent variety among them. From the microscopic protozoa to the giant squid, all invertebrates are alike in that they possess no internal skeleton, yet they are all very different. The awesome diversity of the invertebrates displays God's delight in variety.

**Arthropods,** invertebrates with jointed appendages, make up more than three fourths of all species in the animal kingdom and include such familiar creatures as insects, spiders, crayfish, centipedes, and millipedes. Arthropods range in size from giant *king crabs,* with a leg span of 10 feet, to microscopic *mites* and tiny marine crustaceans. Arthropods are found in all climates and inhabit every conceivable environment—20,000 feet high in the Himalayas, at the surface of the world's oceans, deep within the frozen tundra of the Arctic, on the burning sands of the earth's deserts, and burrowed far underground. Although man uses some arthropods for food, many are parasites and others transmit such deadly diseases as African sleeping sickness, bubonic plague, malaria, Rocky Mountain spotted fever, typhus, and yellow fever. Arthropods compete with man for food, but they also form an essential link in most food chains and provide a virtually inexhaustible supply of food for such creatures as fish, amphibians, reptiles, birds, and some mammals.

The phylum Arthropoda [är·thrŏp′ə·də] includes such diverse creatures as butterflies, ants, shrimp, and horseshoe crabs. Yet these very different creatures have certain features in common. In place of an internal backbone, like the vertebrates, arthropods have a strong, lightweight and flexible **exoskeleton** (external skeleton). The exoskeleton of an arthropod is composed of **chitin** [kīt′n], a substance that is chemically similar to cellulose or starch; when reinforced by special proteins or by hard minerals, chitin can be as strong as bone. The exoskeleton is perhaps an arthropod's most striking feature, and it provides the animal with several advantages—protection, strength, support, lightness, and locomotion. *All arthropods have jointed legs,* as their name implies (Greek: *arthro,* "jointed"+ *pod,* "feet"). In addition, *all arthropods have a segmented body* divided into distinct body regions, a dorsal heart located above the nervous system, and a ventral nervous system with the main nerves located below the digestive system.

## Table 20.1 Major Groups of Arthropods (Kingdom Animalia; phylum Arthropoda)

| Group | Description | Number of living species (approximate) | Examples |
|---|---|---|---|
| **Insects** (class Insecta) | invertebrates with an exoskeleton, three distinct body regions, three pairs of jointed legs, one pair of antennae, compound and simple eyes, and tracheae for breathing | 1 million | honeybee, mosquito, beetle, ant |
| **Crustaceans** (subphylum Crustacea) | invertebrates with an exoskeleton, two distinct body regions, jointed legs, and two pairs of antennae; most have gills for breathing | 45,000 | lobster, shrimp, woodlouse |
| **Arachnids** (class Arachnida) | invertebrates with an exoskeleton, four pairs of jointed legs, simple eyes, and no antennae; most have two distinct body regions | 90,000 | garden spider, daddy longlegs |
| **Centipedes** (class Chilopoda) | invertebrates with an exoskeleton, a flattened body with many segments, one long pair of jointed antennae, one pair of legs per body segment, and one pair of poison claws | 3000 | house centipede |
| **Millipedes** (class Diplopoda) | invertebrates with an exoskeleton, a rounded body with many segments, one short pair of jointed antennae, and two pairs of legs per body segment | 10,000 | common millipede |

The huge phylum Arthropoda includes five major groups, each of which has the characteristics common to all arthropods and its own unique features. The *insects* form the largest class in the animal kingdom; there are four times as many insect species as the species of all other animals combined. The *arachnids* include spiders, scorpions, mites, and ticks. The *centipedes* and the *millipedes* consist of creatures with numerous body segments and legs. The *crustaceans* include crayfish as well as common seafoods such as lobsters, crabs, and shrimp.

## 20.1 Insect Anatomy and Physiology

**Insects,** the most varied and numerous of all God's living creatures, make up more than 70% of all animal species. **Entomologists** (scientists who study insects) have identified about 1 million species of insects and estimate that many more exist. There are over 360,000 species of beetles alone. Yet all of these tremendously varied and numerous creatures have certain common characteristics. All insects have an exoskeleton that is divided into three distinct body regions: the *head,* the *thorax,* and the *abdomen.* All insects have a single pair of *antennae,* or "feelers." As adults, all insects have both *compound eyes* and *simple eyes* and breathe through breathing tubes called *tracheae* [trā′kē·ē′]. All insects have three pairs of jointed *legs,* and most insects have one or two pairs of *wings.*

### Life Cycle of Insects

Most insects do not live for a long time. Generally, their lifespans are months, weeks, or even days; relatively few have lifespans that stretch over years. The lifespan of the common *cockroach* is about 1½ years, making it relatively long-lived among the insects, and some bees and ants can live more than 5 years. Although the *cicada* [sĭ·kā′də], or 17-year locust, lives approximately 17 years (13 years in the southern United States), it spends only a few months as an adult insect. For most of its life, the cicada remains underground in the larval stage.

Although a few species are viviparous or ovoviviparous, most insects begin life as an *egg.* In some wingless insects, such as silverfish and springtails, the young insect that hatches from the egg is a miniature replica of its parents. It simply grows larger each time it molts until it develops into a mature adult; very little change in form takes place. Most insects, however, undergo a series of changes in form as each hatches from an egg and develops into an adult. This wonderful process by which insects mature, called **metamorphosis,** may be either incomplete or complete.

**Incomplete metamorphosis.** In **incomplete metamorphosis,** the insect begins life as an egg that hatches into a **nymph,** an immature form of the insect that looks much like the adult but has different body proportions and lacks wings. About 11 of more than 25 orders undergo incomplete metamorphosis exclusively, including such common insects as dragonflies, mayflies, crickets, grasshoppers, locusts, roaches, and termites. During incomplete metamorphosis, the nymph molts its outgrown exoskeleton several times and secretes a new one. *Grasshoppers* molt five times during incomplete metamorphosis, and each time the nymph looks more and more like the adult grasshopper. After the final molt, the winged adult insect emerges. All during this time of growth, the nymph's major activity is satisfying its voracious appetite.

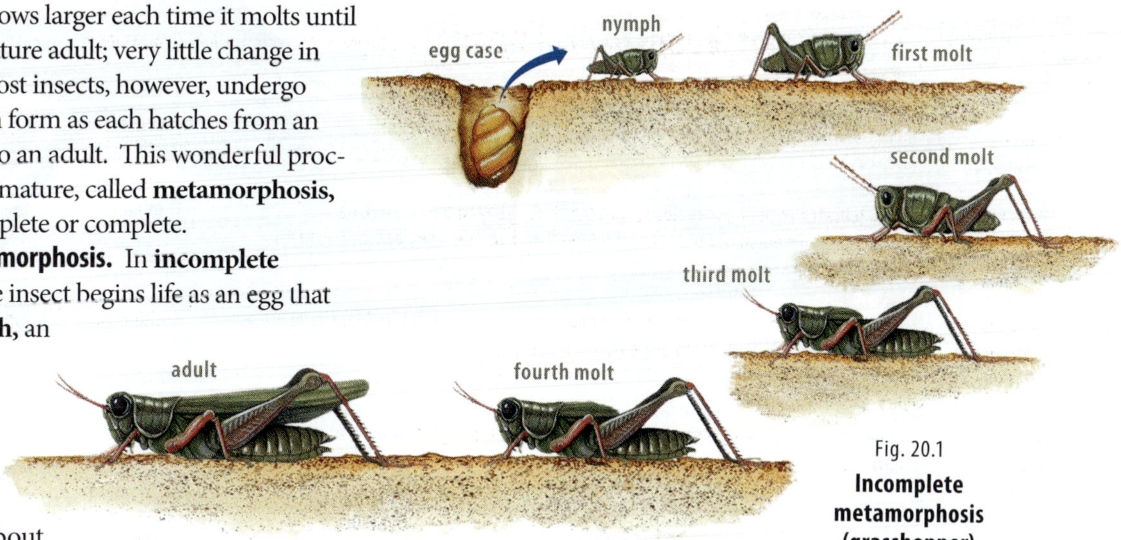


Fig. 20.1
**Incomplete metamorphosis (grasshopper)**

**Complete metamorphosis.** In **complete metamorphosis,** the egg develops into a **larva,** a wormlike eating and growing stage. The larvae of various insects are known by many common names—*grubs* (beetles), *maggots* (flies), *wigglers* (mosquitoes), *caterpillars* (butterflies and moths). Insect larvae have a nearly insatiable appetite, devouring everything in sight and often causing great damage to trees and crops. When the larva attains its full development, it forms either a *cocoon* or a *chrysalis* about itself and enters the resting stage of its life cycle as a **pupa.** During the pupal stage, the insect stops feeding and rests while its body experiences near-miraculous changes that transform it into the adult insect.

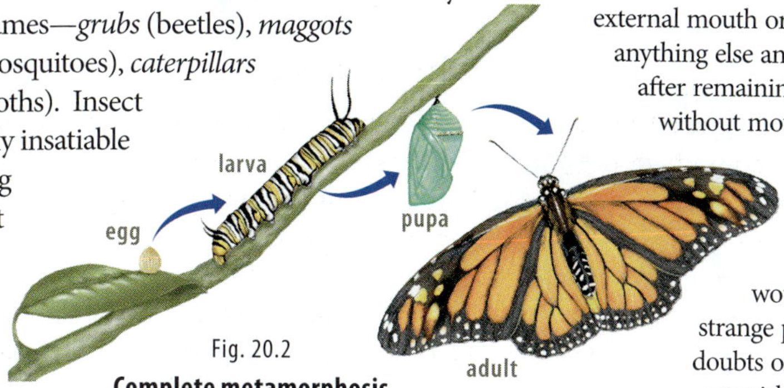


Fig. 20.2
**Complete metamorphosis (monarch butterfly)**

Two of the most familiar insects that undergo complete metamorphosis are the beautiful *monarch butterfly* and the gorgeous *cecropia* [sĭ·krō′pē·ə] *moth.* The monarch larva forms a hard chrysalis about itself, and the cecropia larva spins a silken cocoon; both insects rest as they transform from ungainly caterpillars into magnificent creatures with fluttering wings. William Kirby and William Spence, two pioneer 19th-century entomologists, described this event in these glowing words:

> Were a naturalist to announce to the world the discovery of an animal which for the first . . . of its life existed in the form of a serpent; which then penetrated into the earth, and weaving a shroud of pure silk of the finest texture, contracted itself within this covering into a body without external mouth or limbs, and resembling more than anything else an Egyptian mummy; and which, lastly after remaining in this state without food and without motion . . . should at the end of that period burst its silken cerements, struggle through its earthly covering and start into day a winged bird,—what think you would be the sensation excited by this strange piece of intelligence? After the first doubts of its truth were dispelled, what astonishment would succeed!*

## The Grasshopper: A Typical Insect

The grasshopper is one of the most common and most familiar insects and serves as a typical example of this

**An Introduction to Entomology,* 3rd ed. (London: Longman, Hurst, Rees, Orne, and Brown, 1818), 1:60.

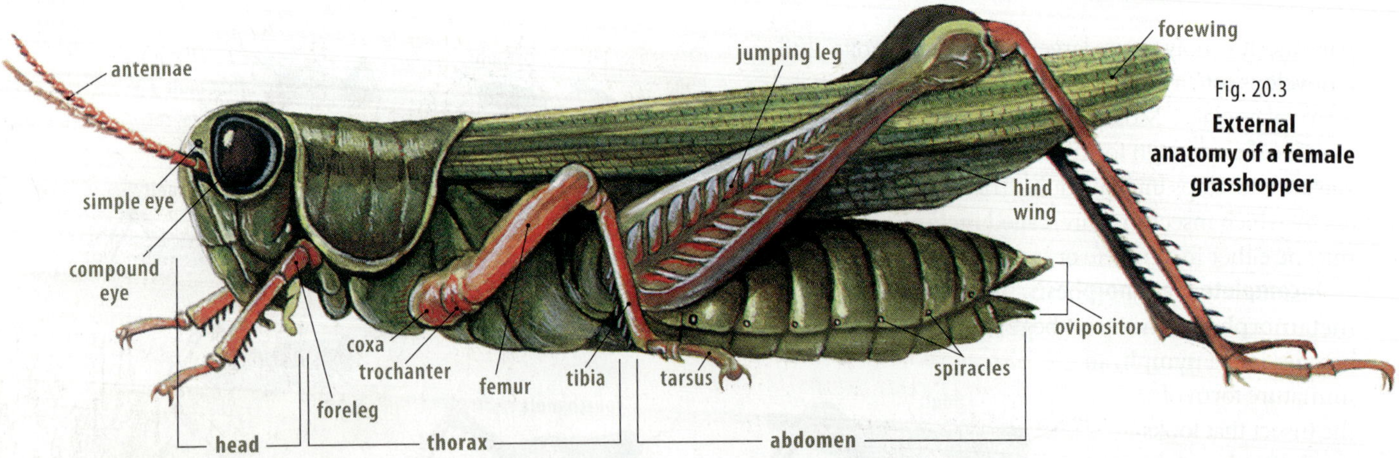


Fig. 20.3
**External anatomy of a female grasshopper**

class of arthropods. Like all arthropods, the grasshopper has jointed appendages and an exoskeleton composed of chitin. Like all insects, *the grasshopper's exoskeleton is divided into three distinct body regions*—the head, the thorax, and the abdomen—and attached to its body are three pairs of legs. Like many other insects, the grasshopper has two pairs of wings and undergoes incomplete metamorphosis in its growth and development.

**The head.** The front segment of the grasshopper's body is the **head.** Attached to the grasshopper's head is a pair of **antennae,** or feelers, which give the grasshopper an excellent sense of smell, touch, and taste. Each antenna is extremely flexible because of its jointed design. The grasshopper has three **simple eyes,** two by the antennae and one between, which can detect only light and shadow. Located on either side of its head are two larger **compound eyes** that are sensitive to shape, color, and movement. A compound eye is an eye that is composed of many independent lenses and "retinas," giving the eye a faceted appearance under a microscope. Images transmitted by the compound eyes to the *brain,* a small mass of nerve tissue located in the insect's head, are assembled into a single image by the brain's optic lobes. Although a grasshopper's eyes are immovable and its head is capable of only limited movement, its marvelously designed compound eyes give the creature a wide angle of vision, allowing it to see much of its surroundings at once.

A grasshopper's mouth is specially designed for its diet of plants and plant parts. Grasshoppers (especially locusts, which are types of grasshoppers) have been infamous throughout man's history for the great damage they do to crops; you can read the descriptions of invasions by locusts in Exodus 10:1–20 and Joel 1:1–12. The components of an insect's mouth are **mouthparts.** The specially designed chewing mouthparts of a grasshopper's mouth are the *labrum* [lā′brəm] and *labium* [lā′bē·əm], which form the upper and lower "lips," and the *mandibles* and *maxillae,* which form the jaws. The labrum and labium are flattened structures that guide food into the insect's jaws and help to hold it there. The main serrated jaws,

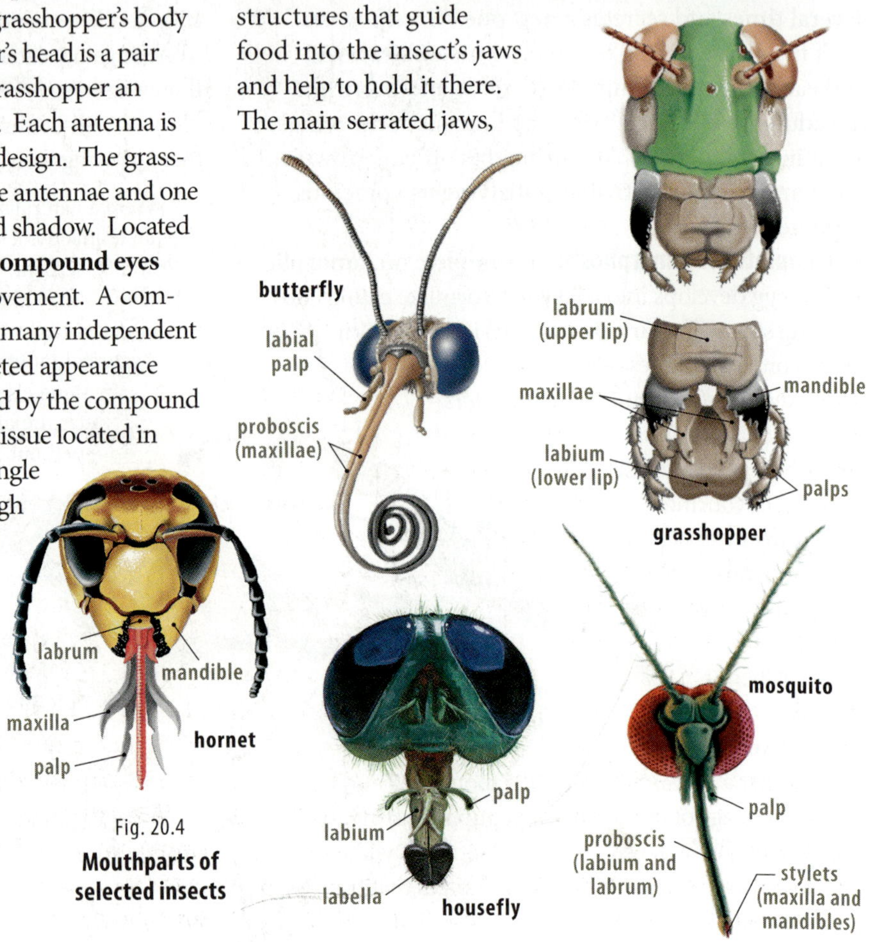


Fig. 20.4
**Mouthparts of selected insects**

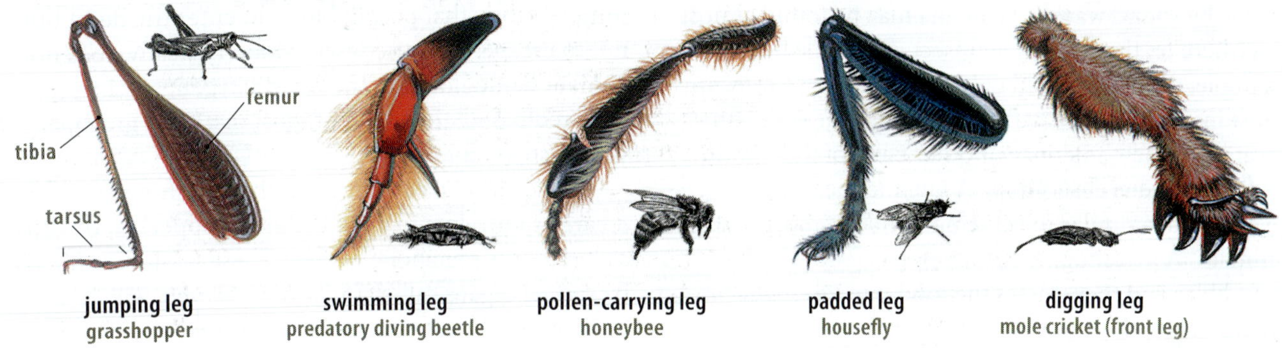


Fig. 20.5 **Legs and feet of selected insects**

or mandibles, are horizontally opposed (move side to side) and function to cut and grind the grasshopper's meal. Maxillae are secondary jaws that work with the mandibles to hold and chew the food. On the labium and on each maxilla is a *palp,* a fingerlike organ that may also taste plants to see if they are suitable for consumption. In other insects, such as flies and mosquitoes, these four mouthparts form sponges or sucking tubes, enabling the insects to drink plant and animal juices.

**The thorax.** The middle part of the grasshopper's body, the **thorax,** is divided into three segments: the *prothorax*, the *mesothorax*, and the *metathorax*. The first section, the prothorax, lies just behind the head; the grasshopper's front pair of legs are attached to this segment. The middle pair of legs and the front pair of wings are attached to the mesothorax, and the rear pair of legs and rear wings are connected to the metathorax.

Because of its attached legs and wings, the thorax is the grasshopper's locomotion center. Within the exoskeleton are sets of powerful muscles that move these appendages. When the wings are not in use, they are folded along the insect's back, the tough forewings serving as protective covers for the membranous hind wings. To fly, the grasshopper extends both sets of wings and flaps its hind wings to provide lift and propulsion. The forewings are held relatively stationary and help provide stability and control in flight.

Stimuli received by the grasshopper's sense organs are relayed by nerves and nerve centers (ganglia) to the proper muscles for a coordinated response. Even within an invertebrate as small and insignificant as a grasshopper, the muscular and nervous systems are of incredible complexity and design. In fact, grasshoppers have more different muscles than do humans. Although man has more than 600 distinct muscles, a grasshopper has about 900.

Each of the grasshopper's six legs has five jointed sections. The first segment, which attaches the leg to the body, is the *coxa*. The *trochanter* is the second segment, and these first two segments function like a ball-and-socket joint to give the grasshopper a wide range of movement. Attached to the trochanter is the long, muscular *femur,* which in turn is connected to the *tibia*. Finally comes the *tarsus*, which serves as the insect's foot and is equipped with small extensions that provide the grasshopper with additional traction. The grasshopper uses its first two pairs of legs for crawling, clinging, and climbing, but its rearmost set of legs are specially designed for jumping. The other insects also have legs that help them fill the niche for which they are designed.

**The abdomen.** The rearmost portion of the grasshopper's body, the **abdomen,** contains many of the insect's vital organs, including the heart, much of the digestive system, and most of the respiratory and excretory systems. The abdomen is divided into 11 segments. Eight of these segments contain minute respiratory openings called *spiracles;* each segment has a pair of spiracles, and there are two additional pairs in the thorax. The spiracles are surrounded by delicate hairs that keep out foreign particles but allow for an efficient exchange of oxygen in the insect's air tubes, or *tracheae*. Within the grasshopper's abdomen, these tracheae form larger air sacs and then divide into smaller and smaller tubes so that all cells of the insect's body are richly supplied with oxygen. The abdomen is flexible, allowing it to expand and contract as the insect breathes. As air is forced through the grasshopper's tracheae, oxygen is carried to the cells and carbon dioxide is removed.

A single eardrum, called a *tympanum* [tĭm′pə·nəm], is located on each side of the first abdominal segment, next to the thorax. The tympanum, a thin, oval-shaped membrane, vibrates much like a drumhead when struck by sound waves. Tympana may be found almost anywhere on the body of an insect except the head; for example, they are located on the front legs of a cricket. Most insects do not have tympana but hear with hairs located on their antennae or other parts of their body.

**Digestion and absorption.** A grasshopper's digestive system is similar in function to that of most other animals. A grasshopper first chews its food with its mandibles and then passes the food into its *esophagus,* where the food is moistened by salivary juices. The softened food is then swallowed and travels to the *crop,* a temporary storage area, before being passed along into the *gizzard.* Inside the gizzard, the food is further broken down by grinding before passing to the *stomach.* The stomach continues the process of digestion with the help of special enzymes produced by glands called *gastric ceca.* Unlike the stomachs of most vertebrates, the stomach of an insect serves as the primary organ for absorbing nutrients. The intestine of a grasshopper is composed solely of the *colon* and *rectum,* which store undigested material until it can be eliminated through the *anus.*

**Circulation and excretion.** A grasshopper's tubelike *heart* is located in its upper abdomen, lying just under the exoskeleton. As the heart contracts, the blood travels forward into the *aorta,* which extends into the insect's head. Insects do not have a closed system of blood vessels like mammals and humans. Instead, the insect's blood exits the aorta at the front of the insect's body and flows rearward through the open body cavity and around the vital organs, supplying them with nutrients. The blood reenters the heart through tiny openings, called *ostia,* along its sides. These openings have minute valves that permit blood to enter the heart but not to escape. Because insect blood typically does not contain hemoglobin to transport oxygen, few insects have red blood; the blood of most insects is greenish, yellowish, or colorless.

As the blood passes through the grasshopper's body, it carries nutrients to all of the grasshopper's body cells and removes metabolic wastes. These wastes are filtered by special organs called *Malpighian tubules,* which serve as "kidneys." Wastes accumulated by the tubules are emptied into the colon to be eliminated.

**Reproduction and life cycle.** The reproductive organs of male and female insects are located in their abdomens. Female grasshoppers have two *ovaries* in which *eggs* develop. An *oviduct* carries the eggs away from each ovary, and the two oviducts unite to form a single tube that opens near the tip of the female's abdomen. Male grasshoppers have a pair of *testes* that produce *sperm.* As in the female, special tubes carry the sperm from the testes and unite to form a single tube that extends to the outside of the abdomen. When the insects mate, the female stores the male's sperm in her *seminal receptacle,* a specialized container where the sperm remain until the female is ready to lay her eggs. A female grasshopper prepares to lay her eggs by digging a hole in the soil

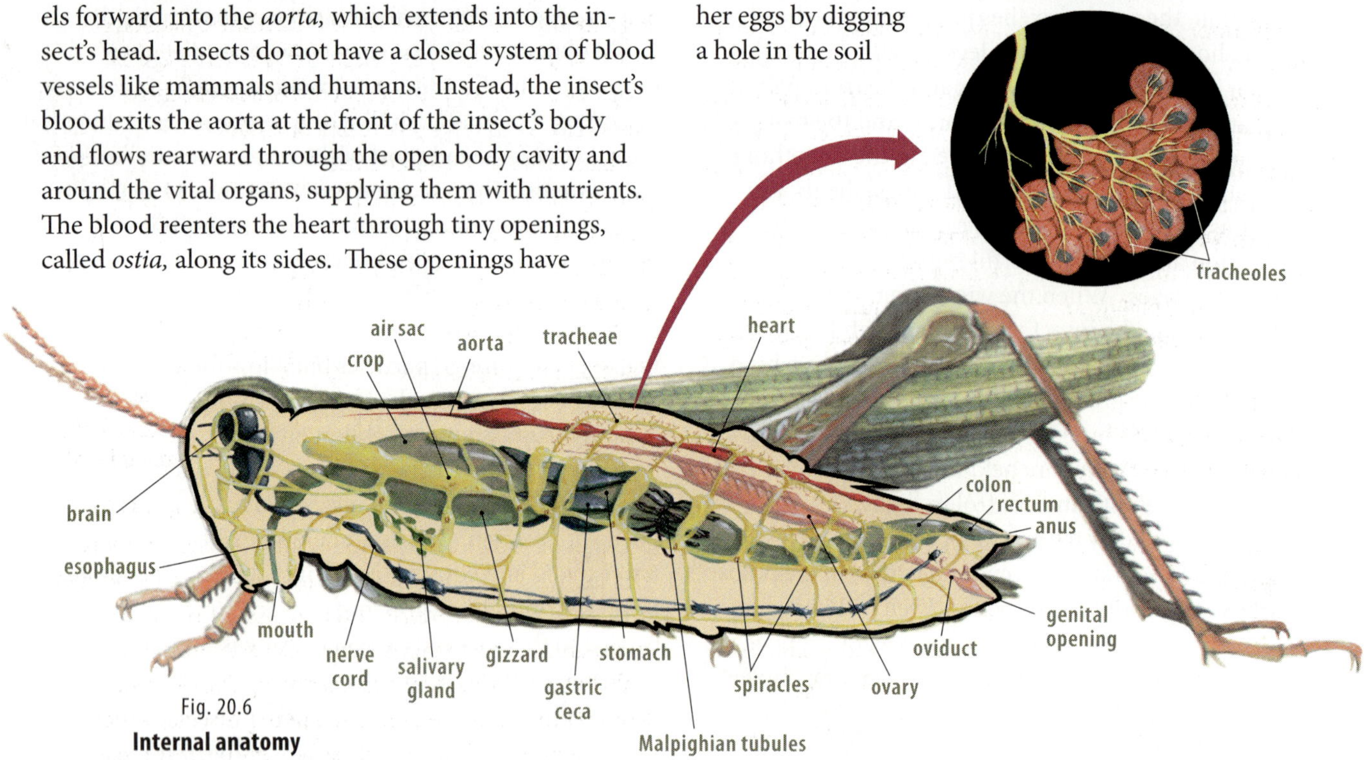


Fig. 20.6
**Internal anatomy of a female grasshopper**

with a hard, sharp-pointed organ called an *ovipositor* at the end of her abdomen. She lays her eggs in the hole, the sperm from the seminal receptacle fertilizing the eggs as they leave her body. Every fall each female grasshopper lays several dozen eggs in the ground. The grasshopper nymphs hatch from the eggs in the spring. Because grasshoppers undergo incomplete metamorphosis, the immature nymphs look like miniatures of their parents. Through the spring and summer, the nymphs grow and develop into mature adult grasshoppers that are ready for reproduction, and a new generation of grasshoppers will come into being.

## Section Review 20.1

1. What characteristics do all arthropods have in common?
2. List the five major groups of arthropods.
3. What characteristics do all insects have in common?
4. What is the difference between complete and incomplete metamorphosis?
5. Which body segment of an insect contains most of the sensory organs?
6. To which body segment of an insect are the insect's legs and wings attached?
7. Which body segment of an insect contains the heart and most of the digestive and excretory systems?

### Identify

invertebrate, arthropod, exoskeleton, chitin, entomologist, nymph, compound eye, mouthparts

### Observe

- Look closely at the insects that live in your area. See if you can identify the three regions of the insects' bodies. Can you find both the compound eyes and the simple eyes?

# 20.2 Orders of Insects and Their Economic Significance

## Important Orders of Insects

Of more than 25 orders of insects, 8 orders include the common insects with which you are probably most familiar. Most of these orders have names that end with the suffix *-ptera,* a Greek word meaning "wing." This reflects the fact that entomologists generally classify insects according to the design and structure of their wings.

**Orthoptera: straight-winged insects.** Crickets, grasshoppers, locusts, mantises, and roaches all belong to the same order, **Orthoptera** [ôr·thŏp′tər·ə: "straight wing"]. These insects get their name from the way they hold their folded wings straight back along the body when not flying.

Fig. 20.7
**Insects of the order Orthoptera**

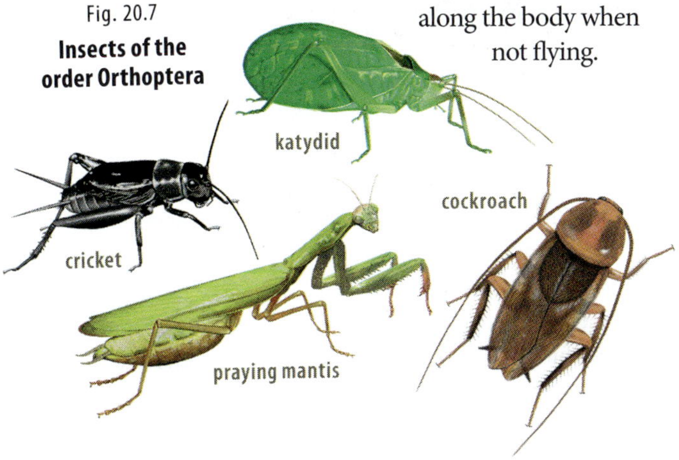


Most orthopterans are plant-eating insects with chewing mouthparts. Many of these insects are considered pests; *locusts* and similar insects do much damage to crops, and some species of *cockroach* are household pests. However, many other orthopterans are harmless or even helpful; the *praying mantis* is extremely beneficial to man because it kills and eats other insects that damage plants.

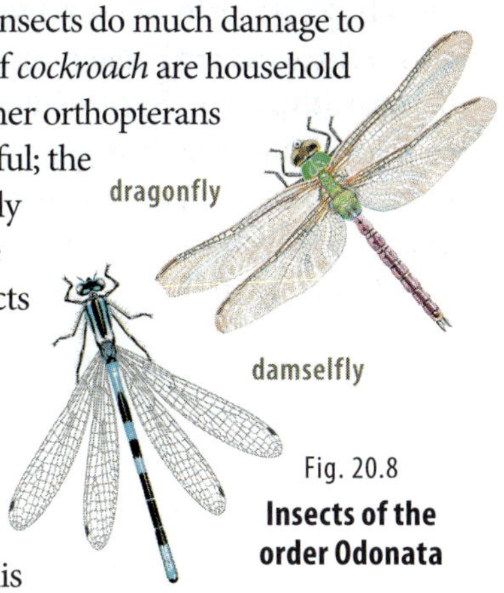


Fig. 20.8
**Insects of the order Odonata**

**Odonata: toothed insects.** *Dragonflies* and *damselflies* are typical members of the order **Odonata** [ō′dō·nā′tə]. This name, meaning "toothed," refers to the sharp, toothlike projections on the chewing mouthparts of these insects. Dragonflies are recognizable by their long, slender bodies, huge eyes, small antennae, and two pairs of equal-sized, transparent, membranous wings. Dragonflies cannot walk but are strong fliers reportedly capable of speeds up to 60 mi/hr. They feed on small flying insects, which they catch in midair. Because they eat harmful insects,

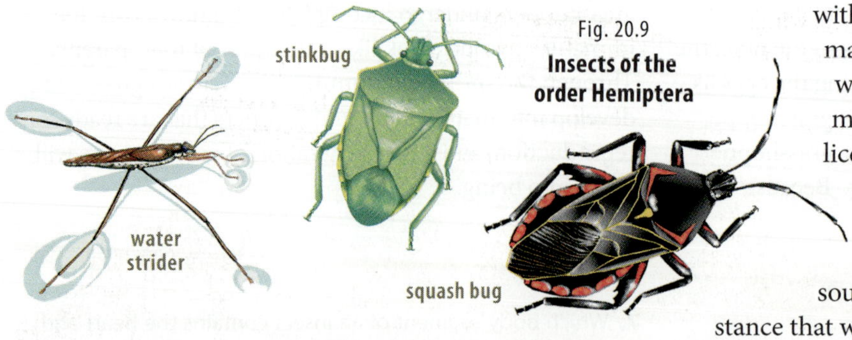


Fig. 20.9
**Insects of the order Hemiptera**

these predators are quite helpful to man. Fossil remains of giant dragonflies with wingspreads of nearly 3 feet have been found, indicating that huge insects apparently inhabited the earth before the Flood.

**Hemiptera: half-winged insects.** The order **Hemiptera** [hĭ·mĭp′tər·ə], whose name means "half wings," is made up of the *true bugs* and includes such common pests as the bedbug, chinch bug, squash bug, and stinkbug. The order derives its name from the "half wing" appearance of the membranous wings of some of its members; but not all bugs have wings. Although many people commonly use the term *bug* to refer to any sort of insect, to an entomologist a **bug** is an insect with a half-wing design and piercing-sucking mouthparts that it uses to suck sap from plants and body fluids from animals.

**Homoptera: same-winged insects.** **Homoptera** [hō·mŏp′tər·ə] means "same wings," and the members of this order that have wings hold them over their bodies in an inverted V, like the roof of a house. You are probably most familiar with the *cicada,* which lives underground for a number of years, depending upon the species, and then tunnels to the surface to spend a week or two as an adult. On hot summer evenings, the treetops ring with the high-pitched, strident notes of the male cicada. Not all homopterans have wings, but all possess piercing-sucking mouthparts. Such homopterans as plant lice, scale insects, mealybugs, and leafhoppers do a great deal of damage to both wild and cultivated plant life.

One homopteran, the *lac insect* of southern Asia, secretes a resinous substance that when processed becomes *shellac,* used all over the world as a base for lacquer and wood finishes.

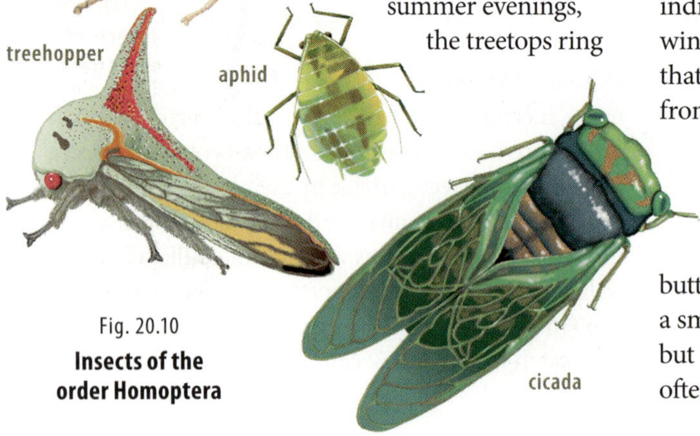


Fig. 20.10
**Insects of the order Homoptera**

**Lepidoptera: scale-winged insects.** The order name **Lepidoptera** [lĕp′ĭ·dŏp′tər·ə] means "scale wings"; *butterflies* and *moths,* the members of this order, have two pairs of large, membrane-like wings that are covered with tiny scales.

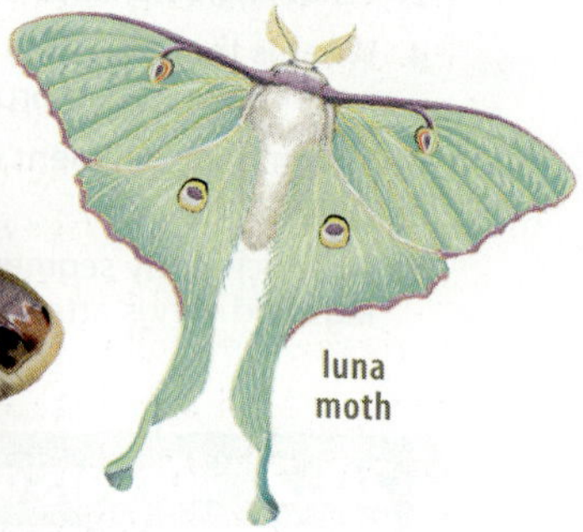

These microscopic scales form a beautiful mosaic pattern and give the wings of such lepidopterans as monarch butterflies and cecropia moths their brilliant colors. Larger butterflies may have as many as one million individual scales on their wings. Butterflies and moths have sucking mouthparts that are shaped like a coiled tube when not in use. Their front wings are usually larger than their hind wings, and they have long antennae and big eyes.

Fig. 20.11
**Insects of the order Lepidoptera**

People often confuse butterflies and moths, but *there are several distinct differences between the two types of insects.* A moth has a heavier body than a butterfly. A moth appears "hairy," whereas a butterfly has a smooth outer covering. Moths have feathery antennae, but the antennae of butterflies are smooth, rodlike, and often knobbed at their ends. A moth holds its wings horizontally when at rest; a butterfly holds its wings vertically. A moth pupa rests in a soft cocoon; a butterfly pupa is always found in a hard chrysalis. Butterflies are active during the day, but moths fly about at night. Finally, butterflies are extremely beneficial to man as pollinators of flowers. Moths, except for nonfeeding adults, are usually harmful. Moth larvae devour food and clothing; the "worm" you might find in an apple is really the larva of the *codling moth.*

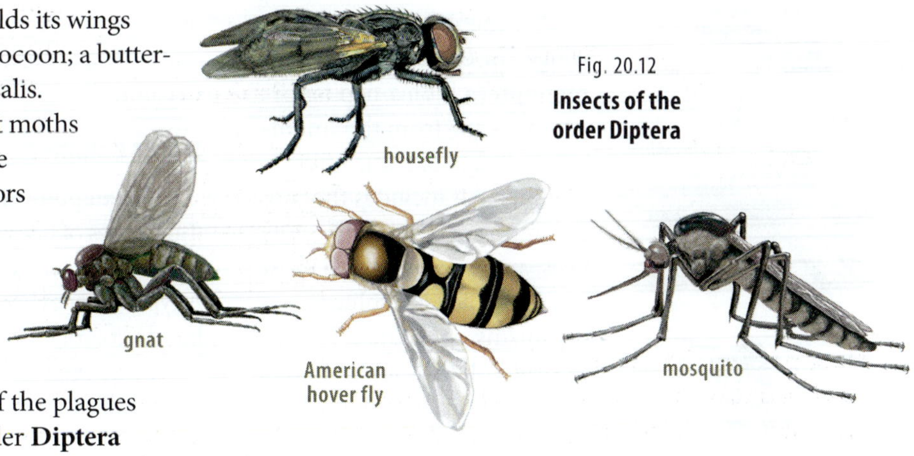


Fig. 20.12
**Insects of the order Diptera**

**Diptera: two-winged insects.** Two of the plagues on Egypt involved members of the order **Diptera** [dĭp′tər·ə], insects with two wings and piercing-sucking or sponging mouthparts. First came the gnats (Exo. 8:16–19; the Hebrew word translated "lice" literally means "gnats") and then the plague of flies (Exo. 8:20–32). The dipterans (flies, gnats, mosquitoes) include some of Earth's most pestilent insects. *Flies* breed in and crawl over all manner of refuse and then carry disease germs, including the microbes responsible for dysentery, sleeping sickness, and typhoid fever, wherever they go. Projections on the legs and sticky hairs on the feet of a housefly make this insect a carrier of refuse. Female flies deposit nearly 200 eggs each time they lay eggs. The eggs soon hatch into *maggots,* the larvae of flies. If all of their descendants survived, a single pair of flies could produce over 2 trillion offspring within three months!

A common *mosquito* in Europe and North America often carries viruses that cause encephalitis; mosquitoes in tropical and subtropical regions transmit the diseases yellow fever (caused by a virus) and malaria (caused by a protozoan). Mosquitoes transmit these deadly microorganisms in their saliva. A mosquito's mouth resembles a funnel; the broadest part is nearest the head, and a tubular proboscis extends downward. At the end of the proboscis are six sharp, lancelike stylets, which the mosquito uses to puncture the skin of an animal or a person. It is the female mosquito that bites; she is in search of animal or human blood, which she needs for the proper development of the eggs within her body. As the stylets pierce the skin, the mosquito's saliva flows into the wound to keep the blood from clotting while the mosquito sips the blood through her proboscis. The saliva contains not only an anticoagulant but also any viruses or protozoa that the mosquito has picked up from the blood of previous victims. Because the saliva causes an allergic reaction, an itchy welt appears on the skin of the victim.

**Coleoptera: sheath-winged insects.** There are over 360,000 species of *beetles.* This vast array of insects constitutes the order **Coleoptera** [kō′lē·ŏp′tər·ə: "sheath wing"]. These insects are easily distinguished by the way the forewings form a protective shell around the body when the beetle is not in flight. Beetles are also characterized by their powerful mandibles and chewing mouthparts. Common beetles include the burying beetle, Colorado potato beetle, firefly, June beetle, six-spotted tiger beetle, stag beetle, striped cucumber beetle, and ladybug.

Fig. 20.13
**Insects of the order Coleoptera**

Fig. 20.14

**Insects of the order Hymenoptera**

**Hymenoptera: membrane-winged insects.** The order **Hymenoptera** [hī′mə·nŏp′tər·ə] gets its name from the "membrane wings" of some of its members. It includes the most common **social insects**—ants, bees, and wasps. Unlike most insects, which are solitary organisms, social insects live in large colonies, and different groups within the colony have specialized tasks. For example, some colony members gather food, others defend their home, and others reproduce. These patterns of insect behavior are quite complex and show the glory and wisdom of the Creator.

## Insect Colonies

**Hymenopterans with wings.** A **honeybee** colony has three types of bees: the *queen* (an egg-laying female), the *drones* (males), and the *workers* (non-egg-laying females). A typical honeybee colony supports one queen, several hundred drones, and perhaps 50,000 worker bees.

The **queen** bee does not rule the colony, but she is responsible for reproduction in the colony. A queen bee mates shortly after she emerges from her pupal stage, making several mating flights over a period of days. The sperm she collects from the drones during these flights supplies the queen for a lifetime. A few days later, she begins her lifetime work of laying eggs for the colony. Although she may lay thousands of eggs in a day, the queen takes the time to put each egg in a separate cell constructed by the worker bees. She may lay unfertilized eggs as well as fertilized eggs. Unfertilized eggs develop into drones; fertilized

### Table 20.2 Common Insect Orders

*(Kingdom Animalia; phylum Arthropoda; class Insecta)*

| Order | Wing characteristics | | Adult mouthparts | Economic significance | Metamorphosis | Examples |
|---|---|---|---|---|---|---|
| | *first wing* | *second wing* | | | | |
| **Orthoptera** "straight wings" | straight, papery | thin, folded, fanlike | chewing | damage crops; pests (except mantises) | incomplete | grasshoppers, crickets, roaches |
| **Odonata** "toothed" | straight, thin | straight, thin | chewing | destroy harmful insects | incomplete | dragonflies, damselflies |
| **Hemiptera** "half wings" (true bugs) | leathery, thin tip | membranous | piercing-sucking | damage plants; pests; carry disease | incomplete | squash bugs, bedbugs, water striders |
| **Homoptera** "same wings" | leathery or thin | membranous | piercing-sucking | damage crops and gardens; source of lac | incomplete | aphids, cicadas, scale insects, lac insects |
| **Lepidoptera** "scale wings" | scaly, large | scaly, smaller | sucking | pollinate flowers; produce silk; damage crops; damage clothing | complete | butterflies, moths |
| **Diptera** "two wings" | straight, thin | none | piercing-sucking or sponging | carry disease; pests | complete | houseflies, mosquitoes, gnats |
| **Coleoptera** "sheath wings" | heavy, shieldlike | membranous, folded | chewing | destroy crops; pests; prey on other insects | complete | beetles |
| **Hymenoptera** "membrane wings" | thin, membranous | thin, much smaller | chewing or chewing-lapping | pollinate flowers; make honey; pests; prey on other pests | complete | bees, ants, wasps |

eggs develop into queens or workers. When a queen bee grows weak and begins to fail in her service to the colony, the workers prepare a new queen to replace her.

**Workers** "manage" the colony. Young workers are sometimes called *house bees* because their major occupation is feeding and caring for the colony's larvae and cleaning the cells. Workers produce a jellylike, protein-rich substance known as *royal jelly* in their mouths; they feed royal jelly to newly hatched larvae. Most larvae are also fed honey and a substance made from pollen known as *beebread.* If the queen bee dies unexpectedly, the workers can cause a worker larva to develop into a queen by continuing to feed it royal jelly and thus alter its development. When the larvae mature into pupae, older house bees produce wax and use it to seal the pupae within the cells of the honeycomb. Pupae become adult honeybees in about 16 to 24 days. Queen bees emerge from their cells in about 16 days; the workers emerge in about 21 days; and the drones emerge from the pupal stage in about 24 days. After two or three weeks of caring for the larvae, cleaning cells, and repairing combs or building new ones, the workers become *foragers,* or *field bees,* and leave the hive in search of nectar and pollen for the colony. Some of the new foragers become *guards* to protect the colony against intruders. If there is a shortage of young workers to feed and care for the brood of larvae, workers remain in the hive longer; some of the foragers may even revert to household duties. How marvelous is God's provision of instinct for bees and other social insects!

Fig. 20.16
**Queen bee surrounded by workers in hive**

*Stingers* are "standard equipment" on queen and worker bees, but drones do not have stingers. Compared to the stinger of the workers, queens have small stingers. When the queen uses her stinger, she does not lose it, and she can use it again. A queen generally uses her stinger to kill rival queens. When a worker uses her stinger on a fleshy animal or on a human, the stinger is ripped from the worker's body along with certain internal organs, and the worker bee dies. (However, worker bees can sting other insects and not lose their stingers.)

Fig. 20.15
**Types of honeybees**

**Drones** are larger than worker bees and have large eyes, strong wings, and broad bodies. The drones are almost helpless and must be fed by the workers. It appears that their only function is to mate with the queen. The drones that successfully mate with the queen soon die, and the others are allowed to return to the colony. In the fall, all the drones are either stung to death or starved to death by workers (which also die in the fall). New drones develop from eggs that are laid the following spring. Worker bees ordinarily live for several months if they hatch in the fall or winter, but they live for only three or four weeks during the working season. Queens live longer than workers or drones; they have an average lifespan of 1½ to 2 years, although some live as long as 6 years.

**Hymenopterans without wings.** **Ants** live in large colonies, with each colony consisting of workers, males, and queens. Large workers, sometimes called *soldier ants,* are equipped with large mandibles, which they may use in fighting for the colony. In some species, soldiers use their jaws to crush seeds for food. Unlike a bee colony, which has only one queen, *an ant colony may have several queens.* Ant queens do not fight each other; they all go about their business of laying eggs. Legless larvae, which hatch from the eggs, mature and form cocoons a few days after hatching. Because ant eggs are extremely small and can hardly be seen by the unaided human eye, the ant pupae in their easily visible white cocoons are often confused with the eggs. When conditions of temperature and moisture become unfavorable for the development of the pupae, the cocoons are transferred to more suitable quarters within the colony. If the ant hill is disturbed, the workers can be seen with cocoons in their mandibles, carrying the pupae to safety.

As in bee colonies, fertilized eggs develop into females and unfertilized eggs develop into males. Male ants are fed throughout their lives by the workers. Worker ants are always wingless, but the males of some species are winged (as are the queens) during mating season. Ants mate only once in their lifetime. After one mating flight, a queen can lay eggs for the remainder of her life. After mating, the males soon die. Because the queens no longer need to fly, they chew or break off their wings.

## Insects and Man

**The plan of God.** *God created the insect world* (as part of the "creeping things" in Gen. 1:24–25) *to benefit mankind.* Such luscious fruits of the earth as apples, blackberries, grapes, oranges, pears, plums, and strawberries depend upon pollinating insects (bees, butterflies, wasps, and others) for the production of seeds. Insect pollinators make it possible for such field crops and vegetables as alfalfa, cabbages, carrots, clover, cotton, onions, and peas to reproduce. (As you learned in chapter 1, legumes such as alfalfa, clover, and peas perform the absolutely essential task of returning vital nitrogen to the soil, without which continued life on earth would be impossible.) Without insects to pollinate them, beautiful flowers, such as carnations, magnolias, morning glories, orchids, and many others, would not continue to bless mankind. In God's providence, insects provide food for numerous other creatures and provide such products as beeswax, honey, shellac, and silk. Insects, one of God's major provisions for His living creation, were designed to be man's friends.

**Man's fall and God's curse.** When God cursed the earth after the fall (Gen. 3:17–19), the relationship between man and some of the insects changed. No longer were all of the insects man's friends; some became enemies and competitors. Sin, disease, suffering, and death entered the world; as a result, some insects became transmitters of disease and causers of suffering. Although certain helpful insects became predators or parasites of harmful insects (ladybugs, for example, devour several kinds of crop-destroying pests), the proliferation of harmful insects would make man's life more difficult.

**Insects versus man.** Insects now attack and injure almost all types of plants, wild and cultivated. Each year such insects as the boll weevil (cotton), the Hessian fly (wheat), the corn earworm (corn), the chinch bug (a variety of plants), and the Colorado potato beetle (potatoes) destroy tons of crops and do billions of dollars' worth of damage to American agriculture. Other insects are common household pests. Crickets, clothes moths, and carpet beetles ruin clothing, carpets, rugs, furniture, upholstery, and furs. Crickets, roaches, and silverfish damage books. Ants, flies, roaches, and many other insects devour or ruin food stored in homes, bakeries, supermarkets, restaurants, and warehouses. Termites destroy furniture and wood used in buildings. Some of the worst insect enemies, however, are those that directly endanger man's health. Certain biting flies and bugs inject a painful and sometimes fatal toxin when they bite. Some fleas and lice cause soreness or damage to skin and other tissues and occasionally even death. Houseflies and blowflies transmit germs that cause such diseases as cholera, dysentery, and typhoid fever. Bloodsucking insects (fleas and mosquitoes) and arachnids (mites and ticks) carry a variety of deadly diseases, including African sleeping sickness, plague, Chagas' disease, encephalitis, malaria, Rocky Mountain spotted fever, Texas fever, typhus, and yellow fever.

**Man versus insect.** Man has responded in a number of ways to the threat posed by harmful insects. Simple methods of control include the use of screening and the destruction of places where insects breed, such as draining swamps to destroy mosquito larvae or burning rubbish to eliminate maggots. Other methods of control are somewhat more difficult. A system of *quarantine control* seeks to keep insect pests from entering a particular country. Many of the most harmful insects in America, such as the boll weevil, cabbage butterfly, gypsy moth, Hessian fly, and Japanese beetle, entered the United States from other countries. Another means of controlling insect pests is *agricultural control,* which involves rotating crops, rescheduling the times of planting and harvesting, and other agricultural techniques to minimize insect damage to particular crops. A third method, *biological control,* uses natural insect diseases, parasites, or predators to stop harmful insects. For example, ladybugs eat certain insect pests; when the California citrus crop was threatened by a certain insect at one time, citrus growers introduced the ladybug, a natural predator of the harmful insect, and saved their crop.

*Chemical control* is another effective form of controlling insect pests. Insect hormones can be used to sterilize particular insects or to prevent the metamorphosis of larvae into adults, thereby preventing the insects from reproducing. *Insecticides* or *pesticides* can be used to kill harmful insects, but they must be used carefully; improper use of insecticides can also kill helpful insects or cause irritation or other harmful effects to people or animals. Often, a combination of different methods is the most effective approach to controlling harmful insects.

In recent decades, advances in genetics have allowed *genetic control* of insect pests. Since the mid-1990s, farmers have planted genetically modified crops that produce pesticides inside their own cells. By 2014, over three fourths of the corn and cotton grown in the United States were genetically modified to produce toxins derived from the bacterium *Bacillus thuringiensis,* which normally infects insects; these toxins kill pestilent insects but do not harm humans or most animals. Another potential method of genetic control involves releasing genetically modified insects that require specific substances (such as the antibiotic tetracycline) to complete their life cycle. These insects compete with normal insects for food and mating, but their offspring (if any are produced) do not survive to adulthood. This technique is being tested for controlling certain crop pests and the mosquito that transmits dengue [dĕng′gē] fever.

### Section Review 20.2

1. What is the meaning of the suffix *-ptera?*
2. Name three insects of the order Orthoptera and two insects of the order Odonata.
3. What is the scientific definition of a bug?
4. List three differences between moths and butterflies.
5. To what order do pestilent insects such as gnats, flies, and mosquitoes belong?
6. What is the common name for a member of the order Coleoptera?
7. To what order do the most common social insects belong?

#### Application

8. In what ways are insects beneficial to man? In what ways can they be harmful?

#### Identify

social insects, queen, drone, worker

#### Observe

- Observe some of the insects around you. Can you identify what orders they belong to?

## 20.3 Arachnids, Centipedes, and Millipedes

### Arachnids: Insectlike Arthropods

The **arachnids**—spiders and daddy longlegs, mites and ticks, and scorpions—are small, insectlike arthropods that constitute the class Arachnida [ə·răk′nĭ·də]. Arachnids range in size from a microscopic mite to a giant South American tarantula with a 5-inch-long body and legs that span 8 to 10 inches when they are extended. Like the insects, arachnids are found almost everywhere on earth—from scorching desert sands to chilly mountain slopes. One tiny spider, the *water spider,* actually spends its entire life underwater in a bubble of air that it carries beneath the surface.

### Spiders: Helpful Arachnids

Zoologists have identified about 43,000 species of **spiders** but estimate that many more exist. Although spiders are often mistaken for insects, they differ from insects in a number of ways. For example, insects have three distinct body regions (head, thorax, and abdomen), but *spiders have only two distinct body regions*—a fused head and thorax called a **cephalothorax** [sĕf′ə·lə·thôr′ăks′] and an *abdomen.* Insects have a pair of antennae and three pairs of legs; spiders have no antennae and four pairs of legs. Insects have both simple and compound eyes, but *spiders have only simple eyes.* Insects usually have one or two pairs of wings; spiders, however, have no wings whatsoever. All spiders are carnivorous, and almost all are venomous; that is, they

Fig. 20.17 **Spiders**

kill their prey by means of a venom they inject into the bodies of victims.

Although many people regard spiders with feelings of fear and revulsion, spiders are among man's best friends. Each year spiders and daddy longlegs devour trillions of pesky insects that infest our homes, our yards, and our gardens. Even the four North American spiders whose bites are harmful to man—the two species of *black widow,* the *brown recluse,* and the *red widow*—attack only when they are hurt or frightened, and even without medical treatment, fewer than 5% of those bitten die from the bite.

## External Structure of a Spider

Like all arthropods, spiders are invertebrates with a tough, protective exoskeleton. A thin waist, or *pedicel* [pĕd′ĭ·səl], separates the cephalothorax and the abdomen of a spider's exoskeleton. Although the number of eyes varies among different species, most spiders have eight simple eyes arranged in two rows of four each. A pair of short appendages called **chelicerae** [kĭ·lĭs′ə·rē′], which the spider uses to seize and to kill or crush its prey, is located below the spider's eyes. At the end of the chelicerae the spider has fangs, which it uses to inject poison into its prey. Spiders are divided into two main groups based on the movement of their fangs. *True spiders* have fangs that point inward and move in and out, crossing in the middle. *Mygalomorphs* [mĭg′ə·lō·môrfs], the best known of which are the *tarantulas,* have fangs that point down and move up and down.

Fig. 20.18 **Wolf spider**

**Mouthparts.** The spider's *mouth,* located beneath the chelicerae, is designed for sucking liquids. Although a spider cannot chew, it eats the solid portions of its prey by spraying or injecting the victim with powerful, enzyme-containing digestive juices. The juices predigest the prey by dissolving its solid tissue; the spider then sucks up the dissolved tissues and the body fluid. A **pedipalp,** an appendage that looks like a small leg, is attached to each side of a spider's mouth. The spider uses its pedipalps to cut and crush its food. A male spider's pedipalps also contain reproductive organs.

**Other appendages.** A spider's four pairs of *legs* are attached to its cephalothorax. Each leg consists of seven segments, and most spiders have two or three claws on the last segment of each leg. This last leg segment is often surrounded by a pad of hairs, the *scopula* [skŏp′yə·lə], which helps the spider cling to walls and ceilings. A spider walks by moving the first and third leg on one side of its body in tandem with the second and fourth leg on the other side. Although spiders have muscles to bend their legs at the joints, they have no muscles with which to extend their legs. The pressure of the blood in a spider's body keeps its legs extended; when its blood pressure drops because of a lack of fluids, its legs draw up under its body, and it cannot walk.

## Internal Structure of a Spider

**Nervous system.** Within the cephalothorax is the spider's nervous system, consisting of its brain and nerves. Although many spiders are nearsighted, *jumping spiders* and *wolf spiders* have excellent eyesight. God has given all spiders an acute sense of touch. Tiny nerve-bearing hairs on the spider's body serve as organs of touch, hearing, and smell; and "strain gauges" in their exoskeleton inform them of body position and muscle tension. Sensations are relayed by nerves to the spider's brain, which allows the spider to quickly react to stimuli in its surroundings.

**Digestive system.** Also within the cephalothorax is the spider's *sucking stomach* and the beginning of a digestive tube that extends the length of its body. Powerful muscles surrounding the boxlike sucking stomach pull outward on its stiff walls, increasing its volume and drawing liquefied food inward through the mouth like a suction pump. In the *intestine,* digestive juices complete the digestion of the spider's liquefied food, and the food is absorbed through the walls of the intestine into the spider's blood.

**Circulatory system.** A spider's *blood* is pale blue and contains many white blood cells. Its *heart,* a long, slender, muscular tube in the abdomen, pumps the blood to all parts of the body. As in the insects, *a spider's blood flows through*

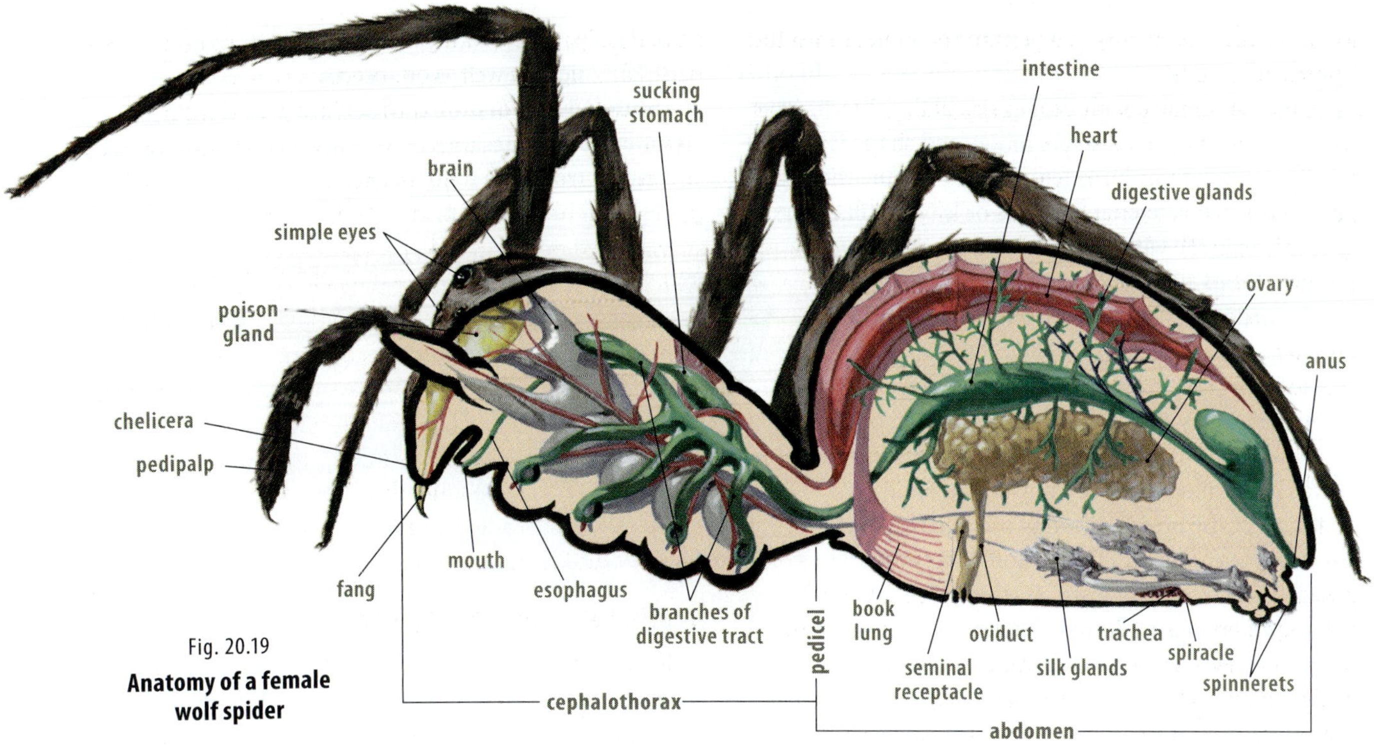


Fig. 20.19

**Anatomy of a female wolf spider**

*open passages in its body rather than through a closed system of blood vessels,* and thus the blood quickly drains from its body if its exoskeleton is damaged.

**Respiratory system.** Closely associated with the circulatory system is the spider's respiratory system, usually consisting of both *tracheae* and **book lungs.** Near the rear of a spider's abdomen are several *spiracles* through which air enters tiny branching tubules called tracheae and is carried throughout its body. The book lungs are located in tiny cavities in the spider's abdomen; mygalomorphs have two pairs of book lungs, while true spiders have only a single pair. Each book lung is made up of at least 15 extremely thin, flat folds of tissue forming parallel air pockets that are arranged like the pages of a book. Air enters the cavities where the book lungs are located through slits in each side of the body wall near the front of the spider's abdomen. As air circulates between the sheets of the book lungs, an exchange of gases takes place and oxygen passes into a blood-filled chamber. A special protein in the spider's blood, *hemocyanin,* carries oxygen in the spider's blood, much like the hemoglobin in vertebrates.

**Spinnerets and silk glands.** At the tip of the abdomen in most spiders are three pairs of short, fingerlike **spinnerets,** the organs that the spider uses to spin silk. Each spinneret contains hundreds of tiny tubes through which liquid silk flows from the spider's silk glands, located inside the abdomen. As the silk passes through the spinnerets, it hardens into filmy threads of silk. The spider uses silk for its webs, cocoons, and nests and as a "safety rope" (dragline) to keep it from falling. A spider normally spins dry silk but can produce sticky silk by coating the thread with a gluelike substance as it is spun; this sticky silk is used to make the insect-snaring portions of a spider web. Spider silk is extremely strong; dragline silk is estimated to be five times as strong as steel cable by weight, yet it is elastic, pliable, and resistant to water and most solvents.

## Reproduction in Spiders

**Mating.** In addition to their food-handling role, the pedipalps of a male spider serve a vital purpose in reproduction. Before mating, the male places sperm in specially designed chambers in his pedipalps. When spiders mate, the male uses his pedipalps to place this sperm in a structure of the female's abdomen called the *seminal receptacle.* After transferring his sperm to the female, the male spider is sometimes eaten by his mate; in most species of spider, the female is larger than the male.

The female stores the sperm until she is ready to lay her eggs and fertilizes them as they are laid. Depending upon the species, the female may lay as few as one egg or as many as 2000 eggs. In most species, the female encloses the fertilized eggs in a silken cocoon. In some species, the female dies after making the egg sac. In other species, she remains with the eggs until they hatch. The egg sac of each species differs in size and shape from that of all other species. Some female spiders hang the egg sac in a web; others attach it to plants or leaves; still others carry the egg sac with them.

**Spiderlings and draglines.** Tiny *spiderlings* hatch inside the silken cocoon and remain inside until the arrival of warm weather. As soon as they leave the egg sac, the spiderlings begin spinning draglines. Many spiderlings climb to the top of a tall object such as a fence post, tilt their spinnerets toward the wind, and spin a fine thread of silk that catches the breeze, allowing them to balloon their way over long distances—sometimes even hundreds of miles out to sea. As they grow into adults, spiderlings molt several times, allowing a new, larger exoskeleton to replace the one that has become too tight. Adult spiders have different life cycles, depending upon the species. Many species live only about one year, while others, such as large *wolf spiders,* live for several years. *Tarantulas are the longest-lived of all spiders;* certain species live over 20 years.

## Other Arachnids

**Daddy longlegs. Daddy longlegs,** or *harvestmen,* are a group of arachnids famous for their long, spindly legs. Although these arachnids are similar to spiders, there are several notable differences. The daddy longlegs has no constriction, or waist, between its cephalothorax and abdomen, and its sense organs are located on its second pair of legs. A spider has poisonous fangs at the end of its chelicerae; the end of the daddy longlegs' chelicerae are pincerlike instead of fanglike. Most spiders have eight simple eyes, but a daddy longlegs has only one pair, located near the back of its cephalothorax. Daddy longlegs lack spinnerets and thus do not spin webs. These rather solitary arachnids may be commonly seen roaming through gardens and fields in search of plant lice and other harmful insect pests. Most daddy longlegs are omnivorous, feeding on plant juices as well as on insects.

Fig. 20.20 **Daddy longlegs**

**Scorpions. Scorpions,** which are found in most temperate climates and in all tropical climates, range in length from less than an inch to about eight inches. These arachnids are most active at night and often crawl into dark places (including empty shoes) during the day to escape bright light. Probably the most prominent feature of a scorpion is its long, segmented abdomen, which is tipped with a venomous stinger capable of inflicting painful, although usually not fatal, wounds to man. As with spiders, female scorpions sometimes devour their mates after mating. *All species of scorpions are either ovoviviparous or viviparous;* the young are born, not hatched. They are cared for and protected by the mother, riding around on her back for 1 to 2 weeks after birth.

Fig. 20.21 **Scorpion**

**Mites and ticks. Mites** and **ticks** make up an order of tiny blood-sucking arachnids that parasitize both man and animals. Ticks transmit such diseases as Rocky Mountain spotted fever (in man) and Texas fever (in cattle). *Harvest mites,* or *chiggers,* suck blood from their host by fastening onto the skin. Usually, their victim does not realize he has been attacked by these tiny mites until an area of his skin reddens, swells painfully, and begins to itch fiercely. Because of the diseases they transmit, mites and ticks join fleas, flies, and mosquitoes as man's worst arthropod enemies.

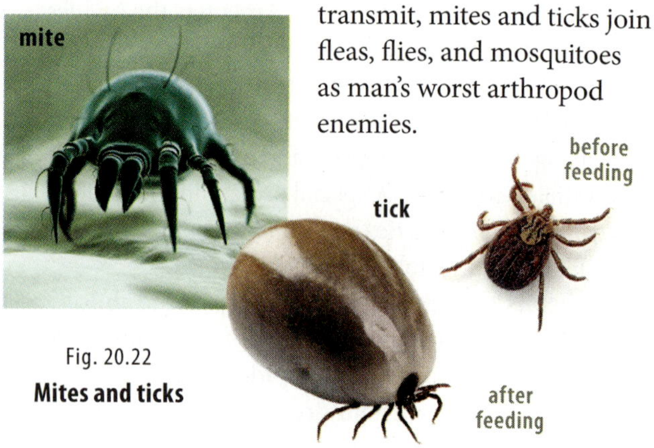


Fig. 20.22 **Mites and ticks**

## Arthropods with Many Legs

**Centipedes** (class Chilopoda [kī·lŏp′ə·də]) and **millipedes** (class Diplopoda [dĭp′lŏp′ə·də]) are caterpillarlike creatures that are often confused with each other, but there are significant differences between the two classes. For example, a centipede has a flattened body; millipedes have rounded bodies. A centipede has a pair of long, jointed antennae; the antennae of the millipede are shorter. Centipedes are poisonous, but millipedes are harmless. Finally, a centipede has a single pair of legs attached to each of its body segments; millipedes have two pairs of legs per segment.

Fig. 20.24
**Millipede**

Although most centipedes are quite small, some tropical species are a foot in length and prey on toads, small snakes, and even small mammals. Centipedes do not necessarily have a hundred feet as their name implies; attached to the elongated, flattened body of a centipede may be as few as 30 or as many as 354 legs (the common *house centipede* has 30; centipedes with more than 50 legs are rare). The first body segment of a centipede has a pair of poison claws (rather than legs), which the centipede uses to prey on insects, spiders, slugs, and earthworms. Just behind the centipede's head are two poison glands that secrete poison to the claws; because of the poison, a centipede bite is quite painful, although North American species are not dangerous to humans. A centipede also lacks legs on its last two body segments. Centipedes have strong mandibles that act as sharp cutters, but they are restricted in their predatory activities by an exoskeleton that does not completely protect them against drying out. For this reason, centipedes hunt at night and are most at home under logs, rocks, and other moist, sheltered places. Many species of centipedes are considered helpful to man because they eat insect pests.

Fig. 20.23
**House centipede**

A millipede is more wormlike than a centipede and has more legs on average. Millipedes may have anywhere from 24 to over 700 legs (although not a thousand legs as its name implies). The millipede has a single pair of legs on each of its first four body segments but two on all the other segments. Millipedes are useful because they are *saprophytes,* feeding mostly on dead and decaying plant materials. Although centipedes are fast movers, millipedes are rather slow creatures that roll into a ball when disturbed. Like centipedes, millipedes prefer dark, moist places in which to live.

### Section Review 20.3

1. How many body regions does an arachnid have? What are they called?
2. What are the two broad categories of spiders?
3. What is the purpose of a spider's chelicerae and pedipalps?
4. What is the most obvious difference between a spider and a daddy longlegs? What are three other differences?
5. What are the differences between centipedes and millipedes?

#### Application

6. Why do the legs of a dead spider curl up under its body?

#### Critical Thinking

7. How is the respiratory system of a spider similar to that of an insect? How is it different?

#### Identify

arachnid, cephalothorax, chelicerae, mygalomorph, pedipalp, book lung, spinnerets

#### Observe

- Examine a piece of spider web with a hand lens.

# 20.4 Crustaceans: Aquatic Arthropods

**Crustaceans** range in size from woodlice and minute water fleas to lobsters and giant crabs (with its legs extended, the *giant spider crab* of Japan measures up to 12 feet across). Even though these creatures vary greatly in size, they all have several things in common: two pairs of antennae on the front of the body; two or more pairs of appendages; a hard, crusty exoskeleton composed of chitin and strengthened by the mineral *calcium carbonate;* and, most often, gills for breathing. Most crustaceans inhabit salt water; however, some live in fresh water, and a few are terrestrial. Some crustaceans—crabs, lobsters, prawns, and shrimp—are relished as seafoods by people, and smaller marine crustaceans are essential components in marine food chains.

## External Structure of a Crustacean

You are probably familiar with **crayfish,** which some people call crawfish or crawdads. They inhabit calcium-containing lakes, ponds, rivers, and streams on every continent except Africa and Antarctica. Crayfish are often studied because they are large and have a body structure similar to that of most other crustaceans. Each crayfish has two major body sections, like the arachnids: the *cephalothorax,* a combination of the head and thorax, and the segmented *abdomen.* Although a crusty skeleton covers the entire animal, an especially hard outer shell called the **carapace** provides the vital organs with an additional defense against hungry predators. A forward extension of the carapace called the **rostrum** protects the well-defined head.

**Appendages of the head.** Two *compound eyes* are attached to the head by flexible stalks. Most adult crustaceans have compound eyes; many species also have additional *simple eyes* that can sense light but cannot form an image. Most young crustaceans have only simple eyes.

Two sets of feelers are also attached to the crayfish's head. The shorter ones are called *antennules* [ăn·tĕn′yo͞olz]; at the base of the antennules are sacs that contain the sense organs of hearing and balance *(statocysts).* Without statocysts, a crayfish could not swim upright. The senses of touch, taste, and smell come from the large *antennae* located behind the antennules. Near the opening of the mouth are the chewing appendages, the larger *mandibles* and the smaller *maxillae.* The crayfish moves these jaws horizontally like an insect rather than vertically like a human.

**Appendages of the thorax.** Associated with the crayfish's thorax are several sets of appendages. The first three sets, designed to help the crayfish keep a firm grip on its food, are called **maxillipeds,** or jaw feet. Behind the maxillipeds lie a pair of large *chelipeds* [kē′lə·pĕdz], appendages with pincers, which the crayfish uses for

Fig. 20.25
**External anatomy of a crayfish**

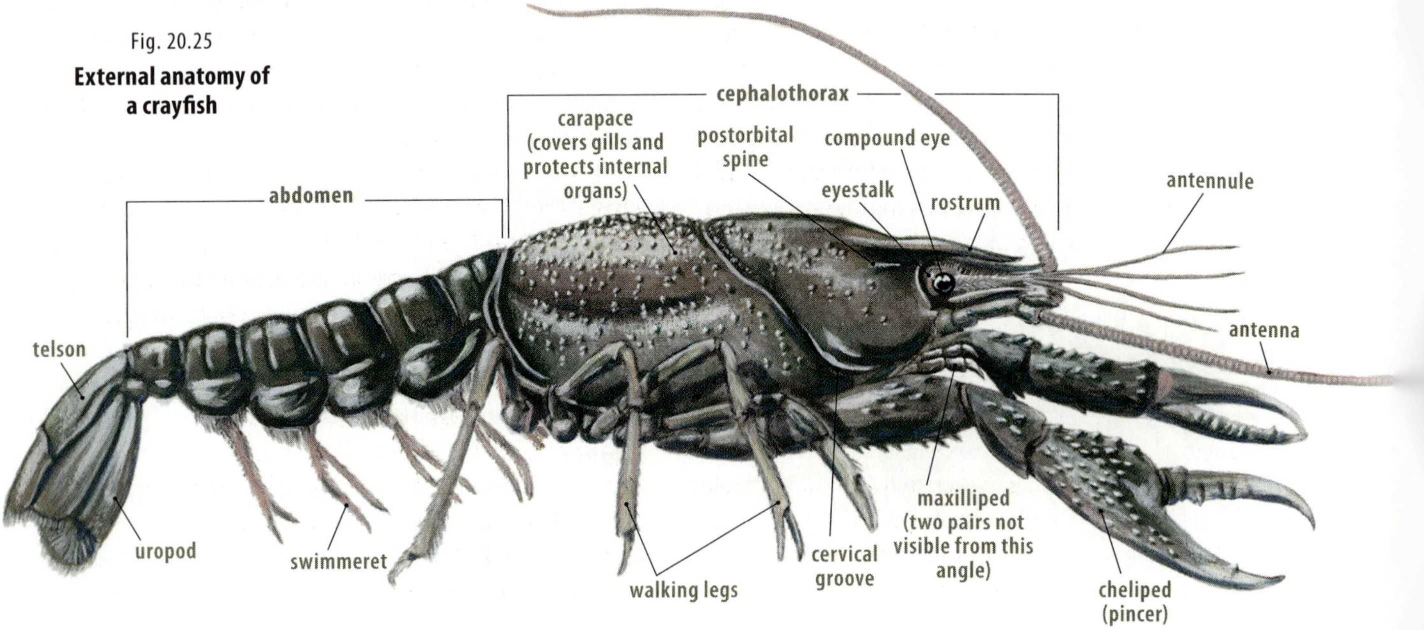

Fig. 20.26
**Internal anatomy of a crayfish**

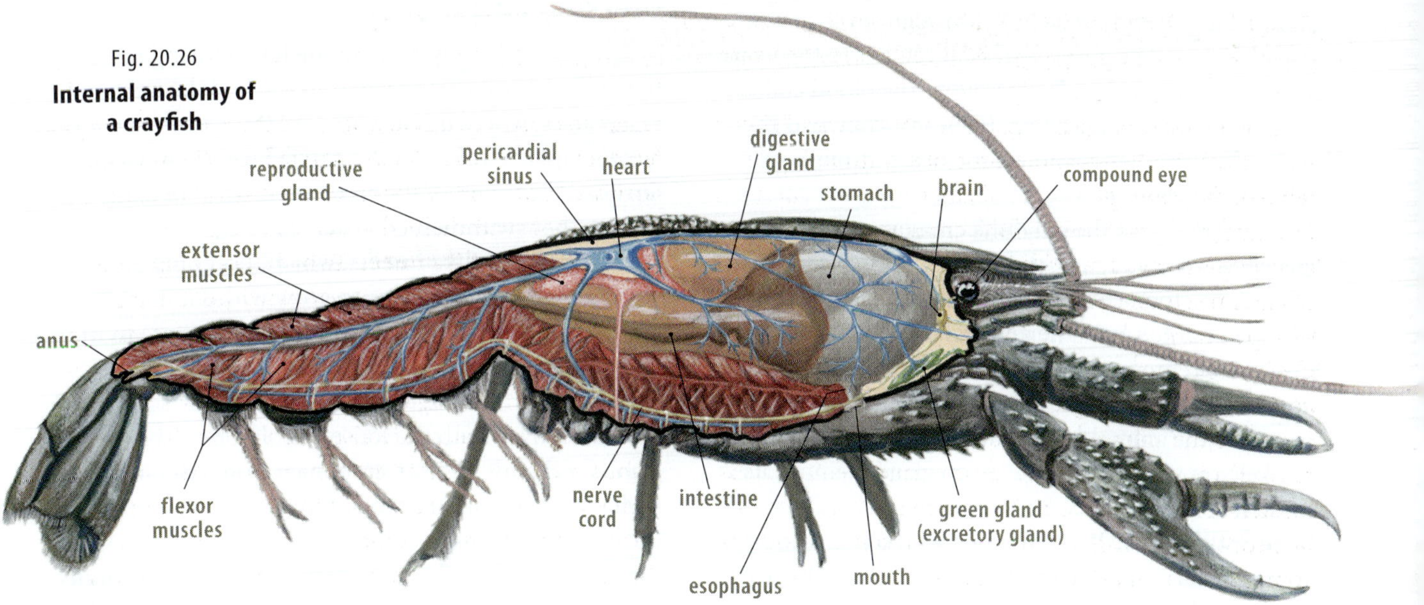


seizing prey and for self-defense. The crayfish uses the remaining thoracic appendages, four pairs of *walking legs,* for locomotion. The first four walking legs are tipped with small pincers, and the rearmost four end in single claws. Beneath the carapace lie several *gills* that are attached to the bases of the crayfish's appendages.

Interestingly, a crayfish can grow back a lost appendage to replace one that has been damaged or severed from the body. Although regenerating a new body part takes time, a lost appendage can eventually be totally restored.

**Appendages of the abdomen.** The crayfish's abdominal appendages, known as **swimmerets,** are found on the body segments that make up the abdomen. The first five pairs of swimmerets are rather small; the female crayfish attaches her eggs to them during reproduction. The swimmerets on the sixth segment form a **uropod** (flipper), which the crayfish uses for swimming. Terminating the sixth segment is a flat, triangular projection **(telson).** The crayfish can dart backward in the water by using its powerful abdominal muscles to whip its uropod and telson forward.

## Internal Structure of a Crayfish

**Digestive system.** Most crayfish are predators, preying on small marine creatures such as snails, small fish, tadpoles, and young insects. However, they are not particular about their diet and may also eat plants or decaying animal matter if necessary. Overall, the digestive system of a crayfish is broadly similar to that of insects, but there are many important differences. The crayfish takes in food through its *mouth,* which is located on the underside of its cephalothorax. Swallowed food moves upward through a short *esophagus* into a section of the *stomach* known as the **gastric mill.** This structure serves as a gizzard, grinding the food into small particles for digestion. The rearmost portion of the gastric mill is guarded by hairlike structures that form a sieve, preventing food from passing into the rest of the stomach until it has been broken down into small particles. The food is then mixed with gastric juice and digestive enzymes, which break down the food into its constituent molecules. Attached to the rear of the stomach is the *digestive gland,* which consists of many small pouches where digestive enzymes are secreted and dissolved nutrients are absorbed. Waste products are then passed along to the *intestine* to be eliminated through the *anus.*

**Respiratory system.** The crayfish, like most other crustaceans, breathes by means of *gills* that are attached to the bases of the legs of the thorax and protected by the carapace. Although most crustaceans obtain oxygen through gills, many small species lack gills and obtain oxygen through the skin instead. Because the gills are attached to the legs, the gills wave through the water every time the crayfish moves, helping them to exchange oxygen and carbon dioxide with the water more efficiently. The crayfish can even survive out of water for a short time because its gills are kept moist beneath the carapace.

**Circulatory system.** Like insects, *crayfish have open circulatory systems.* A crayfish's colorless blood is pumped to the ends of its body through seven arteries (instead of a single aorta) and then returns to the heart by flowing through the body cavity. The blood collects in an open area surrounding the heart (known as the *pericardial sinus*) and reenters the heart through six valve-equipped *ostia*.

Associated with the crayfish's circulatory system are special excretory organs, the **green glands,** which lie in front of the stomach and open to the outside near the base of the antennae. The green glands function like kidneys to remove wastes from the blood. They also store minerals and make them available to the crayfish when the animal is ready to molt. Before the old exoskeleton is discarded, the green glands remove its mineral content.

**Nervous system.** A crayfish's nervous system consists of a brain, nerves, and several sensory structures. Organs that are extremely sensitive to smell, touch, and taste are located in the antennae. Small sensory bristles on the appendages and other parts of the body are sensitive to touch. The crayfish's compound eyes are very sensitive to movement and offer a wide angle of vision.

The **statocysts** at the base of the antennules contain sensory hairs and grains of sand. With every movement of the crayfish, the grains of sand move, stimulating the sensory hairs and thus enabling the animal to adjust its position and maintain its equilibrium. Each time the crayfish molts, the lining of the statocyst is shed with the exoskeleton, and the grains of sand are lost. When the new exoskeleton forms, new grains are picked up. If a crayfish is kept in a tank containing iron filings instead of sand, it will replace the lost sand with filings. A strong magnet held near the crayfish will cause the animal to orient to the magnet as though it were the pull of gravity, and thus the crayfish can be made to swim upside down, because its organs of equilibrium tell the nervous system that down is up.

Fig. 20.27
**Variety among crustaceans**

## Life Cycle of the Crayfish

Crayfish generally mate in the fall. During mating, the male deposits his sperm into the female's seminal receptacles, where it is stored until the spring when the female lays her eggs. As the crayfish lays her eggs, she fertilizes them with the stored sperm and then attaches them to her swimmerets. After six to eight weeks, the young hatch. Unlike insects, which molt only as larvae, most crustaceans continue to grow without reaching a final, definite size; as a result, crayfish continue to molt throughout their lives.

Before a crayfish molts, it begins to grow a new exoskeleton beneath the loosened old one. The green glands reabsorb calcium carbonate from the old exoskeleton to be reused; some of the calcium is stored in the crayfish's stomach walls. When the crayfish is ready to molt, it swells its body by swallowing large amounts of water, causing the carapace to split. The water is then removed, giving the creature room to wriggle out of the old exoskeleton. The creature also sheds its teeth, the plates of its gastric mill, and its statocysts. After shedding the old exoskeleton, the crayfish enlarges the new one by swallowing water while it is still flexible; the new skeleton is then hardened as the stored calcium carbonate is added. A crayfish normally hides while molting and waiting for its new exoskeleton to harden because it is soft and vulnerable during this time. Although the crayfish molts about seven times during its first year, by the end of its life it will average only once or twice a year.

## Other Crustaceans

Although you are probably familiar with crayfish and such crustaceans as *crabs, lobsters,* and *shrimp,* most crustaceans are very small or microscopic. Such tiny creatures as the *fairy shrimp, cyclops, daphnia, Gammarus, woodlouse, Asellus,* and *seed shrimp* are often found in ponds, streams, or moist places on land. The **woodlouse,** commonly called a *sowbug, pillbug,* or *roly-poly,* rolls into a ball whenever it is disturbed. This crustacean keeps its gills moist by living in damp places. **Brine shrimp,** found in salt lakes and marshes, are commonly used as living

Fig. 20.27 (cont.)
**Variety among crustaceans**

food in aquariums. **Barnacles** are marine crustaceans that attach themselves to rocks, piers, and wharves and must be scraped periodically from ships' bottoms. Most other crustaceans are tiny saltwater organisms that form important links in the ocean's major food chains.

## Extinct Arthropods

As with other groups of animals, many interesting arthropods were originally created that have since become extinct; these creatures are presently known only from fossils. Some extinct arthropods were merely larger varieties of present-day creatures, such as dragonflies with 3 ft wingspans. Some other extinct arthropods seem rather bizarre to us today because all the creatures of that kind have apparently died out.

Some of the most familiar of the extinct arthropods are the *trilobites* [trī′lə·bīts′], small aquatic arthropods similar in some ways to horseshoe crabs. Although some trilobites were probably able to swim, most apparently scuttled along the sea floor in search of food.

One of the more bizarre arthropods that has become extinct since Creation was *Anomalocaris,* a two-foot-long aquatic predator thought to have swum gracefully through the water, snaring small marine creatures with two "feeding arms" and placing them in its small, round mouth. Two of the largest known extinct arthropods were a five-foot-long aquatic "sea scorpion" *(Pterygotus)* and a six-foot-long, land-dwelling centipede *(Arthropleura).*

Extinct arthropods to scale (6 ft human shown for comparison)

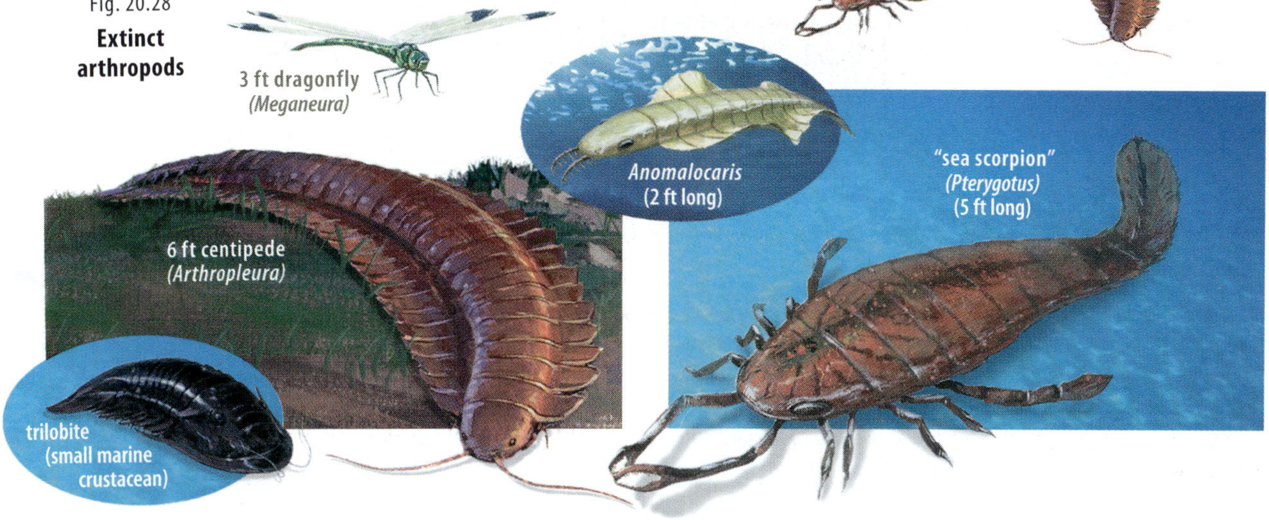


Fig. 20.28
**Extinct arthropods**

## Section Review 20.4

1. What body characteristics do all crustaceans have in common?
2. Name a typical crustacean's two body regions.
3. How does a crayfish's gastric mill serve the same purpose as a gizzard?
4. What are the functions of a crayfish's green glands?
5. Why does a crayfish continue to hide for a while after molting?
6. Name five crustaceans (besides crayfish) that are commonly encountered.

### Identify

carapace, rostrum, maxillipeds, swimmerets, uropod, green glands

### Observe

- Make a terrarium to house a woodlouse. As you place in a potato or carrot slice for it to feed on, observe it roll into a ball. How long does it remain curled? Is the crustacean normally hidden from view? Why?

## Chapter 20 Review

### Define

1. invertebrate
2. arthropod
3. exoskeleton
4. chitin
5. entomologist
6. metamorphosis
7. nymph
8. simple eye, compound eye
9. mouthparts
10. queen, drone, worker
11. arachnid
12. cephalothorax
13. chelicerae
14. pedipalp
15. book lung
16. spinneret
17. carapace
18. rostrum
19. maxilliped
20. swimmeret
21. uropod, telson
22. green glands

### Identify

1. five characteristics of all arthropods
2. five major groups of arthropods
3. the largest group of arthropods
4. five characteristics of insects
5. four characteristics of spiders
6. two broad categories of spiders
7. three types of arachnids that are not spiders
8. two parasitic arachnids mentioned in the text
9. three characteristics that distinguish centipedes and millipedes
10. three characteristics of crustaceans
11. two body regions of a typical crustacean

### Explain

1. What is the difference between complete and incomplete metamorphosis?
2. Which body region of an insect contains the heart and most of the digestive and excretory systems? Which region contains most of the sensory organs? To which region are the legs and wings attached?
3. To what order of insects do cockroaches, crickets, and mantises belong?
4. What do entomologists mean when they use the term *bug?* To what order do the bugs belong?
5. To what orders of insects do butterflies, flies, and beetles belong?
6. What is a social insect? To what order do most of these insects belong?
7. Explain some of the methods man has employed to fight insects.

### Apply

1. Contrast the open circulatory system of insects and other arthropods with the closed circulatory systems of vertebrates.
2. How is the respiratory system of a spider different from that of an insect?
3. Briefly describe some ways in which arthropods are beneficial to mankind and some ways in which they can be harmful.

# Chapter 21 Variety in the World of Invertebrates

**Key Concepts**

- Mollusks
- Spiny-skinned echinoderms
- Microscopic rotifers
- Hollow-intestined coelenterates
- Pore-bearing sponges
- Annelids, platyhelminths, and nematodes
- Protozoa

## 21.1 Mollusks

The largest group of water animals on Earth is the **mollusks,** which includes such soft-bodied, boneless creatures as clams, oysters, slugs, snails, squids, and octopuses. Some snails and slugs live on land, but most mollusks live entirely in water. Mollusks are distinguished by a muscular **foot,** which is used for locomotion; a **visceral hump,** which contains the internal organs; and a **mantle,** which is a skinlike organ that covers the visceral hump and, in most species, produces a calcium carbonate *shell* for protection. The calcium carbonate is produced by special glands in the mantle from minerals provided by the mollusk's food. It is squeezed out of the edges of the mantle as a liquid and is hardened by secretions of other glands. The shell is added to as the mollusk grows. In mollusks that have no shell (such as octopuses and most slugs), the mantle itself serves as a protective wrapper for the body organs. The beautifully colored, gracefully shaped shells that people pick up on the seashore are often mollusk shells. About 1000 new species of mollusks are discovered each year, and there are 100,000 fossil species as well as the 100,000 living species. The shell of each species of mollusk has a distinctive design and shape.

Fig. 21.1 **Bivalves**

Phylum Mollusca [mə·lŭs′kə] is divided into six classes. The most familiar classes are the *bivalves,* such as the clam, mussel, oyster, and scallop; the *gastropods* or *univalves,* such as the snail, slug, and whelk; and the *cephalopods,* such as the squid, octopus, cuttlefish, and chambered nautilus.

### Bivalves: Two-Shelled Mollusks

The two shells (valves) of clams, scallops, oysters, and mussels give these mollusks the name **bivalves.** Edible clams, scallops, oysters, and mussels are often called *shellfish.*

**Shells.** *Each shell of a bivalve consists of three distinct layers:* a hornlike outer layer, a crystalline middle layer, and a pearly inner layer. The inner shell layer of some inedible oysters and clams is called *mother-of-pearl.* This brilliantly colored mineral deposit is used for making buttons, beads, and jewelry. If an irritating foreign substance (usually the larva of a parasitic flatworm) becomes

# Table 21.1 A Survey of the Animal World (Kingdom Animalia)

## Vertebrates

| | | | | | |
|---|---|---|---|---|---|
|  | **Mammals** (class Mammalia) | warm-blooded vertebrates that breathe with lungs and have hair and mammary glands |  | **Amphibians** (class Amphibia) | cold-blooded vertebrates with soft, moist skin; larval forms have gills; most adults breathe with lungs |
|  | **Birds** (class Aves) | warm-blooded vertebrates that breathe with lungs and have wings, feathers, and two feet |  | **Bony fish** (class Osteichthyes) | cold-blooded vertebrates with gills for breathing, fins, and a skeleton of bone |
|  | **Reptiles** (class Reptilia) | cold-blooded vertebrates that breathe with lungs and have dry, scaly skin |  | **Cartilaginous fish** (classes Chondricthyes and Cyclostomata) | cold-blooded vertebrates with gills for breathing, fins, and a skeleton of cartilage |

## Invertebrates

### Arthropods (phylum Arthropoda)

| | | |
|---|---|---|
|  | **Insects** (class Insecta) | invertebrates with an exoskeleton, three distinct body regions, three pairs of jointed legs, one pair of antennae, compound and simple eyes, and tracheae for breathing |
|  | **Crustaceans** (subphylum Crustacea) | invertebrates with an exoskeleton, two distinct body regions, jointed legs, and two pairs of antennae; most have gills for breathing |
|  | **Arachnids** (class Arachnida) | invertebrates with an exoskeleton, four pairs of jointed legs, simple eyes, and no antennae; most have two distinct body parts |
|  | **Centipedes** (class Chilopoda) | invertebrates with an exoskeleton, a flattened body with many segments, one long pair of jointed antennae, one pair of legs per body segment, and one pair of poison claws |
|  | **Millipedes** (class Diplopoda) | invertebrates with an exoskeleton, a rounded body with many segments, one short pair of jointed antennae, and two pairs of legs per body segment |

### Other Invertebrates

| | | |
|---|---|---|
|  | **Mollusks** (phylum Mollusca) | soft-bodied invertebrates with an unsegmented body, one muscular foot, visceral hump, and a mantle; most have a shell (exoskeleton) and gills |
|  | **Echinoderms** (phylum Echinodermata) | marine invertebrates with a spiny endoskeleton and a water-vascular system; most have a five- (or multiple-of-five-) rayed structure |
|  | **Rotifers** (phylum Rotifera) | microscopic invertebrates with one or more rings of cilia that resemble rotating wheels; most inhabit fresh waters |
|  | **Coelenterates** (phylum Coelenterata[1]) | aquatic invertebrates with two body forms, a gastrovascular cavity with a single opening, and tentacles with stinging cells |
|  | **Sponges** (phylum Porifera) | hollow invertebrates that lack a nervous system; their body opening serves only as an outlet for water and wastes |
|  | **Segmented worms** (phylum Annelida) | elongated invertebrates that have a cylindrical body with numerous segments |
|  | **Flatworms** (phylum Platyhelminthes) | elongated invertebrates with a flattened body; many are parasitic |
|  | **Roundworms** (phylum Nematoda) | elongated invertebrates with a cylindrical, unsegmented body; many are parasitic |
|  | **Protozoa** (classification varies)[2] | single-celled creatures that are usually microscopic; some are parasitic |

[1]Or Cnidaria.
[2]Although the protozoa are traditionally studied with animals, they are now usually classified in the kingdom Protista.

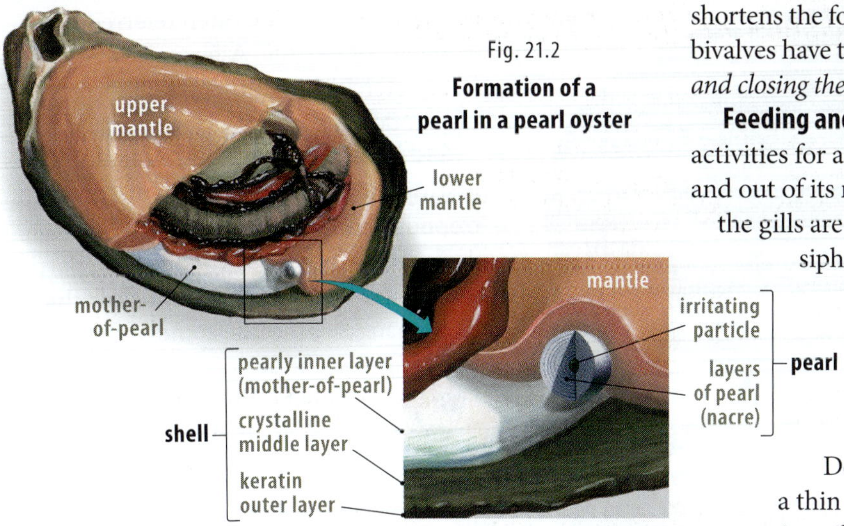


Fig. 21.2 **Formation of a pearl in a pearl oyster**

lodged in the mantle of a bivalve, the animal secretes thin sheets of *nacre* [nā′kər], the substance from which the pearly layer is formed, around the foreign object. The sheets of nacre build up around the irritant and eventually form a **pearl.** Because the pearly layer of edible clams and oysters lacks luster, their pearls are of no value to man. Most of the pearls valued as gems come from *pearl oysters* and other mollusks of the tropical seas. Some valuable pearls may take more than seven years to form.

**Adductor muscles.** The two shells of a bivalve are connected by hinges that look like small teeth and are held together by two very strong **adductor muscles.** These muscles are the main parts of scallops and mussels that are eaten; the entire body of oysters and clams, including the visceral organs, is commonly eaten.

**Locomotion.** *A clam uses its foot for burrowing very quickly into the sand.* The clam pushes its foot into the sand; forces blood to the end of the foot, causing it to swell and form a hatchet-shaped anchor; and then shortens the foot to pull itself down into the sand. Not all bivalves have this foot: *scallops swim by rapidly opening and closing their shells, and oysters are sessile* (immobile).

**Feeding and respiration.** One of the most critical activities for a bivalve is the movement of water into and out of its **mantle cavity,** the open chamber in which the gills are located. This is accomplished with two siphons (tubes) at the back end of the animal. The **incurrent siphon** takes water into the mantle cavity to the *gills,* where oxygen and carbon dioxide are exchanged. Most of this water then passes out through the **excurrent siphon.**

Dead organic particles in the water stick to a thin mucous layer on the gills, and constantly moving cilia on the gills move the mucus to the small mouth, from which it moves through the rest of the digestive system.

**Other systems.** A bivalve has an *open circulatory system,* and its blood, which lacks hemoglobin, is blue or colorless. Its well-developed *nervous system* consists of three pairs of ganglia connected by nerves to the organs of the foot and the body. The sensory cells are sensitive to touch, light, and chemical changes in the water; some bivalves, such as scallops, even have eyes. Reproduction in most bivalves is sexual, with *external fertilization.* Sperm leave the male through his excurrent siphon and enter the female through her incurrent siphon. The eggs stick to the female's gills, where they are fertilized.

**Edible bivalves.** Although clams, oysters, and scallops are popular and nutritious foods, they can be dangerous for man to ingest because they are scavengers, feeding on dead organic matter, which often becomes contaminated. Their beds must be checked closely with modern instruments before the animals can be declared safe for human consumption. It is interesting that God warned the

Fig. 21.3 **Clam burrowing into sand**

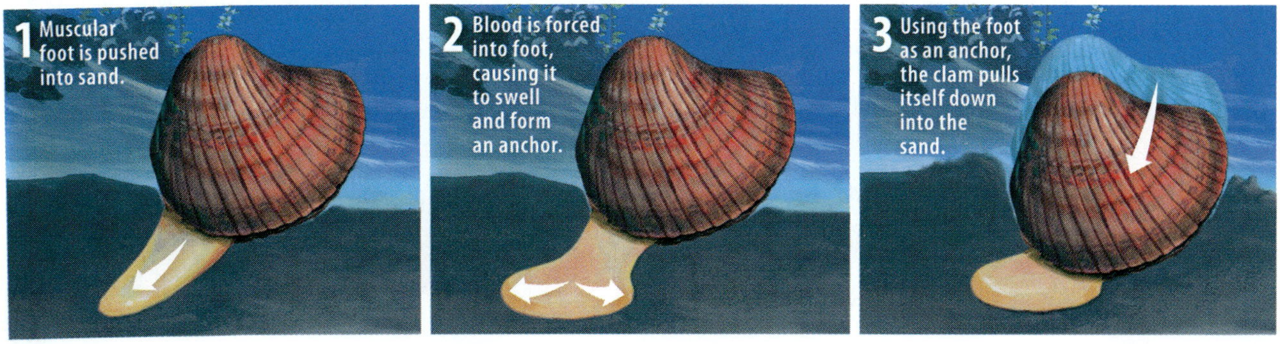

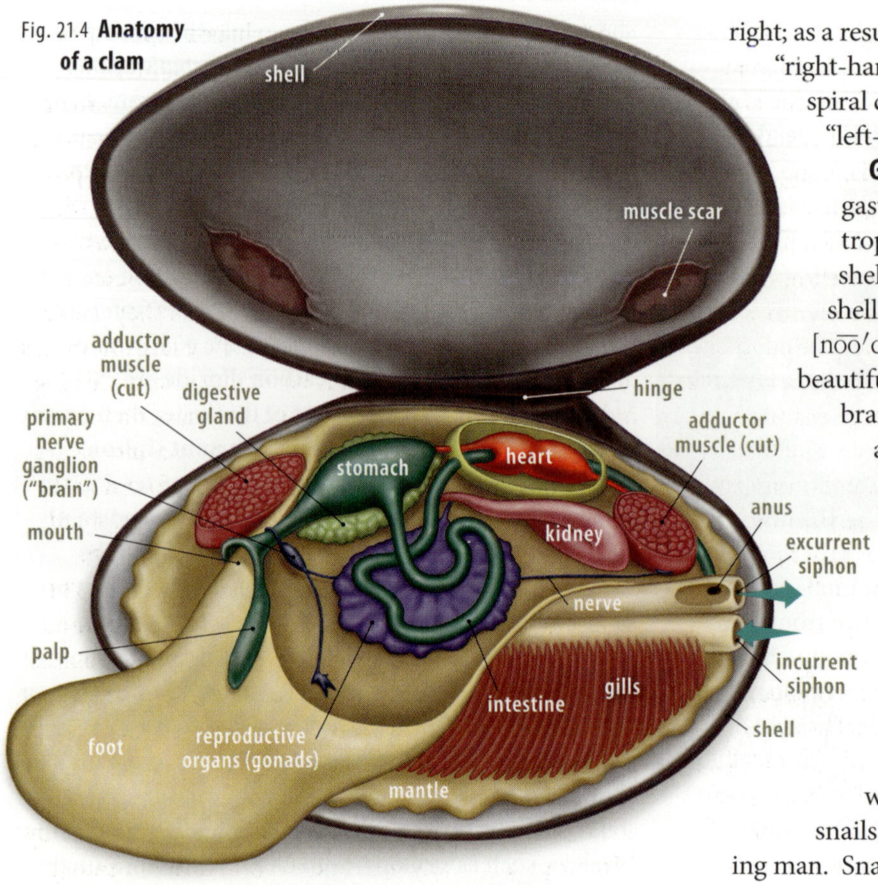

Fig. 21.4 **Anatomy of a clam**

Israelites against eating shellfish or even touching them (Lev. 11:10–11). In this manner, God was protecting His people from dangers beyond their understanding.

## Gastropods: Stomach-Footed Mollusks

**Shells.** **Gastropods** (snails, slugs, whelks, abalones, conchs, cowries, and murexes) are often called *univalves* because most of them have a single shell. This shell is usually very obvious, coiled, and beautifully colored or marked. Because the designs of the shells are unique to each of the 65,000 species, the shell can be used as the main means of identification. Most gastropods have shells that spiral clockwise. When such a shell is held with the spire up and the opening facing the observer, the opening is on the right; as a result, these shells are often referred to as "right-handed," or *dextral.* A few univalve shells spiral counterclockwise and are said to be "left-handed" or *sinistral.*

**Gastropods without shells.** Not all gastropods have shells. **Slugs** are gastropods that look like snails without shells. Some other gastropods have partial shells. *Marine slugs,* called **nudibranchs** [nōō′ də·brăngks′], include some of the most beautiful and graceful sea creatures. Nudibranchs are noted for their brilliant colors and elaborate, plumelike gills extending from their sides and back, giving the meaning of their name "exposed gills" (from Latin *nudi-,* "exposed," + *branchia,* "gill").

**Economic importance.** Gastropods range in size from 2-foot-long aquatic snails to creatures as small as a pinhead. They can be pests, like many land slugs and snails, which destroy crops, or like freshwater snails, which carry parasitic diseases affecting man. Snails are also helpful because they clean up organic debris, and they commonly serve as food for man, fish, and birds.

**Body plan.** Like bivalves, *univalves have a foot, a visceral hump, a mantle, and (usually) a shell.* Unlike bivalves, they also have a *head* that protrudes from the shell. The class name Gastropoda [găs·trŏp′ə·də] comes from the Greek words for "stomach" and "foot." The gastropod's belly *(gastro)* rests on its foot *(pod).*

Fig. 21.6 **Gastropods**

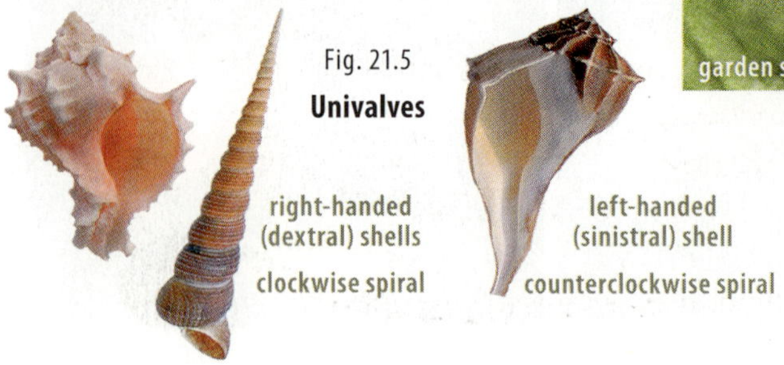

Fig. 21.5 **Univalves**

The animal moves on the foot by gliding over a ribbon of slime produced by special glands in the foot. Many snails have a lidlike *operculum* at the back of the foot. When the creature senses danger, it draws back into its shell, and the operculum closes. Land-dwelling slugs and snails are equipped with simple lungs instead of gills.

**Head structures.** Some univalves have two pairs of *tentacles* on their heads. One pair is for feeling, and the other pair may end in eyes. The mouth contains a **radula** [răj′o͞o·lə: plural radulae, răj′o͞o·lē′], a rough, file-like organ that scrapes food and carries it like a conveyor belt into the digestive tract. If you were to examine the radula under a microscope, you would see that it is covered with tiny teeth; some univalves have thousands of these radula teeth. Most univalves are herbivorous, but a few are carnivorous, preying on fish or other mollusks; carnivorous univalves have larger teeth.

## Cephalopods: Head-Foot Mollusks

The class Cephalopoda [sĕf′ə·lŏp′ə·də] consists of creatures such as the octopus, squid, cuttlefish (very similar to squid), and chambered nautilus. **Cephalopods** are considered by biologists to be mollusks because of similar body structures such as a mantle, visceral hump, and foot, but we would not recognize the similarities quickly. The mantle appears as an outer body wall, the visceral hump is very elongated, and the head and foot are indistinguishably merged, giving them the name "head foot." The octopus has no shell, the squid and the cuttlefish have an internal shell for support, and the nautilus has a beautifully spiraling external shell. The cephalopods are truly uniquely designed animals of the ocean.

**Squids.** The **squid** is a torpedo-shaped cephalopod noted for its ability to move by "jet propulsion" by quickly forcing water out of a *siphon.* It jets forward or backward by directing the siphon in opposite directions. Undulating fins on the sides of its head are used to propel the squid in steady swimming. When threatened, the squid ejects an inky fluid into the water, confusing an attacker and allowing the squid to escape. Squids possess ten arms, on which are suckers that they use to catch and hold fish for food. They can eat large numbers of mackerel and herring, thus becoming pests to fishing industries. The *common squid,* which is about a foot long, is found in large numbers from the marine waters of Nova Scotia to Florida. The *giant squid,* found in the temperate regions of the oceans, and the *colossal squid,* which lives in the oceans around Antarctica, are the largest invertebrates. Giant squids can grow up to 43 ft long and weigh up to 660 lb. The largest colossal squid discovered was only about 30 ft long but weighed almost 1100 lb; however, some incomplete specimens suggest that colossal squids can be much larger than this. The eyes of the colossal squid, which can be 11 inches in diameter, are the largest of any known animal.

Squids possess beaklike jaws that can tear at the body of their prey (usually fish) and radulae to aid in eating the catch. The gills in a squid have no cilia. Water moves in and out of the squid's body by muscle action as the animal moves easily through the water. This rapid water movement allows free exchange of oxygen and carbon dioxide and elimination of wastes.

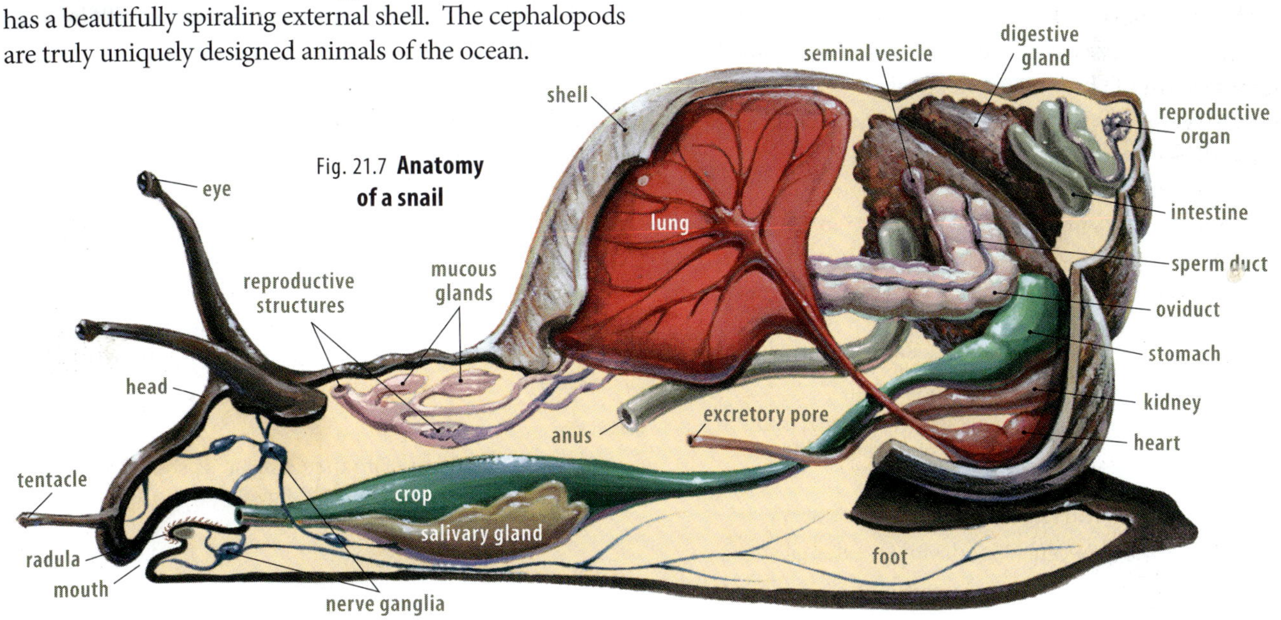


Fig. 21.7 **Anatomy of a snail**

The squid's body is designed to accommodate its active life. It has more arteries and veins than snails or clams and has a number of booster hearts to help drive the blood into the gills. *The circulatory system of cephalopods is not open as in other mollusks* but is a closed system of arteries, veins, and capillaries like that of vertebrates. In a marvelous way, God designed the squid's body to perform its unique purpose in nature.

**Octopuses.** The **octopus** is another cephalopod with unique features. It has eight arms, which are studded with suction cups for catching prey. The *Pacific giant octopus* may attain an arm spread more than 20 ft and a weight of over 150 lb, but most species are smaller. Some species are less than 2 inches in length. The octopus, which is quite timid, crawls along the ocean bottom and hides in rock crevices. It can move quickly when it wants to by a jet propulsion like that of the squid, and it also will eject an inky fluid to confuse its enemy.

The most conspicuous part of the octopus is its head. It has a complexly designed brain enclosed in a cartilage cranium—the most complex brain of all invertebrates. The octopus is thought to be the most intelligent invertebrate. The octopus has two very complex eyes that are remarkably like the eyes of vertebrates, giving them excellent vision; squids have similar eyes.

Fig. 21.8 **Cephalopods**

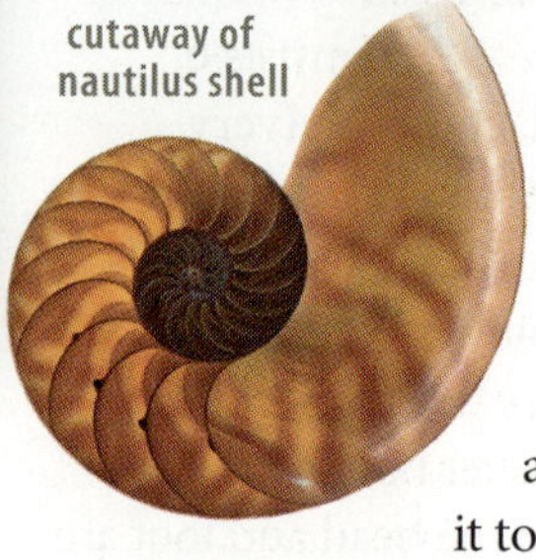


**Nautiluses.** The **nautilus** is the only cephalopod with an external shell, which is a marvel of architecture. Each shell consists of a series of chambers arranged in a spiral. The angles of the chambers are mathematically precise, and they form what mathematicians call a *logarithmic spiral.* New chambers are formed as the nautilus grows; the empty chambers, which form the bulk of the shell, are abandoned living quarters. These hollow chambers are connected by a tube to the animal's body and are used like the ballast chambers of submarines. The nautilus can fill the chambers with water, causing the animal to sink, or it can pump the water out, displacing it with air, to create a buoyancy that allows it to float. The nautilus is usually about 10 inches in diameter and possesses up to 94 tentacles, without suckers, to catch its favorite prey, the shrimp. The nautilus has two eyes, one on either side of its head; because nautiluses' eyes lack lenses, their vision is believed to be poorer than that of other cephalopods.

## Section Review 21.1

1. What body characteristics do mollusks have in common?
2. What are three familiar, broad groups of mollusks mentioned in the text?
3. What is the purpose of a clam's adductor muscles?
4. Why are a bivalve's siphons important?
5. What structure of a clam allows it to burrow into the sand? How is the locomotion of a scallop different from that of a clam?
6. What are the two largest known invertebrates?
7. What is the only cephalopod with an external shell?

### Critical Thinking

8. What are three differences between a squid and an octopus?

### Identify

foot, visceral hump, mantle, pearl, mantle cavity, incurrent siphon, excurrent siphon, univalve, slug, nudibranch, radula

### Observe

- Obtain an oyster or clam that is still in its shell. Open the shell and identify the different parts.

# 21.2 Echinoderms and Rotifers

## Echinoderms: Spiny-Skinned Invertebrates

Starfish, sea urchins, sand dollars, sea cucumbers, and brittle stars (similar to starfish, but with thinner rays) are all spiny-skinned animals, or **echinoderms.** Their phylum name, Echinodermata [ĭ·kī′nə·dûr′mə·tə], literally means "hedgehog skin." All echinoderms are marine animals, and all have spines that grow from a meshlike internal skeleton made of calcium carbonate. The spines may be long, as in the sea urchin, or very short, as in the sand dollar. Most echinoderms have a somewhat flattened **central disk** from which flexible arms, or rays, radiate. As adults, most echinoderms exhibit a five-sided radial symmetry; they come in a variety of shapes based on this one design. All echinoderms exhibit bilateral symmetry in their larval stages. Most are about the size of your hand, but some grow up to a yard across. Many are of a brilliant red, orange, green, or purple hue.

Fig. 21.9 **Echinoderms**

Echinoderms are the only animals with a **water-vascular system,** a complex system of water-filled tubes that extends throughout the body. This system is used in sensory perception, feeding, and respiration. An echinoderm also uses its water-vascular system for movement; it moves its limbs largely by pumping water into and out of the tubes in each limb.

**Starfish,** the best-known echinoderms, have a body that resembles a five-pointed star. Starfish generally have five arms, or **rays,** which extend from a central disk like spokes in a wheel. (Some starfish have as many as twenty rays.) At the end of each ray is a small light-sensitive **eyespot.** The body of a starfish consists of many jointed plates, which allow the animal to bend in many directions. Between the plates are small tufts of skin that act as *gills* for exchanging gases with the body fluid inside and the water outside. Also scattered among the plates are tiny *pincers.* Each pincer has two or three jaws and is attached to the starfish by a ball-and-socket joint. The pincers keep the surface of the starfish clean, preventing the gills from getting clogged with debris.

On the underside of each ray are two double rows of **tube feet,** which are used for holding onto things, for moving, and for opening the shells of clams and oysters. The tube feet of the starfish are connected to *radial canals,* which run the length of each ray from a *ring canal* in the *central disk.* The ring canal, radial canals, and tube feet form the primary parts of a starfish's water-vascular system. Water is taken into the body through the *madreporite* [măd′rə·pôr′īt], or sieve plate, and forced into the tube feet, causing them to expand. When the water is forced out of the tube feet, they contract, acting

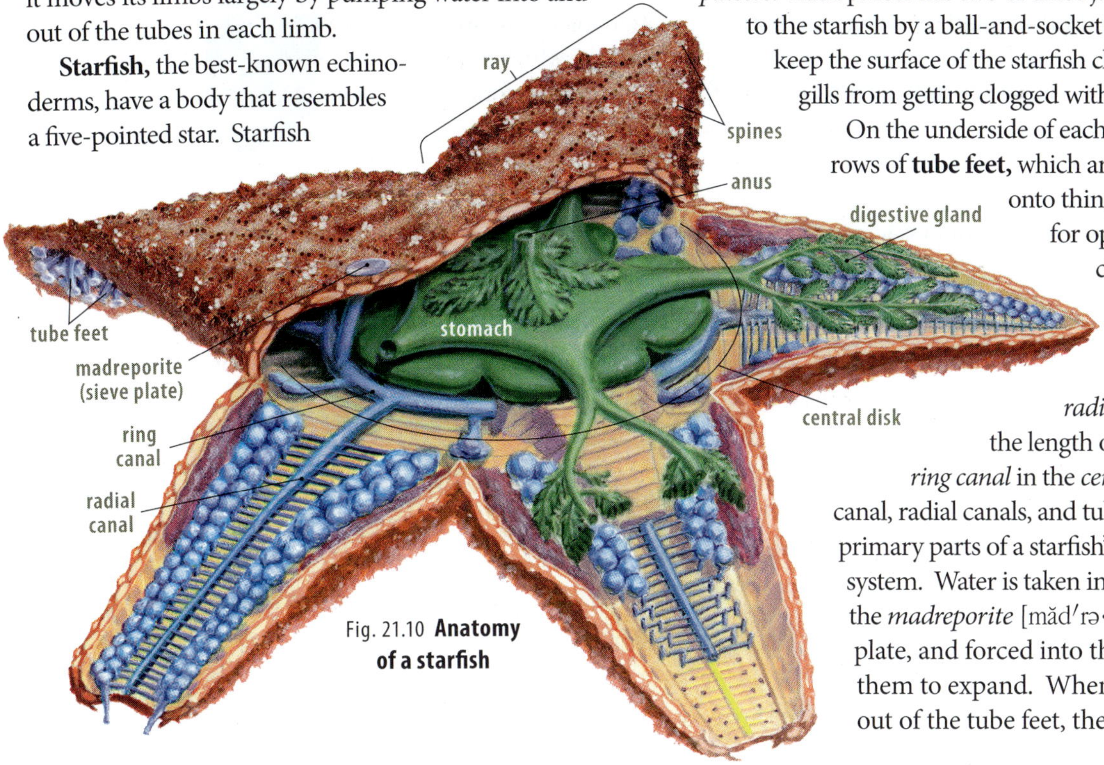


Fig. 21.10 **Anatomy of a starfish**

Fig. 21.11 **Starfish opening a clam**

like suction cups. With its tube feet, a starfish can pull on the two valves of a bivalve shell until the bivalve's muscles fatigue. This process may last for hours; tests have shown that a starfish may exert pressure on a bivalve for as long as 36 hours before giving up. If the bivalve relaxes and opens its shell even a fraction of an inch, the starfish everts (turns inside out) its stomach and digests the bivalve in its own shell. The stomach of a starfish, a baglike structure that can be everted easily through the animal's mouth, practically fills the central disk and extends into the animal's rays as well. After feeding, the starfish retracts its stomach and moves on.

It is estimated that a single starfish can eat a dozen or more clams and oysters in a single day. They are, therefore, a menace to oystermen and clam fishermen. Oyster fishermen once made the mistake of catching starfish and cutting them into many pieces and throwing the pieces overboard. They did not realize that *starfish have the ability to regenerate.* If part of the central disk remains, a torn-off ray will regenerate into a completely new starfish. The fishermen were actually mass-producing more starfish instead of ridding the oyster bed of them. Even without regeneration, starfish reproduce in large numbers, as females may produce 200 million eggs each year.

## Rotifers: "Wheel Bearers"

Perhaps the smallest multicellular animals are the **rotifers,** microscopic invertebrates found in freshwater lakes and ponds. The phylum name Rotifera [rō·tĭf′ər·ə] means "wheel bearer" and aptly describes these microscopic marvels of creation. In 1703, *Anton van Leeuwenhoek* discovered rotifers. Since then, about 2000 species have been identified. The movement of cilia on the front end of these animals resembles the rotation of a wheel.

Some rotifers are sessile, while others swim, creep on the bottom, or float on the surface of the water. These tiny animals are common in pond-culture studies. Even an organism this small, composed of only 500 to 1000 cells, has a complex digestive system consisting of a mouth, pharynx, *mastax* (grinding organ), esophagus, digestive gland, stomach, intestine, and cloaca. Wastes are expelled from its body through the excretory system, which consists of two long tubes connected to a bladder that empties into the cloaca.

A rotifer has a brain that is quite large compared to the total size of the organism. Connected to the brain are large nerves leading to sensory tufts located on the outside anterior (front) portion of the rotifer. The sexes are separate and reproduction is sexual. Some of the female's eggs are **parthenogenic** [pär′thə·nō·jĕn′ĭk]; that is, they require no fertilization. These eggs are laid in the summer. The eggs laid in the winter, on the other hand, require fertilization to be viable. These eggs are different from the summer eggs because they are protected from freezing by a thick shell.

Rotifers are important to the aquatic realm because they clean up natural wastes and provide food for small crustaceans, which in turn provide food for fish and, ultimately, man.

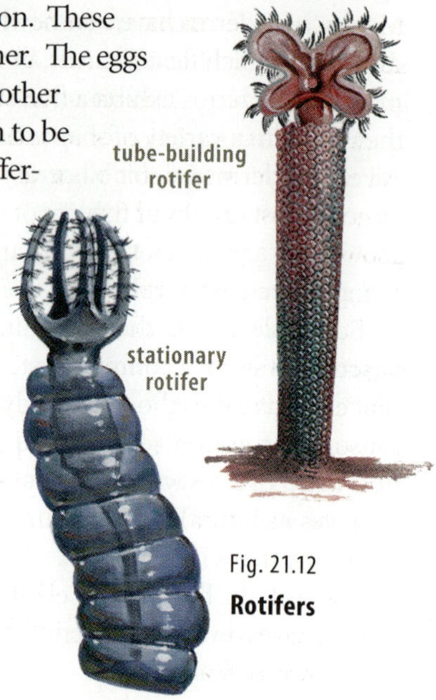

Fig. 21.12 **Rotifers**

## Section Review 21.2

1. Name five characteristics that echinoderms have in common.
2. What purposes does an echinoderm's water-vascular system serve?
3. Name three groups of echinoderms mentioned in the text.

### Application

4. How do rotifers demonstrate that small organisms are not necessarily simple?
5. In what ways are rotifers helpful to mankind?

### Identify

echinoderm, water-vascular system, ray, tube feet, rotifer, parthenogenic

# 21.3 Coelenterates and Sponges

## Coelenterates: Hollow-Intestined Invertebrates

Phylum Coelenterata [sə·lĕn′tə·răt′ə: "hollow intestine"] includes jellyfish, sea anemones [a·nĕm′ə·nēz], corals, and hydras [hī′drəz]. This phylum is also known as Cnidaria [nī·dâr′ē·ə] after the stinging cells, or **cnidocytes,** that the organisms possess. Most coelenterates live in the shallow, warm waters of tropical seas, but hydras live in fresh water.

**Coelenterates** have a saclike body with a central digestive cavity (the *gastrovascular cavity*) that has a single opening, or mouth, through which food enters and wastes are expelled. Extending from the body and surrounding the mouth are a number of slender *tentacles,* which are used for capturing prey. Numerous cnidocytes on the tentacles kill or incapacitate the prey (usually microorganisms, worms, and small fish), and the tentacles draw the prey to the mouth opening. Each cnidocyte contains an organelle called a *nematocyst,* which delivers the sting. The tentacles and stinging cells are perhaps the most characteristic structural features of coelenterates.

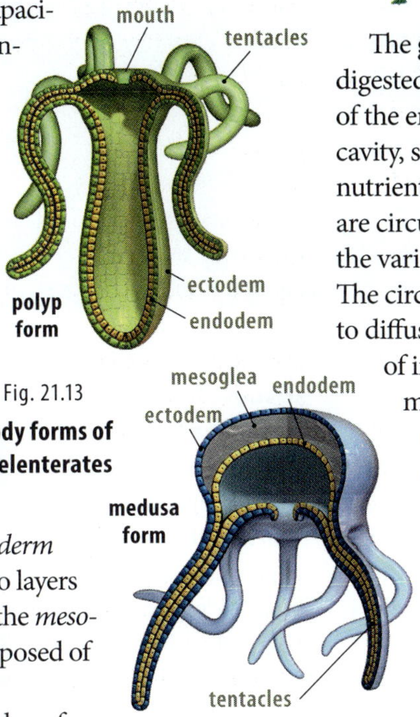


Fig. 21.13 **Body forms of coelenterates**

The body of a coelenterate is made of two layers of cells, the *ectoderm* (outer layer) and the *endoderm* (inner layer). Between these two layers is a jellylike middle layer called the *mesoglea* [mĕz′ə·glē′ə], which is composed of protein fibrils and sugars.

Coelenterates are built on a plan of radial symmetry and have two body forms—the polyp [pŏl′ĭp] and the medusa [mĭ·do͞o′sə]. A **polyp** is a cylindrical body type that is sessile and has its mouth facing upward. Examples of the polyp form are hydras, sea anemones, and corals. A **medusa** is a free-swimming, umbrella-shaped body type that has its mouth facing downward. An adult jellyfish is a medusa. Some coelenterates, including most jellyfish, have a life cycle that alternates between polyp and medusa. The polyp form of these coelenterates reproduces asexually, and the medusa form reproduces sexually.

**Hydras.** The **hydra** is a small (less than half an inch long) freshwater coelenterate of the polyp form. Its body consists of a round base, a cylindrical stalk, and a gastrovascular cavity with a mouth surrounded by tentacles. Although it usually remains attached to some object in the water, a hydra can move about by sliding along on its base or by turning "somersaults" with its armlike tentacles. When it is disturbed, a hydra may contract into a nearly spherical shape.

Fig. 21.14 **Hydra somersaulting**

The gastrovascular cavity digests food and circulates digested food to all parts of the hydra's body. The cells of the endoderm, which form the lining of the body cavity, secrete digestive enzymes, which break down the nutrient materials captured by the hydra. These materials are circulated around the cavity as they are digested, and the various cells of the hydra absorb whatever they need. The circulation of water within the cavity allows oxygen to diffuse into individual cells and wastes to diffuse out of individual cells of the hydra's body. Undigested materials that remain in the gastrovascular cavity are expelled through the mouth.

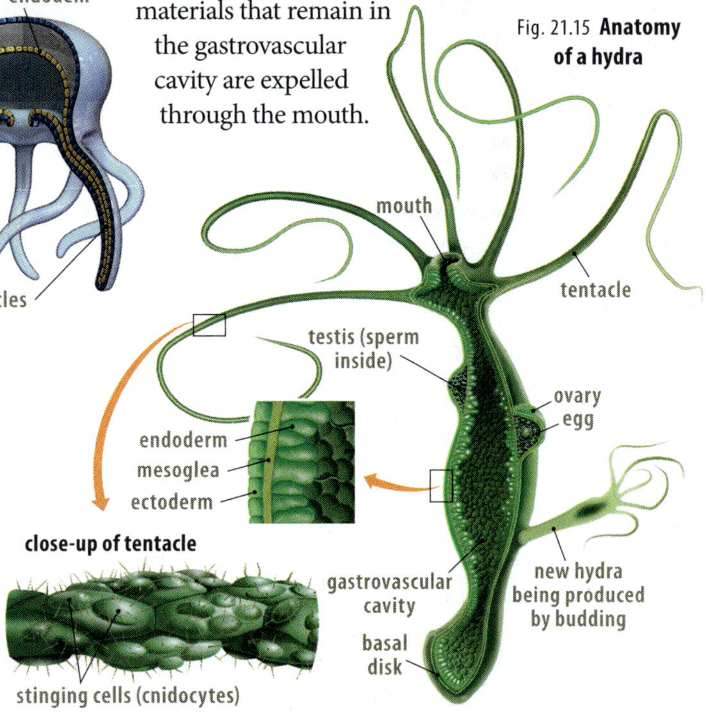


Fig. 21.15 **Anatomy of a hydra**

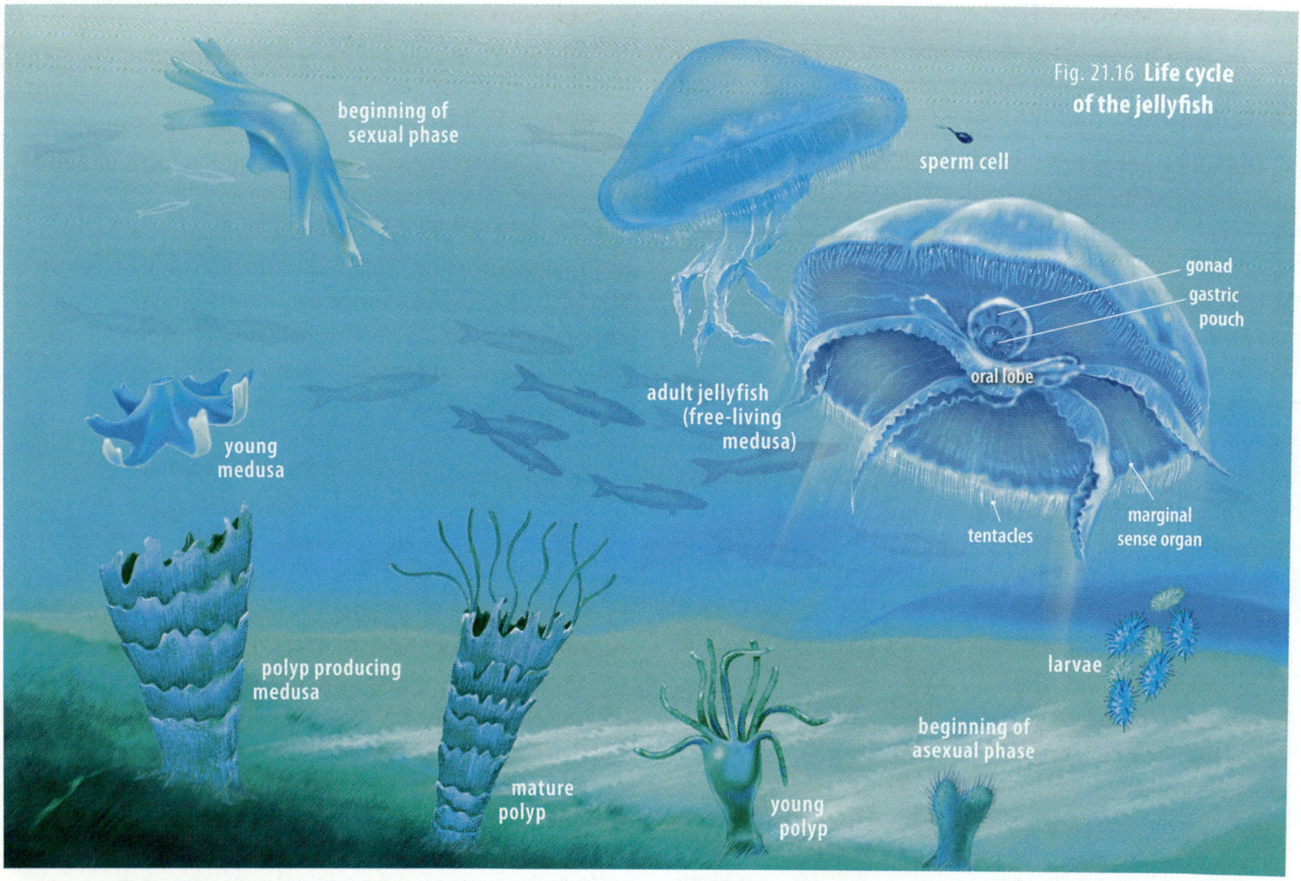


Fig. 21.16 **Life cycle of the jellyfish**

Hydras reproduce asexually by budding and sexually by the formation of sperm and eggs. In some species, both sperm and eggs are produced by the same animal; but in other species, there are definite male and female organisms.

The hydra shows a remarkable ability to regenerate. If a hydra is cut into several pieces, each piece will grow into a new, complete hydra. In a famous regeneration experiment, a hydra was turned inside out. Soon this tiny animal changed its exposed inner layer (endoderm) into an outer layer (ectoderm) and also changed its old ectoderm into endoderm.

**Jellyfish.** The adult **jellyfish** is a free-swimming coelenterate with the medusa type of body structure. Its mouth is located at the center of the "umbrella," and its tentacles hang down around the rim. The tentacles of a jellyfish are equipped with cnidocytes with which it paralyzes its prey or defends itself. Small jellyfish often eat plankton, while larger jellyfish may feed on small fish. A jellyfish's sting can be excruciatingly painful to man, and severe stings by some jellyfish can even be fatal. Muscle cells arrayed around the jellyfish's rim allow the jellyfish to rhythmically expel water from its mouth, allowing it to move by jetlike propulsion. Although most jellyfish are only a few inches in diameter, the *lion's mane jellyfish* of the North Atlantic *(Cyanea capillata)* may occasionally measure more than 6 feet in diameter.

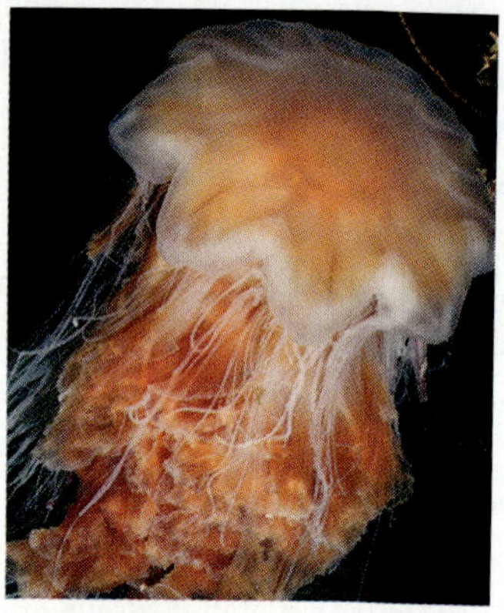

Fig. 21.17 **Lion's mane jellyfish**

Jellyfish undergo *alternation of generations* when they reproduce, meaning they use sexual and asexual methods alternately to reproduce. In the sexual part of the cycle, eggs and sperm are released into the central body cavity, where the eggs are fertilized. After fertilization, the *zygote* attaches itself to the outer edge of its parent's mouth and undergoes a period of development. When the zygote develops into a *larva,* it detaches itself from its parent and swims away.

The asexual part of the cycle begins when the larva stops swimming and attaches itself to some suitable object in the water. At this stage, the larva develops into a polyp. The jellyfish polyp forms buds, which mature and are released one at a time as medusae. The medusae are released when they are still microscopic, but these medusae can grow into adult jellyfish.

**Sea anemones.** Beautiful "flowers" of the ocean, **sea anemones,** are named after the daisylike flower called the anemone. Sea anemones are comparatively large marine coelenterates that have a polyp type of body; the mouth is located at the top of the organism, surrounded by an array of stinging tentacles. Some sea anemones are only a few inches long and perhaps half an inch in diameter, but other species may measure as much as 2 or 3 feet across. The relatively large size of the sea anemone complicates matters when it comes to supplying its body cells with nutrients. Diffusion of molecules alone cannot supply the cells of large organisms, nor can it rid those cells of wastes. To solve this problem, the Creator provided sea anemones with grooves that extend from the slitlike mouth at the top of the organism down into the gastrovascular cavity. Cilia line these grooves, and their motion creates currents that wash water down into the body cavity and between the grooves. In this way, a constant supply of food and oxygen is brought to even the most remote cells of the sea anemone, and these same currents of water sweep wastes away from the animal's cells and out of its gastrovascular cavity.

Fig. 21.18 **Sea anemone**

**Corals.** The **coral** is a colonial coelenterate. In body structure, corals are very much like miniature sea anemones, except that the corals produce a limestone (calcium carbonate) cup at the base of their bodies. This cup is a shelter into which the small and delicate coral polyp can retract when disturbed. Because millions of corals live together in colonies, the limestone cups eventually form vast limestone structures called **coral reefs.**

Corals grow in warm, shallow seas where the water temperature does not drop below 70°F. The warm *Coral Sea,* which is northeast of Australia, is noted for its numerous coral reefs. This is the location of the world's largest coral reef, the *Great Barrier Reef,* which is about 50 miles wide and extends for about 1200 miles along the northern coast of Australia.

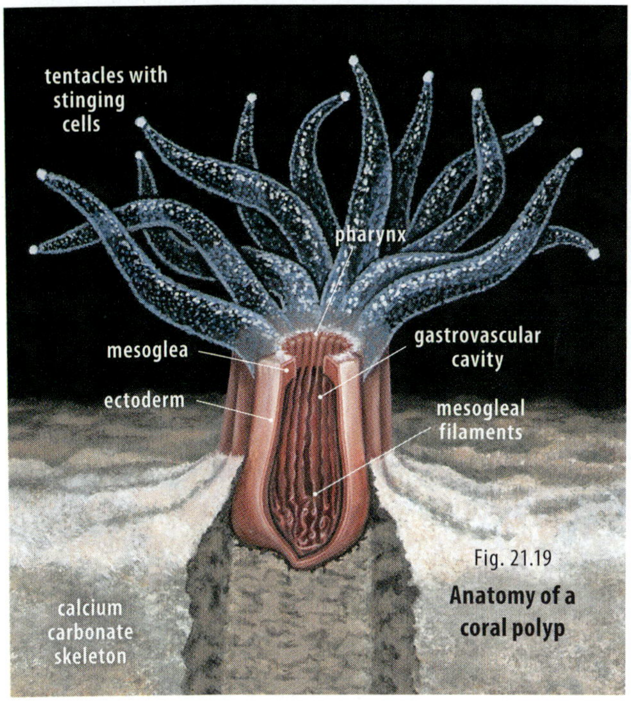


Fig. 21.19 **Anatomy of a coral polyp**

Fig. 21.20 **Coral**

## Porifera: Sponges

There are about 5000 species of **sponges,** which make up the phylum Porifera [pə·rĭf′ər·ə: "pore-bearers"]. Their bodies consist of only two cell layers, and they always grow attached to some object in the water where they are found. Most live in the sea; only one group lives in fresh water. Most sponges prefer warm, shallow waters, but some have been found living successfully in waters that are three miles deep.

Although sponges are multicellular animals, their cells act more or less independently of each other, with no coordination of their activities. For example, if a sponge is touched, only those cells that are actually touched respond to the touch. *Sponges are the only multicellular animals that do not have at least a rudimentary nervous system;* apparently they do not even have nerves. Sponges do have specialized cells that perform special jobs, but the cells are not organized into tissues and organs.

A sponge can be described as a hollow, saclike animal with a large opening opposite its base. The body opening, or mouth, of a sponge is called an **osculum** [ŏs′kyə·ləm]. The osculum is not a mouth as you think of a mouth; it does not take in food for the sponge. In fact, water and waste materials flow out of the sponge through the osculum; an opening of this type through which water leaves an organism is called an *excurrent pore.* The sponge is alone among multicellular animals in possessing an osculum that serves only as an outlet. Water enters the sponge through very small **ostia** *(incurrent pores),* which are located in the sides of the animal.

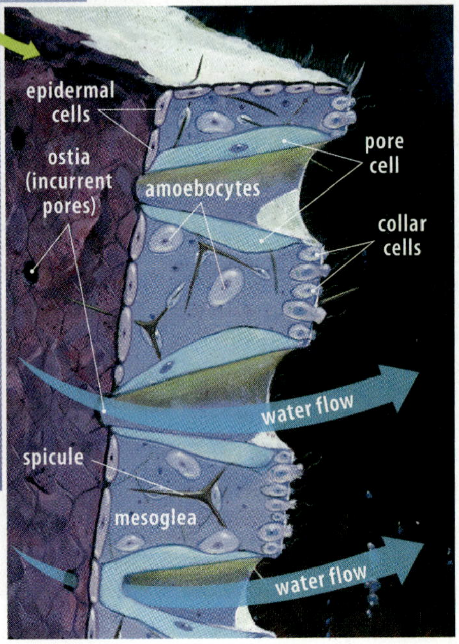

Fig. 21.21 **Anatomy of a simple sponge (asconoid type)**

Fig. 21.22 **Sponges**

Like the coelenterates, the simplest sponges have an inner layer of cells and an outer layer of cells in their walls (Fig. 21.21). The outer layer of cells are the *epidermal cells,* which provide the sponge with its outer covering. The inner layer of cells, which line the body cavity, are called **collar cells** in the sponge because of their unusual shape. These cells have a "collar," and each of them has a flagellum extending from the cell out through the "collar" into the body cavity. More complex sponges have folded body walls with numerous canals branching inward to the central chamber. The most complex sponges are honeycombed with thousands of small chambers lined with collar cells and connected by narrow canals. It is this latter group that was once widely used as household sponges, although they have been largely replaced by synthetic imitations.

A jellylike *mesoglea* fills the space between the collar cells and the pore cells. Situated within the mesoglea are special cells that manufacture **spicules** [spĭk′yo͞olz], the crystalline support structures of the sponge. Special cells called *amoebocytes* (because they resemble amoebas) wander about within the mesoglea and help transport food, oxygen, and wastes to or from all parts of the sponge.

Sponges obtain the nutrient materials they need from the water. The flagella of the collar cells create a continuous current of water that flows into the central cavity through the pore cells and out through the osculum. The collar cells trap, ingest, and digest microorganisms that flow in with the water. Sponges also obtain oxygen from the water flowing through their bodies.

*Sponges may reproduce sexually or asexually.* When a sponge reproduces sexually, a special type of flagellated larva develops from the zygote. After a short period of living independently, the larva attaches itself to an object on the sea floor and develops into a mature sponge. One way that sponges reproduce asexually is by *budding.* Sponges have exceptional powers of regeneration; even the smallest piece of detached sponge may grow into a new sponge. In one experiment, sponge tissue was passed through screen with mesh so fine that only single sponge cells could pass through it. Amazingly, the sponge cells rejoined, and a new sponge soon developed.

### Section Review 21.3

1. List four characteristics of coelenterates.
2. What are three differences between the polyp and medusa forms of coelenterates?
3. What body form (medusa or polyp) characterizes a hydra? An adult jellyfish? An anemone? A coral?
4. How are corals and sea anemones similar? In what ways are they different?

**Identify**

cnidocytes, polyp, medusa, coral reef, osculum, ostia, spicules

## 21.4 Annelids: Segmented Worms

There are several phyla of worms and wormlike animals, of which we shall study three: *segmented worms, flatworms,* and *roundworms.* **Segmented worms** *(annelids),* which include earthworms, sea worms, and leeches, belong to phylum Annelida [ə·nĕl′ĭ·də]. Segmented worms have cylindrical bodies that are divided into many consecutive ringlike segments. On one end of the long body is a head, and on the other end is an anus. Inside most annelids, each segment is separated from the next by a dividing wall called a *septum;* the worm's digestive tract, nerves, and blood vessels pass down the middle of the worm through openings in each septum. Most annelids have rows of external bristles attached to each segment.

### Earthworms

The common **earthworm** is a typical annelid and makes a fine specimen to study. These nocturnal animals are easily captured and preserved for laboratory dissection.

Earthworms have four pairs of **setae** [sē′tē; sing. seta: sē′tə], or bristles, on each body segment. The worm can extend or retract these structures as necessary and can regenerate lost ones. An earthworm moves by extending the setae of its anterior (front) segments into the soil and shortening its body. The worm then extends the

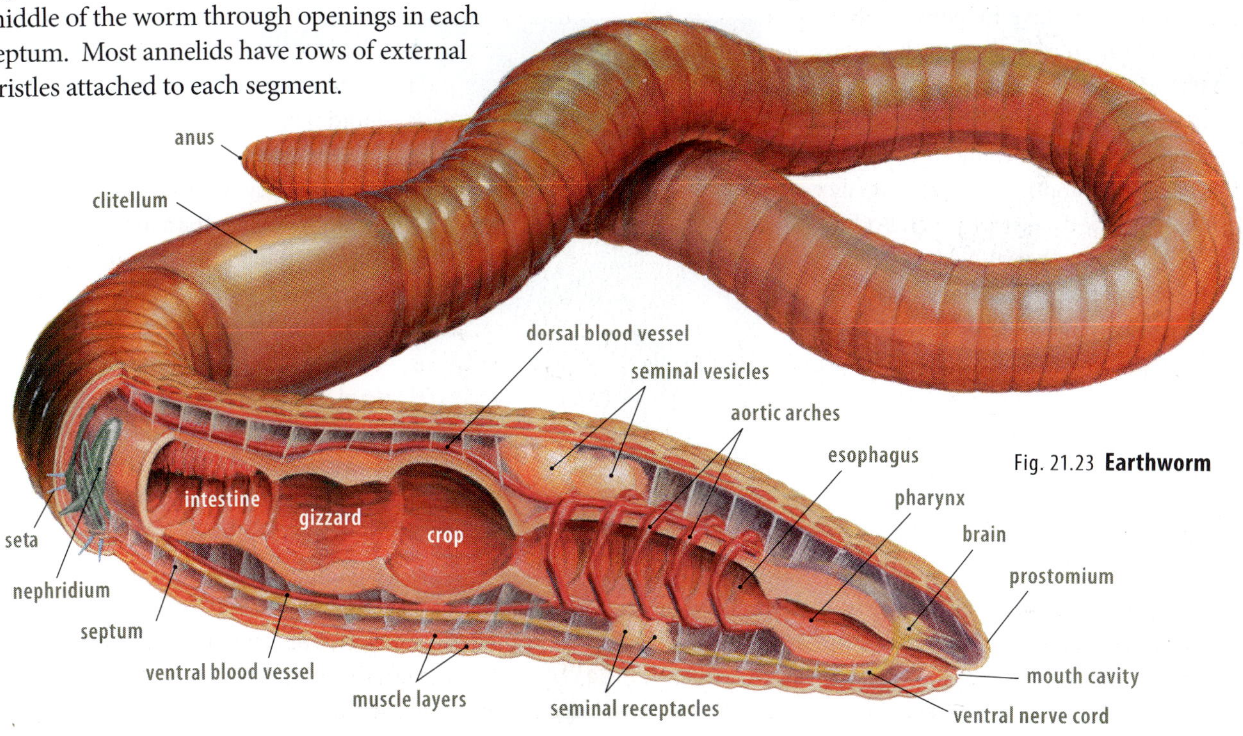


Fig. 21.23 **Earthworm**

setae of its posterior (rear) segments into the soil, retracts the setae of the anterior segments, and pushes its body forward through the soil by stretching its length.

**Respiration.** An earthworm's smooth skin permits direct exchange of gases with the air. Because oxygen must diffuse into the earthworm through its body wall, the worm's outer surface must remain moist. Mucous glands maintain this moist condition, but when the weather is unusually dry or the temperature is low, as in winter, the earthworm burrows deep into the earth to keep from drying out or freezing. During drought conditions, the earthworm can secrete a heavy coating of mucus around itself and become dormant until rain relieves the situation. If an earthworm is dried by the sun, it will die of suffocation because the exchange of oxygen and carbon dioxide cannot take place. When heavy rain saturates the topsoil, earthworms crawl to the surface for air. If they remained in the water-soaked earth, they would drown.

**Sensation.** The *nervous system* of the earthworm consists of nerves, a nerve cord, and a tiny brain. The *brain,* which is about the size of a pinhead, consists of a cluster of nerve cell bodies, or ganglion, and is located in the third body segment. The nerve cord coordinates the muscles of the body and allows the earthworm great mobility. Earthworms do not have eyes and other sensory organs as do certain other annelids, but the earthworm is sensitive to many stimuli, including light, touch, certain chemicals, and electricity.

**Circulation.** The *circulatory system* consists of red blood (containing hemoglobin), two large blood vessels, two smaller connecting blood vessels in each body segment, and ten "hearts," or **aortic arches,** arranged in five pairs near the earthworm's head.

**Digestion.** In structure, the earthworm is like a tube, and its *digestive system* is like a tube within a tube, consisting of a mouth opening, a gullet (esophagus), a crop, a gizzard (stomach), an intestine, an anus, and special digestive glands. Humus (dead and decaying organic matter of the soil) is the food of earthworms. To get this food, the earthworm quite literally eats its way through the soil. The shovel-like upper lip, or **prostomium,** scoops soil into the mouth. The food and soil mixture passes from the mouth into the *pharynx,* the *esophagus,* and then into the *crop,* which is a temporary storage place for food. From the crop, the food passes into a *gizzard,* a thick-walled muscular structure that grinds food into fine particles; fine grains of sand assist the gizzard in the grinding process. After being ground up, the food passes into the *intestine,* where it is digested and absorbed. The earthworm's intestine extends for three quarters of the worm's total length. A fold in the intestine provides a large absorbing surface. Undigested wastes and soil particles pass out of its body through the *anus.*

Because they live in the soil and eat humus, earthworms play an important role as soil builders. Earthworm waste materials help enrich the soil, and earthworm tunnels help aerate the soil and improve water drainage. Thus earthworms condition the soil and improve growing conditions for plants.

**Excretion.** Liquid wastes are eliminated from the earthworm's body by **nephridia** [nə·frĭd′ē·ə], coiled excretory tubes that are open at both ends and function like tiny kidneys. One end opens into the earthworm's body cavity, and the other end opens to the exterior. Liquid wastes drain from the body cavity through these structures. Nephridia are quite complex structures; besides removing liquid wastes from the earthworm, they help maintain a proper balance of salt for the worm and help prevent the loss of excess moisture from its body.

Fig. 21.24 **Sea worm**

## Sea Worms

**Sea worms** *(polychaetes)* are the largest class of segmented worms, but they are not very well known because they live in the sea. They range from less than an inch to over six feet in length. Most hide from predators in tubes that the worms construct, often by cementing together grains of sand. When they feed, the sea worms may come only partially out of their tubes or may come completely out. These worms usually have two pairs of eyes on their heads plus several other sensory appendages. All sea worms have a pair of fleshy lobes called **parapodia** [păr′ə·pō′dē·ə] on each body segment. The parapodia assist in movement and in the exchange of oxygen and carbon dioxide.

## Leeches

**Leeches** do not have setae like the earthworms or parapodia like the sea worms. Leeches have somewhat flattened bodies with one *suction disk,* or sucker, at each end of their body. With these suction disks, leeches cling to their hosts. Some leeches are free-living organisms, but most are parasites. The free-living leeches prey on insect larvae, earthworms, and other small animals. The parasitic leeches feed on the blood of their hosts: snails, fish, reptiles, mammals, and humans. Most leeches are aquatic.

A leech attaches to its host with its posterior sucker. Sharp teeth in the anterior sucker puncture the host's skin so that the leech can suck its blood. An *anticoagulant* (a substance that prevents blood from clotting) in the leech's saliva keeps the blood flowing into the animal. When feeding, a leech gorges itself, sucking several times its weight of blood from the host organism. It stores the blood in its greatly expandable crop until needed. A single feeding can last a leech for several months.

Leeches have been used for medicinal purposes since the Middle Ages. For many years, they were used to remove blood from sick persons (often in the false belief that diseases and fevers were caused by "bad blood"). Even today, however, leeches are used medically to remove excess blood from limbs that have just been surgically reattached following an accident. The anticoagulant found in the saliva of leeches, which seems to work far better than any anticoagulant that man has devised, is also used to prevent and treat blood clots.

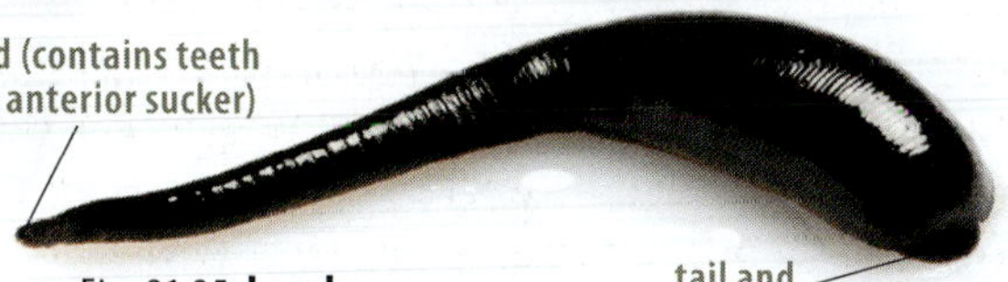

Fig. 21.25 **Leech**

### Section Review 21.4

1. What primary body characteristic do all segmented worms have in common?
2. How does an earthworm obtain oxygen? Why is it essential that an earthworm's skin remain moist?
3. In what ways do earthworms benefit mankind?
4. What unique features do sea worms have on each body segment? What purposes do these structures serve?

**Identify**

setae, aortic arches, nephrida, parapodia

# 21.5 Flatworms and Roundworms

## Platyhelminths: Flatworms

**Flatworms** *(platyhelminths)* belong to phylum Platyhelminthes [plăt′ĭ·hĕl·mĭn′thēz], which is a Greek word meaning "flat worm." From their name, you have no doubt realized that flatworms are worms with flattened bodies. There are two classes of parasitic flatworms (the *flukes* and *tapeworms*) and one class of free-living, or nonparasitic, flatworms (the *planarians*). Flatworms are quite common and are found primarily in aquatic environments, although some are terrestrial, dwelling in soil and in decaying plant matter. Over 20,000 species are known.

**Planarians.** The **planarians** are free-living flatworms that live as scavengers in ponds, streams, and lakes. They can be collected by suspending a piece of liver from a string into a pond. Within a day or so, planarians will be attracted to the liver. Planarians are small animals—most are less than an inch in length. They can be black, brown, or white. Cilia on the ventral (lower) surface allows a planarian to glide smoothly along. When a planarian is on the surface of an object, it secretes a slimelike substance that helps it in its locomotion. If it is touched, muscular contractions permit it to move quickly away from the contact stimulus.

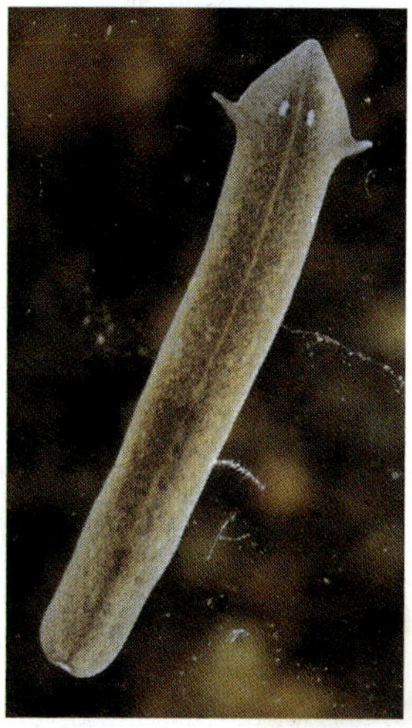
Fig. 21.26 **Planarian**

Planarians have four body systems: a nervous system, a digestive system, an excretory system, and a reproductive system. As part of its *nervous system,* each planarian has two light-sensitive **eyespots.** (The eyespots are responsible for the planarian's nickname, "the cross-eyed worm.") A ladderlike network of sensory

Fig. 21.27
**Anatomy of a planarian**

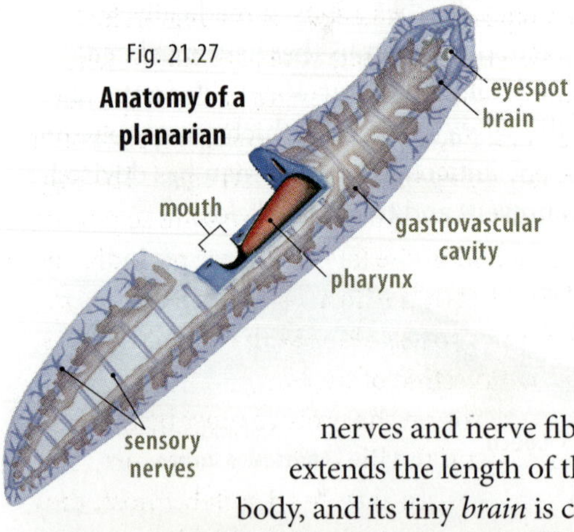


nerves and nerve fibers extends the length of the worm's body, and its tiny *brain* is connected to the eyespots and the nerve network. This nervous system allows a planarian to respond in a coordinated fashion to many types of stimuli.

The *mouth* of a planarian is not in its head, as you might expect, but is located near the middle of the animal's lower (ventral) surface. Attached to the inside of the mouth is a tubelike *pharynx,* or throat, which can be extended out through the mouth. A branched *gastrovascular cavity* is attached to the pharynx, and cells lining this cavity digest food particles that the animal sucks in. Projections of the gastrovascular cavity extend into all parts of the planarian's body so that none of its body cells are very far from a supply of food. Undigested wastes are expelled through the animal's mouth, but cell wastes are collected and excreted through tiny *excretory pores.*

Fig. 21.28
**Regeneration of a planarian**

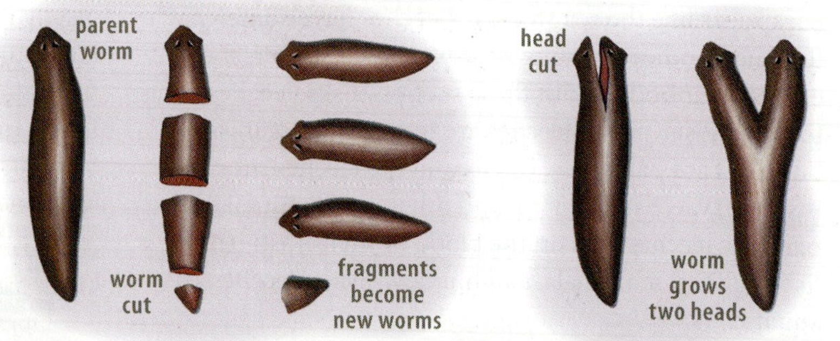


*Most planarians reproduce sexually, but they can also reproduce asexually.* If a planarian is separated into two or more pieces, each of these pieces except the tip of the tail can grow into a new planarian. If the planarian's head is sliced vertically, it will develop two heads.

**Flukes.** **Flukes** [class Trematoda: trĕm′ə·tō′də] are parasitic flatworms that spend a part of their lives as parasites within snails. Unfortunately, many flukes spend another part of their lives as parasites within man or domestic animals, afflicting many millions of people in Asia and Africa. These flukes may infest the liver,

Fig. 21.29 **Life cycle of a sheep liver fluke**

lungs, intestines, or blood of man, depending on the species. Other types of flukes infest various mammals. Flukes have no cilia, but they do have one or more suckers by which they cling to the tissues of their host.

*The life cycle of a fluke is quite complicated.* Most require at least two alternate hosts, and all spend a part of their lives as parasites in snails. The most familiar fluke, the *sheep liver fluke,* has a typical life cycle. As an adult, the sheep liver fluke lives in the liver and then the gallbladder of a sheep; the fluke releases its eggs from the gallbladder into the intestine. If these eggs enter water after being passed by the sheep, they hatch into a larval form of the fluke. The larvae enter a certain type of snail and develop there. The larvae also reproduce asexually before leaving the snail. Outside the snail, the larvae form cysts and remain on the grass. If sheep eat these cysts as they graze, the flukes enter the sheep's body and migrate from the stomach to the liver, beginning the cycle again.

*Clonorchiasis* [klō′nôr·kī′ə·sĭs] is a liver disease prevalent in certain areas of China, Korea, Japan, and other nations of southeast Asia. The *Clonorchis* liver fluke lives in a snail and then moves into freshwater fish. When these fish are eaten raw or undercooked, as they often are in Asian countries, the fluke matures in the human liver, causing medical problems. If eggs released into human feces make their way to bodies of water in which fish live, the cycle continues. This disease cycle can be broken by cooking seafood thoroughly before eating it.

**Tapeworms. Tapeworms** are parasitic flatworms that live in the intestines of their hosts. An adult tapeworm has a long, flat, ribbonlike body divided into numerous segments. In extreme cases, tapeworms have attained lengths of 10–20 feet, although they are only a few inches long in most cases. The small, knoblike head, called the **scolex** [skō′lĕks′], is equipped with hooks and suckers that permit the tapeworm to attach itself firmly to the wall of the host's intestine (Fig. 21.30). These worms have no mouth and no digestive system; the tapeworm simply absorbs whatever nutrients it needs from the digested food in its host's intestine.

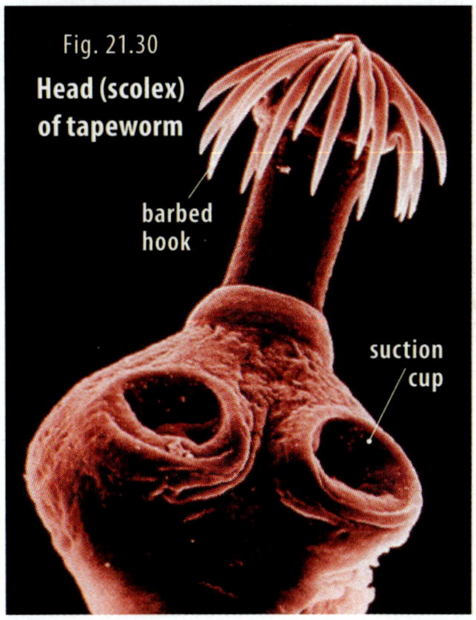

Fig. 21.30
**Head (scolex) of tapeworm**

Growth in a tapeworm takes place in a region just behind the head, where new segments are continually formed; the tapeworm grows in length each time a new segment is formed. Each segment contains reproductive structures for producing both sperm and eggs. The eggs are fertilized within the segments, and mature segments near the posterior end of the tapeworm are periodically detached and pass out of the host. The tapeworm eggs are dispersed in this process along with the solid wastes that pass from the intestine of the host.

Tapeworm embryos on the ground must be taken into an alternate host if the life cycle of a tapeworm is to continue. If these eggs are eaten by a suitable host, they hatch, and the larvae migrate through the bloodstream to the muscles of the alternate host. When the juvenile tapeworms reach these tissues, they form cysts and become dormant. Later, if the meat of the alternate host is eaten, the tapeworms in the cysts are freed, and they then attach themselves to the intestine of their new host and begin their existence as adult tapeworms. The tapeworms that live as adults in the intestines of man use cattle, pigs, fish, and certain other animals as alternate hosts. They can be killed only by freezing the meat for at least a week or by cooking it thoroughly.

For some species of tapeworm, man is the alternate host. These tapeworms are dangerous parasites because they may form cysts in vital organs such as the brain, heart, or eye. But tapeworms that use man as the final host are also serious, because they weaken the host and rob him of nutrients, sometimes block the passage of food through the intestine, and release toxic substances that are harmful to the host.

## Nematodes: Roundworms

**Roundworms** *(nematodes)* belong to phylum Nematoda [nĕm′ə·tō′də], a name meaning "thread worm." Roundworms are nonsegmented animals that have long and slender bodies covered by a thin cuticle. There are about 20,000 known species, and they can be found in practically every environment. Nematodes are especially abundant in soil; many live as parasites in plants or animals, but some are free-living.

The body of a roundworm is typically long, slender, and tapered on both ends. Some are microscopic, and most are no more than a few inches long, but some species that parasitize whales can be almost 30 feet long. Roundworms have a *digestive system* with two openings: a *mouth* at the anterior end and an *anus* at the posterior end. Roundworms also have a *muscular system,* a *nervous system,* and a *reproductive system.* Four parasitic roundworms that affect man are the *filaria worm,* the *hookworm,* the *trichina* [trĭ·kī′nə] *worm,* and the Ascaris [ăs′kə·rĭs] worm.

**Filaria worm.** The **filaria worm** inhabits the vessels of the lymphatic system and produces a serious condition called *elephantiasis* [ĕl′ə·fən·tī′ə·sĭs]. Filaria worms block the lymph vessels, causing tremendous swelling of the legs and other extremities. Filaria worms and elephantiasis are serious problems in Asia, Africa, many of the Pacific islands, and parts of South America. Another type of filaria worm lives in the heart of dogs and cats, producing a condition known as *heartworm,* which can be fatal to the animal. Filaria worm larvae are transmitted to their hosts by mosquitoes or other biting insects.

**Hookworm.** **Hookworms** are parasitic roundworms that live in their host's intestine and suck the host's blood. They are widespread in subtropical and tropical parts of the world. The young hookworm hatches from an egg and develops as a larva in soil. The hookworm larva works its way into a host by burrowing through the skin (often the feet if the host goes barefooted) and entering the bloodstream. The hookworm larva travels to the lungs, where it breaks through a capillary wall into an alveolus. From there, the worm works its way up the host's trachea to the throat. The larva is then swallowed and passes into the intestine. Normally, stomach acid destroys organisms that are swallowed, but the worm's cuticle protects it. Once in the intestine, a hookworm larva attaches itself to the intestinal wall and matures into an adult worm. The adult feeds upon the host's blood from the rich supply of blood vessels in the intestinal wall, causing anemia. Fortunately, hookworms can be controlled by observing good sanitary practices in the disposal of human feces and by wearing shoes.

**Trichina worm.** The **trichina worm** is a parasitic roundworm that causes the disease *trichinosis.* Humans become infected with trichina worms usually through eating poorly cooked pork that is infested with this roundworm's cysts. One ounce of

Fig. 21.31 **Life cycle of a hookworm**

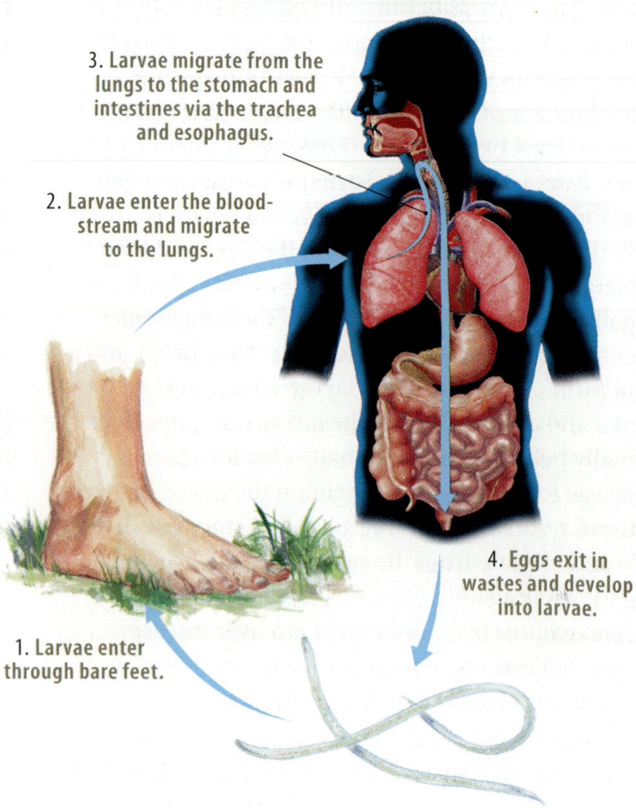


Fig. 21.32 **Life cycle of a trichina worm**

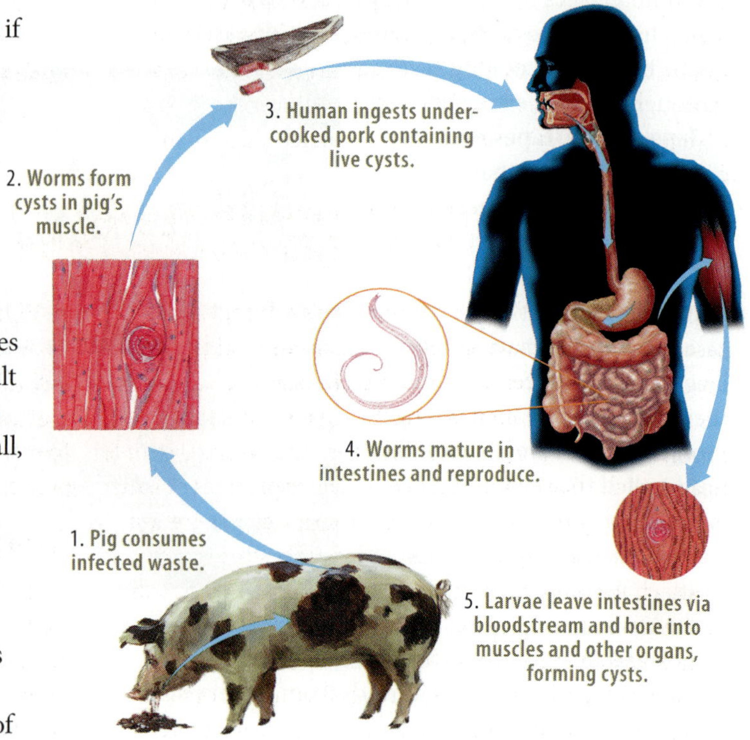

infected pork may contain up to 100,000 encysted worms. The cysts dissolve in the human intestine, and the worms inside grow to adulthood attached to the intestinal wall. The adult female worms produce eggs, which hatch internally. The larvae that emerge from the eggs enter the host's lymphatic system and bloodstream and thus are distributed to all parts of the body. One infection can result in a half billion larvae simultaneously boring into the body, causing pain, fever, anemia, weakness, and swelling. The larvae bore their way into muscle tissue, where they form cysts. They may remain in this dormant stage for years; after the worms move into the muscles and form cysts, there is no way to remove them. The best prevention against trichinosis is to cook meat, especially pork, thoroughly.

***Ascaris* worm.** Because they put soiled hands and other soiled objects into their mouths, children are most commonly affected by ***Ascaris* worms.** These parasitic roundworms, which feed on partially digested food, can live in the intestines of humans and of most mammals. If they become so numerous that they block the intestines, the host could die.

Eggs from these worms pass out of a host's intestine and find their way into new hosts by means of dirty hands, unwashed vegetables, or contaminated water. The eggs hatch in the intestine, but the larvae burrow through the intestinal wall and enter the host's bloodstream. Larval *Ascaris* worms leave the bloodstream at the lungs and crawl up the windpipe into the throat. Like hookworms, these larvae are swallowed and attach themselves to the wall of the intestine. They mature and live as adults attached to their host's intestinal wall. A female may lay as many as 200,000 eggs each day.

*Ascaris* infection is prevented much the same way other parasitic worm infections are prevented: through cleanliness. Disposing of human wastes, washing well before eating, not putting foreign objects into the mouth, and thoroughly cleaning foods can all but eliminate *Ascaris* worm infection. Modern medical science has found drug treatments for these diseases, but they are of little value in stopping the spread of disease in areas where good sanitation is not practiced.

## Section Review 21.5

1. What are three groups of flatworms mentioned in the text? Which of these groups is *not* parasitic?
2. How does a tapeworm obtain nutrients, since it has no digestive system?
3. List four parasitic roundworms (nematodes).
4. How are flatworms and roundworms similar to segmented worms? How are they different?

### Critical Thinking

5. Read God's rules of health and sanitation for Israel in Lev. 11:1–8 and Deut. 23:12–14. How might these rules have protected His people from diseases such as those caused by parasitic worms? Relate your answer to the mode of transmission of many parasitic worms.

### Identify

eyespots, fluke, scolex, filaria worm, trichina worm

## 21.6 Protozoa: Single-Celled Creatures

The last group of organisms that we shall study are the **protozoa,** unicellular animal-like creatures in kingdom Protista. As you would suspect, most of these creatures are microscopic. Even though many people think of the protozoa as the simplest creatures, they are far from simple forms of life. Each individual cell is a complete organism performing functions of digestion, excretion, respiration, movement, and responsiveness. In many respects, protozoa are far more complex than most of the cells in your body. The more biologists discover about the immense complexity of a protozoan, the more we are amazed at God's power and wisdom. It took no less an act of Creation to create an amoeba than it did to create a whale.

There are about 30,000 living species of protozoa. They may be found in nearly every habitat on Earth, including ponds, oceans, mud, soil, and even within the bodies of many plants, animals, and people. To organize their study, biologists have divided the protozoa into four groups according to the mode of locomotion.

### Flagellates: Protozoa with Flagella

**Euglenas.** Members of the genus *Euglena* are especially interesting to scientists because of their unusual resemblance to both plants and animals. **Euglenas** are found in freshwater ponds and streams. Often, they give the water a brilliant green color because, like plants,

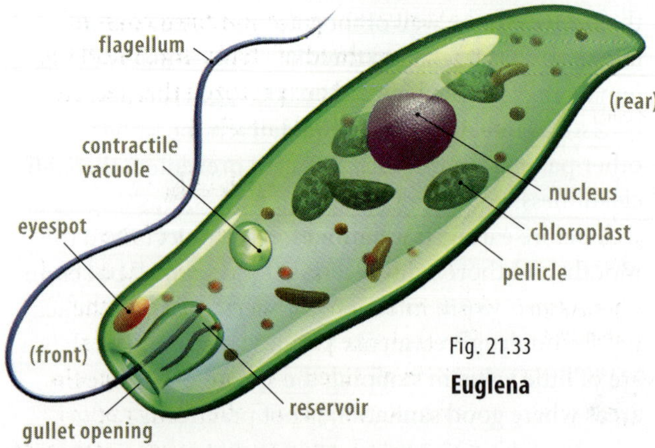

Fig. 21.33
**Euglena**

they contain chlorophyll. They range in size from $\frac{1}{1000}$ to $\frac{1}{100}$ inch and can be seen best with the microscope under high power. A euglena does not have a cell wall like a plant, but its long and somewhat cylindrical body is covered by a thickened membrane called a **pellicle** [pĕl′ĭ·kəl], which acts as a cell wall. A euglena moves about freely by means of its **flagellum,** a whiplike filament located on its anterior end. The flagellum is held straight in front, and the tip is rotated to propel the euglena forward through the water. The euglena has another method of movement, which is found in no other organism, **euglenoid** [yo͞o·glē′noid] **movement.** In this type of movement, the body itself changes shape. The posterior end is drawn forward, giving the body a rounded shape. Then the anterior end is stretched forward, moving the organism and returning it to its original shape.

At the anterior end, near the flagellum, there is a *gullet opening* leading to a reservoir. Beside the reservoir is a *contractile vacuole,* which stores and discharges water and liquid wastes, and small helper vacuoles. A euglena's *nucleus* is located near the center of the cell.

Like a plant, a euglena possesses *chloroplasts* and thus can manufacture its own food. It is aided in the process of photosynthesis by a red, photosensitive *eyespot.* The eyespot, which is near the gullet, directs the euglena toward the light. A euglena can also catch and ingest food like an animal, but it seldom needs to.

Euglenas reproduce chiefly by **binary fission.** In this asexual method of reproduction, a euglena splits lengthwise. One half receives the old flagellum, and the other half grows a new one. When unfavorable conditions occur, a euglena protects itself by losing its flagellum and forming a cyst. When conditions become favorable for growth, a young euglena emerges from the cyst.

**Trypanosome.** The **trypanosome** (genus *Trypanosoma* [trĭp′ə·nə·sō′mə]) is a parasitic flagellate that is transmitted to humans by bites from the *tsetse* [tsĕ′tsē] *fly.* It causes the disease *African sleeping sickness.* The trypanosome has two flagella. One flagellum is free, and the other is attached to the length of the body to form an *undulating membrane,* which aids the organism in its movement through the blood.

## Sarcodines: Protozoa with Pseudopods

The **amoeba** (genus *Amoeba*) is perhaps the best known and most common of the **sarcodines,** protozoa that have flexible bodies and can change shape at will; amoebas use this extraordinary ability both to move about and to feed. To move, an amoeba continually pushes out fingerlike projections of its cell membrane called **pseudopods,** or "false feet," which give the organism an ever-changing, irregular body outline. *The pseudopods are used for locomotion and for ingesting food particles.* Ordinarily, the amoeba has several pseudopods sticking out in several directions at the same time.

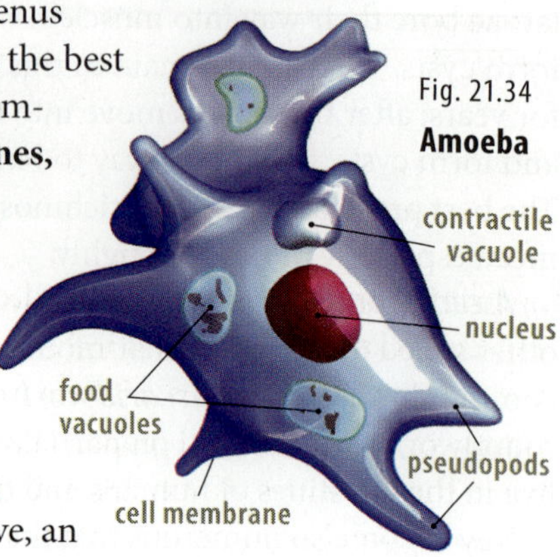

Fig. 21.34
**Amoeba**

If a pseudopod comes into contact with a particle of food or a small organism, the pseudopod changes shape to engulf the particle and take it into the cell. Food taken into the amoeba's body is enclosed within a *food vacuole,* where it is digested by special enzymes. When digestion is complete, the food is absorbed into the cell and any waste materials are expelled from the cell. The food vacuole then disappears.

At least one contractile vacuole is always present in the body of the amoeba. This vacuole regulates the water content of the cell. As the vacuole collects unneeded water, it swells; when it becomes full, the contractile vacuole suddenly contracts, expelling excess water from the amoeba.

When an amoeba reaches its maximum size, the surface area of the membrane can no longer absorb the quantity of materials necessary to support a cell of its volume. At this time, the amoeba reproduces asexually by fission; the nucleus divides in half, and each half migrates to an opposite end of the cell; the mature amoeba

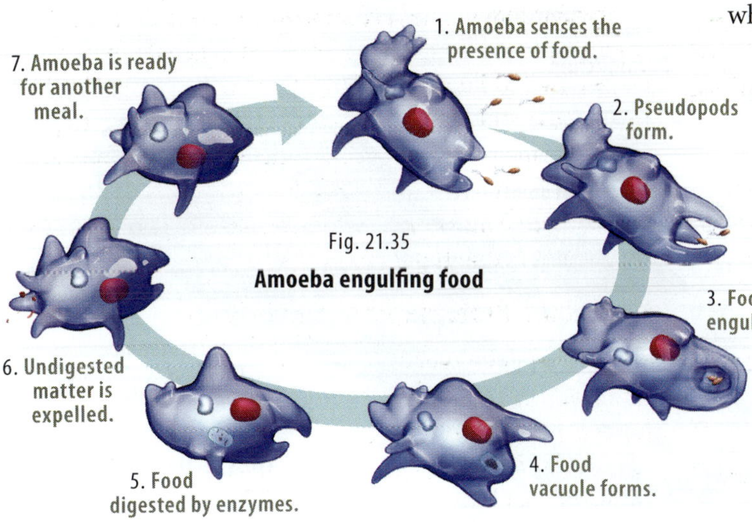


Fig. 21.35 **Amoeba engulfing food**

then simply pinches into two new daughter cells. The amoeba takes about 20 minutes to complete cell division. If ideal conditions exist, the amoeba divides every 3 days. If conditions become unfavorable for active living, the amoeba may form a cyst by secreting a thick wall about itself.

Although it has no eyespots, the amoeba moves away from bright lights, toward food substances, and away from harmful chemicals. Exactly how the amoeba is able to distinguish between these things is not yet known. Amoebas can be obtained by collecting the slime from the bottom of a pond or from the leaves of aquatic plants.

Most amoebas live in water, but some are found in the bodies of humans and animals. One harmful species causes a sickness known as **amoebic dysentery,** which is common in areas of the world where unsanitary conditions prevail. The dysentery amoebas feed on the wall of the human intestine and ingest red blood cells, causing bleeding ulcers.

Some protozoa build shells around themselves that are reminiscent of the shells of diatoms (ch. 5).

Fig. 21.36 **Shells of radiolarians**

**Foraminiferans** [fə·răm′ə·nĭf′ər·ənz] produce shells made of calcium carbonate; large masses of these protozoa may be compressed together under certain conditions to form *chalk* or other types of limestone. The famous white cliffs of Dover along the southern coast of England are composed of chalk. **Radiolarians** [rā′dē·ō·lâr′ē·ənz] are protozoa that secrete intricate shells made of silica.

### Ciliates: Protozoa with Cilia

**Paramecium.** Members of the genus *Paramecium* can be distinguished by their unique slipper shape and the **cilia** (hairlike projections) that cover them on all sides. (Actually, **paramecia** are submarine-shaped; they only appear slipper-shaped when seen in cross section, as when viewed with a light microscope.) Paramecia can be collected by gathering submerged weeds from a stagnant pond and placing them in a jar with some of the pond water. In a few days, this culture should be teeming with paramecia, which can be seen spiraling through the water as a result of the rhythmic motion of their cilia. The body of a paramecium is large enough to be seen with the unaided eye as a white spot, but a microscope is needed for clear viewing.

A paramecium possesses a **macronucleus** (large nucleus) and at least one **micronucleus** (small nucleus). The macronucleus controls the basic activities of the organism, and the micronucleus functions in reproduction. Biologists believe paramecia need multiple nuclei because of their size and complexity.

Biologists often divide the body of a paramecium into two layers. The outer layer is called the *cortex* or *ectoplasm,* while the inner layer is the *endoplasm.* The thick cell membrane of a paramecium is known as the *pellicle;* the pellicle is reinforced by strong fibers running beneath it. The cilia covering the outer body extend through the pellicle. It is also through this pellicle that the paramecium accomplishes respiration by the diffusion of oxygen into the cell and carbon dioxide out of the cell. Embedded in the cortex are **trichocysts** [trĭk′ə·sĭsts′], dartlike structures that the paramecium shoots out when it is attacked or (sometimes) when it is feeding. Paramecia possess food vacuoles and two contractile vacuoles.

Paramecia are the most important consumers of algae in freshwater environments. They also feed on bacteria, other protozoa, and yeast cells. The **oral groove** (mouth cavity) of a paramecium resembles a funnel. Cilia lining the oral groove sweep food into the gullet. Once the food is inside, a food vacuole forms around it. This vacuole breaks away from the gullet,

Fig. 21.37

**Paramecium**

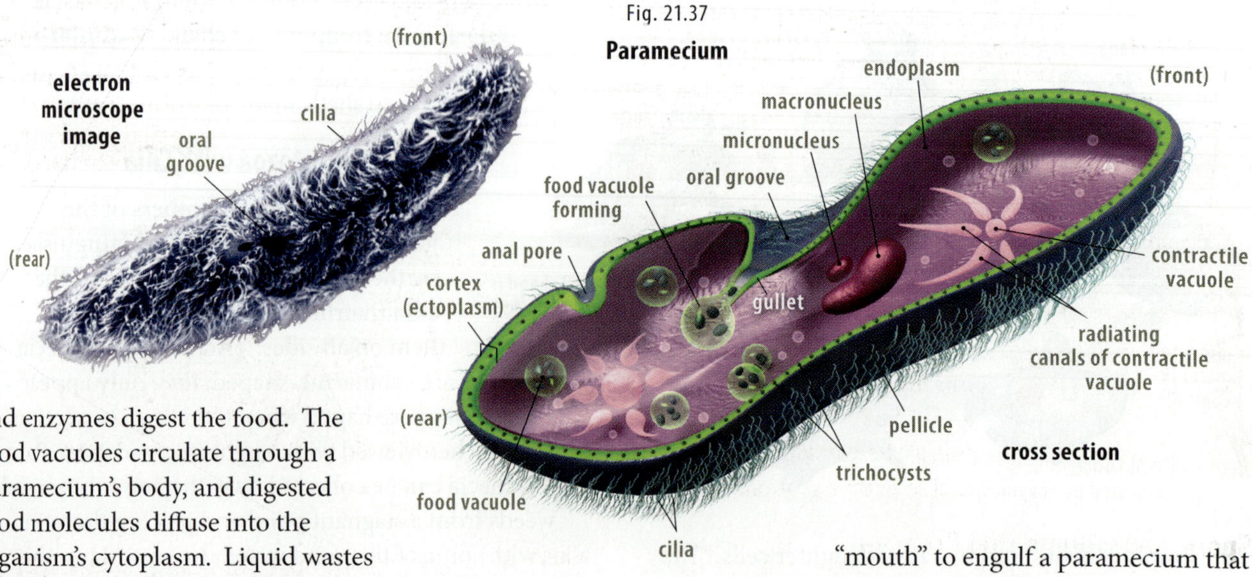


and enzymes digest the food. The food vacuoles circulate through a paramecium's body, and digested food molecules diffuse into the organism's cytoplasm. Liquid wastes are excreted by the two star-shaped contractile vacuoles, and undigested food particles are released at the **anal pore.** This anal pore is difficult to see because it is closed except when in use.

Paramecia do not have nerves, but they do have the ability to respond to their environment. When touched, a paramecium may move either toward or away from the point of contact, depending on what made the contact.

Paramecia reproduce in two ways: by fission and by **conjugation.** In conjugation, two paramecia exchange nuclear materials. After the exchange of nuclear material, each cell undergoes repeated divisions. The offspring of each cell have genetic material from each parent cell.

**Other ciliates.** *Didinium* is a barrel-shaped ciliate that commonly feeds on paramecia. It can expand its "mouth" to engulf a paramecium that is even larger than itself.

**Vorticella** [vôr′tĭ·sĕl′ə] is a stalked ciliate commonly found in stagnant ponds. Its body is shaped like an upside-down bell and is attached to underwater objects by a stalk. The stalk can expand and contract to provide limited movement. Layers of cilia encircle the rim of the "bell" and sweep food particles into the oral cavity. It is thought that vorticella was the first microorganism seen by Anton van Leeuwenhoek.

**Stentor** is a trumpet-shaped ciliate that grows to a length of 1⁄10 of an inch and can be found swimming freely in fresh and salt water. A good place to look for it is in small pools containing dead leaves. Its color varies. Often it is green or blue because of algae that live in its cytoplasm. It feeds by attaching itself to

Fig. 21.38

**Reproduction in paramecia**

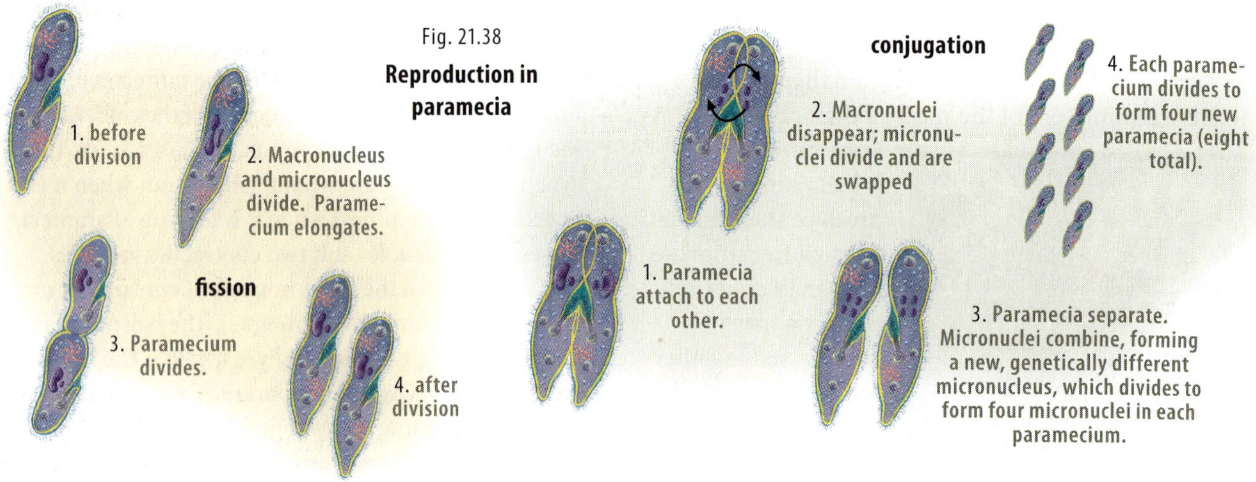

Fig. 21.39
**Other ciliates**

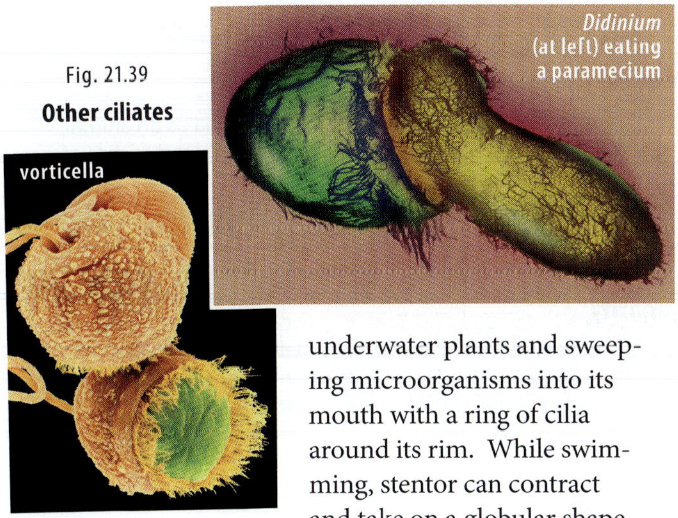


underwater plants and sweeping microorganisms into its mouth with a ring of cilia around its rim. While swimming, stentor can contract and take on a globular shape.

## Sporozoans: Nonmotile Protozoa

The spore-forming protozoa, or **sporozoans,** are almost all parasites requiring two hosts. These organisms are responsible for several diseases and are transmitted from one host to another by mosquitoes, ticks, and flies. Sporozoans are characterized by a complex reproductive cycle that includes both asexual and sexual reproduction.

One of the most common infectious diseases in the world is *malaria,* which infects an estimated 300 million people each year and causes 1.5 million deaths. Malaria is caused by sporozoa of the genus *Plasmodium* [plăz·mō′dē·əm] and is transmitted by the female *anopheles* [ə·nŏf′ə·lēz′] *mosquito.* **Plasmodium** has one of the most complicated reproductive cycles of any protozoan; it reproduces both asexually and sexually and undergoes *alternation of generations* (Fig. 21.40).

**Careers in BIOLOGY**

### Zoology

**Veterinary medicine** – using scientific knowledge to prevent or treat injuries or diseases in animals

**Zoology** – learning how man can be a better steward of God's living creation by studying particular groups, populations, or kinds of animals

**Pest control** – using scientific knowledge to prevent or reduce infestation of homes, crops, etc. by harmful insects or other animals; using scientific knowledge to prevent the spread of diseases transmitted by insect or arthropod vectors

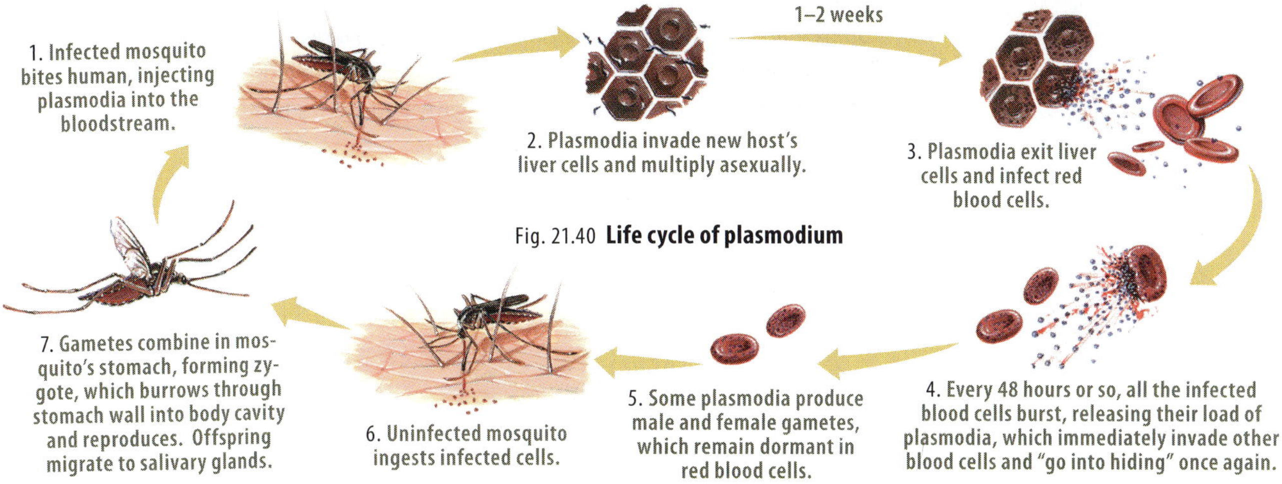


Fig. 21.40 **Life cycle of plasmodium**

## Section Review 21.6

1. What are the four main groups of protozoa? On what major characteristic is this classification based?
2. In what way(s) is the reproduction of amoebas different from that of paramecia?
3. Name three diseases caused by protozoa.

### Application

4. How are foraminiferans and radiolarians similar to diatoms (ch. 5)? How are they different?

### Critical Thinking

5. Many scientists in the past classified euglena as an animal, while others classified it as a plant. What characteristics of euglena could contribute to this confusion?
6. Explain why the protozoa cannot be called simple creatures but are complex organisms that give evidence of the Creator's wisdom, power, and design.

### Identify

protozoa, pellicle, flagellum, euglenoid movement, pseudopods, cilia, macronucleus, micronucleus, trichocysts, oral groove, conjugation, sporozoans

## Chapter 21 Review

### Define

1. mollusk
2. visceral hump, mantle
3. adductor muscle
4. mantle cavity
5. incurrent siphon, excurrent siphon
6. radula
7. echinoderm
8. water-vascular system
9. rotifer
10. parthenogenic
11. osculum, ostia
12. spicules
13. setae
14. aortic arches
15. prostomium
16. nephridia
17. scolex
18. protozoa
19. pellicle
20. macronucleus, micronucleus
21. trichocysts
22. conjugation
23. sporozoans

### Identify

1. the two largest known invertebrates
2. the only cephalopod with an external shell
3. five characteristics common to all echinoderms
4. four characteristics of coelenterates
5. two body forms of coelenterates
6. the only multicellular animal without a nervous system
7. the largest class of segmented worms
8. three groups of flatworms mentioned in the text
9. four parasitic roundworms
10. four broad groups of protozoa
11. three disease-causing protozoa and the diseases they cause

### Explain

1. Describe the formation of a pearl.
2. How does a clam's foot allow it to burrow into the sand?
3. What is the purpose of a bivalve's siphons?
4. Compare and contrast the three broad groups of mollusks discussed in the text.
5. What body system do echinoderms have that is found in no other type of animal?
6. What primary body characteristic do all segmented worms have in common?

### Apply

1. In what ways do earthworms benefit mankind?
2. Explain why protozoa are typically studied with animals instead of being studied with plants (as are other members of kingdom Protista).

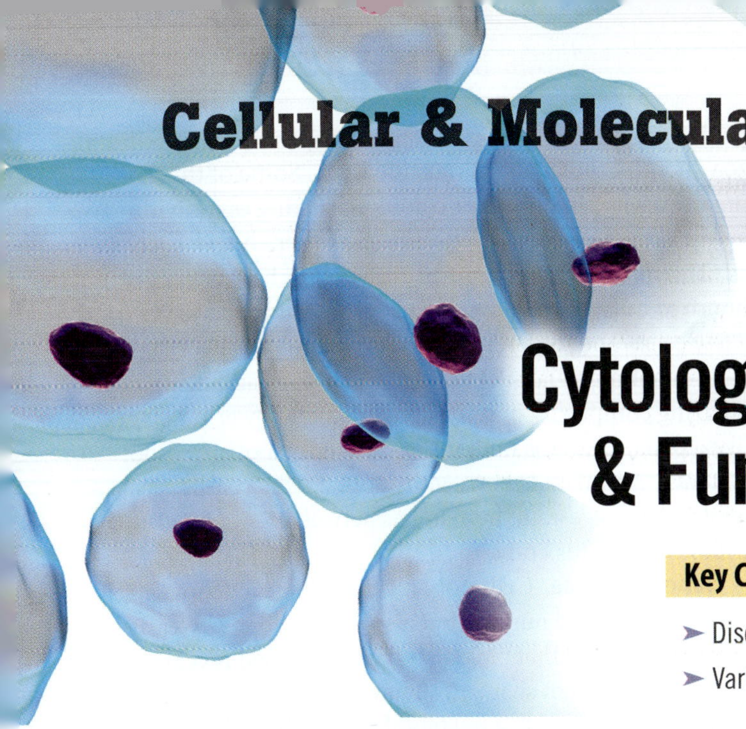

# Cellular & Molecular Biology | UNIT 5

## The Complexity of Living Things

# Cytology: The Design & Function of Cells

CHAPTER 22

**Key Concepts**

- Discovery of cells
- Variety in cells
- Structure of a cell
- Life processes of a cell
- Cell cycle
- Cell death

## 22.1 The Variety and Complexity of Cells

We have studied many aspects of the world around us, including a vast variety of plants and animals, and have looked at the systems of many organisms and at the organs that compose those systems. We have also studied the different types of tissues that help each organ perform certain functions. As we looked more closely at tissues, we found that all tissues are composed of microscopic, individual **cells,** the units of structure and function of living things. The study of cells was slow in coming to man's world. For centuries, man did not even know that cells existed; perhaps people thought of an animal or plant as a sculpture of clay formed by God and given the element of life. It has only been since the late 1800s that God has allowed man to know some of the intricate design and deep secrets hidden within the living cell. The study of cells is known as the science of **cytology** [sī·tŏl′ə·jē] or *cell biology.* Since the 1970s and 1980s, the field of cell biology has exploded with new discoveries, giving us a new appreciation for the design of God's living creation.

### The Discovery of Cells

The idea that God divided living things into smaller parts called cells began in 1665, when **Robert Hooke** observed thin sections of cork with a simple microscope of his own making and wrote a description of his observations. In 1838, after hundreds of naturalists found the same basic design in all the plants and animals they observed, the German botanist *Matthias Schleiden* [shlīd′n: 1804–1881], stated what is now known as the cell theory. One year later, the German physiologist *Theodor Schwann* [shvän: 1810–1882] advanced the same idea. The **cell theory,** which states that all living things are composed of living units called cells and of cell products and that all cells come only from preexisting cells, is universally accepted today as the most fundamental law of biology.

Fig. 22.1
**Robert Hooke observing cork cells with microscope**

*Cells do all the work that is done in an organism.* For example, every physical activity in your body can be traced to the work of individual cells. Cells manufacture all of the materials that are necessary for life and work together to transport substances throughout your body.

Cells communicate with each other and respond to different situations. They grow and reproduce to carry on the work that they were created to do. Scientists have found that, although cells are small, they are anything but simple. Discoveries in the field of cell biology give us a new appreciation for the complex design of God's living creation.

## Variety among Cells

**Variety in size.** You might think that small creatures have small cells and large creatures have large cells, but that is not the case. An important principle of biology states that *the size of an organism is determined by the number of its cells, not the size of its cells.* A typical human or animal cell is 10–20 micrometers across, typical plant cells are 50–70 micrometers across, and bacterial cells average 1 micrometer across.

**Variety in number.** Many cells are independent, living their entire existence as a single cell. Unlike your body, which depends heavily upon specialized cells, each capable of only a few important tasks, the protists (protozoa and unicellular algae) must do it all. Some of these organisms live in groups or colonies, existing as individuals yet providing some needs for each other. A few, such as filamentous algae or fungi, consist of perhaps several hundred cells and can live and grow even if broken into a few cells. The rotifer is composed of about 1000 cells, yet it is not much larger than many protozoa and displays amazing complexity in its design. A small flowering plant like a pansy contains millions of cells in its root and shoot systems. An infant is born with about 2 trillion cells, while an average adult human has between 30 and 100 trillion cells in his body.

**Variety in shape.** Cells not only exhibit an amazing variety in size and number but also vary in shape. Nearly every kind of cell that biologists have examined reveals a new and different shape. You have only to think of the shape of a muscle cell, a red blood cell, a white blood cell, a nerve cell, a paramecium, a bacterium, or a cell in the stem of a plant to see that this is true. The variety of size, shape, and number, combined with the magnificent architecture, both inside and on the surface, make living cells phenomenal creations of God.

Fig. 22.2 **Variety in cell design** (all cells drawn to scale)

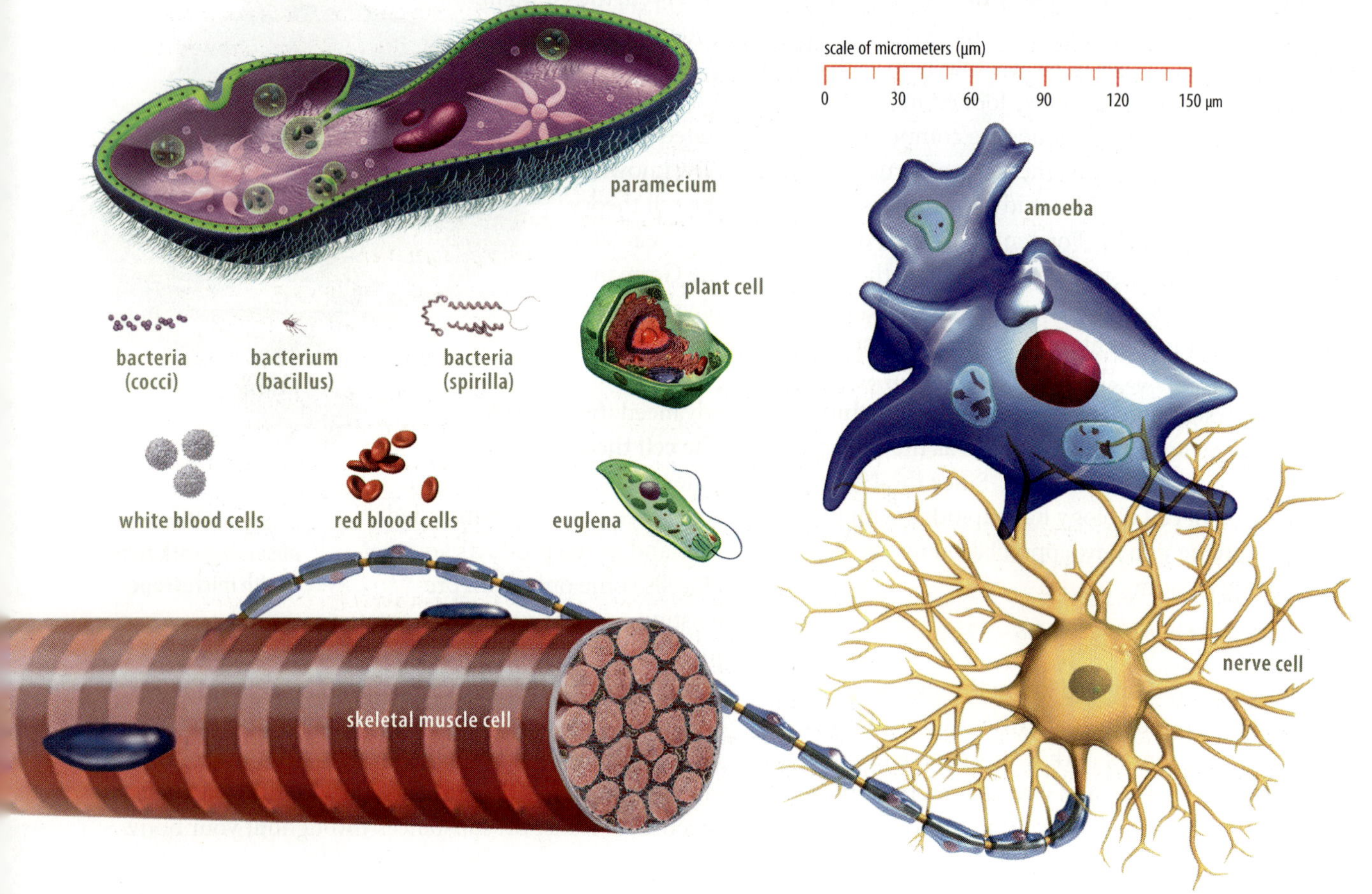

## *The Living Cell* by Dr. Michael Denton

Design in the LIVING CREATION

Viewed down a light microscope at a magnification of some several hundred times, such as would have been possible in Darwin's time, a living cell is a relatively disappointing spectacle appearing only as an ever-changing and apparently disordered pattern of blobs and particles which, under the influence of unseen turbulent forces, are continually tossed haphazardly in all directions. To grasp the reality of life as it has been revealed by molecular biology, we must magnify a cell a thousand million times until it is twenty kilometers in diameter and resembles a giant airship large enough to cover a great city like London or New York. What we would then see would be an object of unparalleled complexity. . . . On the surface of the cell we would see millions of openings, like the port holes of a vast space ship, opening and closing to allow a continual stream of materials to flow in and out. If we were to enter one of these openings we would find ourselves in a world of supreme technology and bewildering complexity. We would see endless highly organized corridors and conduits branching in every direction away from the perimeter of the cell, some leading to the central memory bank in the nucleus and others to assembly plants and processing units. The nucleus itself would be a vast spherical chamber more than a kilometer in diameter, resembling a geodesic dome inside of which we would see, all neatly stacked together in ordered arrays, the miles of coiled chains of the DNA molecules. A huge range of products and raw materials would shuttle along the manifold conduits in a highly ordered fashion to and from all the various assembly plants in the outer regions of the cell.

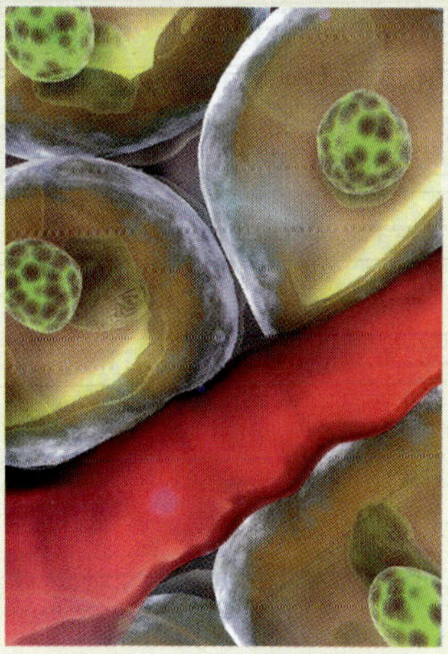

We would wonder at the level of control implicit in the movement of so many objects down so many seemingly endless conduits, all in perfect unison. We would see all around us, in every direction we looked, all sorts of robot-like machines. We would notice that the simplest of the functional components of the cell, the protein molecules, were astonishingly, complex pieces of molecular machinery. . . . We would wonder even more as we watched the strangely purposeful activities of these weird molecular machines, particularly when we realized that, despite all our accumulated knowledge of physics and chemistry, the task of designing one such molecular machine—that is one single functional protein molecule—would be completely beyond our capacity at present. . . . Yet the life of the cell depends on the integrated activities of thousands, certainly tens, and probably hundreds of thousands of different protein molecules.

We would see that nearly every feature of our own advanced machines had its analogue in the cell: artificial languages and their decoding systems, memory banks for information storage and retrieval, elegant control systems regulating the automated assembly of parts and components, error fail-safe and proof-reading devices utilized for quality control, assembly processes involving the principle of prefabrication and modular construction. In fact, so deep would be the feeling of *deja-vu,* so persuasive the analogy, that much of the terminology we would use to describe this fascinating molecular reality would be borrowed from the world of late twentieth-century technology.

What we would be witnessing would be an object resembling an immense automated factory, a factory larger than a city and carrying out almost as many unique functions as all the manufacturing activities of man on earth. However, it would be a factory which would have one capacity not equalled in any of our own most advanced machines, for it would be capable of replicating its entire structure within a matter of a few hours. To witness such an act at a magnification of one thousand million times would be an awe-inspiring spectacle.

*Michael Denton,* Evolution: A Theory in Crisis *(Bethesda, MD: Adler & Adler, 1986), pp. 328–329. Reprinted by permission of Michael Denton.*

### The Complexity of Cells

Before the mid-20th century, living cells were regarded as simple bags of an amorphous gel or slime called **protoplasm,** thought to be a slush of amino acids, protein, fat, and carbohydrates. The processes of life were thought to be simple chemical reactions in the protoplasm, driven by a mysterious but simple *"vital force"* (analogous to electricity) that allowed the protoplasm to grow and reproduce.

Scientists now know that a cell is like a tiny, self-contained city with its own power plants, chemical factories, food warehouses, and waste-disposal facilities. A cell also has an extensive transportation network that it uses to import raw materials for its "factories" and to export products and wastes. All of these activities are directed by programs in the cell's control center. The immense complexity apparent in the living cell is a magnificent testimony to its Creator.

## Section Review 22.1

1. What are the units of structure and function in living things?
2. State the cell theory.
3. Why are the cells of a whale roughly the same size as the cells of a flea?

### Critical Thinking

4. What are the implications of the cell theory for the theory of evolution?
5. Ernst Haeckel, a vocal evolutionist and a contemporary of Darwin, described the cell as "a simple little lump of albuminous combination of carbon." How might this false view of the cell have led Haeckel to view spontaneous generation as a plausible explanation for the origin of life?

### Identify

cell, cytology, cell biology

## 22.2 The Design of Eukaryotic Cells

We have studied the almost limitless variety in cells with regard to their size, number, and shape, but now we will concentrate on the design of the cell itself. Far from being a random congregation of available molecules or a haphazard mass of living matter, cells are built on a plan; they possess a blueprint.

Fig. 22.3 **Main parts of a cell**

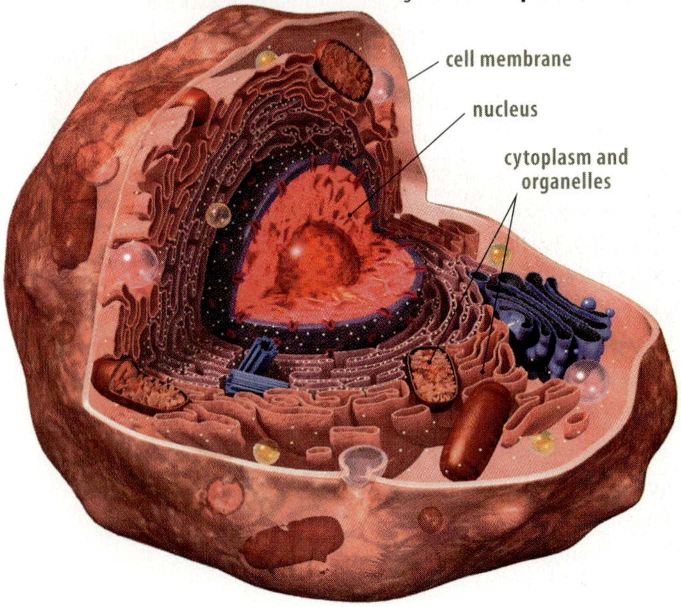


*God designed the cell to be composed primarily of four main elements:* oxygen, carbon, hydrogen, and nitrogen. By mass, about 70% of the cell is *water,* a compound of oxygen and hydrogen. The other major substances in the cell are *carbohydrates* (sugars, starches, etc.), *lipids* (fatty acids, cholesterol, etc.), *nucleic acids,* and *proteins;* all of these substances contain carbon and hydrogen. Many lipids and all carbohydrates also contain oxygen, and all nucleic acids and proteins contain both oxygen and nitrogen. Other elements that are common in cells include sodium, magnesium, phosphorus, sulfur, chlorine, potassium, and calcium.

You may recall that *biologists recognize three main parts of every cell in plants, animals, and man:*

- the *cell membrane,* the outer boundary layer that separates the cell from its environment and controls what enters and leaves the cell
- the *cytoplasm,* which serves as the fluid medium for the many molecules and *organelles* (little organs) the cell contains
- the *nucleus,* a spherical body, often located near the center of the cell, that contains the genetic material of the cell and serves as the cell's master control center

## Cell Membrane: The Covering of a Cell

**Importance of the cell membrane.** The contents of human and animal cells are separated from the outside environment by a very thin, flexible barrier called the **cell membrane** (also known as the *plasma membrane*). The cell membrane is a vital component of the cell, for through it go all substances that enter or leave the cell. The regulation of this flow of materials requires a masterful design, for within the borders of the cell membrane operate the most complicated processes in creation, and failure of the membrane brings instant death.

**Structure of the cell membrane.** A cell membrane is composed of special lipid molecules called **phospholipids** [fŏs′fō·lĭp′ĭdz], with a small amount of *cholesterol* added to maintain the desired flexibility. The phospholipid molecules cluster back to back, forming a two-layered structure referred to as a *lipid bilayer.* The cell membrane is quite flexible and can be easily squashed or deformed without breaking it. Embedded in the membrane are various *membrane proteins* that carry out important functions. All the organelle membranes inside the cell are lipid bilayers with essentially the same structure as the cell membrane. They differ, however, in the types of proteins they contain, which are specific for each organelle's function.

**Support and reinforcement.** Beneath the phospholipid bilayer of the cell membrane lies a network of flexible reinforcement fibers called the **membrane skeleton.** The fibers of the membrane skeleton of red blood cells, for instance, are made from a strong, cablelike protein called *spectrin.* Like the steel belts of an automobile tire, these fibers reinforce the cell membrane, protecting it from rips and tears while still allowing it to be flexible. These fibers also anchor certain membrane proteins in place so that they do not drift around the surface of the cell membrane. Many of the spectrin reinforcement fibers are attached to *actin* filaments, similar to those in the myofibrils of muscle cells, that allow the cell to constrict its cell membrane by tightening selected fibers.

**Machinery of the cell membrane.** The cell membrane is a *semipermeable membrane,* allowing water and some small molecules to cross freely through it. However, to be transported across the cell membrane, most of the larger molecules need special membrane proteins. *Ion pumps* are membrane proteins that allow the cell to pump atoms such as sodium, potassium, calcium, and hydrogen into or out of the cell in order to preserve the right ion balance. Some proteins serve as *portals* to other cells or as *gates* to allow various molecules into or out of the cell.

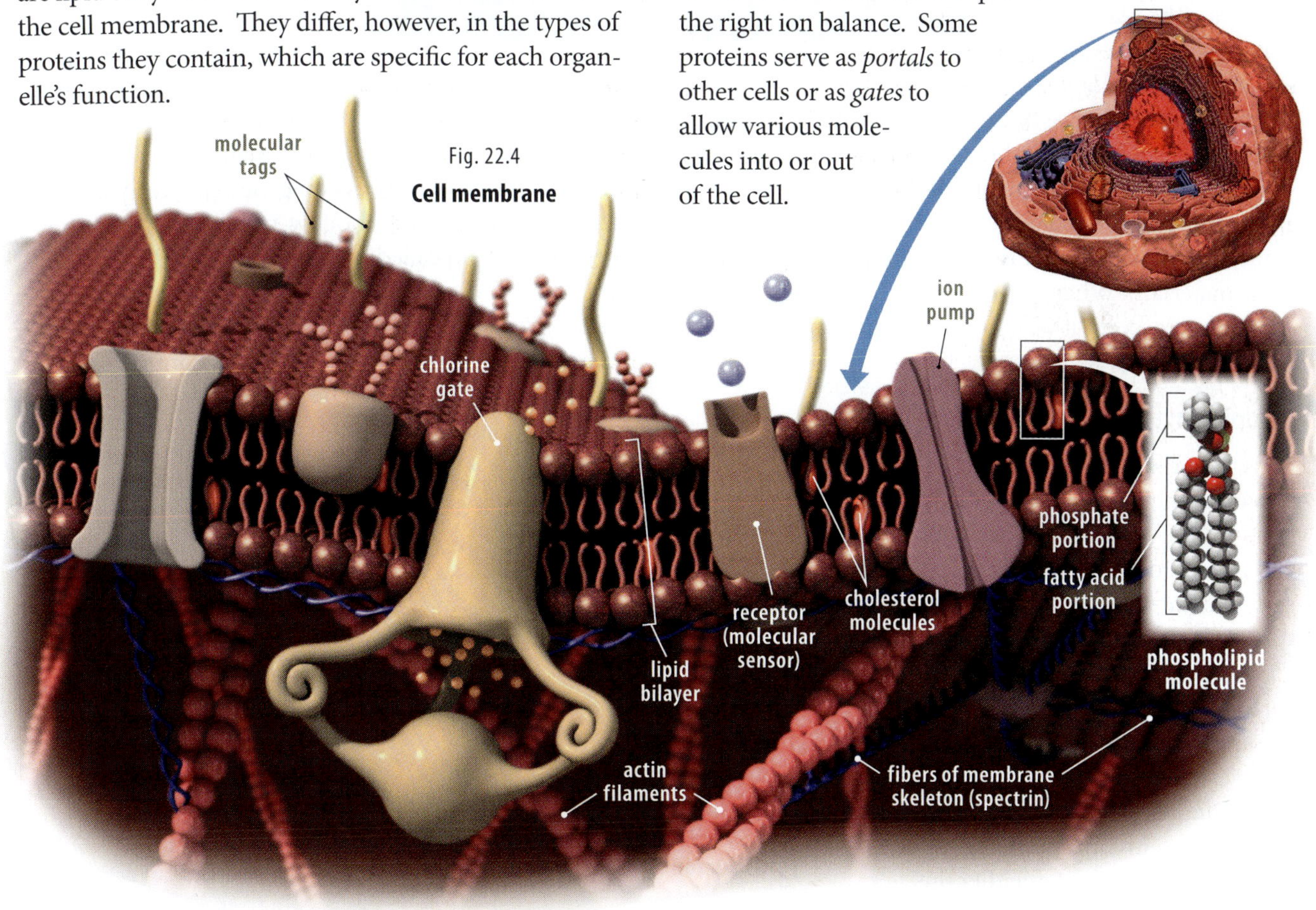


Fig. 22.4
**Cell membrane**

Other proteins, called *receptors*, inform the cell about the outside environment; many of these receptors are designed to detect the chemical messengers known as hormones, which signal the cell to carry out specific tasks. Some membrane molecules are *molecular tags*, serving as identification tags that allow other cells to recognize the cell's identity.

**Cell wall of plants.** Outside the cell membrane, plant cells manufacture a rigid, boxlike structure called the **cell wall.** This layer stiffens the cell, allowing a plant to stand upright even though it lacks a skeleton. Cell walls also protect against many bacteria and other pathogens that would otherwise infect plants.

Fig. 22.5 **Electron microscope photo of plant cell wall**

A cell wall is not a solid mass but rather is composed of millions of threads of cellulose at angles to each other, forming a strong network. The plant cell may also manufacture other materials, such as *lignin,* that make the cell wall stiffer by linking the cellulose threads together more strongly. (Lignin is responsible for the strength and stiffness of wood.) Certain types of plant cells (such as xylem cells) are designed to die when they are mature, leaving an empty box made of cellulose and lignin, which transports materials. When Robert Hooke viewed cork under his microscope, it was the empty boxes of cellulose and lignin that he saw.

## Cytoplasm: Fluid Medium of the Cell

Within the cell membrane is a jellylike fluid known as the **cytoplasm.** The cytoplasm contains an extraordinary diversity of **organelles,** or "little organs," and numerous dissolved chemicals. We will find as we examine the different kinds of cytoplasmic organelles that each has a unique design and purpose.

Cytoplasm's chief role in the life of the cell is to act as a medium, similar to the way that water in a pond acts as a medium for fish and hundreds of other creatures. *Most of the activities of life occur within the cytoplasm and its organelles.* Molecules of proteins, lipids, and carbohydrates abound and act as either the raw materials for the manufacture of cell parts and cell products or as items for export to be packaged and delivered to the cell's surface. Large protein molecules called **enzymes** are found both in the cytoplasm and in the organelles, carrying out various chemical reactions to produce energy, transform raw materials into useful substances, or break down old proteins to be recycled. The diversity and rate of activity within cytoplasm are phenomenal. Within the cytoplasm of one tiny cell occur thousands of simultaneous chemical reactions involving tens of thousands of different compounds. The cytoplasm could be compared to the hustle and bustle of New York City with its thousands of factories and transportation systems all wrapped up in a unit too small to be seen without a microscope.

## The Skeleton of the Cell

**A cellular framework.** To prevent a cell from being squashed or distorted by the slightest external pressures, God designed the cell with an intricate internal skeleton, called the **cytoskeleton,** to help it maintain its shape. The cytoskeleton of a cell is composed primarily of **microtubules,** which are hollow, rodlike filaments composed of a protein called *tubulin* [tōō′byə·lĭn] (Fig. 22.7). Thousands of microtubules are arranged in a complex lattice to give the cell its rigidity. The various organelles of the cell, such as the nucleus (the cell's control center), the mitochondria (the cell's power plants) and

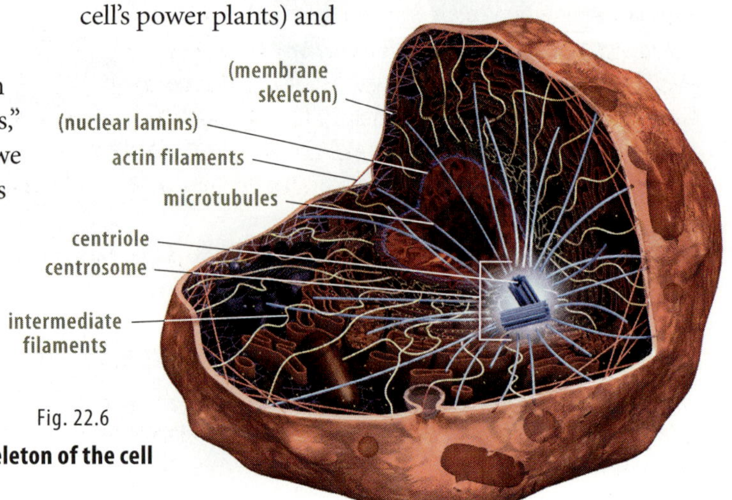


Fig. 22.6
**Cytoskeleton of the cell**

the ribosomes (the cell's factories) are held in place in the cell by the microtubule framework. Microtubules are also an integral part of certain organelles, such as cilia and flagella. The cytoskeleton also contains smaller *actin* filaments and *intermediate filaments,* which act as cables, girders, and winches.

**A "mass-transit" system.** Microtubules also act as the "highway system" for the cell. Proteins, produced in ribosomes (protein factories) can be packaged in special containers called *vesicles* and then transported to other destinations inside the cell or to the cell membrane using the microtubule network. Special motor proteins whisk the vesicle and its contents along the microtubule to its destination. Vesicles travel in opposite directions along the same microtubule, passing by each other like cars on a two-lane highway.

**Microtubule-organizing center.** The microtubules of a living cell can be moved, or even rapidly dissolved and reconstructed, to change the cell's internal arrangement. This allows a cell to move its organelles and structures around the cell as they are needed. This dynamic behavior of the microtubules is coordinated in every cell by the *microtubule-organizing center.* In human and animal cells, the microtubule-organizing center is a structure called the **centrosome,** which is located near the nucleus, at the center of the cell. The centrosome contains the **centrioles,** a pair of cylindrical structures made of tubulin. They play an important role in cell movement and division. Although plant cells do not have centrioles or a centrosome, they do have a microtubule-organizing center with the same functions as the one in human and animal cells.

**Creeping cells.** In addition to transporting materials and organelles within the cell, the cell's power-driven cytoskeleton can allow a cell to move about by changing its shape. For example, the white blood cells of your body can move about by "shape-shifting," squeezing through your body's connective tissue and even entering and leaving blood vessels at will to pursue invading germs. The *amoeba,* a well-known protozoan, also moves about by adjusting its cytoskeleton to change its shape. This movement occurs as *myosin* motor proteins (similar to those in muscle cells) move actin filaments.

## Mitochondria: Power Plants of the Cell

The life and work of the cell requires a great deal of energy. The "power plants" of a cell are organelles known as **mitochondria** (sing. mitochondrion). *A mitochondrion oxidizes or "burns" carbohydrates and fats and uses the energy to produce ATP* (pp. 147–148).

A mitochondrion is a sausage-shaped structure consisting of an outer membrane and an inner membrane. The smooth *outer membrane* contains many pores that allow molecules to pass through, while the *inner membrane* is highly folded and very selective in what it permits to pass through. The inner mitochondrial membrane contains so many transport proteins, ion pumps, and large ATP generators that nearly 80% of its mass is composed of protein (in contrast with a typical cell membrane, which contains about equal amounts of lipids and proteins). Many of the energy-generating reactions take

Fig. 22.7
**Microtubule**

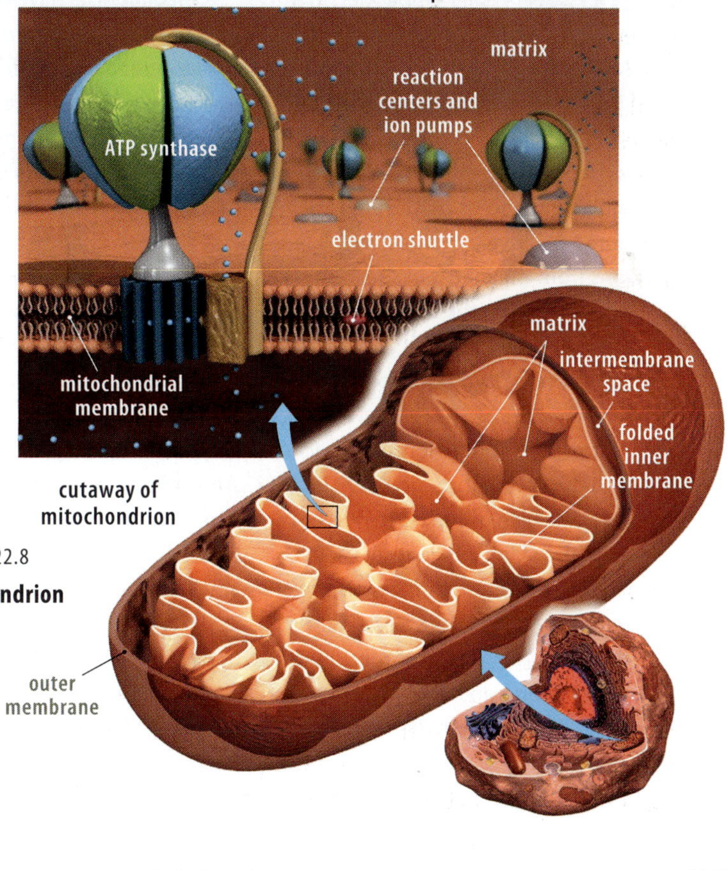


Fig. 22.8
**Mitochondrion**

place on the surface of the inner membrane. The space at the core of the mitochondrion is called the *matrix,* and the space between the outer and inner membranes is called the *intermembrane space.*

Most mammal cells contain between 1000 and 2000 mitochondria; the more active the cell, the more mitochondria it needs. In the human body, cardiac muscle, the muscle of the heart, has very large numbers of mitochondria per cell to supply the energy for pumping, whereas ordinary fat cells contain relatively few mitochondria.

Although the primary purpose of a mitochondrion is the production of ATP for energy, it can also export some of the intermediate byproducts to the rest of the cell as raw materials for the construction of amino acids and nucleic acids. Power generation also results in a certain amount of heat, which an organism can use to maintain its operating temperature.

One of the major differences between plant and animal cells is the presence in plant cells of *chloroplasts,* very complex organelles quite similar in design to mitochondria. Within the folded membranes of the chloroplast occur the complex reactions of *photosynthesis.*

## Ribosomes: Protein Factories

Some of the most important molecules in a cell are *proteins.* Although proteins were once thought to be relatively simple substances, we now know that many proteins are complex molecular machines. In a cell, some proteins serve as pumps, valves,

## Mitochondrial Generators

Understanding the LIVING CREATION

The most important molecular machines in a mitochondrion are the **ATP synthases,** which are embedded in the mitochondrion's inner membrane. A mitochondrion generates power (in the form of ATP) by breaking down food molecules and using the energy to pump positively charged hydrogen ions from the matrix of the mitochondrion into the intermembrane space. When there is a greater concentration of hydrogen ions in the intermembrane space than in the matrix, some of the ions flow back into the matrix through the ATP synthases, which use the energy from this flow of ions to generate ATP. An almost identical process is used to make ATP in chloroplasts during the process of photosynthesis; the primary difference is that the energy required to pump hydrogen ions across the membrane comes from light instead of food.

Cell biologists have discovered that *ATP synthase is a tiny, ion-powered turbogenerator.* As hydrogen ions pass through the enzyme, they turn a turbinelike rotor. The rotor turns a drive shaft that opens and closes the clamshell -like reaction chambers of the synthase. ATP synthase can also be run in reverse (breaking down ATP) to function as a motor-driven ion pump.

Scientists are baffled as to how such a structure might have supposedly evolved; Darwin's hypothesis states that all complex structures arose one tiny, random step at a time, but ATP synthase would be absolutely useless to a cell until all of its parts were present and functioning. In fact, a cell that was still "evolving" ATP synthase would be less likely to survive because it would be wasting energy manufacturing useless proteins that it could have used to gather food and reproduce. ATP synthase, like the living cell itself, testifies of its Creator.

*Model of ATP synthase*

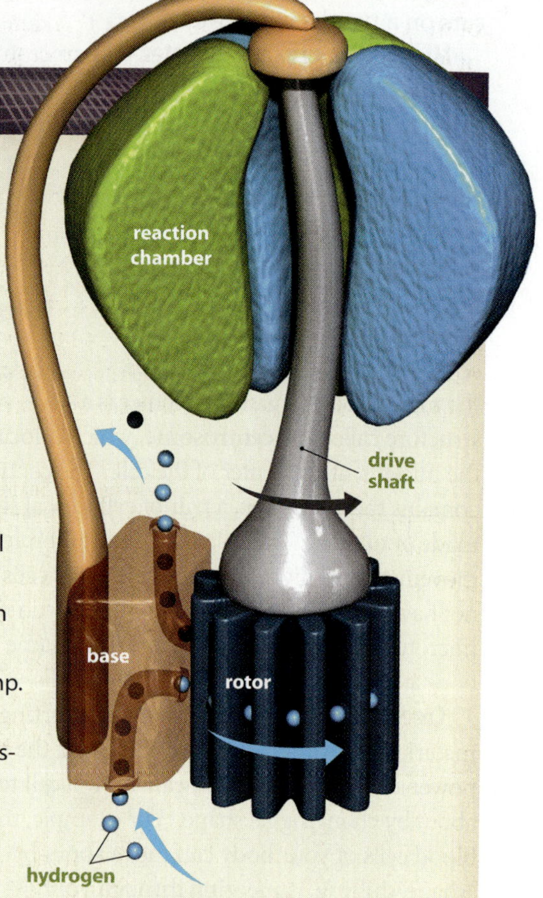

winches and motors, information retrieval devices, and chemical catalysts; others serve as power generators, light and position sensors, and communication relays. Proteins are also used to construct the cell's skeleton and to produce such structures as your hair and fingernails.

In a living cell, proteins are manufactured by special "protein factories" called **ribosomes** [rī′bə·sōmz′]. A single cell may contain tens of thousands of ribosomes. Ribosomes follow "blueprints" from the cell's nucleus to assemble amino acids into complex proteins through a process called *translation;* we will study ribosomes and the process of translation in more detail in chapter 24.

## Endoplasmic Reticulum: A Multipurpose Network

You may have noticed that much of the cytoplasm of the cell in Figure 22.3 is filled with a dense network of membranes. This membrane network is called the **endoplasmic reticulum** (*ER* for short). (The word *reticulum* comes from a Latin word meaning "small net.") The endoplasmic reticulum is a network of interconnected sacs and tubules that is connected to the nucleus and extends throughout much of the cell.

**Purpose of the ER.** When a ribosome is making a protein that is to be secreted outside the cell or attached to one of the cell's many membranes, the ribosome attaches itself to the ER and injects the protein directly into the ER's internal cavity, the **ER lumen.** The protein goes through several chemical modifications inside the ER lumen before it is shipped to the Golgi complex, where additional modifications will be performed to make it suitable for its specific function. The ER also breaks down toxins and wastes and manufactures fat molecules.

**Two types of ER.** The ER near the cell's nucleus is usually studded with thousands of ribosomes busily manufacturing proteins. Because the ribosomes give it a rough appearance, the ER near the nucleus is called *rough ER.* ER farther from the nucleus does not contain ribosomes but instead contains enzymes for fat production and for the breakdown of drugs and waste products. Because this ER has a smoother appearance than that near the nucleus, it is called *smooth ER.* This type is particularly abundant in cells involved in lipid metabolism. In liver cells, most of the ER is smooth.

## The Golgi Apparatus: Shipping Center of the Cell

Proteins that are processed in the rough ER are transported to one of the **Golgi apparatus,** structures that serve as the shipping centers of the cell. A Golgi apparatus consists of a stack of flattened, membrane-enclosed compartments that looks somewhat like a stack of hollow pancakes. The compartments are held together by the cytoskeleton of the cell.

*The purpose of the Golgi apparatus is to receive proteins from the rough ER, modify them if needed, package them in vesicles, and ship them to their final destinations.* Some proteins are shipped to various places within the cell, while others are shipped to the cell membrane. Still other proteins are intended to be secreted outside the cell; these include digestive enzymes, blood proteins, and glandular secretions.

Fig. 22.9
**Endoplasmic reticulum and Golgi apparatus**

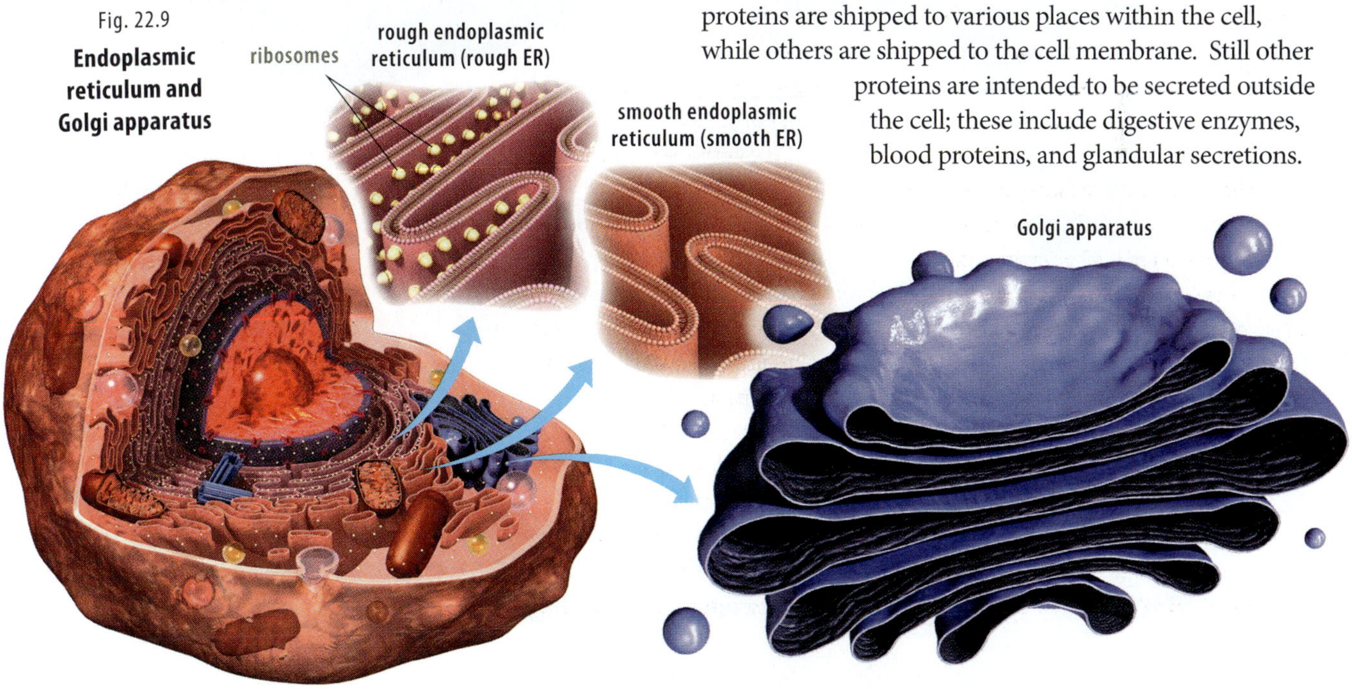

In the rough ER, newly manufactured protiens are labeled with molecules that function as shipping tags to tell where each protein needs to go. When the Golgi apparatus receives a batch of proteins from the ER, it sorts the proteins according to their tags. Proteins that are going to the same destination (such as the cell membrane, for example) are loaded into a container called a **vesicle,** which is attached to a nearby microtubule. A motor protein whisks the vesicle to the final destination, where the proteins are unloaded from the vesicle to do their jobs.

## Lysosomes: Recycling Centers of the Cell

Although we have covered the most important organelles of the cell, other organelles are also found in cells. One such organelle is the **lysosome,** a special type of vesicle that functions as a recycling center.

Fig. 22.10

**Animal and plant cells**

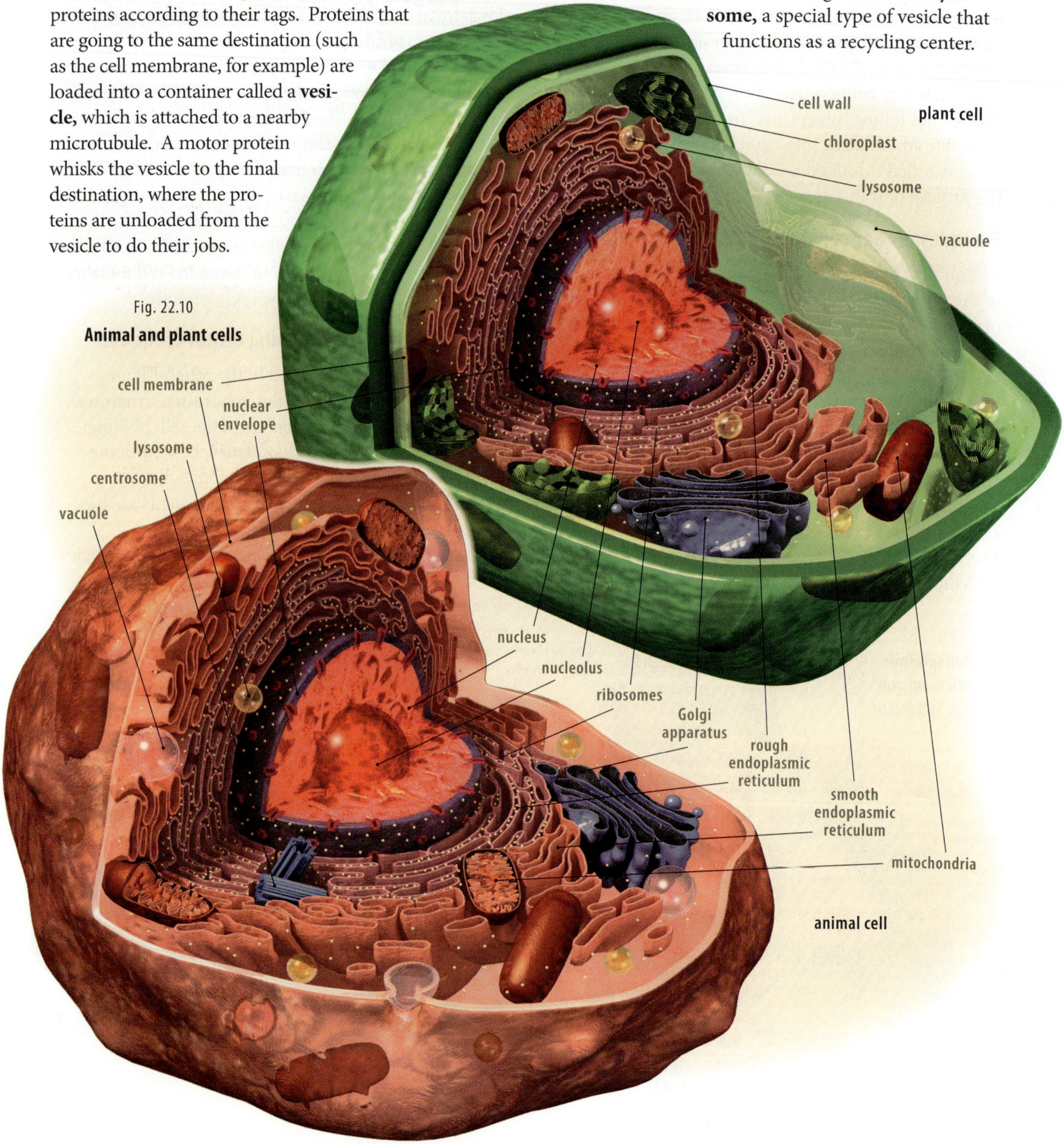

## Table 22.1 The Parts of a Cell and Their Purposes

| Structure | Design | Primary purpose(s) |
|---|---|---|
| **Cell membrane** | flexible, self-repairing membrane that forms the outer boundary of a cell; contains numerous embedded proteins and reinforcement fibers | separates cell from environment; determines what materials can enter or leave the cell; actively transports some materials into and out of the cell; senses chemicals in the surrounding environment; contains molecular tags for recognition by other cells |
| **Cell wall** | rigid, boxlike structure outside the cell membrane of plants; composed of cellulose fibers, lignin, and pectin | forms protective coat and provides firmness and strength to plant cells |
| **Cytoplasm** | the fluid medium of the cell | forms watery environment for cellular organelles; allows small molecules to diffuse throughout the cell |
| **Cytoskeleton** | complex network of hollow, rodlike girders and other fibers that extend throughout the cell | maintains the shape of the cell and provides support for organelles; microtubules serve as the "highways" of the cell; allow the cell to change its shape or to creep from place to place; provides support for cilia and flagella |
| **Centrosome** | pair of microtubule bundles found only in human and animal cells; located close to nucleus; often considered part of cytoskeleton | serves as foundation of cytoskeleton; also helps distribute chromosomes to daughter cells during cell division |
| **Mitochondria** | sausage-shaped structure consisting of a smooth outer membrane and a highly folded inner membrane | "power plants" of the cell; contain ATP synthase and other enzymes that break down food molecules and use the energy released to produce ATP |
| **Chloroplasts** | green, oval-shaped organelles containing hundreds of flattened compartments connected by narrow tubes | capture energy of light and use it to produce sugars |
| **Ribosomes** | small organelles composed of protein and RNA; consist of two subunits that clamp together to form a functioning unit | protein factories of the cell; manufacture proteins from amino acids, guided by "blueprints" provided by the nucleus |
| **Endoplasmic reticulum** (ER) | system of folded membranes that attaches to the outside of the cell's nucleus and extends throughout the cell; rough ER is studded with ribosomes, while smooth ER is not | processes proteins for membranes or for export; breaks down toxins and wastes and manufactures necessary fat molecules |
| **Golgi apparatus** | stack of flattened, membrane-enclosed compartments held together by cell's cytoskeleton | processes and packages proteins and other molecules for shipment to their intended destinations in the cell |
| **Lysosome** | spherical membrane-enclosed compartment with a highly acidic interior | breaks down proteins and other cell molecules for recycling |
| **Vacuoles** | spherical membrane-enclosed compartments; especially large in plant cells | store food and other materials for later use |
| **Nucleus** | large, often spherical body separated from the rest of the cell by a double membrane (the nuclear envelope); surface is studded with numerous gates known as nuclear pores | contains the genetic material of the cell and serves as the cell's control center |
| **Nucleolus** | distinct region within the nucleus; may be irregular in shape | assembles ribosomes |

The wall of a lysosome contains ion pumps that make its interior very acidic; proteins destined for recycling are dumped into the lysosome to be broken down by the acid and special enzymes it contains. Lysosomes can be used to destroy bacteria that invade body cells.

## Vacuoles: Storage Containers

One of the most obvious characteristics of the cytoplasm in plant cells is the presence of a large **vacuole** or storage space. Vacuoles in plant cells perform diverse functions. They recycle proteins and other large molecules, as lysosomes do in animal cells, and they serve as storage compartments; they also allow the cell to regulate its size and water pressure by swelling or shrinking the vacuole. Human and animal cells also have vacuoles, but they are more numerous and much smaller than those of plants.

## The Nucleus: Control Center of the Cell

One of the largest and most important cell parts is the **nucleus,** the "control center" of the cell. Within the nucleus can be found the master program that controls the life of the cell and determines the physical characteristics of the organism. That master program is encoded in molecules of **DNA,** short for *deoxyribonucleic* [dē·ŏk′sē·rī′bō·no͞o·klē′ĭk] *acid.* The DNA program governs the construction and operation of the cell and contains the "blueprints" used to construct the cell's machinery. In chapter 24 we will study the nature of DNA and learn how it controls the cell's activities.

**The nuclear envelope.** The nucleus is separated from the rest of the cell by the **nuclear envelope.** The nuclear envelope, like most of the other membranes in the cell, resembles the cell membrane in its chemical structure, but it consists of two separate layers with a narrow water-filled space between them. Large protein complexes in the envelope, called **nuclear pores,** serve as gates, regulating the transport of large molecules into and

Fig. 22.11 **Nucleus**

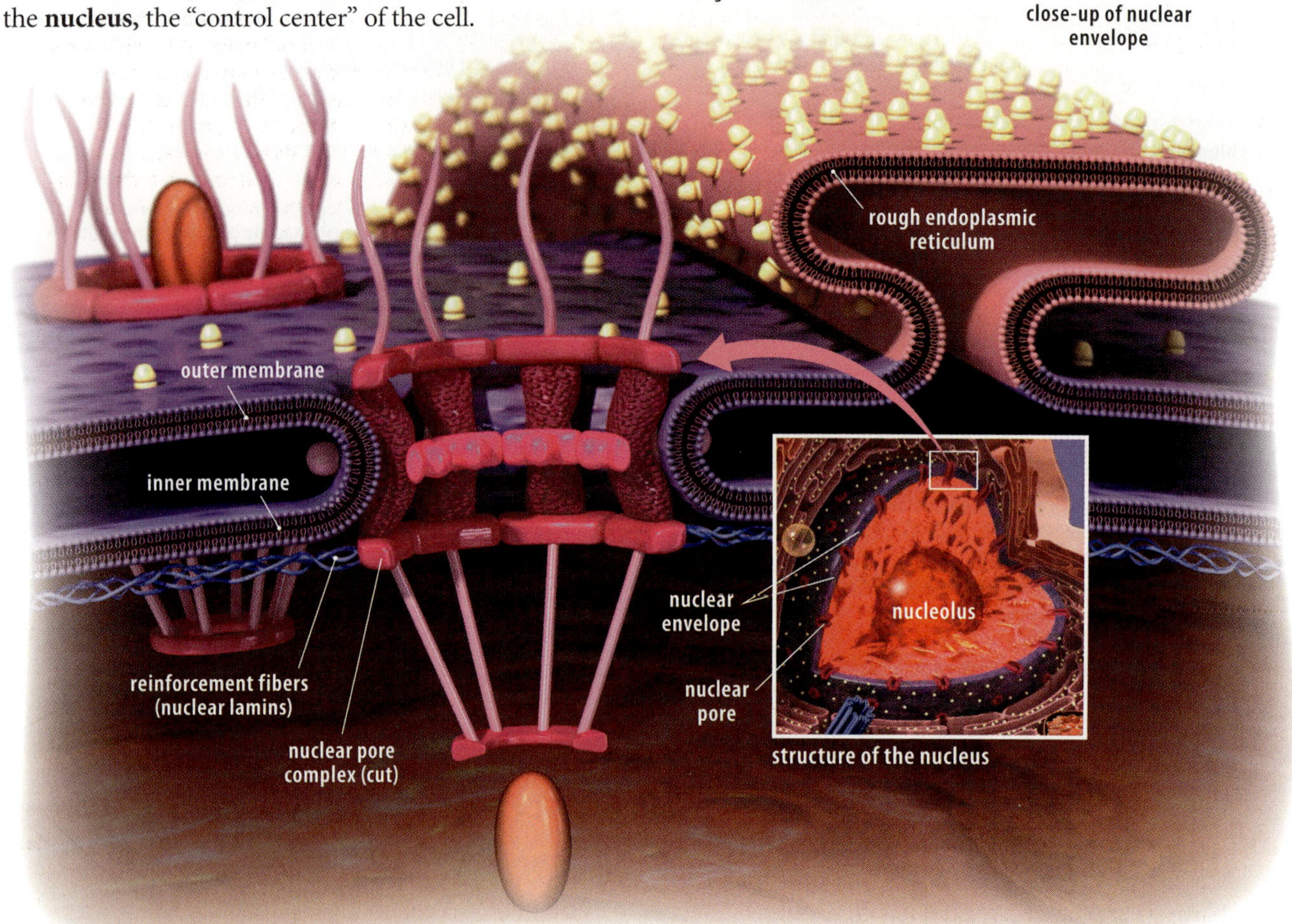

out of the nucleus. Up to one third of the membrane's surface may be seen as pores. The nucleus is very selective in what it allows to cross the membrane. Water and other small molecules are allowed to pass through unhindered, but proteins are blocked by the nuclear pores unless they contain a special identification tag.

**DNA storage.** The main function of the nucleus is to store and maintain the cell's DNA blueprints and to retrieve information from the DNA as the cell needs it. In human cells, the DNA is divided into 46 molecules, each of which would be about 1.5 inches long if stretched out. Thus, a human cell contains about 4 to 6 feet of DNA, yet the average nucleus has a diameter of only a few ten-thousandths of an inch. The DNA molecules are coiled on protein spools and racks in such a way that they can fit inside the tiny nucleus yet still be accessible for information retrieval. The DNA and the proteins associated with it, which fill the bulk of the nucleus, are called *chromatin*.

**The nucleolus.** Embedded within the nucleus is a distinct region called the **nucleolus,** where the ribosomes are assembled. After assembly, the ribosomes are transported out of the nucleus through the pores of the nuclear membrane into the cytoplasm, where they manufacture proteins.

**Conclusion.** Just a few decades ago, much of what you now know about cells was no more than theory. By no means is man's knowledge of living cells complete; we have only just begun to understand the cell's complexity. The age of the "simple cell" is gone forever as man studies a living creation that is far more complex than he ever dreamed.

## Section Review 22.2

1. What are the three main parts of a cell described in this chapter?
2. What are some of the purposes of the cell membrane?
3. What is the purpose of rough and smooth endoplasmic reticulum?
4. How do the Golgi apparatus and the cytoskeleton work together to transport proteins from one region of a cell to another?
5. Briefly state the purpose of mitochondria, chloroplasts, ribosomes, lysosomes, and vacuoles.
6. Briefly describe the overall structure of the nucleus. What purpose does the nucleus serve?

### Critical Thinking

7. In the light of what you have learned about the complex structure and intricate design of living cells, explain the following quotation from the renowned cytologist Edmund B. Wilson: "The study of the cell has on the whole seemed to widen rather than narrow the enormous gap that separates even the lowest forms of life from the inorganic world." Is there such a thing as a "simple" cell? What are the implications of cellular and molecular biology for spontaneous generation and the hypothesis of evolution?

### Identify

organelles, cytoplasm, enzyme, microtubules, centrosome, centriole, ER, vesicle, DNA, nucleolus

## 22.3 The Life and Work of Cells

### Maintaining Life

Cells were designed for one purpose: to do work. Their specialized tasks vary widely, as we have seen throughout our course in biology. For example, skin cells protect the tissues beneath them; gland cells secrete special substances that aid in various body activities; nerve cells carry electrochemical impulses from the brain to the muscles and other parts of the body; muscle cells contract, providing for movement; liver cells act as miniature chemical factories.

All living cells, however, must perform certain tasks as part of daily life. *The most pressing task that any living cell must accomplish is that of simply staying alive*—constantly taking in food and disposing of wastes; pumping ions in and out to balance its osmotic pressure and its pH; and replacing or refurbishing worn-out ribosomes, mitochondria, and other organelles. The cytoskeleton must be rearranged to meet the cell's changing needs, and accumulated poisons must be deactivated. A living cell must remain fantastically busy just to keep functioning and to maintain a stable internal environment (a condition known as **homeostasis**).

**Osmotic pressure.** A major physical requirement of cells is maintaining the proper *osmotic pressure,* which is a property related to the concentration of dissolved substances, or solutes, inside a cell. Because of the semipermeable properties of the cell membrane, water diffuses into and out of the cell. Whether water goes in or out depends upon the concentration of dissolved salts in the cell's cytoplasm in relation to the surrounding environment. For example, if a red blood cell is placed in pure water, water will diffuse into the cell until it bursts; but if the red blood cell is placed in a solution with a high concentration of salts, water will move out of the cell until it shrinks and dies. If it is placed in a solution with the same concentration of dissolved substances as the inside of the cell, the cell will neither swell nor shrink. The cells of your body must constantly regulate their internal osmotic pressures or they will die.

**Proper pH.** All cells need certain chemical substances to maintain life and to balance the delicate world in which they live. Biologists use chemical measurements to describe such an environment. For example, a cell must maintain a certain internal **pH,** a measurement of the concentration of acids and bases. Deviating from the proper pH will alter the delicate molecules in the cell. If the pH shifts, cells can compensate by using their ion pumps to move ions inside or outside the cell as necessary to keep the pH within the proper range.

**Temperature.** The temperature of the cell is critical for the optimal performance of its enzymes, the special protein molecules that carry out chemical reactions in the cell. Our enzymes perform best at normal body temperature (98.6°F or 37°C). At temperatures below normal, the enzyme reactions slow down and are no longer useful. If the body's temperature rises too far above normal, the enzymes' structure is destroyed, and the enzymes can no longer perform their functions. To regulate temperature, the body either sweats or shivers and produces goose bumps.

**Food and wastes.** The cell must carefully regulate the amount of nutrients and wastes in its cytoplasm. It does this by taking in food from the outside, by releasing stored food from its food reservoirs, and by breaking down or excreting waste products.

**Maintenance and repair.** Because of its vast complexity, a cell must devote considerable effort to maintenance and repair of its components. A cell is constantly taking old proteins out of service and replacing them with newly manufactured proteins; the old proteins are broken down into individual amino acids for recycling. On average, housekeeping proteins, which are proteins involved in tasks that are part of the daily life of the cell, are replaced every 2 to 3 days. Proteins involved in regulatory activities are replaced more often, every 5 minutes to 2 hours. The cell also contains systems that can repair damage to DNA molecules in the nucleus.

## Energy Requirements of Cells

To accomplish all of its tasks, a cell requires a great deal of energy. All cells obtain this energy from food. The cells of green plants use the energy of the sun to manufacture their food, but many other cells—including human and animal cells—must consume food from other sources. *The primary task of all cells is to obtain energy,* for it is energy that gives the cell its ability to do work.

## Photosynthesis and Cellular Respiration

**Making and storing energy.** You may recall that *photosynthesis* is the process by which nearly all autotrophs (self-feeding organisms) transform radiant solar energy into the chemical energy of sugar molecules. Photosynthesis forms glucose and other sugars, the basic energy food of the cell, from carbon dioxide and water. All of the various reactions involved with photosynthesis occur in the *chloroplasts,* organelles in green plants that are most abundant in the palisade layer of leaves.

**Releasing energy.** Living cells break down chemical substances, such as glucose, and combine them with oxygen to release energy through the process of **cellular respiration;** because this process requires oxygen, it is also called *aerobic* (with oxygen) *respiration.* The energy released by cellular respiration is used to synthesize the special phosphorus compound *adenosine triphosphate,* or **ATP.** ATP serves as the energy carrier of the cell and as a convenient form for the temporary storage of chemical energy. Nearly all of a cell's life processes are powered by ATP. The energy to produce ATP may come from any of several sources, including carbohydrates (primarily glucose), fats, and proteins.

Cellular respiration involves several steps. The first step, called *glycolysis,* takes place in the cytoplasm and breaks a glucose molecule down into two *pyruvate* [pī·ro͞o′vāt] molecules, producing two molecules of ATP per glucose molecule in the process. In the mitochondria, pyruvate is further broken down and *oxidized* (combined with oxygen) to form carbon dioxide. The ATP produced (about 36 molecules of ATP for each original molecule of glucose) is exported from the mitochondria to all parts of the cell.

Under anaerobic (without oxygen) conditions, cellular respiration cannot take place, so cells rely on **fermentation** (sometimes called *anaerobic respiration*) for energy. In fermentation, glycolysis takes place as in cellular respiration, but the pyruvate produced is converted to alcohol (in yeasts) or lactic acid (in humans and most other organisms) instead of being combined with oxygen. Only the two ATP produced during glycolysis are then available.

Cellular respiration permits an organism to carry on the functions necessary to maintain life; during this process, the organism's cells release energy for their own use and produce carbon dioxide and water as waste products. The pattern of cellular respiration is the same in all organisms because all living things are built on the same general pattern—the pattern designed by the Creator Himself.

Fig. 22.12 **Photosynthesis and respiration**

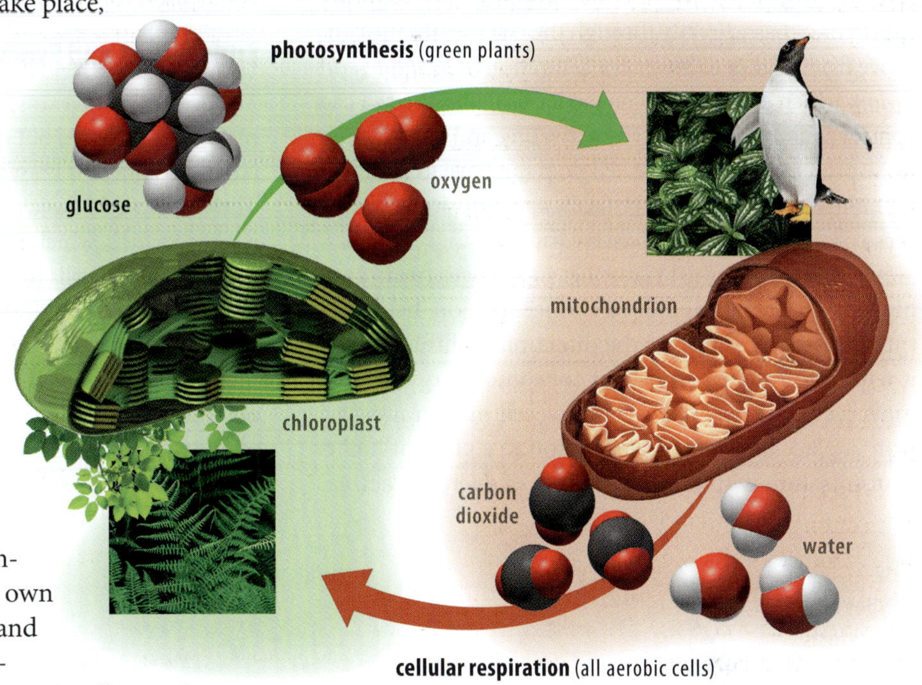


**Opposite processes.** *Both photosynthesis and cellular respiration convert energy into a form that living cells can use.* However, generally speaking, these processes are the opposite of each other. During photosynthesis, water molecules are split, yielding oxygen, hydrogen ions, and electrons. Oxygen is released to the environment, but the hydrogen ions and the electrons are used by the chloroplast to generate ATP and other energy-rich molecules. These molecules provide the energy needed to produce sugar molecules from atmospheric carbon dioxide.

The opposite occurs during cellular respiration. Electrons and hydrogen ions resulting from the breakdown of glucose and other sugar molecules are combined with oxygen in the mitochondria to form water and generate ATP and other energy-rich molecules. All that remains of the sugar molecule is converted to carbon dioxide, which is released to the environment.

## Membrane Transport

Designed into every cell is the ability to regulate what enters it and what leaves it; this regulation is a function of proteins in the cell membrane. A few substances, such as water, oxygen, and lipids, can move in and out of

Fig. 22.13 **Facilitated diffusion (passive transport) by a carrier protein**

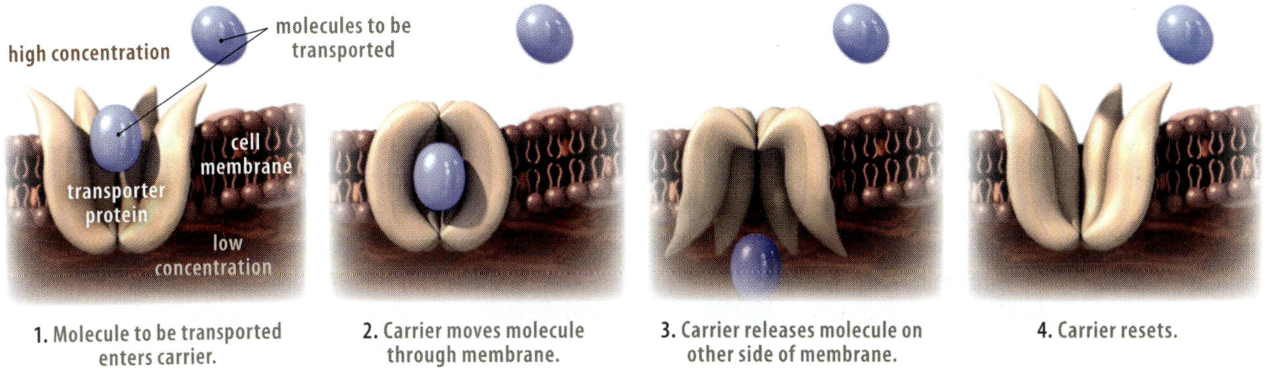

cells by *simple diffusion* from a place of higher concentration to a place of lower concentration. Compounds that cannot pass directly through the cell membrane can move from higher to lower concentration in *facilitated diffusion,* which uses channel or carrier proteins; the cell can control facilitated diffusion by activating and deactivating these proteins. Because the cell does not have to use energy for either simple diffusion or facilitated diffusion to occur, these processes are called **passive transport.**

Often, the cell must transport substances from higher concentration to lower concentration; because this transport is opposite the direction of diffusion, it requires protein pumps that consume energy. The movement of molecules across the cell membrane from a place of lower concentration to a place of higher concentration with the expenditure of energy is **active transport.** About 40% of all the energy released in the cell is used for active transport.

## Endocytosis and Exocytosis

A cell can also transport particles of material larger than molecules into and out of itself. The process of taking in large substances is known as **endocytosis** [ĕn′dō·sī·tō′sĭs]. The cell uses two basic methods to accomplish this task. One is *phagocytosis,* which means "cell eating." It is the process used by certain cells to surround solid particles with the cell membrane and engulf them. In the human body, certain white blood cells engulf bacteria by phagocytosis.

Another form of endocytosis involves the intake of liquids. The process is called **pinocytosis** [pĭn′ə·sī·tō′sĭs], which means "cell drinking." The method of intake is essentially the same as for solids, with the cell membrane allowing an amount of liquid to enter the cell before it pinches off to entrap the engulfed material. Pinocytosis is valuable to cells because most of a cell's environment is liquid containing important dissolved molecules needed by the cell.

Fig. 22.14

**Types of endocytosis**

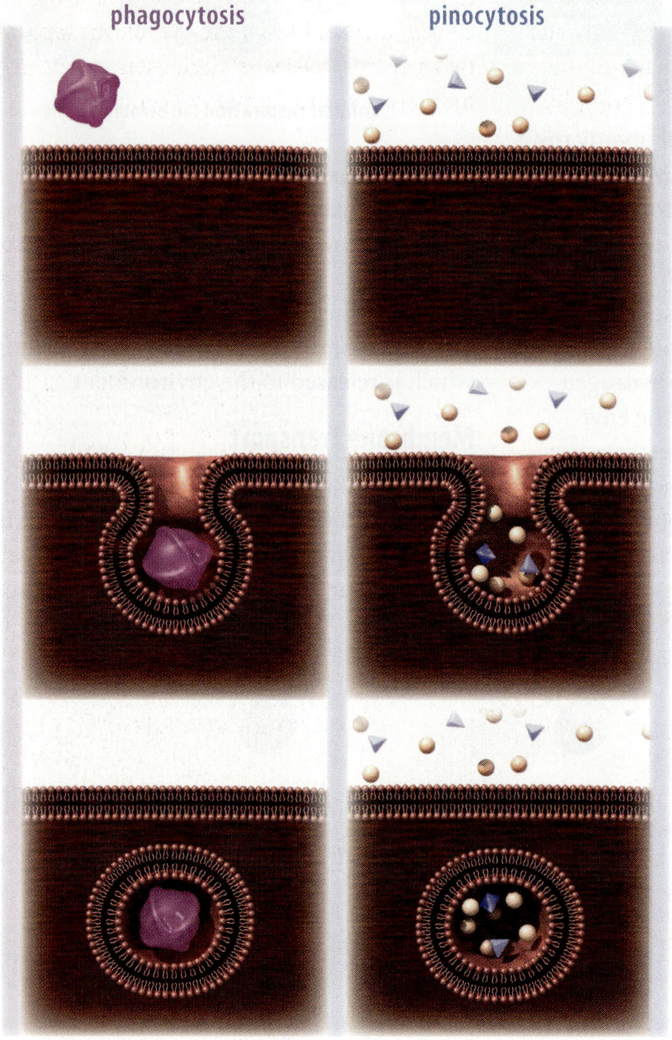


Cellular waste products are eliminated from the cytoplasm by the process of **exocytosis.** In this process, secretion-containing vesicles move to the surface of the cell and open, spilling their contents to the exterior. A similar process is also used to export materials from the Golgi apparatus to the outside of the cell.

## Cell Movement

Many cells in animals and humans (and even in some plants) are equipped with movable appendages. In some cells (such as sperm cells and protozoa), these appendages are used for swimming; in other cells (such as those lining your respiratory passages), they are used to create moving currents of fluids outside the cell.

**Cilia and flagella.** Euglenas, dinoflagellates, and sperm cells are all examples of cells that move about by means of **flagella,** whiplike tails that extend from the cells. Other cells (such as paramecia) move by means of **cilia,** tiny hairlike projections that extend from the cell membrane.

Both cilia and flagella are complex molecular machines consisting of a bundle of several microtubules and motor proteins encased in a flexible membrane. The pull of the motor proteins slides the microtubules past each other, causing the entire assembly to bend. By beating the flagella or cilia rapidly back and forth like a fish's tail or a row of oars, a cell can swim rapidly through the water. Cilia are also found in stationary cells in the linings of our lungs and respiratory tract, where their continual back-and-forth motion sweeps foreign particles out of the lungs and up the windpipe to be swallowed or expelled.

## The Cell Cycle and Mitosis

One of the most amazing aspects of the design of living cells is their ability to reproduce. Most cells reproduce by constructing duplicate cell components and then dividing into two separate cells. It is difficult for us to grasp the idea of a cell, a structure more complex than all of man's creations, constructing a duplicate of itself in only a few hours. Yet during the lifetime of many organisms, this process has occurred billions or trillions of times: one cell doubled to make two, those two doubled to make four, and so forth. Biologists study this reproductive process and summarize it as part of the **cell cycle.**

*The cell cycle has four phases:* $G_1$, S, $G_2$, and M. The first three phases, $G_1$, S, and $G_2$, are collectively called **interphase,** which is the stage between cell divisions. All of the cell's normal activities occur during interphase, when the cell does the work it was designed to do. The "gap" phases ($G_1$ and $G_2$) are times for the cell to grow and get ready to enter either the S phase (after $G_1$) or the M phase (after $G_2$).

During the *S phase* (S for synthesis), each DNA molecule, or **chromosome,** is carefully replicated. A replicated chromosome consists of two identical DNA molecules, each of which is a *chromatid* [krō′mə·tĭd], that are fastened together in the middle by a small structure called the *centromere.* In human and animal cells, the *centrosome,* which controls the formation of microtubules in the cytoskeleton, also replicates during the S phase, forming two centrosomes.

During the *M phase,* the cell divides. The M phase consists of two main processes. In **mitosis** [mī·tō′sĭs], the nucleus of a cell divides to generate two nuclei with the same number of chromosomes as the original nucleus. In **cytokinesis,** the cytoplasm separates to form two separate cells with their own nuclei, organelles, cell membranes, and (in plants) cell walls.

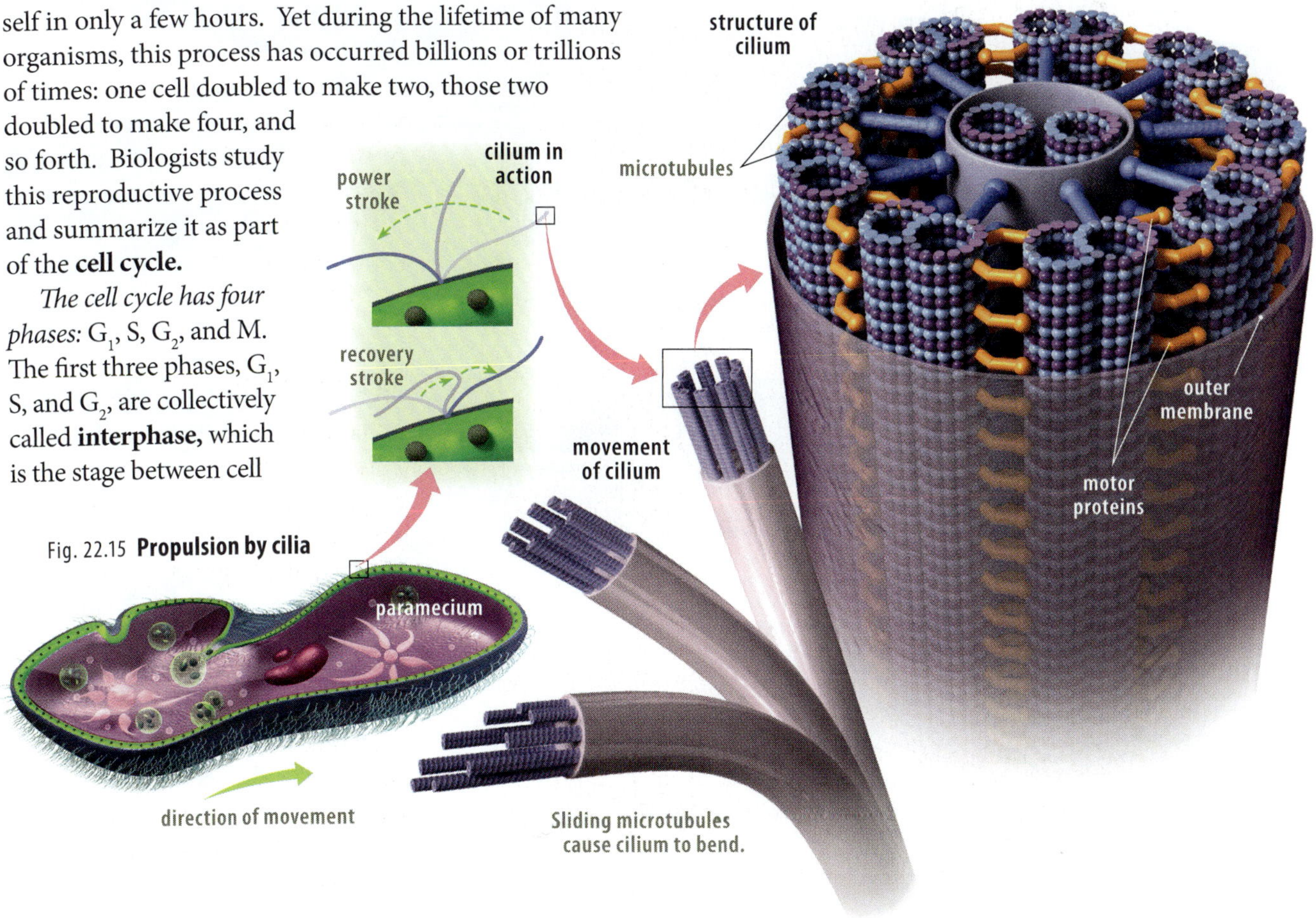


Fig. 22.15 **Propulsion by cilia**

## The Bacterial Motor: A Marvel of Design

Design in the LIVING CREATION

For their size, bacteria are some of the fastest swimmers in all of God's creation. At top speed, a fast bacterium can swim nearly 50 times the length of its body in 1 second—the equivalent of a 6-foot human swimming at 200 mi/hr—while devoting less than 2% of its energy to swimming. Bacteria achieve this remarkable feat using a very special propulsion system—high-speed *electric motors* turning spiral propellers (considered a type of flagellum) at up to 54,000 revolutions per minute. The reversible motors are controlled by a complex feedback system that monitors the concentrations of food and/or toxins outside the bacterium and adjusts the motors accordingly.

Each bacterial motor is composed of dozens of different types of protein components that make up its bearings and power mechanisms. Yet the motor cannot function at all until every part is present and functioning properly. The systems that coordinate the motors would also be useless until all their parts were present and functioning. Such devices constitute powerful evidence that bacteria and other living systems could not have come together by chance.

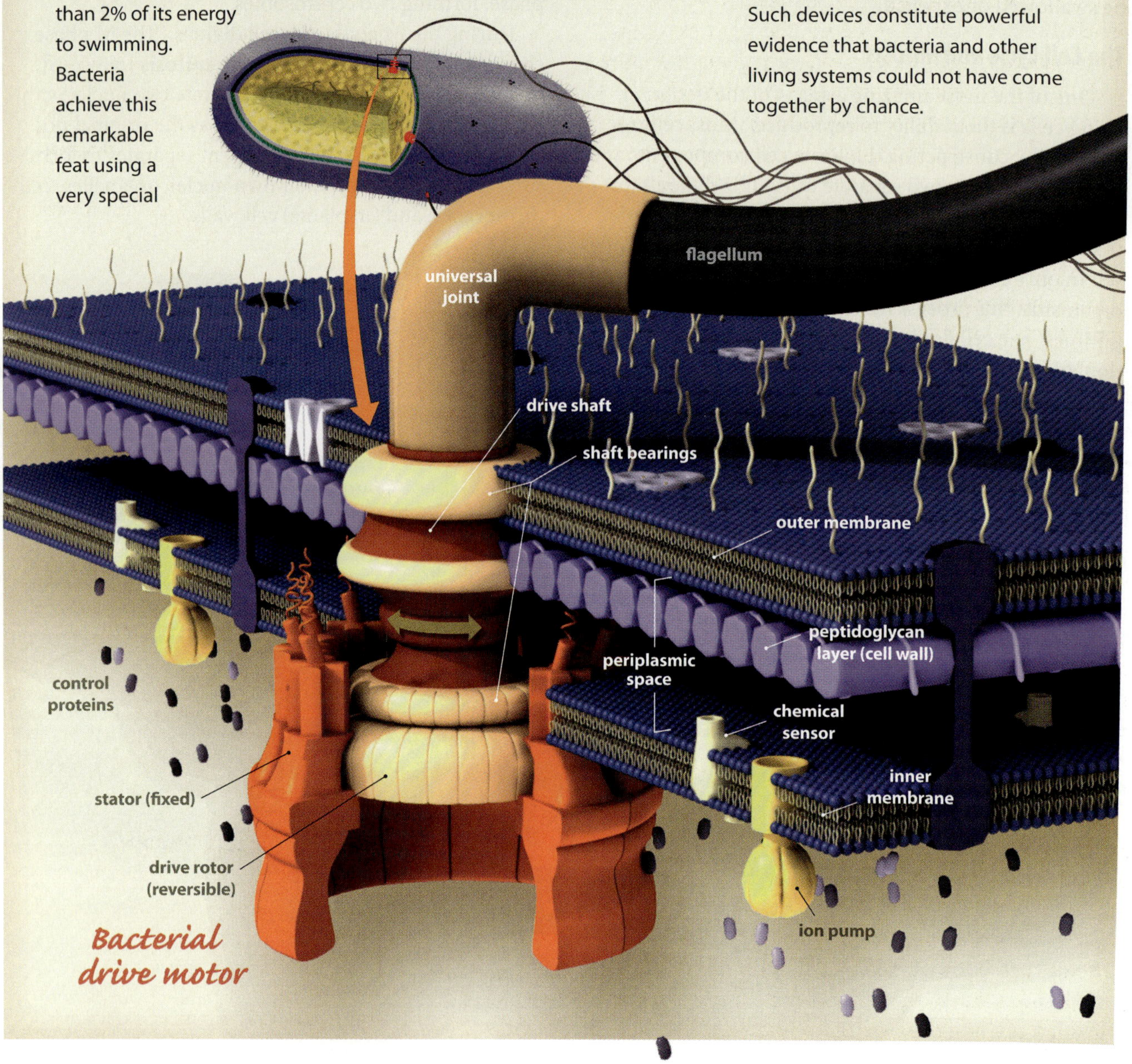


Bacterial drive motor

Fig. 22.16 **Chromosomes during mitosis**

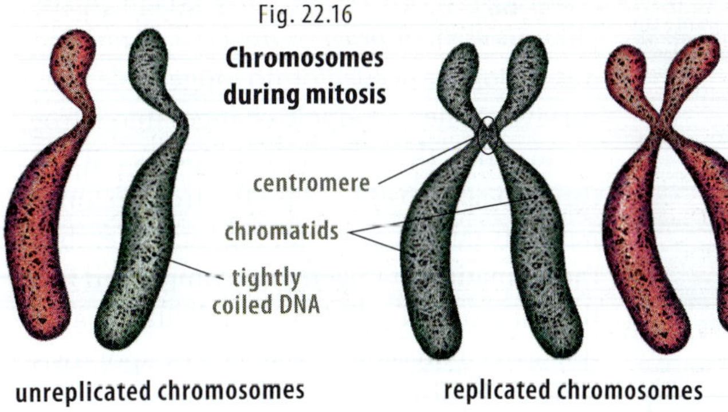


Fig. 22.17 **Mitosis**

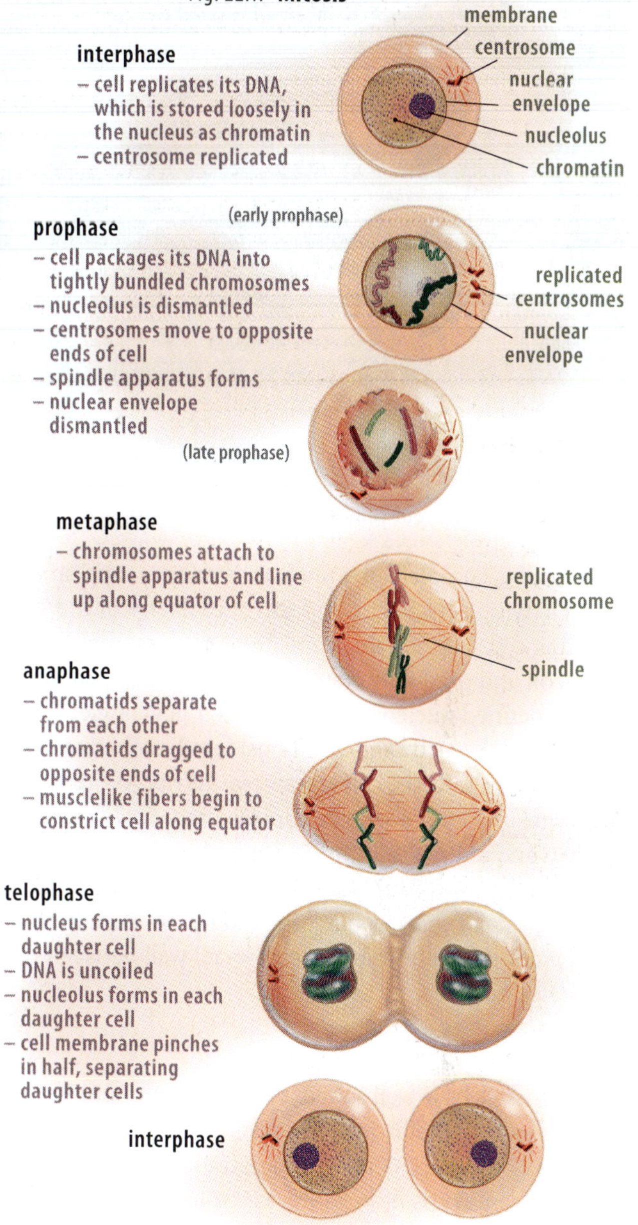


**Mitosis.** *The process of mitosis consists of four stages:* prophase, metaphase, anaphase, and telophase. In **prophase,** the first stage of mitosis, the cell packages its chromosomes, which were spread throughout the entire nucleus as chromatin during interphase, into tightly packed structures that are visible through a microscope. The chromosomes at first appear elongated but become progressively shorter and thicker as prophase progresses. As the chromosomes become more visible, their paired, two-stranded structure becomes clear. While this is occurring, the cell also dismantles its nucleolus.

Also during prophase, the centrosomes move to opposite ends, or poles, of the cell as the cell rearranges its cytoskeleton. In the space between the centrosomes, arcs of microtubules become visible. These fibers and the centrosomes that produce them form the **spindle apparatus,** the structure to which chromosomes attach during metaphase. At the end of prophase, the cell dismantles its nuclear envelope.

As **metaphase** begins, the nuclear envelope has completely disappeared, and the centrosomes are at opposite poles of the cell. The spindle apparatus attaches to the centromere of each replicated chromosome and pulls the chromosomes until they are aligned along the equatorial plane of the cell.

The next stage of mitosis, **anaphase,** begins as the chromatids in each replicated chromosome separate from each other and begin to move apart. The spindle apparatus drags the chromatids toward opposite ends of the cell.

**Telophase** follows anaphase and is in some respects the reverse of prophase. The migration of chromosomes to opposite ends

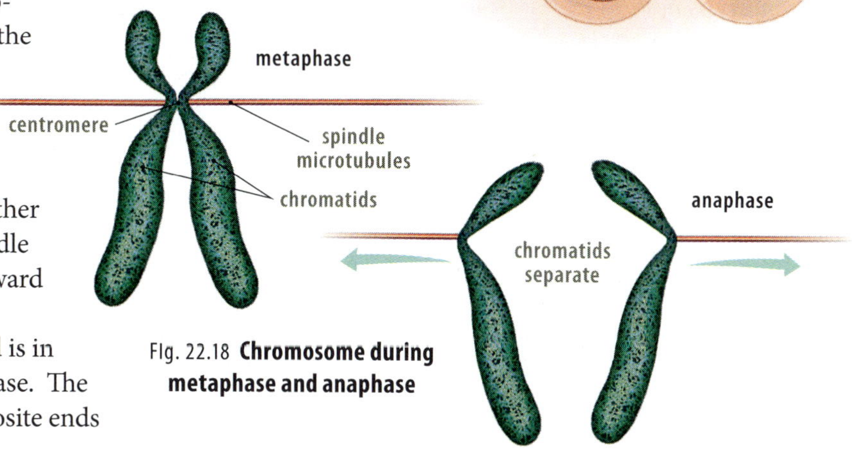


Fig. 22.18 **Chromosome during metaphase and anaphase**

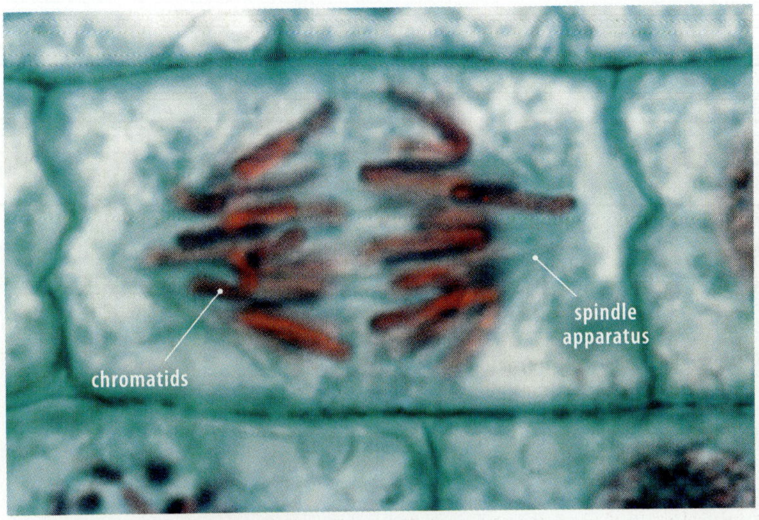


Fig. 22.19 **Onion root cells during early anaphase**

of the cell is complete, the nuclear envelope reappears, and a new nucleus begins to form at each end of the dividing cell. At the same time, the chromosomes begin to uncoil and gradually resume their former appearance. A well-defined nucleolus forms in each new nucleus.

**Cytokinesis.** Although it is considered a separate process from mitosis, *cytokinesis typically begins during anaphase and ends during or shortly after telophase.* In human and animal cells, the cell membrane pinches in half, forming two daughter cells. In plant cells, cytokinesis takes place not by pinching but by the formation of a *cell plate,* which becomes the new cell wall between the two daughter cells.

## Cell Death

No matter how well suited a cell's environment is or how hard the cell works to survive, every cell will eventually die. Cell death through trauma, such as disease or injury, is *necrosis* [nə•krō′sĭs]. But each cell is also designed with a special internal mechanism that can trigger its own death through **apoptosis** [ăp′ep•tō′sĭs], or *programmed cell death,* once it is no longer needed or no longer able to survive. When a tadpole changes into a frog, for instance, the cells in the tadpole's tail die through apoptosis because the frog no longer needs its tail. In a developing human embryo, the formation of fingers is accomplished by the apoptosis of cells between digits. In adult tissues, cell division is balanced by apoptosis to keep the size of the organs approximately the same. As apoptosis begins, the cell activates special destructive enzymes that immediately begin to break down the cell's complex structure. Although these enzymes destroy the cell's components, they leave the cell membrane intact, neatly packaging the cell's raw materials for recycling by other cells. So ends the life of a cell, the most complex design in God's creation.

## Section Review 22.3

1. What are some processes that a cell must carry out continually in order to maintain life?
2. How does cellular respiration allow a cell to produce ATP?
3. What two basic methods of endocytosis does a cell use to take in substances?
4. How do flagella and cilia allow a cell to move?
5. Describe mitosis in detail beginning with the preparations during interphase.

### Identify

homeostasis, pH, cellular respiration, passive transport, active transport, exocytosis, cell cycle, apoptosis

# Chapter 22 Review

## Define

1. cell, cytology
2. phospholipid
3. cytoplasm
4. organelle
5. microtubule
6. centrosome, centriole
7. ribosome
8. vesicle
9. vacuole
10. nucleus, nuclear envelope, nuclear pores
11. homeostasis
12. ATP
13. fermentation
14. endocytosis, exocytosis
15. flagella, cilia
16. chromosome
17. mitosis, cytokinesis
18. apoptosis

## Identify

1. three scientists whose work led to the formulation of the cell theory
2. four main chemical elements of the cell
3. three main parts of every living cell
4. the large protein molecules that carry out the chemical activities of the cell
5. three purposes of the cytoskeleton
6. the organelle in which photosynthesis occurs
7. the purpose of the Golgi apparatus
8. the purpose of lysosomes
9. the type of molecule that stores the cell's master program
10. the purpose of the nucleolus
11. the two forms of passive transport
12. the two forms of endocytosis

## Explain

1. What is the cell theory?
2. Is the cell membrane a passive boundary, or is it an active part of the cell? Explain your answer.
3. What purposes does a cell wall serve in a plant cell?
4. Why are mitochondria vitally important to a cell?
5. What are the two types of endoplasmic reticulum? What is the purpose of each?
6. In what ways must a cell stay busy just to stay alive?
7. In what ways are photosynthesis and cellular respiration opposite processes? What purpose does each serve?
8. Explain the process of mitosis.

## Apply

1. Why is it incorrect to refer to the cell as a "blob of protoplasm"?
2. What would happen to a cell if its cell membrane were removed? Why?
3. The concentration of glucose inside the tube of the small intestine is much lower than the concentration inside the cells lining the intestine. Which type of membrane transport does the intestine use to absorb glucose? Explain your answer.
4. How do ATP synthase, the bacterial motor, and other complex structures of living cells pose problems for Darwin's hypothesis of evolution?

CHAPTER 23

# Heredity: The Continuity of Life

**Key Concepts**

- God's provision of sexual and asexual reproduction
- Cell differentiation
- Meiosis
- Classical genetics
- Incomplete dominance
- Linkage
- Sex-linked traits
- Codominance
- Multiple allele inheritance
- Polygenetic inheritance
- Human genetics

## 23.1 God's Provision for the Continuity of Life

We know from our study of cells that all living things are made of cells and that cells originate from other cells in a process of cell division called mitosis. Since your body is made of cells, it must have grown over the years from a smaller number of cells; in fact, the many trillions of cells now making up your body originated from one single cell. Yet as you grew, your body took on the form of a human, not that of a horse or a hydra. Your body displays a design that reflects in many ways both of your parents, who are only two among the billions of people in the world. How is it that your physical body exhibits a size, shape, and features similar to those of two particular individuals? The answer lies in God's design for the continuity of life.

God ordained that living creatures would reproduce after their own kinds in a continual stream of life that flows from the time of Creation to the very day in which we live. Your physical body is a gift from your parents; the exact size and shape of your ears, the length of your toes, the color of your eyes, the color and texture of your hair, and your blood type are all examples of God's design and providence for your physical frame.

Every living creature follows the principle of the continuity of life. An oak tree is an oak tree because its ancestors were oak trees. The flavor of a strawberry is distinct because its forebears had the same distinct flavor. The bat's ability to fly at night was a gift of God through its parents. All living things on the earth today are the product of this providential and purposeful biological process—the continuity of life—called heredity. **Heredity** is the transmission of characteristics from parent to offspring through **genes,** units of genetic information, in the chromosomes of the cells.

Although heredity controls basic body structure and affects many of the physical traits that make every organism unique, *an organism's environment and other factors also affect genetic traits.* For example, a plant may have genes that would make it grow tall relative to other plants of the same species; but insufficient nutrition and water or an unsuitable climate can stunt its growth regardless of its genes. A person with genes that make him unusually susceptible to heart disease can still reduce his risk of disease through proper diet and exercise. Importantly, *genes do not control a person's character:* every person has the responsibility to make right choices and serve God, regardless of his heredity.

### Asexual Reproduction

*The continuity of life is assured by the miracle of mitosis,* the process by which a cell divides to form two new daughter cells, each having identical characteristics and carrying on the same life functions as the parent cell. Many whole organisms reproduce in much the same manner as single cells do in mitosis. **Asexual reproduction** is God's plan for the continuity of life from one generation to the next that requires only mitotic cell divisions. Some mechanism inside the organism triggers its cells to undergo a rapid series of mitotic divisions that result in a new organism. For instance, a bud of a sponge breaks off from the adult sponge and grows into a complete sponge, and a strawberry plant sends out runners along the ground to grow a new strawberry plant. In each case, only one individual was required for the organism's reproduction.

## Differentiation of Cells

To the casual observer, the process of asexual reproduction seems simple enough. After all, what else would a strawberry stem develop into if not a strawberry plant? But think for a moment about what is required for such a process. When the strawberry runner meets the soil, little roots grow from it and penetrate the earth. Where do the roots come from? As the new shoot reaches into the air, leaves form to begin making food by photosynthesis. Where do the leaves come from? How do the cells of the strawberry runner, dividing by mitosis, know to make strawberry roots and strawberry leaves and then strawberry fruits? The answer to these questions is found in the process of **differentiation.**

A similar phenomenon to that which occurs in the strawberry plant also occurs in the human body. At one time, your body was a single cell, a fertilized egg. As that cell divided and multiplied, a brain and a heart developed, and then arms and legs. How did that one cell know how to make a brain, a heart, arms, and legs? According to God's plan for the continuity of life, all of the information needed to form one particular person's body must be contained in that very first cell, the fertilized egg. It also follows that every cell that results by mitosis from that very first cell must have all of that information as well. As the number of cells increases, certain cells are called out to become specialized components of tissues and organs through differentiation. Blood cells, brain cells, bone cells, skin cells, and all of the other specialized cells of your body resulted from the differentiation of similar cells very early in your embryonic development.

The process of cell differentiation continues even today within your body, as illustrated by blood-cell production. Within your bone marrow are countless cells that are continuously undergoing cell division and differentiation. Some of their offspring become red blood cells, and some become white blood cells of several types.

God's plan for the continuity of life is astounding when you consider that one skin cell has the same heredity as the trillions of other cells in your body and the same heredity as the countless other cell generations before it, all the way back to the moment after conception when your body was only a single fertilized egg cell. It is a skin cell only because various chemical signals triggered it to express the genes needed for skin cells, not the genes for bone or muscle cells. Even more astounding is the fact that the heredity of all human beings extends back through hundreds of generations in the past to the first pair of human beings—Adam and Eve.

Fig. 23.1 **Asexual reproduction**

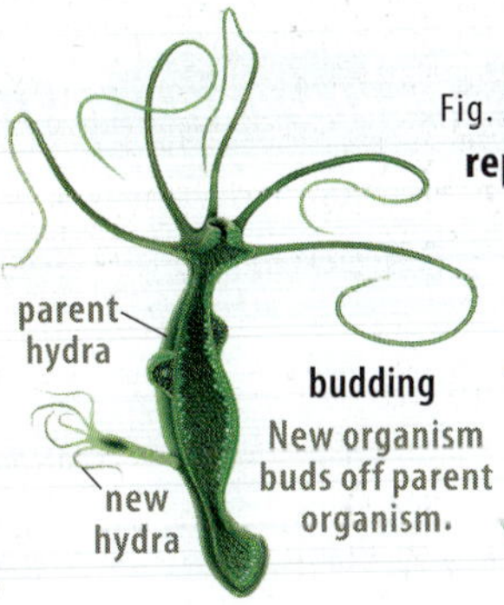


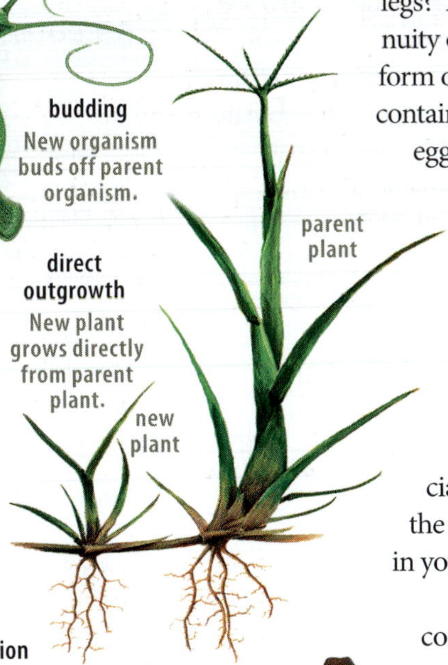


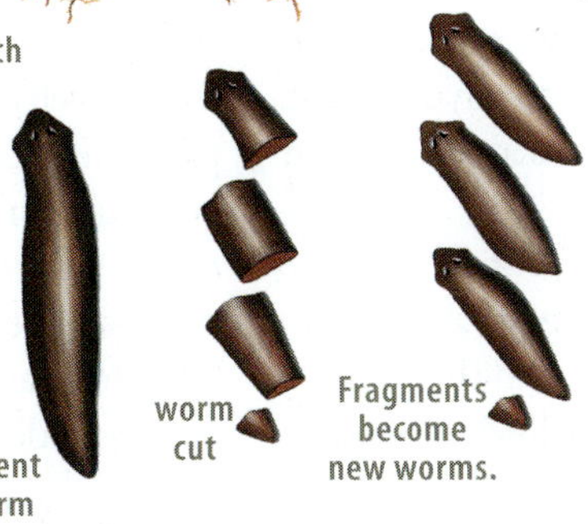


Fig. 23.2 **Development of specialized cells**

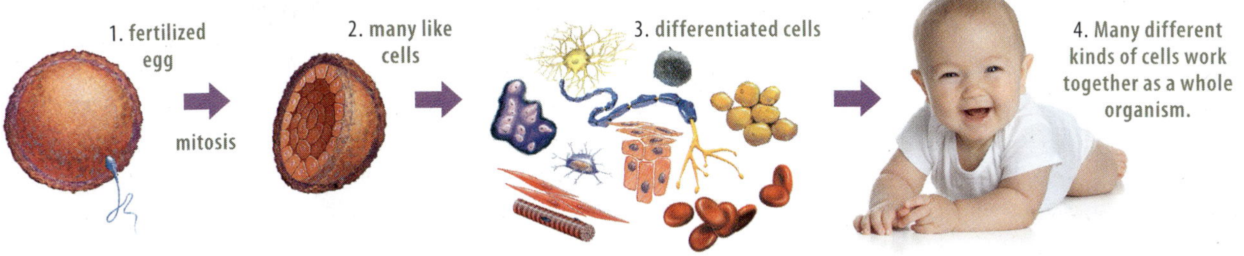

## Sexual Reproduction

Knowing that your heredity can be traced back to a solitary cell, the question is where that cell came from. The answer lies in one of the most remarkable designs in God's plan for the continuity of life—**sexual reproduction.**

Ever since God commanded Adam and Eve to populate the earth with their descendants, man has known that most living things reproduce sexually; that is, a mating must occur between the male and female of the same kind. It was not until fairly recently, however, that biologists began to understand the marvelous design God placed in sexual reproduction.

In organisms that reproduce sexually, the heredity of both parents is combined to provide the heredity of the offspring. The male and the female contribute equal amounts of genetic material in the form of DNA, the chief component of chromosomes.

In humans, each **somatic cell** (body cell) has 46 chromosomes; but the *gametes* (egg and sperm) of the parents each contain only 23 chromosomes (Fig. 23.3). The egg and sperm originate when **germ cells** in the ovaries of females and the testes of males undergo a specially designed type of cell division called **meiosis.** Meiosis also occurs in the germ cells of plants and animals.

When an egg and a sperm unite in the process of *fertilization,* the result is a single fertilized egg cell known as a **zygote.** Since the sperm cell contains half the usual number of chromosomes (known as a **haploid** set) and the egg cell contains half, the zygote contains the complete number of chromosomes (a **diploid** set) for that particular organism. The letter "*n*" designates the haploid number for an organism, and "2*n*" designates the diploid number (see Table 23.1). The single-celled zygote divides by mitosis as its cells differentiate, and a fully formed organism develops.

**Table 23.1 Chromosome Numbers**

| Common name | Scientific name | Number of chromosomes (2n) |
|---|---|---|
| man | *Homo sapiens* | 46 |
| dog | *Canis lupus familiaris* | 78 |
| cat | *Felis catus* | 38 |
| horse | *Equus caballus* | 66 |
| American alligator | *Alligator mississipiensis* | 32 |
| housefly | *Musca domestica* | 12 |
| house cricket | *Acheta domesticus* | 22 |
| common carp | *Cyprinus carpio* | 104 |
| Easter lily | *Lilium longiflorum* | 24 |

Fig. 23.3 **Mitosis, meiosis, and reproduction**

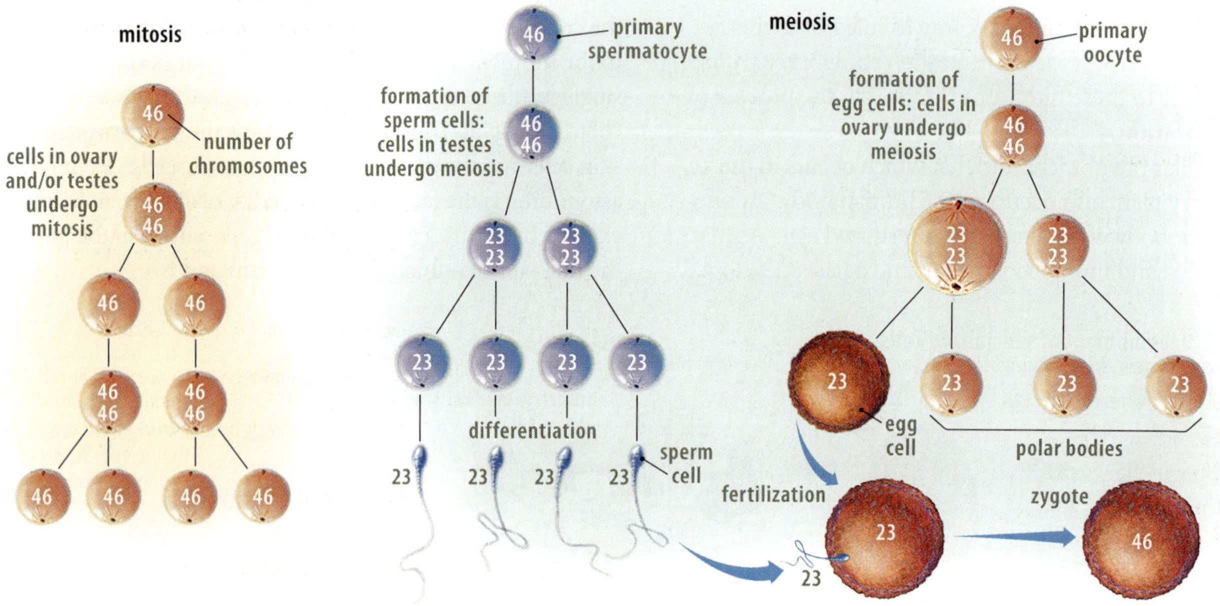

# The Miracle of Meiosis

Understanding the LIVING CREATION

The special type of cell division that makes sexual reproduction possible is meiosis. Because the number of chromosomes is reduced by one half in meiosis, it is also called *reduction division*. In male organisms, meiosis is a major part of *spermatogenesis* [spər·măt′ə·jĕn′ĭ·sĭs], the formation of sperm cells. In female organisms, meiosis occurs as part of *oogenesis* [ō′ə·jĕn′ĭ·sĭs], the production of egg cells. Since the purpose of meiosis is the production of gametes, it occurs only in the germ cells.

Humans have 46 chromosomes, arranged in 23 pairs. The chromosomes in each pair are not identical, but they are *homologous* [hə·mŏl′ə·gəs]—that is, they are similar in size and shape and contain the genetic information for the same functions. One of the homologous chromosomes in each pair comes from the mother and the other from the father. Just before the first meiotic division, the DNA is replicated. As a result of this replication, each individual chromosome contains two chromatids connected through the centromere.

**First meiotic division.** *Meiosis occurs in two successive stages*—a first and a second meiotic division. As the first meiotic division begins, the chromosomes become visible. Homologous chromosomes pair up in a process called *synapsis.* During synapsis, an exchange of genetic information takes place between paired homologous chromosomes. Such exchange of genetic material is called *crossing over.* It results in the **recombination** of genes, an important source of biological variation. The spindle fibers then attach to the centromeres, and the

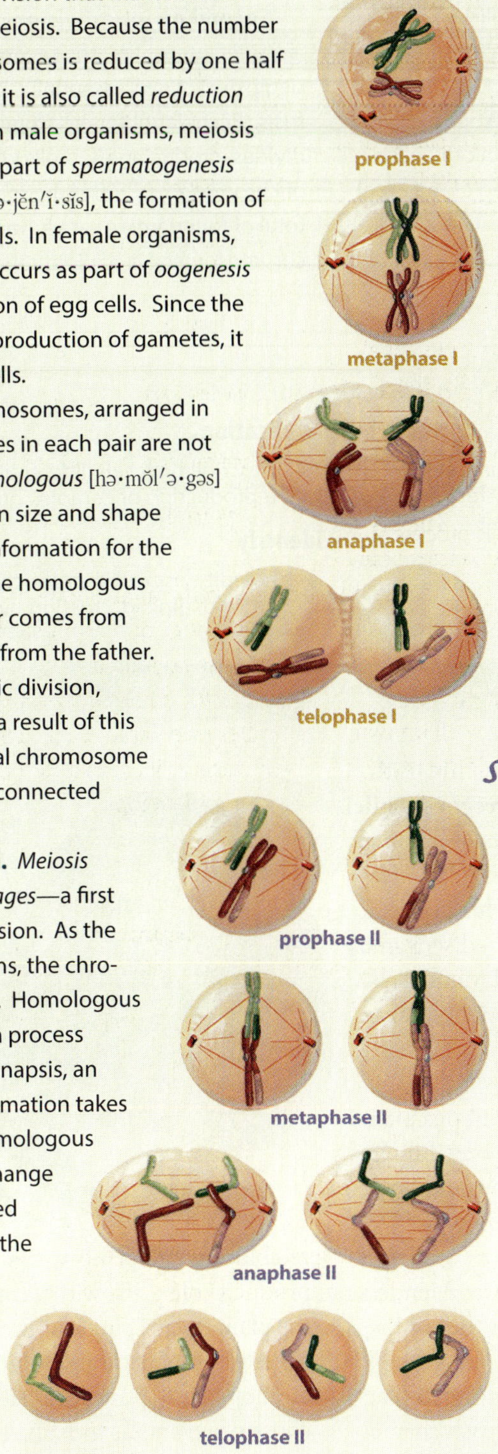


## *First meiotic division*

**interphase**
- cell replicates its DNA, which is stored loosely in the nucleus as chromatin
- centrosome replicated

**prophase I**
- cell packages its DNA into tightly bundled chromosomes
- nucleolus and nuclear envelope dismantled
- homologous chromosomes pair up during synapsis
- crossing over takes place between paired homologous chromosomes, causing recombination
- spindle apparatus formed

**metaphase I**
- paired-up homologous chromosomes line up along the equator of cell

**anaphase I**
- homologous chromosomes become randomly separated, and each moves to a different end of the cell
- musclelike fibers begin to constrict cell along equator

**telophase I**
- cell membrane pinches in half, separating daughter cells
- nucleus and nucleolus form in each daughter cell
- each daughter cell has half the original number of chromosomes

## *Second meiotic division*

**interphase**
- brief interphase, but cell does not replicate its DNA

**prophase II**
- chromosomes are visible
- nucleolus and nuclear envelope dismantled
- spindle apparatus formed

**metaphase II**
- chromosomes attach to spindle apparatus through the centromere and line up along equator of cell

**anaphase II**
- chromatids separate from each other
- chromatids dragged to opposite ends of cell
- musclelike fibers begin to constrict cell along equator

**telophase II**
- cell membrane pinches in half, separating daughter cells
- nucleus and nucleolus form in each daughter cell
- each daughter cell has half the original number of chromosomes and possesses only one copy of each

**differentiation**
- daughter cells become gametes

(cont.)

(cont. from previous page)

homologous pairs line up at the cell's equator (as in mitosis). As the first meiotic division continues, the chromosomes in the homologous pairs separate randomly and move to opposite ends of the cell; unlike in mitosis, the paired chromatids in each replicated chromosome remain attached. The parent cell then divides into two daughter cells, each with one half the number of chromosomes as the parent cell (in human cells, each daughter cell has 23 chromosomes instead of the usual 46). However, each daughter cell contains two identical copies of each chromosome. When division is complete, the nuclear envelopes appear around the haploid sets of chromosomes, nucleoli reappear, and the spindle fibers fade away.

**Second meiotic division.** After *interkinesis,* which is a brief interphase, the second meiotic division begins. It follows essentially the same sequence as mitosis. Each daughter cell from the first meiotic division divides, producing two cells with the same number of chromosomes. After the second meiotic division, a total of four gametes are formed, each containing 23 chromosomes—a haploid set of human chromosomes.

## Section Review 23.1

1. How many chromosomes are in each somatic cell in humans? How many are in a human gamete?
2. What is the purpose of meiosis? How is this process different from mitosis?
3. What is synapsis? Why is it significant?

**Application**

4. Describe how cell differentiation is vital to the reproduction of complex organisms.

**Identify**

heredity, asexual reproduction, meiosis, zygote, haploid, diploid, crossing over, recombination

# 23.2 Classical Genetics

Little was known about heredity until the mid-19th century when an Austrian monk named **Gregor Mendel** (1822–1884) began working with pea plants in his small monastery garden in what is now the Czech Republic. Although Mendel's report on the results of his experiments lay neglected for 34 years, scientists rediscovered it at the turn of the century and recognized Mendel as the Father of **Genetics,** the scientific study of heredity.

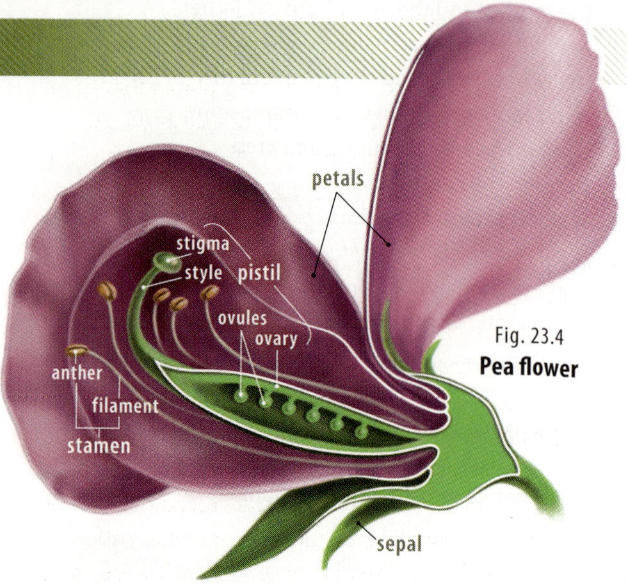


Fig. 23.4
**Pea flower**

## Mendel's Experiments

When Mendel began his experiments with pea plants in 1853, it was already known from animal breeding that characteristics were passed to the offspring from each parent and that these traits were somehow carried by egg and sperm cells. But the laws that governed the mechanism of heredity were unknown. It was Mendel's task to uncover the secrets of God's plan for the continuity of life.

After cultivating and studying pea plants for two years, Mendel established plants that bred true for each characteristic that he was studying. For example, he had plants with red blooms that always produced offspring with red blooms and plants with wrinkled seeds that always produced offspring with wrinkled seeds. He said that the plants were "pure" for these traits. Through many generations of natural self-fertilization, garden peas had developed into pure lines. (It was fortunate that Mendel chose pea plants as the subject of his experimentation, because the flowers of pea plants generally pollinate themselves. Cross-pollination is rare among pea plants

## Table 23.2 Selected Traits Studied by Mendel

| | Seed form | Seed color | Flower color | Pod color | Stem length |
|---|---|---|---|---|---|
| Dominant trait* |  round |  yellow |  red |  green |  tall |
| Recessive trait |  wrinkled |  green |  white |  yellow |  dwarf |
| Alleles | R, r | Y, y | W, w | G, g | T, t |

*Trait exhibited by all the $F_1$ offspring

because of the compact arrangement of the stamen, pistil, and petals.)

Mendel took his pure lines and began to cross pea plants that had different characteristics for the same trait. For example, one set of experiments involved only the texture of the seed coat. He crossed plants that bred true for wrinkled seed coats with those that bred true for round seed coats. He found that all plants in the first generation of offspring (the *first filial* [fĭl′ē·əl: "having to do with offspring"] *generation,* abbreviated $\mathbf{F_1}$) produced only seeds with round seed coats. This pattern held true for all the traits that Mendel studied (see Table 23.2). Whenever true-breeding plants with different traits were crossed, all the offspring in the $F_1$ generation exhibited only one of the two traits, and always the same one.

**The law of dominance.** Mendel concluded that these traits are caused by paired heredity factors that today we call *genes.* Generally, each parent transmits one gene for each trait to the offspring. In modern genetics, different forms of a gene for a particular trait, such as various seed colors, are called **alleles** [ə·lēlz′]. For example, the two alleles for the seed color gene in pea plants are yellow and green. If an organism has two identical alleles for the same gene, it is said to be **homozygous** [hō′mō·zī′gəs]. If one allele is different from the other, the organism is said to be **heterozygous** [hĕt′ər·ə·zī′gəs]. Mendel further concluded that the traits present in the $F_1$ (round seeds in peas, for example) are the dominant traits. **Dominant traits** show up in the offspring even if an allele for a different trait is also present. This fact is known as the **law of dominance.** If a pea plant inherits an allele for wrinkled seeds from one parent and an allele for round seeds from the other parent, the plant will have round seeds because round is dominant over wrinkled. The trait that does not show up when the dominant allele is present is known as the **recessive trait.** For a recessive trait to show up, both alleles must be for that trait. Thus, for a pea plant to produce wrinkled seeds, it must inherit the recessive wrinkled seed allele from each parent.

**Genotype and phenotype.** In genetics, *dominant and recessive alleles are represented by capital and lowercase letters.* For instance, we can use R to represent the dominant allele for round seeds and r for the recessive allele for wrinkled seeds. Then, RR refers to a plant that is homozygous for the dominant trait, Rr refers to a plant that is heterozygous, and rr to a plant that is homozygous for the recessive trait. These letters describe the **genotype** [jĕn′ə·tīp′] of an individual, which is the actual combination of alleles inherited for a particular trait. The outcome or expression of the genes in the physical appearance of the individual is the **phenotype** [fē′nə·tīp′]. In the previous example, both the RR plant and the Rr plant have the same phenotype, round seeds, even though their genotypes are different. We will see many other examples in which the genotype of two individuals may be different but their phenotype the same.

Fig. 23.5 **Punnett square**

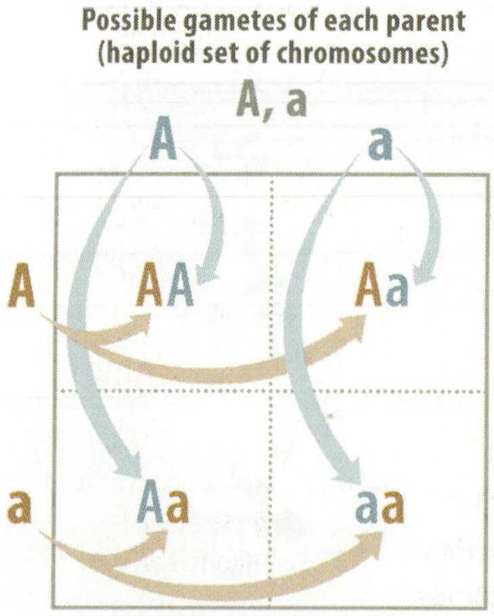


Haploid gametes combine to give offspring a diploid (full) set of chromosomes.

**Steps to making a Punnett square**

1. Assign a letter to each trait. Use a capital for dominant alleles and a lowercase letter for recessive alleles.
2. Write the genotype for each parent (AA, Aa, aa, etc.).
3. Set up the box by placing possible gametes for one parent above the square and gametes for the other parent along the side.
4. Form the crosses by bringing into each box the alleles listed above it and to its side.
5. Analyze the Punnett square by listing the percentage of offspring with each phenotype.

**Punnett squares.** The **Punnett square** is a chart that helps us to visualize all the possible combinations of alleles that could result in the offspring (Fig. 23.5). On top of the Punnett square, you place all the possible combinations of alleles found in the gamete of one parent. Along the side, you place all the possible combinations of alleles in the gamete of the other parent. Remember, gametes can contain only one allele for each trait. By filling in each box with one allele from the parent on top and one allele from the parent on the side, you can determine all the possible combinations of alleles for the offspring and predict the outcome of genetic crosses.

**Hybrids and hybridization.** Suppose pure-breeding plants with round and wrinkled seeds are crossed. Since each parent contributes a different allele for seed-coat texture, the offspring are heterozygous for seed-coat texture. Mendel used the term ***hybrid*** to describe an organism with two different alleles for a particular trait (Fig. 23.6). (Outside genetics, the term *hybrid* can have a different meaning, such as a cross between two different species of organism.) Figure 23.7 illustrates the results of Mendel's crosses based on the color of the seed: dominant yellow (Y) and recessive green (y) color.

**The law of segregation.** When Mendel crossed the hybrid plants with yellow seeds from the $F_1$ generation with each other, he found that there were always some plants with green seeds in the $F_2$ (or *second filial*) **generation** (Fig. 23.8). Based on these results, Mendel formulated a second law known as the **law of segregation.** This law states that alleles for each trait segregate or separate randomly during the formation of gametes. Mendel observed that, because of the segregation of alleles into different gametes, about one fourth of the $F_2$ showed the recessive trait.

**Incomplete dominance.** Since Mendel's time, scientists have observed that *the law of dominance does not hold true in every case.* When Japanese four-o'clock plants with white flowers are crossed with plants having red flowers, the $F_1$ generation consists only of plants with pink flowers. This mixing or blending of traits is called **incomplete dominance.** The red parents in Figure 23.9 (p. 479), with the genotype RR, have red pigment in their flowers. In contrast, the white parents, with the genotype WW, do not have any pigment. Because their cells cannot make any pigment, their flowers are white. In a RW heterozygous $F_1$ plant, one of the alleles, R, makes pigment while the other, W, does not. The dilution of the red pigment gives the appearance of pink. Pink flowers do have red pigment, but much less than

Fig. 23.6 **Round and wrinkled seeds**

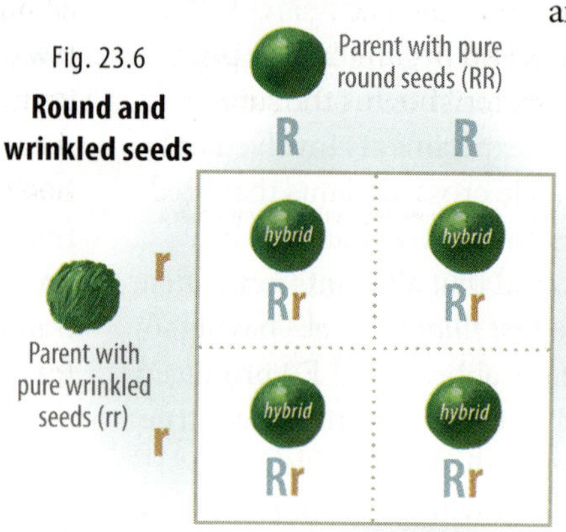


**$F_1$ generation**
100% hybrid round

Fig. 23.7 **Yellow and green seeds**

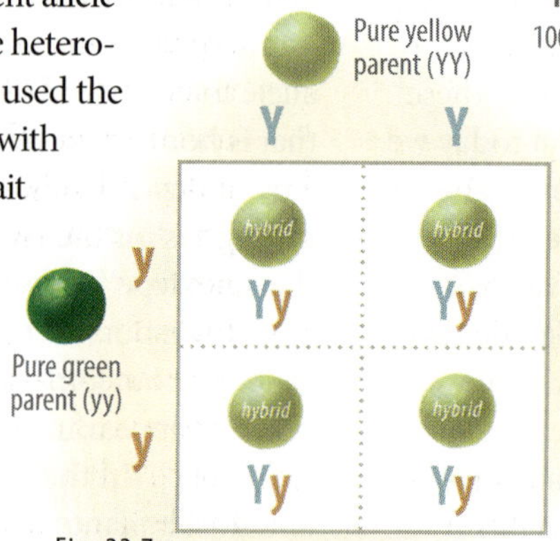


**$F_1$ generation**
100% hybrid yellow

Fig. 23.8 **Segregation of seed color**

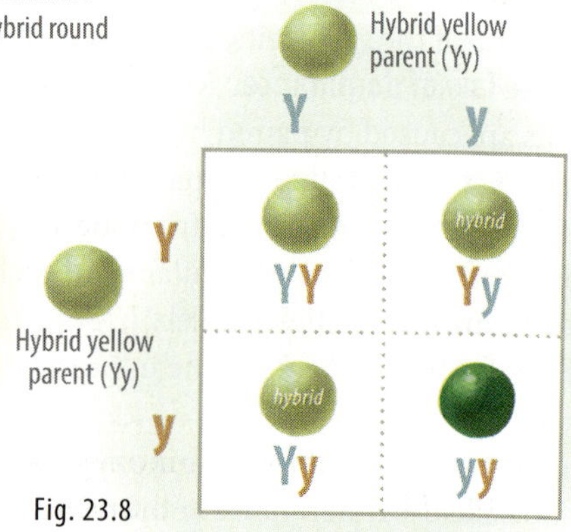


**$F_2$ generation**
25% pure yellow (YY),
50% hybrid yellow (Yy),
25% pure green (yy)

red flowers because only one of the alleles is operating. Now you can understand why all three colors, red, white, and pink, would be present in the $F_2$.

Fig. 23.9

**Incomplete dominance in four-o'clock flowers**

**$F_1$ generation**
100% hybrid pink

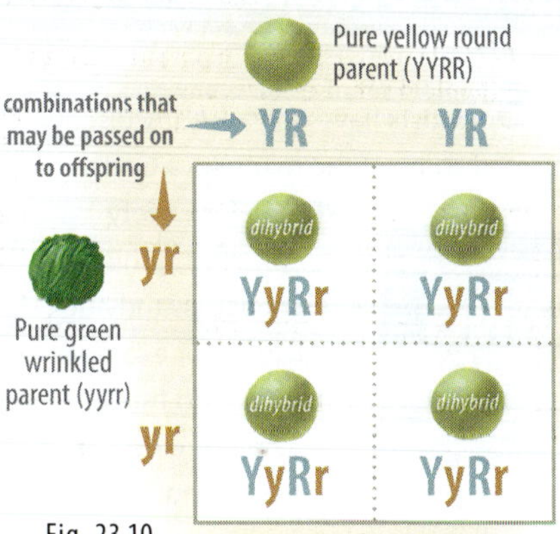


Fig. 23.10

**Production of dihybrids**

**$F_1$ generation**
100% dihybrid yellow round

**The law of independent assortment.** In another set of experiments, Mendel analyzed more than one trait at a time. For instance, he crossed homozygous pure-bred plants having yellow round peas (YYRR) with homozygous pure-bred plants having green wrinkled peas (yyrr). As is evident from Figure 23.10, all the $F_1$ from this cross are heterozygous for both traits and therefore have yellow round peas. An organism that is heterozygous for two traits is called a **dihybrid:** each offspring from the $F_1$ generation is a dihybrid for yellow seeds (Yy) and round seeds (Rr).

From the four available alleles (YyRr) in the $F_1$ generation, there are four possible combinations that can be passed on to the $F_2$ generation from each parent (YR, Yr, yR, yr). When Mendel crossed his dihybrids, he got some interesting results, which are shown in Figure 23.11.

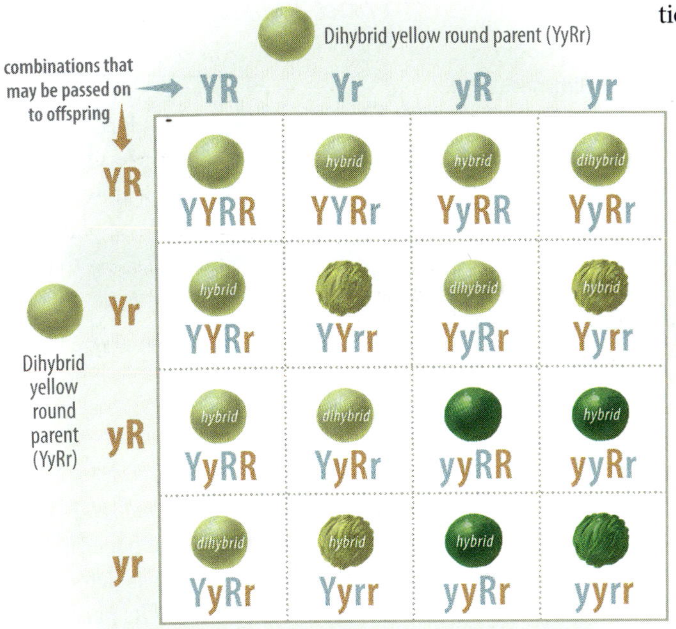


**$F_2$ generation** ratio per every sixteen offspring
9 yellow round, 3 yellow wrinkled,
3 green round, 1 green wrinkled

Fig. 23.11

**Law of independent assortment**

Because the $F_2$ generation contained plants with yellow wrinkled peas and green round peas, Mendel correctly concluded that each trait (yellow or green and round or wrinkled) must be inherited separately and independently of the other traits. Scientists now know that many traits of an organism, such as the seed color and the seed shape in peas, are inherited independently of each other according to the **law of independent assortment.** In other words, the segregation of one gene does not influence the segregation of the other.

**Linkage.** Since Mendel, scientists have found many exceptions to the law of independent assortment. Traits that follow this law have, as a general rule, their corresponding genes located on different chromosomes. But when the two genes are located on the same chromosome, two situations are possible. If they are located far enough from each other, at the extremes of the chromosome, for example, they will segregate in a way resembling the law of independent assortment. But if they are close to each other, both traits may be inherited together and not be sorted independently. This is known as **linkage,** and the genes are said to be linked. Genes that are linked do not follow the law of independent assortment. In humans, for example, the gene for hemophilia A and the gene for a certain type of colorblindness are linked. It is observed that most individuals with hemophilia A are colorblind as well. Molecular studies have shown that both genes are linked because they are found very close to each other on the X chromosome. Because they are linked together, all the genes on a given chromosome are known collectively as a **linkage group.**

## God's Design for Variety

Design in the LIVING CREATION

Scientists have only recently discovered some of the marvels of God's design for variety among living things. Geneticists studying meiosis, for example, have found that chromosomes crossing over during synapsis exchange alleles. In this way, a chromosome, which goes on to become part of a gamete, either keeps its original alleles or exchanges them for the alleles on its partner (homolog). This exchange is called **recombination.** After recombination, the gamete has a different combination of alleles than does its parent cell. This is why you are different in many respects from your parents or siblings.

God has provided every plant or animal that reproduces sexually with this marvelous means of diversity; at the same time, there is a continuity of life from generation to generation according to a highly organized scheme that we see all around us in the living world.

*Another source of genetic variation is the varying assortment of whole chromosomes during meiosis.* Any given pair of homologous chromosomes in an organism contains one chromosome derived from the organism's father and one derived from the organism's mother. For example, a human egg cell might receive paternal copies of chromosomes 1 and 2, maternal copies of chromosomes 3–5, and so on. This also allows a great deal of variety from generation to generation and helps explain how a child can have features that resemble multiple grandparents.

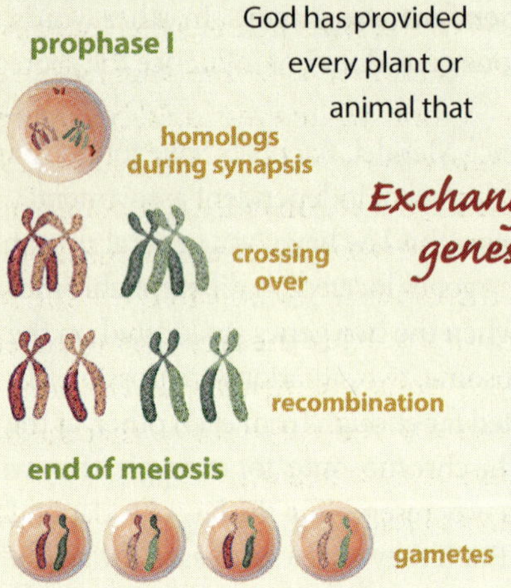


Man, in thinking God's thoughts after Him, has used these principles of variety (along with the laws of independent assortment and segregation) to develop "improved" strains of wheat and other agricultural products. Horticulturists and animal breeders selectively breed plants and animals that possess certain qualities that they want to preserve or enhance.

Other domestic animals, such as different breeds of cows, hogs, chickens, dogs, and cats, are the products of man's systematized hybridization programs, selectively breeding for certain desirable characteristics (such as more meat and less fat in hogs). An example of plant hybridization is found in corn. If a strain of corn with exceptional root systems and another strain with exceptional ear qualities are developed, they may be crossed to produce a hybrid. These hybrids may grow into top-yielding crops but are often sterile because they are genetic "dead ends." (As mentioned before, the word *hybrid* is used in this case to indicate corn plants produced by crossing two different corn varieties.)

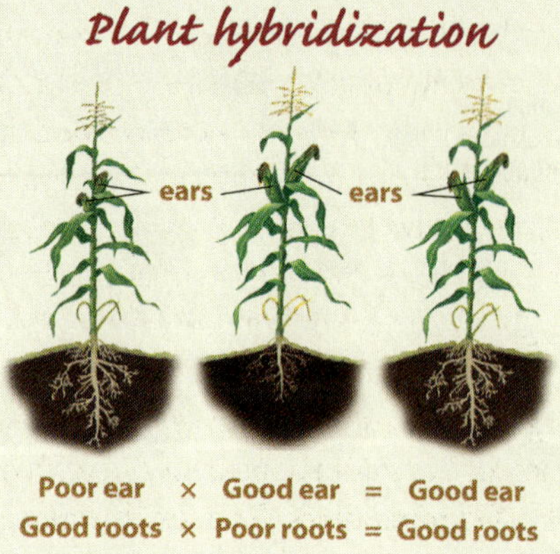


### Genetic Research in the Twentieth Century

Interest in Mendel's pioneering work revived at the turn of the 20th century when biologists rediscovered his long-neglected papers. The next four decades, 1900 to 1940, are often referred to as the "golden age" of genetics because they witnessed great advances in this science. Following Mendel's lead, geneticists began to uncover the secrets of God's plan for the continuity of life.

**Sutton's hypothesis.** In 1902, *Walter S. Sutton* (1877–1916), a young American scientist, published a paper in which he theorized that an organism's heredity is determined by chromosomes within the sex cells of its parents. He theorized that the chromosomes carried "hereditary factors" (now called genes). Sutton explained Mendel's dihybrid experiments with pea plants by saying that the "factors" for green or yellow seeds were on one chromosome and that the "factors" for round or wrinkled seeds

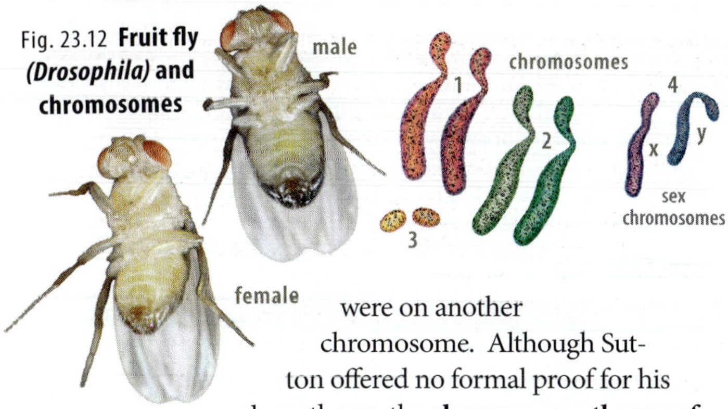


Fig. 23.12 **Fruit fly (*Drosophila*) and chromosomes**

were on another chromosome. Although Sutton offered no formal proof for his hypotheses, the **chromosome theory of inheritance,** as his hypotheses came to be called, soon laid the foundation for all modern genetic research.

**Morgan's research.** In 1904, *Thomas Hunt Morgan* (1866–1945), an eminent American biologist, began a quarter century of genetic research that would earn him a Nobel prize in 1933. Although Morgan initially doubted Mendel's theories and Sutton's hypotheses, his work soon convinced him that both men were right. For his research, Morgan chose the common fruit fly, *Drosophila melanogaster,* because the tiny insect is easy to handle and store and has an extremely short life cycle of two weeks. Thus, in only a few months, Morgan would be able to study several different generations of fruit flies.

**Sex chromosomes.** Sutton and other biologists had observed that the chromosomes in cells from male and female fruit flies were identical except for one pair. They suspected that this pair of chromosomes determined the organism's sex. Morgan's research confirmed this idea; he discovered that fruit flies have four homologous pairs of chromosomes, with one pair being different in males and females. These homologous pairs are the **sex chromosomes,** the chromosomes that determine an organism's sex.

Morgan observed that both of a female fly's sex chromosomes are similar in appearance to each other; he called them the *X chromosomes.* The sex chromosomes of a male fly, however, are quite different from each other. Because one of the male's sex chromosomes was similar to the female's chromosomes, Morgan also called it an X chromosome; but its counterpart, which was unlike the other chromosomes, he termed the *Y chromosome.* Thus, the female's sex chromosomes are denoted XX, and the male's are denoted XY. As you can see in Figure 23.13, it is the male who determines the sex of the offspring. The same basic principle also determines the sex of humans, mammals, and most other animals; but in a few organisms, such as birds, butterflies, and some fish, the male has two identical sex chromosomes, and the female's sex chromosomes are different.

Fig. 23.13 **Chromosomes determine sex of offspring**

**Sex-linked traits.** One day in 1909, an unexpected event occurred. A male fruit fly with white eyes instead of the normal brick-red eyes appeared in one of Morgan's culture bottles. Morgan assigned the symbol w for the white allele and the symbol $w^+$ for the red allele, which is the wild type allele. (The *wild type* allele refers to the most frequent allele for that gene in the population and is represented by the plus (+) sign; sometimes the wild type allele is called "normal" because that is how the majority of the fruit flies look.)

Morgan wondered how white eyes could be inherited, so he crossed the lone white-eyed male with red-eyed females. All the the $F_1$ generation of this cross had red eyes. Morgan then interbred a male and a female fly from the $F_1$ generation, and he found white-eyed flies among the $F_2$ generation. All of the flies with white eyes were males. Morgan thereby demonstrated not only that white eyes are recessive to red eyes but also that the trait is **sex-linked** because it is located on the X chromosome, as illustrated by Figure 23.14.

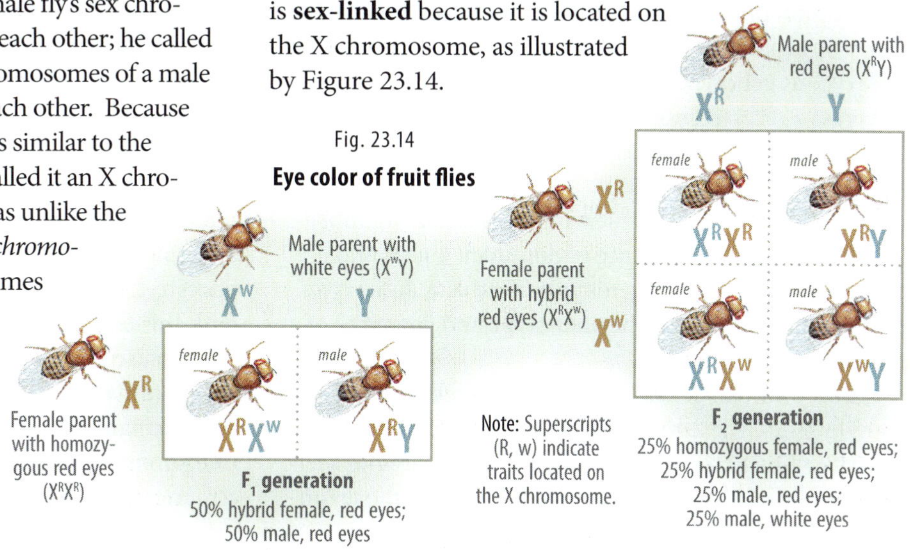


Fig. 23.14 **Eye color of fruit flies**

## Section Review 23.2

1. Who is considered the Father of Genetics?
2. Why is cross-pollination rare among pea plants?
3. Explain the difference between genotype and phenotype.
4. Briefly state the law of dominance, the law of segregation, and the law of independent assortment.
5. Why did Morgan choose the common fruit fly for his experiments?

### Application

6. Use Punnett squares to determine the possible genetic combinations if you crossed a pure tall (TT) pea plant with a pure dwarf (tt) pea plant. What are the possible genetic combinations between a pure tall plant and a hybrid tall plant (Tt)? What if both plants are hybrids?
7. Draw a Punnett square and determine the possible genetic combinations if you crossed a yellow wrinkled pea plant with a recessive green allele (Yyrr) and a green round pea plant with a recessive wrinkled allele (yyRr). What outcomes are possible?

### Critical Thinking

8. Explain why rearrangements of preexisting genes during reproduction can increase variety within a kind of organism but cannot create new kinds or new genetic information.

### Identify

genetics, allele, homozygous, heterozygous, dominant trait, recessive trait, genotype, phenotype, incomplete dominance, dihybrid, linkage group, sex chromosomes

## 23.3 Human Genetics

Nearly all of the pioneering geneticists worked with plants and animals because these organisms can be experimentally bred. Genetics in humans, however, is largely confined to observation and documentation of inherited characteristics. Today we know of thousands of human traits that are inherited, many of which are responsible for disease; hundreds of genetic disorders have been identified. Recent genetic research has enabled geneticists to understand more about these inherited traits and diseases.

Fig. 23.15 **Short and long eyelashes**

### Dominant Allele Inheritance

A noticeable human characteristic that demonstrates inheritance from a single gene with two alleles is the length of eyelashes (Fig. 23.15). Long eyelashes are the expression of the dominant allele (we may call it E) and may be caused by a homozygous genotype (EE) or a heterozygous genotype (Ee). Persons with short eyelashes express a recessive allele (e); they must be homozygous (ee), because the presence of a dominant allele would express that dominant trait. You can study your family history of this gene by observing each family member and recording the presence of dominant alleles and recessive homozygous genotypes. In this manner, you may be able to predict the phenotype and genotype of a new baby in the family.

Many human genetic diseases are determined by a single gene. These mutations can be dominant or recessive. Most human dominant mutations are not seen very frequently because there are very few of those mutant alleles in the population. We say therefore that their frequency is low. **Allele frequency** is the extent which a certain allele exists in a population. For instance, *polydactyly* [pŏl′ē·dăk′tə·lē: having more fingers or toes than usual] is caused by a dominant allele; but how many people do you know with six fingers on each hand? We must conclude that very few alleles for six fingers exist in the human gene pool.

**Cystic fibrosis** (CF) is one of the most common hereditary diseases among children in the United States, affecting about 30,000 Americans, mostly of European ancestry. One in every 3500 white Americans is born with this disease; the median age of survival is 33 years. CF is caused by a recessive allele. A child must receive the recessive CF allele from both parents before this disease manifests itself. Receiving a CF allele from only one parent makes the child a **carrier.** About 12 million Americans are carriers of CF.

## Table 23.3 Selected Human Genetic Characteristics

| Dominant characteristics (homozygous or heterozygous) | Recessive characteristics (homozygous) |
|---|---|
| thin lips | full lips |
| dark hair | light hair |
| brown eyes | blue eyes |
| susceptibility to nearsightedness | normal vision |
| astigmatism | normal vision |
| long eyelashes | short eyelashes |
| bone disease of the ear with hearing loss | normal hearing |
| polydactyly | five fingers or toes per limb |
| freckles | lack of freckles |
| can roll tongue into a U shape | lack of ability to roll tongue |
| feet with normal arches | flat feet |

Note: Some of these traits are complicated by having multiple genes, more than two possible alleles for the same gene, or influence from nongentic factors.

### Codominance

**Codominance** occurs when both alleles for a trait are equally expressed in the phenotype of a heterozygous individual. Codominance should not be confused with *incomplete dominance,* in which the heterozygous phenotype is a blend of the homozygous phenotypes but is distinct from either. For example, the red and white flower-color alleles in four-o'clock flowers are incompletely dominant, producing pink flowers in a heterozygote (pp. 478–479). If these alleles were instead codominant, the heterozygote's flowers would be both red and white.

The production of *hemoglobin* in red blood cells is an example of codominance in humans. The shape of part of the hemoglobin protein is specified by a gene called *Hbβ* (*β* is the lowercase Greek letter *beta*); this gene has two alleles, which we will designate A and S. (Other genes specify the rest of the hemoglobin protein.) A normal person is AA and produces normal hemoglobin only. A person who is SS has **sickle cell anemia** and makes only abnormal hemoglobin molecules that clump together under certain conditions, causing the red blood cells to become sickle shaped and possibly puncturing the cells. If a normal person (AA) marries a person with sickle cell anemia (SS), their children must be heterozygous (AS). Because these alleles are codominant, both express themselves: some of the hemoglobin molecules are normal and some are abnormal. This is **called sickle cell trait.** Studies have shown that the sickle cell allele, when present in a heterozygous individual, protects the carrier against malaria, although the exact mechanism of protection is not understood. This protection is advantageous in places, like central Africa, where malaria is abundant. About 2 million Americans, mostly of African ancestry, possess this allele; and approximately 1 in every 500 black American children is born with sickle cell anemia.

### Multiple Allele Inheritance

So far in our discussion of genetics we have talked about two possible alleles that a gene can possess. **Multiple allele inheritance** occurs when there are more than two different alleles of a gene for a particular trait.

One important example of multiple allele inheritance is the *ABO blood group.* The blood types A, B, AB, and O are determined by different combinations of three alleles. There is an allele $I^A$, which will express A type blood; an allele $I^B$, which will express B type blood; and an allele i, which will express neither A nor B blood. Alleles $I^A$ and $I^B$ are codominant; that is, they are both expressed when present. $I^A$ and $I^B$ are both dominant to allele i.

Red blood cells have *antigens,* carbohydrate chains that serve as identification tags, attached to their cell membrane proteins. A particular terminal sugar molecule is added by the gene $I^A$ and a different sugar molecule by the gene $I^B$. Thus, the only difference between blood types A and B is the kind of sugar at the end of a long carbohydrate chain on the cell membrane. Type AB adds both sugar molecules, and type O adds neither.

Knowing a person's blood type may tell you his genotype. A person with type O or AB blood has only one possible genotype. A person with type A or B blood may have either of two possible genotypes. Suppose you have type A blood. Your possible genotypes are $I^AI^A$ or $I^Ai$. If your father has type O blood, his genotype must be ii. That means the allele contributed by your father must have been i, making your genotype $I^Ai$. Your mother must have either type A ($I^AI^A$ or $I^Ai$) blood or type AB ($I^AI^B$) for you to get the $I^A$ allele.

## Table 23.4 Genes and Blood Types

| Genotypes | Phenotypes |
|---|---|
| $I^AI^A$ or $I^Ai$ | type A blood |
| $I^BI^B$ or $I^Bi$ | type B blood |
| $I^AI^B$ | type AB blood |
| ii | type O blood |

Fig. 23.16

**Inheritance of hemophilia**

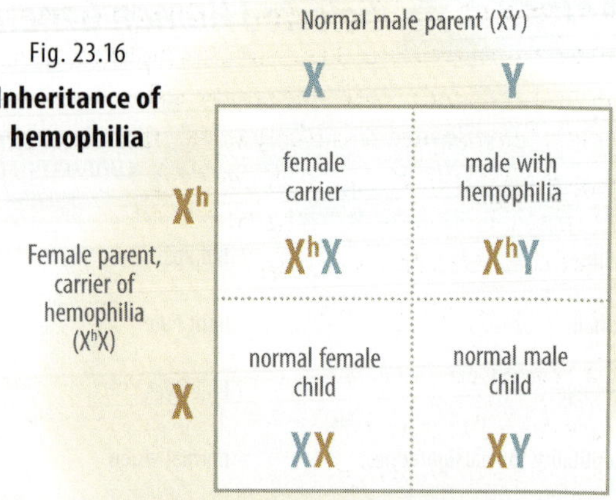


**$F_1$ generation**

25% female carrier, 25% male with hemophilia, 25% normal female, 25% normal male

## Pleiotropy and Polygenic Inheritance

**Pleiotropy.** Most genes influence not just one trait but several seemingly unrelated traits; this is known as **pleiotropy** [plī·ŏt′rə·pē]. One example of pleiotropy is the gene for sickle cell anemia, which causes the body to produce abnormal hemoglobin molecules, as explained before. In homozygous individuals, this gene causes anemia and other complications. On the other hand, in heterozygous individuals, the gene provides protection against malaria because the organism that causes malaria does not thrive in red blood cells of heterozygous individuals.

**Polygenic inheritance.** Many of the characteristics that make you unique from either your mother or your father are influenced by a number of different genes. **Polygenic** [pŏl′ē·jĕn′ĭk] **inheritance** involves traits controlled by many different genes. Thus, instead of one gene controlling an expression, many genes are involved. Skin color, for example, is influenced by at least four genes. Height, another example of polygenic inheritance, also involves at least four genes. Wherever there seems to be a great variety of possible phenotypes, polygenic factors are involved. These factors help provide the many varied traits that make you like no other person on earth.

## Sex-Linked Disorders

**Hemophilia.** Humans inherit certain characteristics along with the sex chromosomes in much the same manner as the fruit fly. A well-known sex-linked characteristic is **hemophilia,** a blood-clotting disorder caused by recessive abnormal alleles for the clotting mechanism. Because the genes for clotting are carried on the X chromosome, both males and females can possess hemophilia alleles. Males express the disorder more readily, however, because they lack another X chromosome on which the dominant allele for normal clotting may be found. A male with hemophilia would have a genotype of $X^hY$ as opposed to a normal male XY. However, a female with hemophilia would have to have the genotype $X^hX^h$ (a normal female is XX). Although a female with the genotype $X^hX$ shows no signs of hemophilia, she carries the recessive allele, which can be transmitted to her offspring, and thus she is a carrier.

If a normal male (XY) marries a female carrier ($X^hX$), what will be the genotypes and phenotypes of any children that may be born? The gametes for the male will be X or Y. For the female, they can be either $X^h$ or X. Using a Punnett square (Fig. 23.16), we see that half of the boys will be normal (XY) while the other half will be hemophiliac ($X^hY$). Also, half of the girls will be normal (XX) while half will be carriers ($X^hX$). Although the genetics is accurate, it might take dozens, hundreds, or even thousands of children for these ratios to hold true precisely. Therefore, geneticists use *probabilities* to express potential results in these situations. In our case, if the baby is a boy, there is a 50% probability that he will have hemophilia. If the baby is a girl, on the other hand, there is a 100% certainty that she will not have hemophilia. However, her chances of being a carrier are 50%. Probabilities are often expressed as fractions of 1, so 50% will be represented as ½ or 0.5.

**Colorblindness.** Another disorder with sex-linked inheritance is **colorblindness.** The allele for a certain type of colorblindness is carried by a gene on the X chromosome and is recessive. A colorblind male has the genotype $X^cY$; a colorblind female must be $X^cX^c$; and a carrier female $X^cX$. The recessive gene operates with the same pattern of inheritance as hemophilia. Because a female will be colorblind only if she inherits two recessive alleles, females are less likely to express this sex-linked recessive trait than are males (Fig. 23.17). In the United States, the frequency of colorblindness is 1 in 12 in males but 1 in 250 in females.

Fig. 23.17
**Inheritance of colorblindness**

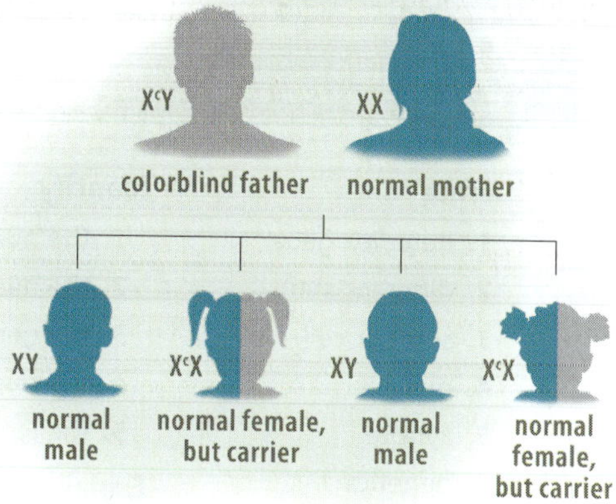


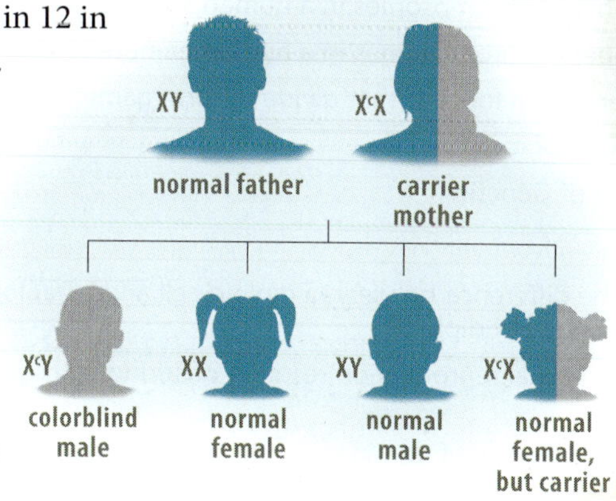


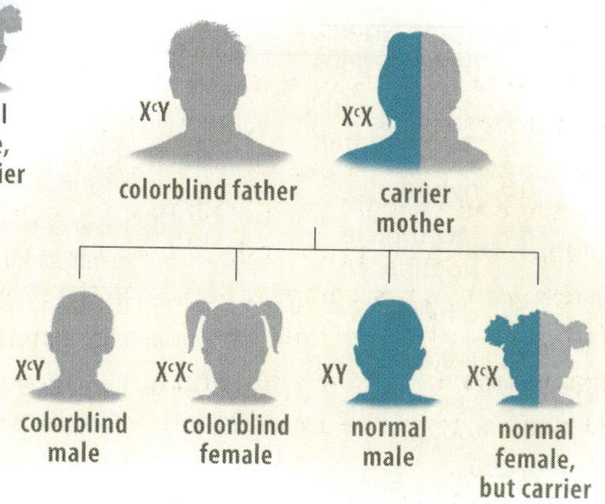


## Genetic Advances

As a result of recent research, man's knowledge of genetics has increased remarkably, enabling medical scientists to detect various genetic abnormalities. Through early detection of the genetic disorder *PKU* (phenylketonuria [fĕn′əl·kēt′n·ŏŏr′ē·ə]), which can cause skin disorders, seizures, and severe mental retardation, the effects of the disorder can be controlled. Geneticists can also determine if parents have a high risk of passing on a serious genetic disease.

Man's advanced knowledge of genetics can also be harmful. Although a child developing in the womb can now be tested for many genetic diseases, most of these diseases have no known cure. Parents are often advised to have the developing child put to death by means of a "therapeutic" abortion. In the United States, for instance, 7 out of 10 children with *Down syndrome* are aborted instead of being allowed to live. Some people have even advocated the practice of **eugenics,** an attempt to improve the human race through the control of hereditary factors. The sequencing of the human genome and recent developments in biotechnology and genetics seem to offer the possibility of modifying the genetic constitution of human beings before birth. However, the world's problems resulting from the fall of man into sin cannot be solved by developing a new "master race."

## Section Review 23.3

1. Genetically, what is the difference between a person with sickle cell trait and one with sickle cell anemia?
2. How is polygenic inheritance different from pleiotropy?

### Application

3. Using a Punnett square, find the possible genotypes and phenotypes of the children if the blood of the father is homozygous type A and the blood of the mother is heterozygous type B.
4. Draw a Punnett square and determine the possible genetic combinations if a colorblind male ($X^cY$) marries a carrier female ($X^cX$). Could any of their daughters be colorblind? Could any of their sons have normal vision?
5. What are some ways that man's advancing genetic knowledge can be used for good? What are some ways that it can be misused?

### Critical Thinking

6. Why do males more readily express recessive sex-linked genetic characteristics than females?

### Identify

allele frequency, carrier, codominance, multiple allele inheritance, pleiotropy, polygenic inheritance, eugenics

# Chapter 23 Review

### Define

1. heredity, gene
2. differentiation
3. somatic cell
4. meiosis
5. zygote
6. genetics
7. allele
8. homozygous, heterozygous
9. dominant trait, recessive trait
10. hybrid
11. incomplete dominance
12. dihybrid
13. linkage group
14. sex chromosomes
15. allele frequency
16. carrier
17. codominance
18. sickle cell anemia
19. eugenics

### Identify

1. the number of chromosomes in a human somatic cell
2. the number of chromosomes in a human gamete
3. the general term for cells that divide to form gametes
4. the purpose of meiosis
5. the Father of Genetics

### Explain

1. What is the difference between a diploid cell and a haploid cell? Why are gametes haploid instead of diploid?
2. How does sexual reproduction result in offspring with genetic information from both parents?
3. Briefly state the law of dominance, the law of independent assortment, and the law of segregation.
4. How can the process of recombination contribute to additional genetic variety within created kinds?
5. Explain the difference between genotype and phenotype. Why are all the characteristics in the genotype not necessarily expressed in the phenotype?
6. Explain the difference between multiple allele inheritance, pleiotropy, and polygenic inheritance.
7. What determines whether a genetic characteristic is a sex-linked trait?

### Apply

1. What are three differences between mitosis and meiosis?
2. A man heterozygous for long eyelashes marries a woman with short eyelashes. Draw a Punnett square to determine the possible genotypes of their children. What is the likelihood of their child having short eyelashes?
3. Using a Punnett square, state the genotypes and the phenotypes of the possible offspring if both parents are carriers of cystic fibrosis.
4. One of the most famous carriers of hemophilia was Queen Victoria of England; her husband, Prince Albert, was not a hemophiliac. Use a Punnett square to predict the possible genotypes and phenotypes of their children and the probability of each. Their oldest son, Edward VII, was not a hemophiliac; explain why he could not have passed the hemophilia allele to his children.

CHAPTER 24

# DNA: The Regulation of Life

**Key Concepts**

- Discovery of DNA
- How DNA is stored
- Structure of DNA
- DNA replication
- Structure and types of RNA
- Transcription
- Protein manufacture
- Mutations

## 24.1 DNA: Master Program of the Cell

### The Master Plan of Life

The work of Gregor Mendel and other geneticists revealed the general patterns in which hereditary traits are passed from one generation to the next. Mendel and other pioneers of genetics knew that something in the gametes from the parent organisms determined the physical characteristics of the offspring. However, they had no idea how hereditary information was transferred. By the mid-1940s, scientists had begun to realize that a mysterious substance in the nucleus of a cell, known as **deoxyribonucleic acid** *(DNA),* somehow controls heredity. (The DNA molecule is classified as a *nucleic acid* because it was first identified in the cell's nucleus and it has certain acidic properties.) Somehow, they suspected, DNA contains the genetic information of an organism.

Although the importance of DNA was recognized, not until 1953 was the real nature of DNA discovered. The greatest breakthrough came when two biochemists, American **James D. Watson** and Englishman **Francis H. C. Crick** (1916–2004), deduced the actual structure of DNA. These discoveries revealed that *each gene of an organism is the part of the DNA molecule that controls a particular trait.* This genetic information is encoded in the DNA molecule.

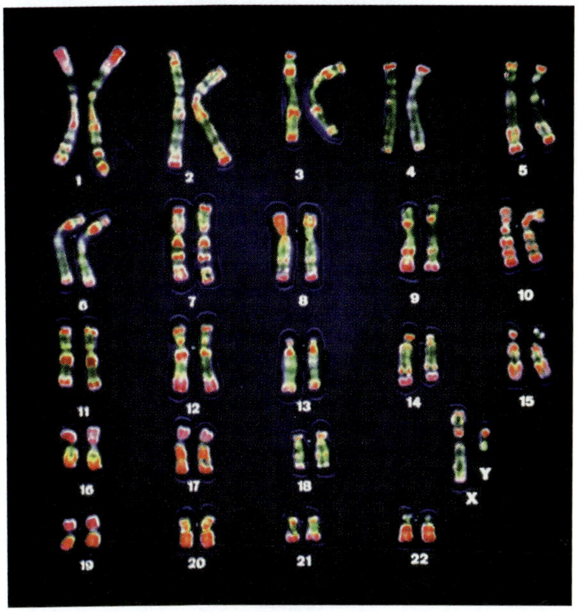


Fig. 24.1
**Complete set of male human chromosomes**

### DNA and You

Within your body, *every somatic cell contains 46 chromosomes in the nucleus* (Fig. 24.1). Each chromosome contains one molecule of DNA. Therefore, every somatic cell in your body contains 46 molecules of DNA. Each cell contains all of the information necessary to manufacture all of the unique substances that make up your body.

It has been estimated that if the information carried by the DNA in one cell were to be spelled out in English, it would require somc 400,000 pages of print averaging 500 words per page. That is to say that *each cell of your body contains enough information in its DNA molecules to fill a library of about one thousand books.* What is even more amazing is that your body contains between 60 and 100 trillion copies of your genetic material, yet your DNA composes only about 0.0007% of your body weight.

## How DNA Is Stored

To fit several feet of DNA into a nucleus that is less than $\frac{1}{10,000}$ of an inch across, each DNA molecule is tightly coiled. Ordinarily, short segments of each DNA molecule are coiled on protein "spools," each made of eight *histones.* Each spool and the DNA wrapped around it form a *nucleosome* (Fig. 24.2). The millions of nucleosomes filling the nucleus give the nucleus a grainy texture when viewed under a powerful microscope. Scientists often refer to DNA stored in this manner as **chromatin.** The long, loose loops of DNA allow easy access to the molecules that interact with DNA.

When the cell prepares to divide, the nucleus coils the chromatin fibers into tight coils, which are then bound together into a large bundle. *These bundles of tightly coiled DNA are the chromosomes that are visible during mitosis and meiosis.* Bundling the DNA allows the cell to easily distribute DNA copies to its daughter cells. Once the cell has divided, each daughter cell "unpacks" the DNA and stores it as chromatin.

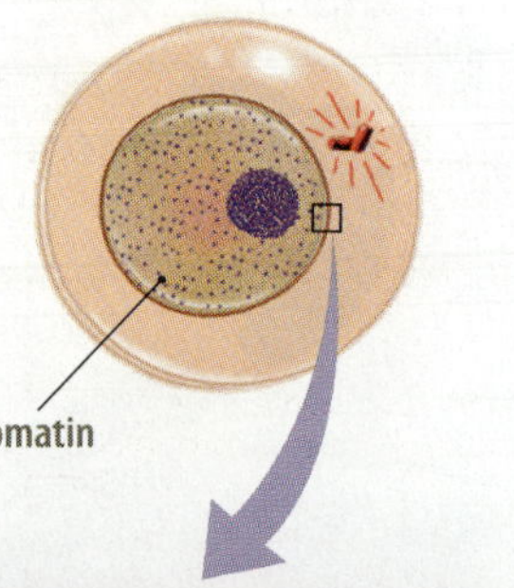


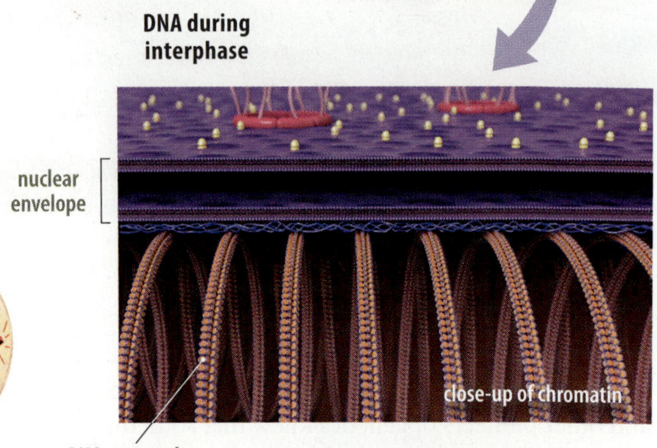


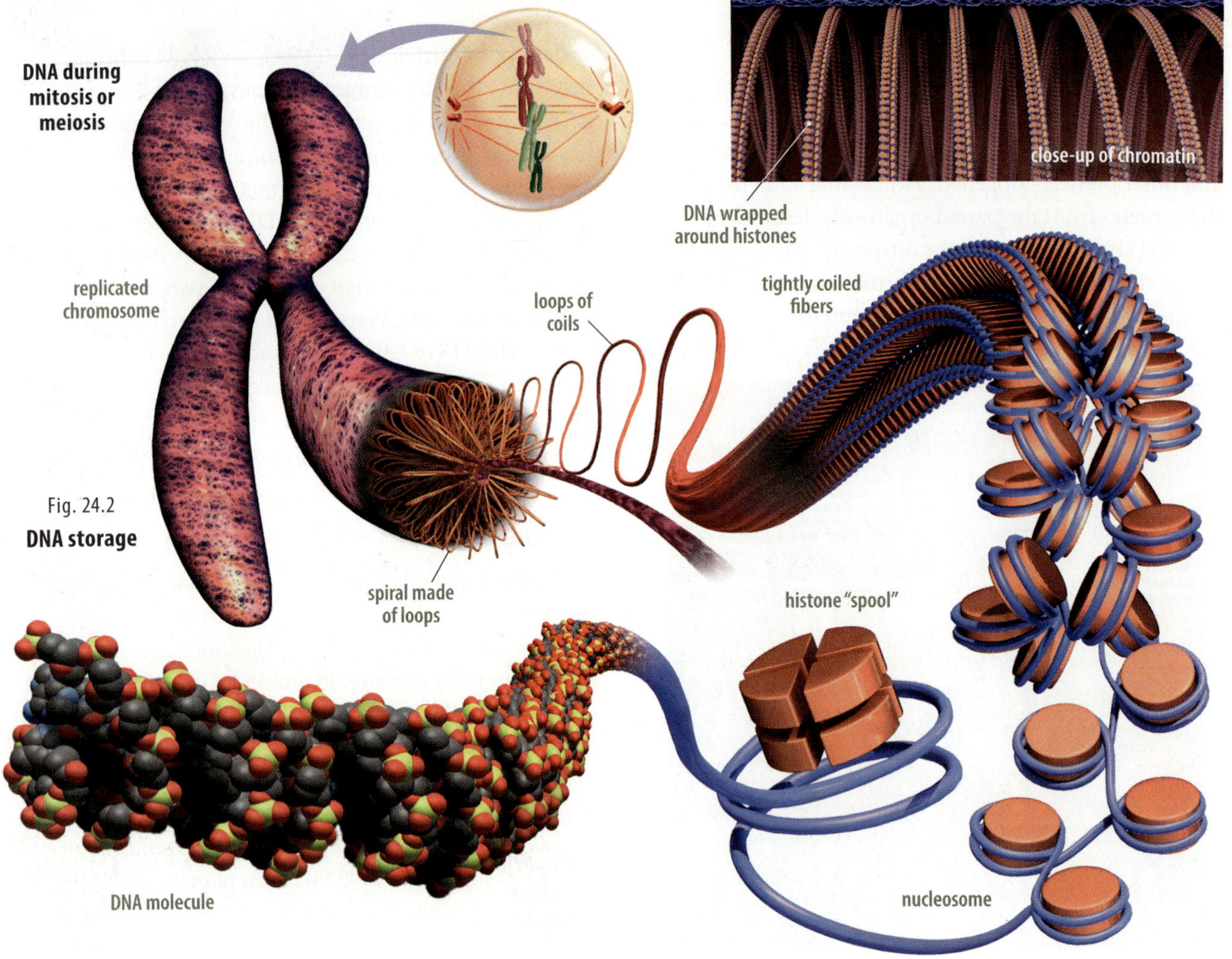


Fig. 24.2
**DNA storage**

## Structure of Nucleic Acids

Besides DNA, the other important nucleic acid in the cell is **ribonucleic** [rī′bō·no͞o·klē′ĭk] **acid** *(RNA)*. Both nucleic acids, DNA and RNA, are composed of special molecules called **nucleotides.** Each nucleotide is composed of three parts (Fig. 24.3): a sugar molecule (deoxyribose in DNA and ribose in RNA), a phosphate group, and a nitrogen-containing **base.**

**DNA structure.** Each DNA molecule consists of two long nucleotide chains that are twisted about one another, giving the molecule the shape of a twisted ladder. Scientists refer to this double-spiral structure as a **double helix** (*helix* means "spiral").

The strands forming the "legs" *(backbone)* of the DNA ladder consist of nucleotide chains connected by bonds between sugar and phosphate molecules. The "rungs," which connect the legs of the ladder, consist of paired nitrogen-containing bases, or **base pairs.** The DNA molecules in human cells average about 65 million base pairs per molecule.

Fig. 24.3
**Structure and composition of DNA**

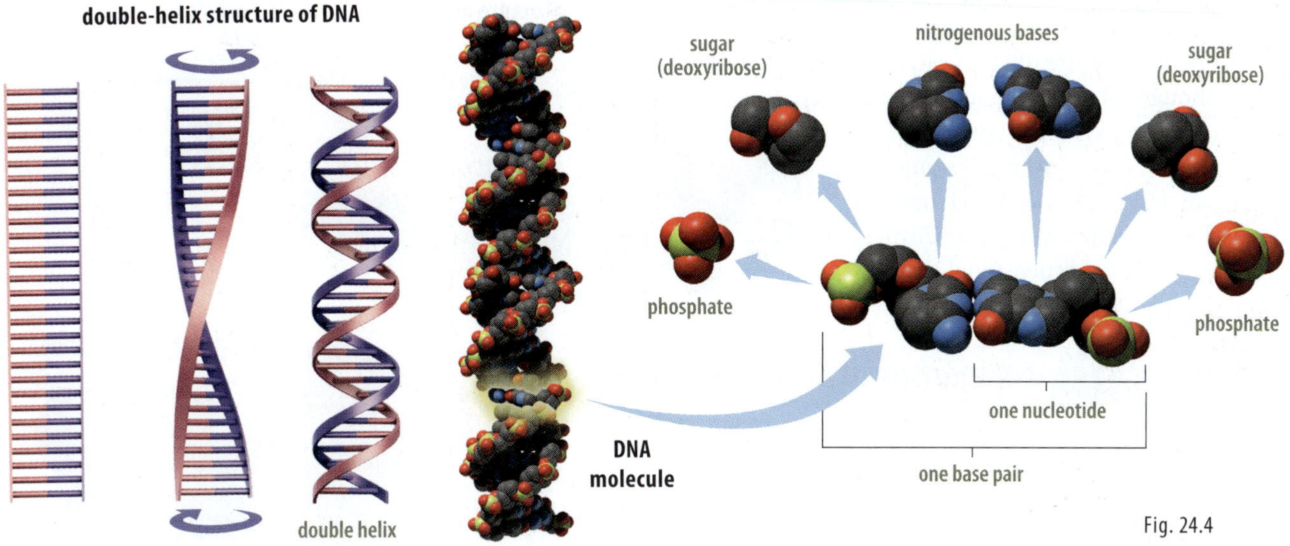


Fig. 24.4
**Base pairing in DNA**

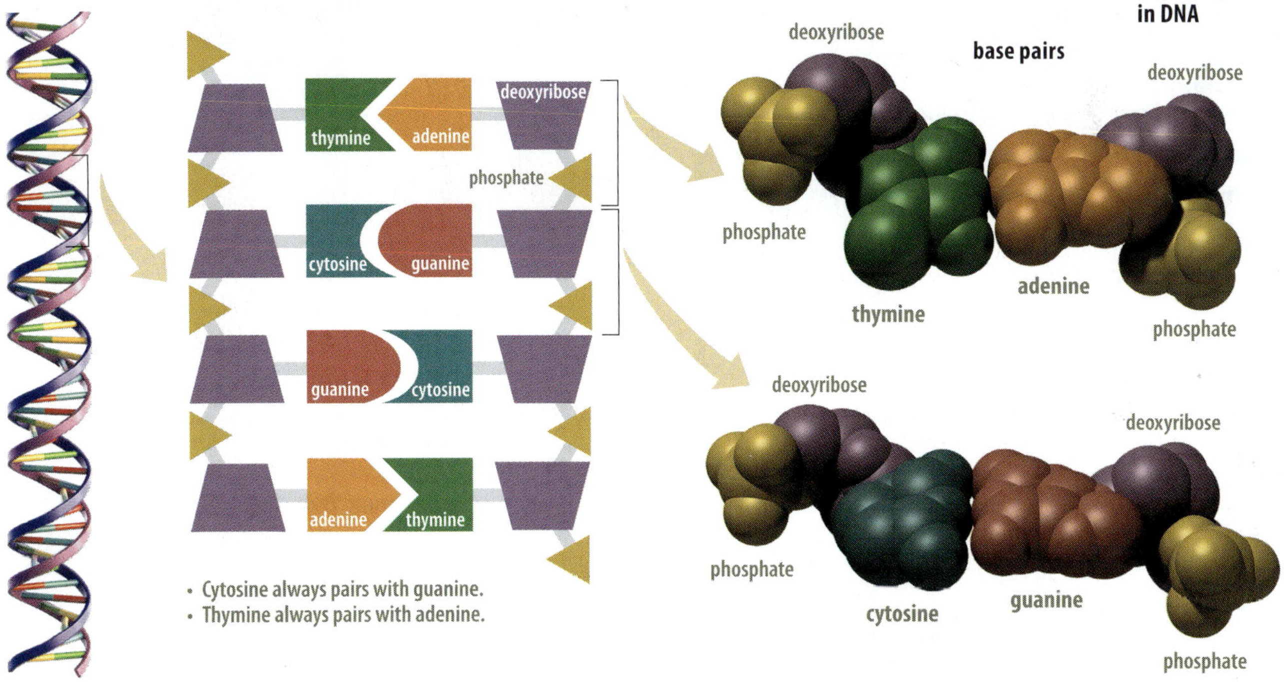

Four different types of bases are found in a DNA molecule. These bases are *guanine* [gwä′nēn′], *cytosine* [sī′tə·sēn′], *adenine* [ăd′n·ēn′], and *thymine* [thī′mēn′], often abbreviated *G, C, A,* and *T,* respectively. The two bases of each pair attract each other strongly. *Only one type of base will fit another.* Where there is a thymine, there must be an adenine opposite it; where there is a guanine, there must be a cytosine opposite it. This process in which only one type of base will fit another is called **base pairing.**

**RNA structure.** The structure of an RNA molecule is similar to that of DNA, except that *RNA consists of only one strand instead of two.* As a result, RNA lacks the double-helix structure of DNA. RNA molecules should not be thought of as simple straight strands, however; although some types of RNA molecules are indeed straight, some others are designed to fold back upon themselves and latch into place, forming complex three-dimensional shapes.

Like DNA, *RNA may contain four different types of nitrogen-containing bases.* Three of these bases (guanine, cytosine, and adenine) are identical to those of DNA. However, RNA uses a base called *uracil* (abbreviated *U*) instead of thymine. Although uracil is very similar to thymine, it requires much less energy to manufacture, allowing the cell to save energy when making large amounts of RNA. Like thymine, uracil pairs up with adenine but will not bond to guanine or cytosine.

Fig. 24.5
**Structure of RNA**

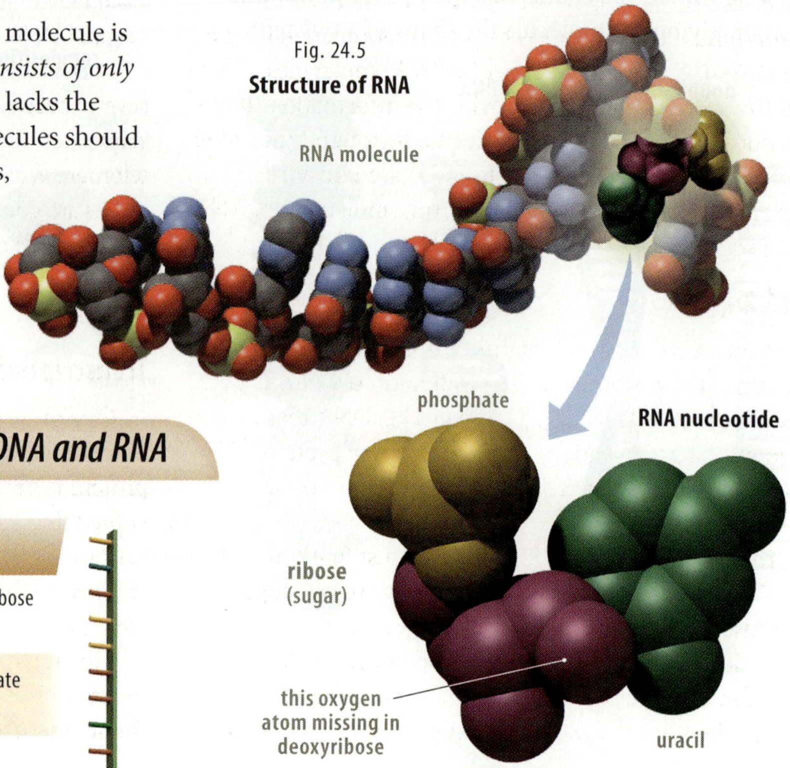


## Table 24.1 Comparison of DNA and RNA

| DNA | RNA |
| --- | --- |
| Contains sugar deoxyribose | Contains sugar ribose |
| Contains phosphate groups | Contains phosphate groups |
| Has nitrogen-containing bases called guanine, cytosine, adenine, and thymine | Has nitrogen-containing bases called guanine, cytosine, adenine, and uracil |
| Has double-helix structure | Has single-stranded structure; shape varies |
| Found only in the nucleus | Found in the nucleus, the cytoplasm, and as part of the structure of the ribosomes |
| Only one type | Several types |
| Replicated from DNA | Transcribed from DNA |

## Section Review 24.1

1. Where in a cell is the DNA found?
2. What two scientists are credited with discovering the structure of the DNA molecule?
3. Briefly describe the structure of a DNA molecule.
4. What are the four types of bases found in a DNA molecule? Which type pairs only with adenine? With guanine?
5. Describe five structural differences between DNA and RNA.

**Identify**

DNA, gene, chromatin, nucleotide, double helix

# 24.2 The Activities of DNA

## The Central Dogma of Molecular Biology

The flow of biological information from DNA to RNA and then from RNA to protein is known as the **central dogma of molecular biology.** This principle is explained by the processes of replication, transcription, and translation. The process by which DNA makes a copy of itself before cell division is called *replication.* Then, through *transcription,* the cell copies the information contained in the DNA into messenger RNA (mRNA) molecules. These mRNA molecules in turn, provide the information to the ribosomes to make protein molecules through *translation.* More recently, scientists have found that many viruses, like HIV, can make DNA starting from molecules of RNA in a process called *reverse transcription.*

## DNA Replication

Before a cell divides, it makes a copy of its DNA by a process called **replication.** Replication occurs during the *S phase* of the cell's life cycle (chapter 22). DNA replication requires the participation of a complex set of proteins. This process is another example of God's marvelous design in the cell.

To copy a DNA molecule, the cell must first unwind the DNA and "unzip" it so that it consists of two separate strands. The enzyme *DNA polymerase* then attaches individual nucleotides to each strand of the DNA molecule, forming new complementary strands (Fig. 24.6). The nucleotides always pair: adenines with thymines and guanines with cytosines. During replication, the new DNA molecules are also checked for errors by proofreading enzymes that look for mismatched or damaged base pairs, helping ensure that both of the new copies are identical to the original molecule. *Replication results in two identical DNA molecules where there had been only one.* Each new DNA molecule consists of one strand from the original DNA molecule and one newly constructed strand.

Because DNA polymerase cannot replicate the last several base pairs of a chromosome, DNA has special sequences, called *telomeres,* on its ends; a small part of the telomere is used during each replication. After the telomeres are gone, DNA can no longer replicate and the cell dies through *apoptosis.* Some cells, such as skin cells and germ-line cells, have their telomeres regularly extended by special enzymes, allowing them to divide indefinitely.

## Transcription

One of the important purposes of RNA is its role as a working copy of the DNA blueprint. When a new protein is needed, the DNA coding for that protein is copied into a **messenger RNA** molecule *(mRNA).* This mRNA produced by **transcription** in the nucleus is then shipped to the cytoplasm, where it is used by the ribosomes as a template to make the protein.

Like replication, transcription is a carefully ordered process. The cell's transcription machinery first locates the segment of DNA that contains the information for

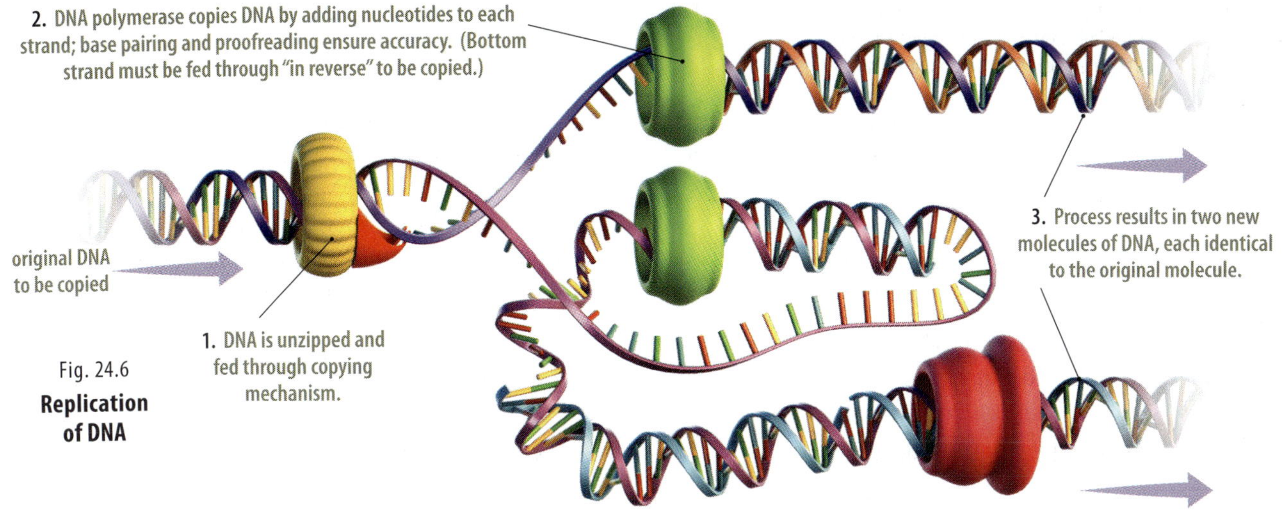


Fig. 24.6
**Replication of DNA**

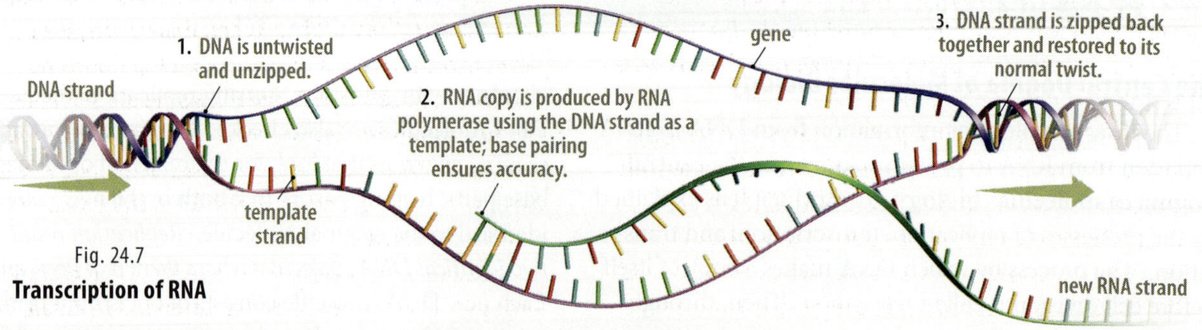

Fig. 24.7
**Transcription of RNA**

the desired protein. That segment of the DNA molecule is then untwisted and unzipped, allowing an RNA copy of the gene to be made by the enzyme *RNA polymerase.* After the segment is copied, the DNA is zipped back together and twisted back into its normal shape. Transcription, like replication, depends on the strict rules governing base pairing: uracil or thymine pairs with adenine, and guanine pairs with cytosine. In humans and other eukaryotes, transcription does not produce mRNA directly; rather it produces a *pre-mRNA* molecule that must be processed to prepare it for translation.

## Translation

**The genetic code.** It took years of careful genetic experiments to understand how a sequence of nucleotides in the mRNA can be translated into a sequence of amino acids in a protein. Scientists discovered that the information is coded in the mRNA in the form of codons. A **codon** is a consecutive sequence of three nucleotides in the mRNA that translates into a specific amino acid. The equivalence of these codons to amino acids is given by the **genetic code** (Table 24.2), a table with all the possible combinations of nucleotides and the amino acids they generate. For instance, the GAA codon (guanine-adenine-adenine) in the mRNA codes for the amino acid glutamic acid. The codon GAU (guanine-adenine-uracil), although similar, codes for a different amino acid, aspartic acid. Not all combinations produce amino acids. Three *stop codons,*

**Table 24.2 Genetic Code**

| First Nucleotide | Second Nucleotide: U | C | A | G | Third Nucleotide |
|---|---|---|---|---|---|
| U | UUU phenylalanine | UCU serine | UAU tyrosine | UGU cysteine | U |
| | UUC phenylalanine | UCC serine | UAC tyrosine | UGC cysteine | C |
| | UUA leucine | UCA serine | UAA *stop* | UGA *stop* | A |
| | UUG leucine | UCG serine | UAG *stop* | UGG tryptophan | G |
| C | CUU leucine | CCU proline | CAU histidine | CGU arginine | U |
| | CUC leucine | CCC proline | CAC histidine | CGC arginine | C |
| | CUA leucine | CCA proline | CAA glutamine | CGA arginine | A |
| | CUG leucine | CCG proline | CAG glutamine | CGG arginine | G |
| A | AUU isoleucine | ACU threonine | AAU asparagine | AGU serine | U |
| | AUC isoleucine | ACC threonine | AAC asparagine | AGC serine | C |
| | AUA isoleucine | ACA threonine | AAA lysine | AGA arginine | A |
| | AUG *start* methionine | ACG threonine | AAG lysine | AGG arginine | G |
| G | GUU valine | GCU alanine | GAU aspartic acid | GGU glycine | U |
| | GUC valine | GCC alanine | GAC aspartic acid | GGC glycine | C |
| | GUA valine | GCA alanine | GAA glutamic acid | GGA glycine | A |
| | GUG valine | GCG alanine | GAG glutamic acid | GGG glycine | G |

UAA, UAG, and UGA, signal the end of the protein and instruct the ribosome to stop translation. A *start codon,* AUG, indicates the beginning of a protein and also codes for the amino acid methionine.

A simple example will help you understand how the genetic code works. Consider an mRNA molecule with the sequence AUGAAAAGUGGCAGUGUUUGUC-CCUGA. Divide the sequence into groups of consecutive triplets (codons): AUG AAA AGU GGC AGU GUU UGU CCC UGA. Then, using the genetic code, you will see that this mRNA codes for a small protein with amino acids bonded in the following order: methionine-lysine-serine-glycine-serine-valine-cysteine-proline.

| AUG | AAA | AGU | GGC | AGU | GUU | UGU | CCC | UGA |
|---|---|---|---|---|---|---|---|---|
| methionine | lysine | serine | glycine | serine | valine | cysteine | proline | stop |

**Construction of proteins.** Proteins are manufactured by the **ribosomes.** A ribosome consists of two subunits, one large and one small. These subunits are normally separate in the cytoplasm but come together to make proteins. Ribosomes are made of both protein and **ribosomal RNA** *(rRNA).* Just like mRNA, rRNA is produced in the nucleus by transcription; but rRNA serves as an RNA "enzyme" instead of containing the information to produce proteins. Ribosomal RNA is the most abundant type of RNA in the cell.

When an mRNA "blueprint" reaches the cytoplasm, a small ribosomal subunit and a large ribosomal subunit clamp over it to form a functional ribosome. Amino acids are brought to the ribosome by molecules of **transfer RNA** *(tRNA),* which enter the ribosome through a special slot. On one end of each tRNA is an *anticodon* that base-pairs with a codon on the mRNA, and on the other end is the corresponding amino acid (as determined by the genetic code). As each codon of the mRNA is fed through the ribosome, a tRNA molecule brings an amino acid that the ribosome adds to the growing protein chain.

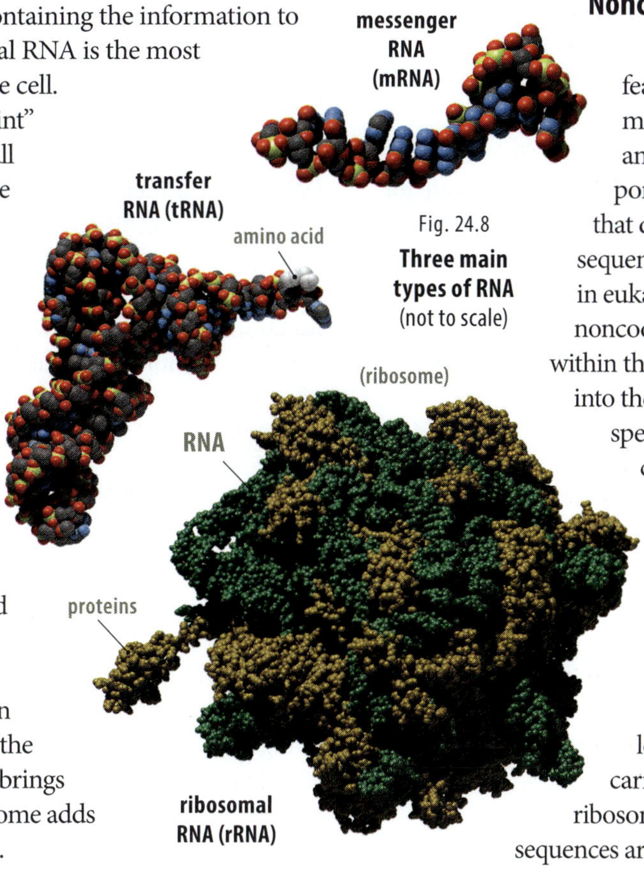


Fig. 24.8
**Three main types of RNA** (not to scale)

This process continues until the ribosome encounters a stop codon in the mRNA. When this occurs, the completed protein chain is released from the ribosome. The mRNA is also released as the ribosome disassembles back into free subunits. The process of synthesizing proteins from the RNA pattern is called **translation** because the cell has, in a sense, translated the RNA information into a functional protein.

Translation may seem to be a very slow and cumbersome process when each step is described in detail, but all of these complex steps are amazingly coordinated so that translation occurs at high speed. Bacteria are especially fast at protein manufacture; a bacterial ribosome assembles amino acids at a rate of about 15 amino acids per second, so an average-sized protein may be assembled in only 20 to 30 seconds. (Eukaryotic cells, such as those of your body, are slower, assembling a comparable protein in 3 minutes or so, but this is still fast considering the complexity of even a "simple" protein.) *Random processes could never result in such a marvelous and intricate mechanism.* The more we learn about molecular biology, the more insight we have into God's creative mind.

## Noncoding DNA

One of the most fascinating features of the cell's genetic information is the presence of large amounts of **noncoding DNA,** portions of the chromosomes that do not specify the amino-acid sequence of proteins. Almost all genes in eukaryotic cells include **introns,** noncoding DNA sequences located within the genes. Introns are transcribed into the pre-mRNA molecules, but special mechanisms in the nucleus carefully *splice* them out of the pre-mRNA, resulting in a mature mRNA composed of only the coding sequences, called **exons.** For example, the pre-mRNA for human beta globin, a portion of hemoglobin, is 1608 nucleotides long, but the processed mRNA carries only 628 nucleotides to the ribosome. Other noncoding DNA sequences are located between genes.

Fig. 24.9

**Protein synthesis in a cell**

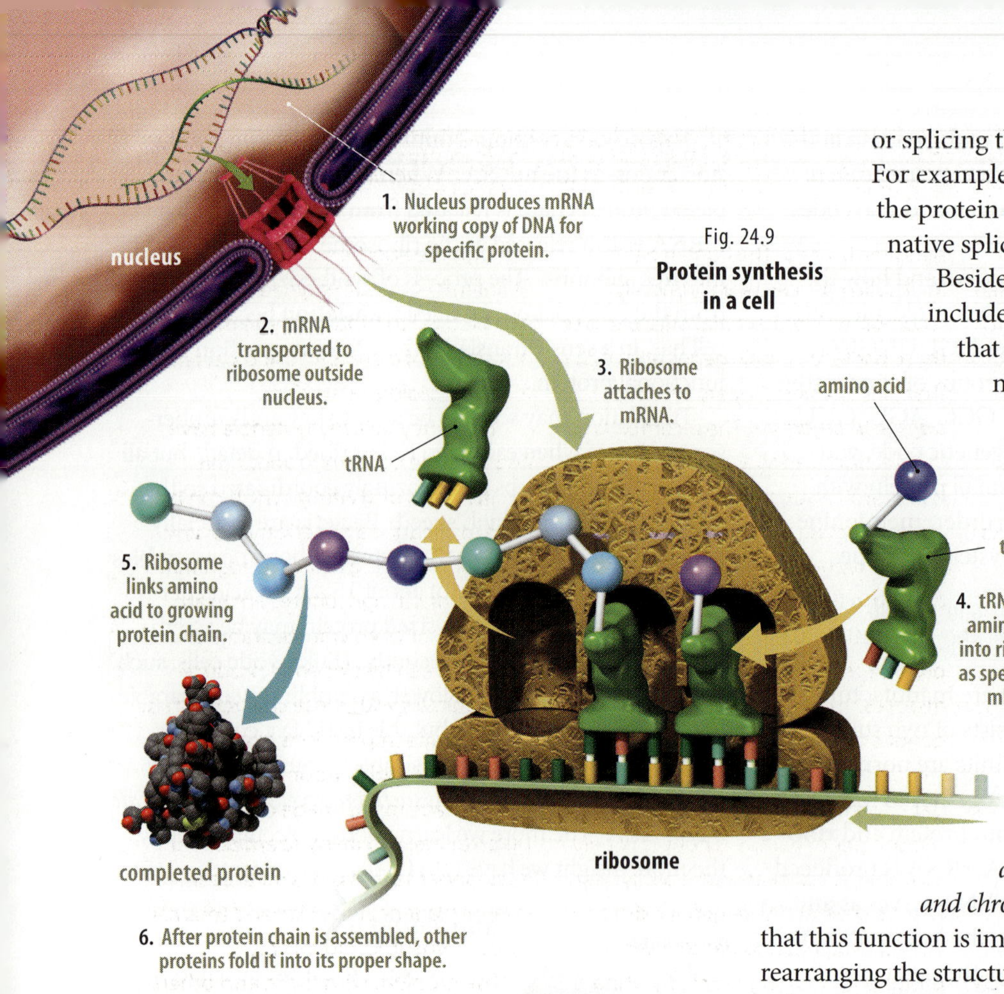


Several functions of noncoding DNA have been discovered, mostly since the early 1990s. In about 60% of human genes, introns allow a single gene to code for several proteins. Through *alternative splicing,* different mRNA molecules can be produced by leaving in some introns, excluding some exons, or splicing the pre-mRNA at different points. For example, at least forty different forms of the protein tropomyosin are formed by alternative splicing of just four genes.

Besides mRNA, tRNA, and rRNA, the cell includes many types of RNA molecules that are collectively called **small RNA;** much of the noncoding DNA, including some introns, are used to produce these RNA molecules. Small RNA's perform many functions. Some are RNA "enzymes," or *ribozymes,* that perform reactions such as splicing and processing pre-mRNA. Others regulate gene expression by binding to DNA or to mRNA.

*Noncoding DNA also helps determine the structure of chromatin and chromosomes.* Scientists believe that this function is important in cell differentiation: rearranging the structure of the chromatin blocks RNA polymerase from reaching certain genes, effectively turning off those genes in cells that do not need them.

Evolutionists sometimes call noncoding DNA *"junk" DNA* and consider it to be a vestigial leftover of evolution; but much as new discoveries showed that "vestigial" organs are important to the body, recent research has shown that noncoding DNA is important to the cell. Still, many functions of noncoding DNA are poorly understood. Continued research is needed to clarify and modify our understanding of the life processes occurring in the cell.

Fig. 24.10

**Alternative splicing**

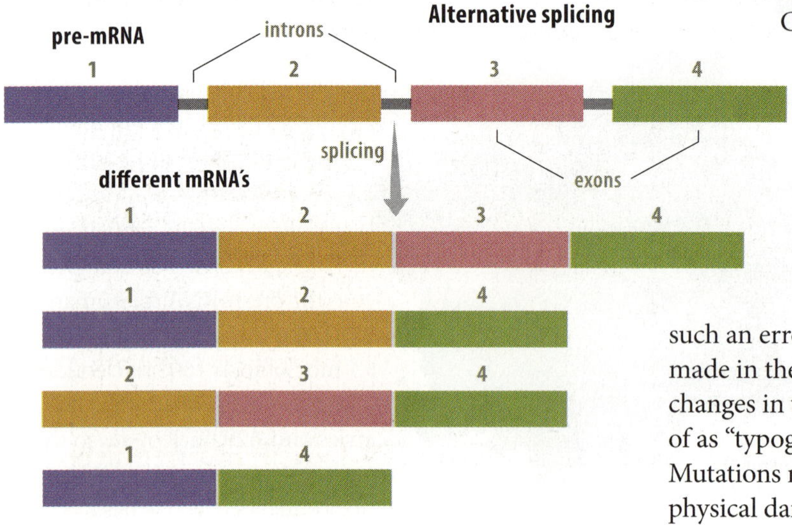


## Errors in Replication

The mechanism of DNA replication is amazingly accurate. But because of the curse of sin upon creation, the cell occasionally makes mistakes. When such an error in replication occurs, small changes are made in the DNA sequence. These small, random changes in the DNA, called **mutations,** can be thought of as "typographical errors" in the genetic material. Mutations result from improper base pairings or from physical damage.

## Genetic Engineering

Using the LIVING CREATION

The manipulation of genes in organisms is not new—man has used selective breeding to manipulate genetic characteristics in plants and animals for centuries. However, **genetic engineering** allows man to make genetic changes in organisms more quickly and more precisely. It has also made possible genetic changes that would not have been possible by selective breeding.

Genetic engineering has many beneficial uses. Besides various applications in agriculture, genetic engineering has tremendous potential in the field of medicine. For example, the *insulin* that diabetics use to control their blood sugar is usually produced by genetically engineered organisms. *Type 1 diabetes* is caused by the failure of the pancreas to produce insulin. Patients with this type of diabetes depend on daily injections of insulin to maintain healthy levels of sugar in their blood. Before genetic engineering, insulin for diabetics was extracted from the pancreases of cows and pigs obtained from slaughterhouses. The human gene for insulin has now been artificially recombined into bacterial cells; because these bacteria contain genes from more than one species, they are a form of *transgenic organism.* The recombinant insulin produced by transgenic bacteria is identical in structure to "real" human insulin. Today, most diabetics depend on recombinant human insulin and not on animal insulin.

Genetic engineering may also provide cures for previously incurable diseases. In principle, a person with a genetic disease can be partially cured through *gene therapy* by replacing the defective gene with a normal one. Although gene therapy is not yet widely available, it has been successfully used to treat several severe genetic diseases. Genetic engineering has also shown promise for treating cancer by using engineered viruses that only attack cancer cells and by modifying some of the body's white blood cells to specifically seek and destroy the cancer cells.

As with any technology, genetic engineering can be used for good or evil. One of the saddest results of our increasing knowledge of genetics is an increasing emphasis on *eugenics* through the abortion of "genetically unfit" infants. Genetic engineering could also be used to produce more dangerous strains of disease-causing bacteria for use in biological warfare.

In recent years, many people have also become concerned about the ramifications of **cloning,** which can be used to produce a genetic twin of any organism. Cloning has some legitimate applications in agriculture. For example, sheep or cows with desirable characteristics can be cloned to produce more animals with those characteristics. However, the implications of cloning a human are highly troubling because man is made in the image of God. Unfortunately, to many scientists man is no longer sacred but is on the same level as plant or animal life and thus can be experimented with at will.

The problem with these and other distortions of genetic engineering is not with the technology itself but with man's depraved nature (see Jer. 17:9). In the hands of moral, ethical scientists, genetic engineering may be used to benefit mankind in numerous ways and bring glory to God by increasing man's knowledge of His master plan for heredity.

### Cloning Sheep

Improper base pairs occur when a base is added, deleted, or substituted in the original DNA sequence, and the results are generally very serious. Physical damage to the DNA molecule can result from oxidation, radiation, or carcinogens. Mutations can also involve the accidental deletion of entire genes, or even groups of genes, from the chromosome. All of these errors, of course, are serious; just as a damaged part can destroy the functioning of a fine watch, a damaged gene can cause the intricate machinery of the cell to break down. If the damage occurs in a less vital gene, however, the cell may live, although it may be hindered or deformed and may malfunction. If the mutation occurs in the DNA of an organism's reproductive (germ) cells, the damaged gene can be passed along to the organism's offspring, and even to the offspring's offspring.

Mutations are usually harmful to an organism because they specify the wrong amino-acid sequence for the proteins they code for. (Some mutations are considered *silent* because they change a codon to a different codon for the same amino acid.) When mutations damage the genes that control the cell cycle, for example, cancerous tumors may result as the cells divide uncontrollably. Many scientists think that accumulated mutations in the DNA of mitochondria, the "power plants" of the cell, cause many of the symptoms of aging. According to this idea, body cells gradually lose power over time, causing the body as a whole to gradually degenerate.

Mutations that are passed from generation to generation also have harmful effects. Relatively minor mutations have been responsible for causing albinism, or lack of skin pigment, in people and animals; a near absence of feathers on chickens; extra fingers or toes on people; wrinkled skin and near hairlessness on mice; and other harmful effects. More serious hereditary mutations cause such diseases as cystic fibrosis, hemophilia, muscular dystrophy, phenylketonuria (PKU), and sickle

## Careers in BIOLOGY

### Cellular and Molecular Biology

**Cell biology** – determining the anatomy and physiology of cells for application in fields such as biotechnology and pharmacology

**Biochemistry** – determining the chemical processes of life for application in fields such as biotechnology and pharmacology

**Biotechnology/industrial microbiology** – using knowledge of microbiology, cell biology, molecular biology, and biochemistry to produce useful commercial products

**Genetics** – using knowledge of DNA and genetics for man's benefit

- plant breeding
- livestock and pet breeding
- forensic science/law enforcement
- genetic engineering

**Pharmacology/biomedical research** – using knowledge of cell biology, molecular biology, and biochemistry to design or produce medical treatments, medicines, etc.

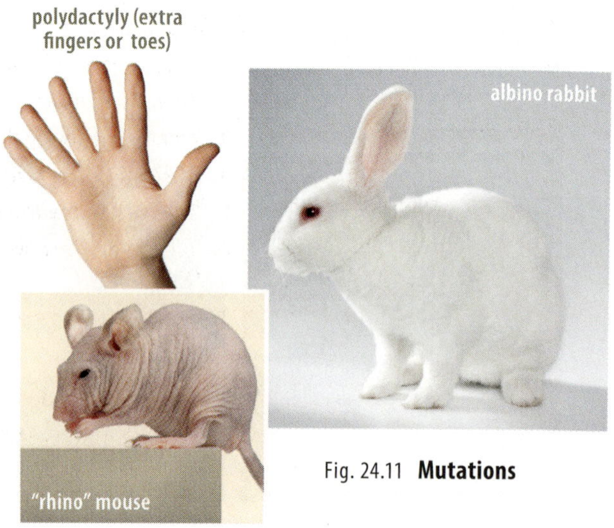


Fig. 24.11 **Mutations**

cell anemia. Mutations are part of the meaning of the scriptural observation that "the whole creation groaneth and travaileth in pain together until now" (Rom. 8:22).

DNA, the master plan of life, is the Master's plan for controlling and perpetuating His living creation on Earth. Surely we can see that the cell and the processes occurring in it are far more complex than man could have imagined even a few years ago, and they are more complex than we can imagine today. This points out the greatness and wisdom of God and shows us that scientists' old idea of the "simple" cell was far too simple to make life possible. "For my thoughts are not your thoughts . . . saith the LORD. For as the heavens are higher than the earth, so are my ways higher than your ways, and my thoughts than your thoughts" (Isa. 55:8–9). There is much still to be learned about God's living creation.

## Section Review 24.2

1. Why is DNA replication necessary in a cell? Why must it be very accurate?
2. Explain the roles of DNA and the three main types of RNA in protein synthesis.

### Application

3. How does the process of base pairing help ensure that DNA is replicated and transcribed correctly?
4. What is one way mentioned in the text that accumulated mutations might lead to the degenerative diseases that accompany aging? In what other ways might accumulated mutations contribute to aging?

### Critical Thinking

5. Why are mutations in somatic (body) cells not transmitted from generation to generation?

### Identify

replication, transcription, mRNA, genetic code, rRNA, tRNA, translation, noncoding DNA, intron, mutation

## Chapter 24 Review

### Define

1. chromatin
2. nucleotide
3. base pair
4. replication
5. transcription
6. codon, genetic code
7. translation
8. transgenic organism, gene therapy, cloning
9. mutation

### Identify

1. the type of molecule that stores a cell's genetic information
2. the two scientists who are credited with deducing the structure of the DNA molecule
3. the four types of bases that are found in a DNA molecule
4. five differences between DNA and RNA
5. the base found in RNA that is not found in DNA
6. four types of RNA and their functions

### Explain

1. Why does a cell bundle its chromosomes into compact structures during mitosis?
2. Why is the structure of DNA considered a double helix?
3. What is meant by the term *base pairing?*
4. Briefly describe DNA replication.
5. Briefly describe the process by which RNA is transcribed from DNA.
6. Briefly describe how a cell produces a protein using the information contained in the DNA.
7. What circumstances can cause mutations in a cell's DNA?

### Apply

1. State the central dogma of molecular biology and explain its importance to all life.
2. How do introns and other noncoding DNA demonstrate God's creative design?

# Credits

**Photo credits are listed from top to bottom, left to right on each page. Locations are noted t–top, b–bottom, l–left, c–center. Other abbreviations used for some sources are as follows: AA–Animals Animals; CB–Corbis Corporation; CC–Corel Corporation; GR–Granger, NYC; IS–©iStockphoto.com; JI–Jupiterimages; ScSo–Science Source; SS–SuperStock; VU–Visuals Unlimited, Inc. Materials provided by the publisher (AB) are credited only to clarify location of other images. If all photos on a page are from the same source, the source is given only once.**

Cover–FarukUlay/IS, helen-light/IS, Yakobchuk/IS, RhettOlson/IS, j_markow/IS, AB; images throughout–aamorim/IS, FarukUlay/IS; title page–FarukUlay/IS, AB; iv–RhettOlson/IS, Firina/IS, tanatat/IS, Beboy_ltd/IS, maxhomand/IS, brozova/IS; v–nickp37/IS, Milan Zeremski/IS, Sergey Nivens/IS, kzenon/IS, Sashkinw/IS, Petit Format/Nestle/ScSo, Ted Kinsman/ScSo; vi–8213erika/IS, Kantapatp/IS, hiphunter/IS, Chameleon335/IS, Serega/IS, GlobalP/IS; vii–FrankyDeMeyer/IS, benjamint444/IS, Brendan Hunter/IS, Eraxion/IS, hotshotsworldwide/IS, Sergey Nivens/IS; 2–Firina/IS, CC, AB, AB, CC, CC, CC, AB, CC, CC, AB, CC, CC, AB, AB, AB; 4–AB, Cimmerian/IS, AVTG/IS, Cararsa/IS, Paija/IS, HelpingHandPhotos/IS, AB, AB; 5–PicturePartners/IS, Elenathewise/IS, DonNichols/IS, Fiery_Phoenix/IS, Hsvrs/IS, Smileus/IS; 6–stocksnapper/IS, MariaBrzostowska/IS, Stacey Newman/IS, naran/IS, Gala_Kan/IS; 7–Vilor/IS, Hope Milam/IS, Dan70/IS, Lisa Thornberg/IS, levifoto/IS, LocusArt/IS, Nao/IS, Bosca78/IS, L_amica/IS, NickR/IS, njnightsky/IS; 9–Arpad Benedek/IS, schnuddel/IS, Art Wager/IS, CharlesGibson/IS, InCommunicado/IS, flammulated/IS, John M. Burnley/ScSo, Rosta/IS, GeorgHanf/IS, Vladimir Salman/IS, Vtls/IS; 11–anafcsousa/IS, Purdue9394/IS, EcoPic/IS, OtmarWinterleitner/IS; 12–baianliang/IS, alxpin/IS; 14–MorePixels/IS (background), CC (silver maple leaf); 16–Ed Young/Corbis (walnut catkins), Patrick Johns/Corbis (willow catkins), Dr. Ken Wagner/VU (red oak catkins); 19–tanatat/IS, photo by Andrew Massyn-Kirstenbosch Gardens Cape Town, andipantz/IS; 25–AB, AB, CC, CC; 28–Michal Krakowiak/IS, Remains/IS, jtyler/IS, AB; 30–AB, photosbyash/IS; 31–zulufriend/IS, Marcus Lindstrom/IS, goodaftermoon/IS; 33–Beboy_ltd/IS, double_p/IS; 34–fstockfoto/IS, Lisa Thornberg/IS, AB, AB, AB, AB, 35–Olaf Simon/IS, Smileus/IS, Ainis/IS, Vilor/IS, Dole08/IS, hbak/IS, AB, Vidok/IS, Vilor/IS, CC, SondraP/IS, StockPhotosArt/IS, Whirler/IS, AB; 36–SteveByland/IS, Epitavi/IS, AB, AB; 37–erperlstrom/IS, Dennis Kunkel Microscopy, Inc./Phototake, SPL/ScSo, SPL/ScSo; 40–CC, CC, CC, CC, CC, CC, CC, AB, AB, jojobob/IS; 41–lepas2004/IS, PicturePartners/IS, Diana Talium/IS, toto8888/IS, Somus/IS, subjug/IS, Mny-Jhee/IS, Anna Kucherova/IS, ValentynVolkov/IS; 43–Vitalii Hulai/IS, Suljo/IS, Charles E. Mohr/ScSo; 47–maxhomand/IS (t), 51–"The Power to Change Matter" from *The Wonders of Creation* by Alfred M. Rehwinkel Copyright © 1974, Bethany House Publishers, a division of Baker Publishing Group, Used by permission, AB, Magnified growth rings courtesy of the estate of Winfred Côté; 57–Dr. Jeremy Burgess/ScSo; 62–brozova/IS (sago); 67–AB, lightpix/IS; 68–gmnicholas/IS (c), CB (b); 69–"Variety of Plants" from *The Wonders of Creation* by Alfred M. Rehwinkel Copyright © 1974, Bethany House Publishers, a division of Baker Publishing Group, Used by permission, marcoregalia/IS, keiichihiki/IS; 70–IrinaKorshunova/IS, jopelka/IS, Joe_Potato/IS; 72–Michael Gadomski/ScSo; 73–AB, Kathy Merrifield/ScSo, Mantonature/IS; 74–Rod Planck/ScSo, samaro/IS; 75–Carolina Biological Supply Company/Phototake, Perennou et Nuridsany/ScSo, SPL/ScSo, Frank Fox/ScSo, Andrew Syred/ScSo; 76–Kent Wood/ScSo, Asa Thoresen/ScSo, AB, paule858/IS; 77–Snowshill/IS, Mathieu Foulquié/ScSo, Dr. Keith Wheeler/ScSo, Dr. Paul Zahl/ScSo; 78–StudioAsp/IS, Michael Abbey/ScSo; 79–MLiberra/IS, kellylovesphotography/IS, Whiteway/IS; 80–Nigel Cattlin/ScSo, ©William Ferguson, Patrick Grace/ScSo, AB, AB; 81–Nigel Cattlin/ScSo, AB, epantha/IS; 82–IS, cpsnell/IS, Alexei Cruglicov/IS, CreativeNature_nl/IS; 84–nickp37/IS; 87–"The Wonders of Man's Body" from *The Wonders of Creation* by Alfred M. Rehwinkel Copyright © 1974, Bethany House Publishers, a division of Baker Publishing Group, Used by permission, Petit Format/Nestle/ScSo; 90–"Cells" taken from *Fearfully and Wonderfully Made* by Philip D. Yancy, Est. of Paul W. Brand. Copyright 1980. Used by permission of Zondervan; 92–Milan Zeremski/IS (t); 98–AB, AB, huseyintuncer/IS, "The Bones of the Human Foot" from *The Wonders of Creation* by Alfred M. Rehwinkel Copyright © 1974, Bethany House Publishers, a division of Baker Publishing Group, Used by permission; 110–"The Human Hand" from *The Wonders of Creation* by Alfred M. Rehwinkel Copyright © 1974, Bethany House Publishers, a division of Baker Publishing Group, Used by permission, MiguelMalo/IS; 114–mirefa/IS, RapidEye/IS, JMichl/IS, lbrakovic/IS, miljko/IS; 116–Sergey Nivens/IS; 117–Eraxion/IS, AB; 122–SPL/ScSo (tr), pixologicstudio/IS (bl); 126–Robertus/IS (green/silver brain); 127–CNRI/Phototake (tc); 130–"Science Can't Explain "Who Am I? Why Am I Here? Making room for the 'mystery of existence'" by Sir John Eccles, quoted in *U.S. News and World Report,* Dec. 10, 1984. Copyrighted 2015 U.S. News & World Report 115414:0215BC; 132–monkeybusinessimages/IS; 137–doram/IS (c); 138–coloroftime/IS; 143–ScSo; 145–kzenon/IS, CB; 149–Sergey Peterman/IS; 160–CNRI/Phototake, SPL/ScSo; 161–Steve Gschmeissner/ScSo, Biophoto Associates/ScSo; 163–Sashkinw/IS; 166–SPL/ScSo, "Soldier Cells" taken from *Fearfully and Wonderfully Made* by Philip D. Yancy, Est. of Paul W. Brand. Copyright 1980. Used by permission of Zondervan; 168–Dr. Dennis Kunkel Microscopy, Inc./Phototake (c); 176–courtesy of Thoratec Corporation; 181–lisafx/IS; 188–Petit Format/Nestle/ScSo, Dario Egidi/IS; 189–SPL/ScSo (bc); 190–Nikada/IS, Viktor_Gladkov/IS, szefei/IS, Maica/IS, encrier/IS, miljko/IS, jrphoto6/IS; 191–SPL/ScSo, CC (hammer); 192–SPL/ScSo (pore); 196–Image Copyright gopixa, Used under license from Shutterstock.com; 203–Motta & Blerkom/SPL/ScSo, Motta & Blerkom/SPL/ScSo, Motta & Blerkom/SPL/ScSo, Petit Format/Nestle/ScSo, Petit Format/Nestle/ScSo, Petit Format/Nestle/ScSo, ScSo, Petit Format/Nestle/ScSo, Petit Format/Nestle/ScSo, Petit Format/Nestle/ScSo; 205–"The Miracle of New Life" from *The Wonders of Creation* by Alfred M. Rehwinkel Copyright © 1974, Bethany House Publishers, a division of Baker Publishing Group, Used by permission; 207–kevinruss/IS (baby); 209–Ted Kinsman/ScSo, Zoran Kolundzija/IS; 217–CC, SPL/ScSo, SPL/ScSo, SPL/ScSo; 219–GR; 223–GR; 225–Meckes/Ottawa/ScSo, Dr. Dennis Kunkel Microscopy, Inc./Phototake; 231–Michael Freeman/Phototake; 232–TommL/IS; 233–londoneye/IS; 235–8213erika/IS; 236–Scala/Art Resource, NY; 240-241–Adapted from "Bible-Believing

Scientists of the Past" by Henry M. Morris, published in the January 1982 issue of *Acts & Facts,* Used by permission of Institute of Creation Research; 243–*Leeuwenhoek and the "Little Animals"* Robert Thom (1915–1979) Collection of the University of Michigan Health System, UMHS.15, Used by permission; 246–Lenar Musin/IS; 249–Kantapatp/IS, Logray-2008/IS; 250–GR; 259–Peter Scoones/ScSo; 261–SPL/ScSo; 262–AB, Institute for Creation Research. Used by permission, CC; 264–National Museums of Kenya, Nairobi © David L. Brill; 268–ScSo; 269–CC, CB, CC, CC, CC, CB, CC, CC, CC, CC, CC, AB, CC, LorenFFile/IS, Frank Leung/IS, GlobalP/IS, GlobalP/IS, GlobalP/IS, teekaygee/IS, Frank Leung/IS, AB, Nikada/IS, Viktor_Gladkov/IS, szefei/IS, encrier/IS, miljko/IS, Maica/IS; 270–Michael Willmer Forbes Tweedie/ScSo; 271–Eraxion/IS; 277–FrankRamspott/IS; 279–hiphunter/IS; 280–Galen Rowell/Corbis (l); 283–CB; 284–Bepsimage/IS, chatursunil/IS, michael1959/IS, CC, Adobe; 286–CB, CB, CB, JI, Mik122/IS, tenback/IS, samaro/IS, ErikKarits/IS, Achim Prill/IS, CB; 287–Anest/IS, JI, Andyworks/IS, CB, Tajarja/IS, Vtls/IS, RTimages/IS, CB, CB, Jeya-kumar/IS, Torchuck/IS, D.P. Wilson/FLPA/ScSo, pilipenkoD/IS, Byronsdad/IS, Kent Wood/ScSo; 289–Elena Alyuko-Sergeeva/IS, CC; 290–IS, Martin Harvey/ScSo; 291–stephaniki2/IS, Waynerd/IS, John Shepherd/IS; 294–Hansen, M., DeFries, J.R.G. Townshed, and R. Sohlberg (1998) UMD Global Land Cover Classification, 1 Kilometer, 1.0, Department of Geography, University of Maryland, College Park, Maryland, 1981–1994, JamesTung/IS, AlbyDeTweede/IS, RichardSeeley/IS; 295–Mike Cherim/IS, JI, AB; 296–JI, impr2003/IS, bookwyrmm/IS, YinYang/IS, Eyezaya/IS, Adobe; 297–Cookie1974/IS, Adobe, vlynder/IS, twildlife/IS, johnaudrey/IS, altmarkfoto/IS, tilo/IS, CB; 298–IBorisoff/IS, crisod/IS, AB; 301–Photodisc, AB, CB, JI, IPGGutenbergUKLtd/IS, JI, CB; 302–BartCo/IS; 304–Chameleon335/IS, IPGGutenbergUKLtd/IS; 305–Jasmin Awad/IS, Gwenvidig/IS, CB, AB, mcsilvey/IS, CC, olyniteowl/IS, Mirostaw Kijewski/IS, GlobalP/IS; 310–astonerattnet/IS; 311–GlobalP/IS; 312–Kosin_Sukhum/IS, Johannes Gerhardus Swanepoel/IS, Nigel Pavitt/JAI/Corbis, mpwoodib/IS; 313–s_white/IS, AndreaBovolo/IS; 314–Saddako/IS, CC, CC; 315–ross1248/IS, erniedecker/IS, J Dennis Nigel/Getty Images; 316–mpot/IS, eeqmcc/IS, keiichihiki/IS; 317–GlobalP/IS, NHPA/Photoshot; 318–Tramper2/IS, CC; 319–AB, CB, CC, CC, AB; 320–jonmccormackphoto/IS, aldorado10/IS; 321–ksumano/IS, kevdog818/IS, Adobe; 322–Antenore/IS, Frans Lanting/MINT Images/ScSo, martin_33/IS; 323–fotojagodka/IS, N. Bergkessel, Jr./ScSo, MikeLane45/IS, MikeLane45/IS; 324–dimj/IS, Anna39/IS; 325–OndagoArts/IS, Natures Display/IS, BrianLasenby/IS, abadonian/IS; 326–112flieger/IS, Jan Gottwald/IS; 327–EMPPhotography/IS, Coldimages/IS, Jose Angel Astor Rocha/IS; 328–Michael Madsen/IS, EEI_Tony/IS, savoia/IS, Kyslynskyy/IS, Whiteway/IS, Sportstock/IS; 329–tankbmb/IS, eAlisa/IS, CB, ElsvanderGun/IS, Martin Wimmer/IS, PaulTessier/IS; 330–CC; 331–bgsmith/IS, EEI_Tony/IS; 332–jim-kruger/IS, kmwphotography/IS; 333–zanskar/IS, Alex Hibbert/Robert Harding World Imagery/Corbis, Ray Hems/IS, JasonPrince/IS, mthaler/IS, Wayne Lynch/All Canada Photos/Corbis; 335–Jared Hobbs/All Canada Photos/Corbis; 337–Serega/IS; 338–AB, BrianLasenby/IS, John Alves/IS, MichaelStubblefield/IS, rmarnold/IS, AB, AB, AB, Frank Leung/IS; 339–CC, fbosse/IS, AB, SteveByland/IS, SteveByland/IS, SteveByland/IS, Frank Leung/IS, Debra Lee Wiseberg/IS, Bill Coster/FLPA/ScSo, AB, AB, WilliamSherman/IS, MikeLane45/IS; 341–JillLang/IS, johnandersonphoto/IS, bgsmith/IS, ca2hill/IS; 342–CB, DaddyBit/IS, Marcus Lindstrom/IS, mkaminski/IS, KristianBell/IS, leekris/IS, Andyworks/IS, CC, twildlife/IS, HinelineDesign/IS, scanrail/IS, stokaji/IS; 343–AB, AB, G.I. Bernard/ScSo, AB, AB, CC, CC, KeithSzafanski/IS, GlobalP/IS, GlobalP/IS; 344–Andrew_Howe/IS; 346–Asa C. Thoresen, Reprinted with adaptations from *Creation: Nature's Designs and Designer,* by Richard Utt copyright 1971 by Pacific Press Publishing Association, Inc. Used by permission of the publisher, ArtBoyMB/IS (seagulls); 347–MichaelStubblefield/IS; 348–NaturesThumbPrint/IS; 358–Frans Lanting/MINT Images/ScSo, amrishwad/IS, Mark Newman/ScSo; 359–Whiteway/IS; 360–La_Corivo/IS, NaturesDisplay/IS; 362–GlobalP/IS; 364–MikeLane45/IS, JillLang/IS, Bobbushphoto/IS; 365–Stephen Dalton/ScSo, fremme/IS, GlobalP/IS, VibeImages/IS, StephanHoerold/IS, ShaneMyersPhoto/Getty Images; 369–Karl Switak/ScSo, bugphai/IS, JasonOndreicka/IS, Stas_K/Depositphotos.com, amwu/IS, huronphoto/IS, Austin J. Stevens/AA; 370–National Geographic/SS, Minden Pictures/SS, skynavin/IS, Dr. William Weber/VU, AB; 371–the4js/IS, Joe McDonald/VU, Inc., Image Copyright Tom Reichner, Used under license from Shutterstock.com, Zigmund Leszczynski/AA; 373–ConstanceMcGuire/IS, Joseph Pulitano/IS, ccahill/IS; 374–j4r3k/IS, cookelma/IS, Jan Gottwald/IS, ralphp/IS, vusta/IS, chris2766/IS; 375–VMJones/IS; 376–CB (horse), CC (vehicle), CC (elephant); 378–tzooka/IS, shaunl/IS, DougVonGausig/IS; 382–Oktay Ortakcioglu/IS; 385–Photo 24/Exactstock-1555/SS, Gerold & Cynthia Merker/VU, DesignPicsInc/Depositphotos.com; 386–epantha/IS, Nature's Images/ScSo, ScienceSource/Getty Images, Danté Fenolio/ScSo; 388–FrankyDeMeyer/IS, Tom McHugh/ScSo, cynoclub/IS, Stephen Dalton/AA, IS, RbbrDckyBK/IS, GlobalP/IS; 391–Exactostock-1527/SS; 392–AB, CoreyFord/IS; 398–Andrew J. Martinez/ScSo, Minden Pictures/SS, Frank Leung/IS; 399–Winfred Wisniewski/FLPA/Corbis, Lynda Richardson/Corbis; 401–thomasmales/IS, cdascher/IS, shihina/IS, Whitepointer/IS, Dray van Beeck/IS; 402–Paul Souders/Corbis, crisod/IS, Image Copyright Jill Lang, Used under license from Shutterstock.com; 403–AB, age fotostock/SS, bit245/IS, Tom McHugh/ScSo; 405–benjamint444/IS; 412–Darrell Gulin/Corbis (Cecropia moth); 415–Ikonoklast_Fotografie/IS (b); 417–CathyKeifer/IS, GarysFRP/IS, JasonOndreicka/IS; 418–CathyKeifer/IS; 420–johnaudrey/IS, Holly Kuchera/IS, Eraxion/IS, ajafoto/IS, Lezh/IS; 421–jeridu/IS, Emmanouil Filippou/IS; 424–micro_photo/IS, Miyuki Satake/IS; 425–FernandoAH/IS, Jeya-kumar/IS, David Wrobel/VU; 427–Brendan Hunter/IS, Andrew J. Martinez/ScSo, PicturePartners/IS, E.R. Degginger/AA; 430–AB, AB, jianghongyan/Depositphotos.com, joebelanger/IS, NunoLopes/IS; 432–hdere/IS, Jman78/IS, Bluexhand/IS, FlamingPumpkin/IS; 433–tiburonstudios/IS, CC, Rainer von Brandis/IS, nature-diver/IS, AB; 434–Jacana/ScSo, AB, AB; 436–AB, NaturePL/SS; 437–Brendan Hunter/IS, AB, johnandersonphoto/IS, Steven David Miller/AA, johnandersonphoto/IS; 438–KevinPanizza/IS, AB; 440–CharlesWhiting/IS; 441–Antagain/IS, Breck P. Kent/AA; 443–SPL/ScSo; 447–AB, Dennis Kunkel Microscopy, Inc./VU; 448–ScSo, AB, 449–K.H. Kjeldsen/SPL/ScSo, Greg Antipa/Biophoto Associates/ScSo, humonia/IS, AB; 451–Eraxion/IS, AB; 453–Ugreen/IS; 456–AB, Biophoto Associates/ScSo, AB; 465–CC (leaves), javarman3/IS (penguin), tanatat/IS (leaves), Jupiterimages (ferns); 470–Carolina Biological Supply/Phototake; 472–hotshotsworldwide/IS; 473–DaydreamsGirl/IS (baby); 481–Biology Pics/ScSo; 485–sx70/IS, Leontura/IS; 487–Sergey Nivens/IS, CNRI/ScSo; 495–Look at Sciences/ScSo, Roslin Institute/Phototake; 496–AlexRaths/IS, stocksnapper/IS, antsmarching/IS, lifeonwhite/Depositphotos.com.

# Glossary

Section numbers are given at the end of each entry.

## Pronunciation Key

| Symbol | Example | Symbol | Example |
|---|---|---|---|
| ā | āte | ŏ | nŏt |
| â | dâre | oi | boil |
| ă | făt | o͞o | fo͞od |
| ä | fäther | o͝o | bo͝ok |
| ə | ago (ə·gō′) | ou | out |
| ē | ēven | th | thin |
| ĕ | ĕgg | t̶h̶ | t̶h̶ere |
| ẽ (ər) | pondẽr | tû | pictûre |
| ī | īce | ū | ūnit |
| ĭ | ĭt | û | hûrt |
| ō | ōver | ŭ | ŭp |
| ô | côrd, taught, saw | zh | measure |

## A

**aardvark:** ant-eating African mammal that is the sole member of order Tubulidentata (16.2)

**abdomen:** rearmost portion of an insect's body (20.1)

**abdominal cavity:** body cavity located below the diaphragm; contains liver, stomach, pancreas, spleen, kidneys, urinary bladder, intestines, and gonads; also called the abdominopelvic cavity (6.1)

**abiotic factor:** nonliving influence in a habitat (15.1)

**ABO blood group:** red-blood-cell antigen group consisting of antigens A and B (10.1)

**abortion:** medical term for the death of a developing child by either natural or artificial means; in common use, reserved for a deliberate action taken to kill a child before birth (11.4)

**abscisic acid:** hormone that signals plant tissues to remain dormant; also called ABA (4.4)

**abscission layer:** special cell layer that forms to cut a leaf or fruit from a stem (2.4, 3.2)

**absorption:** movement of water, digested food, and other dissolved substances from the small intestine into the bloodstream (9.3)

**accommodation:** increase in the curvature of the eye's lens to focus on nearby objects (8.5)

**achene:** simple fruit consisting of a seed and a shell, much thinner than the shell of a nut (3.3)

**acquired immunity:** immunity that results from circulating antibodies or memory cells (12.3)

**acromegaly:** condition in which the bones of the head, hands, and feet become enlarged due to excessive somatotropin production after normal childhood development (11.3)

**action potential:** brief pulse of positive charge that travels down the axon of a neuron that has been triggered to fire (8.2)

**active transport:** movement of molecules across the cell membrane from a place of lower concentration to a place of higher concentration with the expenditure of energy (22.3)

**acute:** describes a disease that is severe but lasts only a short time (12.1)

**acute confusion:** disturbance of the brain caused by physical illness elsewhere in the body (8.4)

**adductor muscles:** pair of muscles in a bivalve that hold the shells together (21.1)

**adenoids:** accumulations of lymph nodes in the throat that, with the tonsils, protect against pathogens entering through the throat (12.3)

**ADH:** pituitary hormone that signals the kidneys to reabsorb more water and produce less urine; short for *antidiuretic hormone* (11.2)

**adipose cell:** fat cell (9.1)

**adolescence:** time of physical transition from child to adult (11.4)

**adrenal gland:** endocrine gland, located on top of the kidney, that produces epinephrine and steroid hormones (11.3)

**adventitious root:** root that grows from an unexpected region of a plant (4.4)

**aerobic bacteria:** bacteria that live where oxygen is present and obtain oxygen from the air (12.2)

**agar:** gelling agent, derived from red algae, used as a culture medium for growing bacteria in the laboratory (5.5)

**agent dispersal:** seed dispersal in which an outside agent carries the seeds (3.3)

**aggregate fruit:** fruit that forms from a flower that has several pistils (3.3)

**AIDS:** serious immune deficiency disease caused by HIV; short for *acquired immune deficiency syndrome* (12.3)

**air sac:** sac in a bird's body through which air passes during respiration (17.2)

**albino:** individual that cannot produce melanin (11.1)

**albumen:** egg white (17.3)

**albumin:** blood protein that helps regulate the water content of blood (10.1)

**aldosterone:** steroid hormone, produced by the adrenal glands, that stimulates the kidneys to conserve sodium and excrete potassium (11.3)

**alga:** member of the group containing the smallest green, chlorophyll-containing organisms (5.5)

**algin:** gummy substance derived from kelp and used to hold liquids in an emulsion (5.5)

**alimentary canal:** tube leading from the mouth to the anus that forms the main portion of the digestive system (6.1, 9.2)

**allele:** any of the different forms of a particular gene (23.2)

**allele frequency:** extent to which a certain allele exists in a population (23.3)

**allergy:** disease that occurs when the immune system overreacts to foreign substances such as pollen, dust, or certain foods (12.1)

**all-or-none principle:** principle stating that when a muscle fiber is stimulated to contract it does so completely and then relaxes completely (7.6)

**alpine tundra:** mountain biome that has vegetation and climatic conditions similar to those of Arctic tundra but lacks permafrost (15.5)

**alternate:** leaf pattern in which only one leaf grows from each node in an alternating pattern up the branch (2.1)

**alternation of generations:** life cycle, found in ferns, mosses, and some other plants, that involves both asexual and sexual reproduction in alternating generations (5.3)

**altricial:** describes a bird that is naked and helpless at hatching (17.3)

**alula:** retractable group of wing feathers supported by a bird's "thumb" bones (17.2)

**alveolus:** one of the microscopic air sacs of the lungs in which gas exchange takes place; plural alveoli (10.4)

**Alzheimer's disease:** disease similar to dementia but more severe and sometimes occurring much earlier in life (8.4)

**amensalism:** relationship in which one organism inhibits another (15.4)

**amino acids:** group of about 20 types of simple molecules that combine to produce proteins (9.1)

**amnesia:** memory loss (8.4)

**amniotic fluid:** fluid that protects the unborn young of a placental mammal from shocks caused by the mother running or being bumped (16.1)

**amniotic sac:** membranous sac that protects an unborn child as he develops (11.4)

**amoeba:** sarcodine that uses pseudopods for locomotion and for ingesting food particles (21.6)

**amoebic dysentery:** disease in which a certain species of amoeba feeds on the wall of the human intestine, causing bleeding ulcers (21.6)

**amphibian:** smooth-skinned, cold-blooded vertebrate that has a three-chambered heart and generally spends part of its life in the water and part on land (18.4)

**amplexus:** process in which a male frog clasps a female frog to help her expel her eggs (18.4)

**anaerobic bacteria:** bacteria that live in the absence of oxygen (12.2)

**anal fin:** fin that projects downward from the rear of a fish's trunk (19.1)

**anal pore:** pore from which a paramecium releases undigested food particles (21.6)

**analogy:** similarity among organisms designed for the same environment but different in internal structure (14.3)

**anaphase:** third stage of mitosis, in which the chromatids in each replicated chromosome separate and are dragged toward opposite ends of the cell (22.3)

**anemia:** disorder resulting from too little oxygen being transported to body cells (10.1)

**angiosperm:** plant with flowers that produce seeds covered by fruit (1.3)

**animal:** living organism that moves from place to place in its environment (16.1)

**Animalia:** animal kingdom; consists of multicellular eukaryotes without cell walls (5.1)

**annual:** plant that develops from a seed and produces new seeds in a single growing season (1.3)

**annual growth ring:** distinct layer in the xylem of a woody stem formed by annual variation in the rate of secondary growth (4.2)

**antenna:** feeler that serves as a sensory organ in an arthropod (20.1)

**anther:** enlarged pollen-producing structure at the tip of a stamen (3.1)

**anthocyanin:** pigment in leaves that produces bright red, blue, and purple colors (2.4)

**Anthophyta:** phylum consisting of angiosperms (5.2)

**antibiotic:** substance produced by bacteria, mold, or other organisms that is effective in stopping the growth of microorganisms (12.3)

**antibody:** Y-shaped protein molecule that binds to pathogen antigens to inactivate pathogens and mark them for destruction (12.3)

**antigen:** carbohydrate molecule on the surface of a cell that functions as an identification tag (10.1)

**antioxidant:** substance that neutralizes harmful free radicals (9.1)

**antiserum:** blood extract that contains antibodies to provide passive immunity (12.3)

**antitoxin:** antibody that attacks toxin molecules to neutralize their effects (18.2)

**antler:** solid, annually shed, bony elevation on the skull of a deer (16.4)

**anuran:** frog or toad (18.4)

**anus:** valve that closes the rectum and serves as the digestive tract's second opening to the exterior (9.3)

**aorta:** the body's largest artery, from which all other systemic arteries branch (10.2)

**aortic arch:** earthworm circulatory structure that is one of its ten "hearts" (21.4)

**aortic semilunar valve:** heart valve that allows blood to exit the left ventricle and enter the aorta (10.2)

**ape:** primate without a tail (16.3)

**apoptosis:** process through which a cell can trigger its own death when it is no longer needed or no longer able to survive; also called programmed cell death (22.3)

**appendicitis:** condition caused by infection of the vermiform appendix (9.3)

**appendicular skeleton:** division of the skeletal system that includes the shoulders, hips, and appendages (7.2)

**appendix:** small, fingerlike structure attached to the cecum (9.3)

**aquatic:** refers to water (15.1)

**aquatic biome:** biome that is located in water (15.5)

**aqueous humor:** clear fluid in the eye that fills the space between the cornea and the iris and lens (8.5)

**arachnid:** arthropod with two main body regions and eight legs (20.3)

**Archaebacteria:** kingdom consisting of prokaryotes that do not contain peptidoglycan in their cell walls (5.1)

***Archaeopteryx:*** unusual fossil bird often presented as an evolutionary link between dinosaurs and modern birds (14.2)

**Arctic tundra:** treeless northern biome that has long, harsh winters and permafrost (15.5)

**Aristotle:** disciple of Plato who developed the doctrine of intellect (13.1)

**arrector pili:** smooth muscle attached to the hair follicle (11.1)

**arteriole:** small blood vessel connecting an artery to capillaries (10.3)

**arteriosclerotic dementia:** dementia caused by death of brain cells due to blockage in the arteries (8.4)

**artery:** thick-walled blood vessel that transports blood away from the heart (10.3)

**arthritis:** inflammation of the joints (7.4)

**arthropod:** invertebrate with jointed appendages (20.1)

**artiodactyl:** even-toed hoofed mammal of order Artiodactyla (16.4)

**Artiodactyla:** order of even-toed hoofed mammals (16.4)

***Ascaris* worm:** parasitic roundworm that lives in the host's intestine and eats partially digested food (21.5)

**asexual reproduction:** reproduction without the union of sex cells, requiring only mitotic cell divisions (4.4, 23.1)

**astigmatism:** focusing problem in which the light rays entering the eye are bent incorrectly, usually caused by an incorrectly curved cornea (8.5)

**asymmetric:** without symmetry (16.1)

**atherosclerosis:** form of arteriosclerosis in which a buildup of plaque and scar tissue narrow the inside of an artery (10.3)

**atmospheric cycle:** any biogeochemical cycle that involves nutrients found mainly in gaseous forms in the atmosphere (15.4)

**ATP:** special phosphorus compound that serves as the energy carrier of the cell; short for *adenosine triphosphate* (9.1, 22.3)

**ATP synthase:** molecular machine that generates ATP using power provided by the flow of hydrogen ions (22.2)

**atrium:** either of the two thin-walled upper chambers of the heart; plural atria (10.2)

**atrophy:** muscle degeneration (7.7)

**auditory nerve:** nerve connecting the hairlike nerve cells of the cochlea to the brain (8.5)

***Australopithecus afarensis:*** small extinct ape represented by the fossil nicknamed "Lucy" (14.2)

***Australopithecus africanus:*** small extinct ape, represented by the "Tuang child" fossil, that was once considered a human ancestor (14.2)

***Australopithecus boisei:*** fossil ape discovered in 1959 that was originally considered a human ancestor (14.2)

**autoimmune diseases:** immunological diseases in which the immune system attacks the body (12.1)

**autonomic nervous system:** portion of the peripheral nervous system that controls the heart and other internal organs (8.1)

**autotroph:** organism that makes its own food using an inorganic energy source; also called a producer (2.3, 15.3)

**auxin:** any of a group of plant growth hormones that are involved in many tropisms, cause the apical stem to remain dominant, and stimulate root growth and fruit formation (4.4)

**A-V valve:** either of the two heart valves that allow blood to flow from an atrium into a ventricle (10.2)

**axial skeleton:** division of the skeletal system that includes the head, spine, and rib cage (7.1)

**axolotl:** aquatic salamander of Mexico and the southwestern United States that normally becomes sexually mature and breeds in the larval form but can metamorphose in dry conditions (18.5)

**axon:** long extension of a neuron that relays nerve impulses from the cell body to other neurons (8.1)

## B

**bacillus:** rod-shaped bacterium; plural bacilli (12.2)

**bacteria:** group of unicellular prokaryotes, some of which are pathogens; sing. bacterium (12.1)

**baleen:** comblike plate in the mouth of a baleen whale; also called whalebone (16.3)

**baleen whale:** any whale whose mouth has rows of comblike baleen plates that are used to filter small prey out of the water (16.3)

**ball-and-socket joint:** joint in which the rounded head of a bone fits into a hollow socket in another bone, allowing the bone to move in two planes and rotate in place (7.4)

**barb:** filament that helps form the vane of a feather (17.2)

**barbule:** small structure attached to a barb that holds barbs together to form a vane (17.2)

**bark:** outer section of a woody stem, which provides protection and helps transport food (4.2)

**barnacle:** marine crustacean that attaches to rocks, piers, wharves, and ships (20.4)

**base:** nitrogen-containing portion of a nucleotide whose sequence stores the information in DNA or RNA (24.1)

**base pair:** "rung" of the DNA "ladder," composed of two matching bases (24.1)

**base pairing:** process in which only one type of base will fit another to form a base pair (24.1)

**basidium:** microscopic, club-shaped, spore-producing structure of a club fungus; plural basidia (5.6)

**bat:** flying mammal of order Chiroptera (16.3)

**behaviorism:** philosophy that the behavior of a person is determined by his environment, not his mind (8.3)

**benign:** describes a tumor that is not cancerous and shows no tendency to spread throughout the body (12.1)

**berry:** simple fruit in which the entire ovary is fleshy and juicy throughout (3.3)

**biceps brachii:** muscle on the front of the upper arm that bends the forearm; also called biceps (7.5)

**bicuspid:** any of the four light-grinding teeth next to the cuspids (one pair on each side) on each jaw; also called premolar (9.2)

**bicuspid valve:** heart valve connecting the left atrium and left ventricle; also called mitral valve (10.2)

**biennial:** plant that lives through two growing seasons to complete its life cycle (1.3)

**bilateral symmetry:** form of symmetry in which an object can be cut into identical halves along only one plane (16.1)

**bile:** greenish substance secreted from the liver that helps in the digestion of fats (9.3)

**bile salts:** group of substances in bile that emulsify fats (9.3)

**binary fission:** asexual method of reproduction in which a unicellular organism splits into two new cells (12.2, 21.6)

**binomial nomenclature:** system of assigning scientific names to organisms by using two words (genus and species) for each scientific name (5.1)

**biodiversity:** number and variety of species living within an ecosystem (15.2)

**biogeochemical cycle:** cycle that recycles nutrients in the environment; also called nutrient cycle (15.4)

**biology:** study of living things (1.1)

**biomass:** total mass of living matter per unit area (15.3)

**biomass pyramid:** diagram that represents the biomass at each trophic level as a pyramid (15.3)

**biome:** large geographic region with characteristic plant and animal life (15.1)

**biosphere:** relatively thin layer of the earth's surface in which life exists (15.1)

**biotic factor:** living influence in a habitat (15.1)

**bipedal:** two-footed (17.1)

**bird:** vertebrate with feathers, wings, and scale-covered legs (17.1)

**bird of prey:** any of the birds known for their fierce and precise hunting skills (17.1)

**bivalve:** mollusk that has two hinged shells (21.1)

**blade:** flat, green portion of a leaf (2.1)

**blastocyst:** hollow ball of cells formed by repeated division of a zygote (11.4)

**blind snake:** any of several small, harmless, wormlike burrowing snakes (18.2)

**blind spot:** spot on the retina where the optic nerve exits the eye and there are no light-sensitive cells (8.5)

**blood:** thick fluid that transports gases, nutrients, and wastes as part of the circulatory system (10.1)

**blood pressure:** force that blood exerts upon the walls of the blood vessels (10.3)

**blood-brain barrier:** system of tightly sealed capillary walls that prevents bacteria, viruses, and harmful molecules from entering the brain and spinal cord (8.1)

**boid:** any member of a family of nonvenomous snakes that overcome their prey by constriction (18.2)

**bolus:** small ball into which food is formed by the tongue (9.2)

**bone conduction:** transmission of vibrations (caused by speech, etc.) from the jawbone to the inner ear (8.5)

**bony fish:** fish whose skeleton is made either partially or wholly of bone (19.1)

**book lung:** arachnid respiratory structure consisting of flat folds of tissue forming parallel air pockets (20.3)

**botany:** study of plants (1.1)

**botulism:** severe food poisoning caused by the bacterium *Clostridium botulinum* (9.4)

**bovid:** any ruminant of family Bovidae (16.4)

**Bowman's capsule:** spherical container surrounding the glomerulus that collects water and dissolved substances (11.2)

**bract:** special leaf that looks like a flower petal (1.4, 3.1)

**brain:** principal organ of the nervous system; the chief organ of the mind (8.1)

**brain stem:** portion of the brain that connects the rest of the brain to the spinal cord; includes three regions responsible for most automatic body functions (8.3)

**bran:** shell-like fruit covering of a wheat kernel (3.3)

**broadleaf tree:** tree that has broad, flat leaves (1.4)

**bronchial asthma:** allergic condition characterized by spasms of the bronchial tube muscles (10.4)

**bronchial tube:** a bronchus or any of its branches (10.4)

**bronchiole:** small tube, branching from a secondary bronchus, that ends in alveoli (10.4)

**bronchitis:** inflammation of the lining of the bronchial tubes (10.4)

**bronchus:** either of the two tubes into which the trachea divides, one going to each lung; plural bronchi (10.4)

**brood patch:** exposed patch of skin that a parent bird uses to transfer heat to the eggs (17.3)

**Brunfels, Otto:** Protestant pastor and schoolmaster who wrote *Herbarum Vivae Eicones* (13.2)

**bryophyte:** spore-bearing nonvascular plant (5.4)

**buccal respiration:** respiration through the membranes lining the oral cavity (18.4)

**bud:** part of a stem that allows it to grow in length or develop new stems, flowers, or leaves (4.1)

**bud scale:** any of the scales that cover a bud in winter to protect it from frost injury and drying out (4.1)

**budding:** form of grafting in which the scion is a bud (4.4)

**bud-scale scar:** ring that circles a stem, indicating where last year's growth started (4.1)

**bug:** insect of the order Hemiptera (20.2)

**bulb:** structure made of a special stem and leaves that stores food underground (4.4)

**bundle scar:** marking on a leaf scar that indicates a passage point of vascular tissue from the stem to the leaf petiole (4.1)

**bursa of Fabricius:** organ in young birds that lies close to the wall of the cloaca and produces lymphocytes (17.2)

## C

**caecilian:** wormlike burrowing amphibian (18.5)

**Calorie:** unit used to measure the energy content of food; equal to the amount of energy needed to raise the temperature of one kilogram (or one liter) of water by one degree Celsius (9.1)

**calyx:** collective term for a flower's sepals (3.1)

**cambium:** internal layer of meristematic tissue that enables stems and roots to grow in diameter (2.2)

**Cambrian explosion:** sudden appearance of a great variety of complex organisms in the Cambrian rocks, with no evidence of ancestors in the Precambrian rocks (14.2)

**cancer:** condition that results from the runaway growth of the body's own cells (12.1)

**canine:** dog, wolf, or other similar mammal of family Canidae (16.4)

**capillarity:** upward movement of liquids through a very narrow tube due to differences in adhesion and cohesion; also called capillary action (4.5)

**capillary:** microscopic blood vessel, linking an artery and vein, in which the exchanges between body cells and blood take place (10.3)

**capsule:** spore case of a moss plant (5.4)

**carapace:** upper shell of a turtle (18.3)

**carapace:** hard shell that provides extra protection to the vital organs of a crayfish (20.4)

**carbohydrate:** organic compound consisting of the elements carbon, hydrogen, and oxygen in about a 1:2:1 ratio (9.1)

**carbon-oxygen cycle:** atmospheric cycle that recycles carbon and oxygen (15.4)

**carcinogen:** substance that is thought to be capable of causing cancer (12.1)

**cardiac conduction system:** system of special cardiac muscle fibers that generate electrical impulses and distribute them to all parts of the heart (10.2)

**cardiac muscle:** self-stimulating striated muscle tissue found in the heart (7.5)

**cardiac veins:** system of veins that collect blood from the heart tissues and carry it to the right atrium (10.2)

**cardiologist:** doctor specializing in the heart and its disorders (10.2)

**cardiovascular system:** another name for the circulatory system (10.1)

**carnivore:** general term for any consumer that feeds on other animals; also specifically refers to any meat-eating mammal of order Carnivora (15.3, 16.4)

**carotene:** pigment in leaves that produces yellowish-orange colors (2.4)

**carpal:** any of the eight bones of the wrist (7.2)

**carrageenan:** gelatinous substance that comes from the alga Irish moss and is used in many food products (5.5)

**carrier:** person who displays no symptoms of an infectious disease but harbors the pathogens in his body and can pass them on to infect others; person who has only one allele for a recessive genetic disease (12.2, 23.3)

**carrying capacity:** ability of an ecosystem to support biodiversity (15.2)

**cartilage:** tough, rubbery connective tissue that cushions the joints between bones (7.1)

**cartilaginous fish:** fish whose skeleton is made of cartilage rather than bone (19.3)

**cat:** any member of the large group of land-dwelling carnivores that includes tigers, lions, and housecats (16.4)

**catalyst:** substance that speeds up a chemical change but is not consumed in the process (9.2)

**catkin:** long, tassel-like cluster formed by the flowers of some trees (1.4)

**caudal fin:** tail fin of a fish (19.1)

**cecum:** pouch in the large intestine at the junction with the small intestine; similar structure in horses and rabbits where fermentation of cellulose occurs (9.3, 16.4)

**cell:** basic unit of structure and function of living things (2.2, 22.1)

**cell biology:** another name for cytology, the study of cells (22.1)

**cell body:** central portion of neuron, which contains the nucleus and most of the cytoplasm (8.1)

**cell cycle:** life cycle of a cell (22.3)

**cell membrane:** thin, flexible barrier that surrounds a cell, separates it from its environment, and regulates what enters and leaves the cell; also called plasma membrane (2.2, 22.2)

**cell theory:** idea that all living things are composed of living units called cells and of cell products and that all cells come only from preexisting cells (13.2, 22.1)

**cell wall:** rigid, boxlike structure outside the cell membrane of a plant cell that stiffens, supports, and protects the cell (2.2, 22.2)

**cellular respiration:** process through which a cell breaks down chemical substances such as sugars and combines them with oxygen (oxidizes them) to release energy; also called aerobic respiration (2.3, 22.3)

**cellulase:** enzyme that weakens the cell walls of the abscission layer (2.4)

**cellulose:** complex carbohydrate that forms the cell wall of plants and is a form of insoluble fiber (2.2, 9.1)

**cementum:** bonelike substance that fastens the root of a tooth to the jaw (9.2)

**centipede:** caterpillarlike arthropod with a long, flattened body; long, jointed antennae; poison claws; and a single pair of legs attached to each body segment (20.3)

**central disk:** somewhat flattened region from which the arms of an echinoderm radiate (21.2)

**central dogma of molecular biology:** flow of biological information from DNA to RNA and from RNA to protein (24.2)

**central nervous system:** portion of the nervous system that consists of the brain and spinal cord (8.1)

**centrioles:** pair of cylindrical structures in the centrosome (22.2)

**centrosome:** microtubule-organizing center of a human or animal cell (22.2)

**cephalopod:** aquatic mollusk that has an elongated visceral hump and a merged head and foot (21.1)

**cephalothorax:** combined head and thorax of arachnids and some crustaceans (20.3)

**ceratopsian:** any member of the group of dinosaurs that included *Triceratops* (18.3)

**cereal:** any of the types of grass used by man for food; the grain of such a plant (1.3, 3.3)

**cerebellum:** portion of the brain that coordinates skeletal muscle activity (8.3)

**cerebral cortex:** outer layer of the cerebrum, which contains much of the brain's gray matter (8.3)

**cerebral palsy:** damage to the cerebral motor area, resulting in reduced motor abilities without affecting intelligence (8.3)

**cerebrospinal fluid:** clear fluid that circulates through the fibers of the arachnoid mater (8.1)

**cerebrum:** portion of the brain responsible for consciousness, memory, voluntary actions, thinking, and intelligence (8.3)

**cervical:** of or related to the neck (7.1)

**cervical vertebra:** any of the seven vertebrae of the neck (7.1)

**chaff:** husk of a wheat kernel (3.3)

**chalaza:** dense, twisted cord that supports the yolk of a bird egg and keeps the yolk upright (17.3)

**chamber:** any of the four internal hollow spaces of the heart (10.2)

**cheese mold:** any of the molds used to make some types of cheese (5.6)

**chelicerae:** pair of short, fanged appendages used by a spider to seize and to kill or crush prey (20.3)

**chemical sense:** any of the senses that result from chemical stimuli (8.5)

**chemoreceptor:** receptor for a chemical sense (8.5)

**chemotherapy:** any procedure in which chemical compounds are used to treat a disease (12.3)

**chimaera:** medium-size cartilaginous fish with huge eyes; a long, slender, pointed tail; and opercula (19.3)

**chitin:** substance, chemically similar to cellulose or starch, of which an arthropod's exoskeleton is made (20.1)

**chlorophyll:** green pigment of plants that enables them to capture the energy of light (2.2)

**chloroplast:** organelle, found in the cells of green plants, that is a tiny chemical "factory" that uses light to manufacture food (2.2)

**cholesterol:** lipid used by the body to make bile, vitamin D, and important hormones and used in cell membranes (9.1)

**Chordata:** phylum that includes vertebrates (subphylum Vertebrata) and a few invertebrates that have notochords but lack some of the other distinguishing features of true vertebrates (16.1)

**choroid:** layer of tissue within the uvea that provides most of the eye's nourishment (8.5)

**chromatin:** loosely coiled DNA in a cell nucleus (24.1)

**chromatophore:** branched pigment cell found in the skin of reptiles, amphibians, and fish (18.4)

**chromosome:** individual DNA molecule within a cell (22.3)

**chromosome theory of inheritance:** theory stating that genes are stored on chromosomes (23.2)

**chronic:** describes a disease that is long lasting or recurring (12.1)

**chyme:** thick liquid formed in the stomach, consisting of digestive juices and partially digested food (9.3)

**ciliary body:** muscular structure near the front of the uvea that adjusts the lens of the eye to focus at varying distances (8.5)

**cilium:** tiny hairlike projection extending from the cell membrane of some cells (21.6, 22.3)

**circular reasoning:** logic error of basing an argument on the assumption it attempts to prove (14.2)

**circulatory shock:** potentially lethal condition in which the body's blood pressure falls to a critical low (10.1)

**circulatory system:** body system that brings food, water, and oxygen to the cells and carries away carbon dioxide and waste products; also called cardiovascular system (6.1)

**cirrhosis:** buildup of fatty deposits and fibrous tissue in the liver (10.3)

**class:** level of taxonomy between phylum and order (5.1)

**clavicle:** collarbone (7.2)

**climax community:** combination of a region's climax vegetation and its animal populations (15.5)

**climax vegetation:** species of plants that dominate a region and characterize a biome (15.1)

**cloaca:** organ of a bird where wastes are readied for elimination and gametes are held to be released to the outside (17.2)

**cloning:** genetic engineering process that produces a genetic twin of another organism (24.2)

**club fungi:** group of fungi that produce spores in club-shaped basidia (5.6)

**club moss:** creeping vascular plant that has an erect stem and bears spores in club-shaped, conelike structures; also called ground pine (5.3)

**cnidocyte:** stinging cell of a coelenterate (21.3)

**coccus:** spherical bacterium that usually grows in clusters or chains; plural cocci (12.2)

**coccyx:** bone, consisting of four separate vertebrae that are fused in adults, that is an attachment for muscles involved in standing and the elimination of wastes; also called tailbone (7.1)

**cochlea:** coiled tube in the inner ear that detects vibrations and sends them to the brain (8.5)

**codominance:** form of inheritance in which two alleles for a trait are equally expressed in the phenotype of a heterozygous individual (23.3)

**codon:** consecutive sequence of three mRNA nucleotides that translates into a specific amino acid (24.2)

**coelacanth:** fish of the order Crossopterygii (14.2)

**coelenterate:** aquatic invertebrate that has a saclike body, a central digestive cavity with a single opening, and tentacles containing cnidocytes (21.3)

**coenzyme:** molecule that assists a cell's enzymes in performing their jobs (9.1)

**cold-blooded:** regulating body temperature by external factors (18.1)

**Coleoptera:** order of insects that undergo complete metamorphosis and have forewings that form a protective shell when the insect is not in flight (20.2)

**collagen:** tough, resilient protein fiber that reinforces bone and, with elastin, gives skin its stretchability and elasticity (7.3, 11.1)

**collar cell:** any of the cells that line a sponge's body cavity (21.3)

**colon:** long, tubular section of the large intestine that absorbs water from undigested food (9.3)

**colony:** group of single-celled algae clumped together and living as a group (5.5)

**colorblind:** unable to distinguish one or more of the primary colors of light (8.5)

**colubrid:** member of the largest family of living snakes, most of which are nonvenomous (18.2)

**coma:** prolonged unconscious state that may not be reversible (8.1)

**commensalism:** symbiotic relationship in which one organism is benefited and the other is neither harmed nor helped (15.4)

**common bile duct:** duct through which bile enters the small intestine from the gallbladder (9.3)

**common cold:** most common respiratory disease, caused by any of over a hundred different types of viruses (10.4)

**communicable:** describes an infectious disease in which the invading microbe can be transferred from person to person (12.2)

**community:** collective term for all the living things in an ecosystem (15.1)

**compact bone:** strong, dense tissue composing the outer layer of a bone (7.3)

**comparative anatomy:** study of similarities and differences in the body structure of organisms (14.3)

**competition:** relationship in which two organisms compete for the same limited resource (15.4)

**complete flower:** flower that has all four basic flower parts (sepals, petals, stamens, and pistils) (3.1)

**complete metamorphosis:** process in which an insect metamorphoses from a larva into a pupa and then into an adult (20.1)

**complex carbohydrate:** carbohydrate formed from a long chain of simple carbohydrates; also called polysaccharide (9.1)

**compound eye:** arthropod eye that is composed of many independent lenses and "retinas" (20.1)

**compound fracture:** fracture in which the broken bone pierces through the skin; also called open fracture (7.3)

**compound leaf:** leaf that has more than one blade on every petiole (2.1)

**conception:** fertilization (11.4)

**concertina movement:** movement in which a snake moves the front part of its body forward, coils it to anchor it, and then pulls the back part forward (18.2)

**concussion:** period of paralysis of the central nervous system accompanied by a short period of unconsciousness (8.4)

**condensation:** process in which water vapor forms liquid water (15.4)

**cone:** type of cell in the retina that detects colors (8.5)

**congenital disease:** disease that occurs during development in the mother's womb and is present at birth (12.1)

**conifer:** cone-bearing gymnosperm (5.2)

**conjugation:** method of reproduction in which two paramecia exchange nuclear material before undergoing repeated divisions (21.6)

**connective tissue:** any of the tissues that connect parts of the body (6.2)

**conservation:** using resources in a responsible way that allows their continued use (15.6)

**constipation:** condition that occurs when feces are so dry that defecation is difficult (9.4)

**constrictor:** snake that overcomes its prey by constriction (18.2)

**contour feather:** strong feather that points toward a bird's tail to aid in streamlining (17.2)

**control group:** group in a controlled experiment that forms the standard for comparison (13.3)

**controlled experiment:** experiment in which all factors are identical except the one being tested (13.3)

**coral:** colonial coelenterate that resembles a miniature sea anemone but produces a limestone cup at the base of its body (21.3)

**coral reef:** marine ecosystem formed from the remains of stony corals and coralline algae; vast limestone structure made of the limestone cups of a coral colony (15.5, 21.3)

**cork:** plant structural tissue that forms the bark of woody plants (2.2, 4.2)

**cork cambium:** cambium tissue that produces new cork tissue (2.2)

**corm:** swollen stem that stores food underground; distinguished from a bulb by a thicker stem and thinner leaves (4.4)

**corn smut:** smut that attacks the tassels and ears of corn plants, forming large, shiny blisters (5.6)

**cornea:** transparent structure that replaces the sclera at the front of the eye (8.5)

**corolla:** collective term for a flower's petals (3.1)

**coronary arteries:** two arteries that branch from the base of the aorta to supply blood to the cardiac muscle cells (10.2)

**coronary circulation:** portion of systemic circulation in which blood flows to the heart muscle and back to the right atrium (10.3)

**corpus callosum:** mass of nerve fibers through which the cerebral hemispheres communicate (8.3)

**cortex:** food-storage tissue in a herbaceous stem in which the vascular bundles are embedded; similar tissue in a root or young woody stem (4.3)

**cortisol:** steroid hormone, produced by the adrenal glands, that stimulates the body to repair itself following an injury or stress (11.3)

**cotyledon:** portion of an angiosperm embryo that contains stored food used to supply the energy for sprouting (1.3, 3.4)

**cranial cavity:** body cavity containing the brain (6.1)

**cranial nerve:** any of the twelve pairs of nerves that branch directly from the brain stem (8.1)

**cranium:** portion of the skull that protects the brain; also called braincase (7.1)

**cretinism:** condition in which a child does not grow and develop properly due to a thyroxine deficiency during childhood (11.3)

**Crick, Francis H.:** British scientist who, with James Watson, discovered the double-helix structure of DNA (24.1)

**crocodilian:** large, thick-skinned, long-tailed aquatic reptile (18.3)

**Cro-Magnon man:** group of fully human fossils, first discovered in a cave in southwest France, that were originally considered subhuman by evolutionists (14.2)

**crop:** special storage sac in the esophagus of a bird (17.2)

**crop milk:** protein- and fat-rich substance produced by doves and pigeons to feed their young (17.3)

**crossing over:** exchange of genetic material that results in recombination (23.1)

**cross-pollination:** pollination in which the pollen from one plant is transferred to another plant (3.2)

**crown:** bushy head of a tree (1.4)

**crown:** part of a tooth that is exposed above the gums (9.2)

**crustacean:** aquatic arthropod that has two pairs of antennae; two or more pairs of appendages; a hard, crusty exoskeleton composed of chitin and strengthened by calcium carbonate; and usually gills (20.4)

**cud:** small balls of partially digested food returned to a ruminant's mouth from the rumen for further chewing (16.4)

**cuspid:** either of two tearing teeth flanking the incisors on each jaw; also called canine (9.2)

**cutaneous membrane:** another term for the skin (6.2)

**cuticle:** waxy layer that usually covers the epidermal tissue of a plant's shoot system (2.2)

**cutting:** piece of stem or root that can grow into a new plant; also called a slip (4.4)

**Cuvier, Georges:** French scientist who founded the science of comparative anatomy and showed that the bodies of men and animals have certain basic similarities in their overall designs (14.3)

**cyanobacteria:** group of bacteria that form colonies resembling those of unicellular algae; also called blue-green algae (5.5)

**cycad:** palmlike tropical gymnosperm (5.2)

**cystic fibrosis:** recessive genetic disease caused by a defect that affects the chlorine gates of lung cell membranes; abbreviated *CF* (10.4, 23.3)

**cytokinesis:** process in which the cytoplasm of a cell separates to form two separate cells (22.3)

**cytokinin:** any of a group of plant growth hormones that promote cell division and stimulate lateral buds to grow into new shoots (4.4)

**cytology:** study of cells; also called cell biology (22.1)

**cytoplasm:** jellylike fluid medium of the cell, containing organelles and numerous dissolved chemicals (2.2, 22.2)

**cytoskeleton:** intricate internal skeleton that helps a cell maintain its shape (22.2)

## D

**daddy longlegs:** long-legged arachnid similar to spiders but having only two eyes and lacking a constriction between cephalothorax and abdomen, fangs, and spinnerets; also called harvestman (20.3)

**Darwin, Charles:** British naturalist who wrote *The Origin of Species* and *The Descent of Man* and popularized evolution (14.1)

**deciduous:** describes a tree that loses its leaves each fall (1.4)

**decomposer:** bacterium or fungus that feeds on detritus by secreting enzymes that break it down into simpler molecules (15.3)

**deer:** any member of a ruminant family characterized by the presence of antlers (16.4)

**deficiency disease:** nutritional disease caused by a diet low in vitamins, minerals, or certain other nutrients (12.1)

**degenerative disease:** disease in which organs and tissues gradually lose their ability to function properly (12.1)

**deltoid:** muscle that forms the curve of the shoulder; lifts the upper arm away from the body (7.5)

**dendrite:** short, branched extension of a nerve cell that receives nerve impulses from other neurons and conducts them toward the cell body (8.1)

**denitrifying bacteria:** bacteria in the soil that convert unused nitrates back into atmospheric nitrogen (1.3)

**dental caries:** disease in which acid attacks the teeth, penetrating the enamel and dentin; also called tooth decay and cavities (9.2)

**denticle:** toothlike scale of a shark, ray, or chimaera (19.3)

**dentin:** bonelike layer that surrounds the pulp of a tooth (9.2)

**deoxyribonucleic acid:** full name of DNA (24.1)

**depressant:** substance that slows the central nervous system (8.4)

**dermis:** layer of the skin below the epidermis (11.1)

***Descent of Man, The:*** book in which Charles Darwin applied his ideas of evolution to the origin of mankind (14.2)

**desert:** hot, dry biome where more water is lost through evaporation than is gained through rain (15.5)

**desmid:** any of a large group of unicellular freshwater green algae characterized by a beltlike constriction that divides cells into two identical semicells (5.5)

**detritivore:** organism that feeds on detritus but cannot actually convert the detritus into a soil component (15.3)

**diabetes mellitus:** disease in which sugar accumulates in the bloodstream instead of being stored as glycogen, caused by either inadequate insulin production or an improper response to the insulin that is produced (11.3)

**dialysis:** process of separating molecules of different sizes using a membrane, used by artificial kidneys (11.2)

**diaphragm:** thick sheet of muscle that separates the thoracic cavity from the abdominal cavity (10.4)

**diaphysis:** shaft of a long bone (7.3)

**diarrhea:** condition that occurs when insufficient water is removed from feces, resulting in soft and watery feces and frequent defecation (9.4)

**diastole:** relaxing and filling phase of heart action (10.2)

**diatom:** any of a group of yellow algae characterized by rigid cell walls composed of silica (5.5)

**diatomaceous earth:** substance formed from packed layers of diatom shells; also called diatomite (5.5)

**dicot:** plant with two cotyledons per seed; typically characterized by flower petals that are arranged in groups of four or five, leaves that are usually broad and flat with branching venation, and a taproot system (1.3)

***Didinium:*** barrel-shaped ciliate that commonly feeds on paramecia (21.6)

**dietary fiber:** group of complex carbohydrates that are not digested or incorporated into the body (9.1)

**differentiation:** process in which unspecialized cells become specialized (23.1)

**diffusion:** process of mixing molecules of one substance through another by random molecular motion (4.5)

**digestion:** process that changes food substances with large, complex chemical molecules into substances with smaller, less complex molecules (9.2)

**digestive enzyme:** any of the enzymes involved in digestion (9.2)

**digestive gland:** secreting body that produces digestive enzymes and digestive juices (9.2)

**digestive system:** body system that breaks down food into a form that can be absorbed by the bloodstream (6.1)

**dihybrid:** organism heterozygous for two traits (23.2)

**dilation and evacuation:** abortion technique in which forceps inserted into the mother's womb are used to pull the child out of the womb; abbreviated *D&E;* also called sharp curettage (11.4)

**dinoflagellate:** unicellular aquatic alga that has two unlike flagella used for locomotion (5.5)

**dinosaur:** any of a large group of extinct land-dwelling creatures that resemble reptiles in some ways (18.3)

**dioecious:** having staminate and pistillate flowers borne by separate plants (3.1)

**diploid:** containing the complete number of chromosomes for a particular organism (23.1)

**Diptera:** order of insects that undergo complete metamorphosis and have two wings and either piercing-sucking or sponging mouthparts (20.2)

**disease:** any condition that interferes with the normal functioning of the body (12.1)

**disk flower:** one of the flowers in the center of a composite flower head (1.3)

**dislocation:** injury in which a joint is overstressed to the extent that a bone pops out of alignment (7.4)

**diurnal:** active during the day (16.2)

**DNA:** chemical, found in the nucleus, that stores the genetic information needed for cellular function; short for *deoxyribonucleic acid* (22.2, 24.1)

**doctrine of ideas:** system of thought, developed by Plato, based on the idea that the physical world is not genuinely real (13.1)

**doctrine of intellect:** system of thought, developed by Aristotle, based on the idea that intellectual speculation is the highest form of reality (13.1)

**dominant trait:** trait that appears in the phenotype even if an allele for a different trait is present (23.2)

**dominion:** man's God-given authority over nature (15.6)

**dormancy:** period of inactivity, which helps prevent seeds from sprouting when conditions are unfavorable for the growth of young plants (3.4)

**dorsal fin:** fin that projects upward from a fish's back (19.1)

**double helix:** double-spiral structure of DNA (24.1)

**down feather:** soft, fluffy feather close to a bird's body that provides insulation (17.2)

**drone:** male social insect whose purpose is to mate with the queen (20.2)

**droplet infection:** infection carried by droplets, such as those discharged during a sneeze or cough (12.2)

**drupe:** simple fruit that has an outer fleshy layer and an inner woody layer (3.3)

**dry fruit:** fruit that has a dry ovary (3.3)

**duodenum:** first and shortest section of the small intestine (9.3)

**dysentery:** painful infection of the colon that causes severe diarrhea and cramps accompanied by blood in the feces (9.4)

**dyspepsia:** gastric indigestion (9.4)

## E

**eardrum:** membrane stretched across the ear canal that vibrates in response to sound waves (8.5)

**echinoderm:** spiny-skinned invertebrate with a water-vascular system (21.2)

**echolocation:** system, found in bats and dolphins, of producing sounds and listening for the echoes to detect objects (16.3)

**ecological succession:** replacement of early pioneer species in a disturbed ecosystem by later species in an orderly progression until the climax community is established (15.6)

**ecology:** study of how living things interact with one another and with their physical environment (15.1)

**ecosystem:** small area within a biome that has its own specific environmental conditions and living organisms (15.1)

**edema:** condition in which the body tissues absorb extra fluids and cannot maintain proper water balance (9.1)

**eft:** term referring to the terrestrial juvenile forms of certain salamanders, whose adult forms are called newts (18.5)

**egg cell:** female reproductive cell (3.1)

**elapid:** any of a family of venomous snakes with immovable fangs in front of the upper jaw (18.2)

**elastin:** protein fiber in the dermis that, with collagen, gives skin its ability to stretch and its elasticity (11.1)

**electrocardiogram:** medical procedure in which the tiny electrical currents in the heart are detected by electrodes taped to the patient's skin; abbreviated *EKG* or *ECG* (10.2)

**elephant:** large, trunked mammal of order Proboscidea (16.2)

**ellipsoid joint:** joint that consists of a bone with a convex surface that fits into a concave portion of another bone and that allows movement in two planes but does not permit rotation; also called condyloid joint (7.4)

**embryo:** living part within a seed that develops into a new plant (3.4)

**embryo:** scientific term for an unborn child during the first eight weeks after conception, when the basic body structure and major organs and systems are forming (11.4)

**embryonic recapitulation:** discredited hypothesis stating that an unborn baby, while developing in its mother's womb, goes through stages of development that "replay" man's supposed evolution from a simple invertebrate to his present form (14.3)

**emphysema:** progressive deterioration of the alveoli caused by malfunctioning enzymes (10.4)

**emulsification:** process of breaking large droplets into smaller droplets (9.3)

**enamel:** mineral substance that covers the crown of a tooth (9.2)

**endangered:** in danger of extinction (16.4)

**endocardium:** inner layer of the heart, which provides a smooth coating for the heart's interior (10.2)

**endocrine gland:** any of the glands within the endocrine system, which secrete hormones into the bloodstream (6.1, 11.3)

**endocrine system:** body system containing hormone-producing glands (6.1, 11.3)

**endocytosis:** process by which a cell takes in large substances (22.3)

**endoplasmic reticulum:** network of interconnected sacs and tubules that is connected to the nucleus and extends through much of the cell; abbreviated *ER* (22.2)

**endoskeleton:** internal skeleton, as found in vertebrates (16.1)

**endosperm:** part of a seed that provides nutrition to the growing embryo, such as the starchy part of a wheat kernel (3.4)

**endospore:** thick, protective internal capsule that a bacterium forms to protect its nucleoid and other key machinery during unfavorable conditions (12.2)

**energy pyramid:** diagram that represents the energy transferred from one trophic level to the next as a pyramid (15.3)

**entomologist:** scientist who studies insects (20.1)

**enzyme:** large protein molecule that is a catalyst to carry out chemical reactions (9.2, 22.2)

**epicardium:** layer of slippery tissue covering the outside of the heart to allow the heart to move easily within the pericardial sac (10.2)

**epidemic:** condition in which a disease spreads rapidly and extensively among a population (12.2)

**epidermal tissue:** plant structural tissue that protects and covers leaves, roots, stems, and other exposed areas and prevents excessive water loss and injury by outside agents (2.2)

**epidermis:** outer layer that provides covering and protection to plant structures (2.2)

**epidermis:** outermost layer of the skin (11.1)

**epididymis:** small organ, attached to the testis, in which sperm mature and are stored; plural epididymides (11.4)

**epiglottis:** small flap of cartilage in the pharynx that closes the top of the trachea during swallowing (9.2)

**epilepsy:** disorder in which neurons malfunction and begin to fire together over and over, disrupting ordinary cerebral processing and causing unconsciousness (8.4)

**epinephrine:** hormone, produced by the adrenal glands, that prepares the body to respond to emergency situations during times of excitement or alarm (11.3)

**epiphysis:** bulged end of a long bone (7.3)

**epithelial tissue:** any of the tissues that line body parts, cover and protect organs, regulate temperature, and regulate what enters and exits organs (6.2)

**equid:** any member of the equine family, which includes horses and other similar perissodactyls (16.4)

**ER:** abbreviation for *endoplasmic reticulum* (22.2)

**ER lumen:** internal cavity of the endoplasmic reticulum (22.2)

**erector spinae:** muscle that extends along the spinal column from the upper back to the pelvis; straightens the back (7.5)

**esophagus:** foot-long muscular tube connecting the pharynx to the stomach (9.2)

**estivation:** state of summer dormancy or limited activity (18.3)

**estrogen:** any of the group of chemicals that are the primary female reproductive hormones (11.4)

**estuary:** marine ecosystem that occurs where nutrient-rich fresh water from rivers and streams meets and mixes with seawater, producing an environment with intermediate salinity (15.5)

**ethylene:** hydrocarbon that functions as a plant hormone to signal leaf abscission and to trigger green fruits to ripen (4.4)

**Eubacteria:** bacteria kingdom; consists of prokaryotes with cell walls made of peptidoglycan (5.1)

**eugenics:** "science" that attempts to improve the human species by controlling hereditary factors and selectively breeding humans to produce a "master race" (14.1, 23.3)

**euglena:** unusual protozoan that has several plantlike characteristics (21.6)

**euglenoid movement:** unique method of movement in which a euglena draws its posterior end forward and then stretches the anterior end forward (21.6)

**eukaryote:** organism that has a membrane-bound nucleus and membrane-enclosed organelles (5.1)

**eustachian tube:** tube connecting the throat to the middle ear (8.5)

**evaporation:** process in which liquid water enters the atmosphere as water vapor (15.4)

**evergreen:** describes a tree that retains its leaves through the winter (1.4)

**evolution:** teaching that all living things developed from a common ancestor (13.1)

**excretion:** removal of waste substances from the body (11.2)

**excretory system:** body system that excretes waste materials from the body (6.1)

**excurrent siphon:** tube that expels water from a bivalve's mantle cavity (21.1)

**exhalation:** process in breathing in which air rushes from the lungs due to increased pressure (10.4)

**exocrine gland:** gland that secretes its products to the outside of the body (11.3)

**exocytosis:** process by which waste products are eliminated from the cytoplasm (22.3)

**exon:** coding sequence of a gene that is present in mature mRNA (24.2)

**exoskeleton:** external skeleton (20.1)

**experimentation:** special kind of observation that involves testing theoretical speculations in a controlled and systematic way (14.4)

**external oblique:** muscle that connects the ribs to the pelvis but more to the side than does the rectus abdominis (7.5)

**external respiration:** process of breathing (10.4)

**extinct:** describes a species no longer found alive on the earth (16.4)

**extracellular fluid:** fluid that surrounds the body's cells (9.1)

**extrinsic muscles:** set of six muscles that move the eye (8.5)

**eye:** node consisting of buds and small leaves, found on a tuber (4.4)

**eyelid:** protective skin structure that blinks to keep foreign objects from entering the eye and cleans and moistens the eyeball (8.5)

**eyespot:** light-sensitive spot found on invertebrates such as starfish and planarians (21.2)

## F

**$F_1$:** first generation of offspring in a genetics experiment; short for *first filial* (23.2)

**$F_2$:** second generation of offspring in a genetics experiment; short for *second filial* (23.2)

***Fabrica:*** book by Andreas Vesalius that laid the foundation for the modern study of human anatomy (13.2)

**facial skeleton:** bones that are the framework of the face and jaw (7.1)

**fallopian tube:** tube connecting an ovary to the uterus; also called oviduct (11.4)

**family:** level of classification between order and genus (5.1)

**fang:** hollow or grooved venom-injecting tooth (18.2)

**farsightedness:** the ability to see distant objects better than near objects due to the eye being too short or the lens being too flat; also called hyperopia (8.5)

**fascia:** tough, translucent sheath encasing a muscle; also called epimysium (7.6)

**fat:** molecule consisting of three fatty acid molecules attached to a single glycerol molecule; also called triglyceride (9.1)

**fat-soluble vitamin:** any of the vitamins that are insoluble in water and can accumulate in the body if ingested in excess (9.1)

**fatty acid:** simplest type of lipid molecule, made of carbon and oxygen atoms attached to a long chain of carbon and hydrogen atoms (9.1)

**feces:** undigested waste materials from the digestive process (9.3)

**feedback mechanism:** body mechanism that helps maintain homeostasis by detecting changes in the body and directing the body to respond appropriately (6.1)

**femur:** thigh bone (7.2)

**fermentation:** process used by microorganisms to break down cellulose (16.4)

**fermentation:** process by which cells produce ATP when the oxygen required for cellular respiration is not available; also called anaerobic respiration (22.3)

**fern:** nonflowering vascular plant with spore-bearing leaves and horizontal underground stems (5.3)

**fertilization:** process in which a sperm cell fuses with an egg cell to form a new organism (3.2)

**fetus:** scientific term for an unborn child from the ninth week of development to birth, whose basic body structure and major organs are formed (11.4)

**fever:** elevated body temperature (12.3)

**fibrinogen:** blood protein that helps in blood clotting (10.1)

**fibrous root:** type of root system in which the primary root remains small and many slender secondary roots grow from it in all directions (1.2, 4.5)

**fibula:** thin bone of the lower leg (7.2)

**filament:** slender, elongated part of a stamen (3.1)

**filament:** hairlike structure formed by green algae that join end to end, forming a long chain of cells (5.5)

**filaria worm:** parasitic roundworm that blocks lymph vessels, causing tremendous swelling of the extremities (21.5)

**fin:** structure of a fish's trunk or tail that helps provide control and propulsion while swimming (19.1)

**fish:** cold-blooded, water-dwelling vertebrate with gills and scales (19.1)

**flagellum:** whiplike tail that extends from a cell and is used for propulsion; plural flagella (5.5, 21.6, 22.3)

**flat bone:** bone that has a flattened shape (7.3)

**flatworm:** any of a group of worms with flattened bodies; also called platyhelminth (21.5)

**fleshy fruit:** fruit that has a fleshy ovary (3.3)

**fleshy root:** taproot that enlarges to store large quantities of food (4.5)

**flight feather:** contour feather on a bird's wings that helps provide the necessary wing shape for flight (17.2)

**flightless bird:** any bird that has wings or winglike structures but cannot fly (17.1)

**flock:** gathering of birds in a large group (17.3)

**flower:** reproductive structure of an angiosperm (3.1)

**fluke:** parasitic flatworm that spends part of its life as a parasite within a snail (21.5)

**flying lemur:** gliding mammal of order Dermoptera that resembles a flying squirrel but lacks the continually growing incisors characteristic of rodents (16.3)

**flyway:** route followed year after year by migrating birds (17.3)

**fontanel:** tough connective-tissue membrane connecting the bones of an infant's cranium (7.1)

**food chain:** one-way series of steps showing the transfer of food and nutrients from one trophic level to another (15.3)

**food poisoning:** gastrointestinal disease caused by consumption of food contaminated with bacteria or bacterial toxins (9.4)

**food web:** model that combines food chains to show all possible feeding relationships at each trophic level (15.3)

**foot:** muscular organ used by a mollusk for locomotion (21.1)

**forage grass:** any of the types of grass used to feed livestock (1.3)

**foraminiferan:** sarcodine that produces a shell made of calcium carbonate (21.6)

**fossil:** remains or impression of a plant, animal, or human preserved in sedimentary rock (14.2)

**fovea:** small depression on the retina where the light-sensitive cells are most densely packed (8.5)

**fracture:** crack or break in a bone (7.3)

**freely movable joint:** type of joint that allows a wide range of motion; also called synovial joint (7.4)

**French Academy of Sciences:** French scientific society founded in 1666 (13.2)

**freshwater ecosystem:** aquatic ecosystem characterized by flowing or standing water with low salinity (15.5)

**frond:** fern leaf (5.3)

**frontal lobe:** lobe of the cerebrum corresponding to the frontal bone; associated with personality, judgment, and self-control; includes motor area (8.3)

**fruit:** ripened ovary of a flower (3.2)

**fruiting body:** umbrella-shaped reproductive structure formed by mushrooms; similar reproductive structure formed by another club fungus (5.6)

**Fuchs, Leonhard:** 16th-century German naturalist who wrote *The Natural History of Plants* (13.2)

**Fungi:** fungus kingdom; consists of multicellular eukaryotes that have cell walls and do not conduct photosynthesis (5.1)

**fungus:** nonvascular plantlike organism that lacks chlorophyll and is heterotrophic; plural fungi (5.6)

## G

**Gaia hypothesis:** quasi-religious idea that the earth and all the living things that inhabit it constitute a single living thing (14.4)

**Galen:** 2nd-century Greek physician who was considered the absolute authority in human anatomy for over a thousand years (13.1)

**gallbladder:** saclike organ that stores bile (9.3)

**game bird:** bird that is hunted for food and sport (17.1)

**gamete:** reproductive cell (sperm cell or egg cell) (3.2, 11.4)

**gametophyte:** sexually reproducing generation in alternation of generations (5.3)

**ganglion:** mass of neuron cell bodies; plural ganglia (8.1)

**gastric juice:** digestive juice formed by the secretions of the stomach (9.3)

**gastric mill:** portion of a crayfish's stomach that serves as a gizzard to grind food (20.4)

**gastrocnemius:** muscle that forms the bulge on the upper calf; extends the foot downward (7.5)

**gastropod:** mollusk that has a head and glides over a ribbon of slime produced by glands in its foot (21.1)

**gene:** unit of genetic information, consisting of the portion of a DNA molecule that controls a particular trait (23.1, 24.1)

**gene therapy:** process of completely or partially curing a genetic disease by replacing a defective gene with a normal one (24.2)

**genetic code:** table with all the possible combinations of nucleotides in a codon and the amino acids they generate (24.2)

**genetic disease:** disease that results from inherited characteristics (12.1)

**genetic engineering:** using knowledge of genes and DNA to make genetic changes in organisms (24.2)

**genetics:** scientific study of heredity (14.3, 23.2)

**genotype:** actual combination of alleles inherited for a particular trait (23.2)

**genus:** level of classification between family and species; plural genera (5.1)

**geologic column:** simple-to-complex sequence of fossils used by evolutionists to present the fossil record as a straightforward evolutionary progression; also called geologic time chart (14.2)

**geotropism:** growth of a plant in response to gravity (4.4)

**germ:** embryo of a wheat kernel (3.3)

**germ cell:** cell that undergoes meiosis to form egg or sperm (23.1)

**germ concept of disease:** idea that every infectious disease is caused by a specific germ (12.2)

**germinal spot:** location in the yolk where a bird embryo develops (17.3)

**germination:** process in which a seed sprouts (3.4)

**Gesner, Konrad:** 16th-century Swiss naturalist who wrote *Historae Animalium* and *Opera Botanica* (13.2)

**gestation:** in humans, the 38- to 40-week period from conception to birth; corresponding period in a placental mammal (11.4, 16.1)

**gibberellin:** any of a group of plant growth hormones that help regulate normal plant growth and trigger germination of dormant seeds (4.4)

**gill:** basidia-containing "spoke" on the underside of a mushroom cap (5.6)

**gill:** special organ that allows oxygen and carbon dioxide to pass between the water and the blood of a fish or other aquatic animal (19.2)

**gill arch:** curved arch of cartilage that supports a fish's gill filaments (19.2)

**gill chamber:** chamber beneath the opercula that contains a fish's gills (19.2)

**gill filament:** long, narrow portion of a fish's gill in which oxygen and carbon dioxide are exchanged (19.2)

**gill raker:** vanelike projection from the front of a fish's gill arch that helps strain food particles and other debris out of the water (19.2)

**gingivitis:** periodontal disease in which the gums become tender and inflamed due to accumulation of plaque and tartar (9.2)

**ginkgo:** unique oriental gymnosperm with fan-shaped leaves; also called maidenhair tree (5.2)

**gizzard:** muscular rear portion of a bird's stomach that has horny plates to grind food (17.2)

**glaucoma:** condition in which the pressure of the fluid inside the eye becomes much higher than normal, causing permanent damage to the retina (8.5)

**glial cell:** cell that helps support and insulate nerve tissue (8.1)

**gliding joint:** joint in which one bone merely slides across the surface of the other (7.4)

**globulin:** any of a broad group of blood proteins that help transport fat throughout the body and fight infections (10.1)

**glomerulus:** sievelike structure in a nephron that allows water and small molecules to escape the blood vessel (11.2)

**glucagon:** pancreatic hormone that signals the liver to convert glycogen to glucose and release the glucose into the bloodstream (9.3, 11.3)

**glucose:** simple sugar produced by photosynthesis ($C_6H_{12}O_6$) (2.3)

**gluteus maximus:** muscle that pulls the upper leg downward (7.5)

**Golgi apparatus:** organelle that serves as a shipping center of a cell (22.2)

**gonad:** reproductive organ of a human or animal that produces gametes (5.6, 11.4)

**gonadotropin:** any of a group of pituitary hormones that affect the development of the reproductive organs and influence their secretions (11.3)

**gout:** condition in which uric acid accumulates in the joints instead of being filtered properly by the kidneys (11.2)

**grafting:** process of transplanting living tissue from one plant to another (4.4)

**grain:** fruit of a grass plant (1.3, 3.3)

**grain:** light and dark streaks in wood formed by differences in the size of springwood and summerwood vessels (4.2)

**graminoid:** general term for a member of the grass family (1.3)

**grassland:** biome dominated by grasses, with occasional trees (15.5)

**gray matter:** portion of the brain and spinal cord that consists largely of neuron cell bodies (8.1)

**Greeks:** ancient civilization whose people developed many ideas about nature but were blinded by superstition (13.1)

**green gland:** crayfish excretory organ that functions like a kidney (20.4)

**guard cell:** either of a pair of crescent-shaped cells that expand and contract to open and close a stoma (2.2)

**gulfweed:** common name for the brown algae of the genus *Sargassum* (5.5)

**gymnosperm:** plant that does not produce flowers but does produce seeds (5.2)

## H

**habitat:** region where a particular organism normally lives; the organism's environment or home (1.1, 15.1)

**Haeckel, Ernst:** German evolutionist who developed the theory of embryonic recapitulation (14.3)

**hagfish:** jawless cartilaginous fish with a slitlike mouth and no sucking organ (19.3)

**hair follicle:** vertical tube in the dermis designed to manufacture hair (11.1)

**hamstrings:** group of three muscles in the back of the thigh that bend the leg at the knee (7.5)

**haploid:** containing only half the normal number of chromosomes for a particular organism (23.1)

**hardwood:** another term for a broadleaf tree (1.4)

**Harvey, William:** English physiologist known for his study of blood circulation (13.2)

**head:** flower cluster of a member of the composite family, which superficially resembles a single flower (1.3)

**head:** front portion of an insect's body (20.1)

**heart:** pump that propels blood through the circulatory system (10.2)

**heart attack:** type of heart failure caused by a blockage of blood flow to the heart muscle (10.2)

**heart murmur:** abnormal rushing or blowing heart sound produced by valve damage (10.2)

**heartbeat:** "lub-dup" sounds produced by the heart during the cardiac cycle (10.2)

**heartwood:** inner portion of a woody stem, made of older xylem tubes that have become clogged and been sealed off (4.2)

**Hebrews:** people who had a distinct advantage over other ancient peoples in the study of the natural world because they possessed God's truth as preserved in the Scriptures (13.1)

**Hemiptera:** order of insects that undergo incomplete metamorphosis and have piercing-sucking mouthparts and membranous wings with a half-wing design (20.2)

**hemisphere:** right or left half of the cerebrum or cerebellum (8.3)

**hemoglobin:** iron-containing blood protein, found in red blood cells, that carries oxygen in the blood (10.1)

**hemophilia:** recessive sex-linked disease in which clots cannot form due to a hereditary defect in any of several blood-clotting proteins (10.1, 23.3)

**hemotoxin:** blood poison (18.2)

**herb:** herbaceous plant (1.2)

**herbaceous:** nonwoody (1.2)

**herbivore:** consumer that feeds on plants (15.3)

**herbivory:** relationship in which grazing animals feed on plants by cropping portions of the plant without killing it (15.4)

**heredity:** transmission of characteristics from parent to offspring through genes (23.1)

**herpetologist:** zoologist who studies reptiles and amphibians (18.2)

**heteroecious:** refers to a parasite that must have two hosts to complete its life cycle (5.6)

**heterotroph:** organism that cannot produce its own food and must obtain energy from organic sources; also called a consumer (2.3, 15.3)

**heterozygous:** possessing two different alleles for a particular gene (23.2)

**hibernation:** form of dormancy in which a warm-blooded animal's body temperature falls to near the freezing point and respiration and heart rate slow down (16.1)

**hinge joint:** joint that allows a bone to move back and forth in a single plane (7.4)

**histamine:** chemical messenger, produced by white blood cells, that causes blood vessels to become more permeable, causing inflammation (12.3)

**HIV:** virus that causes AIDS; short for *human immunodeficiency virus* (12.3)

**holdfast:** special type of anchoring cell found in certain types of algae (5.5)

**homeostasis:** condition of having and maintaining a stable internal environment, as in an individual cell or in the body as a whole (6.1, 11.1, 22.3)

**homeothermic:** maintaining the same internal body temperature regardless of environment (16.1)

***Homo erectus:*** fossil species that is considered a human ancestor by evolutionists but most likely represents an extinct race of true humans (14.2)

***Homo habilis:*** group of "human ancestors" that includes small members that are extinct apes and large members that are fully human (14.2)

**homology:** similarity between organisms that have similar function and internal structure (14.3)

**Homoptera:** order of insects that undergo incomplete metamorphosis, have piercing-sucking mouthparts, and hold their wings over their bodies in an inverted V (20.2)

**homozygous:** possessing identical alleles for a particular gene (23.2)

**hoof:** enlarged and thickened toenail found on some mammals (16.4)

**Hooke, Robert:** 17th-century naturalist who discovered cells in cork and was the first scientist to devote an entire book exclusively to microscopic observation (13.2, 22.1)

**hookworm:** parasitic roundworm that lives in the host's intestine and sucks the host's blood (21.5)

**hormonal disease:** disease that occurs because of malfunctions of the endocrine system (12.1)

**hormone:** chemical produced by humans, plants, or animals to control or stimulate specific processes and regulate body activities; produced by the endocrine glands in humans and animals (3.2, 6.1, 11.3)

**horn:** usually permanent structure, consisting of a hollow sheath of keratinized epidermal cells surrounding a bony core, found on the heads of some ruminants (16.4)

**horse series:** hypothetical arrangement of horses and similar fossils that supposedly demonstrates how the horse evolved from a small, tapirlike creature (14.2)

**horsetail:** spore-producing vascular plant with unique hollow, jointed stems and very small leaves (5.3)

**horticulturist:** agricultural technician who specializes in growing flowers, fruits, vegetables, and shrubs (3.1)

**Huguenots:** group of French Protestants who supported the French Academy of Sciences (13.2)

**human anatomy and physiology:** study of the physical structure and function of the human body (1.1)

**humerus:** upper-arm bone (7.2)

**husk:** outer layer of a wheat kernel; also called chaff (3.3)

**hybrid:** organism resulting from a cross between two different species of the same kind of plant or animal (3.2)

**hybrid:** organism with two different alleles for a particular trait (23.2)

**hydra:** small freshwater coelenterate of the polyp form (21.3)

**hydrochloric acid:** component of gastric juice that softens foods, kills bacteria, reacts with certain minerals to make them soluble, and provides the acid medium required for the enzyme pepsin (9.3)

**hydrogenated oil:** vegetable oil with hydrogen added to change it from liquid to solid or semisolid (9.1)

**hydrologic cycle:** biogeochemical cycle in which water moves from the soil and from the water surfaces of the earth, through the atmosphere, and back to the earth again (15.4)

**hydroxyapatite:** rigid mineral that gives bone its hardness $[Ca_{10}(PO_4)_6(OH)_2]$ (7.3)

**Hymenoptera:** order of social insects that undergo complete metamorphosis and are either wingless or have membranous wings (20.2)

**hyoid bone:** U-shaped bone that is the foundation of many of the tongue muscles and of certain muscles involved in swallowing (7.1)

**hypertension:** condition in which resting blood pressure exceeds either 140 mmHg systolic or 90 mmHg diastolic; also called high blood pressure (10.3)

**hyperthyroidism:** condition caused by an overactive thyroid gland (11.3)

**hypertrophy:** enlargement of muscles through use (7.7)

**hyphae:** threadlike filaments that form the body of a fungus (5.6)

**hypothalamus:** portion of the limbic system that controls the autonomic nervous system and endocrine system and is responsible for the physical effect of emotions (8.3)

**hypothesis:** tentative answer to a scientific problem; plural hypotheses (13.3)

**hypothyroidism:** condition caused by deficient thyroid function in adulthood (11.3)

**hyrax:** rabbit-sized herbivorous mammal of order Hyracoidae; also called coney (16.3)

## I

**ichthyologist:** scientist who studies fish (19.1)

**ileum:** third and longest section of the small intestine (9.3)

**immune deficiency:** condition in which a person's immune system breaks down, leaving him susceptible to infections to which he would normally be immune (12.3)

**immune system:** body system that defends against bacteria, viruses, and other invaders (6.1)

**immune therapy:** treatment and prevention of disease using vaccines and antiserums (12.3)

**immunity:** condition of resistance to a pathogen in which the pathogen will be destroyed before symptoms of infection develop (12.3)

**immunological disease:** any disease caused by malfunctions of the immune system (12.1)

**impacted:** describes a tooth (usually a wisdom tooth) that has erupted too close to the adjacent tooth and become stuck (9.2)

**implantation:** process in which a blastocyst attaches to the uterine wall (11.4)

**inbreeding:** breeding within a small, isolated group (14.3)

**incisor:** any of the four front cutting teeth on each jaw (9.2)

**incomplete dominance:** form of inheritance in which the phenotype of a heterozygous individual is a blend of the homozygous phenotypes (23.2)

**incomplete flower:** flower that lacks one or more of the four basic flower parts (3.1)

**incomplete metamorphosis:** process in which an insect metamorphoses from a wingless nymph into an adult (20.1)

**incubation:** process of keeping eggs warm so the young birds will develop properly (17.3)

**incubation period:** period of pathogen development in which a person has an infection but does not show signs or symptoms of disease (12.2)

**incurrent siphon:** tube that takes water into a bivalve's mantle cavity (21.1)

**incus:** second bone of the middle ear; also called anvil (8.5)

**index fossil:** fossil considered by evolutionists to be characteristic of a certain time period and therefore used to identify and date rock layers (14.2)

**inductive reasoning:** reasoning from specific events in nature to their general causes (13.3)

**infection:** state of having pathogens present and growing in the body (12.2)

**infectious:** describes a disease caused by invading pathogens (11.1)

**inferior vena cava:** vein that returns blood from the lower part of the body to the heart (10.2)

**Inflorescence:** flower cluster (3.1)

**influenza:** viral respiratory disease that causes fever, aching muscles, headache, and chest pain (10.4)

**inhalation:** process in breathing in which air rushes into the lungs due to decreased pressure (10.4)

**innate immunity:** immunity that results from factors other than the presence of antibodies and memory cells (12.3)

**inner ear:** portion of the ear that is the actual organ of hearing (8.5)

**insect:** arthropod with three distinct body regions, simple and compound eyes, tracheae, and three pairs of legs (20.1)

**insectivores:** common name for insect-eating mammals of order Insectivora (16.3)

**insectivorous:** insect eating (16.3)

**insectivorous plant:** plant that uses special leaves to trap and digest small insects (2.4)

**instinct:** unlearned knowledge that an animal possesses from birth (17.3)

**insulin:** pancreatic hormone that signals the liver and muscles to remove glucose from the blood and store it as glycogen (9.3, 11.3)

**integumentary system:** body system that provides communication with the world and protects the body from its surroundings; consists of the skin (6.1, 11.1)

**intercostal muscles:** group of muscles located between the ribs; lift the rib cage to aid breathing (7.5, 10.4)

**interferon:** chemical messenger produced by virus-infected cells to limit the spread of the virus (12.3)

**internal respiration:** exchange of oxygen and carbon dioxide between blood cells and tissue cells (10.4)

**interneuron:** neuron that relays signals between neurons or groups of neurons and is involved in information processing (8.1)

**internode:** portion of a stem between nodes (4.1)

**interphase:** stage of the cell cycle between cell divisions (22.3)

**interstitial fluid:** fluid that fills the area around and between tissues (6.2)

**intertidal zone:** area between the highest and lowest tide along a coast; also called littoral zone (15.5)

**intervertebral disk:** disk of flexible cartilage between vertebrae (7.1)

**intestinal glands:** digestive glands in the small intestine that produce intestinal juice (9.3)

**intracellular fluid:** fluid inside the body's cells (9.1)

**intron:** noncoding DNA sequence within a gene that is spliced out of the pre-mRNA (24.2)

**invertebrate:** animal without a backbone (16.1, 20.1)

**involuntary muscles:** cardiac and smooth muscles (7.5)

**iris:** front-most muscular portion of the uvea, which forms the colored portion of the eye and controls the size of the pupil (8.5)

**irregular bone:** bone that cannot be easily classified as a long bone, short bone, or flat bone (7.3)

**islets of Langerhans:** clumps of endocrine glands in the pancreas that produce insulin and glucagon (11.3)

## J

**Jacobson's organ:** two tiny cavities in the roof of a snake's mouth that contain odor-sensitive nerve endings (18.2)

**Jansenists:** group of Calvinist French Catholics who emphasized salvation through God's love and grace (13.2)

**Java man:** "human ancestor" fossil discovered in 1893 on the Indonesian island of Java (14.2)

**jejunum:** middle section of the small intestine (9.3)

**jellyfish:** free-swimming coelenterate that has a medusa body form and undergoes alternation of generations (21.3)

**joint:** place where bones join (7.4)

## K

**kelp:** any of the group containing the largest brown algae (5.5)

**keratin:** tough, waterproof protein fiber that strengthens and hardens skin cells (11.1)

**kidney:** excretory organ that cleanses the blood of wastes to produce urine (11.2)

**kidney stone:** crystallized particle of calcium and other substances that forms in the kidneys due to abnormal function (11.2)

**kilocalorie:** unit of energy sometimes used in physics, equal to one food Calorie (9.1)

**kind:** group of similar organisms that are all descended from a single group of originally created organisms (5.1)

**kingdom:** one of the main groups into which all organisms are divided in the Linnaean system of classification (5.1)

**Koch, Robert:** German physician who developed techniques used to study bacteria safely and formulated guidelines for determining if a particular bacterium is the cause of a specific disease (12.2)

**Koch's postulates:** set of guidelines developed by Robert Koch for determining if a particular bacterium is the cause of a specific disease (12.2)

## L

**labor:** process of birth (11.4)

**lacrimal gland:** tear gland (8.5)

**lacteal:** tubelike lymph vessel that carries absorbed fats out of a villus (9.3)

**lagomorph:** any mammal of order Lagomorpha, which has teeth similar to a rodent's but has four upper incisors (16.3)

**lamprey:** jawless cartilaginous fish whose mouth is a round sucking organ with a toothed tongue (19.3)

**large intestine:** tubelike organ that prepares undigested food for expulsion from the body (9.3)

**larva:** immature wormlike eating and growing stage of an insect (20.1)

**laryngeal muscles:** muscles in the larynx that move the vocal cords (10.4)

**larynx:** special enlargement of the trachea in which speech originates; also called voice box (10.4)

**lateral bud:** bud that allows growth of a new stem from the side of the main stem; also called axillary bud (4.1)

**lateral line:** system of special nerve endings that extends over a fish's head and the sides of its body (19.2)

**lateral root:** root that originates from the taproot; also called secondary root (4.5)

**lateral undulation:** movement in which a snake bends into a series of curves while keeping its body flat on the ground (18.2)

**latissimus dorsi:** muscle that connects the humerus to the lumbar region of the spine and pulls the arms to the rear (7.5)

**law of biogenesis:** scientific law stating that living things can come only from other living things (13.3)

**law of dominance:** law stating that dominant traits appear in the offspring even if an allele for a different trait is also present (23.2)

**law of independent assortment:** law stating that the segregation of one gene does not influence the segregation of another (23.2)

**law of segregation:** law stating that alleles for each trait separate randomly during the formation of gametes (23.2)

**layering:** method of vegetative propagation that involves causing an existing plant to generate adventitious roots at a node (4.4)

**leaf hair:** one of the hairlike structures that cover the epidermis of certain plant leaves (2.2)

**leaf scar:** scar left on a stem where a leaf was previously attached (4.1)

**leaflet:** one of the small blades of a compound leaf (2.1)

**leech:** segmented worm that has a suction disk on each end of its body and does not have setae or parapodia (21.4)

**Leeuwenhoek, Anton van:** Dutch naturalist who was the first person to devote his whole life to studies with the microscope (13.2)

**legume:** plant of the pea family; a simple fruit consisting of a pod enclosing several seeds, which is characteristic of members of this family (1.3, 3.3)

**lens:** portion of the eye that focuses the rays of light (8.5)

**lenticel:** opening in the bark of a stem that allows air to enter the stem (4.1)

**Lepidoptera:** order of insects that undergo complete metamorphosis and have large, membranelike wings covered with tiny scales (20.2)

**leukemia:** form of cancer in which white blood cells are produced in uncontrolled numbers (10.1)

**lichen:** single unit formed by a fungus and an alga living in close relationship (5.6)

**ligament:** tissue that joins bones to other bones (7.1)

**limbic system:** collective term for the brain structures, clustered around the brain stem, that coordinate the activity of the parts of the brain and seem to be involved in emotions (8.3)

**limiting factor:** environmental factor outside an organism's tolerance range (15.2)

**linkage:** situation in which genes do not obey the law of independent assortment because they are located close to each other on the same chromosome (23.2)

**linkage group:** collective term for all the genes on a given chromosome (23.2)

**Linnaeus, Carolus:** Swedish naturalist who devised the modern system of biological classification (5.1)

**lipid:** any of a group of macronutrient compounds that are characterized by their insolubility in water (9.1)

**liver:** large abdominal organ that functions as the body's chemical factory; produces bile; filters blood; and stores glycogen, vitamins, minerals, and reserve blood (9.3, 10.3)

**liverwort:** bryophyte that grows along the surface of the soil and often looks like tiny leaves (5.4)

**lizard:** elongated, four-legged reptile with a tapered tail (18.1)

**lobe:** any of the main regions into which the hemispheres of the cerebrum are divided (8.3)

**long bone:** bone longer than it is wide (7.3)

**lumbar:** of or related to the lower back (7.1)

**lumbar vertebra:** any of the five vertebrae of the lower back (7.1)

**lung:** either of two large, spongy organs that are the primary organs of the respiratory system (10.4)

**Lyell, Charles:** geologist who wrote *Principles of Geology* and developed the false doctrine of uniformitarianism (14.1)

**lymph:** name for tissue fluid that has entered the lymph vessels (12.3)

**lymph duct:** either of two large ducts that empty lymph from the lymph vessels into the subclavian veins (12.3)

**lymph node:** special enlargement of lymphatic tissue that filters tissue fluid (12.3)

**lymph vessel:** vessel that drains tissue fluid from the body's tissues (12.3)

**lymphatic system:** portion of the immune system that is a drainage and sanitation system for the tissue fluid (12.3)

**lymphocyte:** white blood cell that produces antibodies (12.3)

**lysosome:** special type of vesicle that functions as a recycling center (22.2)

**lysozyme:** enzyme, found in tears, that kills bacteria by destroying their cell walls (8.5, 12.3)

## M

**macroevolution:** hypothetical process by which new kinds of creatures emerge from existing kinds over time; often simply called evolution (14.2)

**macronucleus:** large nucleus of a paramecium (21.6)

**macronutrient:** any of the group of nutrients, consisting of carbohydrates, proteins, and fats, that are needed in relatively large amounts (9.1)

**malaria:** disease caused by protozoa of the genus *Plasmodium* and transmitted by the female anopheles mosquito (12.2)

**malignant:** describes a cancerous tumor, which has cells that can leave the tumor and spread throughout the body (12.1)

**malleus:** first bone of the middle ear; also called hammer (8.5)

**mammal:** vertebrate that is warm-blooded and has hair and mammary glands (16.1)

**mammary gland:** milk-producing gland of a female mammal (16.1)

**mantle:** skinlike organ that covers a mollusk's visceral hump (21.1)

**mantle cavity:** open chamber in which a bivalve's gills are located (21.1)

**margin:** edge of a leaf blade (2.1)

**marine ecosystem:** any aquatic ecosystem associated with the ocean and having high salinity (15.5)

**marsupial:** mammal that gives birth to immature young, which are nourished within a pouch; member of order Marsupialia (16.2)

**marsupium:** pouch in which the young of a marsupial are nourished by the mother's milk (16.2)

**masseter:** muscle that, with the temporalis, connects the mandible to the cranium and helps close the jaw (7.5)

**mastication:** chewing (9.2)

**materialism:** idea that the universe consists of nothing but matter and energy and has no spiritual or supernatural aspects (13.2, 14.1)

**matrix:** material surrounding cells in a tissue (6.2)

**maturation region:** region of a root where cells organize into the vascular cylinder (4.5)

**maxilliped:** jaw foot of crayfish's head (20.4)

**mechanical dispersal:** seed dispersal in which fruits scatter their seeds by bursting open and catapulting the seeds to a new location (3.3)

**mechanoreceptor:** any of a group of sensory receptors responsible for sensations of touch and pressure and for muscle sense (8.5)

**medical abortion:** method of abortion in which the mother is given drugs that kill the developing child (11.4)

**medulla oblongata:** lowest part of the brain stem; monitors and regulates various vital body functions (8.3)

**medullary cavity:** cavity in the diaphysis of a long bone that stores red marrow (in infants and children) or yellow marrow (in adults); also called marrow cavity (7.3)

**medusa:** free-swimming, umbrella-shaped coelenterate body form with a downward-facing mouth (21.3)

**meiosis:** special type of cell division in which a diploid germ cell divides into four haploid cells (23.1)

**melanin:** dark pigment that is the main pigment responsible for skin color (11.1)

**melatonin:** pineal hormone that triggers sleepiness (11.3)

**melon:** fatty organ in a dolphin's head through which sounds produced for echolocation are sent out (16.3)

**membrane skeleton:** network of flexible reinforcement fibers found beneath a cell membrane (22.2)

**Mendel, Gregor:** 19th-century Austrian monk who founded the modern science of genetics (23.2)

**meninges:** triple layer of protective tissues surrounding the brain and spinal cord (8.1)

**meningitis:** inflammation and swelling inside the skull, usually caused by an infection (8.1)

**meristematic region:** region near the tip of a root where cell division takes place (4.5)

**meristematic tissue:** plant tissue responsible for the growth and repair of plants (2.2)

**mesentery:** special membrane that holds an anuran's pancreas in place (18.4)

**mesophyll:** middle portion of a leaf, where most of the photosynthesis takes place (2.2)

**messenger RNA:** type of RNA that carries the DNA "blueprint" for a protein from the nucleus to the ribosome; abbreviated *mRNA* (24.2)

**metabolism:** general term for the chemical processes of life (9.1)

**metacarpal:** any of the five bones that form the base of the hand and the palm (7.2)

**metamorphosis:** transformation of an immature form such as a nymph or larva into an adult; characteristic of amphibians and invertebrates such as insects (18.4, 20.1)

**metaphase:** second stage of mitosis, in which the spindle apparatus attaches to the replicated chromosomes and pulls them to the equatorial plane of the cell (22.3)

**metatarsal:** any of the five bones that form the framework of the instep and the ball of the foot (7.2)

**microbe:** microscopic living creature; also called microorganism (12.1)

**micronucleus:** small nucleus of a paramecium (21.6)

**micronutrient:** any of the group of nutrients, consisting of vitamins and minerals, that are needed in relatively small amounts (9.1)

**microtubule:** hollow, rodlike filament composed of tubulin (22.2)

**microvilli:** fine extensions of the cell membrane on epithelial cells of the small intestine (9.3)

**midbrain:** region of the brain stem above the pons; helps coordinate eye movements, adjust pupil size, and operate the lens muscles (8.3)

**middle ear:** chamber containing the malleus, incus, and stapes (8.5)

**midrib:** large vein running down the center of a leaf blade; major vein of a leaf with pinnate venation (2.1, 2.2)

**migration:** annual movement between different regions (17.3)

**mildew:** sac fungus that appears as a white, powdery substance on plants (5.6)

**millipede:** caterpillarlike arthropod with a long, rounded body; short antennae; and two pairs of legs attached to each body segment (20.3)

**milt:** sperm-containing fluid released by a male fish (19.2)

**mineral:** inorganic nutrient (9.1)

**mint family:** family of plants that has square stems and whose members are often very aromatic (1.3)

**mite:** small, parasitic arachnid in the same order as ticks (20.3)

**mitochondrion:** organelle that serves as the power plant of a cell; plural mitochondria (22.2)

**mitosis:** process in which the nucleus of a cell divides to generate two nuclei with the same number of chromosomes as the original nucleus (22.3)

**mixed nerve:** nerve that includes both sensory and motor fibers (8.1)

**molar:** any of the six heavy grinding teeth (three to a side) at the back of each jaw (9.2)

**molecular biology:** branch of science that seeks to discover how the mechanisms of living cells work (14.3)

**mollusk:** any of the soft-bodied invertebrates distinguished by a muscular foot, a visceral hump, and a mantle (21.1)

**monocot:** plant with only one cotyledon per seed; typically characterized by flower petals that are arranged in groups of three, leaves that are usually long and narrow with parallel venation, and a fibrous root system (1.3)

**monoecious:** having staminate and pistillate flowers produced by the same plant (3.1)

**moss:** nonvascular spore plant that undergoes alternation of generations (5.4)

**motile:** able to move from place to place (16.1)

**motor area:** portion of the frontal lobe that controls voluntary movement of skeletal muscles (8.3)

**motor nerve fiber:** axon that carries impulses from the brain and spinal cord to muscles and organs (8.1)

**motor neuron:** neuron that relays signals from the central nervous system to the other parts of the body (8.1)

**motor unit:** motor neuron and the group of muscle cells it controls (7.6)

**mouthparts:** components of an arthropod's mouth (20.1)

**mRNA:** abbreviation for *messenger RNA* (24.2)

**mucous membrane:** type of membrane that lines openings to the outside (6.2)

**mucus:** slimy substance that lubricates the mucous membrane of the oral cavity (9.2)

**multiple allele inheritance:** form of inheritance in which a single gene has more than two possible alleles (23.3)

**multiple fruit:** fruit that forms from several flowers (3.3)

**multiple sclerosis:** disease that occurs when the body's immune system attacks the glial cells that provide myelin sheaths for central-nervous-system axons (8.2)

**muscle fiber:** individual muscle cell of a skeletal muscle (7.5)

**muscle sense:** the brain's total awareness of body movements (7.7)

**muscle tissue:** tissue that moves the body and its parts (6.2)

**muscular system:** body system that functions together with the skeletal and nervous systems to make body movements possible (6.1, 7.1)

**mushroom:** club fungus that consists of a mass of subterranean hyphae and, at reproduction, produces an umbrella-shaped fruiting body (5.6)

**mutation:** error in an organism's genetic material caused by a small, random change in the DNA (14.3, 24.2)

**mutualism:** symbiotic relationship in which both organisms benefit (15.4)

**mycelium:** mass of hyphae forming the main body of a fungus (5.6)

**myelin:** white, specialized covering found on axons of nerve cells (8.1)

**mygalomorph:** spider whose fangs point down and move vertically (20.3)

**myocardium:** middle layer of the heart, containing the actual heart muscle tissue (10.2)

**myofibril:** long cylinder inside a muscle fiber that contains the contraction machinery (7.6)

**myomere:** W-shaped band of skeletal muscle in a fish (19.1)

## N

**nasal cavity:** cavity inside the nose that warms, moistens, and filters air (10.4)

**nasal meatus:** irregularly shaped pocket in the nasal cavity (10.4)

***Natural History:*** work by Pliny the Elder that touched on many scientific subjects (13.1)

**natural selection:** idea that the fittest and strongest of each species were more likely to survive and reproduce than weaker, poorly adapted animals (14.1)

**naturalism:** idea that nature is all that exists (14.1)

**nautilus:** cephalopod with an external shell (21.1)

**Neanderthal man:** group of fully human fossils, first discovered in Germany's Neander Valley, that were once considered by evolutionists to be subhuman, apelike brutes (14.2)

**nearsightedness:** ability to see only near objects clearly due to the lens being too thick or the eye being too long; also called myopia (8.5)

**Nebraska man:** "human ancestor" fossil that was discovered in Nebraska and later proved to be an extinct pig (14.2)

**neck:** part of a tooth at the gum line (9.2)

**nectar:** sweet-tasting, watery liquid produced by plants (3.2)

**nectar guide:** striking pattern on a flower that is visible only in the ultraviolet region of the spectrum and helps attract birds to the flower (17.2)

**nephridia:** coiled excretory tubes of an earthworm that function like tiny kidneys (21.4)

**nephron:** microscopic network of blood vessels and renal tubules that forms the working unit of the kidney (11.2)

**neritic zone:** region from the low-tide line to the edge of the continental shelf; also called coastal zone (15.5)

**nerve:** bundle of nerve fibers (axons) branching from the brain and spinal cord (8.1)

**nerve center:** group of nerve cell bodies in the brain or spinal cord (8.1)

**nerve tissue:** tissue that controls organs and coordinates their functions (6.2)

**nervous system:** body system that coordinates the activities of the body (6.1)

**Nestorians:** group of Greek Christians who carried ancient natural knowledge to Persia and helped preserve this knowledge during the Middle Ages (13.1)

**neuritis:** degeneration and inflammation of nerves (8.4)

**neurobiologist:** biologist who studies the nervous system (8.1)

**neurologist:** physician who specializes in disorders of the nervous system (8.1)

**neuromuscular junction:** point at which a neuron connects to a muscle cell (7.6)

**neuron:** nerve cell (8.1)

**neurotoxin:** nerve poison (18.2)

**neurotransmitter:** chemical released into a synapse by a neuron to cause an action in the receiving cell (e.g., making a muscle contract or a neuron fire) (7.6, 8.2)

**neutralism:** indirect relationship between organisms (15.4)

**New World monkey:** any monkey that lives in Central and South America and has widely spaced nostrils and a prehensile tail (16.3)

**newt:** term referring to the aquatic adult forms of certain salamanders, whose juvenile forms are called efts (18.5)

**niche:** function or "occupation" of a living thing (15.3)

**nictitating membrane:** transparent membrane that allows a bird to "blink" and moisten its cornea without obstructing its vision; a similar membrane in a reptile or anuran (17.2, 18.1, 18.4)

**night blindness:** condition in which the eyes do not become adjusted to darkness (8.5)

**nitrate:** any of the special compounds of nitrogen and oxygen that plants can use to make proteins (1.3)

**nitrifying bacteria:** bacteria that provide nitrates to the soil from decomposing plants and animals (1.3)

**nitrogen cycle:** biogeochemical cycle involving the movement of nitrogen from the air, into the soil, and back into the air (1.3)

**nitrogen-fixing bacteria:** bacteria that grow on the roots of legumes and convert nitrogen in the air into ammonia and nitrates (1.3)

**nocturnal:** active at night (16.2)

**node:** one of the points at which leaves grow from the stem (2.1)

**noncoding DNA:** portions of chromosomes that do not specify the amino-acid sequence of proteins (24.2)

**noncommunicable:** describes an infectious disease caused by pathogens that are not transferred from one person to another (12.2)

**noninfectious:** describes a disease that is not caused by pathogens (12.1)

**nonrenewable resource:** natural resource that regenerates slowly, if at all (15.6)

**nonseed plant:** any vascular plant that does not reproduce using seeds (5.2)

**nonvascular plant:** any nonseed plant that does not have vascular tissue (5.2)

**northern coniferous forest:** northern biome that has long, harsh winters and is dominated by conifers; also called boreal forest and taiga (15.5)

**nuclear envelope:** double-layered structure that separates the nucleus from the rest of the cell (22.2)

**nuclear pore:** large protein complex in the nuclear envelope that serves as a gate to regulate the transport of large molecules into and out of the nucleus (22.2)

**nucleolus:** distinct region within the nucleus that assembles ribosomes (22.2)

**nucleotide:** type of molecule, composed of a phosphate, a sugar, and a nitrogenous base, of which DNA and RNA are made (24.1)

**nucleus:** "control center" of a cell, which stores the DNA (2.2, 22.2)

**nudibranch:** marine slug (21.1)

**number pyramid:** diagram that represents the number of organisms at each trophic level as a pyramid (15.3)

**nut:** simple dry fruit consisting of a seed enclosed in a hard shell (3.3)

**nutritional disease:** any disease caused by improper diet (12.1)

**nymph:** immature form of an insect that resembles the adult but has different body proportions and lacks wings (20.1)

## O

**obesity:** being severely overweight due to excess body fat (9.1)

**observation:** looking very closely at the physical universe for the purpose of gathering scientific knowledge about it (14.4)

**occipital lobe:** lobe of the cerebrum corresponding to the occipital bone; associated with vision (8.3)

**oceanic zone:** region extending from edge of the continental shelf into the open ocean (15.5)

**octopus:** cephalopod with conspicuous head and eight arms (21.1)

**Odonata:** order of insects that undergo incomplete metamorphosis and have toothlike projections on their chewing mouthparts (20.2)

**oil:** fat that is liquid at room temperature (9.1)

**Old World monkey:** any monkey that lives in the Old World and has close nostrils and a nonprehensile tail (16.3)

**olfactory nerve:** nerve connecting the nose's sensory receptors to the brain (8.5)

**omnivore:** consumer that feeds on both plants and animals (15.3)

**operculum:** gill cover of a fish; plural opercula (19.1)

**opposite:** leaf pattern in which two leaves grow from each node (2.1)

**optic nerve:** nerve that connects the eye to the brain (8.5)

**optimum range:** ideal range of values for an environmental factor in which an organism can thrive (15.2)

**oral cavity:** inside of the mouth (9.2)

**oral groove:** funnel-like mouth cavity of a paramecium (21.6)

**orca:** strikingly patterned black and white toothed whale also known as a killer whale (16.3)

**order:** level of classification between class and family (5.1)

**organ:** structure within a system that has a definite form and performs a definite function or functions for the system (2.1, 6.1)

**organelle:** "little organ" within the cytoplasm of a cell (2.2, 22.2)

**organism:** individual living thing (1.1, 15.1)

***Origin of Species, The:*** book in which Charles Darwin presented his theory of evolution by natural selection (14.1)

**ornithologist:** zoologist who studies birds (17.1)

**Orthoptera:** order of insects that undergo incomplete metamorphosis and hold their wings straight back along their body when not in flight (20.2)

**osculum:** large body opening through which a sponge expels water and waste materials (21.3)

**osmosis:** one-way diffusion through a semipermeable membrane (4.5)

**ossification:** process by which cartilage is replaced with bone (7.3)

**osteoblast:** cell that moves through a Haversian canal, constructing new collagen fibers and hydroxyapatite crystals (7.3)

**osteoclast:** cell that continually moves through a bone removing old material (7.3)

**ostia:** small body openings through which water enters a sponge (21.3)

**outer ear:** portion of the ear consisting of the visible sound-collecting structure and the ear canal (8.5)

**ovary:** swollen base of a pistil; contains ovules (3.1)

**ovary:** female gonad (11.4)

**oviduct:** tube connecting the ovary of a placental mammal to the uterus, where fertilization takes place (16.1)

**oviparous:** egg laying (16.1)

**ovoviviparous:** giving birth through eggs that hatch within the mother's body (18.1)

**ovulate cone:** cone that produces seeds; also called seed cone (5.2)

**ovulation:** process in which an egg cell is released from an ovarian follicle (11.4)

**ovule:** one of a plant's future seeds, which contain the female reproductive cells (3.1)

## P

**pain receptor:** any of a group of sensory receptors that consist of a bare dendrite activated by extreme heat or cold, excessive mechanical stress, damage to body cells, etc. (8.5)

**palate:** structure that forms the roof of the mouth (9.2)

**paleontology:** study of fossils (14.2)

**Paley, William:** 18th-century British theologian who argued that the intricate mechanisms of living things exhibit evidence of design by their Creator (14.3)

**palmate venation:** leaf venation in which two or more major veins extend outward from one point (2.2)

**pancreas:** organ that produces pancreatic juice and contains the islets of Langerhans (9.3, 11.3)

**pancreatic juice:** digestive juice, produced by the pancreas, that contains digestive enzymes needed in the small intestine (9.3)

**pangolin:** ant-eating mammal of order Pholidota, having scales made from fused bundles of hair; also called scaly anteater (16.2)

**papillae:** taste-bud-containing projections on the tongue (9.2)

**parallel venation:** leaf venation in which the veins are usually parallel to each other along the length of the leaf (2.2)

**paralysis:** inability of muscles to move (8.2)

**paramecium:** ciliate that has a unique slipper shape when seen in cross section (21.6)

**parapodia:** paired fleshy lobes found on sea worms (21.4)

**parasite:** organism that lives on or in another living organism and derives its nutrition from that organism (5.6)

**parasitism:** symbiotic relationship in which one organism benefits while the other is harmed (15.4)

**parasympathetic division:** portion of the autonomic nervous system that counteracts the sympathetic division (8.1)

**parenchyma:** plant structural tissue that makes and stores food (2.2)

**parietal eye:** "third eye" of a tuatara (18.3)

**parietal lobe:** lobe of the cerebrum corresponding to the parietal bones; analyzes senses to feel temperature, pressure, pain, shape, and texture (8.3)

**Parkinson's disease:** disease characterized by tremors and stiffness of the limbs that appears to result from a lack of dopamine (8.2)

**parthenogenic:** referring to reproduction through unfertilized eggs (21.2)

**passive transport:** movement of molecules across the cell membrane from a place of higher concentration to a place of lower concentration without the expenditure of energy (22.3)

**Pasteur, Louis:** French scientist who developed the germ concept of disease, developed a vaccine for rabies, and demonstrated that not even the simplest organisms can develop from nonliving matter (12.2, 13.3)

**patella:** kneecap (7.2)

**pathogen:** any organism that causes disease (12.1)

**pearl:** sheets of nacre built up around an irritating substance lodged in a bivalve's mantle (21.1)

**peat:** compressed accumulation of various partially decayed plant materials that collect in swamps and marshes (5.4)

**peat moss:** large, branching type of swamp-dwelling moss that helps prevent soil erosion and has antiseptic qualities (5.4)

**pectoral fin:** either of a pair of fish fins located just behind the head on each side of the body (19.1)

**pectoral girdle:** collective term for the shoulder bones, which attach the arms to the axial skeleton (7.2)

**pectoralis major:** largest chest muscle; pulls arm forward (7.5)

**pedicel:** special stem on which a flower forms; flower stalk (3.1)

**pedipalp:** leglike appendage attached to the side of a spider's mouth (20.3)

**pedomorphosis:** condition in which an animal retains larval characteristics even after becoming sexually mature (18.5)

**Peking man:** "human ancestor" fossil discovered near Beijing, China that was very similar to Java man (14.2)

**pellicle:** thickened membrane of a euglena that acts as a cell wall (21.6)

**pelvic bone:** either of the two large, heavy bones that attach to the sacrum and form most of the pelvic girdle (7.2)

**pelvic fin:** either of a pair of fish fins located to the rear of the pectoral fins and lower on the trunk (19.1)

**pelvic girdle:** collective term for the bones that form the hips and attach the legs to the axial skeleton (7.2)

**pelvis:** another name for the pelvic girdle (7.2)

**peppered moth:** English moth whose coloration demonstrates variation within a kind due to natural selection (14.3)

**peptic ulcer:** ulcer caused by the digestive action of pepsin; includes both gastric and duodenal ulcers (9.4)

**perching bird:** any of the birds that have three toes pointing forward and one pointing back, allowing them to perch even while sleeping (17.1)

**perennial:** plant that lives from year to year and blooms each season (1.3)

**pericardium:** tough, white sac in which the heart is encased (10.2)

**periodontitis:** periodontal disease in which the gums detach from the teeth and eventually the bones supporting the teeth are destroyed (9.2)

**peripheral nervous system:** portion of the nervous system that consists of nerves (8.1)

**Perissodactyla:** order of odd-toed hoofed mammals (16.4)

**peristalsis:** rhythmic wave of muscular contractions that moves food through the digestive tract (9.2)

**permafrost:** permanently frozen layer of soil 2 to 3 feet below the surface, found in Arctic tundra (15.5)

**permanent wilting:** wilting that occurs when there is a drought and there is not sufficient soil water to replace that lost by transpiration; results in the loss of leaves and possibly the death of the plant (2.4)

**persistence of vision:** phenomenon in which an object's image on the retina persists for about $1/10$ second after the object is gone (8.5)

**petal:** most conspicuous part of a flower (3.1)

**petiole:** part of a leaf that attaches the blade to the stem (2.1)

**Peyer's patches:** lymph node groups in the wall of the ileum that protect that digestive area from pathogens (12.3)

**pH:** measurement of the concentration of acids and bases (22.3)

**phagocyte:** any of the white blood cells that roam through the tissues and body cavities, engulfing and disposing of any bacteria, viruses, and cellular debris they encounter (12.3)

**phalanx:** any of the bones of the fingers and toes; plural phalanges (7.2)

**pharynx:** throat (9.2)

**phenotype:** outcome or expression of genes in an organism's physical appearance (23.2)

**phloem:** plant vascular tissue that transports food manufactured in the leaves downward (2.2)

**phospholipid:** special type of lipid molecule of which the cell membrane is composed (22.2)

**phosphorus cycle:** sedimentary cycle that recycles phosphorus (15.4)

**photoperiodism:** phenomenon in which a plant requires a definite period of light and darkness before it will flower (3.1)

**photosynthesis:** process whereby a plant's chloroplasts capture the radiant energy of light and convert it into the chemical energy of food (2.3)

**phototropism:** growth response of a plant to move toward light (2.1, 4.4)

**phylum:** one of the main groups into which a kingdom is divided; also called a division in botany; plural phyla (5.1)

**Piltdown man:** "early human" fossil that was discovered near Piltdown, England, and proved to be a hoax (14.2)

**pineal gland:** endocrine gland, located between the cerebral hemispheres of the brain, that regulates the body's sleeping and waking cycles (11.3)

**pinnate venation:** leaf venation in which the veins are in a branching pattern characterized by a midrib with smaller veins extending out from it (2.2)

**pinocytosis:** form of endocytosis involving the intake of liquids (22.3)

**pioneer species:** first organisms to colonize an ecosystem after a disturbance (15.6)

**pistil:** elongated, vase-shaped structure of a flower that contains the female reproductive structures (3.1)

**pistillate:** describes a female flower, which lacks stamens and bears only pistils (3.1)

**pit viper:** viperid that has a heat-sensitive pit organ on each side of its head (18.2)

**pith:** water-storage region at the center of a young woody stem (4.2)

**pituitary dwarfism:** condition in which a person does not grow to normal height due to a somatotropin deficiency during childhood (11.3)

**pituitary gigantism:** condition in which a person grows to a great height because of excessive somatotropin production during the growing years (11.3)

**pituitary gland:** endocrine gland located just above the nasal cavity; produces hormones that regulate growth and trigger the development and function of other glands (11.3)

**pivot joint:** joint in which one bone rotates in place against another bone (7.4)

**placenta:** sac of blood vessels that connects a mother to an unborn child; similar organ in a placental mammal (11.4, 16.1)

**placental mammal:** mammal whose developing young are nourished by a placenta until birth (16.1)

**planarian:** free-living flatworm (21.5)

**plankton:** collective term for microorganisms that float near the surface of the water and provide food for larger organisms (5.5)

**Plantae:** plant kingdom; consists of multicellular eukaryotes that have cell walls and usually perform photosynthesis (green plants) (5.1)

**plaque:** sticky, colorless film of bacteria that forms on and between teeth (9.2)

**plaque:** deposit of fatty material and calcium formed on the inside of an artery (10.3)

**plasma:** straw-colored liquid part of the blood (10.1)

**plasmid:** small DNA circle that can be exchanged between bacteria (12.2)

**plasmodium:** sporozoan that causes malaria (21.6)

**plastron:** lower shell of a turtle (18.3)

**platelet:** tiny disk-shaped cell fragment that helps patch blood vessels during clotting (10.1)

**Plato:** early Greek philosopher who developed the doctrine of ideas (13.1)

**pleiotropy:** form of inheritance in which a single gene influences several seemingly unrelated traits (23.3)

**pleura:** double membrane that provides lubrication between the ribs and the surface of the lungs (10.4)

**pleurisy:** inflammation of the pleura (10.4)

**plexus:** large mass of ganglia grouped together (8.1)

**Pliny the Elder:** Roman general who wrote *Natural History* (13.1)

**plumule:** tiny shoot within a seed that will develop into the shoot system (3.4)

**pneumonia:** bacterial or viral infection of the linings in the lungs that cause the lungs to secrete large quantities of fluid (10.4)

**poikilotherm:** animal whose body temperature varies with the temperature of its surroundings (18.1)

**poliomyelitis:** disease caused by a virus that attacks motor neurons of the spinal cord, paralyzing muscles connected to the brain by these neurons (8.1)

**pollen:** plant reproductive structure, produced by the anther of the stamen, that contains the sperm cells (3.1)

**pollen tube:** tube formed by a pollen grain through which the sperm cells reach the ovule (3.2)

**pollination:** transfer of pollen from an anther to the stigma of a pistil (3.2)

**polygenic inheritance:** form of inheritance in which a single trait is controlled by multiple genes (23.3)

**polyp:** sessile, cylindrical coelenterate body form with an upward-facing mouth (21.3)

**pome:** simple fruit with an outer fleshy layer and an inner papery core (3.3)

**pons:** portion of the brain stem that links the cerebrum and cerebellum; assists the medulla oblongata in regulating breathing and helps coordinate some eye movements and facial expressions (8.3)

**population:** group of organisms of the same species living in the same ecosystem (15.1)

**pore:** opening through which sweat reaches the skin (11.1)

**portal circulation:** portion of systemic circulation in which blood moves from the digestive organs and spleen to the liver (10.3)

**precipitation:** process in which small droplets in clouds return to earth as rain, snow, or ice (15.4)

**precocial:** describes a bird that is covered with down and able to run or swim almost immediately after hatching (17.3)

**predation:** relationship in which one organism hunts, kills, and eats another (15.4)

**prehensile tail:** tail designed for grasping objects (16.3)

**presbyopia:** inability to focus on near objects as the lens becomes less elastic with age (8.5)

**primary color:** any of the three colors corresponding to those absorbed most strongly by the three different types of cone cells (8.5)

**primary consumer:** animal that eats plants directly; herbivore (15.3)

**primary germ layer:** any of the three distinct cell layers that form in the blastocyst, each of which develops into specific parts of the body (11.4)

**primary growth:** growth of a stem in length (4.1)

**primary odor:** any of the seven basic odors of which all other odors are thought to be composed (8.5)

**primary root:** root that grows directly from the plant embryo; taproot (4.5)

**primary succession:** ecological succession that starts with barren ground (15.6)

**primate:** arboreal mammal of order Primates (16.3)

**prokaryote:** organism that does not have a membrane-bound nucleus (5.1)

**prophase:** first stage of mitosis, in which the cell packages the chromosomes, forms the spindle apparatus, and dismantles the nucleolus and nuclear envelope (22.3)

**prostomium:** shovel-like upper lip of an earthworm (21.4)

**protein:** any of a group of complex organic molecules used to build and maintain living cells; made of a long chain of amino acids (9.1)

**Protista:** kingdom consisting of eukaryotes that are not green plants, fungi, or multicellular animals (5.1)

***Protococcus:*** type of spherical, unicellular green alga that often grows on the trunks of trees and on unpainted buildings and fenceposts (5.5)

**protoplasm:** gel or slime of which cells were thought to consist before modern discoveries of cell complexity (22.1)

**protozoa:** unicellular animal-like creatures in kingdom Protista; sing. protozoan (12.2, 21.6)

**pseudopod:** any of the "false feet" of an amoeba or other sarcodine (21.6)

**psoriasis:** skin condition that occurs when the immune system attacks epidermal cells, causing them to be formed and discarded too rapidly (11.1)

**pterosaur:** any of the flying dinosaurlike creatures characterized by long wings, large heads, and (apparently) fur-covered bodies (18.3)

**puberty:** stage of development in which the reproductive organs mature and the person becomes physically able to reproduce; the start of adolescence (11.4)

**pulmonary artery:** either of two blood vessels that carry deoxygenated blood from the heart to the lungs (10.2)

**pulmonary circulation:** movement of blood between the heart and the lungs (10.3)

**pulmonary semilunar valve:** heart valve that allows blood to exit the right ventricle and enter the pulmonary arteries (10.2)

**pulmonary vein:** either of two pairs of blood vessels that carry oxygenated blood from the lungs back to the heart (10.2)

**pulp:** inner part of a tooth consisting mainly of blood vessels and nerves (9.2)

**pulse:** rhythmic blood-pressure changes in the arteries as the heart beats (10.3)

**punctuated equilibrium hypothesis:** idea that evolution occurs in sudden spurts, followed by long periods without noticeable change (14.2)

**Punnett square:** chart used to visualize all possible combinations of alleles in the offspring of two organisms (23.2)

**pupa:** form of an insect in which it stops feeding and rests while its body transforms from a larva into an adult (20.1)

**pupil:** opening through which light enters the eye (8.5)

**pyloric sphincter:** muscle structure that opens and closes to control the movement of food from the stomach to the small intestine (9.3)

## Q

**quadrate bone:** bone that loosely attaches a snake's lower jaw to the skull (18.2)

**quadriceps femoris:** group of four muscles in the front of the thigh that straighten the leg at the knee (7.5)

**queen:** female social insect that lays eggs for the colony (20.2)

## R

**rachis:** stiff quill of a feather (17.2)

**radial symmetry:** form of symmetry in which an object can be cut into identical halves along several planes (16.1)

**radicle:** portion of a plant embryo that develops into the root system (3.4)

**radiolarian:** sarcodine that secretes an intricate shell made of silica (21.6)

**radiometric dating:** technique that assigns hypothetical ages to rocks and fossils based on the decay rates of radioactive elements (14.2)

**radius:** bone of the forearm on the same side as the thumb (7.2)

**radula:** rough, filelike gastropod organ that scrapes food and carries it into the digestive tract; plural radulae (21.1)

***Ramapithecus:*** medium-size ape fossil once considered humanlike by evolutionists but now known to be similar to modern orangutans (14.2)

**ray:** cartilaginous fish with a broad, flat body and generally a long, whiplike tail (19.3)

**ray:** arm of a starfish (21.2)

**Ray, John:** 17th-century English clergyman and naturalist who introduced a precise concept of species, wrote books on plants, and described how man can learn of God's power by studying His creation (5.1, 15.1)

**ray flower:** one of the petal-like flowers arranged around the edge of a composite flower head (1.3)

**reabsorption:** return of glucose, hormones, and other useful substances from the fluid in the renal tubule to the blood (11.2)

**receptacle:** enlarged end of a pedicel, designed to hold the developing seeds (3.1)

**recessive trait:** trait that appears in the phenotype only if both alleles are for that trait (23.2)

**recombination:** process in which homologous chromosomes exchange genetic material during meiosis (23.1)

**rectilinear movement:** movement in which a snake alternately stretches and shortens segments of its body to creep forward in a straight line (18.2)

**rectum:** section of the large intestine that stores feces until they are expelled from the body (9.3)

**rectus abdominis:** group of muscles extending from the bottom of the sternum to the front of the pelvic girdle (7.5)

**red blood cell:** blood cell that carries oxygen in hemoglobin molecules; also called erythrocyte (10.1)

**red fiber:** muscle fiber that contains many mitochondria and large amounts of myoglobin (7.7)

**red marrow:** special tissue, found in bones, that manufactures red blood cells and white blood cells (7.3)

**red tide:** phenomenon that occurs when a certain dinoflagellate blooms, producing great quantities of a highly toxic nerve poison (5.5)

**Redi, Francesco:** Italian physician whose experiments with flies provided strong evidence against the idea of spontaneous generation (13.3)

**reflex:** quick, automatic response (8.2)

**reflex arc:** nerve pathway controlling a reflex, which allows nerve impulses to bypass the brain for a speedy response (8.2)

**region of elongation:** region in a root where cells increase in length (4.5)

**REM sleep:** period of sleep in which the eyes move rapidly as if they were watching the events of a dream (8.4)

**renal circulation:** portion of systemic circulation in which blood moves through the kidneys (10.3)

**renal tubule:** small tube through which fluid flows after exiting the Bowman's capsule (11.2)

**renewable resource:** natural resource that regenerates very quickly (15.6)

**renin:** enzyme released by the kidneys to increase blood pressure (11.2)

**replication:** process in which a cell makes a copy of its DNA (24.2)

**reproductive system:** body system that carries out the process of reproduction and is responsible for the continuation of the human race (6.1, 11.4)

**reptile:** cold-blooded animal that has dry, tough skin covered by scales; breathes with lungs; and has a three- or four-chambered heart (18.1)

**respiratory system:** body system that takes oxygen into the body and releases carbon dioxide from the body (6.1, 10.4)

**reticular formation:** intricate network of neurons within the brain stem that serves as the "master switch" of the cerebrum (8.3)

**retina:** light-sensitive innermost layer of the eye (8.5)

**retrovirus:** RNA virus that makes DNA copies of its genetic material and inserts those copies into a host cell's DNA (12.3)

**Rh blood group:** red-blood-cell antigen group whose most important member is the Rh factor (10.1)

**Rh factor:** most important antigen in the Rh blood group (10.1)

**rhizoid:** hairlike thread that acts as a bryophyte "root" (5.4)

**rhizome:** thick stem that grows horizontally underground to provide food storage and cause vegetative reproduction, such as the underground stem of a fern (4.4, 5.3)

**rhodopsin:** light-absorbing substance in rod cells (8.5)

**rib:** long, curved bone that attaches to the thoracic vertebrae in back and to the sternum in front (7.1)

**ribonucleic acid:** full name of RNA (24.1)

**ribosomal RNA:** type of RNA that serves as an RNA "enzyme" as part of a ribosome; abbreviated *rRNA* (24.2)

**ribosome:** cell structure that performs translation to manufacture proteins based on mRNA "blueprints" from the nucleus (22.2, 24.2)

**rickets:** childhood disease in which the bones are weak and deformed due to vitamin D deficiency (7.3)

**rickettsia:** any of a group of small bacteria that reproduce only inside other living cells; plural rickettsiae (12.2)

**rind:** outer covering of a monocot stem (4.3)

**ripening:** process in which a flower ovary grows larger and develops into a fruit (3.2)

**RNA:** nucleic acid that is similar to DNA but contains ribose instead of deoxyribose; short for *ribonucleic acid* (24.1)

**rockweed:** brown alga that lives attached to rocks along the seashore and has gas-filled air bladders that serve as floats (5.5)

**rod:** type of cell in the retina that detects light of all colors (8.5)

**rodent:** any of the small gnawing mammals that have two pairs of sharp, chisel-like incisor teeth and are in order Rodentia (16.3)

**roe:** fish eggs (19.2)

**Romans:** ancient civilization that compiled and spread the natural science of the Greeks and other earlier civilizations (13.1)

**root:** part of a plant that is ordinarily underground, anchoring the plant, and that functions in absorption of nutrients and storage of food (2.1)

**root:** part of a tooth that anchors it below the gum (9.2)

**root cap:** cone-shaped structure that covers and protects the tip of a plant root (4.5)

**root cortex:** region of a plant root where food is stored (4.5)

**root hair:** fingerlike projection of epidermal cells that increases the water-absorbing surface area of a plant root (4.5)

**rosette:** leaf pattern in which a cluster of leaves grows around the base of the plant; also called a basal arrangement (2.1)

**rostrum:** extension of the carapace that protects the head of a crayfish (20.4)

**rotifer:** microscopic freshwater invertebrate that has wheel-like cilia (21.2)

**roundworm:** nonsegmented invertebrate with a long, slender body covered by a thin cuticle; also called nematode (21.5)

**Royal Society of London for Improving Natural Knowledge:** English scientific society founded in the 17th century to study science and repudiate the unbelieving ideas of materialistic philosophers (13.2)

**rRNA:** abbreviation for *ribosomal RNA* (24.2)

**rumen:** first and largest portion of a ruminant's stomach, where digestion begins and food is formed into small balls called cud (16.4)

**ruminant:** artiodactyl that chews the cud (16.4)

**rust:** any of a group of parasitic club fungi that leave rusty-colored spores on their hosts (5.6)

## S

**sac fungi:** group of fungi with saclike spore-producing structures (5.6)

**sacral vertebra:** any of the five vertebrae (fused into the sacrum in adults) that are joined rigidly to the hip bones; also called pelvic vertebra (7.1)

**saddle joint:** joint in which a saddle-shaped portion of one bone nestles into a saddle-shaped portion of another bone (7.4)

**salamander:** any of the amphibians that superficially resemble lizards (18.5)

**saliva:** sticky digestive juice produced by the salivary glands (9.2)

**salivary amylase:** digestive enzyme produced by the salivary glands that begins the digestion of starch (9.2)

**salivary gland:** any of the six saliva-secreting digestive glands of the mouth (9.2)

**salmonellosis:** common food infection caused by bacteria of the genus *Salmonella* (9.4)

**salt gland:** excretory gland located above the eye of marine birds that concentrates and eliminates excess salt (17.2)

**salt poisoning:** abortion technique in which a solution of salt or urea is injected into the amniotic sac to poison the unborn child; medically called intrauterine instillation (11.4)

**samara:** fruit consisting of a small dry seed with one or more winglike structures attached to it (1.4, 3.3)

**sap:** sweet liquid found within the vascular tissue of plants (2.2)

**sap stream:** upward flow of water and minerals through the xylem (4.5)

**saprophyte:** organism that obtains nutrition from dead organisms (5.6)

**sapwood:** younger, functioning xylem tissue found in the outer portion of a woody stem (4.2)

**sarcodine:** protozoan that has a flexible body and can change shape at will (21.6)

**sarcomere:** repeated unit within a myofibril, consisting of overlapping protein fibers that give skeletal muscle its striated appearance (7.6)

**sartorius:** muscle that extends diagonally downward from the front of the pelvic bone and passes around the side of the thigh to connect to the tibia; lifts the lower leg and turns it inward (7.5)

**saturated fatty acid:** fatty acid in which every possible bond to hydrogen is occupied (9.1)

**sauropod:** any of the large, herbivorous dinosaurs characterized by long necks and columnlike tails (18.3)

**savanna:** tropical grassland biome (15.5)

**scab:** fibrin and platelet structure formed by a blood clot on the surface of the skin (10.1)

**scale:** any of the bony plates located in the dermis of a fish's skin (19.1)

**scapula:** shoulder blade (7.2)

**scavenger:** carnivore that does not kill its own food but feeds on dead animals instead (15.3)

**Schwann cell:** type of glial cell that produces layers of myelin sheathing on nerves in the peripheral nervous system (8.2)

**sciatica:** condition, characterized by pain in the lower back and the back of the thigh, caused by a pinching of the sciatic nerve (8.4)

**scientific law:** theory that stands the test of time and is verified by experiment after experiment (13.3)

**scientific method:** systematic thought process used by scientists to collect data and assemble that data to confirm or discard a hypothesis (13.3)

**scientism:** idea that science can find answers for all the problems in life (13.3)

**scion:** branch that will be grafted onto a stem growing on another tree (4.4)

**sclera:** tough outer layer of the eye; the white of the eye (8.5)

**scolex:** small, knoblike head of a tapeworm (21.5)

**scoliosis:** severe lateral curvature of the spine (7.1)

**scorpion:** arachnid whose long, segmented abdomen is tipped with a venomous stinger (20.3)

**sea anemone:** relatively large marine coelenterate with a polyp body form (21.3)

**sea cow:** aquatic grazing mammal of order Sirenia (16.2)

**sea worm:** marine segmented worm with a pair of parapodia on each body segment; also called polychaete (21.4)

**seaweed:** large saltwater alga (5.5)

**sebaceous gland:** oil gland in the skin (11.1)

**sebum:** oil that waterproofs the skin and keeps it soft and flexible (11.1)

**secondary bronchi:** branches into which a bronchus divides after entering the lungs (10.4)

**secondary consumer:** animal that eats herbivores (15.3)

**secondary growth:** growth of a stem in width (4.2)

**secondary sex characteristic:** any of the body characteristics that develop during adolescence and are not directly related to reproduction (11.4)

**secondary succession:** ecological succession that starts with ready soil and pioneer species (15.6)

**sedimentary cycle:** any of the biogeochemical cycles that recycle the nutrients found in the soil and rocks of the earth's surface (15.4)

**seed:** fully developed and mature ovule that can produce a new plant (3.4)

**seed coat:** protective covering of a seed (3.4)

**seed plant:** any vascular plant that produces seeds (an angiosperm or gymnosperm) (5.2)

**segmented worm:** invertebrate with a cylindrical body divided into many consecutive ringlike segments; also called annelid (21.4)

**self-pollination:** pollination that occurs within a flower (3.2)

**semicircular canals:** assembly of fluid-filled tubes in the inner ear that serve as balance sensors (8.5)

**semilunar valve:** either of the two heart valves that allow blood to flow to the body from the ventricles (10.2)

**senile dementia:** disorder that results from the slow process of natural nerve cell loss (8.4)

**sense organ:** intricately designed organ for a special sense (8.5)

**sensorineural deafness:** partial or complete hearing loss caused by repeated or prolonged exposure to excessively loud sounds (8.4)

**sensory nerve fiber:** axon that carries sensory impulses to the spinal cord and brain (8.1)

**sensory neuron:** neuron that transmits information from the senses to the central nervous system (8.1)

**sensory receptor:** specialized neuron that obtains sensory information and relays it to the brain and spinal cord (8.5)

**sepal:** leaflike structure attached to the edge of a flower's receptacle (3.1)

**septum:** thick vertical wall dividing the left and right sides of the heart (10.2)

**serous membrane:** type of membrane that lines enclosed body cavities that do not open to the exterior (6.2)

**sessile:** describes a leaf that lacks a petiole (2.1)

**sessile:** unable to move from place to place (16.1)

**seta:** bristle on a body segment of an earthworm; plural setae (21.4)

**sex chromosome:** either of a pair of chromosomes that determine an organism's sex (23.2)

**sex-linked:** describes traits located on the sex chromosomes (23.2)

**sexual reproduction:** form of reproduction in which genetic material is contributed by two parents through the union of sperm and egg (3.2, 23.1)

**sexually transmitted disease:** disease that is usually spread by sexual relations, especially those outside of marriage; abbreviated *STD;* also called venereal disease (12.2)

**shark:** cartilaginous fish with a body similar to that of a bony fish, an asymmetrical caudal fin, a complex reproductive system, and no swim bladder (19.3)

**sheath:** bottom portion of a grass leaf, which wraps partially around the stem (1.3)

**shelf fungi:** group of shelflike club fungi that grow slowly and do not have gills; also called bracket fungi (5.6)

**shoot system:** the part of the plant that is usually found above ground, holding the leaves toward the sun for the manufacturing of food and providing for the production of flowers, fruits, and seeds (2.1)

**short bone:** bone that is roughly cube shaped (7.3)

**shrub:** plant that has several low, woody stems near the ground and can stand without support (1.2)

**sickle cell anemia:** genetic disease that causes hemoglobin molecules to clump together, making red blood cells brittle and sickle shaped under certain conditions (10.1, 23.3)

**sickle cell trait:** condition in which a person is heterozygous for the allele that causes sickle cell anemia (23.3)

**sidewinding movement:** movement in which a snake supports its body with a few points of contact while throwing the rest of its body sideways (18.2)

**sieve plate:** perforated plate at the end of a woody-stem phloem cell (4.2)

**sieve tube:** phloem cells connected end to end at the sieve plates to conduct food through a woody plant (4.2)

**silica:** hard substance in sand, glass, and diatom cell walls; chemical formula $SiO_2$ (5.5)

**simple carbohydrate:** either a monosaccharide or disaccharide (9.1)

**simple eye:** arthropod eye that has only one lens and "retina" (20.1)

**simple fracture:** fracture in which the bone cracks or breaks but does not pierce through the skin; also called closed fracture (7.3)

**simple fruit:** fruit that forms from a flower that has only one pistil (3.3)

**simple goiter:** swelling in the neck caused by insufficient iodine (11.3)

**simple leaf:** leaf that has one blade on every petiole (2.1)

**sinoatrial node:** portion of the cardiac conduction system that functions as the heart's pacemaker; also called S-A node (10.2)

**sinus:** hollow space within the design of a skull bone (7.1)

**siren:** salamander that has a long, slender, eel-like body; minute forelimbs; and no hind limbs and never completely metamorphoses (18.5)

**skate:** long-nosed nonvenomous ray that, unlike true rays, is oviparous (19.3)

**skeletal muscle:** any muscle whose primary function is to move the parts of the skeleton (7.5)

**skeletal system:** body system that provides support and rigidity and serves as the framework of the body (6.1, 7.1)

**skin:** the organ of the integumentary system (11.1)

**skull:** bones forming the structure of the head (7.1)

**slime mold:** jellylike organism that oozes into its environment to engulf bacteria and other particles (5.6)

**slug:** gastropod that lacks a shell (21.1)

**small intestine:** long, tubular organ that is the primary organ of digestion and absorption (9.3)

**small RNA:** type of RNA that performs various functions within the nucleus (24.2)

**smooth muscle:** involuntary muscle tissue used to perform functions such as operating sphincters, adjusting tension in blood-vessel walls, squeezing food through the alimentary canal, and adjusting the focus of the lenses in the eyes (7.5)

**smut:** any of a group of parasitic club fungi that produce masses of black spores on their hosts (5.6)

**snake:** legless reptile with a long, cylindrical body covered with dry scales (18.2)

**social insect:** any of the insects that live in large colonies whose members perform different tasks (20.2)

**socket:** portion of the skull that protects the eye by surrounding it with heavy bone; also called orbit (8.5)

**sodium bicarbonate:** chemical produced by the pancreas to neutralize stomach acid in chyme that enters the small intestine (9.3)

**Solomon:** king of Israel who studied natural history (13.1)

**somatic cell:** body cell (23.1)

**somatic sense:** any of the senses that detect touch, temperature, pain, etc. (8.5)

**somatotropin:** pituitary hormone that regulates skeletal growth; also called growth hormone (11.3)

**sorus:** small, brownish dot on the underside of a fern frond that contains spore cases; plural sori (5.3)

**spawn:** for a fish to lay eggs (19.2)

**special leaf:** leaf that has a special design for a special task; also called a modified leaf (2.4)

**special sense:** any of the senses that are provided by neurons in sense organs (8.5)

**speciation:** changes within a particular kind of organism because of variations in the gene pool within a population; also called microevolution (14.2)

**species:** group of very similar organisms that can interbreed freely in nature; lowest of the seven main levels of the Linnaean classification system (5.1)

**sperm cell:** male reproductive cell (3.1)

**spermatophore:** gelatinous sperm-containing structure deposited by a male salamander for fertilization (18.5)

***Sphagnum:*** genus of peat moss (5.4)

**spicule:** crystalline support structure of a sponge (21.3)

**spider:** arachnid with a constriction between cephalothorax and abdomen, a venomous bite, and both tracheae and book lungs (20.3)

**spinal cavity:** body cavity containing the spinal cord (6.1)

**spinal cord:** thick bundle of nerve fibers within the spinal cavity by which the brain communicates with most of the body (8.1)

**spinal nerve:** any of the thirty-one pairs of nerves that branch from the spinal cord (8.1)

**spindle apparatus:** structure, consisting of centrosomes and microtubules, that pulls chromatids apart during mitosis (22.3)

**spine:** special protective leaves that have no chlorophyll (2.4)

**spinneret:** organ used by a spider to spin silk (20.3)

**spirelike branching:** type of branching in which a tree grows tall and straight with branches coming from a large central stem (4.1)

**spirillum:** spiral-shaped bacterium; plural spirilla (12.2)

**spirogyra:** filamentous green alga characterized by ribbonlike chloroplasts arranged in a spiral (5.5)

**spleen:** hand-sized lymphatic organ that holds a large emergency supply of blood and red blood cells; cleans the blood of microbes; and works with the liver to destroy old, worn-out red blood cells (10.1, 12.3)

**sponge:** hollow-bodied, sessile, aquatic invertebrate whose body consists of only two cell layers and lacks a nervous system (21.3)

**spongy bone:** lightweight, porous tissue composing the interior of a bone's epiphyses (7.3)

**spontaneous generation:** idea that living things can arise from nonliving things (13.1)

**sporangium:** spore case of a fern; plural sporangia (5.3)

**spore:** tiny, one-celled reproductive structure that can grow into a distinct or independent organism under the proper conditions (5.3)

**sporophyte:** asexually reproducing generation in alternation of generations (5.3)

**sporozoan:** nonmotile, spore-forming protozoan (21.6)

**sprain:** injury that occurs when the ligaments of a joint are overstrained (7.4)

**spreading branching:** type of branching in which a tree develops several main branches, usually close to the ground (4.1)

**squid:** torpedo-shaped cephalopod that has ten arms (21.1)

**stability:** ability of an ecosystem to withstand and recover from changes (15.1)

**stamen:** pollen-producing organ within a flower (3.1)

**staminate:** describes a male plant reproductive structure that produces only pollen, such as a flower that lacks a pistil and has only stamens or a pollen-producing cone (3.1, 5.2)

**stapes:** second bone of the middle ear; also called stirrup (8.5)

**starch:** mixture of long, chainlike molecules that consist of thousands of glucose molecules linked together, manufactured by plants for glucose storage (2.3, 9.1)

**starfish:** echinoderm with a body that resembles a five-pointed star (21.2)

**statocyst:** saclike sensory organ located at the base of a crayfish's antenna (20.4)

**STD:** abbreviation for *sexually transmitted disease* (12.2)

**stentor:** free-swimming, trumpet-shaped ciliate (21.6)

**sternocleidomastoid:** muscle that connects the temporal bone to the sternum and clavicles; moves head forward and helps turn the head from side to side (7.5)

**sternum:** breastbone (7.1)

**steroid:** any of a group of special lipid molecules that includes the hormones secreted by the outer shell of the adrenal glands (11.3)

**stewardship:** man's responsibility to manage creation in a way that balances the wise use of natural resources and their conservation (15.6)

**stigma:** tip of the pistil, which receives pollen grains during pollination (3.1)

**stigma:** eyespot of certain green algae (5.5)

**stock:** tree that receives a new stem by grafting (4.4)

**stolon:** creeping special stem that grows above the ground and produces new plants wherever a node touches the soil; common in grass plants; also called runner (1.3, 4.4)

**stoma:** tiny opening in the surface of a leaf that allows gases to enter and leave the leaf; also called leaf pore; plural stomata (2.2)

**stomach:** muscular digestive organ that thoroughly mixes food with digestive juices (9.3)

**stone:** inner woody layer of a drupe (3.3)

**striated muscle:** muscle tissue that has a striated (striped) appearance under the microscope; skeletal and cardiac muscle tissue (7.5)

**stroke:** death of brain tissue caused by internal blockage or rupture of a blood vessel in the brain (8.4)

**structural tissue:** type of tissue that composes most of the body of a plant (2.2)

**style:** stalklike portion of a pistil, which connects the ovary to the stigma (3.1)

**subcutaneous layer:** fatty layer of loose connective tissue beneath the dermis (11.1)

**sucrose:** table sugar (2.3)

**suction method:** abortion method in which a tube connected to a suction device is used to draw an unborn child out of the uterus and into a container for disposal; also called suction curettage and vacuum aspiration (11.4)

**sugar:** any of a group of small carbohydrates (9.1)

**superior vena cava:** vein that returns blood from the upper part of the body to the heart (10.2)

**support and strengthening tissue:** plant structural tissue consisting of fibers that run through roots, stems, and large veins of leaves (2.2)

**suture:** jagged joint line that connects cranial bones tightly (7.1)

**sweat duct:** duct connecting a skin pore to a sweat gland (11.1)

**sweat gland:** skin gland that helps regulate body temperature by producing sweat (11.1)

**swim bladder:** hollow, gas-filled chamber that allows a fish to adjust its buoyancy (19.2)

**swimmerets:** abdominal appendages of a crayfish (20.4)

**symbiosis:** close relationship between two different species over a period of time, especially one in which they mutually benefit (5.6, 15.4)

**symmetry:** property of being able to be cut into identical halves (16.1)

**sympathetic division:** portion of the autonomic nervous system that responds to increased activity and emergency situations (8.1)

**synapse:** enclosed junction between two neurons or a neuron and another cell (8.2)

**synovial fluid:** lubricating fluid inside a freely movable joint (7.4)

**synovial membrane:** type of membrane that lines the joint cavities (6.2)

**syrinx:** voice box of a bird (17.2)

**system:** group of structures designed to function together as a unit to perform a particular job for an organism's body (2.1, 6.1)

**systemic circulation:** movement of blood through all parts of the body except the lungs (10.3)

**systole:** contracting and pumping phase of heart action (10.2)

## T

**tanning:** temporary increase in melanin production caused by exposure to sunlight (11.1)

**tapeworm:** parasitic flatworm that has a long, flat, ribbonlike body divided into numerous segments (21.5)

**taproot:** root system in which the primary root grows straight down and remains larger than the secondary roots that branch off from it, penetrating the soil with relatively little branching (1.2, 4.5)

**tarsal:** any of the seven bones of the ankle and heel (7.2)

**tartar:** hard deposit on teeth formed by plaque not removed by daily brushing and flossing; medically called calculus (9.2)

**taste bud:** organ of taste, consisting of chemoreceptors for different types of taste (8.5)

**taxonomy:** process of arranging organisms into groups based on similarities (5.1)

**tear:** fluid secreted by the lacrimal glands to clean and moisten the eye (8.5)

**telophase:** fourth and final stage of mitosis, in which the chromosomes are unpacked and the nuclear envelope and nucleolus reform (22.3)

**telson:** flat, triangular projection on the sixth abdominal segment of a crayfish (20.4)

**temperate deciduous forest:** biome that has well-defined seasons and is dominated by deciduous broadleaf trees (15.5)

**temporal lobe:** lobe of the cerebrum corresponding to the temporal bones; associated with senses of hearing, taste, and smell (8.3)

**temporalis:** muscle that, with the masseter, connects the mandible to the cranium and helps close the jaw (7.5)

**temporary wilting:** wilting that occurs when transpiration takes place faster than the roots can absorb soil water; plant recovers at night, when transpiration is reduced (2.4)

**tendon:** strong "cable" of fibers that attaches a muscle to a bone (7.2)

**tendril:** special leaf or stem that has the unique function of holding a plant upright by coiling around some support structure (2.4, 4.4)

**terminal bud:** bud that allows a stem to grow in length; also called apical bud (4.1)

**terrapin:** any of a group of hard-shelled freshwater turtles that are the only turtles to live in brackish waters (18.3)

**terrestrial:** refers to land (15.1)

**territoriality:** behavior in which an animal maintains and guards its territory (17.3)

**testis:** male gonad; plural testes (11.4)

**testosterone:** primary male reproductive hormone, produced by the testes (11.4)

**tetanus:** disease caused by the soil bacterium *Clostridium tetani,* whose toxin damages motor nerves, resulting in death if untreated; also called lockjaw (8.4)

**tetrapod:** animal with four appendages or limbs (16.1)

**thalamus:** structure in the limbic system that acts as a switchboard, routing signals to the appropriate area of the cerebral cortex (8.3)

**theoretical speculation:** thinking seriously and rigorously about the phenomena of the physical universe and forming general principles to explain them (14.4)

**theory:** hypothesis that has passed the test of many well-designed experiments and has the support of other scientists (13.3)

**thermoreceptor:** sensory receptor that responds to temperature changes (8.5)

**theropod:** any of the bipedal dinosaurs characterized by powerful lower legs used for walking and smaller "arms" used for grasping (18.3)

**thoracic:** of or related to the chest (7.1)

**thoracic cage:** cagelike bone structure in the chest that protects the organs of the thoracic cavity (7.1)

**thoracic cavity:** chest cavity, containing the heart, lungs, esophagus, trachea, thymus, and aorta (6.1)

**thoracic vertebra:** any of the twelve vertebrae that are the attachments for the thoracic cage (7.1)

**thorax:** middle portion of an insect's body (20.1)

**thorn:** long, sharp special stem that provides protection to a plant (4.4)

**thylakoid:** one of the special chlorophyll-containing disks inside a chloroplast (2.3)

**thymus:** lymphatic organ where immature T cells are matured and "trained" after being produced in the bone marrow (12.3)

**thyroid gland:** endocrine gland, located in the neck, that regulates the rate of chemical activity in the body's cells (11.3)

**thyroxine:** thyroid hormone that regulates general activities related to metabolism (11.3)

**tibia:** main weight-bearing bone in the lower leg; the shin bone (7.2)

**tick:** any of the relatively large parasitic arachnids in the same order as mites (20.3)

**tidal volume:** volume of air inhaled and exhaled during normal breathing (10.4)

**tinnitus:** ringing in the ears (8.5)

**tissue:** group of similar cells that work together to perform a particular task for an organ (2.2, 6.2)

**tissue culturing:** method of vegetative propagation in which seedlings are produced from individual plant cells grown in the laboratory (4.4)

**tissue fluid:** fluid that originates in blood plasma and surrounds the body's cells (12.3)

**tolerance range:** range of values for an environmental factor in which an organism can survive (15.2)

**tongue:** muscular organ that pushes food between the teeth, mixes it with mucus and saliva, and prepares chewed food for swallowing (9.2)

**tonsils:** accumulations of lymph nodes in the throat that, with the adenoids, protect against pathogens entering through the throat (12.3)

**tooth:** structure in the mouth that cuts, breaks, or crushes food into smaller pieces (9.2)

**toothed whale:** any whale that has teeth instead of baleen (16.3)

**top carnivore:** carnivore at the top of a food chain (15.3)

**tortoise:** land turtle with feet and claws suitable for terrestrial life (18.3)

**trace element:** any of the mineral nutrients that are needed only in extremely small quantities (9.1)

**trachea:** tube that carries air from the pharynx to the lungs; also called windpipe (10.4)

**tracheid:** long, narrow, rod-shaped cell that helps conduct water in the xylem of a woody stem (4.2)

**transcription:** process in which a cell makes an mRNA copy of a particular gene (24.2)

**transfer RNA:** type of RNA that carries amino acids to a ribosome during translation; abbreviated *tRNA* (24.2)

**transgenic organism:** organism that has been genetically engineered to contain genes from more than one species (24.2)

**transitional form:** hypothetical fossil that connects one form of organism with another by tiny steps (14.2)

**translation:** process of synthesizing proteins from an mRNA pattern (24.2)

**transpiration:** loss of water vapor from leaves and stems of a plant (2.2)

**transpiration pull:** force that draws sap up a stem as leaf cells pull water into themselves (4.5)

**trapezius:** muscle in the back of the shoulder; pulls the shoulders and head back and helps turn the head from side to side (7.5)

**tree:** plant with a single tall, woody stem capable of standing erect without support (1.2)

**triceps brachii:** muscle located in the back of the upper arm that extends the forearms; also called triceps (7.5)

**trichina worm:** parasitic roundworm that forms larval cysts in muscles (21.5)

**trichocyst:** dartlike structure embedded in a paramecium's cortex (21.6)

**tricuspid valve:** heart valve connecting the right atrium and right ventricle (10.2)

**triglyceride:** another name for a fat (9.1)

**tRNA:** abbreviation for *transfer RNA* (24.2)

**trophic level:** classification that describes an organism's feeding relationship to other organisms in its ecosystem; step in a food chain (15.3)

**tropical bird:** bird native to tropical climates (17.1)

**tropical rain forest:** tropical biome characterized by abundant rain and a year-round growing season (15.5)

**tropism:** response in which an organism moves either toward or away from a stimulus (4.4)

**trunk:** main stem of a tree (1.4)

**trypanosome:** parasitic flagellate that causes African sleeping sickness (21.6)

**tuatara:** lizardlike reptile in the genus *Sphenodon* (18.3)

**tube foot:** water-vascular structure on the underside of a starfish that is used for holding on to things, for moving, and for opening the shells of clams and oysters (21.2)

**tuber:** bud-containing special stem that grows underground to store food (4.4)

**tuberculosis:** severe respiratory disease caused by infection with *Mycobacterium tuberculosis* (10.4)

**tumor:** mass of rapidly reproducing cells caused by uncontrolled reproduction (12.1)

**turf grass:** any of the grasses used as coverings for lawns, athletic fields, golf courses, and playgrounds (1.3)

**turgid:** swollen, distended (2.4)

**turgor pressure:** pressure produced by water within guard cells; causes stomata to open and close (2.4)

**turtle:** reptile with a shell and toothless jaws (18.3)

**tusk:** enlarged incisor of an elephant (16.2)

**typhoid fever:** serious illness acquired from food or water contaminated with the bacteria *Salmonella* Typhi (9.4)

## U

**ulcer:** open sore in the mucous membrane of the digestive tract (9.4)

**ulna:** bone of the forearm on the same side as the little finger (7.2)

**umbilical cord:** cord, consisting of three blood vessels and supporting tissue, that connects the placenta to an unborn baby; similar structure in a placental mammal (11.4, 16.1)

**umbilicus:** scar formed where the umbilical cord was attached to a person during prenatal development; also called navel (11.4)

**ungulate:** hoofed mammal (16.4)

**uniformitarianism:** false doctrine that states that the present is the only key to the past and that all things continue by natural processes at the same rate as they always have (14.1)

**univalve:** alternate name for a gastropod (21.1)

**unsaturated fat:** fat containing unsaturated fatty acids (9.1)

**unsaturated fatty acid:** fatty acid in which the hydrogen-carbon chain can form one or more additional pairs of bonds to hydrogen (9.1)

**urea:** waste substance made from excess amino acids (9.3)

**ureter:** tube that transfers urine from a kidney to the urinary bladder (11.2)

**urethra:** tube that carries urine from the urinary bladder to the outside of the body (11.2)

**urinary bladder:** organ that stores urine until it is discharged to the outside (11.2)

**urinary system:** portion of the excretory system that produces and excretes urine (11.2)

**urinary tract infection:** potentially severe infection of the urinary system (11.2)

**urine:** fluid, consisting of water and dissolved wastes, produced by the nephron after filtering and reabsorption (11.2)

**urogenital opening:** opening through which a fish releases gametes (19.2)

**uropod:** flipper formed by the swimmerets on the sixth abdominal segment of a crayfish (20.4)

**uterus:** hollow muscular organ in a woman's lower abdomen that provides a place for an unborn child to develop; also called womb (6.1, 11.4)

**uvea:** middle layer of the eye (8.5)

## V

**vaccination:** process of using a vaccine to prevent the spread of a disease (12.2)

**vaccine:** substance that stimulates the body to develop an acquired immunity to a disease (12.2)

**vacuole:** structure in a cell that stores food materials, fluid substances, and minerals (2.2, 22.2)

**valve:** half of a diatom's cell wall (5.5)

**valve:** structure in the heart that prevents blood from flowing in the wrong direction (10.2)

**vane:** flat part of a feather (17.2)

**variable:** factor in a controlled experiment that is being tested and will not be identical in all experiments (13.3)

**vas deferens:** either of two ducts that transfer sperm from the epididymides to the urethra; plural vasa deferentia (11.4)

**vascular bundle:** bundle of xylem and phloem cells found in a herbaceous stem (4.3)

**vascular cambium:** cambium tissue that produces new vascular tissue (2.2)

**vascular cylinder:** innermost portion of a root, containing xylem and phloem; also called stele (4.5)

**vascular plant:** any of the plants that have vascular tissue; also called tracheophyte (5.2)

**vascular ray:** structure found in most mature trees that helps transport materials between the center and edges of the tree (4.2)

**vascular tissue:** sap-conducting tissue of plants, composed of elongated, tubular cells (2.2)

**vector:** animal that transmits infection (12.2)

**vegetative propagation:** process of deliberately using vegetative reproduction to start a new plant (4.4)

**vegetative reproduction:** formation of a new plant from a stem, leaf, or root (4.4)

**vein:** one of the tubes running through the mesophyll of a leaf that contain the leaf's vascular tissue (2.2)

**vein:** relatively thin-walled blood vessel that returns blood to the heart (10.3)

**velvet:** skin covering over the growing antlers of a deer (16.4)

**venae cavae:** two large veins that return deoxygenated blood from the body to the heart; sing. vena cava (10.2)

**venation:** pattern of the veins within a leaf (2.2)

**ventricle:** thick-walled lower chamber of the heart (10.2)

**ventricular fibrillation:** type of heart failure in which the heart twitches randomly instead of beating (10.2)

**venule:** small blood vessel connecting capillaries and veins (10.3)

**vertebra:** bone of the vertebral column; plural vertebrae (7.1)

**vertebral column:** chief structural member of the body, made of 33 vertebrae; the support to which all the other parts of the skeleton are attached; also called backbone, spinal column, and spine (7.1)

**Vertebrata:** subphylum within phylum Chordata in which vertebrates are typically classified (16.1)

**vertebrate:** animal with a backbone (16.1)

**Vesalius, Andreas:** 16th-century physician who wrote *Fabrica* and is considered the Father of Anatomy (13.2)

**vesicle:** container within a cell that serves purposes such as transporting proteins (22.2)

**vessel:** large water-conducting tube in the xylem of a woody stem (4.2)

**vestigial:** describes organs that allegedly serve no useful purpose today (6.2)

**viable:** able to germinate and grow (3.3)

**villus:** tiny, hairlike projection of the small intestine that absorbs digested food; plural villi (9.3)

**vine:** any plant with a long, thin stem that grows along the ground or has tendrils that twine around a wall or other support (1.2)

**viperid:** any of a family of venomous snakes whose long fangs can be folded into a sheath on the roof of the mouth (18.2)

**viral hepatitis:** liver infection that can be acquired by skin contact, contaminated food, blood transfusions, and intravenous drug use (10.3)

**virus:** pathogen that is a molecule of genetic material enclosed in a geometric protein capsid (12.2)

**viscera:** general term for the soft internal organs within the thoracic and abdominal cavities (6.1)

**visceral hump:** mollusk body structure that contains the internal organs (21.1)

**vital capacity:** maximum volume of air that can be inhaled and exhaled (10.4)

**vitamin:** any of a group of organic substances needed in small amounts for proper function of metabolism (9.1)

**vitreous humor:** transparent, jellylike fluid that fills the space between the lens and the rear of the eye (8.5)

**viviparous:** giving birth to live young (16.1)

**vocal cords:** structures in the larynx that vibrate to produce sound (10.4)

**voluntary muscle:** any of the muscles that are generally under conscious control; a skeletal muscle (7.5)

**volvox:** colonial green alga that lives in pale-green globular colonies containing hundreds or thousands of individual cells (5.5)

**vomerine teeth:** two teeth that extend from the roof of an anuran's mouth (18.4)

**vorticella:** ciliate shaped like an upside-down bell attached to an object by a stalk (21.6)

## W

**warm-blooded:** regulating body temperature by internal mechanisms (16.1)

**water bird:** any bird that commonly lives or feeds in the water (17.1)

**water-soluble vitamin:** any of the vitamins that are soluble in water and can be easily eliminated from the body (9.1)

**water-vascular system:** complex system of water-filled tubes that extends throughout an echinoderm's body (21.2)

**Watson, James D.:** American scientist who, with Francis Crick, discovered the double-helix structure of DNA (24.1)

**whale:** any of the marine mammals of order Cetacea (16.3)

**wheat rust:** widespread rust fungus that lives on wheat and barberry bushes (5.6)

**whiplash:** injury, caused by the body being jolted, to the spinal nerves that branch from the spinal cord in the neck (8.4)

**white blood cell:** blood cell that functions as part of the immune system, helping the body fight disease; also called leukocyte (10.1, 12.3)

**white fiber:** muscle fiber containing fewer mitochondria and less myoglobin than red fibers (7.7)

**white matter:** portion of the brain and spinal cord composed of axons and glial cells that are white because of their myelin content (8.1)

**whorled:** leaf pattern in which three or more leaves grow from each node (2.1)

**wilting:** effect of a high rate of transpiration on a plant (2.4)

**wisdom tooth:** last molar on each side of the jaw (9.2)

**Wolff's law:** principle that bones adjust their shapes to the physical stress placed upon them (7.3)

**Wollemi pine:** unusual Australian conifer only known from fossils until a grove of living trees was discovered in 1994 (5.2)

**wood:** strong, resilient inner section of a woody stem (4.2)

**woody plant:** plant that produces wood (1.2)

**worker:** female social insect that performs tasks such as maintaining the colony and caring for larvae (20.2)

## X

**xanthophyll:** pigment in leaves that produces yellowish colors (2.4)

**xylem:** plant vascular tissue that transports water and dissolved minerals upward from the roots to the leaves (2.2)

## Y

**yeast:** small, colorless, single-celled sac fungus that feeds on foods containing sugar, converting the sugar into alcohol and carbon dioxide (5.6)

**yellow marrow:** fat-storing tissue found in the diaphyses of adult bones (7.3)

**yolk:** center part of a bird egg that serves as a fat-rich food source (17.3)

## Z

**zoology:** study of animals (1.1)

**zygote:** fertilized egg cell (11.4, 23.1)

# Glossary of Medical Meanings

**Note:** For many prefixes ending in a vowel, the final vowel is often dropped when forming words.

**a-, an-:** without
**aero-:** air
**ana-:** up, positive
**ante-:** before
**arterio-:** artery
**auditory:** related to hearing
**auto-:** self
**bi-:** two
**brachi-:** arm
**carcino-:** cancer
**cardio-:** heart
**cephalo-:** head
**cervical:** related to the neck
**chemo-:** chemical
**-cide:** killing
**con-:** with
**coronary:** related to the heart
**cranio-:** skull
**cryo-:** cold, freezing
**cut-:** skin
**-cyte:** cell
**dent-:** tooth
**derm-, derma-:** skin
**di-:** two
**digit-:** finger, toe
**dont-:** tooth
**dorso-:** back, dorsal
**ecto-:** outside
**-emia:** condition of the blood
**en-, end-, endo-:** in, inside, within
**entero-:** intestines
**epi-:** on, over, above
**exo-:** outside
**fore-:** before, the front part of
**gastro-:** stomach
**-gen:** something that produces
**-genous:** produced by
**hema-, hemo-:** blood
**-hemia:** condition of the blood
**hepar, hepato-:** liver
**histo-:** tissue
**hydro-:** water (or sometimes hydrogen)
**hyper-:** over, above, excessive
**hypo-:** less than, beneath, below
**-ia:** condition
**-iasis:** disease
**im-, in-:** not; in, into
**infarction:** death of tissue due to insufficient blood supply
**-itis:** inflammation
**kinesis, -kinesis:** movement
**lacri-, lacrim-:** tears
**leuko-:** white
**lingu-:** tongue
**lipo-:** fat, fatty
**-logy:** science, study
**lumbar:** related to the lower back
**-lytic:** causing decomposition
**macro-:** large
**mal-:** bad, abnormal
**mechano-:** mechanical, having to do with physical forces
**medio-:** middle
**melano-:** dark, black
**meso-:** middle
**meta-:** after, beyond, accompanying
**micro-:** small
**mid-:** middle
**mono-:** one
**mot-:** move
**multi-:** many
**myelo-:** bone marrow; spinal cord
**myo-:** muscle
**naso-:** nose
**nat-:** birth
**nephro-:** kidney
**neuro-:** nerve, nervous system
**oculo-:** eye
**olfact-:** smell
**-oma:** tumor
**-opia, -ops:** eye, vision
**-osis:** disease
**osteo-:** bone
**oto-:** ear
**para-:** beside, near
**patho-, -pathy:** disease
**pell-:** skin
**peri-:** surrounding, enclosing
**phago-:** eat, eater
**phlebo-:** vein
**pleuro-:** side; pleura
**pneumo-:** lungs
**poly-:** many, more than one
**poro-:** passage, pore (opening)
**post-:** after
**pre-, pro-:** before
**prox-, proximo-:** next
**psycho-:** mind
**pulmo-:** lungs
**quad-:** four
**ren-:** kidney
**rhino-:** nose
**sarco-:** flesh, striated muscle
**sclero-:** hard
**semi-:** half
**soma-, somato-:** body
**sub-:** under, beneath, below
**super-, supra-:** above
**thermo-:** heat
**thoraco-:** chest, thorax
**toxi-, toxico-:** poison, poisonous
**trans-:** across
**tri-:** three
**vaso-:** duct, vessel (e.g., blood vessels)
**viscero-:** internal organs (viscera)

# Index

Page numbers for photos and illustrations are in *italic* type. Page numbers for tables are followed by the letter *T*.

## A

## F



## G

## H

## I

## N

## O

## Q



## R

## S